Ben Ryan

FOUNDATIONS OF RADIATION HYDRODYNAMICS

Dimitri Mihalas
and
Barbara Weibel-Mihalas

DOVER PUBLICATIONS, INC.
Mineola, New York

Bibliographical Note

This Dover edition, first published in 1999, is an unabridged reprint, with corrections and a new preface, of the first edition of *Foundations of Radiation Hydrodynamics*, published by Oxford University Press in 1984.

Library of Congress Cataloging-in-Publication Data

Mihalas, Dimitri, 1939–
Foundations of radiation hydrodynamics / Dimitri Mihalas and Barbara Weibel-Mihalas.
p. cm.
An unabridged reprint with corrections and a new preface of the first edition published by Oxford University Press in 1984.
Includes bibliographical references and index.
ISBN 0-486-40925-2 (pbk.)
1. Radiative transfer. 2. Hydrodynamics. 3. Astrophysics. I. Title. II. Weibel-Mihalas, Barbara.

QC175.25.R3 M54 1999
523.01'92 21—dc21 99-042987

Manufactured in the United States by Courier Corporation
40925204
www.doverpublications.com

To the memory of our fathers, Emmanuel Demetrious Mihalas and Emil Edwin Weibel, who helped us set our standards, and with homage to Thomas Stearns Eliot, whose inspiring words:

All men are ready to invest their money
But most expect dividends.
I say to you: *Make perfect your will.*
I say: take no thought of the harvest,
But only of proper sowing.

have given us the courage to try to attain them.

Preface to the Dover Edition

The Oxford edition of this book, published fifteen years ago, has been out of print for several years. Nevertheless, interest in, and the relevance of the book to, astrophysical and laboratory applications has risen steadily. Our students and coworkers in this field have often been frustrated by the unavailability of the text. Therefore, on behalf of both authors of the book I would like to express my sincere gratitude to Dover Publications, Inc., for bringing out a new edition of the book in an inexpensive form that will make it readily available to all potential users. In this connection I would like particularly to thank Dr. Randy M. Roberts of the Transport Group, the Applied Theoretical and Computational Physics Division, Los Alamos National Laboratory for insisting that the time had come to issue a new edition, and for recommending the book to the Dover editors.

In addition, I would like to thank the many readers who have pointed out typographical errors. Colleagues at Los Alamos deserve my gratitude for many illuminating discussions that have helped improve my understanding of radiation hydrodynamics over the past twenty years. They are too numerous to name here, but my gratitude to them is real.

I also would like to thank again the Laboratory's administration for its generous support during the time the book was being written, and in subsequent years.

Dimitri Mihalas
Los Alamos, New Mexico
January, 1999

Preface

This book is the result of our attempt, over the past few years, to gather the basic tools required to do research on radiating flows in astrophysics. The subject of radiation hydrodynamics is very large and cuts across many disciplines in physics and astronomy: fluid dynamics, thermodynamics, statistical mechanics, kinetic theory, and radiative transfer, to name a few. The relevant material is scattered widely among a large number of books, journal papers, and technical reports; indeed, some of it exists only as folklore among practitioners in the field. As a result it has been difficult for both students and research scientists to do productive work in this area with a clear understanding of the full significance of the assumptions they have made, of how their work relates to other problems, and without having to reinvent techniques already in use.

In writing this book our primary goal has been to expose the great foundation-stones of the subject, and to erect upon them solid, if incomplete, walls of methodology on which others can later build. Accordingly, we have quite deliberately concentrated on fundamentals, and have limited severely the discussion of applications to only a few examples whose purpose is to instruct, to illustrate a point, or to provoke deeper thought.

The book divides naturally into three parts; throughout we have attempted to keep the discussion self-contained. In the first part, comprising Chapters 1 to 5, we focus on the dynamics of *nonradiating fluids*, both ideal and real, classical and relativistic, and then consider applications to a few astrophysically interesting problems: waves, shocks, and stellar winds. As an illustration of numerical methods we outline the basic von Neumann–Richtmyer technique for one-dimensional Lagrangean hydrodynamics. While many of these topics are covered in other books, it is nevertheless necessary to develop them here to the level of completeness, and with the particular emphasis, required to make a meaningful connection to the theory of *radiating fluids*.

The second part of the book, Chapters 6 to 8, deals with the physics of radiation, radiation transport, and the dynamics of radiating fluids. Here we have attempted to emphasize the very close relationship of radiation hydrodynamics to ordinary fluid dynamics, and to display the underlying unity and strong parallelism of the two formalisms. We therefore approach radiation hydrodynamics as the study of a composite fluid, consisting of material particles and photons. We develop both the continuum and kinetic-theory views for both matter and radiation, exploiting the conceptual advantages of each in order to paint a complete picture of the physics of the composite radiating fluid. An essential difference between the dynamics of radiating and nonradiating fluids is that because photons typically have much longer mean free paths than their material counterparts (perhaps approaching or exceeding the physical size of the entire flow), they can introduce a fundamental global coupling between widely separated parts of the flow, which must be treated by a full transport theory. We have attempted to counter what seems to be a commonly held opinion that radiation transport theory is an arcane art, accessible only to specialists, by arguing that conceptually it is simply a nonlocal kinetic theory for a special class of particles (photons) that do not experience body forces but interact strongly with the material component of the fluid locally, while being responsive to the global properties of the flow. We have found this paradigm to be extremely fruitful for our own thinking. Furthermore, we have attempted to show how radiation transport fits naturally into fluid-dynamical computations, in particular how a fully Lagrangean treatment of radiation transport can be incorporated into numerical calculations of one-dimensional flows in both the diffusion and transport regimes. Finally, we discuss a few illustrative examples of astrophysical flows in which radiation plays an important role.

The third part of the book is a short appendix on tensor calculus. We have found that many astronomers and physicists working on radiation hydrodynamics problems are unfamiliar with tensor techniques, and therefore cannot appreciate the power, beauty, and deep physical insight they afford. In the text we exploit tensor concepts to write equations that are covariant by inspection, an approach that allows one to make the transition from ordinary fluids, to relativistic fluids, to radiation almost automatically. The appendix summarizes only the basic material used in the text, and we assume that our readers have this minimum background. Those who do not should read the appendix first; the effort will be amply repaid.

The theory developed in this book has a wide range of application, including such diverse astrophysical phenomena as waves and oscillations in stellar atmospheres and envelopes, nonlinear stellar pulsation, stellar winds, supernova explosions, accretion flows onto compact objects, the initial phases of the cosmic expansion, and many others. It also has direct application in other areas, for example to the physics of laser fusion and of reentry vehicles. Since our professional backgrounds are those of solar/stellar astronomers we have focused almost exclusively on stellar-oriented applications. We have limited the discussion by choice to relatively low-energy phenomena, and by necessity mainly to one-dimensional flows. Even with these restrictions the essential physics of the problem emerges clearly. We hope that this book will serve as a useful starting point and as a guide to a larger literature.

We wish to acknowledge with sincere thanks the helpful contributions of many colleagues, too numerous to list here. We particularly thank Dr. John I. Castor, who understands radiation hydrodynamics better than anyone else we know, for sharing his insights with us, and for helping us learn much of what we know about the subject. We thank Dr. Robert P. Weaver for offering constructive suggestions and asking penetrating questions, which have clarified both our thinking and our writing; we are especially grateful for his unrelenting insistence that we always keep the physics in the foreground. We also thank Dr. J. Robert Buchler for several stimulating and illuminating exchanges of ideas. In addition, one of us (D. M.) thanks Drs. Robert W. Selden and Keith A. Taggart of Los Alamos National Laboratory for hospitality and financial support during the summers of 1981 and 1982 when portions of this book were written and revised.

Finally, we wish to thank Kathleen Welch and Paulina Franz for conscientiously producing a beautiful typescript.

Boulder, Colorado D.M.
January, 1984 B.W.M.

Contents

1

Microphysics of Gases

The material component of the astrophysical fluids that we consider in this book is typically a dilute gas composed of single atoms, ions, and free electrons. Insofar as we are interested in the large-scale dynamical behavior of this compressible fluid in response to imposed external forces (both gravitational and radiative), it often suffices to view the gas as a continuum and to describe it in terms of macroscopic properties such as pressure, density, temperature, etc., along with certain macroscopic transport coefficients that specify energy and momentum transport within the gas itself (i.e., thermal conduction and viscosity effects). But when we seek actually to calculate these macroscopic properties and their relationships with one another, particularly when the gas is not in equilibrium or when it interacts with a radiation field, we must examine its microscopic properties in detail and develop appropriate models on an atomic scale. In this chapter we therefore discuss the microphysics of gases from three distinct points of view, each of which complements the others, and all of which yield useful information. Throughout this chapter we ignore radiation, returning to its effects in Chapter 6.

A very basic macroscopic description of gases is afforded by the theory of *thermodynamics*. Although it is possible to develop such a theory within a purely axiomatic framework [see, e.g., (**C1,** Chap. 1)] we shall instead regard it as the formal expression of empirical results obtained from astute experimentation. An alternative description is provided by *kinetic theory*, in which one develops a model of the gas as a system of individual particles interacting according to prescribed laws of force. In the end the results of kinetic theory can be no better than our knowledge of the laws of interaction among the constituent particles (and our ability to solve the equations that result from these laws), and are, in this sense, model dependent. But, in practice, this theory gives a remarkably accurate account of the behavior of real gases even for manifestly crude atomic models. A rather different approach, in some ways more powerful than the preceding ones, is taken by classical *statistical mechanics*, by which we can calculate the most probable state of a gas under prescribed external conditions and evaluate all of its properties quite completely in terms of fundamental atomic constants and an irreducible set of thermodynamic parameters. We obtain this complete picture, however, only for gases in *equilibrium*. In contrast, despite its dependence on a definite (and usually oversimplified) atomic model, kinetic theory permits one to treat *non-equilibrium* gases in regimes where both thermodynamics and statistical mechanics provide little, if any, information. Thus it is apparent that it will repay our efforts to consider each of these approaches in turn.

Even though most astrophysical fluids are composed of several chemical species, for the purposes of developing the basic theory we shall confine attention to a pure hydrogen gas and consider mixtures of elements only where required in specific applications. By doing so we can treat all the essential phenomena (e.g., ionization) but still enjoy expository simplicity while obtaining equations in their least complicated form. Furthermore, we shall show that it is usually adequate to describe a gas by a single set of macroscopic parameters (e.g., temperature) despite the fact that the gas may be composed of at least three rather different constituents (atoms, ions, and electrons). Finally, on the spatial scales of interest here we can usually ignore *plasma* properties and view the gas as an electrically neutral, single-component (in the sense just mentioned) material. Extensive discussions of multicomponent gases can be found in (**B1**), (**C4**), and (**H1**), while plasma properties are treated in some detail in (**C4,** Chap. 19), (**K1**), and (**S2**).

1.1 Thermodynamics

1. *Equation of State of a Perfect Gas*

The force exerted by a gas on the walls of a container is directed along the outward normal to the containing surface. In the absence of external forces acting on the gas, it will exert the same force per unit area, the *pressure* p, on all points of the walls. Experiment shows that the pressure exerted by a gas at constant temperature is inversely proportional to its *volume* V (Boyle's law), and that the pressure exerted by a fixed volume of gas is directly proportional to its thermodynamic *temperature* T (the law of Charles and Gay-Lussac). These results are combined into an *equation of state* for a *perfect gas*, which states that

$$pV = n\mathcal{R}T, \tag{1.1}$$

where n is the number of *moles* of gas present (i.e., the mass of the gas divided by the atomic mass of its constituent particles), and $\mathcal{R}$ is the *universal gas constant*. Equation (1.1) provides an excellent approximation to the behavior of *dilute* (i.e., low-density), gases, the main case we consider in our work; more accurate expressions for imperfect (i.e., real) gases are given in (**H1,** Chaps. 3 and 4).

It has been established experimentally that the number of particles in a mole of gas is a universal constant, *Avogadro's number* $\mathcal{A}_0$; hence n moles contain $n\mathcal{A}_0$ particles. We can therefore rewrite (1.1) as

$$p = NkT \tag{1.2}$$

where N is the number density of particles per unit volume, and $k = \mathcal{R}/\mathcal{A}_0$ is Boltzmann's constant. If the particles in the gas have atomic weight A, then the *density* of the material is $\rho = NAm_H$, where m_H is the mass of a hydrogen atom. Thus we can write (1.2) alternatively as

$$p = \rho kT/Am_H = \rho RT, \tag{1.3}$$

where $R = k/Am_H$ is the gas constant for the particular gas being considered. In some applications it is convenient to use the *specific volume* $v = 1/\rho$, the volume per unit mass; (1.3) then becomes

$$pv = RT. \tag{1.4}$$

Equations (1.1) to (1.4) can be used interchangeably, as convenient.

The *state* of a gas is described by *state variables* such as p, ρ, T, etc. One finds empirically (and can show theoretically, cf. §§1.2 and 1.3) that all thermodynamic properties of a gas are specified when the values of any two state variables are given.

2. *First Law of Thermodynamics*

Experiments show that a gas may exchange energy with its environment by absorbing or releasing heat and by performing mechanical work, and that in such processes an energy conservation principle applies. Consider a *reversible* process, in which the gas is taken infinitely slowly through a sequence of *equilibrium* states, each differing only infinitesimally from its antecedent. Then if $\mathscr{E}$ is the *internal energy* in a volume V of the gas (which we will later see can be identified with the energy of microscopic motions and internal excitation of the particles in the gas), the *first law of thermodynamics* states that

$$d\mathscr{E} = dQ - dW. \tag{2.1}$$

Here dQ is the amount of heat gained or lost (counted positive for gains) and dW is the work done by the gas (counted positive when the gas delivers work to its surroundings) in an infinitesimal process.

We can write dW in terms of state variables because the force exerted by the gas on an element of the container's surface is $\delta\mathbf{F} = p\mathbf{n}\,\delta A$, and thus the work done if the surface moves by an infinitesimal displacement $d\mathbf{x}$ is $\delta(dW) = \delta\mathbf{F}\cdot d\mathbf{x} = p(\mathbf{n}\cdot d\mathbf{x})\,\delta A = p\,\delta V$ where δV is the volume element swept out by δA. Hence, summing over all surface elements,

$$d\mathscr{E} = dQ - p\,dV. \tag{2.2}$$

In terms of the *specific internal energy* e (i.e., internal energy per unit mass), we have

$$de = dq - p\,d(1/\rho), \tag{2.3}$$

where dq is the heat input per unit mass.

If in some process the volume changes by a finite amount, then the total work done by the gas is $\Delta W = \int p(V)\,dV$, which clearly depends on the nature of the process, that is, on how p varies with V. We therefore say that dW is an *inexact* differential. In contrast, it is found experimentally that $\mathscr{E}$ for a gas in equilibrium depends only on the state of the gas as defined in §1; hence $\Delta\mathscr{E}$ in any finite process depends only on the values of the state variables in the initial and final states but not on the details of the process between them. We therefore say that $d\mathscr{E}$ is an *exact* differential. It follows that because dW is an inexact differential, dQ must also be an inexact differential.

An *adiabatic* process is one in which the gas exchanges no heat with its surroundings ($dQ \equiv 0$); such processes can be achieved by thermally insulated systems. Consider now the *Joule–Kelvin experiment* in which a perfect gas at temperature T_1, confined to a volume V_1 within a thermally insulated container, is allowed to expand adiabatically into an additional volume V_2 initially containing a vacuum. It is found empirically that the final temperature of the gas is again T_1. In this process $\Delta Q \equiv 0$ (adiabatic) and $\Delta W \equiv 0$ (because the gas meets no resistance in its expansion and therefore does no work); therefore by the first law $\Delta\mathscr{E} \equiv 0$. In general, we can write $\mathscr{E} = \mathscr{E}(T, V)$; but we have just shown that $\Delta\mathscr{E}$ is zero for arbitrary ΔV at fixed T, hence we reach the important conclusion that, for a perfect gas, $\mathscr{E}$ is a function of T only [see also equation (5.5)].

If an input of heat dQ into a system induces a temperature-change dT, we define the *heat capacity* C as

$$C \equiv (dQ/dT). \tag{2.4}$$

This quantity depends on the amount of material present; it is more useful to work with the *specific heat* c, the heat capacity per unit mass, given by

$$c \equiv (dq/dT). \tag{2.5}$$

If the volume of the gas is unchanged during the delivery of dq, we have the *specific heat at constant volume* c_v; similarly, if the pressure is unchanged we have the *specific heat at constant pressure* c_p.

A large number of useful relations can be deduced directly from the first law (2.3) by choosing different sets of state variables. Because any thermodynamic variable can be expressed in terms of any other two, in calculating derivatives it is helpful to employ a notation of the form $(\partial z/\partial x)_y$ to indicate that (x, y) are the independent variables chosen, and that y is being held constant in the calculation of the derivative of z with respect to x. Now suppose that we have any three variables (x, y, z) connected by a functional relation $F(x, y, z) = 0$, in which any two of the variables may be considered to be independent. Then it is straightforward to derive [see, e.g., (**S1,** §3–3)] the important relations

$$(\partial x/\partial y)_z = 1/(\partial y/\partial x)_z \tag{2.6}$$

and

$$(\partial x/\partial y)_z(\partial y/\partial z)_x(\partial z/\partial x)_y = -1. \tag{2.7}$$

Furthermore, if w is a function of any two of (x, y, z), then

$$(\partial x/\partial y)_w(\partial y/\partial z)_w = (\partial x/\partial z)_w. \tag{2.8}$$

Suppose now that we choose (T, ρ) as the independent variables. Then

$$de = (\partial e/\partial T)_\rho \, dT + (\partial e/\partial \rho)_T \, d\rho, \tag{2.9}$$

hence from (2.3),

$$dq = (\partial e/\partial T)_\rho \, dT + [(\partial e/\partial \rho)_T - (p/\rho^2)] \, d\rho. \tag{2.10}$$

In a process at constant volume $d\rho \equiv 0$ and $dq = c_v \, dT$; therefore

$$c_v = (\partial e/\partial T)_\rho. \tag{2.11}$$

This result is completely general. For a process at constant pressure $dq = c_p \, dT$, and in view of (2.11), we have

$$c_p = c_v + [(\partial e/\partial \rho)_T - (p/\rho^2)](\partial \rho/\partial T)_p. \tag{2.12}$$

For adiabatic processes $dq \equiv 0$, therefore

$$c_v(\partial T/\partial \rho)_s = [(p/\rho^2) - (\partial e/\partial \rho)_T], \tag{2.13}$$

where the subscript s indicates the derivative at constant entropy (cf. §3).

These results can be cast into a standard form by using the following definitions. Write the *coefficient of thermal expansion* of the gas as

$$\beta \equiv (\partial \ln v/\partial T)_p = -(\partial \ln \rho/\partial T)_p, \tag{2.14}$$

its *coefficient of isothermal compressibility* as

$$\kappa_T \equiv -(\partial \ln v/\partial p)_T = (\partial \ln \rho/\partial p)_T, \tag{2.15}$$

and its *coefficient of adiabatic compressibility* as

$$\kappa_s \equiv -(\partial \ln v/\partial p)_s = (\partial \ln \rho/\partial p)_s. \tag{2.16}$$

Then (2.9) and (2.10) become

$$(\partial e/\partial \rho)_T = (p/\rho^2) - (c_p - c_v)/\beta\rho \tag{2.17}$$

and

$$(\partial T/\partial \rho)_s = (c_p - c_v)/\beta\rho c_v. \tag{2.18}$$

Suppose now we take (p, T) as the independent variables. Then

$$de = (\partial e/\partial T)_p \, dT + (\partial e/\partial p)_T \, dp \tag{2.19}$$

and

$$d\rho = (\partial\rho/\partial T)_p\, dT + (\partial\rho/\partial p)_T\, dp, \tag{2.20}$$

so that (2.3) becomes

$$dq = [(\partial e/\partial T)_p - (p/\rho^2)(\partial\rho/\partial T)_p]\, dT + [(\partial e/\partial p)_T - (p/\rho^2)(\partial\rho/\partial p)_T]\, dp. \tag{2.21}$$

By considering a process at constant pressure we find

$$c_p = (\partial e/\partial T)_p - (p/\rho^2)(\partial\rho/\partial T)_p; \tag{2.22}$$

a process at constant volume implies

$$c_v = c_p + [(\partial e/\partial p)_T - (p/\rho^2)(\partial\rho/\partial p)_T](\partial p/\partial T)_\rho; \tag{2.23}$$

and an adiabatic process implies

$$c_p(\partial T/\partial p)_s = (p/\rho^2)(\partial\rho/\partial p)_T - (\partial e/\partial p)_T. \tag{2.24}$$

Using (2.6) and (2.7) we can rewrite (2.22) to (2.24) into the standard forms

$$(\partial e/\partial T)_p = c_p - \beta p/\rho, \tag{2.25}$$

$$(\partial e/\partial p)_T = \kappa_T[(p/\rho) - (c_p - c_v)/\beta] \tag{2.26}$$

and

$$(\partial T/\partial p)_s = \kappa_T(c_p - c_v)/\beta c_p. \tag{2.27}$$

Finally, choosing (p, ρ) as independent variables we find by a similar analysis

$$(\partial e/\partial p)_\rho = \kappa_T c_v/\beta, \tag{2.28}$$

$$(\partial e/\partial \rho)_p = (p/\rho^2) - c_p/\beta\rho, \tag{2.29}$$

and

$$(\partial p/\partial \rho)_s = c_p/\kappa_T \rho c_v. \tag{2.30}$$

The *specific enthalpy* of a substance is defined as

$$h = e + pv = e + (p/\rho). \tag{2.31}$$

Thus

$$dh = de + p\, d(1/\rho) + dp/\rho, \tag{2.32}$$

so from the first law (2.3) we have

$$dq = dh - dp/\rho. \tag{2.33}$$

We then see that for a process at constant pressure we have, quite generally,

$$c_p = (\partial h/\partial T)_p. \tag{2.34}$$

Comparing (2.34) with (2.11), we see that enthalpy plays the same role in isobaric processes that internal energy does in processes at constant volume.

3. Second Law of Thermodynamics

The first law of thermodynamics is insufficient by itself to provide a complete theory for thermodynamics. From practical experience, it is found that certain physical processes cannot actually be realized despite the fact that they conserve energy. In general terms, one finds that in certain processes energy may be channeled into forms in which it becomes effectively unrecoverable from the gas as useful work or as heat that can be transferred to its surroundings. In a sense the energy has been *degraded*; in fact, the energy has been dissipated at the molecular level by processes that result in a more highly disordered system. We return to this point in §11.

These empirical findings are summarized into a *second law of thermodynamics*, which may be stated in several equivalent forms. For our purposes it is most direct to introduce a state function S, called the *entropy* of the system, defined such that if the system exchanges an amount of heat dQ with a reservoir at a temperature T in a reversible process, then

$$dS \equiv dQ/T. \tag{3.1}$$

The second law is equivalent to the statement that in *any* cyclic transformation, the following inequality holds:

$$\oint \frac{dQ}{T} \leq 0 \tag{3.2}$$

where the integral is evaluated over the entire cycle.

Consider now a reversible cycle. Let the system traverse the cycle in one direction, and let $dQ_1(T)$ denote the heat received in the process. Now reverse the cycle, and consider a new cycle with $dQ_2(T) \equiv -dQ_1(T)$. In both cases the material starts and ends in the same state. Now because (3.2) holds for *any* cycle we must simultaneously have

$$\oint \frac{dQ_1(T)}{T} \leq 0 \tag{3.3a}$$

and

$$\oint \frac{dQ_2(T)}{T} = -\oint \frac{dQ_1(T)}{T} \leq 0, \tag{3.3b}$$

whence we see that for a reversible cycle

$$\oint_{\text{rev}} \frac{dQ}{T} = 0. \tag{3.4}$$

We can now show that dS is a perfect differential, and hence that the entropy difference between two states is independent of the nature of the reversible processes connecting those states. Thus let A and B be two definite states at chosen points on a reversible cycle, and let path 1 denote a process leading from A to B and path 2 a different process leading from B back to A. Then by (3.4),

$$\oint \frac{dQ}{T} = \left(\int_A^B \frac{dQ}{T}\right)_1 + \left(\int_B^A \frac{dQ}{T}\right)_2 = 0. \tag{3.5}$$

But by (3.1),

$$\left(\int_A^B \frac{dQ}{T}\right)_1 = S(B) - S(A), \tag{3.6}$$

hence from (3.5)

$$S(B) - S(A) = -\left(\int_B^A \frac{dQ}{T}\right)_2 = \left(\int_A^B \frac{dQ}{T}\right)_2, \tag{3.7}$$

which shows that the entropy difference is path independent, as stated.

Consider now an *irreversible* process I joining states A and B. Follow this process by a reversible process R that returns the system from B to A in a cycle. From (3.2) we then have

$$\int_I \frac{dQ}{T} + \int_R \frac{dQ}{T} \le 0, \tag{3.8}$$

or

$$\int_I \frac{dQ}{T} \le \left(-\int_B^A \frac{dQ}{T}\right)_R = \left(\int_A^B \frac{dQ}{T}\right)_R = S(B) - S(A), \tag{3.9}$$

the equality holding only if I were reversible.

We can now prove that the entropy of an isolated system must always increase. In such a system $dQ \equiv 0$ (because the system is isolated from its surroundings), and hence from (3.9) we immediately have

$$S(B) \ge S(A), \tag{3.10}$$

the equality holding only if all processes occurring were to be reversible. But all natural processes are actually irreversible, for they always take place at a finite rate with finite departures from a sequence of perfect equilibrium states. Thus the entropy content of a natural system tends always to increase, and a thermally isolated system will therefore ultimately find itself in equilibrium in the state of maximum entropy consistent with imposed external constraints.

In terms of entropy we can rewrite the first law for a reversible process as

$$T\,dS = d\mathscr{E} + p\,dV \tag{3.11}$$

or

$$T\,ds = de + p\,d(1/\rho) \tag{3.12}$$

where s is the *specific entropy*.

4. *Thermal Properties of a Perfect Gas*

For a perfect gas we can derive explicit expressions for the thermodynamic variables; in practice these are extremely useful because we can often approximate astrophysical fluids as perfect gases. We recall that the Joule–Kelvin experiment implies that for a perfect gas the internal energy $e = e(T)$, hence from (2.11) we have an exact relation

$$e = \int c_v \, dT. \tag{4.1}$$

From kinetic theory we will find that for a perfect gas c_v is a constant, hence

$$e = c_v T, \tag{4.2}$$

where we suppress an additive constant.

Now, using (1.3) in (2.14) and (2.15), we find the thermal expansion coefficient is

$$\beta = 1/T \tag{4.3}$$

and the isothermal compressibility is

$$\kappa_T = 1/p. \tag{4.4}$$

Then from (2.17) we immediately obtain the important relation

$$c_p = c_v + (\beta p/\rho) = c_v + R. \tag{4.5}$$

The *ratio of specific heats* for a perfect gas is the constant

$$\gamma \equiv (c_p/c_v) = 1 + (R/c_v). \tag{4.6}$$

From kinetic theory we shall find $\gamma = \frac{5}{3}$ for a perfect monatomic gas.

The specific enthalpy for a perfect gas follows from (2.31), (4.2), and (4.5) as

$$h = c_v T + RT = c_p T. \tag{4.7}$$

The specific entropy of a perfect gas can be calculated directly from (4.2), (3.12), (2.11), and (1.3), which imply

$$ds = c_v (dT/T) - R(d\rho/\rho). \tag{4.8}$$

Or, in view of (4.5),

$$ds = c_p (dT/T) - R(dp/p). \tag{4.9}$$

Thus

$$s = s_0 + c_v \ln T - R \ln \rho \tag{4.10}$$

or

$$s = s_0' + c_p \ln T - R \ln p. \tag{4.11}$$

We cannot evaluate the constants in (4.10) and (4.11) from thermodynamic considerations alone; they can be calculated explicitly using statistical mechanics, cf. §12.

Now consider an adiabatic change for a perfect gas. From (4.8) and (4.5) we have

$$c_v\, d(\ln T) = (c_p - c_v)\, d(\ln \rho), \tag{4.12}$$

or

$$T = T_0(\rho/\rho_0)^{\gamma-1}. \tag{4.13}$$

Equivalent forms of (4.13) are

$$p = p_0(\rho/\rho_0)^{\gamma} \tag{4.14}$$

and

$$p = p_0(T/T_0)^{\gamma/(\gamma-1)}. \tag{4.15}$$

From (4.13) to (4.15) we see why γ is called the *adiabatic exponent.* Relations of the form (4.13) to (4.15) are called *polytropic laws;* they can be generalized to cases other than perfect gases [see §14 and §56 below, and (**C5**, Chaps. 9 and 12)].

Finally, from (2.16) and (4.14), we have, for the adiabatic compressibility of a perfect gas,

$$\kappa_s = 1/\gamma p. \tag{4.16}$$

5. *Some Consequences of the Combined First and Second Laws*

A great many useful results can be derived from the combined first and second laws for reversible processes as expressed by (3.12), along with the fact that dS is an exact differential. Thus suppose we expand de as in (2.9) to obtain

$$ds = T^{-1}(\partial e/\partial T)_\rho\, dT + T^{-1}[(\partial e/\partial \rho)_T - (p/\rho^2)]\, d\rho. \tag{5.1}$$

Now because ds is exact we know that we can also write

$$ds = (\partial s/\partial T)_\rho\, dT + (\partial s/\partial \rho)_T\, d\rho, \tag{5.2}$$

and we thereby conclude that

$$(\partial s/\partial T)_\rho = T^{-1}(\partial e/\partial T)_\rho \tag{5.3}$$

and

$$(\partial s/\partial \rho)_T = T^{-1}[(\partial e/\partial \rho)_T - (p/\rho^2)]. \tag{5.4}$$

Moreover, $(\partial^2 s/\partial T\, \partial \rho) \equiv (\partial^2 s/\partial \rho\, \partial T)$; hence from (5.3) and (5.4) we obtain

$$(\partial e/\partial \rho)_T = [p - T(\partial p/\partial T)_\rho]/\rho^2 = [p - (\beta T/\kappa_T)]/\rho^2, \tag{5.5}$$

where we have used (2.6), (2.7), (2.14), and (2.15). In particular, for a perfect gas, we now see from (4.3) and (4.4) that $(\partial e/\partial \rho)_T \equiv 0$, hence $e \equiv e(T)$, in agreement with the Joule–Kelvin experiment.

From the first law alone we were able to derive (2.12), giving a relation between c_v and c_p; but now in light of (5.5) we can rewrite this result as

$$c_p - c_v = -(T/\rho^2)(\partial p/\partial T)_\rho(\partial \rho/\partial T)_p. \tag{5.6}$$

Or, using (2.6), (2.7), (2.14), and (2.15), we have the general expression

$$c_p - c_v = \beta^2 T/\kappa_T \rho. \tag{5.7}$$

From experiment we find that $\kappa_T > 0$ for most substances, hence (5.7) implies that $(c_p - c_v) \geq 0$ for most materials, a result that follows directly from kinetic theory and statistical mechanics.

From (5.3), (5.4), and (5.5) we see that by taking (ρ, T) as independent we have

$$(\partial s/\partial T)_\rho = c_v/T \tag{5.8}$$

and

$$(\partial s/\partial \rho)_T = -\beta/\kappa_T \rho^2, \tag{5.9}$$

and thus

$$T\, ds = c_v\, dT - (\beta T/\kappa_T \rho^2)\, d\rho. \tag{5.10}$$

By a similar analysis, one finds by taking (p, T) as independent

$$(\partial e/\partial p)_T = (\kappa_T p - \beta T)/\rho, \tag{5.11}$$

$$(\partial s/\partial T)_p = c_p/T, \tag{5.12}$$

$$(\partial s/\partial p)_T = -\beta/\rho, \tag{5.13}$$

and

$$T\, ds = c_p\, dT - (\beta T/\rho)\, dp. \tag{5.14}$$

Finally, by taking (ρ, p) as independent we find

$$(\partial s/\partial p)_\rho = \kappa_T c_v/\beta T, \tag{5.15}$$

$$(\partial s/\partial \rho)_p = -c_p/\beta \rho T, \tag{5.16}$$

and

$$T\, ds = (\kappa_T c_v/\beta)\, dp - (c_p/\beta\rho)\, d\rho. \tag{5.17}$$

1.2 Kinetic Theory

6. The Distribution Function and Boltzmann's Equation

We can gain a much deeper insight into the physical properties of gases by forsaking the macroscopic picture of thermodynamics and developing in its place a microscopic kinetic theory. We restrict attention to a gas composed of a single species of particle. A macroscopic sample of gas typically contains an enormous number of particles; for example, in a stellar atmosphere the characteristic particle density is $N = 10^{16}\ \text{cm}^{-3}$. Such large numbers immediately show that there is no hope (nor any point) in our attempting to develop an exact particle-by-particle description of the system, but instead that we require a *statistical* picture that gives the distribution of particles in space and over velocity.

The large particle density quoted above implies that the average distance between particles is quite small: $d_0 = (4\pi N/3)^{-1/3} \approx 3 \times 10^{-6}$ cm for $N \sim 10^{16}$. This spacing is very large, however, compared to a typical particle size, which we can estimate to be of the order of a Bohr radius $a_0 = 5 \times 10^{-9}$ cm; these numbers imply that the particles occupy only about a part in 10^8 of the volume available to them. The interparticle spacing is also very much larger than the de Broglie wavelength $\lambda = h/p \approx h/\sqrt{3mkT}$ associated with each particle; for example, for atomic hydrogen at $T = 10^4$ K, $\lambda \approx 2 \times 10^{-9}$ cm, so that $\lambda \ll a_0 \ll d_0$.

Furthermore, if density fluctuations in the medium are random, the fractional root-mean-square fluctuation in a macroscopic volume, say 1 cm^3, is very small: $\delta N/N = N^{1/2}/N = N^{-1/2}$. For example, in a stellar atmosphere, a characteristic scale of interest for fluid flow is of the order of 100 km (a pressure scale height) so that even a 1 cm^3 volume with its fractional rms fluctuation of 10^{-8} would be considered minuscule. These numbers show clearly why it is usually reasonable to consider the gas to be a continuum: the smallness of d_0 implies that the material is exceedingly fine grained from a macroscopic view, and the smallness of $N^{-1/2}$ implies that it is extremely smooth.

The considerations outlined above show that to a very high degree of approximation we can consider the gas to be a dilute collection of *classical point particles* (because their wave packets are so highly localized). For electrically neutral particles the interparticle forces are very short range, typically falling off as a large power of the interparticle separation; therefore we can well describe the motion of a typical particle as a sequence of straight-line paths, each interrupted by a brief collision with another particle. Because we view the particles classically, the collision, which is characterized by a collision cross section, can be described by classical mechanics. Furthermore, because the probability of collisions is so small, we can neglect the possibility that three or more particles may collide simultaneously, and can consider only *binary collisions*. As we shall see in §10, the situation is rather different for charged particles interacting by Coulomb forces, which are long range; in that case the dominant contribution comes from large numbers of overlapping weak collisions.

To describe the physical state of the gas statistically, we introduce the *distribution function* $f(\mathbf{x}, \mathbf{u}, t)$ defined such that the average number of particles contained, at time t, in a volume element d^3x about $\mathbf{x}$ and a velocity-space element d^3u about $\mathbf{u}$ is $f\, d^3x\, d^3u$. We demand that $f \geq 0$ everywhere (no negative particle densities), and that as $u_i \to \pm\infty$, $f \to 0$ sufficiently rapidly to guarantee that a finite number of particles has a finite energy.

Macroscopic properties of the gas are computed from the distribution function. For example, the particle density is

$$N(\mathbf{x}, t) = \iiint_{-\infty}^{\infty} f(\mathbf{x}, \mathbf{u}, t)\, du_1\, du_2\, du_3, \tag{6.1}$$

and the mass density of the material is

$$\rho(\mathbf{x}, t) = Am_{\rm H}N(\mathbf{x}, t), \tag{6.2}$$

where A is the atomic weight of each particle. Similarly the average velocity of an element of gas (and hence its macroscopic flow velocity) is

$$\mathbf{v}(\mathbf{x}, t) = \langle \mathbf{u} \rangle \equiv N^{-1} \iiint_{-\infty}^{\infty} f(\mathbf{x}, \mathbf{u}, t)\mathbf{u}\, du_1\, du_2\, du_3. \tag{6.3}$$

To study the microscopic properties of the gas it is helpful to decompose the particle velocity $\mathbf{u}$ into

$$\mathbf{u} = \mathbf{v} + \mathbf{U}, \tag{6.4}$$

where now $\mathbf{U}$ is the *random velocity* of the particle relative to the mean flow; notice that $\langle \mathbf{U} \rangle \equiv 0$.

We now seek an equation that determines how the distribution function changes in time. For the moment, ignore collisions and assume that the gas particles do not interact. Suppose that at a time t_0 we choose a definite group of particles located in the phase-space volume element (dx_0, dy_0, dz_0, du_0, dv_0, dw_0) around the point (x_0, y_0, z_0, u_0, v_0, w_0). Let the external force acting on these particles be $\mathbf{F}(\mathbf{x}, t)$ so that they experience accelerations $\mathbf{a}(\mathbf{x}, t) = \mathbf{F}(\mathbf{x}, t)/m$, where m is the particle mass. We assume that $\mathbf{a}(\mathbf{x}, t)$ is a smooth function, so that the phase-space element will evolve into a new element (dx, dy, dz, du, dv, dw) centered on the point $\mathbf{x} = \mathbf{x}_0 + \mathbf{u}_0\, dt$ and $\mathbf{u} = \mathbf{u}_0 + \mathbf{a}\, dt$. The phase volume of this new element is related to that of the original by

$$d^3x\, d^3u = J(x, y, z, u, v, w/x_0, y_0, z_0, u_0, v_0, w_0)(d^3x)_0(d^3u)_0 \tag{6.5}$$

where J is the Jacobian of the transformation [cf. equation (A3.3)]. Bearing in mind that (x, y, z) and (u, v, w) are independent sets of variables, we see that

$$J = \begin{vmatrix} 1 & 0 & 0 & dt & 0 & 0 \\ 0 & 1 & 0 & 0 & dt & 0 \\ 0 & 0 & 1 & 0 & 0 & dt \\ (\partial a_x/\partial x)\, dt & (\partial a_x/\partial y)\, dt & (\partial a_x/\partial z)\, dt & 1 & 0 & 0 \\ (\partial a_y/\partial x)\, dt & (\partial a_y/\partial y)\, dt & (\partial a_y/\partial z)\, dt & 0 & 1 & 0 \\ (\partial a_z/\partial x)\, dt & (\partial a_z/\partial y)\, dt & (\partial a_z/\partial z)\, dt & 0 & 0 & 1 \end{vmatrix}. \tag{6.6}$$

Thus $J=1+O(dt^2)$, and hence to first order in dt the volume of the phase-space element remains constant, that is,

$$d^3x\, d^3u=(d^3x)_0(d^3u)_0. \tag{6.7}$$

Now the number of particles inside the original phase-space element is

$$\delta N_0=f(\mathbf{x}_0, \mathbf{u}_0, t_0)(d^3x)_0(d^3u)_0; \tag{6.8}$$

in the absence of collisions all of these particles end up in the new element, and thus δN_0 must equal δN, where

$$\delta N=f(\mathbf{x}_0+\mathbf{u}_0\, dt, \mathbf{u}_0+\mathbf{a}\, dt, t_0+dt)\, d^3x\, d^3u. \tag{6.9}$$

By virtue of (6.7) we therefore find that in the absence of collisions the phase-space density of a group of particles is invariant, that is,

$$f(\mathbf{x}_0+\mathbf{u}_0\, dt, \mathbf{u}_0+\mathbf{a}\, dt, t_0+dt)=f(\mathbf{x}_0, \mathbf{u}_0, t_0). \tag{6.10}$$

By expanding to first order in dt we then find

$$(\partial f/\partial t)+u^i(\partial f/\partial x^i)+a^i(\partial f/\partial u^i)=0, \tag{6.11}$$

which is known as the *collisionless Boltzmann equation*, or *Vlasov's equation.*

Equation (6.11) holds, of course, only in the absence of collisions and is thus not yet the equation we seek. When collisions occur, their effect is to shuffle particles around in phase space. Thus at $t_1=t_0+dt$, some particular particle that was in the original group, but which suffered a collision during the interval dt, will not, in general, find itself within the velocity element d^3u centered around $\mathbf{u}=\mathbf{u}_0+\mathbf{a}\, dt$, but rather in some other velocity element $(d^3u)'$ centered around some other $\mathbf{u}'\neq\mathbf{u}$. Similarly, other particles not originally in the velocity element $(d^3u)_0$ may suffer collisions during dt that leave them with a final velocity within d^3u. Therefore, when collisions occur, f is no longer invariant, and we must add a source term on the right-hand side of (6.11) which gives the net rate at which particles are shuffled into the phase-space element under consideration. We write this net rate symbolically as $(Df/Dt)_{\text{coll}}$, where we use the Lagrangean time derivative (D/Dt) because we have followed the motion of a particular group of particles in the fluid (cf. §15). Thus the desired equation is

$$(\partial f/\partial t)+u^i(\partial f/\partial x^i)+a^i(\partial f/\partial u^i)=(Df/Dt)_{\text{coll}}, \tag{6.12}$$

which is known as the *Boltzmann transport equation.*

We must next obtain an explicit expression for $(Df/Dt)_{\text{coll}}$; as will be seen in §7 it can be expressed in terms of an integral over the distribution functions of the collision partners. But before doing that it is well to emphasize that (6.12) provides only an approximation to the physics of real gases. In particular, f is only a *one-particle distribution function*, which is based on the tacit assumption that the probability of finding a particle at a particular point in phase space is independent of the coordinates of all other particles in phase space. Such a description is inadequate for anything but a dilute gas, and for a system of $\mathcal{N}$ particles the most general description is given by the $\mathcal{N}$-particle distribution function $f_{\mathcal{N}}$, which is defined such that $f_{\mathcal{N}}(\mathbf{x}_1, \ldots, \mathbf{x}_{\mathcal{N}}, \mathbf{u}_1, \ldots, \mathbf{u}_{\mathcal{N}}, t)\, d\mathbf{x}_1 \ldots d\mathbf{x}_{\mathcal{N}}\, d\mathbf{u}_1 \ldots d\mathbf{u}_{\mathcal{N}}$ gives the joint probability of finding the $\mathcal{N}$ particles in phase-space elements

$$(\mathbf{x}_1, \mathbf{x}_1 + d\mathbf{x}_1), \ldots, (\mathbf{x}_{\mathcal{N}}, \mathbf{x}_{\mathcal{N}} + d\mathbf{x}_{\mathcal{N}}), \qquad (\mathbf{u}_1, \mathbf{u}_1 + d\mathbf{u}_1), \ldots, (\mathbf{u}_{\mathcal{N}}, \mathbf{u}_{\mathcal{N}} + d\mathbf{u}_{\mathcal{N}}).$$

Notice that $f_{\mathcal{N}}$ depends explicitly on the coordinates of all the particles.

The time evolution of $f_{\mathcal{N}}$ is described exactly by *Liouville's equation*, which determines the continuous trajectory of a point representing the system in an abstract $6\mathcal{N}$-dimensional phase space. In practice it is too difficult to handle the problem with this degree of generality, so Liouville's equation is reduced to a less complicated system of equations known as the *BBGKY hierarchy*. These equations provide a systematic way of treating correlations among particle positions, velocities, and interactions, and thus allow one to describe to various degrees of approximation how a given particle perturbs all others in its vicinity. One can allow for correlations between two, three, or more particles. The BBGKY equations permit one to construct a kinetic theory for dense gases; for further details the reader can consult (**B2,** §10.2) and (**C4,** §16.7). We shall not pursue these approaches further because we will deal exclusively with dilute gases. Boltzmann's transport equation can also be derived directly from Liouville's equation; see (**H1,** 449–452).

7. *The Collision Integral*

DYNAMICS OF BINARY COLLISIONS

Consider two particles of mass m_1 and m_2 starting from infinity with initial velocities $\mathbf{u}_1$ and $\mathbf{u}_2$, which suffer a collision and move away to infinity with velocities $\mathbf{u}_1'$ and $\mathbf{u}_2'$. Then, because the total momentum of the system comprising both particles is conserved, we have

$$m_1\mathbf{u}_1 + m_2\mathbf{u}_2 = m_1\mathbf{u}_1' + m_2\mathbf{u}_2' = (m_1 + m_2)\mathbf{G} \tag{7.1}$$

where $\mathbf{G}$ is the velocity of the center of mass of the system in the laboratory frame. Let $\mathbf{g}_{12}$ and $\mathbf{g}_{12}'$ be the initial and final velocity of particle 1 relative to particle 2, and $\mathbf{g}_{21}$ and $\mathbf{g}_{21}'$ the velocities of particle 2 relative to particle 1, that is,

$$\mathbf{g}_{12} = \mathbf{u}_1 - \mathbf{u}_2 = -\mathbf{g}_{21} \tag{7.2a}$$

and

$$\mathbf{g}_{12}' = \mathbf{u}_1' - \mathbf{u}_2' = -\mathbf{g}_{21}'. \tag{7.2b}$$

Then clearly

$$g_{12} = |\mathbf{g}_{12}| = g_{21} \equiv g \tag{7.3a}$$

and

$$g_{12}' = g_{21}' = g'. \tag{7.3b}$$

Using (7.2) in (7.1) we can express the laboratory velocities of the particles in terms of their relative velocities and the center-of-mass velocity:

$$\mathbf{u}_1 = \mathbf{G} - m_2\mathbf{g}_{21}/(m_1+m_2), \tag{7.4a}$$

$$\mathbf{u}_2 = \mathbf{G} + m_1\mathbf{g}_{21}/(m_1+m_2), \tag{7.4b}$$

$$\mathbf{u}_1' = \mathbf{G} - m_2\mathbf{g}_{21}'/(m_1+m_2), \tag{7.4c}$$

$$\mathbf{u}_2' = \mathbf{G} + m_1\mathbf{g}_{21}'/(m_1+m_2). \tag{7.4d}$$

Thus knowledge of $\mathbf{G}$, $\mathbf{g}_{12}$, and $\mathbf{g}_{12}'$ is equivalent to knowledge of $\mathbf{u}_1$, $\mathbf{u}_2$, $\mathbf{u}_1'$, and $\mathbf{u}_2'$, and vice versa.

In the collision the total energy is conserved, and because the interparticle potential can be set to zero at infinite separation, this is equivalent to the statement that the sum of the initial kinetic energies equals the sum of the final kinetic energies:

$$\tfrac{1}{2}(m_1u_1^2 + m_2u_2^2) = \tfrac{1}{2}(m_1u_1'^2 + m_2u_2'^2). \tag{7.5}$$

Equations (7.1) and (7.5), along with the statement of conservation of mass ($m_1 + m_2 = m_1' + m_2'$) are known as the *summational invariants* of the collision.

Using (7.4) it is easy to show that

$$\tfrac{1}{2}(m_1u_1^2 + m_2u_2^2) = \tfrac{1}{2}(MG^2 + \tilde{m}g^2) \tag{7.6a}$$

and

$$\tfrac{1}{2}(m_1u_1'^2 + m_2u_2'^2) = \tfrac{1}{2}(MG^2 + \tilde{m}g'^2) \tag{7.6b}$$

where M is the total mass of the system

$$M = m_1 + m_2, \tag{7.7a}$$

and $\tilde{m}$ is the *reduced mass*

$$\tilde{m} \equiv m_1m_2/(m_1+m_2). \tag{7.7b}$$

From (7.5) and (7.6) we immediately see that

$$g = g', \tag{7.8}$$

which shows that the relative velocity of the two particles is changed only in direction but not in magnitude by the collision.

A specification of the full three-dimensional trajectories of the particles is a bit complicated [see, e.g., (**V1,** 351–353)], but we do not require this much information and can consider the collision in the plane of the relative orbit of the two collision partners. Assume that the interaction between particles is a *central force*, that is, one that depends only on the magnitude of their separation $x_{12} = |\mathbf{x}_1 - \mathbf{x}_2|$. Let the force exerted by particle 1 on particle 2 be $\mathbf{F}$, and by 2 on 1 be $-\mathbf{F}$; then

$$m_1\ddot{\mathbf{x}}_1 = -\mathbf{F} \tag{7.9a}$$

and

$$m_2\ddot{\mathbf{x}}_2 = \mathbf{F}, \tag{7.9b}$$

so that

$$m_1 m_2 \ddot{\mathbf{x}} \equiv m_1 m_2(\ddot{\mathbf{x}}_2 - \ddot{\mathbf{x}}_1) = (m_1 + m_2)\mathbf{F}. \tag{7.10}$$

Thus the relative orbit of particle 2 with respect to 1 is identical to the motion of an equivalent particle of mass $\tilde{m}$ around a fixed center of force **F**.

Decomposing (7.10) into radial and tangential components we find

$$\ddot{r} - r\dot{\theta}^2 = F/\tilde{m} \tag{7.11a}$$

and

$$r\ddot{\theta} + 2\dot{r}\dot{\theta} = 0. \tag{7.11b}$$

We can integrate (7.11b) straightaway to find *conservation of angular momentum*

$$r^2\dot{\theta} = \text{constant} = gb, \tag{7.12}$$

where b is the *impact parameter* of the collision, that is, the perpendicular distance of particle 1 from the straight-line extension of the incident relative velocity vector $\mathbf{g}$ of particle 2 (see Figure 7.1).

Using (7.12) in (7.11a), and integrating with respect to t after multiplying through by $\dot{r}$, we obtain the energy integral

$$\tfrac{1}{2}(\dot{r}^2 + r^2\dot{\theta}^2) + \phi(r)/\tilde{m} = \text{constant} = \tfrac{1}{2}g^2 \tag{7.13}$$

where $\phi(r)$ is the potential energy of the interaction. Suppose now that the interaction can be represented by a power-law potential

$$\phi(r) = C_\alpha / r^\alpha. \tag{7.14}$$

Then using (7.14) in (7.13) and eliminating dt in favor of $d\theta$ via (7.12) we

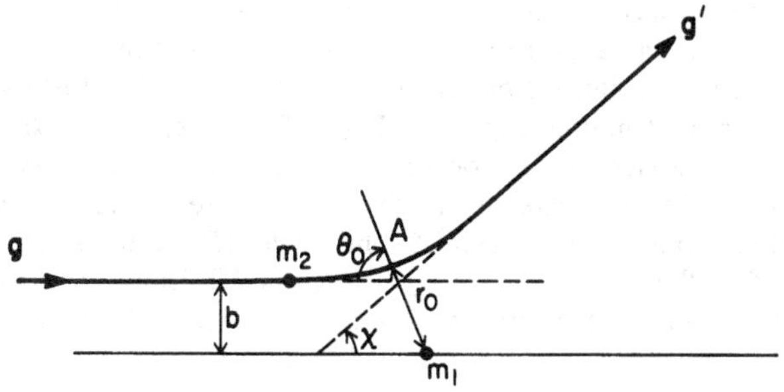

Fig. 7.1 Geometry of binary collision.

obtain the integral

$$\begin{aligned}\theta &= \int_R^\infty \left(1 - \frac{2C_\alpha}{\tilde{m}g^2 r^\alpha} - \frac{b^2}{r^2}\right)^{-1/2} \frac{b\,dr}{r^2} \\ &= \int_0^{(b/R)} \left[1 - \zeta^2 - \frac{2}{\alpha}\left(\frac{\zeta}{\mathscr{b}}\right)^\alpha\right]^{-1/2} d\zeta\end{aligned} \tag{7.15}$$

where $\zeta \equiv (b/r)$ and

$$\mathscr{b} \equiv b(\tilde{m}g^2/\alpha C_\alpha)^{1/\alpha}. \tag{7.16}$$

Here θ is measured from an axis parallel to the initial asymptote of the orbit. The apse A of the orbit is the point at which $(dr/d\theta) = 0$, which occurs when $R = R_0 = (b/\zeta_0)$ where ζ_0 is the largest root of the equation

$$1 - \zeta_0^2 - (2/\alpha)(\zeta_0/\mathscr{b})^\alpha = 0. \tag{7.17}$$

When $R = R_0$, θ achieves the value θ_0 shown in Figure 7.1; the angle of deflection χ between the asymptotes is the supplement of twice θ_0, hence

$$\chi(\mathscr{b}) = \pi - 2\int_0^{\zeta_0} [1 - \zeta^2 - (2/\alpha)(\zeta/\mathscr{b})^\alpha]^{-1/2}\, d\zeta. \tag{7.18}$$

Equations (7.16) to (7.18) show that χ depends on b, g, and the inter-particle potential only through the parameters α and $\mathscr{b}$.

THE COLLISION CROSS SECTION

The rate at which collisions occur can be expressed in terms of a collision cross section. Suppose the initial velocities of the particles are $\mathbf{u}_1$ and $\mathbf{u}_2$; these suffice to fix the center of mass velocity $\mathbf{G}$, the relative velocity $\mathbf{g}$, and the orientation of the plane of the relative orbit. The initial velocities do not completely fix the properties of the collision, however, because they do not determine the impact parameter b, which must therefore be taken as an independent variable. When b is also given, the collision is uniquely determined.

Choose a particle to act as a collision center and bombard it with an incident flux of particles; let $\mathscr{I}$ be the number of particles in the incident beam crossing a unit area normal to the beam in a unit time. Then the rate of collisions having impact parameters on the range $(b, b+db)$ within an increment $d\varepsilon$ of azimuthal angle (see Figure 7.2) is $R_1 = \mathscr{I}b\,db\,d\varepsilon$. Alternatively we can assign to the process a *differential cross section* $\sigma(\mathrm{g}, \chi)$, defined such that the rate at which particles are scattered out of the incident beam into an increment of solid angle $d\Omega$ around some direction $\mathbf{n}'$ specified by the angles (χ, ε) is $R_2 = \mathscr{I}\sigma(\mathrm{g}, \chi)\,d\Omega$. But R_1 and R_2 refer to the same particles. Therefore noting that $d\Omega = \sin\chi\,d\chi\,d\varepsilon$ we conclude that

$$\sigma(\mathrm{g}, \chi) = b\,|\partial b/\partial\chi|/\sin\chi. \tag{7.19}$$

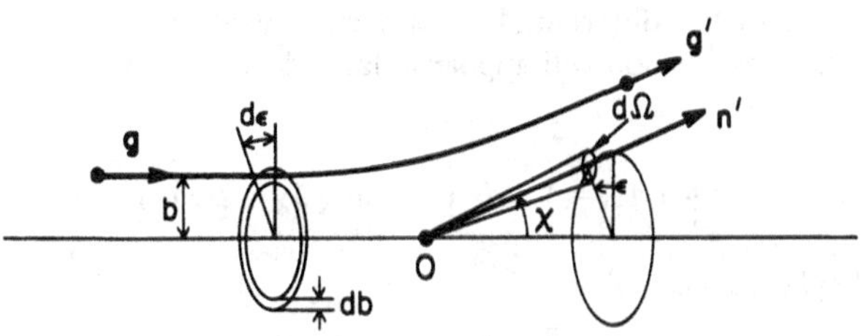

Fig. 7.2 Scattering from a collision center.

Another useful notation for the cross section is $\sigma = \sigma(\mathbf{u}_1, \mathbf{u}_2; \mathbf{u}_1', \mathbf{u}_2')$ because the initial and final velocities together are sufficient to determine the collision uniquely. In calculating collision rates we may write either $\sigma\, d\Omega$ or the differential target area $b\, db\, d\varepsilon$, as convenient.

For the important case of power-law interactions we find from (7.16) and (7.19) that

$$\sigma(g, \chi)\, d\Omega = (\alpha C_\alpha / \tilde{m})^{2/\alpha} g^{-4/\alpha} \beta\, d\beta\, d\varepsilon. \tag{7.20}$$

Another extremely useful model is to imagine that the particles are *rigid elastic spheres* of diameter d that interact only on contact. In this case, if θ_0 is the angle of the point of contact relative to the incident path, then $b = d \sin \theta_0$, and $\chi = \pi - 2\theta_0$ (see Figure 7.3) so that from (7.19)

$$\sigma = \tfrac{1}{4} d^2, \tag{7.21}$$

which is independent of both the relative speed g and the deflection angle χ of the collision partner. Despite the extreme simplicity of this model, it actually yields useful estimates of collision rates, mean-free-paths, and transport coefficients, and we shall employ it in later work.

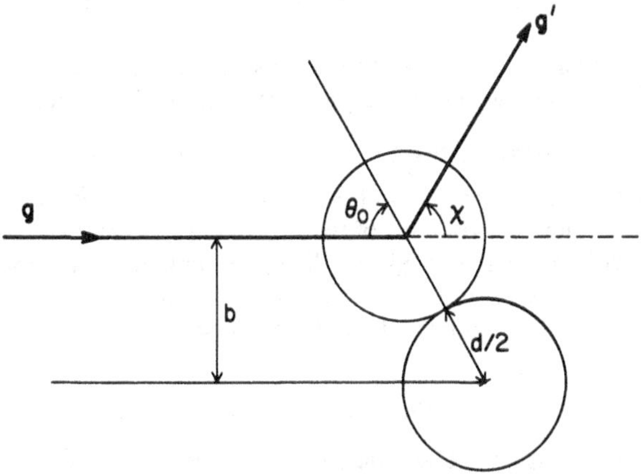

Fig. 7.3 Collision of two rigid spheres.

In addition to the differential cross section itself, various angular moments of the cross section will appear in later developments. The *total cross section* is

$$\sigma_T = \sigma_{(0)} \equiv \oint \sigma \, d\Omega = \int_0^{2\pi} d\varepsilon \int_0^{\pi} \sigma \sin \chi \, d\chi = 2\pi \int_0^{\pi} \sigma \sin \chi \, d\chi. \quad (7.22)$$

For rigid spheres we find

$$\sigma_T(\text{rigid sphere}) = \pi d^2, \quad (7.23)$$

which is clearly just the geometrical cross section of the two colliding particles. For power-law interactions (7.22) diverges, hence classically the total cross section is infinite; convergence of the integral is, however, obtained when quantum-mechanical effects are taken into account (**W2,** 9).

The transport coefficients for viscosity and thermal conductivity are related to the moment

$$\sigma_{(2)} \equiv \oint \sigma \sin^2 \chi \, d\Omega = 2\pi \int_0^{\pi} \sigma \sin^3 \chi \, d\chi. \quad (7.24)$$

For rigid spheres

$$\sigma_{(2)} = \tfrac{2}{3}\pi d^2. \quad (7.25)$$

For power-law interactions

$$\sigma_{(2)} = 2\pi(\alpha C_\alpha/\tilde{m})^{2/\alpha} g^{-4/\alpha} A_2(\alpha), \quad (7.26)$$

where

$$A_2(\alpha) \equiv \int_0^{\infty} \sin^2 \chi(\mathscr{b}) \mathscr{b} \, d\mathscr{b} \quad (7.27)$$

is a pure number tabulated in (**C4,** 172) for various values of $\nu \equiv \alpha + 1$.

The differential collision cross section has certain important invariance properties:

(a) Time reversal. If we reverse the sense of time (i.e., run the collision backward) each particle must merely retrace its original trajectory. Therefore

$$\sigma(\mathbf{u}_1, \mathbf{u}_2; \mathbf{u}_1', \mathbf{u}_2') = \sigma(-\mathbf{u}_1', -\mathbf{u}_2'; -\mathbf{u}_1, -\mathbf{u}_2). \quad (7.28)$$

(b) Rotation and reflection. Let $\mathbf{u}^\dagger$ denote the vector obtained from $\mathbf{u}$ by transformation under a rotation of the coordinate axes in space, or under a reflection with respect to a given plane, or both. Because the collision event depends only on the magnitudes and *relative* orientations of $\mathbf{u}_1$, $\mathbf{u}_2$, $\mathbf{u}_1'$, $\mathbf{u}_2'$ it is obviously unaffected by these transformations, hence we must have

$$\sigma(\mathbf{u}_1, \mathbf{u}_2; \mathbf{u}_1', \mathbf{u}_2') = \sigma(\mathbf{u}_1^\dagger, \mathbf{u}_2^\dagger; \mathbf{u}_1'^\dagger, \mathbf{u}_2'^\dagger). \quad (7.29)$$

(c) Inverse collisions. The *inverse* of a collision is defined to be the collision obtained by interchanging $(\mathbf{u}_1, \mathbf{u}_2)$ with $(\mathbf{u}_1', \mathbf{u}_2')$. As shown in

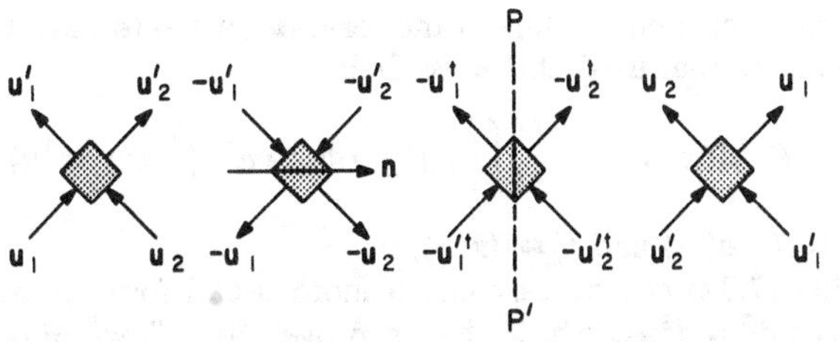

Fig. 7.4 Invariance properties of binary collisions.

Figure 7.4, the inverse collision can be obtained from the direct collision by a combination of a time reversal, followed by a 180° rotation about an axis $\mathbf{n}$ perpendicular to the center-of-mass momentum $M\mathbf{G}$, and then by a reflection through a plane PP' perpendicular to $\mathbf{n}$. It follows from (7.28) and (7.29) that direct and inverse collisions necessarily have the same cross section:

$$\sigma(\mathbf{u}_1, \mathbf{u}_2; \mathbf{u}_1', \mathbf{u}_2') = \sigma(\mathbf{u}_1', \mathbf{u}_2'; \mathbf{u}_1, \mathbf{u}_2). \tag{7.30}$$

BASIC FORM OF THE COLLISION INTEGRAL

We are now in a position to evaluate the collision term $(Df/Dt)_{\text{coll}}$ in (6.12). Label target particles in the phase-element considered with the subscript "1" and their collision partners with the subscript "2". We calculate first R_{out}, the rate at which particles of type 1 are scattered out of the phase-space element $d^3x_1\,d^3u_1$. For each particle of type 1, the number of particles of type 2 moving with velocities on the range $(\mathbf{u}_2, \mathbf{u}_2 + d\mathbf{u}_2)$, incident within a range of impact parameters $(b, b+db)$ and within azimuth range $d\varepsilon$ in a unit time, is $(f_2\,d^3u_2)gb\,db\,d\varepsilon$, where g is the incident speed of particles 2 relative to particle 1, and for brevity we have written $f_2 \equiv f_2(\mathbf{x}, \mathbf{u}_2, t)$. The total number of collisions is obtained by summing over all impact parameters, azimuths, and incident velocities, and then multiplying by the number of particles of type 1 present, namely $f_1\,d^3x_1\,d^3u_1$. Thus

$$R_{\text{out}}\,d^3x_1\,d^3u_1 = -\left(\iiint f_1 f_2 g b\,db\,d\varepsilon\,d^3u_2\right) d^3x_1\,d^3u_1. \tag{7.31}$$

In calculating R_{out} we have tacitly made the assumption of *molecular chaos*, that is, that both sets of particles are independently distributed according to f without any correlation between velocity and position or location of other particles.

Next we calculate R_{in}, the rate at which particles not originally within the phase-space element $d^3x_1\,d^3u_1$ are scattered into that element. We now wish to consider only those collisions whose result is a particle moving with velocity $\mathbf{u}_1$ within d^3u_1 *after* the collision. This is most simply done by considering inverse encounters of the form $(\mathbf{u}_1', \mathbf{u}_2') \to (\mathbf{u}_1, \mathbf{u}_2)$. By an argument similar to that used above we find

$$R_{\rm in}\, d^3x_1\, d^3u_1 = \left(\iiint f_1' f_2' g' b\, db\, d\varepsilon\, d^3u_2'\right) d^3x_1\, d^3u_1' \qquad (7.32)$$

where $f_1' \equiv f(\mathbf{x}, \mathbf{u}_1', t)$ and $f_2' \equiv f(\mathbf{x}, \mathbf{u}_2', t)$.

Equation (7.32) can be cast into a more useful form by noticing that $d^3u_1\, d^3u_2 \equiv d^3u_1'\, d^3u_2'$, which can be proved as follows. From (7.4) we know that we can express $(\mathbf{u}_1, \mathbf{u}_2)$ in terms of $(\mathbf{G}, \mathbf{g})$; hence we can write

$$d^3u_1\, d^3u_2 = J(\mathbf{u}_1, \mathbf{u}_2/\mathbf{G}, \mathbf{g})\, d^3G\, d^3g, \qquad (7.33)$$

where J is the Jacobian of the transformation. From (7.4) we find

$$J = \begin{vmatrix} 1 & -m_1/(m_1+m_2) \\ 1 & m_2/(m_1+m_2) \end{vmatrix} \equiv 1. \qquad (7.34)$$

Therefore, $d^3u_1\, d^3u_2 = d^3G\, d^3g$. Similarly we can express $(\mathbf{u}_1', \mathbf{u}_2')$ in terms of $(\mathbf{G}, \mathbf{g}')$ and write

$$d^3u_1'\, d^3u_2' = J'\, d^3G\, d^3g' = d^3G\, d^3g' \qquad (7.35)$$

because one finds $J' \equiv 1$. But $\mathbf{G}$ is unchanged by the collision, and from (7.8) $g = g'$ because the relative velocity changes only in direction. Therefore $d^3G\, d^3g' \equiv d^3G\, d^3g$, which in turn implies from (7.33) and (7.35) that $d^3u_1'\, d^3u_2' = d^3u_1\, d^3u_2$. Using this result and (7.8) we can thus rewrite (7.32) as

$$R_{\rm in}\, d^3x_1\, d^3u_1 = \left(\iiint f_1' f_2' g b\, db\, d\varepsilon\, d^3u_2\right) d^3x_1\, d^3u_1 \qquad (7.36)$$

where now in f_1' and f_2' we consider $\mathbf{u}_1' = \mathbf{u}_1'(\mathbf{u}_1, \mathbf{u}_2)$ and $\mathbf{u}_2' = \mathbf{u}_2'(\mathbf{u}_1, \mathbf{u}_2)$.

The *net* rate of scattering into $d^3x_1\, d^3u_1$ is simply $(Df_1/Dt)_{\rm coll} = R_{\rm in} + R_{\rm out}$, so that the Boltzmann equation for f_1, accounting for binary collisions, becomes

$$(\partial f_1/\partial t) + u^i(\partial f_1/\partial x^i) + a^i(\partial f_1/\partial u^i) = \iiint (f_1' f_2' - f_1 f_2) g b\, db\, d\varepsilon\, d^3u_2, \qquad (7.37)$$

or, in terms of the differential cross section,

$$(\partial f_1/\partial t) + u^i(\partial f_1/\partial x^i) + a^i(\partial f_1/\partial u^i) = \iint (f_1' f_2' - f_1 f_2) g \sigma(\mathbf{\Omega})\, d\Omega\, d^3u_2. \qquad (7.38)$$

In $\sigma(\mathbf{\Omega})$, $\mathbf{\Omega}$ denotes the angles between $\mathbf{g}$ and $\mathbf{g}'$. From (7.37) and (7.38) we see that the Boltzmann transport equation is a nonlinear integrodifferential equation for the distribution function f; its complicated mathematical nature makes the equation difficult to solve.

ALTERNATE FORMS OF THE COLLISION INTEGRAL

It is convenient to develop here several equivalent forms of moments of the collision integral for later use. Let $Q(\mathbf{u}_1)$ be any function of the particle velocity $\mathbf{u}_1$, and define

$$I \equiv \iiint Q(\mathbf{u}_1)(f_1'f_2' - f_1f_2) g\sigma(\mathbf{\Omega})\, d\Omega\, d^3u_1\, d^3u_2. \tag{7.39}$$

Suppose now that we merely interchange the labeling 1 and 2 on all particles; obviously the same value of I must result. Adding the two expressions, we see that an equivalent expression for I is

$$I = \tfrac{1}{2} \iiint [Q(\mathbf{u}_1) + Q(\mathbf{u}_2)](f_1'f_2' - f_1f_2) g\sigma(\mathbf{\Omega})\, d\Omega\, d^3u_1\, d^3u_2. \tag{7.40}$$

Suppose now in (7.40) we replace the collision by its inverse; the integral must have the same value because for every collision there is an inverse collision with the same cross section, and we are thus merely writing two completely equivalent expressions for the sum over all collisions of the value, before the collision, of the function Q. Thus

$$\begin{aligned} I &= \tfrac{1}{2} \iiint [Q(\mathbf{u}_1') + Q(\mathbf{u}_2')](f_1f_2 - f_1'f_2') g'\sigma(\mathbf{\Omega}')\, d\Omega'\, d^3u_1' d^3u_2' \\ &= \tfrac{1}{2} \iiint [Q(\mathbf{u}_1') + Q(\mathbf{u}_2')](f_1f_2 - f_1'f_2') g\sigma(\mathbf{\Omega})\, d\Omega\, d^3u_1\, d^3u_2, \end{aligned} \tag{7.41}$$

where we have used the facts, proven earlier, that $\sigma(\mathbf{\Omega}') = \sigma(\mathbf{\Omega})$, $d\Omega' = d\Omega$, $g' = g$, and $d^3u_1'\, d^3u_2' = d^3u_1\, d^3u_2$. Adding (7.40) and (7.41) we have

$$I = -\tfrac{1}{4} \iiint [\delta Q(\mathbf{u}_1) + \delta Q(\mathbf{u}_2)](f_1'f_2' - f_1f_2) g\sigma(\mathbf{\Omega})\, d\Omega\, d^3u_1\, d^3u_2, \tag{7.42}$$

where

$$\delta Q(\mathbf{u}) \equiv Q(\mathbf{u}') - Q(\mathbf{u}). \tag{7.43}$$

By the same line of argument leading to (7.41), we see that

$$\begin{aligned} \iiint Q(\mathbf{u}_1')f_1f_2 g\sigma(\mathbf{\Omega})\, d\Omega\, d^3u_1\, d^3u_2 &\equiv \iiint Q(\mathbf{u}_1)f_1'f_2' g'\sigma(\mathbf{\Omega}')\, d\Omega'\, d^3u_1'\, d^3u_2' \\ &\equiv \iiint Q(\mathbf{u}_1)f_1'f_2' g\sigma(\mathbf{\Omega})\, d\Omega\, d^3u_1\, d^3u_2. \end{aligned} \tag{7.44}$$

Then using (7.44) in (7.39) we see immediately that

$$I = \iiint \delta Q(\mathbf{u}_1) f_1f_2 g\sigma(\mathbf{\Omega})\, d\Omega\, d^3u_1\, d^3u_2. \tag{7.45}$$

We have shown above that I is unchanged under interchange of $\mathbf{u}_1$ and $\mathbf{u}_2$;

thus, from (7.45), we see that

$$I=\tfrac{1}{2}\iiint[\delta Q(\mathbf{u}_1)+\delta Q(\mathbf{u}_2)]f_1f_2g\sigma(\mathbf{\Omega})\,d\Omega\,d^3u_1\,d^3u_2. \qquad (7.46)$$

8. *The Maxwellian Velocity Distribution*

Boltzmann's equation is quite general and can be applied to a wide variety of physical situations. First, let us suppose that the gas is in a state of *equilibrium*: the material is homogeneous, isotropic, and at rest. The distribution function will then be independent of both position and time, and therefore the number of particles in each velocity class must be constant. If this is to be true when there is a reshuffling of particles by collisions, then it must be possible to pair each collision with its inverse, and we will have *detailed balance.* That is, in equilibrium we must have $(Df/Dt)_{\text{coll}}\equiv 0$. We see from (7.37) and (7.38) that this will be the case if

$$f_0(\mathbf{u}_1)f_0(\mathbf{u}_2)-f_0(\mathbf{u}_1')f_0(\mathbf{u}_2')\equiv 0, \qquad (8.1)$$

which is thus a *sufficient* condition for equilibrium; here the subscript zero denotes the equilibrium distribution function. Taking logarithms of (8.1) we have

$$\ln f_0(\mathbf{u}_1)+\ln f_0(\mathbf{u}_2)=\ln f_0(\mathbf{u}_1')+\ln f_0(\mathbf{u}_2'), \qquad (8.2)$$

which states that $\ln f_0$ is a function such that the sum of that function for the two collision partners is conserved in the collision for *any* possible collision. Hence $\ln f_0$ must be expressible as a linear combination of summational invariants.

We saw in §7 that the total mass, momentum, and energy of the collision partners are summational invariants; it is possible to show that these are the *only* linearly independent summational invariants that exist for structureless point particles (**C4,** 50). Therefore we can write f_0 as

$$\ln f_0=\alpha_1+\boldsymbol{\alpha}_2\cdot\mathbf{u}+\alpha_3u^2=-\tfrac{1}{2}\beta m(\mathbf{u}-\mathbf{v})^2+\gamma, \qquad (8.3)$$

where β must be positive to guarantee that $f\to 0$ as $|\mathbf{u}|\to\infty$; the factor $\tfrac{1}{2}m$ was chosen to simplify later results. In (8.3), $\mathbf{v}$ must be the mean velocity defined by (6.3) because the distribution function must be isotropic in the frame in which the material is at rest. Hence, in terms of random velocities $\mathbf{U}$, we have

$$f_0(\mathbf{U})=A\exp(-\tfrac{1}{2}\beta mU^2) \qquad (8.4)$$

We can determine the normalization factor in (8.4) by invoking (6.1):

$$N=\iiint_{-\infty}^{\infty}f_0(\mathbf{U})\,d^3U$$
$$=A\int_0^{2\pi}d\phi\int_0^{\pi}\sin\theta\,d\theta\int_{-\infty}^{\infty}\exp(-\tfrac{1}{2}\beta mU^2)U^2\,dU=A(2\pi/\beta m)^{3/2}, \qquad (8.5)$$

or

$$A = (\beta m/2\pi)^{3/2} N. \tag{8.6}$$

To evaluate β we use $f_0(\mathbf{U})$ to calculate a directly measurable quantity—the pressure. By definition, p equals the momentum transfer from the gas to a wall per unit area per unit time. Erect a perfectly reflecting wall in the (y, z) plane, and confine the gas to the region $x \le 0$ so that particles hit the wall only if $U_x > 0$. If an incoming particle has velocity (U_x, U_y, U_z), after hitting the wall its velocity is $(-U_x, U_y, U_z)$, and the momentum transferred to the wall is $2mU_x$; the flux of these particles per unit area and time is $U_x f_0(\mathbf{U})$. Therefore the pressure is

$$\begin{aligned} p &= \int_{-\infty}^{\infty} dU_z \int_{-\infty}^{\infty} dU_y \int_0^{\infty} (2mU_x) U_x A e^{-(\beta m U^2/2)}\, dU_x \\ &= mA \iiint_{-\infty}^{\infty} U_x^2 e^{-(\beta m U^2/2)}\, d^3U \\ &= \tfrac{1}{3} mA \iiint_{-\infty}^{\infty} U^2 e^{-(\beta m U^2/2)}\, d^3U \\ &= \tfrac{4}{3}\pi m A \int_{-\infty}^{\infty} e^{-(\beta m U^2/2)} U^4\, dU = \frac{mA}{2\pi}\left(\frac{2\pi}{\beta m}\right)^{5/2}. \end{aligned} \tag{8.7}$$

Here we noticed that, by symmetry, the averages $\langle U_x^2\rangle$, $\langle U_y^2\rangle$, and $\langle U_z^2\rangle$ are identical and hence equal to $\frac{1}{3}\langle U^2\rangle$. Now combining (1.2), (8.6), and (8.7) we find $NkT = N/\beta$; therefore

$$\beta = 1/kT. \tag{8.8}$$

The equilibrium velocity distribution, the *Maxwellian distribution*, is thus

$$f_0(\mathbf{U}) = N(m/2\pi kT)^{3/2} \exp(-mU^2/2kT). \tag{8.9}$$

Equations (8.7) to (8.9) can be viewed as giving the definition of the *kinetic temperature*. In equilibrium, the kinetic temperature is identical to the absolute temperature of thermodynamics. But, more important, a kinetic temperature can be uniquely defined by an approach similar to the one above even in situations where the thermodynamic definition can no longer be applied (see §30). We shall therefore take the kinetic temperature as the fundamental operational definition of T in our later work.

The distribution of particles in speed (magnitude of $\mathbf{U}$) is

$$f_0(U)\, dU = N(m/2\pi kT)^{3/2} \exp(-mU^2/2kT) 4\pi U^2\, dU. \tag{8.10}$$

The *most probable speed*, at which the maximum of f_0 occurs, is

$$U_{mp} = (2kT/m)^{1/2}; \tag{8.11}$$

the *average speed* is

$$\langle U\rangle = N^{-1}\int_0^\infty U f_0(U)\,dU = (8kT/\pi m)^{1/2}; \tag{8.12}$$

and the *root-mean-square speed* is

$$\langle U^2\rangle^{1/2} = \left[N^{-1}\int_0^\infty U^2 f_0(U)\,dU\right]^{1/2} = (3kT/m)^{1/2}. \tag{8.13}$$

The distribution function $N^{-1}f_0(U)$ can be factored into the product

$$N^{-1}f_0(U)\,dU_x\,dU_y\,dU_z = [\Phi(U_x)\,dU_x][\Phi(U_y)\,dU_y][\Phi(U_z)\,dU_z] \tag{8.14}$$

where

$$\Phi(U_x) = (m/2\pi kT)^{1/2}\exp(-mU_x^2/2kT), \tag{8.15}$$

and similarly for U_y and U_z. In contrast to f_0, Φ peaks at $U_x = 0$ and by symmetry $\langle U_x\rangle = 0$ (the range being $-\infty \leq U_x \leq \infty$); as remarked earlier $\langle U_x^2\rangle^{1/2} = (\langle U^2\rangle/3)^{1/2} = (kT/m)^{1/2}$.

The distribution function gives a complete description of the microscopic properties of the gas, hence a knowledge of f_0 is sufficient to derive all macroscopic properties of a gas. For instance, the total translational energy of $\mathcal{N}$ particles is

$$\mathscr{E}_{\text{trans}} = \mathcal{N}(\tfrac{1}{2}m\langle U^2\rangle) = (\tfrac{3}{2}NkT)V, \tag{8.16}$$

whence we have the important result that for a perfect gas

$$p = \tfrac{2}{3}(\mathscr{E}_{\text{trans}}/V) \equiv \tfrac{2}{3}\hat{e}_{\text{trans}}, \tag{8.17}$$

where $\hat{e}_{\text{trans}}$ denotes the translational energy per unit volume. The translational energy per unit mass is

$$e_{\text{trans}} = \tfrac{3}{2}RT, \tag{8.18}$$

and the energy per particle is $\tilde{e}_{\text{trans}} = \tfrac{3}{2}kT$. The specific enthalpy (per unit mass) is

$$h = e_{\text{trans}} + p = \tfrac{5}{2}RT, \tag{8.19}$$

hence the enthalpy per particle is $\tilde{h} = \tfrac{5}{2}kT$. For a gas of point particles without internal structure (our present model), the total internal energy $\mathscr{E} \equiv \mathscr{E}_{\text{trans}}$ because there are no other modes of energy storage available to the particles; as we shall show in §12 the situation is different for a gas of real atoms.

From (8.18) we see that e_{trans} is a function of T only, as expected from the Joule–Kelvin experiment, and using (2.11) we have

$$c_v = \tfrac{3}{2}R = \tfrac{3}{2}(k/m). \tag{8.20}$$

Hence from (4.5) or (4.11)

$$c_p = \tfrac{5}{2}R = \tfrac{5}{2}(k/m), \tag{8.21}$$

and therefore

$$\gamma = \tfrac{5}{3}. \tag{8.22}$$

The above results justify the statements made in §4 that c_v, c_p, and γ for a perfect gas are all constants.

Finally, using (8.20) and (8.21) in (4.10) and (4.11), we find that the entropy of a perfect gas is

$$s = s_0 + \tfrac{3}{2}R \ln T - R \ln \rho = s_0' + R \ln T - R \ln p. \tag{8.23}$$

As was true for thermodynamics, kinetic theory is unable to evaluate the additive constants.

Suppose now that we have a *mixture* of gases containing several *non-interacting* chemical species (i.e., particles whose identities do not change when the state of the gas changes); as before, assume the particles have no internal structure. Then

$$N = \sum_s n_s \tag{8.24}$$

where n_s is the number density of particle species s, and

$$\rho = \sum_s \rho_s = m_H \sum_s n_s A_s \tag{8.25}$$

where A_s is the atomic weight of species s. If the mixture is in equilibrium at a given temperature T, then experiments show that

$$p = \sum_s p_s = \sum_s n_s kT = NkT \tag{8.26}$$

which is *Dalton's law of partial pressures.* Now from (8.16) and (8.17)

$$n_s kT = p_s = \tfrac{2}{3}\hat{e}_s = \tfrac{1}{3}n_s m_s \langle U_s^2 \rangle \tag{8.27}$$

which implies that $\tfrac{1}{2}m_s\langle U_s^2\rangle = \tfrac{3}{2}kT$, that is, that all particles have the same amount of kinetic energy (the principle of *equipartition of energy*). Furthermore, we again find that (8.20) to (8.22) hold for c_v, c_p, and γ, and also that the velocity distribution for each species is Maxwellian:

$$f_{0s}(U_s) = (m_s/2\pi kT)^{3/2} \exp(-m_s U_s^2/2kT). \tag{8.28}$$

It must be stressed that (8.26) to (8.28) apply only in equilibrium where all particles have the same kinetic temperature. In general, different species of particles may have different kinetic temperatures as defined via equations (8.7) and (8.8); we discuss this question further in §10 where we shall show that the use of a single kinetic temperature is a good approximation in the cases of primary interest to us. Furthermore, when the species can interact "chemically" (e.g., in a gas of atoms, ions, and electrons, where atoms can ionize to ions plus electrons, and vice versa, or in a relatively cool gas where atoms can form molecules), the internal energy of the mixture contains terms in addition to translational energy, and the expression for e, c_v, c_p, and the adiabatic exponent must be modified (cf. §14).

9. *Boltzmann's H-Theorem*

We have seen above that (8.1) is a sufficient condition for equilibrium; we shall now prove that it is also a *necessary* condition, and hence that the Maxwellian velocity distribution is the unique equilibrium distribution function. Suppose that the sample of gas is thermally isolated, is homogeneous so that f is independent of $\mathbf{x}$, and that external forces are absent. Consider the functional

$$H(t) \equiv \int f(\mathbf{u}, t) \ln [f(\mathbf{u}, t)]\, d^3u \tag{9.1}$$

introduced by Boltzmann. Differentiating with respect to time we have

$$(dH/dt) = \int (1 + \ln f)(\partial f/\partial t)\, d^3u. \tag{9.2}$$

Clearly, if $(\partial f/\partial t) \equiv 0$, as it must be for equilibrium, then $(dH/dt) = 0$, which is therefore a necessary condition for equilibrium. We shall show that $(dH/dt) = 0$ implies (8.1), which is thus both necessary and sufficient for equilibrium.

For a uniform gas $(\partial f/\partial x^i) \equiv 0$, and with no external forces $a^i \equiv 0$, so from (7.38)

$$(\partial f/\partial t) \equiv (Df/Dt)_{\text{coll}} = \iint (f_1' f_2' - f_1 f_2) g\sigma(\mathbf{\Omega})\, d^3u_2, \tag{9.3}$$

hence

$$(dH/dt) = \iiint (1 + \ln f_1)(f_1' f_2' - f_1 f_2) g\sigma(\mathbf{\Omega})\, d\Omega\, d^3u_1\, d^3u_2. \tag{9.4}$$

Then, by virtue of (7.42), we immediately have

$$(dH/dt) = -\tfrac{1}{4} \iiint [\ln (f_1' f_2'/f_1 f_2)](f_1' f_2' - f_1 f_2) g\sigma(\mathbf{\Omega})\, d\Omega\, d^3u_1\, d^3u_2. \tag{9.5}$$

The integrand in (9.5) is clearly greater than zero whether $(f_1' f_2') < (f_1 f_2)$ or $(f_1' f_2') > (f_1 f_2)$. Therefore,

$$(dH/dt) \leq 0, \tag{9.6}$$

the equality holding only when $f_1' f_2' - f_1 f_2 \equiv 0$.

It can easily be shown that $H(t)$ is bounded from below [see (**C4,** 67) or (**H1,** 465)] because the total kinetic energy of an assembly of particles must be finite. Given this fact and (9.6) we conclude that $H(t)$ continually decreases to some final value, at which point $(dH/dt)=0$ and we recover (8.1); this proves the necessity and sufficiency of this condition.

Although we have treated $H(t)$ as an arbitrary functional for our present purposes, it has a much deeper physical meaning: it is proportional to the negative of the entropy. We return to this point in §12.

10. The Time of Relaxation

We have shown in §§8 and 9 that the Maxwellian velocity distribution is the unique velocity distribution function for a gas (or mixture of gases) of point particles in equilibrium, and have seen that the physical properties of such gases are easy to calculate. How quickly can this equilibrium state actually be reached by a gas starting with a non-Maxwellian distribution function? This is a question of central importance to our work because when there is fluid flow in the medium, any particular element of material, initially in equilibrium with its surroundings, can be transported to some other position where the ambient conditions differ from its initial conditions. If the characteristic time required to reequilibrate the velocity distribution to a Maxwellian distribution is very short compared to a characteristic flow time, then the gas may be considered to remain locally in equilibrium at each point in the flow because it rapidly accommodates to its slowly varying environment.

From elementary mechanics, we know that, in elastic collisions, energy exchange is most efficient among particles of equal masses, and is inefficient among particles of widely differing masses. It thus comes as no surprise that in a typical astrophysical fluid the longest time scale in the equilibration toward a single Maxwellian velocity distribution is set by energy exchange between the lightest particles (free electrons) and the much heavier atoms and ions. In keeping with our choice of a pure hydrogen gas, let us therefore consider collisions between pairs of charged particles such as electrons with electrons, protons with protons, and electrons with protons.

COULOMB FORCES

Particles with charges Z_1e and Z_2e interact by the Coulomb force

$$F=Z_1Z_2e^2/r^2, \tag{10.1}$$

which, in contrast to forces among neutral particles [where typical values of α in (7.14) are ≥ 5], are extremely long range. A consequence of this long range is that while the force exerted on a test particle falls as r^{-2}, the number of field particles (in a uniform medium) on a range $(r, r+dr)$ rises as r^2, and therefore there will be roughly equal contributions to the collisional interaction from particles at *all* distances. In fact, we shall shortly see that the dominant effect comes from a multitude of weak collisions with particles at large distances. Because a characteristic collision time is of the order of (b/g), these weak collisions at large b inevitably overlap one another. Therefore the assumptions underlying the binary collision integral in (7.37) and (7.38) break down, and a different approach is indicated.

An alternative expression for the collision integral can be developed by considering changes in a test particle's velocity to be a *Markov process* characterized by a function $P(\mathbf{u}, \Delta\mathbf{u})$, which gives the probability that a particle changes its velocity from $\mathbf{u}$ to $\mathbf{u}+\Delta\mathbf{u}$ in a time interval Δt. On the assumption that weak collisions dominate, $(Df/Dt)_{\text{coll}}$ can then be written in terms of a differential operator containing P and the distribution function f, yielding what is known as the *Fokker–Planck equation*. Derivations of the Fokker–Planck equation from this statistical point of view can be found in (**B2,** §10-9), (**C3**), and (**K1,** §6.3); an in-depth analysis of this equation for an inverse-square force is given in (**R1**). Curiously enough, it is also possible to derive the Fokker–Planck equation directly from the usual Boltzmann collision integral (**C4,** §19.7), despite the radically different physical picture employed in the derivation of that integral.

The Fokker–Planck equation provides a rather complete picture of such important processes as *dynamical friction* (the systematic slowing down of fast-moving particles by drag) and the *diffusion* of particles in phase space (from which the rate of equilibration can be determined). The solution of this equation is, however, complicated, and we shall follow instead a different route by estimating characteristic *times of relaxation* for the equilibration process; a check on these estimates is provided by direct calculations from the Fokker–Planck equation itself (**M1**).

ROUGH CALCULATION OF THE DEFLECTION TIME

A test particle colliding with another particle in the field experiences a change ΔE in its energy and is deflected by an angle χ from its original direction. Because the changes are *random*, we expect that over a long time interval $\langle \Delta E \rangle_T = 0$ and $\langle \chi \rangle_T = 0$. Nevertheless the particle's motion suffers an ever-growing, random-walk departure from the original trajectory, and eventually the particle will be moving on a path essentially unrelated to its original path. We can characterize these cumulative departures by calculating the values of $\sum (\Delta E)^2$ and $\sum \chi^2$ summed over all collisions; because the process is random we expect these sums to grow linearly with time. We can then define an *energy-exchange time* t_E as

$$t_E = E^2/\{d[\sum (\Delta E)^2]/dt\} \qquad (10.2)$$

and a *deflection time* t_D as

$$t_D = 1/(\sum \chi^2)/dt]. \qquad (10.3)$$

In the time t_E the energy of a test particle can no longer be considered to have been even approximately conserved, and thus t_E is a representative time for the velocity distribution to be thermalized towards a Maxwellian. Similarly t_D gives the time required for $\sum (\chi^2)^{1/2}$ to accumulate to about a radian and hence is a representative time over which the velocity distribution becomes isotropized.

Given these definitions, both t_D and t_E can be computed with considerable mathematical precision, but the analysis is intricate [see (**C2**, Chap. 2)]. We therefore give only a heuristic derivation of t_D, which is simple mathematically but retains all the essential physics. Consider a test particle of mass m_1 and charge $Z_1 e$ moving through a field of particles with mass m_2 and charge $Z_2 e$. The integral (7.18) can be evaluated exactly for $\alpha = 1$; one finds

$$\sin \tfrac{1}{2}\chi = [1 + (\tilde{m} g^2 b / Z_1 Z_2 e^2)^2]^{-1/2} \tag{10.4}$$

or

$$\tan \tfrac{1}{2}\chi = Z_1 Z_2 e^2 / \tilde{m} g^2 b \tag{10.5}$$

where, as in §7, g is the relative speed of the collision partners, b the impact parameter, and $\tilde{m}$ the reduced mass defined in (7.7b). Suppose we consider electrons and protons so that $Z = \pm 1$, $\tilde{m} \approx m_e$, and $g^2 \approx (3kT/m_e)$, and choose b to be the interparticle spacing $d_0 \sim 3 \times 10^{-6}$ cm quoted in §6. Then, for $T = 10^4$ K, we find that $\chi \approx 4 \times 10^{-2}$ radians. Therefore, a typical encounter results in a small deflection and we can make the approximation

$$\chi(g, b) \approx 2 Z_1 Z_2 e^2 / \tilde{m} g^2 b. \tag{10.6}$$

Suppose now that the test particle is an electron and the field particles are ions of charge Z_i; then because $m_e \ll m_i$ we can ignore the motions of the field particles, and can also set $\tilde{m} = m_e$. If the density of field particles is n_i, then the number of collisions suffered by the test particle in a time t with field particles having impact parameters on the range $(b, b + db)$ is $2\pi n_i g t b\, db$; hence using (10.6) we have

$$\begin{aligned} \sum \chi^2 &\approx 2\pi n_i g t \int_{b_{\min}}^{b_{\max}} \chi^2(g, b) b\, db = (8\pi Z_1^2 Z_2^2 e^4 n_i t / m_e^2 g^3) \int_{b_{\min}}^{b_{\max}} b^{-1}\, db \\ &= (8\pi Z_i^2 e^4 n_i t / m_e^2 g^3) \ln (b_{\max}/b_{\min}). \end{aligned} \tag{10.7}$$

In what follows we shall write $\Lambda \equiv (b_{\max}/b_{\min})$.

CUTOFF PROCEDURE

The integral over impact parameters in (10.7) diverges at both large and small values of b and therefore two cutoffs must be specified. The divergence for small b is spurious and results from our use of (10.6), which allows $\chi \to \infty$ as $b \to 0$, instead of (10.5), which guarantees that χ remains bounded. A reasonable choice for $b_{\min}$ is the value that results in a 90° deflection, that is,

$$b_{\min} = Z_i e^2 / m_e g^2. \tag{10.8}$$

The divergence for large b is physical and arises from the long range of the Coulomb interaction, which implies, as described at the beginning of this section, equal contributions from radial shells at *all* distances. But in a real plasma we must have overall charge neutrality, which means that charges of opposite signs surround each other. Therefore, individual field particles at large distances will be partially shielded by charges of opposite sign and will have a diminished effect on the test particle.

We can describe this shielding quantitatively with a theory developed by P. Debye. The probability that a charged particle will be found in a volume element dV is not just $n\,dV$ (where n is the particle density), but depends also on the electrostatic potential ϕ in dV. For example, if $\phi>0$ at some position, electrons will tend to migrate toward that position while ions will tend to migrate away. We can account for this effect by introducing a Boltzmann factor (cf. §12) which depends on $\psi \equiv (e\phi/kT)$. Then, for electrons, the probability of being located at a point is proportional to

$$\pi_e = \exp(\psi) \approx (1+\psi), \tag{10.9}$$

and for ions

$$\pi_i = \exp(-Z_i\psi) \approx (1-Z_i\psi), \tag{10.10}$$

where we have assumed weak interactions so that $\psi \ll 1$. The average charge density, summing over all ion types, is

$$\rho = e\left(\sum_i n_i Z_i \pi_i - n_e \pi_e\right) = -e\psi\left(n_e + \sum_i Z_i^2 n_i\right). \tag{10.11}$$

Here we have demanded large-scale charge neutrality, which implies that

$$n_e = \sum_i Z_i n_i. \tag{10.12}$$

To calculate the potential around any particular ion we combine (10.11) with *Poisson's equation*

$$\nabla^2\phi = -4\pi\rho, \tag{10.13}$$

and write

$$\nabla^2\phi = (\phi/D^2) \tag{10.14}$$

where

$$D \equiv \left\{kT\Big/\left[4\pi e^2\left(n_e + \sum_i Z_i^2 n_i\right)\right]\right\}^{1/2} \tag{10.15}$$

is the *Debye length.* Solving (10.14) for ϕ one obtains $\phi = r^{-1}(Ae^{-r/D} + Be^{r/D})$. Demanding that (1) $\phi \to 0$ as $r \to \infty$, we set $B \equiv 0$; and (2) $\phi \to Z_i e/r$ (the potential of the ion itself) as $r \to 0$, we set $A = Z_i e$. Thus the potential produced by a shielded ion is

$$\phi(r) = (Z_i e/r)\exp(-r/D), \tag{10.16}$$

which shows that charges are effectively completely screened at distances larger than D. Thus, in (10.7), it is reasonable to set $b_{max} = D$. Numerically,

$$D = 4.8(T/n_e)^{1/2} \text{ cm} \tag{10.17}$$

for a pure hydrogen plasma. For example, in a stellar atmosphere with $T = 10^4$ K, $n_e = 10^{14}$ cm^{-3}, one finds that $D \approx 5 \times 10^{-5}$ cm, which is only about four times r_0, the interparticle spacing.

Collecting results from above we have, for a hydrogen plasma,

$$\Lambda = 3e^{-3}(k^3T^3/8\pi n_e)^{1/2}. \tag{10.18}$$

Here we have set g^2 in b_{min} equal to $\langle U^2 \rangle = 3kT/m_e$. A table of $\ln \Lambda$ for a hydrogen plasma is given in (**S2,** 73); a typical value for a stellar atmosphere is $\ln \Lambda \approx 10$. Furthermore, for a hydrogen plasma, we obtain

$$t_D \approx m_e^2 g^3/(8\pi e^4 n_p \ln \Lambda). \tag{10.19}$$

We can now justify the claim made earlier that the effects of weak collisions dominate those of strong collisions. The characteristic time between strong collisions (i.e., a 90° deflection) is

$$t_s = 1/(\pi b_{min}^2 n_p g) = m_e^2 g^3/(\pi e^4 n_p). \tag{10.20}$$

Therefore,

$$t_s/t_D \approx 8 \ln \Lambda \tag{10.21}$$

which shows that the weak collisions dominate by two orders of magnitude.

PRECISE EXPRESSIONS FOR THE DEFLECTION AND ENERGY-EXCHANGE TIMES

Equation (10.19) is not rigorous because we have not performed the (complicated) integrations over all angular variables, allowing for the motion of the center of mass, correctly. Exact calculations by Chandrasekhar (**C2,** Chap. 2) show that for a test particle of charge Z_1 and mass m_1, moving with a speed U_1 relative to the velocity centroid of a uniform field of particles having charge Z_2, mass m_2, and an isotropic Maxwellian velocity distribution with dispersion $\langle U_2^2 \rangle = (3kT/m_2)$, the deflection time is

$$t_D = m_1^2 U_1^3/[8\pi e^4 Z_1^2 Z_2^2 n_2 H(x_0) \ln \Lambda] \tag{10.22}$$

and the energy-exchange time is

$$t_E = m_1^2 U_1^3/[32\pi e^4 Z_1^2 Z_2^2 n_2 G(x_0) \ln \Lambda] \tag{10.23}$$

where

$$x_0 \equiv (3/2\langle U_2^2 \rangle)^{1/2} U_1 = (m_2/2kT)^{1/2} U_1. \tag{10.24}$$

The functions $G(x_0)$ and $H(x_0)$ depend on the error function and its derivative; numerical values for $G(x_0)$, $H(x_0)$, and the ratio

$$(t_D/t_E) = 4G(x_0)/H(x_0) \tag{10.25}$$

are listed in Table 10.1. An important point to notice from the table is that

Table 10.1. The functions $G(x_0)$, $H(x_0)$, and the Ratio of Relaxation Times (t_D/t_E)

x_0	$G(x_0)$	$H(x_0)$	(t_D/t_E)	x_0	$G(x_0)$	$H(x_0)$	(t_D/t_E)
0.0	0.0	0.0	2.00	1.5	0.175	0.791	0.89
0.1	0.037	0.075	1.99	1.6	0.163	0.813	0.80
0.2	0.073	0.149	1.98	1.7	0.152	0.832	0.73
0.3	0.107	0.221	1.93	1.8	0.140	0.849	0.66
0.4	0.137	0.292	1.88	1.9	0.129	0.863	0.60
0.5	0.162	0.358	1.81	2.0	0.119	0.876	0.54
0.6	0.183	0.421	1.74	2.5	0.080	0.920	0.35
0.7	0.198	0.480	1.65	3.0	0.056	0.944	0.24
0.8	0.208	0.534	1.56	3.5	0.041	0.959	0.17
0.9	0.213	0.584	1.46	4.0	0.031	0.969	0.13
1.0	0.214	0.629	1.36	5.0	0.020	0.980	0.082
1.1	0.211	0.669	1.26	6.0	0.014	0.986	0.057
1.2	0.205	0.706	1.16	7.0	0.010	0.990	0.041
1.3	0.196	0.738	1.06	8.0	0.008	0.992	0.032
1.4	0.186	0.766	0.97	10.0	0.005	0.995	0.020

t_E is a minimum for $x_0 = 1$, and becomes much larger for particles moving at speeds much above or much below the average speed of the field particles.

THE SELF-COLLISION TIME AND THE EXCHANGE TIME BETWEEN SPECIES

A case of special interest is a group of particles interacting with themselves, for which Spitzer (**S2**) has defined the extremely useful *self-collision time.* For particles having a speed U_1 equal to the rms speed $\langle U_2^2\rangle^{1/2}$ of their collision partners, $x_0 = (1.5)^{1/2} = 1.225$, and from Table 10.1 we see that $t_D/t_E = 1.14$. Therefore in this case t_D is about equal to t_E and thus provides a good estimate both of the time required to isotropize the velocity distribution and of the time needed for the distribution over energy to approach a Maxwellian. Spitzer denotes this special value of t_D as

$$t_c = m^{1/2}(3kT)^{3/2}/[(8\pi e^4 Z^4 n)\times 0.714 \ln \Lambda] = 11.4(AT^3)^{1/2}/(nZ^4 \ln \Lambda) \tag{10.26}$$

where A is the atomic weight of the particles in units of the hydrogen mass. A detailed numerical solution of the Fokker–Planck equation shows (**M1**) that in the neighborhood of the average energy the time required for relaxation of a distribution that initially is strongly non-Maxwellian is quite close to t_c as predicted from (10.26). However, the time needed for the distribution to become Maxwellian for energies ranging from zero to several times the average energy is about $10t_c$. Thus t_c provides a semi-quantitative estimate only.

From (10.26) it is immediately obvious that $t_c(\text{protons})/t_c(\text{electrons}) = (m_p/m_e)^{1/2} \approx 43$. The energy-exchange time between electrons and protons follows from (10.23) by choosing $U_1 = (3kT/m_e)^{1/2}$, $m_1 = m_e$, $m_2 = m_p$, $x_0 = (3m_p/2m_e)^{1/2}$, and using the asymptotic formula $G(x_0) \to \frac{1}{2}x_0^2$ for $x_0 \gg 1$. One then finds that the electron–proton energy-exchange time is a factor $(m_p/m_e)^{1/2}$ greater than $t_c(\text{protons})$. Thus

$$t_c(e\text{-}e):t_c(p\text{-}p):t_E(e\text{-}p) = 1:(m_p/m_e)^{1/2}:(m_p/m_e). \tag{10.27}$$

For a stellar atmosphere with $T = 10^4$ K and $n_e = n_p = 10^{14}\ \text{cm}^{-3}$ one finds from (10.26) and (10.27) that $t_E(e\text{-}p) \approx 10^{-6}$ s. This time may be compared with a characteristic flow time $t_f \sim (l/v)$; in a stellar atmosphere we can take the characteristic length l equal to a scale height $H \approx 100$ km, and the characteristic velocity v equal to the sound speed $\approx 10\ \text{km s}^{-1}$, so that $t_f \sim 10\ \text{s} \sim 10^7 t_E(e\text{-}p)$. Thus, in the absence of any other processes perturbing equilibrium, we conclude that the velocity distribution of all particles in the atmosphere must be quite precisely a Maxwellian at a single kinetic temperature. We shall reexamine this conclusion in §84 where we allow for the effects of recombinations and inelastic collisions between electrons, ions, and atoms.

1.3 Classical Statistical Mechanics

11. Thermodynamic Probability and Entropy

In classical statistical mechanics, as in kinetic theory, we again deal with an atomic model of the gas (which can now include internal structure of the particles). However, we do not now concern ourselves with the detailed mechanics of collisions among the particles, but instead appeal to certain powerful statistical principles. As before, we make no attempt to follow the time history of particles, but consider instead only the distribution of a large number of particles in the six-dimensional phase space (x, y, z, u_x, u_y, u_z). For the present we assume all the particles are identical and ignore internal structure; we shall relax these restrictions later.

The macroscopic (i.e., thermodynamic) state of the system is completely determined when we specify the total number $\mathcal{N}$ of particles in the system, the volume V they occupy, and their total internal energy $\mathscr{E}$. Associated with this single *macrostate* there will, in general, be an enormous number of *microstates* (i.e., arrangements of particles within phase space), all of which yield the same macroscopic properties. We define the *thermodynamic probability* W to be directly proportional to the number of distinct microstates by which a given macroscopic state can be realized consistent with the constraints implied by specified values of $\mathcal{N}$, V, and $\mathscr{E}$. A fundamental assumption of statistical mechanics is that all microstates are equally probable and that, as a result of collisions, the system continuously evolves from one microstate to another. It then follows that the macrostate in which we will most probably find the gas is the one that has the largest number of microstates associated with it. As we shall see, there is one particular macrostate for which there are many more microstates than any other, and even small departures from this state are extremely improbable.

A major task of statistical mechanics is the calculation of W. In practice this is done by dividing the available phase volume into elementary *cells*, and then counting the number of ways particles can be distributed into these cells. The details of this procedure will be discussed in §12; here we merely note that the total number of cells, and hence W, is proportional to the total phase volume available.

In principle, statistical mechanics can be constructed without reference to thermodynamics, and a correspondence between the two theories can be made after the fact by an analysis of the relationships among the mathematical expressions that emerge from the development of statistical mechanics. It is simpler, however, to invoke the fundamental connection between thermodynamics and statistical mechanics provided by *Boltzmann's relation*, which states that the entropy S and thermodynamic probability W are related by the expression

$$S = k \ln W \tag{11.1}$$

where k is Boltzmann's constant.

A complete justification of (11.1) would take us too deeply into the foundations of statistical mechanics and thermodynamics. We can, however, render it plausible by means of a few simple examples and gain useful insight in the process. First, we know from thermodynamics that an isolated system tends toward an equilibrium state of maximum entropy [cf. equation (3.10)]. In statistical terms we expect the system to tend to its most probable state. Boltzmann's relation is consistent with these two statements. It is also in this vein that we can associate entropy with the degree of *disorder* in a system. We can consider a system to be completely ordered if all the particles occupy a single cell in phase space, that is, if all are in the same volume element and move with identical velocities, and hence constitute a unique, perfectly structured, initial configuration. In time, as collisions among particles spread them out in phase space, the entropy rises, the thermodynamic probability increases (because more cells in phase space can be occupied), and at the same time the system becomes more disordered. We may therefore consider entropy to be a quantitative measure of the degree of disorder of a system. Second, suppose we consider the free expansion of a gas into a vacuum (the Joule–Kelvin experiment). We know from experiment that T remains constant in this process, and from (4.10) we have the change in entropy $\Delta S \propto \ln (V_f/V_i)$ where i and f denote "initial" and "final". Thus not only is the process accompanied by an increased uncertainty in our knowledge of the system (because each particle has a larger volume available to it) and hence in its degree of disorder, as expected qualitatively, but also, because $W \propto V$ (i.e., the number of available cells), we see that (11.1) is quantitatively just right to describe the process. Third, consider two different gases at the same temperature and pressure, originally separated by a partition in a container. If we remove the partition, T and p remain unaltered, but the gases intermingle, eventually producing a homogeneous mixture through the whole container. Obviously, the degree of disorder of the system has increased. The entropy increase of the system as a whole equals the *sum* of the entropy increases in the two subsystems, $\Delta S = \Delta S_1 + \Delta S_2$. On the other hand, the thermodynamic probability W assigned to the whole system must be proportional to the *product* of the (independent) thermodynamic probabilities of the two subsystems, $W = W_1 \cdot W_2$ (i.e., particles of each gas can independently be distributed through the larger volume). Again one sees that (11.1) provides the correct relation.

Many other examples of the types discussed above can be constructed, but we need not pursue them further because in the final analysis the justification of (11.1) is that it leads to an internally consistent theoretical structure that is in excellent agreement with experiment.

12. Boltzmann Statistics

COUNTING PROCEDURE

Classical statistical mechanics, which is based on *Boltzmann statistics*, can be derived as limiting case of quantum statistics, valid when the number of phase-space cells greatly exceeds the number of particles to be placed in them (i.e., a dilute gas). While this approach affords deep insight [see, e.g., (**H2,** Chaps. 9–12), (**S1,** Chap. 16), (**V1,** Chap. 4)] it is more direct to proceed purely on classical grounds, and to appeal to quantum mechanical concepts only when required.

Our task is now to calculate the number of microstates associated with a given macrostate.

Suppose that we have a fixed number $\mathcal{N}$ of particles within a volume V. We consider these particles to be distributed among cells, each of which has a definite energy ε_i. The number of particles in the ith cell is its *occupation number* ν_i. We agree that all *acceptable* sets of occupation numbers $\{\nu_i\}$ corresponding to the desired macrostate are those $\{\nu_i\}$ such that the total number of particles $\sum \nu_i$ equals the given total number $\mathcal{N}$, and the total energy $\sum \nu_i \varepsilon_i$ equals the given energy $\mathscr{E}$. To obtain the total number of microstates associated with a macrostate we must therefore (1) compute the number of microstates associated with each acceptable set of occupation numbers $\{\nu_i\}$ and then (2) sum over all such sets.

Each energy cell may correspond to several physically different states, all having the same ε_i; such states are called *degenerate.* The number of degenerate states associated with the ith cell is called the *statistical weight* or *degree of degeneracy* g_i of that cell. Having chosen some definite set $\{\nu_i\}$ of occupation numbers that produces the desired macrostate, for purposes of distributing particles we now discard all empty cells ($\nu_i = 0$), and arrange those with nonzero occupations into a consecutive sequence. We view this sequence as a consecutive set of *partitions* or *boxes*, each of which contains g_i *slots* into which particles can be put.

If we were to assume (as is acceptable classically) that the particles are *distinguishable*, then there are $\mathcal{N}!$ distinct ways we can sort $\mathcal{N}$ particles consecutively into the sequence of partitions, no matter how they are distributed over the slots within each partition. In addition, for each of these $\mathcal{N}!$ sortings, any of the ν_i particles in the ith box can be in any one of the g_i slots available in that box, hence there are $(g_i)^{\nu_i}$ possible ways to arrange ν_i within the ith box. However, each of the $\nu_i!$ permutations of the *order* in which the ν_i particles are inserted into a definite set of slots within the ith box are actually identical, because the particles are, in fact, physically identical (even if we assume that they are distinguishable).

Thus the total number of physically distinct microstates associated with the particular set $\{\nu_i\}$ of occupation numbers is

$$W(\{\nu_i\}) = \mathcal{N}! \prod_i (g_i)^{\nu_i} \Big/ \prod_i \nu_i! \tag{12.1}$$

Here the products now extend over *all* cells [including empty cells, for which $\nu_i! = 0! = 1$ and $(g_i)^{\nu_i} = (g_i)^0 = 1$].

Using (12.1) one can develop statistical mechanics along the lines described below, and one arrives finally at an expression for the entropy. This expression correctly predicts an increase in entropy if two different gases, originally in volumes V_1 and V_2, respectively, and at the same temperature and pressure, are allowed to mix throughout the entire volume $V = V_1 + V_2$. Unfortunately, it also predicts an entropy increase if the two gases are identical [see (**H1**, §7.6)], which is absurd. This catastrophic result, known as the *Gibbs paradox*, would imply that the entropy of a single homogeneous gas depends on its history (i.e., how many stages of "mixing" have occurred) instead of being a function of the thermodynamic state variables alone. Gibbs showed that the paradox is resolved if we postulate that in (12.1) we have made a counting error of a factor of $\mathcal{N}!$, and that the thermodynamic probability given by a "correct Boltzmann counting" is really

$$W\{\nu_i\} = \prod_i (g_i)^{\nu_i} \Big/ \prod_i \nu_i! \tag{12.2}$$

Gibbs's ad hoc correction is vindicated by the realization that quantum mechanically the particles are actually indistinguishable; in fact (12.2) is recovered directly from quantum statistics in the limit described at the beginning of this section. We therefore use (12.2) henceforth.

THE EQUILIBRIUM DISTRIBUTION FUNCTION

Summing over all acceptable sets of occupation numbers, we find that the thermodynamic probability associated with a definite macrostate is

$$W = \sum W\{\nu_i\}. \tag{12.3}$$

As described above, the sum extends over all sets of occupation numbers $\{\nu_i\}$ for which

$$\sum_i \nu_i = \mathcal{N} \tag{12.4}$$

and

$$\sum_i \nu_i \varepsilon_i = \mathscr{E} \tag{12.5}$$

where $\mathscr{E}$ is the total energy of the system. Ideally we should evaluate the sum (12.3), but this turns out to be algebraically quite difficult. However for large $\mathcal{N}$ it is possible to obtain the same results as are given by a rigorous analysis (**F1**) based on (12.3) by means of the following approximation: we assume that there is a single term in the sum, W_{max}, which dominates over all others, and we then analyze only this one partitioning of particles. We shall see below that this assumption is actually justified. For expository simplicity in what follows we shall refer to W_{max} simply as W.

From (12.2) we have

$$\begin{aligned} \ln W &= \sum_i (\nu_i \ln g_i - \ln \nu_i!) \approx \sum_i [\nu_i - \nu_i \ln (\nu_i/g_i)] \\ &= \mathcal{N} - \sum_i \nu_i \ln (\nu_i/g_i), \end{aligned} \tag{12.6}$$

where we have made use of the dominant terms in *Stirling's formula*

$$\ln x! \approx \tfrac{1}{2} \ln 2\pi + \tfrac{1}{2} \ln x + x \ln x - x, \tag{12.7}$$

which is valid for $x \gg 1$, and we also have used the constraint (12.4). We now assume that as the system evolves, the ν_i vary as a result of collisional shuffling of particles, and approach a distribution for which W is maximum. Once this state is attained, the first variation δW resulting from small fluctuations $\delta\nu_i$ in individual occupation numbers will be zero. Thus when W has its maximum value

$$\delta W = \delta\mathcal{N} - \sum_i \delta\nu_i - \sum_i \delta\nu_i \ln (\nu_i/g_i) = -\sum_i \delta\nu_i \ln (\nu_i/g_i) = 0, \tag{12.8}$$

where we have noted from (12.4) that

$$\delta\mathcal{N} = \sum_i \delta\nu_i = 0. \tag{12.9}$$

The variations $\delta\nu_i$ cannot be arbitrary, but must satisfy both (12.9) and the constraint, implied by (12.5), that

$$\delta\mathscr{E} = \sum_i \varepsilon_i \, \delta\nu_i = 0. \tag{12.10}$$

The standard method for solving a variational problem subject to constraints is to use the *method of Lagrange multipliers* [see, e.g., (**M2**, Appendix VI)] in which one considers the variation of a linear combination (with as-yet-undetermined coefficients) of the original equation and the constraint equations. That is, in the equation

$$\sum_i \delta\nu_i[\ln(\nu_i/g_i) - \ln\alpha + \beta\varepsilon_i] = 0 \tag{12.11}$$

we can now consider the $\delta\nu_i$'s to be, in effect, independent; the form of the coefficients in (12.11) is chosen for later convenience. The only solution of (12.11) for arbitrary $\delta\nu_i$ is

$$\nu_i = \alpha g_i e^{-\beta\varepsilon_i}, \tag{12.12}$$

and this is the distribution of occupation numbers that maximizes W.

Summing (12.12) over all i we obtain

$$\mathcal{N} = \alpha \sum_i g_i e^{-\beta\varepsilon_i} = \alpha Z \tag{12.13}$$

where

$$Z \equiv \sum_i g_i e^{-\beta\varepsilon_i} \tag{12.14}$$

is the *partition function* associated with the distribution. Combining (12.12) to (12.14) we thus have

$$(\nu_i/\mathcal{N}) = (g_i e^{-\beta\varepsilon_i})/Z \tag{12.15}$$

and hence from (12.5)

$$\mathscr{E} = (\mathcal{N}/Z) \sum_i \varepsilon_i g_i^{-\beta\varepsilon_i}. \tag{12.16}$$

Using (12.15) and (12.16) in (12.6) we find

$$\ln W = \mathcal{N}[1 + \ln(Z/\mathcal{N})] + \beta\mathscr{E} \tag{12.17}$$

whence

$$S = \mathcal{N}k[1 + \ln(Z/\mathcal{N})] + \beta k\mathscr{E}. \tag{12.18}$$

If we write $dS = (\partial S/\partial\mathscr{E})_v \, d\mathscr{E} + (\partial S/\partial V)_{\mathscr{E}} \, dV$, then we see from the first law of thermodynamics that

$$(\partial S/\partial\mathscr{E})_v = 1/T. \tag{12.19}$$

Now differentiating (12.18) with respect to $\mathscr{E}$ we find

$$(\partial S/\partial\mathscr{E})_v = k[\mathscr{E} + (\mathcal{N}/Z)(\partial Z/\partial\beta)](\partial\beta/\partial\mathscr{E}) + \beta k. \tag{12.20}$$

But from (12.14) and (12.16)

$$(\partial Z/\partial\beta) = -\sum_i \varepsilon_i g_i e^{-\beta\varepsilon_i} = -(Z/\mathcal{N})\mathscr{E}, \tag{12.21}$$

hence (12.20) reduces to

$$(\partial S/\partial\mathscr{E})_V = \beta k, \tag{12.22}$$

and therefore from (12.19) we have finally

$$\beta = 1/kT. \tag{12.23}$$

Having evaluated β we can now rewrite the above results in the usual forms

$$(\nu_i/\mathcal{N}) = (n_i/N) = [g_i \exp(-\varepsilon_i/kT)]/Z, \tag{12.24}$$

$$Z = \sum_i g_i \exp(-\varepsilon_i/kT), \tag{12.25}$$

$$\mathscr{E} = (\mathcal{N}/Z)\sum_i \varepsilon_i g_i \exp(-\varepsilon_i/kT), \tag{12.26}$$

$$S = \mathcal{N}k[1+\ln(Z/\mathcal{N})]+(\mathscr{E}/T), \tag{12.27}$$

and

$$(\nu_m/\nu_l) = (n_m/n_l) = (g_m/g_l)\exp[-(\varepsilon_m-\varepsilon_l)/kT]. \tag{12.28}$$

Equation (12.28) is known as the *Boltzmann excitation formula*. In (12.24) and (12.28) the quantities n and N denote, as before, particle *densities* per cm^3.

FLUCTUATIONS AROUND EQUILIBRIUM

Having determined the equilibrium distribution function, we are now in a position to study fluctuations around equilibrium. First, let us calculate the probability of any other distribution of occupation numbers relative to the probability of the equilibrium state. Let $\nu_i = \nu_i^0 + \Delta\nu_i$ where ν_i^0 denotes the equilibrium populations given by (12.24), and where the $\Delta\nu_i$ are fluctuations subject to the constraint (12.9). For simplicity we consider each elementary state separately, that is, $g_i \equiv 1$. Assuming $|\Delta\nu_i/\nu_i^0| \ll 1$, we have, from (12.6),

$$\begin{aligned}\ln W &= \mathcal{N} - \sum_i (\nu_i^0+\Delta\nu_i)\ln(\nu_i^0+\Delta\nu_i)\\ &\approx \mathcal{N} - \sum_i (\nu_i^0+\Delta\nu_i)[\ln\nu_i^0 + (\Delta\nu_i/\nu_i^0) - \tfrac{1}{2}(\Delta\nu_i/\nu_i^0)^2]\\ &= \ln W^0 - \tfrac{1}{2}\sum_i (\Delta\nu_i)^2/\nu_i^0 - \sum_i \Delta\nu_i - \sum_i \Delta\nu_i \ln\nu_i^0.\end{aligned} \tag{12.29}$$

The last two terms are zero by virtue of (12.8) and (12.9), so we find

$$\ln(W/W^0) \approx -\tfrac{1}{2}\sum_i (\Delta\nu_i)^2/\nu_i^0 \tag{12.30}$$

which explicitly shows that the extremum W^0 is in fact a maximum.

Furthermore, the maximum is very sharp. To illustrate, consider a volume of 1 cm^3 containing $\mathcal{N}=10^{16}$ particles, and take a very small fluctuation, say $|\Delta\nu_i|/\nu_i^0=10^{-6}$ for all i. Then $\ln(W/W^0)=-5\times10^3$, or $W/W_0\approx10^{-2000}$, which is a very small number indeed! Thus even an extremely small departure from the equilibrium distribution implies an enormous reduction in the thermodynamic probability, and this implies that we will almost never observe a state that differs even slightly from the equilibrium state (unless, of course, the system is *driven* to that non-equilibrium state).

Next, let us inquire how large the fluctuations in the total energy of a system are likely to be. Consider a sequence of microstates $\{\nu_i^{(\lambda)}\}$, each of which satisfies (12.4) and (12.5). Let W_0 be the thermodynamic probability of the corresponding macrostate; from Boltzmann's relation $W_0=\exp(S/k)$, therefore for any one of these microstates

$$\exp[(S/k)-\beta\mathscr{E}]=W_0\exp\left[-\beta\sum_i \nu_i^{(\lambda)}\varepsilon_i\right]=\sum_\lambda \exp\left[-\beta\sum_i \nu_i^{(\lambda)}\varepsilon_i\right] \tag{12.31}$$

where the last equality takes advantage of the fact that all microstates have the same energy $\mathscr{E}$. Rewrite (12.31) as

$$W=\exp(S/k)=\sum_\lambda \exp\left[-\beta\sum_i \nu_i^{(\lambda)}\varepsilon_i+\beta\mathscr{E}\right]=\sum_\lambda \exp[-\beta(H_\lambda-\mathscr{E})]. \tag{12.32}$$

Now relax the requirement (12.5) and suppose that the energy H_α of the αth microstate differs slightly from $\mathscr{E}$. Interpreting the right-hand side of (12.32) as the sum of probabilities contributed to the total by the individual microstates λ, we conclude that the probability that the system has a total energy H_λ is proportional to $\exp(-\beta H_\lambda)$. We therefore can calculate the average energy of a group of macrostates with energies clustered around $\mathscr{E}$ as

$$\mathscr{E}=\langle H\rangle\equiv\sum_\lambda H_\lambda e^{-\beta H_\lambda}\Big/\sum_\lambda e^{-\beta H_\lambda}. \tag{12.33}$$

Differentiating (12.33) with respect to β we find

$$\frac{\partial\mathscr{E}}{\partial\beta}=\frac{\left(\sum_\lambda H_\lambda e^{-\beta H_\lambda}\right)^2}{\left(\sum_\lambda e^{-\beta H_\lambda}\right)^2}-\frac{\sum_\lambda H_\lambda^2 e^{-\beta H_\lambda}}{\sum_\lambda e^{-\beta H_\lambda}}=\mathscr{E}^2-\langle H^2\rangle. \tag{12.34}$$

Now

$$\langle H^2\rangle=\langle(\mathscr{E}+\Delta\mathscr{E})^2\rangle=\mathscr{E}^2+2\mathscr{E}\langle\Delta\mathscr{E}\rangle+\langle(\Delta\mathscr{E})^2\rangle=\mathscr{E}^2+\langle(\Delta\mathscr{E})^2\rangle \tag{12.35}$$

where we noted that $\langle \Delta \mathscr{E} \rangle = 0$ because $\langle H \rangle = \mathscr{E}$. Combining (12.34) and (12.35) we have

$$\langle (\Delta \mathscr{E})^2 \rangle = -(\partial \mathscr{E}/\partial \beta) = kT^2(\partial \mathscr{E}/\partial T) = kT^2 C_v, \tag{12.36}$$

where C_v is the heat capacity at constant volume. But $\mathscr{E} \propto \mathcal{N}$ and $C_v \propto \mathcal{N}$, hence

$$\langle (\Delta \mathscr{E})^2 \rangle^{1/2} / \mathscr{E} \propto \mathcal{N}^{-1/2}. \tag{12.37}$$

Thus as $\mathcal{N} \to \infty$, almost all realizations of the physical system will have an energy very nearly equal to $\mathscr{E}$. In fact it can be shown (**H2**, 159–160) that the distribution in energy around $\mathscr{E}$ is a Gaussian with a halfwidth

$$(\Delta \mathscr{E})_{1/2} = (2kT^2 C_v)^{1/2}. \tag{12.38}$$

Again $(\Delta \mathscr{E})_{1/2}/\mathscr{E} \propto \mathcal{N}^{-1/2}$, so as $\mathcal{N} \to \infty$, the distribution function approaches a δ-function centered on $\mathscr{E}$.

Finally, we return to the argument advanced earlier that we can replace the sum in (12.3) by a single term W_{max}. It can be shown rigorously that $\ln W$ as defined by (12.3) can differ from $\ln W_{max}$ only by terms of order $\ln \mathcal{N}$ [see e.g., (**C1**, 370–375), (**D1**, 302–308), (**H2**, §7.2), or (**W1**, 90–94)]. But $\ln W_{max} \propto \mathcal{N}$, hence

$$\ln (W/W_{max}) = 1 + O[(\ln \mathcal{N})/\mathcal{N}], \tag{12.39}$$

which shows that as $\mathcal{N} \to \infty$, W_{max} provides an extremely good estimate of W.

RELATION TO THERMODYNAMICS

We can derive all thermodynamic properties of a gas from the distribution function (12.24). A particularly compact set of formulae can be obtained in terms of the partition function. Thus using (12.25) and (12.26) we have

$$\mathscr{E} = \mathcal{N}kT^2(\partial \ln Z/\partial T)_v. \tag{12.40}$$

Therefore (12.27) can be rewritten as

$$S = \mathcal{N}k[1 + \ln (Z/\mathcal{N}) + T(\partial \ln Z/\partial T)_v], \tag{12.41a}$$

whence

$$s = R[1 + \ln (Z/\mathcal{N}) + T(\partial \ln Z/\partial T)_v]. \tag{12.41b}$$

Furthermore, by expanding dS and $d\mathscr{E}$ in the first law of thermodynamics in terms of dT and dV, one easily finds that

$$p = T(\partial S/\partial V)_T - (\partial \mathscr{E}/\partial V)_T \tag{12.42}$$

and hence, from (12.40) and (12.41a),

$$p = \mathcal{N}kT(\partial \ln Z/\partial V)_T. \tag{12.43}$$

And so on for other macroscopic variables. Thus once we know $Z(T, V)$ we can derive explicit expressions for all state variables.

THERMODYNAMIC PROPERTIES OF A PERFECT GAS

To illustrate the results derived above, let us consider a gas of $\mathcal{N}$ structureless point particles in equilibrium in a volume V. The only energy associated with such particles is their translational kinetic energy $\varepsilon = \frac{1}{2}mu^2$. Consider those particles within the phase-space volume element $dV\,d^3u = dx\,dy\,dz\,du_x\,du_y\,du_z$ centered on a position $\mathbf{x}$ and velocity $\mathbf{u}$. The distribution function f for these particles can be derived from (12.24) if we identify ν_i with $d^6\mathcal{N} = f\,dV\,d^3u$, the number of particles in the phase-space element; write $\varepsilon_i = \frac{1}{2}mu^2$; and accept from quantum statistics the *Pauli exclusion principle*, which implies that the number of states available within the phase-space element is

$$g_i = (m/h)^3\,dx\,dy\,dz\,du_x\,du_y\,du_z = (4\pi m^3/h^3)\,dx\,dy\,dz\,u^2\,du. \quad (12.44)$$

Thus

$$d^6\mathcal{N} = (4\pi m^3\mathcal{N}/h^3Z)\exp(-mu^2/2kT)\,dx\,dy\,dz\,u^2\,du. \quad (12.45)$$

Integrating over V we have the velocity distribution function

$$d^3\mathcal{N} = (4\pi m^3\mathcal{N}V/h^3Z)\exp(-mu^2/2kT)u^2\,du. \quad (12.46)$$

Integrating over all velocities we have

$$\mathcal{N} = (m^3\mathcal{N}V/h^3Z)(2\pi kT/m)^{3/2}, \quad (12.47)$$

whence we obtain the partition function for translational motion

$$Z_{\text{trans}} = (2\pi mkT/h^2)^{3/2}V. \quad (12.48)$$

Using (12.48) in (12.45) we see that the distribution function is none other than the Maxwellian distribution

$$f(u)\,d^3u\,dV = N(m/2\pi kT)^{3/2}\exp(-mu^2/2kT)\,d^3u\,dV \quad (12.49)$$

as expected; here $N \equiv \mathcal{N}/V$ is the particle density per cm^3.

Given the partition function (12.48) we can calculate all the thermodynamic properties of the gas. Thus from (12.43)

$$p = \mathcal{N}kT/V = NkT, \quad (12.50)$$

which is identical to (1.2). Similarly, using (12.40), we have

$$\mathscr{E}_{\text{trans}} = \tfrac{3}{2}\mathcal{N}kT = (\tfrac{3}{2}NkT)V \quad (12.51)$$

or

$$e_{\text{trans}} = \tfrac{3}{2}RT, \quad (12.52)$$

in complete agreement with the results of kinetic theory. Then from (2.11) we recover $c_v = \frac{3}{2}R$, and from (4.5) $c_p = \frac{5}{2}R$, hence $\gamma = \frac{5}{3}$, again in agreement with kinetic theory. Finally, from (12.48) and (12.41), we have

$$S_{\text{trans}} = \mathcal{N}k[\tfrac{5}{2} + \ln(V/\mathcal{N}) + \ln(2\pi mkT/h^2)^{3/2}], \quad (12.53)$$

which is the *Sackur–Tetrode equation* for the entropy of translational motions. Dividing (12.53) by $\mathcal{N}m$ and using the perfect gas law we find for the specific entropy

$$s_{\text{trans}} = \tfrac{5}{2}R \ln T - R \ln p + R[\tfrac{5}{2} + \tfrac{3}{2} \ln (2\pi m/h^2) + \tfrac{5}{2} \ln k]. \quad (12.54)$$

Comparing (12.54) with (8.23) and (4.11) we see that none of the coefficients in the equation were determined by thermodynamics; with kinetic theory we can evaluate explicitly only the coefficients of ln T and ln p; but with statistical mechanics we obtain explicit results for all coefficients including the additive constant.

THERMODYNAMIC PROPERTIES OF A GAS WITH INTERNAL STRUCTURE

Consider now a gas composed of atoms that have a sequence of bound eigenstates with statistical weights g_i, lying at energies ε_i above the ground state, to which the atom's electrons can be excited. Let ε'_j be the translational energy of an atom whose velocity lies within phase-space cell j, which has statistical weight g'_j. Then the total energy of an atom excited to state i and moving in phase-space cell j is $\varepsilon = \varepsilon_i + \varepsilon'_j$, the combined statistical weight associated with this condition is $g = g_i g'_j$, and the partition function is

$$Z = \sum_i \sum_j g_i g'_j \exp[-(\varepsilon_i + \varepsilon'_j)/kT]. \quad (12.55)$$

But the distributions over translational and internal states are statistically independent of each other because atoms in a particular excited state can move at any velocity, and indeed in equilibrium must necessarily have the same Maxwellian velocity distribution as atoms in all other excitation states. For each term in the sum over i there will therefore be a complete sum over j, hence

$$Z = \sum_j g'_j e^{-\varepsilon'_j/kT} \sum_i g_i e^{-\varepsilon_i/kT} = Z_{\text{trans}} Z_{\text{elec}}. \quad (12.56)$$

Thus the partition function can be factorized.

For dilute gases, the independence of translational and internal energies is essentially perfect, and the factorization (12.56) is quite accurate. If the particles in the gas are molecules instead of atoms then there are additional energy states produced by molecular vibration and rotation, and the complete partition function becomes $Z = Z_{\text{trans}} Z_{\text{rot}} Z_{\text{vib}} Z_{\text{elec}}$. In this case the factorization is not as accurate, however, because rapid rotation of a molecule distorts its shape and changes its vibrational potential well, while vibration changes a molecule's moments of inertia and hence affects its rotation; thus vibrational and rotational modes are explicitly coupled, contrary to the assumption used to derive (12.56). We do not consider molecules further in what follows.

The translational part of (12.56) is again given by (12.48), while the electronic excitation part is

$$Z_{\text{elec}} = \sum_i g_i e^{-\varepsilon_{i0}/kT} \tag{12.57}$$

where the subscript zero indicates that ε is measured relative to the ground state. Formally this sum extends over an infinite number of eigenstates, and because ε approaches a limiting value ε_{I0}, the ionization potential, the individual terms remain finite and therefore (12.57) diverges. This divergence is unphysical, however, because for very large quantum numbers the eigenstates fill so large a volume that they overlap adjacent atoms in the gas; these states are so strongly perturbed that they are effectively destroyed. The sum in (12.57) can therefore be truncated after a finite number of terms [cf. (**M5**, 111 and 295)].

Because Z_{elec} is independent of V it contributes nothing to $(\partial \ln Z/\partial V)_T$, hence application of (12.43) again yields the perfect gas law. The pressure in a gas is thus independent of its internal excitation, and reflects only "external" translational motion, as would be expected from its basic definition in terms of rate of momentum transport per unit area (cf. §30). Using (12.40) to calculate the internal energy we have

$$\begin{aligned} \mathscr{E} &= \tfrac{3}{2}\mathscr{N}kT + (\mathscr{N}/Z_{\text{elec}}) \sum_i \varepsilon_{i0} g_i e^{-\beta\varepsilon_{i0}} \\ &= \tfrac{3}{2}\mathscr{N}kT + \sum_i \nu_i \varepsilon_{i0} = \mathscr{E}_{\text{trans}} + \mathscr{E}_{\text{elec}}, \end{aligned} \tag{12.58}$$

where in the second term ν_i denotes the total number of atoms in eigenstate i without regard to velocity, and, as before, $\beta = 1/kT$. The internal energy per particle is

$$\tilde{e} = \tfrac{3}{2}kT + \langle \varepsilon_{i0} \rangle, \tag{12.59}$$

where $\langle \varepsilon_{i0} \rangle \equiv \sum (\nu_i/\mathscr{N})\varepsilon_{i0}$; per gram $e = \tilde{e}/m$ where m is the mass per particle.

From (12.59) we immediately find the specific heat at constant volume:

$$c_v = \tfrac{3}{2}R + R\beta^2(\langle \varepsilon_{i0}^2 \rangle - \langle \varepsilon_{i0} \rangle^2). \tag{12.60}$$

For example, suppose the atom has only a ground state and one excited state; (12.60) then becomes

$$c_v = \tfrac{3}{2}R + R[G(\beta\varepsilon_{10})^2 e^{-\beta\varepsilon_{10}}/(1 + Ge^{-\beta\varepsilon_{10}})^2] = (c_v)_{\text{trans}} + (c_v)_{\text{elec}}, \tag{12.61}$$

where $G \equiv (g_1/g_0)$. Notice that c_v now varies and depends explicitly on T. One sees that $(c_v)_{\text{elec}}$ vanishes both as $T \to 0$ $(\beta \to \infty)$ and as $T \to \infty$ $(\beta \to 0)$, that is, when the gas is unexcited or when the excitation has "saturated" so that $(\nu_i/\nu_0) = (n_1/n_0) = (g_1/g_0)$; in both these limits heat added to the system goes directly into translational motions, and c_v reduces to the value appropriate to a gas composed of structureless particles.

Calculations using (12.61) show that c_v has a strong maximum (whose amplitude depends on the ratio G) at the temperature at which the gas first becomes appreciably excited, typically near $\beta\varepsilon_{10} \sim 2$; for $G=1$, $(c_v)_{\text{elec}}$ reaches a maximum of about $0.45R$. Of course, the behavior of c_v for a real gas having several excitation states is more complex and in general must be computed numerically.

Because the pressure, as we saw above, is the same as for a gas of structureless particles, (2.34) again leads to $c_p = c_v + R$ where c_v is now given by (12.61). Clearly c_p is also a function of temperature, and the ratio $\gamma \equiv (c_p/c_v)$ is not constant. In this case we can no longer write equations (4.13) to (4.15) but must instead develop more general relations, as discussed in §14.

Finally, for the specific entropy we have, from (12.41b),

$$\begin{aligned} s &= s_{\text{trans}} + R[\ln Z_{\text{elec}} + T(\partial \ln Z_{\text{elec}}/\partial T)_v] \\ &= s_{\text{trans}} + R\left[\ln\left(\sum_i g_i e^{-\beta\varepsilon_{i0}}\right) + (\langle\varepsilon_{i0}\rangle/kT)\right] \\ &= s_{\text{trans}} + s_{\text{elec}}, \end{aligned} \tag{12.62}$$

where s_{trans} is given by (12.54). As was true for e and c_v, a major contribution to s comes from the internal excitation term s_{elec}.

INTERPRETATION OF THE BOLTZMANN *H*-THEOREM

The results of statistical mechanics afford us the means by which we can make a physical interpretation of Boltzmann's H-theorem. Again consider an isolated, homogeneous sample of a gas composed of structureless particles, with no external forces applied. An expression for the entropy in terms of the distribution function f can be derived directly from (12.6) by (1) writing $\nu_i = f(\mathbf{u}_i)\, d^3u\, dV$, (2) using (12.44) for g_i, and (3) replacing the sum by integrations over $d^3u\, dV$. We find

$$S = k \ln W = k\mathcal{N} - kV \int \ln (h^3 f/m^3) f\, d^3u. \tag{12.63}$$

Differentiating (12.63) with respect to time, and recognizing that $(\partial f/\partial t)$ under the restrictions stated above is $(Df/Dt)_{\text{coll}}$, we obtain

$$\begin{aligned} (dS/dt) &= -kV \int [1 + \ln (h^3 f/m^3)](Df/Dt)_{\text{coll}}\, d^3u \\ &= -kV \iiint [1 + \ln (h^3 f/m^3)](f_1' f_2' - f_1 f_2) g\sigma(\mathbf{\Omega})\, d\Omega\, d^3u_1\, d^3u_2. \end{aligned} \tag{12.64}$$

and hence that $(dS/dt) \geq 0$, the equality holding in equilibrium. Thus the physical content of Boltzmann's H-theorem is that the entropy of an isolated system continually increases to its equilibrium value (at which point the distribution function is Maxwellian), in agreement with the conclusions reached in §3 from thermodynamic considerations.

13. *Ionization*

In the process of *ionization* a bound electron is removed from one of the discrete eigenstates of an atom into a *continuum* of levels in which it is unbound and has a finite kinetic energy at infinite distance from the atom whence it came. This continuum begins at an energy ε_{I0}, the ionization potential of the atom above the ground state. Every species of atom has a sequence of ionization stages, each with a progressively higher ionization potential as successive electrons are removed, leading ultimately to a completely stripped nucleus. The relative number of atoms and ions in successive ionization stages can be computed using the *Saha ionization formula*, which can be derived as an extension of the Boltzmann excitation formula to the continuum.

Consider a process in which an atom in its ground level is ionized, yielding an ion in its ground level plus a free electron moving with speed u in the continuum. The total energy of the final state of the system relative to its initial state is $\varepsilon = \varepsilon_{I0} + \frac{1}{2}mu^2$ where m is the mass of an electron. Let the statistical weight of the atom be $g_{0,0}$ and of the ion $g_{0,1}$ where the first subscript denotes the atomic level (in this case the ground level) and the second indicates the ionization stage. Using (12.44) we assign a statistical weight

$$g_{\text{electron}} = 2(m/h)^3\, dV(4\pi u^2)\, du \tag{13.1}$$

to an electron in a volume element dV moving with a speed on the range $(u, u+du)$; the factor of 2 has a quantum-mechanical origin and accounts for the two possible *spin* states of the free electron. The statistical weight of the combined electron + ion system is $g = g_{0,1}g_{\text{electron}}$.

Integrate over a volume element large enough to contain exactly one electron (i.e., we consider a unit process), so that we can replace dV by $(1/n_e)$ where n_e is the number density of free electrons. We can then apply (12.28) to the final system with electrons having velocities $(u, u+du)$ to obtain

$$[n_{0,1}(u)/n_{0,0}] = 8\pi(m/h)^3(g_{0,1}/g_{0,0})(1/n_e)\exp[-(\varepsilon_{I0}+\tfrac{1}{2}mu^2)/kT]u^2\, du. \tag{13.2}$$

Then summing over all possible final states by integrating over all electron velocities we have

$$(n_{0,1}/n_{0,0}) = (8\pi m^3 g_{0,1}/h^3 g_{0,0} n_e)(2kT/m)^{3/2} e^{-\varepsilon_{I0}/kT} \int_0^\infty x^2 e^{-x^2}\, dx. \tag{13.3}$$

Or, evaluating the integral,

$$n_{0,0} = n_{0,1} n_e \tfrac{1}{2}(h^2/2\pi mkT)^{3/2}(g_{0,1}/g_{0,0})\exp(\varepsilon_{I0}/kT). \tag{13.4}$$

Notice that in (13.2) to (13.4), and in the equations that follow, the n's are particle *densities* (per cm^3).

In the above derivation, no essential use is made of information about which initial ionization stage is being considered, and (13.4) can actually be applied to any two successive stages $(j, j+1)$, yielding

$$n_{0,j} = n_{0,j+1} n_e \tfrac{1}{2}(h^2/2\pi mkT)^{3/2}(g_{0,j}/g_{0,j+1}) \exp(\varepsilon_{Ij}/kT). \qquad (13.5)$$

By using (12.28) in (13.5) we can obtain the occupation number of any excitation level of ion j in terms of the temperature, electron density, and the ground-state population of ion $(j+1)$:

$$\begin{aligned} n_{i,j} &= C_I n_{0,j+1} n_e (g_{i,j}/g_{0,j+1}) T^{-3/2} \exp[(\varepsilon_{Ij} - \varepsilon_{ij})/kT] \\ &\equiv n_{0,j+1} n_e \Phi_{ij}(T). \end{aligned} \qquad (13.6)$$

In (13.6), ε_{ij} is the energy of excitation state i of ion j relative to that ion's ground state (hence $\varepsilon_{Ij} - \varepsilon_{ij}$ is the ionization potential from level i of ion state j to the ground state of ion $j+1$), and the numerical constant is $C_I = 2.07 \times 10^{-16}$ in cgs units. This is perhaps the most useful form of Saha's equation for our later work; it is sometimes called the *Saha–Boltzmann formula.*

Using (12.24) in (13.6) for ionization stage $j+1$ we find

$$n_{i,j} = N_{j+1} n_e (g_{ij}/Z_{j+1}) \Phi_{ij}(T), \qquad (13.7)$$

and again, for ionization stage j, we have

$$N_j = C_I N_{j+1} n_e (Z_j/Z_{j+1}) T^{-3/2} \exp(\varepsilon_{Ij}/kT) \equiv N_{j+1} n_e \Phi_j(T). \qquad (13.8)$$

Equation (13.8) gives the relation between the total densities of atoms in the two ionization stages, irrespective of their excitation states.

In practically all of our later work we shall restrict attention to pure hydrogen for which only one ionization transition is possible. For hydrogen, the statistical weight of quantum level i is $g_i = 2i^2$, and the statistical weight of the ion is $g_p = 1$. The energy of level i below the continuum is ε_H/i^2 where ε_H equals one rydberg (13.6 eV). Thus (13.6) becomes

$$n_i = 2C_I i^2 n_e n_p T^{-3/2} \exp(\varepsilon_H/i^2 kT) \equiv n_e n_p \Phi_{iH}(T). \qquad (13.9)$$

Similarly (13.8) becomes

$$n_H = C_I n_e n_p Z_H T^{-3/2} \exp(\varepsilon_H/kT) \equiv n_e n_p \Phi_H(T) \qquad (13.10)$$

where

$$Z_H \equiv \sum_{i=1} 2i^2 \exp(-\varepsilon_{iH}/kT). \qquad (13.11)$$

Here

$$\varepsilon_{iH} \equiv (1 - i^{-2})\varepsilon_H. \qquad (13.12)$$

14. Thermodynamic Properties of Ionizing Hydrogen

Let us now examine the effects of ionization on the thermodynamic properties of a gas. We restrict attention to pure hydrogen, and thereby account for the dominant process in astrophysical plasmas with only a minimum of complication. The density of a pure hydrogen gas is

$$\rho = n_H m_H + n_p m_p + n_e m_e. \tag{14.1}$$

But, by charge conservation,

$$n_e = n_p, \tag{14.2}$$

hence

$$\rho = n_H m_H + n_p(m_p + m_e) = (n_H + n_p)m_H \equiv N_H m_H. \tag{14.3}$$

Here N_H denotes the total number density of hydrogen atoms and ions in all forms.

As was true for internal excitation (cf. §12), in an ionizing gas only the translational degrees of freedom contribute to the pressure, so

$$p = NkT \tag{14.4}$$

where now

$$N = n_H + n_p + n_e = N_H + n_e. \tag{14.5}$$

Let x denote the *degree of ionization* of the material,

$$x \equiv n_p/(n_p + n_H) = n_p/N_H. \tag{14.6}$$

In terms of the degree of ionization, Saha's formula (13.10) becomes

$$x^2/(1-x) = \text{Const. } VT^{3/2}e^{-\varepsilon_H/kT}/Z_H \tag{14.7a}$$

or, equivalently,

$$x^2p/(1-x^2) = \text{Const. } T^{5/2}e^{-\varepsilon_H/kT}/Z_H. \tag{14.7b}$$

Furthermore, the equation of state can be written

$$p = (1+x)N_H kT = (1+x)\rho\mathcal{R}T. \tag{14.8}$$

If we regard (T, p) [or equivalently (T, N)] as given, then using (13.10) in (14.5) we obtain a quadratic for n_e:

$$n_e^2\Phi_H + 2n_e = N, \tag{14.9}$$

which yields

$$n_e = [(N\Phi_H + 1)^{1/2} - 1]/\Phi_H. \tag{14.10}$$

Once we know n_e we can compute all the n_i immediately from (13.9).

The specific internal energy is simply the sum of the translational, excitation, and ionization energies of the particles per gram of material:

$$\begin{aligned} e &= \left[\tfrac{3}{2}(N_H+n_e)kT+\sum_i n_i\varepsilon_{iH}+n_p\varepsilon_H\right]\Big/ N_H m_H \\ &= \tfrac{3}{2}(1+x)\mathscr{R}T+\left[(1-x)\sum_i (n_i/n_H)\varepsilon_{iH}+x\varepsilon_H\right]\Big/ m_H \\ &= \tfrac{3}{2}(1+x)\mathscr{R}T+[(1-x)\langle\varepsilon_{iH}\rangle+x\varepsilon_H]/m_H \end{aligned} \tag{14.11a}$$

or per heavy particle (i.e., atom or proton), which is more convenient in what follows,

$$\tilde{e}=\tfrac{3}{2}(1+x)kT+(1-x)\langle\varepsilon_{iH}\rangle+x\varepsilon_H. \tag{14.11b}$$

Equation (14.11) was written directly from its physical meaning; it can of course also be derived from (12.40) if Z_{elec} is extended to include ionization. The specific enthalpy per particle follows immediately from (14.4) and (14.11b):

$$\tilde{h}=\tfrac{5}{2}(1+x)kT+(1-x)\langle\varepsilon_{iH}\rangle+x\varepsilon_H. \tag{14.12}$$

Computations with the Saha equation show that hydrogen in stellar ionization zones is strongly ionized even when $\varepsilon_H/kT \sim 10$; thus the ionization term $x\varepsilon_H$ makes a very large contribution to $\tilde{e}$, and represents a large energy reservoir in the material.

In practice, because the first excited state of hydrogen lies at an energy of $\frac{3}{4}\varepsilon_H$ above ground, one finds from the Boltzmann and Saha formulae that whenever hydrogen has even a small population in any excited state, it is already strongly ionized and therefore $(1-x) \ll 1$. Hence in (14.11) and (14.12) it is a good approximation simply to neglect the term in $\langle\varepsilon_{iH}\rangle$, which we drop henceforth. It is also for this reason that we can set $Z_H = g_{0,0}$ and ignore the temperature dependence of the partition function in (14.7).

With these simplifications it is now relatively easy to calculate the specific heats of ionizing hydrogen. Thus the specific heat at constant volume, per heavy particle, is

$$\tilde{c}_v = c_v m_H = (\partial\tilde{e}/\partial T)_v = \tfrac{3}{2}k(1+x)+(\tfrac{3}{2}kT+\varepsilon_H)(\partial x/\partial T)_v. \tag{14.13}$$

Differentiating (14.7a) logarithmically, we obtain

$$\frac{1}{x}\left(\frac{2-x}{1-x}\right)\left(\frac{\partial x}{\partial T}\right)_v=\frac{1}{T}\left(\frac{3}{2}+\frac{\varepsilon_H}{kT}\right), \tag{14.14}$$

whence

$$(\tilde{c}_v/k)=(c_v/\mathscr{R})=\tfrac{3}{2}(1+x)+\frac{x(1-x)}{(2-x)}\left(\frac{3}{2}+\frac{\varepsilon_H}{kT}\right)^2. \tag{14.15}$$

Clearly, as $x \to 0$ or $x \to 1$, $\tilde{c}_v$ reduces to the contribution from translational energy only; when the material is partially ionized (e.g., when

$x \approx \frac{1}{2}$), $\tilde{c}_v$ can greatly exceed this value. This large increase reflects the fact that when the gas is ionizing, a large fraction of the heat input to the gas is consumed in further ionizing the material rather than in raising its temperature.

Similarly, the specific heat at constant pressure per heavy particle is

$$\tilde{c}_p = (\partial \tilde{h}/\partial T)_p = \tfrac{5}{2}k(1+x) + (\tfrac{5}{2}kT + \varepsilon_H)(\partial x/\partial T)_p. \tag{14.16}$$

Differentiating (14.7b) logarithmically, we obtain

$$(\partial x/\partial T)_p = \tfrac{1}{2}x(1-x^2)[\tfrac{5}{2} + (\varepsilon_H/kT)]/T, \tag{14.17}$$

whence

$$(\tilde{c}_p/k) = (c_p/\mathscr{R}) = \tfrac{5}{2}(1+x) + \tfrac{1}{2}x(1-x^2)[\tfrac{5}{2} + (\varepsilon_H/kT)]^2. \tag{14.18}$$

Again, $\tilde{c}_p$ approaches its pure translational value as $x \to 0$ or $x \to 1$ and is much larger than that value for partially ionized material.

From the results obtained above, it is obvious that the ratio $\gamma \equiv (c_p/c_v)$ is no longer a constant, hence for adiabatic changes we cannot recover relations (4.13) to (4.15). But, following Chandrasekhar (**C1,** 121), it is natural to introduce the *generalized adiabatic exponents* Γ_1, Γ_2, Γ_3, defined to be those values for which

$$(dp/p) - \Gamma_1(d\rho/\rho) = 0, \tag{14.19}$$

$$(dp/p) - [\Gamma_2/(\Gamma_2 - 1)](dT/T) = 0, \tag{14.20}$$

and

$$(dT/T) - (\Gamma_3 - 1)(d\rho/\rho) = 0 \tag{14.21}$$

for adiabatic processes in an ionizing gas. Clearly these gammas are variable (functions of T, x, etc.). It is obvious that only two of the gammas are independent, and that one has a general identity

$$\Gamma_1/(\Gamma_3 - 1) \equiv \Gamma_2/(\Gamma_2 - 1). \tag{14.22}$$

For an adiabatic change, we know from (2.3) that

$$de = (p/\rho^2)\, d\rho, \tag{14.23}$$

and from (14.8) we have

$$(dp/p) = [dx/(1+x)] + (d\rho/\rho) + (dT/T). \tag{14.24}$$

Also, from (14.11a),

$$de = \tfrac{3}{2}(1+x)\mathscr{R}\, dT + [\tfrac{3}{2}\mathscr{R}T + (\varepsilon_H/m_H)]\, dx. \tag{14.25}$$

Thus, using (14.25) in (14.23), and eliminating $(d\rho/\rho)$ via (14.24), we find

$$\frac{dp}{p} = \frac{5}{2}\frac{dT}{T} + \left(\frac{5}{2} + \frac{\varepsilon_H}{kT}\right)\frac{dx}{1+x}. \tag{14.26}$$

But, from (14.7b),

$$\frac{2\,dx}{x(1-x^2)}=\left(\frac{5}{2}+\frac{\varepsilon_H}{kT}\right)\frac{dT}{T}-\frac{dp}{p}. \tag{14.27}$$

Therefore, using (14.27) to eliminate $dx/(1+x)$ from (14.26) we find, in view of (14.20),

$$\left(\frac{\Gamma_2-1}{\Gamma_2}\right)\equiv\left(\frac{\partial \ln T}{\partial \ln p}\right)_s=\frac{2+x(1-x)[\frac{5}{2}+(\varepsilon_H/kT)]}{5+x(1-x)[\frac{5}{2}+(\varepsilon_H/kT)]^2}. \tag{14.28}$$

Clearly, in the limits $x\to 0$ or $x\to 1$, $(\Gamma_2-1)/\Gamma_2\to\frac{2}{5}$, which equals $(\gamma-1)/\gamma$ for a perfect gas. When hydrogen is partially ionized, $(\Gamma_2-1)/\Gamma_2$ can become as small as 0.1.

By similar analyses it is fairly easy to show that

$$\Gamma_1=\left(\frac{\partial \ln p}{\partial \ln \rho}\right)_s=\frac{5+x(1-x)[\frac{5}{2}+(\varepsilon_H/kT)]^2}{3+x(1-x)\{\frac{3}{2}+[\frac{3}{2}+(\varepsilon_H/kT)]^2\}}, \tag{14.29}$$

and

$$\Gamma_3-1=\left(\frac{\partial \ln T}{\partial \ln \rho}\right)_s=\frac{2+x(1-x)[\frac{5}{2}+(\varepsilon_H/kT)]}{3+x(1-x)\{\frac{3}{2}+[\frac{3}{2}+(\varepsilon_H/kT)]^2\}}, \tag{14.30}$$

which clearly satisfy (14.22).

In many applications it is convenient to write the equation of state as

$$p=\rho kT/\mu m_H, \tag{14.31}$$

where μ is the *mean molecular weight*, which may be variable and hence a function of state variables. In the present case we have, from (14.8),

$$\mu=1/(1+x). \tag{14.32}$$

When one considers small-amplitude perturbations around an ambient state, the quantity

$$Q\equiv 1-(\partial \ln \mu/\partial \ln T)_p \tag{14.33}$$

often appears. In the present case we can use (14.17) to evaluate Q as

$$Q=1+\tfrac{1}{2}x(1-x)[\tfrac{5}{2}+(\varepsilon_H/kT)]. \tag{14.34}$$

All of the results derived in this section apply to a gas composed of a single element. Results for mixtures of ionizing gases can be found in (**C5**, §§9.13–9.18), (**K2**), (**M3**, §5.4), (**M4**), and (**U1**, §§56 and 57).

References

(B1) Bond, J. W., Watson, K. M., and Welch, J. A. (1965) *Atomic Theory of Gas Dynamics.* Reading: Addison-Wesley.

(B2) Boyd, T. J. M. and Sanderson, J. J. (1969) *Plasma Dynamics.* New York: Barnes and Noble.

(C1) Chandrasekhar, S. (1939) *An Introduction to the Study of Stellar Structure.* Chicago: University of Chicago Press.
(C2) Chandrasekhar, S. (1942) *Principles of Stellar Dynamics.* Chicago: University of Chicago Press.
(C3) Chandrasekhar, S. (1943) *Rev. Mod. Phys.*, **15,** 1.
(C4) Chapman, S. and Cowling, T. G. (1970) *The Mathematical Theory of Non-Uniform Gases.* (3rd ed.) Cambridge: Cambridge University Press.
(C5) Cox, J. P. and Giuli, R. T. (1968) *Principles of Stellar Structure.* New York: Gordon and Breach.
(D1) Davidson, N. (1962) *Statistical Mechanics.* New York: Wiley.
(F1) Fowler, R. and Guggenheim, E. A. (1956) *Statistical Thermodynamics.* Cambridge: Cambridge University Press.
(H1) Hirschfelder, J. O., Curtiss, C. F., and Bird, R. B. (1954) *Molecular Theory of Gases and Liquids.* New York: Wiley.
(H2) Huang, K. (1963) *Statistical Mechanics.* New York: Wiley.
(K1) Krall, N. A. and Trivelpiece, A. W. (1973) *Principles of Plasma Physics.* New York: McGraw-Hill.
(K2) Krishna-Swamy, K. S. (1961) *Astrophys. J.*, **134,** 1017.
(M1) MacDonald, W. M., Rosenbluth, M. N., and Chuck, W. (1957) *Phys. Rev.*, **107,** 350.
(M2) Mayer, J. E. and Mayer, M. G. (1940) *Statistical Mechanics.* New York: Wiley.
(M3) Menzel, D. H., Sen, H. K., and Bhatnagar, P. L. (1963) *Stellar Interiors.* London: Chapman and Hall.
(M4) Mihalas, D. (1965) *Astrophys. J.*, **141,** 564.
(M5) Mihalas, D. (1978) *Stellar Atmospheres.* (2nd ed.) San Francisco: Freeman.
(R1) Rosenbluth, M. N., MacDonald, W. M., and Judd, D. L. (1957) *Phys. Rev.*, **107,** 1.
(S1) Sears, F. W. (1953) *An Introduction to Thermodynamics, Kinetic Theory of Gases, and Statistical Mechanics.* (2nd ed.) Reading: Addison-Wesley.
(S2) Spitzer, L. (1956) *Physics of Fully Ionized Gases.* New York: Interscience.
(U1) Unsöld, A. (1968) *Physik der Sternatmosphären.* Berlin: Springer.
(V1) Vincenti, W. G. and Kruger, C. H. (1965) *Introduction to Physical Gas Dynamics.* New York: Wiley.
(W1) Wilson, A. H. (1957) *Thermodynamics and Statistical Mechanics.* Cambridge: Cambridge University Press.
(W2) Wu, T.-Y. and Ohmura, T. (1962) *Quantum Theory of Scattering.* Englewood Cliffs: Prentice-Hall.

2

Dynamics of Ideal Fluids

The basic goal of any fluid-dynamical study is to provide (1) a complete description of the motion of the fluid at any instant of time, and hence of the *kinematics* of the flow, and (2) a description of how the motion changes in time in response to applied forces, and hence of the *dynamics* of the flow. We begin our study of astrophysical fluid dynamics by analyzing the motion of a compressible *ideal fluid* (i.e., a nonviscous and nonconducting gas); this allows us to formulate very simply both the basic conservation laws for the mass, momentum, and energy of a fluid parcel (which govern its dynamics) and the essentially geometrical relationships that specify its kinematics. Because we are concerned here with the macroscopic properties of the flow of a physically uncomplicated medium, it is both natural and advantageous to adopt a purely continuum point of view. In the next chapter, where we seek to understand the important role played by internal processes of the gas in transporting energy and momentum within the fluid, we must carry out a deeper analysis based on a kinetic-theory view; even then we shall see that the continuum approach yields useful results and insights. We pursue this line of inquiry even further in Chapters 6 and 7, where we extend the analysis to include the interaction between radiation and both the internal state, and the macroscopic dynamics, of the material.

2.1 Kinematics

15. Velocity and Acceleration

In developing descriptions of fluid motion it is fruitful to work in two different frames of reference, each of which has distinct advantages in certain situations. On the one hand we can view the flow from a fixed *laboratory frame*, and consider any property of the fluid, say α, to be a function of the position $\mathbf{x}$ in this frame and of time t, that is, $\alpha = \alpha(\mathbf{x}, t)$. The time and space variation of α are then described using a time derivative $(\partial/\partial t)$ computed at fixed $\mathbf{x}$ and space derivatives $(\partial/\partial x^i)$ evaluated at fixed t. This approach is generally known as the *Eulerian description*. In particular, in this scheme we describe the *velocity* of the fluid by a vector field $\mathbf{v}(\mathbf{x}, t)$, which gives the rate and direction of flow of the material as a function of position and time, as seen in the laboratory frame.

Alternatively, we may choose a particular fluid parcel and study the time variation of its properties while following the motion of that parcel; this approach is generally referred to as the *Lagrangean description.* The time variation of the properties of a Lagrangean fluid element (also called a *material element*) is described in terms of the *fluid-frame time derivative* (D/Dt) (also known as the *comoving,* or *Lagrangean,* or *material,* or *substantial derivative*). For example, the velocity of an element of fluid is, by definition, the time rate of change of the position of that particular fluid parcel; hence in the Lagrangean scheme we have

$$\mathbf{v} = (D\mathbf{x}/Dt). \tag{15.1}$$

Similarly the *acceleration* $\mathbf{a}$ of a fluid element is the rate of change of its velocity during the course of its motion, hence

$$\mathbf{a} = (D\mathbf{v}/Dt). \tag{15.2}$$

For nonrelativistic flows, the choice between the Eulerian or Lagrangean points of view is usually made purely on the basis of the convenience of one or the other frame for formulating the physics of the situation under study; as we show below, derivatives in the two frames are simply related. In relativistic flows, however, the difference between these two frames is much more fundamental and has a deep physical significance; we return to this point in Chapter 4.

To relate the Lagrangean time derivative to derivatives in the Eulerian frame, we notice that $(D\alpha/Dt)$ is defined as

$$(D\alpha/Dt) = \lim_{\Delta t \to 0} [\alpha(\mathbf{x}+\Delta\mathbf{x}, t+\Delta t) - \alpha(\mathbf{x}, t)]/\Delta t \tag{15.3}$$

where α is measured in a definite fluid parcel at two different times, t and $t+\Delta t$, and also, as a result of the motion of that parcel, at two different positions, $\mathbf{x}$ and $\mathbf{x}+\Delta\mathbf{x} = \mathbf{x}+\mathbf{v}\,\Delta t$, as seen in the laboratory frame. Expanding to first order in Δt, we have

$$\begin{aligned} \alpha(\mathbf{x}+\Delta\mathbf{x}, t+\Delta t) &= \alpha(\mathbf{x}, t) + (\partial\alpha/\partial t)\,\Delta t + (\partial\alpha/\partial x^i)\,\Delta x^i \\ &= \alpha(\mathbf{x}, t) + [(\partial\alpha/\partial t) + v^i(\partial\alpha/\partial x^i)]\,\Delta t \\ &= \alpha(\mathbf{x}, t) + (\alpha_{,t} + v^i\alpha_{,i})\,\Delta t, \end{aligned} \tag{15.4}$$

(cf. §A1 for notation) and therefore

$$(D\alpha/Dt) = \alpha_{,t} + v^i\alpha_{,i}. \tag{15.5}$$

Equation (15.5) holds for any α: scalar, vector, or tensor. In more familiar notation,

$$(D\alpha/Dt) = (\partial\alpha/\partial t) + (\mathbf{v}\cdot\nabla)\alpha. \tag{15.6}$$

The covariant generalization of (15.5), valid in curvilinear coordinates, is

$$(D\alpha/Dt) = \alpha_{,t} + v^i\alpha_{;i} \tag{15.7}$$

where $\alpha_{;i}$ is the *covariant derivative* of α with respect to x^i (cf. §A3.10). By comparing (15.7) with equation (A3.81) we see that, in general, $(D\alpha/Dt)$ is to be identified with the *intrinsic derivative* $(\delta\alpha/\delta t)$, a remark that will assume greater significance in our discussion of relativistic kinematics. It should be noted that here, and in Chapter 3, by the "covariant form" of an equation we mean covariant only with respect to changes of the spatial coordinates in a three-dimensional Euclidian space, treating time as absolute and independent. In Chapters 4 and 7 we use this expression to mean completely covariant with respect to the full four-dimensional spacetime of special relativity.

The value of the formalism provided by (15.7) is seen when we derive vector and tensor expressions in curvilinear coordinates. Consider, for example, the acceleration, which in Cartesian coordinates is simply

$$a^i = v^i_{,t} + v^j v^i_{,j}. \tag{15.8}$$

In curvilinear coordinates we have, from equation (A3.83),

$$a^i = v^i_{,t} + v^j v^i_{;j} = v^i_{,t} + v^j v^i_{,j} + \left\{ \begin{matrix} i \\ j\ k \end{matrix} \right\} v^j v^k. \tag{15.9}$$

In particular, for spherical coordinates we find from equation (A3.63) that the contravariant components of **a** are

$$a^{(1)} = v^{(1)}_{,t} + v^j v^{(1)}_{,j} - r(v^{(2)})^2 - r\sin^2\theta (v^{(3)})^2, \tag{15.10a}$$

$$a^{(2)} = v^{(2)}_{,t} + v^j v^{(2)}_{,j} + 2v^{(1)}v^{(2)}/r - \sin\theta\cos\theta (v^{(3)})^2, \tag{15.10b}$$

and

$$a^{(3)} = v^{(3)}_{,t} + v^j v^{(3)}_{,j} + 2v^{(1)}v^{(3)}/r + 2\cot\theta\, v^{(2)}v^{(3)}. \tag{15.10c}$$

Now replacing the contravariant components of both **a** and **v** with their equivalent physical components via equation (A3.46a), we find

$$a_r = \frac{\partial v_r}{\partial t} + v_r \frac{\partial v_r}{\partial r} + \frac{v_\theta}{r}\frac{\partial v_r}{\partial \theta} + \frac{v_\phi}{r\sin\theta}\frac{\partial v_r}{\partial \phi} - \frac{1}{r}(v_\theta^2 + v_\phi^2), \tag{15.11a}$$

$$a_\theta = \frac{\partial v_\theta}{\partial t} + v_r \frac{\partial v_\theta}{\partial r} + \frac{v_\theta}{r}\frac{\partial v_\theta}{\partial \theta} + \frac{v_\phi}{r\sin\theta}\frac{\partial v_\theta}{\partial \phi} + \frac{v_r v_\theta}{r} - \frac{\cot\theta}{r} v_\phi^2, \tag{15.11b}$$

$$a_\phi = \frac{\partial v_\phi}{\partial t} + v_r \frac{\partial v_\phi}{\partial r} + \frac{v_\theta}{r}\frac{\partial v_\phi}{\partial \theta} + \frac{v_\phi}{r\sin\theta}\frac{\partial v_\phi}{\partial \phi} + \frac{v_r v_\phi}{r} + \frac{\cot\theta}{r} v_\theta v_\phi. \tag{15.11c}$$

Similar expressions for any other coordinate system are easily derived from (15.9).

16. *Particle Paths, Streamlines, and Streaklines*

In a moving fluid, the *particle path* of a particular fluid element is simply the three-dimensional path traced out in time by that element. If we label each element by its coordinates $\boldsymbol{\xi}$ at some reference time $t = 0$, then its particle path is

$$\mathbf{x}(\boldsymbol{\xi}, t) = \boldsymbol{\xi} + \int_0^t \mathbf{v}(\boldsymbol{\xi}, t')\, dt' \tag{16.1}$$

where $\mathbf{v}(\boldsymbol{\xi}, t)$ is the velocity, as a function of time, of the element specified by $\boldsymbol{\xi}$. Equation (16.1) is the parametric equation of a curve in space, with t as parameter.

The *streamlines* in a flow are defined as those curves that, at a *given instant of time*, are tangent at each point to the velocity of the fluid at that position (and time). Thus a streamline can be written as the parametric curve

$$(d\mathbf{x}/ds) = \mathbf{v}(\mathbf{x}, t) \tag{16.2}$$

where t is fixed and s is a path-length parameter. Alternatively, along a streamline,

$$(dx/v_x) = (dy/v_y) = (dz/v_z) = ds. \tag{16.3}$$

Notice also that if $d\mathbf{x}$ lies along a streamline, then $\mathbf{v} \times d\mathbf{x} = 0$. A *stream tube* is a tube, filled with flowing fluid, whose surface is composed of all the streamlines passing through some closed curve C.

A *streakline* is the curve traced out in time by all fluid particles that pass through a given fixed point in the flow field; streaklines may be made visible in a flow, for example, by injection of a dye or colored smoke at some point in the flow.

For *steady* flows, where $\mathbf{v}(\mathbf{x}, t)$ is independent of time, the particle paths, streamlines, and streaklines are all identical; in time-dependent flows all three sets of curves will, in general, be distinct.

17. The Euler Expansion Formula

Suppose we choose a fluid element located, at $t = 0$, within the volume element $dV_0 = d\xi^{(1)}\, d\xi^{(2)}\, d\xi^{(3)}$ around some point $\boldsymbol{\xi}$. As the fluid flows, in time this element will, in general, move to some other position $\mathbf{x}(\boldsymbol{\xi}, t)$, and will occupy some new volume element $dV = dx^{(1)}\, dx^{(2)}\, dx^{(3)}$. These volume elements are related by the expression

$$\begin{aligned} dV &= J(x^{(1)}, x^{(2)}, x^{(3)}/\xi^{(1)}, \xi^{(2)}, \xi^{(3)})\, d\xi^{(1)}\, d\xi^{(2)}\, d\xi^{(3)} \\ &= J\, d\xi^{(1)}\, d\xi^{(2)}\, d\xi^{(3)} = J\, dV_0, \end{aligned} \tag{17.1}$$

where J is the Jacobian of the transformation $\mathbf{x}(\boldsymbol{\xi}, t)$. The ratio

$$J = (dV/dV_0) \tag{17.2}$$

is called the *expansion* of the fluid.

In what follows, we require an expression for the time rate of change of the expansion of a fluid element as we follow its motion in the flow. From equation (A2.23),

$$J = \left|\frac{\partial x^i}{\partial \xi^j}\right| = e^{ijk}\frac{\partial x^{(1)}}{\partial \xi^i}\frac{\partial x^{(2)}}{\partial \xi^j}\frac{\partial x^{(3)}}{\partial \xi^k}, \tag{17.3}$$

hence

$$\begin{aligned}\frac{DJ}{Dt} &= \frac{D}{Dt}\left(e^{ijk}\frac{\partial x^{(1)}}{\partial \xi^i}\frac{\partial x^{(2)}}{\partial \xi^j}\frac{\partial x^{(3)}}{\partial \xi^k}\right)\\ &= e^{ijk}\frac{\partial v^{(1)}}{\partial \xi^i}\frac{\partial x^{(2)}}{\partial \xi^j}\frac{\partial x^{(3)}}{\partial \xi^k} + e^{ijk}\frac{\partial x^{(1)}}{\partial \xi^i}\frac{\partial v^{(2)}}{\partial \xi^j}\frac{\partial x^{(3)}}{\partial \xi^k}\\ &\quad + e^{ijk}\frac{\partial x^{(1)}}{\partial \xi^i}\frac{\partial x^{(2)}}{\partial \xi^j}\frac{\partial v^{(3)}}{\partial \xi^k}.\end{aligned} \tag{17.4}$$

We can expand $(\partial v^i/\partial \xi^j)$ as

$$(\partial v^i/\partial \xi^j) = (\partial v^i/\partial x^l)(\partial x^l/\partial \xi^j). \tag{17.5}$$

Therefore,

$$\begin{aligned}\frac{DJ}{Dt} &= e^{ijk}\frac{\partial v^{(1)}}{\partial x^l}\frac{\partial x^l}{\partial \xi^i}\frac{\partial x^{(2)}}{\partial \xi^j}\frac{\partial x^{(3)}}{\partial \xi^k} + e^{ijk}\frac{\partial x^{(1)}}{\partial \xi^i}\frac{\partial v^{(2)}}{\partial x^l}\frac{\partial x^l}{\partial \xi^j}\frac{\partial x^{(3)}}{\partial \xi^k}\\ &\quad + e^{ijk}\frac{\partial x^{(1)}}{\partial \xi^i}\frac{\partial x^{(2)}}{\partial \xi^j}\frac{\partial v^{(3)}}{\partial x^l}\frac{\partial x^l}{\partial \xi^k}.\end{aligned} \tag{17.6}$$

Writing out the first term on the right-hand side in full, we have

$$\begin{aligned}e^{ijk}\frac{\partial v^{(1)}}{\partial x^l}\frac{\partial x^l}{\partial \xi^i}\frac{\partial x^{(2)}}{\partial \xi^j}\frac{\partial x^{(3)}}{\partial \xi^k} &= \frac{\partial v^{(1)}}{\partial x^{(1)}}e^{ijk}\frac{\partial x^{(1)}}{\partial \xi^i}\frac{\partial x^{(2)}}{\partial \xi^j}\frac{\partial x^{(3)}}{\partial \xi^k}\\ &\quad + \frac{\partial v^{(1)}}{\partial x^{(2)}}e^{ijk}\frac{\partial x^{(2)}}{\partial \xi^i}\frac{\partial x^{(2)}}{\partial \xi^j}\frac{\partial x^{(3)}}{\partial \xi^k}\\ &\quad + \frac{\partial v^{(1)}}{\partial x^{(3)}}e^{ijk}\frac{\partial x^{(3)}}{\partial \xi^i}\frac{\partial x^{(2)}}{\partial \xi^j}\frac{\partial x^{(3)}}{\partial \xi^k}.\end{aligned} \tag{17.7}$$

The last two terms on the right-hand side of (17.7) are obviously identically zero because the permutation symbol is antisymmetric in all indices, while the first equals $(\partial v^{(1)}/\partial x^{(1)})J$. By expanding the second and third terms on the right-hand side of (17.6) in a similar way, we obtain finally

$$\frac{DJ}{Dt} = \left(\frac{\partial v^{(1)}}{\partial x^{(1)}} + \frac{\partial v^{(2)}}{\partial x^{(2)}} + \frac{\partial v^{(3)}}{\partial x^{(3)}}\right)J = (\nabla \cdot \mathbf{v})J, \tag{17.8}$$

which is known as *Euler's expansion formula.*

The covariant generalization of (17.8) to curvilinear coordinates, taking account of equation (A3.86), is

$$(D \ln J/Dt) = v^i_{;i} = g^{-1/2}(g^{1/2}v^i)_{,i} \tag{17.9}$$

where g is the determinant of the metric tensor of the coordinate system (cf. §A3.4).

18. The Reynolds Transport Theorem

With the help of Euler's expansion formula, we can now calculate the time rate of change of integrals of physical quantities within material volumes. Thus let $F(\mathbf{x}, t)$ be any single-valued scalar, vector, or tensor field, and choose $\mathcal{V}(t)$ to be some finite *material* volume composed of a definite set of fluid particles. Then, clearly,

$$\mathscr{F}(t) = \int_{\mathcal{V}} F(\mathbf{x}, t)\, dV \tag{18.1}$$

is a definite function of time. At $t = 0$, the material element dV is identical to a *fixed* element dV_0; at later times dV is related to dV_0 by means of (17.7). The fluid-frame time derivative of $\mathscr{F}$ is then

$$\frac{D\mathscr{F}}{Dt} = \frac{D}{Dt}\int_{V_0} F(\mathbf{x}, t) J\, dV_0 = \int_{V_0} \left(\frac{DF}{Dt} J + F\frac{DJ}{Dt}\right) dV_0$$
$$= \int_{V_0} \left(\frac{DF}{Dt} + F\,\nabla\cdot\mathbf{v}\right) J\, dV_0,$$

hence

$$\frac{D\mathscr{F}}{Dt} = \int_{\mathcal{V}} \left(\frac{DF}{Dt} + F\,\nabla\cdot\mathbf{v}\right) dV. \tag{18.2}$$

In view of (15.6) and equation (A2.48) we can also write

$$\frac{D}{Dt}\int_{\mathcal{V}} F(\mathbf{x}, t)\, dV = \int_{\mathcal{V}} \left[\frac{\partial F}{\partial t} + \nabla\cdot(F\mathbf{v})\right] dV. \tag{18.3}$$

Furthermore, by applying the divergence theorem [cf. (A2.68)], we also have

$$\frac{D}{Dt}\int_{\mathcal{V}} F(\mathbf{x}, t)\, dV = \int_{\mathcal{V}} \frac{\partial F}{\partial t}\, dV + \int_{S} F\mathbf{v}\cdot d\mathbf{S} \tag{18.4}$$

where S is the surface bounding $\mathcal{V}$.

Equations (18.2) through (18.4) are known as the *Reynolds transport theorem*; they hold for any scalar field or the components of a vector or tensor field. We shall use these important results repeatedly in what follows. The physical interpretation of (18.4) is quite clear: the rate of change of the integral of F within a material volume equals the integral of the time rate of change of F within the fixed volume V that instantaneously coincides with the material volume $\mathcal{V}(t)$, plus the net flux of F through the bounding surface of the (moving) material volume.

19. The Equation of Continuity

By definition the fluid in a material volume is always composed of exactly the same particles. The mass contained within a material volume must therefore always be the same; hence

$$\frac{D}{Dt}\int_{\mathcal{V}} \rho \, dV \equiv 0. \tag{19.1}$$

Equation (19.1) is a mathematical statement of the law of *conservation of mass* for the fluid. Now applying the Reynolds theorem (18.2) for $F = \rho$, we find

$$\int_{\mathcal{V}} [(D\rho/Dt) + \rho(\nabla \cdot \mathbf{v})] \, dV = 0. \tag{19.2}$$

But the material volume $\mathcal{V}$ is arbitrary, so in general we can guarantee that the integral will vanish only if the integrand vanishes at all points in the flow field. We thus obtain the *equation of continuity*

$$(D\rho/Dt) + \rho(\nabla \cdot \mathbf{v}) = 0, \tag{19.3}$$

or, from (15.6),

$$(\partial\rho/\partial t) + \nabla \cdot (\rho\mathbf{v}) = \rho_{,t} + (\rho v^i)_{,i} = 0. \tag{19.4}$$

For *steady flow* $(\partial\rho/\partial t) \equiv 0$, hence

$$\nabla \cdot (\rho\mathbf{v}) = 0. \tag{19.5}$$

For a one-dimensional steady flow in planar geometry we then have $d(\rho v_z)/dz = 0$ or

$$\rho v_z = \text{constant} = \dot{m}, \tag{19.6}$$

where $\dot{m}$ is the *mass flux* per unit area in the flow.

For *incompressible flow* $\rho \equiv$ constant, hence $(\partial\rho/\partial t) \equiv 0$ and from (19.5) $\nabla \cdot \mathbf{v} \equiv 0$.

The covariant generalization of (19.4), valid in curvilinear coordinate systems, is

$$\rho_{,t} + (\rho v^i)_{;i} = 0. \tag{19.7}$$

For example, in spherical coordinates [for which $g^{1/2} = r^2 \sin\theta$ and the connection between contravariant and physical components is given by equation (A3.46a)], we find, by using equation (A3.86), that

$$\frac{\partial\rho}{\partial t} + \frac{1}{r^2}\frac{\partial}{\partial r}(r^2\rho v_r) + \frac{1}{r\sin\theta}\frac{\partial(\rho v_\theta \sin\theta)}{\partial\theta} + \frac{1}{r\sin\theta}\frac{\partial(\rho v_\phi)}{\partial\phi} = 0. \tag{19.8}$$

For one-dimensional, spherically symmetric flow, (19.8) simplifies to

$$(\partial\rho/\partial t) + r^{-2}[\partial(r^2\rho v_r)/\partial r] = 0, \tag{19.9}$$

which, for steady flow, implies that

$$4\pi r^2 \rho v_r = \text{constant} = \dot{\mathcal{M}}, \tag{19.10}$$

where $\dot{\mathcal{M}}$ is the mass flux through a spherical shell surrounding the origin.

Finally, by applying the Reynolds transport theorem (18.2) to the function $F=\rho\alpha$, we see that

$$\frac{D}{Dt}\int_{\mathcal{V}} \rho\alpha\, dV + \int_{\mathcal{V}} \left[\frac{D(\rho\alpha)}{Dt} + \rho\alpha(\nabla\cdot\mathbf{v})\right] dV$$
$$= \int_{\mathcal{V}} \left\{\rho\frac{D\alpha}{Dt} + \alpha\left[\frac{D\rho}{Dt} + \rho(\nabla\cdot\mathbf{v})\right]\right\} dV, \quad (19.11)$$

which, in view of the equation of continuity (19.3), yields the useful identity

$$\frac{D}{Dt}\int_{\mathcal{V}} \rho\alpha\, dV = \int_{\mathcal{V}} \rho\frac{D\alpha}{Dt}\, dV. \quad (19.12)$$

Similarly, by combining (19.12) with (18.3), we find the useful result

$$\rho(D\alpha/Dt) = [\partial(\rho\alpha)/\partial t] + \nabla\cdot(\rho\alpha\mathbf{v}). \quad (19.13)$$

In (19.12) and (19.13), α may be a scalar or a component of a vector or tensor.

20. *Vorticity and Circulation*

The *vorticity* $\boldsymbol{\omega}$ of a fluid flow is defined to be the curl of the velocity field:

$$\boldsymbol{\omega} \equiv \nabla\times\mathbf{v}. \quad (20.1)$$

In Cartesian coordinates

$$\omega^i = e^{ijk}v_{k,j}. \quad (20.2)$$

We shall see in §21 that $\boldsymbol{\omega}$ provides a local measure of the rate of rotation of the fluid at each point in the flow. Therefore we define an *irrotational flow* to be one for which $\boldsymbol{\omega}\equiv 0$. Such flows can be described by a scalar *velocity potential* ϕ, defined to be such that $\mathbf{v}=\nabla\phi$, because then by equation (A2.56) we have $\boldsymbol{\omega}=\nabla\times(\nabla\phi)\equiv 0$.

A *vortex line* is a curve that at each point is tangent to the vortex vector at that point. Thus if $d\mathbf{x}$ lies along a vortex line we have

$$(dx/\omega_x) = (dy/\omega_y) = (dz/\omega_z) = ds, \quad (20.3)$$

which generates a parametric curve with s as the parameter. Clearly we must also have $\boldsymbol{\omega}\times d\mathbf{x}=0$ along a vortex line. A *vortex tube* is the surface generated by all vortex lines passing through some closed curve C in the fluid.

The *circulation* in the flow is defined to be

$$\Gamma \equiv \oint_C \mathbf{v}\cdot d\mathbf{x} \quad (20.4)$$

where C is a closed curve. Suppose, in fact, that C is a closed material curve, that is, a closed curve composed of a definite set of fluid particles.

Then the time rate of change of the circulation around that material curve is

$$\frac{D\Gamma}{Dt}=\frac{D}{Dt}\oint_C \mathbf{v}\cdot d\mathbf{x}=\oint_C \frac{D\mathbf{v}}{Dt}\cdot d\mathbf{x}+\oint_C \mathbf{v}\cdot\frac{D}{Dt}(d\mathbf{x}). \tag{20.5}$$

But, by (15.1), on a material curve we have $[D(d\mathbf{x})/Dt]=d(D\mathbf{x}/Dt)=d\mathbf{v}$, hence

$$(D\Gamma/Dt)=\oint_C (D\mathbf{v}/Dt)\cdot d\mathbf{x}+\oint_C \mathbf{v}\cdot d\mathbf{v}=\oint_C (D\mathbf{v}/Dt)\cdot d\mathbf{x}+\oint_C d(\tfrac{1}{2}v^2). \tag{20.6}$$

However v^2 is a single-valued scalar function, and its line integral around a closed path must be identically zero. Hence we obtain *Kelvin's equation*

$$(D\Gamma/Dt)=\oint_C (D\mathbf{v}/Dt)\cdot d\mathbf{x}. \tag{20.7}$$

The *strength* of a vortex tube (or the *flux of vorticity*) is defined to be

$$\Sigma=\int_S \boldsymbol{\omega}\cdot d\mathbf{S} \tag{20.8}$$

where S is the cross section enclosed by a closed curve C lying in the surface of the tube. By Stokes's theorem (A2.70), we then see that the strength of a vortex tube equals the circulation in the flow:

$$\Sigma=\int_S (\boldsymbol{\nabla}\times\mathbf{v})\cdot d\mathbf{S}=\oint_C \mathbf{v}\cdot d\mathbf{x}=\Gamma. \tag{20.9}$$

Because the divergence of the vorticity is identically zero [cf. (A2.54)], it follows from the divergence theorem (A2.68) that the flux of vorticity $\boldsymbol{\omega}\cdot d\mathbf{S}$ integrated over a closed surface S, which is composed of two cross sections S_1 and S_2 of a vortex tube (bounded by closed curves C_1 and C_2) and the segment of the tube joining them, must be zero. Furthermore, because $\boldsymbol{\omega}$ by definition lies along the tube, it is obvious that $\boldsymbol{\omega}\cdot d\mathbf{S}$ must be identically zero on that part of the closed surface. We must therefore have

$$\int_{S_1} (\boldsymbol{\nabla}\times\mathbf{v})\cdot\mathbf{n}_1\, dS+\int_{S_2} (\boldsymbol{\nabla}\times\mathbf{v})\cdot\mathbf{n}_2\, dS=\Sigma_1+\Sigma_2=\Gamma_1-\Gamma_2=0, \tag{20.10}$$

where in choosing the sign of Γ_2 we have recognized that $\mathbf{n}_2$, the outward normal on S_2, is opposite in direction to the positive normal (in the sense of being right handed) defined by the circuit of C_2. Thus we have *Helmholtz's vortex theorem*: the flux of vorticity across any cross-section of a vortex tube is constant, or, equivalently, the circulation around any closed surface lying on the surface of a vortex tube is constant. It should be noted explicitly that the fluid particles that define the surface of a vortex tube at one time will not in general lie on its surface at a later time. Thus (20.10) does not automatically imply that Γ around a *material* curve is constant; we return to the question of when $(D\Gamma/Dt) = 0$ in §23.

The covariant generalization of (20.2) is

$$\omega^i = \varepsilon^{ijk} v_{k;j} = g^{-1/2}(v_{k,j} - v_{j,k}), \tag{20.11}$$

where ε^{ijk} is the Levi-Civita tensor, and the second equality is proved in equation (A3.95). For example, in spherical coordinates we find from equation (A3.96)

$$\omega_r = \frac{1}{r \sin\theta}\left[\frac{\partial(v_\phi \sin\theta)}{\partial\theta} - \frac{\partial v_\theta}{\partial\phi}\right], \tag{20.12a}$$

$$\omega_\theta = \frac{1}{r}\left[\frac{1}{\sin\theta}\frac{\partial v_r}{\partial\phi} - \frac{\partial(r v_\phi)}{\partial r}\right], \tag{20.12b}$$

and

$$\omega_\phi = \frac{1}{r}\left[\frac{\partial(r v_\theta)}{\partial r} - \frac{\partial v_r}{\partial\theta}\right]. \tag{20.12c}$$

Finally, we note that the vorticity $\boldsymbol{\omega}$ is in reality a pseudovector, and has associated with it (cf. §A2.11) an antisymmetric second-rank tensor

$$\Omega_{ij} = e_{ijk}\omega^k = v_{j,i} - v_{i,j}, \tag{20.13}$$

where we used equation (A2.20). This tensor plays a prominent role in fluid kinematics, as we will now see.

21. The Cauchy–Stokes Decomposition Theorem

Let us now analyze in detail the instantaneous motion of a fluid. Consider the nature of the flow field in the neighborhood of some point O, which is moving with velocity $\mathbf{v}^0$, and let $\boldsymbol{\xi}$ be a small displacement away from O. Then to first order

$$v_i(\boldsymbol{\xi}) = v_i^0 + v_{i,j}\xi^j, \tag{21.1}$$

which shows that the relative velocity can be expressed in terms of the *velocity gradient tensor* $v_{i,j}$. Decomposing this tensor into an antisymmetric and symmetric part (cf. §A2.5), we have

$$v_i(\xi) = v_i^0 + \tfrac{1}{2}(v_{i,j} - v_{j,i})\xi^j + \tfrac{1}{2}(v_{i,j} + v_{j,i})\xi^j. \tag{21.2}$$

Each of the terms on the right-hand side of (21.2) admits of a direct physical interpretation. First, we see that the element undergoes a *translation* at a velocity $\mathbf{v}^0$. To interpret the second term, consider first the rotation of a rigid body with an angular rate ω_R around some axis $\mathbf{n}$ through O; write $\boldsymbol{\omega}_R = \omega_R \mathbf{n}$. Then the linear velocity of any point at a position $\boldsymbol{\xi}$ in the body is

$$\mathbf{v}_R = \boldsymbol{\omega}_R \times \boldsymbol{\xi}. \tag{21.3}$$

Now the second term of (21.2) is a sum of the form $\tilde{\Omega}_{ji}\xi^i$ where $\tilde{\Omega}_{ji}$ is the antisymmetric tensor

$$\tilde{\Omega}_{ji} \equiv \tfrac{1}{2}(v_{i,j} - v_{j,i}). \tag{21.4}$$

As is shown in equation (A2.63), this particular sum can be rewritten as $\tilde{\boldsymbol{\omega}} \times \boldsymbol{\xi}$, where $\tilde{\boldsymbol{\omega}}$ is the vector dual of the tensor $\tilde{\boldsymbol{\Omega}}$. From (21.3) we then see that the physical interpretation of the second term of (21.2) is that in the vicinity of O the fluid *rotates* with an effective angular velocity $\tilde{\boldsymbol{\omega}}$. To calculate the rotation rate we use (21.4) and equation (A2.60) to find

$$\tilde{\omega}^i = \tfrac{1}{2}e^{ijk}\tilde{\Omega}_{jk} = \tfrac{1}{2}e^{ijk}v_{k,j} = \tfrac{1}{2}(\text{curl } \mathbf{v})^i = \tfrac{1}{2}\omega^i, \tag{21.5}$$

which shows that the local angular velocity $\tilde{\boldsymbol{\omega}}$ is equal to one half the vorticity of the flow.

The symmetric tensor

$$E_{ij} = \tfrac{1}{2}(v_{i,j} + v_{j,i}) \tag{21.6}$$

appearing in the third term of (21.2) is known as the *rate of strain* tensor, for, as we shall now see, it describes how a fluid element in the neighborhood of O is *deformed* by the flow. In Cartesian coordinates

$$\mathsf{E} = \begin{pmatrix} (\partial v_x/\partial x) & \tfrac{1}{2}[(\partial v_x/\partial y) + (\partial v_y/\partial x)] & \tfrac{1}{2}[(\partial v_x/\partial z) + (\partial v_z/\partial x)] \\ \tfrac{1}{2}[(\partial v_y/\partial x) + (\partial v_x/\partial y)] & (\partial v_y/\partial y) & \tfrac{1}{2}[(\partial v_y/\partial z) + (\partial v_z/\partial y)] \\ \tfrac{1}{2}[(\partial v_z/\partial x) + (\partial v_x/\partial z)] & \tfrac{1}{2}[(\partial v_z/\partial y) + (\partial v_y/\partial z)] & (\partial v_z/\partial z) \end{pmatrix} \tag{21.7}$$

The diagonal elements $E_{(i)(i)}$ are called the *normal rates of strain*. As shown clearly in Figure 21.1a for the velocity field $\mathbf{v} = (x, 0, 0)$, these components represent the rates of *stretching* or *contraction* along the (x, y, z) axes. The off-diagonal terms E_{ij} equal half the *rate of shear deformation*, that is, half the rate of decrease of the angles between the edges of a parallelepiped of fluid originally having its edges along the orthogonal coordinate axes. For example, Figure 21.2 shows that the rate of change of the angle α between two line elements dx and dy originally along the x and y axes is $(d\alpha/dt) = -[(\partial v_x/\partial y) + (\partial v_y/\partial x)]$.

From fundamental matrix theory we know that any real symmetric matrix can be *diagonalized* by a suitable rotation of axes, and that when this is done the diagonal elements of the transformed matrix are equal to the (real) *eigenvalues* of the original matrix. The eigenvalues λ of E are the roots of the *secular equation* (or *characteristic equation*) obtained by setting to zero the determinant

$$|\mathsf{E} - \lambda \mathsf{I}| = |E_{ij} - \lambda\, \delta_{ij}| = 0. \tag{21.8}$$

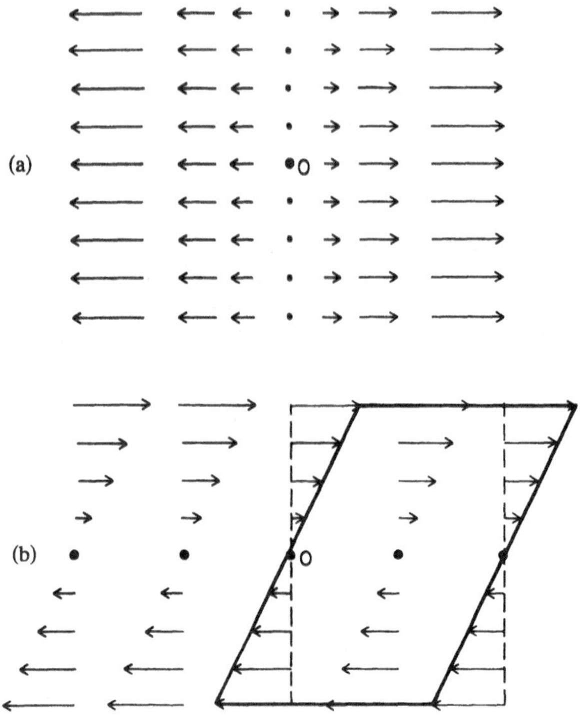

Fig. 21.1 (a) Stretching flow. (b) Flow with shear and vorticity.

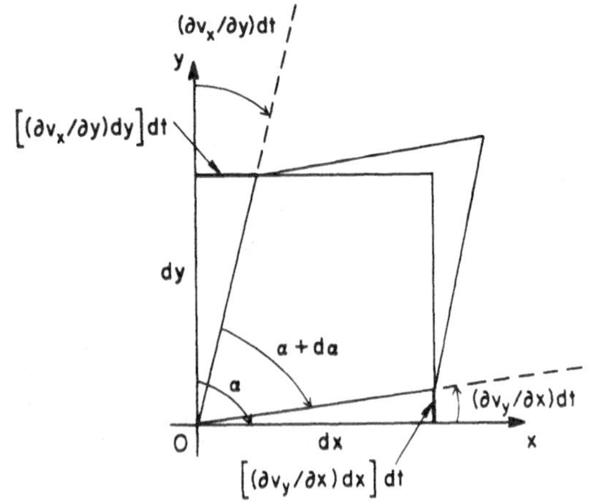

Fig. 21.2 Angular deformation of fluid element in shearing flow.

By expanding the determinant we find the cubic equation

$$\lambda^3 - A\lambda^2 + B\lambda - C = 0. \tag{21.9}$$

where A, B, and C are the three *invariants* (under rotation)

$$A = E_{11} + E_{22} + E_{33} = E_{ii} = \nabla \cdot \mathbf{v}, \tag{21.10}$$

$$B = E_{11}E_{22} + E_{22}E_{33} + E_{33}E_{11} - E_{12}^2 - E_{23}^2 - E_{31}^2 = \tfrac{1}{2}e_{abc}e_{ajk}E_{bj}E_{ck}, \tag{21.11}$$

and

$$C = |E_{ij}| = e_{ijk}E_{i1}E_{j2}E_{k3}. \tag{21.12}$$

The invariant A is clearly the *trace* of E; in this particular case it is equal to the divergence of the velocity field and is called the *dilatation* of the flow. Equation (21.9) yields three eigenvalues.

To show that A, B, and C are, in fact, invariant under rotations of coordinates we can proceed in two different ways. First, we can argue that because E is a definite physical entity regardless of the coordinate system chosen, the matrix E_{ij} must have unique eigenvalues; this will be true only if the secular equation is unique and hence its coefficients A, B, and C, are unique. Alternatively we can directly perform a transformation between two coordinate systems as described in §A2.4 and §A2.8. We find

$$\bar{A} = \bar{E}_{ii} = l_{pi}l_{qi}E_{pq} = \delta_{pq}E_{pq} = E_{pp} \equiv A, \tag{21.13}$$

$$\begin{aligned}\bar{B} &= \tfrac{1}{2}\bar{e}_{ijk}\bar{e}_{ipq}\bar{E}_{jp}\bar{E}_{kq} = \tfrac{1}{2}l_{ai}l_{bj}l_{ck}e_{abc}l_{di}l_{ep}l_{fq}e_{def}l_{sj}l_{tp}E_{st}l_{uk}l_{vq}E_{uv}\\ &= \tfrac{1}{2}\,\delta_{ad}\,\delta_{bs}\,\delta_{cu}\,\delta_{et}\,\delta_{fv}e_{abc}e_{def}E_{st}E_{uv} = \tfrac{1}{2}e_{abc}e_{aef}E_{be}E_{cf} \equiv B,\end{aligned} \tag{21.14}$$

and

$$\bar{C} = |\overline{E_{ij}}| = |l_{pi}l_{qj}E_{pq}| = |l_{pi}|\,|l_{qj}|\,|E_{pq}| = |E_{pq}| \equiv C, \tag{21.15}$$

where we have used equations (A2.12), (A2.29), and (A2.30).

Associated with each of the three eigenvalues of E is an eigenvector. If the eigenvalues are distinct, the eigenvectors are orthogonal; if not, the three eigenvectors can be orthogonalized. These three orthogonal vectors define the *principal axes* (or *principal directions*) relative to which the rate of strain tensor becomes diagonal. The eigenvalues themselves give the *principal rates of strain*, that is, the rates of stretching or contraction along the principal axes. Because the principal rates of strain are, in general, unequal, a spherical element will be deformed into an ellipsoid, as we will now show.

Consider the expression $f = \tfrac{1}{2}(v_{k,j}x^jx^k)$ where $\mathbf{x}$ is an infinitesimal displacement from O; this expression is quadratic in the coordinates x^j, and hence in general represents an ellipsoid. The vector normal to the surface f = constant is given by ∇f. Consider a region so small that $v_{k,j}$ can be taken to be constant; then the components of ∇f are

$$\tfrac{1}{2}(v_{k,j}x^jx^k)_{,i} = \tfrac{1}{2}v_{k,j}(\delta_i^jx^k + \delta_i^kx^j) = \tfrac{1}{2}(v_{i,j} + v_{j,i})x^j = E_{ij}x^j. \tag{21.16}$$

But this expression is just the contribution from the rate of strain tensor to the velocity of $\mathbf{x}$ relative to O. Thus we conclude that the velocity associated with strain of the fluid is normal to the level surfaces

$$v_{k,j}x^j x^k = \text{constant} \tag{21.17}$$

which define the *rate of strain ellipsoid.* Along the principal axes, which are normal to the surface of this ellipsoid, the fluid moves in the same direction as the axes, and therefore the principal axes remain mutually perpendicular during the deformation. Thus in these directions, and only in these directions, the fluid undergoes a pure expansion or contraction.

We can summarize our analysis of the motion of a fluid by the *Cauchy-Stokes decomposition theorem*: at each point in the flow the instantaneous state of motion of a fluid can be resolved into a translation, plus a dilatation along three mutually perpendicular axes, plus a rigid rotation of these axes.

2.2 Equations of Motion and Energy

22. The Stress Tensor

The forces acting on a fluid may be divided into two types. First, *body forces,* arising from external agents such as gravitation, act throughout the whole volume of an element of fluid. Second, *surface forces* act on a volume element of fluid at its boundary surface; we describe these surface forces in terms of fluid *stresses* acting on the surface. As we will see in Chapter 3, fluid stress results from the transport of momentum within the fluid by molecular motions; an example is the pressure within a gas.

Consider a planar material surface (i.e., composed of a definite set of particles) dS located at position $\mathbf{x}$ in the fluid, and oriented with normal $\mathbf{n}$. Let $\mathbf{t}$ be the surface force across this element exerted by the fluid on one side (the side containing $\mathbf{n}$) on the fluid on the other side. In general, this force will depend not only on the surface element's position $\mathbf{x}$, but also on its orientation $\mathbf{n}$, so we write the total force that the surface element experiences as $\mathbf{t}(\mathbf{x}, \mathbf{n})\, dS$. Furthermore, a moment's reflection shows that the force exerted across the surface element by the fluid on the side away from $\mathbf{n}$ on the fluid on the other side is $-\mathbf{t}(\mathbf{x}, \mathbf{n})\, dS$. But we could also regard this force as $\mathbf{t}(\mathbf{x}, -\mathbf{n})\, dS$; hence we conclude that $\mathbf{t}(\mathbf{x}, -\mathbf{n}) = -\mathbf{t}(\mathbf{x}, \mathbf{n})$, that is, that $\mathbf{t}$ is an antisymmetric function of $\mathbf{n}$.

Because the directions of $\mathbf{t}$ and $\mathbf{n}$ do not generally coincide, we infer that the fluid stresses that give rise to $\mathbf{t}$ must have associated with them two independent sets of directional information, hence we conjecture that stress must be a second-rank tensor. To demonstrate that this is so, suppose that an elementary tetrahedron located at a point P finds itself in equilibrium under the action of surface forces in the fluid. (For the moment, the requirement of equilibrium is arbitrary, but in §23 we shall see that the stresses at a point in a fluid are, in fact, always in equilibrium.)

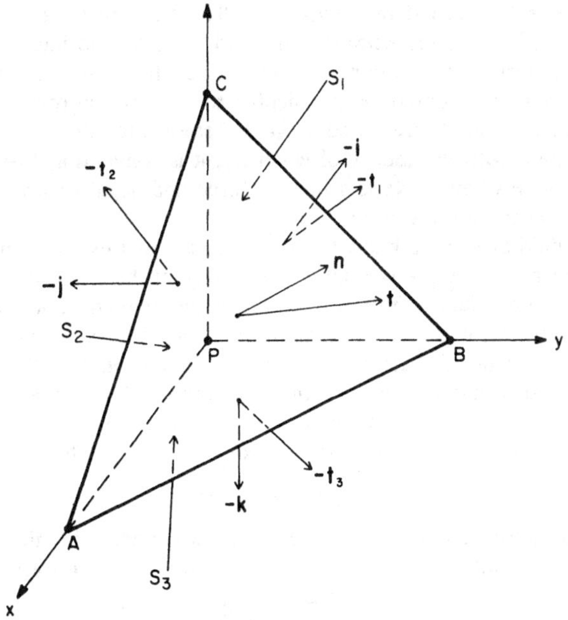

Fig. 22.1 Stresses acting on elementary tetrahedron.

As shown in Figure 22.1, choose three faces of the tetrahedron to lie in the coordinate planes, and let the slant face ABC with normal $\mathbf{n}$ have surface area S. The surface areas of the other three faces are evidently $S_1 = n_1 S$, $S_2 = n_2 S$, and $S_3 = n_3 S$, where n_l is the lth component of $\mathbf{n}$. Next, as above, let the surface force acting on a face with normal $\mathbf{m}$ be $\mathbf{t}(\mathbf{m})$. In particular, for notational convenience, write $\mathbf{t}(\mathbf{i}) \equiv \boldsymbol{t}_1$, $\mathbf{t}(\mathbf{j}) \equiv \boldsymbol{t}_2$, and $\mathbf{t}(\mathbf{k}) \equiv \boldsymbol{t}_3$. Then, noticing that S_1, S_2, and S_3 are oriented along $-\mathbf{i}$, $-\mathbf{j}$, and $-\mathbf{k}$, respectively, we see that to achieve equilibrium we must have

$$\mathbf{t}(\mathbf{n})S + \mathbf{t}(-\mathbf{i})S_1 + \mathbf{t}(-\mathbf{j})S_2 + \mathbf{t}(-\mathbf{k})S_3 \equiv 0, \tag{22.1a}$$

or, using the fact that $\mathbf{t}(\mathbf{n})$ is an antisymmetric function of $\mathbf{n}$,

$$\mathbf{t}(\mathbf{n})S - \boldsymbol{t}_1 S_1 - \boldsymbol{t}_2 S_2 - \boldsymbol{t}_3 S_3 \equiv 0. \tag{22.1b}$$

Expressing S_i in terms of n_i and S we then have

$$\mathbf{t}(\mathbf{n}) = n_1 \boldsymbol{t}_1 + n_2 \boldsymbol{t}_2 + n_3 \boldsymbol{t}_3. \tag{22.2}$$

Now let t_i denote the ith component of $\mathbf{t}(\mathbf{n})$, and let T_{ji} denote the ith component of $\boldsymbol{t}_j$. Then (22.2) states that

$$t_i = n^j T_{ji}. \tag{22.3}$$

Because $\mathbf{t}$ and $\mathbf{n}$ are independent vectors, we conclude that T_{ji} must be the components of a second-rank tensor T called the *stress tensor.* The diagonal elements of T are called *normal stresses,* which is appropriate because $T_{(i)(i)}$ gives the normal component of the surface force acting on a surface element that is oriented perpendicular to the ith coordinate axis. The off-diagonal elements are called *tangential stresses* (or *shearing stresses*). To be consistent with the theory of elasticity, it is conventional to take stress to be positive when it exerts a tension force and negative when it exerts a compressive force on a body.

For a fluid at rest we know from experiment that every element of area experiences only a force normal to the surface of the element, and that this force is independent of the orientation of the element. Such an isotropic normal stress with zero tangential stress is called a *hydrostatic stress.* The surface force $\mathbf{t}$ in this case must be proportional to $\mathbf{n}$, with the constant of proportionality independent of the orientation of $\mathbf{n}$; because we know empirically that the force in a fluid is always compressive (i.e., fluids do not support tension forces), we write the constant as $-p$. Then

$$t_j = T_{ij}n^i = -pn_j, \tag{22.4}$$

from which we recognize that p is to be identified with the hydrostatic pressure. The only form for T that can guarantee (22.4) for arbitrary $\mathbf{n}$ is

$$T_{ij} = -p\,\delta_{ij} \tag{22.5}$$

or

$$\mathsf{T} = -p\mathsf{I} \tag{22.6}$$

which is indeed an isotropic stress. Equation (22.6) applies to all fluids *at rest.*

If the fluid is *ideal,* then it is nonviscous and will not support tangential stress even when the fluid is in motion. The stress acting across a surface in an ideal fluid is thus always normal to the surface; hence (22.5) or (22.6) gives the stress tensor in an ideal fluid whether it is in motion or not. A *nonideal fluid* (i.e., a *real* fluid) will not support tangential stress when it is at rest, but can do so when it is in motion; for such fluids the stress tensor takes a more general form:

$$\mathsf{T} = -p\mathsf{I} + \boldsymbol{\sigma} \tag{22.7}$$

or

$$T_{ij} = -p\,\delta_{ij} + \sigma_{ij}. \tag{22.8}$$

The tensor $\boldsymbol{\sigma}$ is called the *viscous stress tensor*; we discuss it in some detail in §§25, 30, and 32.

23. The Momentum Equation

CAUCHY'S EQUATION OF MOTION

Consider a material volume $\mathscr{V}$ whose boundary surface lies entirely within the fluid. Then the principle of *conservation of linear momentum* asserts that the time rate of change of the momentum associated with the material element equals the total force acting on it; that is,

$$\frac{D}{Dt}\int_{\mathcal{V}} \rho \mathbf{v}\, dV = \int_{\mathcal{V}} \mathbf{f}\, dV + \int_{S} \mathbf{t}\, dS. \tag{23.1}$$

On the right-hand side of (23.1) the first term accounts for body forces and the second for surface forces. Now using the identity (19.12) we have

$$\int_{\mathcal{V}} \rho(Dv^i/Dt)\, dV = \int_{\mathcal{V}} f^i\, dV + \int_{S} t^i\, dS, \tag{23.2}$$

and thus from (22.3) and the divergence theorem (A2.67),

$$\int_{\mathcal{V}} \rho(Dv^i/Dt)\, dV = \int_{\mathcal{V}} f^i\, dV + \int_{S} T^{ji} n_j\, dS = \int_{\mathcal{V}} (f^i + T^{ji}_{\ ,j})\, dV. \tag{23.3}$$

Because the volume element $\mathcal{V}$ is arbitrary, equality of these integrals can be guaranteed only if their integrands are identical. We thus obtain *Cauchy's equation of motion*

$$\rho(Dv^i/Dt) = f^i + T^{ji}_{\ ,j} \tag{23.4}$$

or

$$\rho \mathbf{a} = \rho(D\mathbf{v}/Dt) = \mathbf{f} + \nabla \cdot \mathsf{T}. \tag{23.5}$$

In deriving (23.5) we have made no special assumptions about the physical mechanisms producing the stresses, and therefore the equation is quite general.

For an ideal fluid, (23.5) reduces to

$$\rho(D\mathbf{v}/Dt) = \mathbf{f} - \nabla p \tag{23.6}$$

or

$$\rho(Dv_i/Dt) = \rho(v_{i,t} + v^j v_{i,j}) = f_i - p_{,i}, \tag{23.7}$$

which is known as *Euler's equation of motion* or *Euler's momentum equation*. It is worth noticing here that the familiar pressure gradient term on the right-hand side of (23.6) or (23.7) is not, strictly speaking, the gradient of a scalar, but is in reality the divergence of an isotropic diagonal stress tensor.

SYMMETRY PROPERTIES OF THE STRESS TENSOR

Armed with (23.2) we are now in a position to prove that the stresses in a fluid are in equilibrium. Suppose a volume element has a characteristic dimension l; then its volume V scales as l^3, while its surface area S scales as l^2. Thus we see that

$$l^{-2}\int_{S} \mathbf{t}\, dS = O(l), \tag{23.8}$$

and hence in the limit as $l \to 0$ the integral vanishes. As this integral equals the sum of the surface forces over the surface, our assertion is proved.

Furthermore, by invoking the principle of *conservation of angular momentum*, we can prove that *the stress tensor is symmetric*. Thus, demanding that the time rate of change of the angular momentum of a material element equal the total applied torque we find

$$\frac{D}{Dt}\int_{\mathcal{V}} \rho(\mathbf{x}\times\mathbf{v})\,dV = \int_{\mathcal{V}} \rho\left(\mathbf{x}\times\frac{D\mathbf{v}}{Dt}\right)dV = \int_{\mathcal{V}} (\mathbf{x}\times\mathbf{f})\,dV + \int_S (\mathbf{x}\times\mathbf{t})\,dS, \tag{23.9}$$

where we used (19.12) and the fact that $(\mathbf{v}\times\mathbf{v})\equiv 0$. In (23.9), $\mathbf{x}$ is the position vector from the fixed origin of coordinates. Now writing (23.9) in component form, and applying the divergence theorem (A2.67) to the surface integral, we find

$$\int_{\mathcal{V}} e_{ijk}x^j[\rho(Dv^k/Dt)-f^k]\,dV = \int_S e_{ijk}x^jt^k\,dS = \int_S e_{ijk}x^jT^{lk}n_l\,dS \tag{23.10}$$
$$= \int_{\mathcal{V}} (e_{ijk}x^jT^{lk})_{,l}\,dV.$$

Because $\mathcal{V}$ is arbitrary, (23.10) implies that the integrands must be equal, hence

$$e_{ijk}x^j[\rho(Dv^k/Dt)-f^k-T^{lk}_{,l}]-e_{ijk}x^j_{,l}T^{lk}=0. \tag{23.11}$$

But the first term is identically zero by virtue of Cauchy's equation. Hence

$$e_{ijk}x^j_{,l}T^{lk} = e_{ijk}\,\delta^j_lT^{lk} = e_{ijk}T^{jk} = 0. \tag{23.12}$$

Writing out the last equality in (23.12) in components we have $(T_{12}-T_{21})=0$, $(T_{23}-T_{32})=0$, $(T_{31}-T_{13})=0$, whence it is obvious that

$$T_{ij} = T_{ji}, \tag{23.13}$$

which proves that T is symmetric.

The validity of the result just obtained hinges on the tacit assumption that the fluid is incapable of transporting angular momentum (and hence stress couples) across a surface by some internal process on the microscopic level; such fluids are called *nonpolar*, and include most gases. Certain non-Newtonian fluids (cf. §25) and polyatomic gases *can* transport angular momentum on the microscopic scale; these are called *polar* fluids. Such fluids have both a symmetric and an antisymmetric part to their stress tensors [see e.g., (**A1**, §5.11 and §5.13) for further discussion of these points]. Henceforth in this book we will deal exclusively with nonpolar fluids.

THE MOMENTUM FLUX DENSITY

The momentum equation can be cast into another extremely useful form. Taking the sum of the equation of motion

$$\rho v^i_{,t} = f^i - \rho v^j v^i_{,j} + T^{ij}_{,j} \tag{23.14}$$

and the product of the velocity with the equation of continuity

$$v^i\rho_{,t} = -v^i(\rho v^j)_{,j} \tag{23.15}$$

we find

$$(\rho v^i)_{,t} = f^i - (\rho v^i v^j)_{,j} + T^{ij}_{,j} = f^i - \Pi^{ij}_{,j} \tag{23.16}$$

where

$$\Pi^{ij} \equiv \rho v^i v^j - T^{ij}. \tag{23.17}$$

When integrated over a fixed volume in space, equation (23.16) gives

$$\frac{\partial}{\partial t}\int_V \rho v^i \, dV = \int_V f^i \, dV - \int_V \Pi^{ij}_{,j} \, dV = \int_V f^i \, dV - \int_S \Pi^{ij} n_j \, dS. \tag{23.18}$$

The left-hand side is the rate of change of the ith component of momentum contained in the fixed volume. The first term on the right-hand side accounts for the rate of momentum transfer by external forces, and the second must be interpreted as the rate of momentum flow out through the bounding surface (obvious for $f^i = 0$).

Thus Π_{ij} gives the rate of flow of the ith component of the momentum through a unit area oriented normal to the jth coordinate axis; it is therefore called the *momentum flux-density tensor.* In particular, for an ideal fluid

$$\Pi_{ij} = \rho v_i v_j + p\,\delta_{ij}, \tag{23.19}$$

so that

$$\Pi_{ij} n^j = \rho v_i v_j n^j + p n_i = [\rho \mathbf{v}(\mathbf{v} \cdot \mathbf{n}) + p\mathbf{n}]_i \tag{23.20}$$

which shows that Π_{ij} accounts for momentum transport in the fluid by both macroscopic flow and microscopic particle motions. We will see in Chapter 3 that the same results can be derived from kinetic theory and also can be extended to include viscous terms. In Chapter 4 we will see that Π_{ij} has a natural relativistic generalization.

CURVILINEAR COORDINATES

The covariant generalization of Cauchy's equation of motion is

$$\rho a^i = f^i + T^{ij}_{;j}. \tag{23.21}$$

The covariant generalization of Euler's equation of motion is simply

$$\rho a_i = f_i - p_{,i}. \tag{23.22}$$

We cannot proceed further with Cauchy's equation until we specify in detail the form of the stress tensor; we therefore defer discussion of (23.21) until §26. In Euler's equation we need only use the correct expressions for the acceleration (derived in §15) and for ∇p. For example, in spherical coordinates we have

$$\rho\left[\frac{\partial v_r}{\partial t}+v_r\frac{\partial v_r}{\partial r}+\frac{v_\theta}{r}\frac{\partial v_r}{\partial \theta}+\frac{v_\phi}{r\sin\theta}\frac{\partial v_r}{\partial \phi}-\frac{1}{r}(v_\theta^2+v_\phi^2)\right]=f_r-\frac{\partial p}{\partial r}, \quad (23.23a)$$

$$\rho\left[\frac{\partial v_\theta}{\partial t}+v_r\frac{\partial v_\theta}{\partial r}+\frac{v_\theta}{r}\frac{\partial v_\theta}{\partial \theta}+\frac{v_\phi}{r\sin\theta}\frac{\partial v_\theta}{\partial \phi}+\frac{1}{r}(v_r v_\theta-v_\phi^2\cot\theta)\right]=f_\theta-\frac{1}{r}\frac{\partial p}{\partial \theta}, \quad (23.23b)$$

$$\rho\left[\frac{\partial v_\phi}{\partial t}+v_r\frac{\partial v_\phi}{\partial r}+\frac{v_\theta}{r}\frac{\partial v_\phi}{\partial \theta}+\frac{v_\phi}{r\sin\theta}\frac{\partial v_\phi}{\partial \phi}+\frac{v_\phi}{r}(v_r+v_\theta\cot\theta)\right]=f_\phi-\frac{1}{r\sin\theta}\frac{\partial p}{\partial \phi}. \quad (23.23c)$$

For one-dimensional, spherically symmetric flow under a gravitational force $f_r=-G\mathcal{M}\rho/r^2$, we have simply

$$(\partial v_r/\partial t)+v_r(\partial v_r/\partial r)=-\rho^{-1}(\partial p/\partial r)-G\mathcal{M}/r^2. \quad (23.24)$$

For steady flow, the first term on the left-hand side is identically zero, and $(\partial/\partial r)$ can be replaced by (d/dr); we can then integrate (23.24) explicitly, [see (23.37) below].

HYDROSTATIC EQUILIBRIUM

In a *static* medium $\mathbf{v}=\mathbf{a}\equiv 0$, hence Euler's equation reduces to

$$\nabla p=\mathbf{f}, \quad (23.25)$$

from which we can determine the pressure stratification in the fluid. For instance, consider a plane-parallel atmosphere with homogeneous layers parallel to the (x, y) plane, stratified under a force $\mathbf{f}=-\rho g\mathbf{k}$ where g is the (constant) acceleration of gravity. Then (23.25) implies that $p=p(z)$ and

$$(dp/dz)=-\rho g. \quad (23.26)$$

Suppose the gas has a mean molecular weight μ and obeys the perfect gas law. Then in an *isothermal atmosphere* we can write

$$(dp/dz)=-(g\mu m_H/kT)p\equiv -p/H, \quad (23.27)$$

where H is the *pressure scale height.* In the atmosphere of the Sun H is of the order of 100 km; for the atmosphere of an early-type star it is a few thousand kilometers. Integrating (23.27) we have

$$p=p_0\exp[-(z-z_0)/H] \quad (23.28a)$$

or

$$\rho=\rho_0\exp[-(z-z_0)/H]. \quad (23.28b)$$

Equation (23.28) is valid only for an isothermal atmosphere, for which the scale height is constant; but it can be generalized to an atmosphere with a temperature gradient by writing

$$p = p_0 \exp\left\{-\int_{z_0}^{z} [g\mu m_H/kT(z)]\,dz\right\} = p_0 \exp\left[-\int_{z_0}^{z} dz/H(z)\right]. \tag{23.29}$$

In this case the density is most easily computed from the perfect gas law, given $p(z)$ and $T(z)$.

Notice that if we use the *column-mass* $dm \equiv -\rho\,dz$ as the independent variable, equation (23.26) becomes

$$(dp/dm) = g \tag{23.30}$$

whence

$$p = gm + p_0. \tag{23.31}$$

Here m is measured downward into the atmosphere (i.e., in the direction of gravity). Equation (23.31) holds whether or not the material is isothermal and shows why the column mass is the natural choice of coordinate for problems of hydrostatic equilibrium.

BERNOULLI'S EQUATION

From the vector identity (A2.59) with $\mathbf{a} = \mathbf{b} = \mathbf{v}$ we find

$$(\mathbf{v}\cdot\nabla)\mathbf{v} = \tfrac{1}{2}\nabla v^2 - \mathbf{v}\times(\nabla\times\mathbf{v}), \tag{23.32}$$

hence

$$(D\mathbf{v}/Dt) = (\partial\mathbf{v}/\partial t) - \mathbf{v}\times(\nabla\times\mathbf{v}) + \nabla(\tfrac{1}{2}v^2), \tag{23.33}$$

which is known as *Lagrange's acceleration formula.* Suppose now that the external force $\mathbf{f}$ acting on the fluid is conservative so that it can be written as the gradient of a potential Φ, that is, $\mathbf{f} = -\rho\,\nabla\Phi$. Then Euler's equation of motion for an ideal fluid becomes

$$(\partial\mathbf{v}/\partial t) - \mathbf{v}\times(\nabla\times\mathbf{v}) + \nabla(\tfrac{1}{2}v^2) + \rho^{-1}\nabla p + \nabla\Phi = 0. \tag{23.34}$$

We can derive an important integral of this equation by integrating *along a streamline.* Let $d\mathbf{s}$ be an element of length along any path. Then for any quantity α,

$$d\mathbf{s}\cdot\nabla\alpha = \alpha_{,i}\,dx^i \equiv d\alpha \tag{23.35}$$

where $d\alpha$ is the change in α along $d\mathbf{s}$. Furthermore, if $d\mathbf{s}$ lies along a streamline, it is parallel to $\mathbf{v}$, hence $d\mathbf{s}\cdot[\mathbf{v}\times(\nabla\times\mathbf{v})]$ must be identically zero. Using these results in (23.24) we see that in a conservative field of force the integral of Euler's equation along a streamline is

$$\int (\partial\mathbf{v}/\partial t)\cdot d\mathbf{s} + \tfrac{1}{2}v^2 + \int dp/\rho + \Phi = \text{constant} = C(t). \tag{23.36}$$

This equation is valid for unsteady flow, but applies only instant by instant because, in general, a given streamline in unsteady flows is composed of different sets of fluid particles at different times.

For *steady flow* (23.36) simplifies to

$$\tfrac{1}{2}v^2+\int dp/\rho+\Phi=C \tag{23.37}$$

where now C is a constant for all time along each streamline, but may differ from streamline to streamline. For *incompressible flow* ($\rho=$ constant) we obtain *Bernoulli's equation*

$$\tfrac{1}{2}v^2+(p/\rho)+\Phi=C. \tag{23.38}$$

Explicit forms of (23.37) can also be obtained for *barotropic flow*, in which ρ is a function of p only. For example, for a polytropic gas $p\propto\rho^\gamma$ and we find

$$\tfrac{1}{2}v^2+[\gamma/(\gamma-1)](p/\rho)+\Phi=C. \tag{23.39}$$

For an isothermal flow $p\propto\rho$ and (23.37) yields

$$\tfrac{1}{2}v^2+(kT/\mu m_{\rm H})\ln\rho+\Phi=C \tag{23.40a}$$

or

$$\tfrac{1}{2}v^2+(kT/\mu m_{\rm H})\ln p+\Phi=C. \tag{23.40b}$$

Another general integral of the momentum equation can be written for the case of *irrotational flow*, for which $\nabla\times\mathbf{v}=0$ and therefore $\mathbf{v}=\nabla\phi$. Then, for barotropic flow (23.34) becomes

$$\nabla\left[(\partial\phi/\partial t)+\tfrac{1}{2}v^2+\int dp/\rho+\Phi\right]=0. \tag{23.41}$$

We can now integrate (23.41) along an *arbitrary flow line* to obtain

$$(\partial\phi/\partial t)+\tfrac{1}{2}v^2+\int dp/\rho+\Phi=C(t), \tag{23.42}$$

where $C(t)$ is a function of time, but has the same value everywhere in the flow field. Equation (23.42) is known as the *Bernoulli-Euler equation* for potential flow. It is worth emphasizing that (23.37) holds along a streamline for steady flow, while (23.42) holds along an arbitrary line for a time-dependent but irrotational flow.

KELVIN'S CIRCULATION THEOREM

In §20 we derived Kelvin's equation for the time rate of change of the circulation

$$(D\Gamma/Dt)=\oint_C (D\mathbf{v}/Dt)\cdot d\mathbf{x}, \tag{23.43}$$

where C is a closed material curve. Substituting for $(D\mathbf{v}/Dt)$ from Euler's equation in a conservative force field we have

$$(D\Gamma/Dt)=-\oint_C(\nabla\Phi)\cdot d\mathbf{x}-\oint_C\rho^{-1}(\nabla p)\cdot d\mathbf{x}=-\oint_C d\Phi-\oint_C dp/\rho. \tag{23.44}$$

The first integral on the right-hand side is always zero because the potential is a single-valued point function. The second integral will be zero if the fluid is incompressible or is barotropic so that $(dp/\rho) = df(p)$. From (23.44) we thus obtain *Kelvin's circulation theorem* (or the *law of conservation of circulation*), which states that the circulation around any closed material path in a barotropic (or isothermal) flow of a perfect fluid moving in a conservative force field is independent of time. Such flows are called *circulation preserving.*

A consequence of Kelvin's theorem is that if the flow around some material path is initially irrotational, it remains irrotational. For flows of viscous fluids, particularly in the presence of material boundary surfaces or bodies moving through the fluid, this statement is not generally true, and vortices can be formed in thin boundary layers and then shed into the fluid [see, e.g., (**O1,** 70–72), (**S1**), and (**Y1**, 342–344)].

24. *The Energy Equation*

In addition to conservation laws for the mass and momentum of a fluid element, we can formulate a law of *conservation of energy*, which says that the rate at which the energy of a material element increases equals the rate at which heat is delivered to that element minus the rate at which it does work against its surroundings. Clearly this is merely a restatement of the first law of thermodynamics. In an ideal fluid there are no microscopic processes of energy dissipation from internal friction (viscosity), or of energy transport from one set of particles to another (thermal conduction). Moreover, at present we are disregarding heat exchange with external sources or by other physical mechanisms (e.g., radiation), hence there is a total absence of heat transfer between different parts of the fluid. We therefore conclude that *the motion of an ideal fluid must necessarily be adiabatic.*

TOTAL ENERGY AND MECHANICAL ENERGY EQUATIONS

For a material volume $\mathcal{V}$, the mathematical formulation of the energy conservation principle stated above is

$$\frac{D}{Dt}\int_{\mathcal{V}} \rho(e+\tfrac{1}{2}v^2)\,dV = \int_{\mathcal{V}} \mathbf{f}\cdot\mathbf{v}\,dV + \int_{S} \mathbf{t}\cdot\mathbf{v}\,dS. \tag{24.1}$$

The term on the left-hand side of (24.1) is the rate of change of the internal plus kinetic energy of the material element, and the terms on the right-hand side are the rates at which work is done by external forces and fluid stresses respectively. From (22.4) and the divergence theorem the rightmost term becomes, for an ideal fluid,

$$\int_{S} \mathbf{t}\cdot\mathbf{v}\,dS = -\int_{S} p\mathbf{v}\cdot\mathbf{n}\,dS = -\int_{\mathcal{V}} \nabla\cdot(p\mathbf{v})\,dV. \tag{24.2}$$

Then using (19.12) to transform the left-hand side of (24.1), we have

$$\int_{\mathcal{V}} \{\rho[D(e+\tfrac{1}{2}v^2)/Dt]+\nabla\cdot(p\mathbf{v})-\mathbf{f}\cdot\mathbf{v}\}\,dV=0; \tag{24.3}$$

because $\mathcal{V}$ is arbitrary this implies

$$\rho[D(e+\tfrac{1}{2}v^2)/Dt]+\nabla\cdot(p\mathbf{v})=\mathbf{f}\cdot\mathbf{v}. \tag{24.4}$$

This is the *total energy equation* for an ideal fluid.

Now applying (19.13) to (24.4) we find the alternative forms

$$\partial(\rho e+\tfrac{1}{2}\rho v^2)/\partial t+\nabla\cdot[(\rho e+p+\tfrac{1}{2}\rho v^2)\mathbf{v}]=\mathbf{f}\cdot\mathbf{v}, \tag{24.5}$$

or

$$\partial(\rho e+\tfrac{1}{2}\rho v^2)/\partial t+\nabla\cdot[(h+\tfrac{1}{2}v^2)\rho\mathbf{v}]=\mathbf{f}\cdot\mathbf{v}, \tag{24.6}$$

where h is the specific enthalpy. Equation (24.6) can be integrated over a fixed volume V to obtain

$$\frac{\partial}{\partial t}\int_V \rho(e+\tfrac{1}{2}v^2)\,dV=-\int_S (h+\tfrac{1}{2}v^2)\rho\mathbf{v}\cdot d\mathbf{S}+\int_V \mathbf{f}\cdot\mathbf{v}\,dV. \tag{24.7}$$

In physical terms, the left-hand side of (24.7) is the rate of change of the total fluid energy in the fixed volume. The right-hand side is the rate of work being done by external forces, minus a term that must be interpreted as the rate of energy flow through the boundary surface of the volume. Hence $(h+\tfrac{1}{2}v^2)\rho\mathbf{v}$ must be the *energy flux density vector*; we will obtain a relativistic generalization of this expression in Chapter 4.

By forming the dot product of $\mathbf{v}$ with (23.6), Euler's equation, we find

$$\rho\mathbf{v}\cdot(D\mathbf{v}/Dt)=\tfrac{1}{2}\rho[D(\mathbf{v}\cdot\mathbf{v})/Dt]=\rho[D(\tfrac{1}{2}v^2)/Dt]=\mathbf{v}\cdot\mathbf{f}-(\mathbf{v}\cdot\nabla)p \tag{24.8}$$

which is the *mechanical energy equation* for an ideal fluid. Physically, this equation states that the time rate of change of the kinetic energy of a fluid element equals the rate of work done on that element by applied forces (both external and pressure).

GAS ENERGY EQUATION; FIRST AND SECOND LAWS OF THERMODYNAMICS

Subtracting the mechanical energy equation (24.8) from the total energy equation (24.4) we obtain the *gas energy equation*

$$\rho(De/Dt)+p(\nabla\cdot\mathbf{v})=0, \tag{24.9}$$

or, using the continuity equation (19.3),

$$(De/Dt)+p[D(1/\rho)/Dt]=0. \tag{24.10}$$

An alternative form of (24.10) is

$$\rho(Dh/Dt)-(Dp/Dt)=0. \tag{24.11}$$

We immediately recognize that (24.10) is merely a restatement of the first law of thermodynamics

$$(De/Dt)+p[D(1/\rho)/Dt]=T(Ds/Dt) \tag{24.12}$$

in the case that

$$(Ds/Dt)=(\partial s/\partial t)+(\mathbf{v}\cdot\boldsymbol{\nabla})s\equiv 0. \tag{24.13}$$

This result is completely consistent with our expectation, stated earlier, that the flow of an ideal fluid must be adiabatic. Combining (24.13) with the equation of continuity we have an *equation of entropy conservation* for ideal fluid flow:

$$\partial(\rho s)/\partial t+\boldsymbol{\nabla}\cdot(\rho s\mathbf{v})=0. \tag{24.14}$$

We can therefore interpret $\rho s\mathbf{v}$ as an *entropy flux density*.

If the entropy is initially constant throughout some volume of a perfect fluid then it remains constant for all times within that material volume during its subsequent motion, and (24.13) reduces to $s=\text{constant}$; such a flow is called *isentropic*. A flow in which the entropy is constant and is uniform throughout the entire region under consideration is called *homentropic*.

The second law of thermodynamics for a material volume implies that

$$\frac{D}{Dt}\int_{\mathscr{V}}\rho s\,dV\geq\int_{\mathscr{V}}\frac{\rho}{T}\left(\frac{Dq}{Dt}\right)dV \tag{24.15}$$

where (Dq/Dt) is the rate of heat exchange of the fluid element with its surroundings. For an ideal fluid having no heat exchange with external sources this implies

$$\frac{D}{Dt}\int_{\mathscr{V}}\rho s\,dV\geq 0. \tag{24.16}$$

From what has been said thus far we would conclude that the equality sign must hold in (24.16) for the flow of an ideal fluid. However as we will see in Chapter 5, if we allow discontinuities (e.g., shocks) in the flow, then there can be an increase in entropy across a discontinuity even in an ideal fluid. Even though the entropy increase is correctly predicted mathematically by the equations of fluid flow for such *weak solutions*, it can be interpreted consistently, from a physical point of view, only when we admit the possibility of dissipative processes in the fluid.

CURVILINEAR COORDINATES

The covariant generalization of (24.6) is

$$[\rho(e+\tfrac{1}{2}v^2)]_{,t}+[\rho(h+\tfrac{1}{2}v^2)v^i]_{;i}=f_iv^i. \tag{24.17}$$

For example, in spherical coordinates we have

$$\frac{\partial}{\partial t}[\rho(e+\tfrac{1}{2}v^2)]+\frac{1}{r^2}\frac{\partial}{\partial r}[r^2\rho v_r(h+\tfrac{1}{2}v^2)]$$
$$+\frac{1}{r\sin\theta}\frac{\partial}{\partial\theta}[\rho v_\theta \sin\theta(h+\tfrac{1}{2}v^2)] \tag{24.18}$$
$$+\frac{1}{r\sin\theta}\frac{\partial}{\partial\phi}[\rho v_\phi(h+\tfrac{1}{2}v^2)]=\mathbf{f}\cdot\mathbf{v}.$$

For a one-dimensional, spherically symmetric flow the terms in $(\partial/\partial\theta)$ and $(\partial/\partial\phi)$ vanish identically.

STEADY FLOW

For one-dimensional steady flow we can write an explicit integral of the energy equation. Consider, for example, planar geometry with a gravity force $\mathbf{f}=-\rho g\mathbf{k}$ where g is constant. Then (24.6) becomes

$$d[\rho v_z(h+\tfrac{1}{2}v_z^2)]/dz=-\rho v_z g \tag{24.19}$$

which, recalling from (19.6) that $\dot{m}=\rho v_z=\text{constant}$, integrates to

$$\dot{m}(h+\tfrac{1}{2}v_z^2+gz)=\text{constant}. \tag{24.20}$$

Similarly, in spherical geometry, if $\mathbf{f}=-(G\mathcal{M}\rho/r^2)\hat{\mathbf{r}}$ we have

$$d[r^2\rho v_r(h+\tfrac{1}{2}v_r^2)]/dr=-(G\mathcal{M}/r^2)(r^2\rho v_r). \tag{24.21}$$

which, in view of (19.10), integrates to

$$\dot{\mathcal{M}}[h+\tfrac{1}{2}v_r^2-(G\mathcal{M}/r)]=\text{constant}. \tag{24.22}$$

Both (24.20) and (24.22) have a simple physical meaning: in a steady flow of an ideal fluid the *energy flux*, which equals the mass flux times the total energy (enthalpy plus kinetic plus potential) per unit mass, is a constant. We generalize these results to include the effects of viscous dissipation and thermal conduction in Chapter 3.

MATHEMATICAL STRUCTURE OF THE EQUATIONS OF FLUID DYNAMICS

We have formulated a total of five partial differential equations governing the flow of an ideal fluid: the continuity equation, three components of the momentum equation, and the energy equation. These relate six dependent variables: ρ, p, e, and the components of $\mathbf{v}$. To close the system we require *constitutive relations* that specify the thermodynamic properties of the material. We know that in general any thermodynamic property can be expressed as a function of two state variables. Thus we might choose an equation of state $p=p(\rho, T)$ and an equation for the internal energy, $e=e(\rho, T)$ sometimes called the *caloric equation of state.* These relations may be those appropriate to a perfect monatomic gas of structureless point particles having only translational degrees of freedom, as described in §§1,

4, and 8, or may include the effects of ionization and internal excitation as described in §§12 and 14.

We then have a total of seven equations in seven unknowns. This system of equations can be solved for the spatial variation of all unknowns as a function of time once we are given a set of *initial conditions* that specify the state and motion of the fluid at a particular time, plus a set of *boundary conditions* where constraints are placed on the flow. There are many techniques for solving these nonlinear equations. Analytical methods can yield solutions for some simplified problems, for example, certain incompressible flows, or low-amplitude flows for which the equations can be linearized. But, in general, this approach is too restrictive, and recourse must be had to numerical methods. In the linear regime numerical methods provide flexibility and generality in treating, for example, the depth-variation of the physical properties of the medium in which the flow occurs. In the nonlinear regime we encounter essential new physical phenomena: among them the formation of *shocks*. Here one can employ the *method of characteristics* or solve *difference-equation* representations of the fluid equations. The latter approach is both powerful and convenient, and is easy to explain. We defer further discussion of these methods until Chapter 5 where we discuss the additional physics needed in the context of specific examples.

References

(A1) Aris, R. (1962) *Vectors, Tensors, and the Basic Equations of Fluid Mechanics.* Englewood Cliffs: Prentice-Hall.
(B1) Batchelor, G. K. (1967) *An Introduction to Fluid Dynamics.* Cambridge: Cambridge University Press.
(B2) Becker, E. (1968) *Gas Dynamics.* New York: Academic Press.
(L1) Landau, L. D. and Lifshitz, E. M. (1959) *Fluid Mechanics.* Reading: Addison-Wesley.
(O1) Owczarek, J. A. (1964) *Fundamentals of Gas Dynamics.* Scranton: International Textbook Co.
(S1) Schlichting, H. (1960) *Boundary Layer Theory.* New York: McGraw-Hill.
(Y1) Yuan, S. W. (1967) *Foundations of Fluid Mechanics.* Englewood Cliffs: Prentice-Hall.

3

Dynamics of Viscous and Heat-Conducting Fluids

It is known from experiment that in all real fluids there are internal processes that result in transport of momentum and energy from one fluid parcel to another on a microscopic level. The momentum transport mechanisms give rise to *internal frictional forces* (*viscous forces*) that enter directly into the equations of motion, and that also produce frictional energy *dissipation* in the flow. The energy transport mechanisms lead to energy *conduction* from one point in the flow to another.

In this chapter we derive equations of fluid flow that explicitly account for the processes described above. To begin, we adopt a continuum view, which permits us to derive the mathematical form of the equations from quite general reasoning, drawing on heuristic arguments, symmetry considerations, and empirical facts. We then reexamine the problem from a microscopic kinetic-theory view, from which we recover essentially the same set of equations, but now with a much clearer understanding of the underlying physics. This approach also allows us to evaluate explicitly (for a given molecular model) the transport coefficients that are introduced on empirical grounds in the macroscopic equations.

3.1 Equations of Motion and Energy: The Continuum View

25. The Stress Tensor for a Newtonian Fluid

In §22 we showed that internal forces in a fluid can be described in terms of a stress tensor T_{ij}, which we showed to be symmetric, but otherwise left unspecified. We now wish to derive an explicit expression for the stress tensor in terms of the physical properties of the fluid and its state of motion. We can deduce the form of T_{ij} from the following physical considerations. (1) We expect internal frictional forces to exist only when one element of fluid moves relative to another; hence viscous terms must depend on the space derivatives of the velocity field, $v_{i,j}$. (2) We demand that the stress tensor reduce to its hydrostatic form when the fluid is at rest or translates uniformly (in which case it is at rest for an observer moving with the translation velocity). We therefore write

$$T_{ij} = -p\,\delta_{ij} + \sigma_{ij} \qquad (25.1)$$

where σ_{ij} is the *viscous stress tensor*, which accounts for the internal frictional forces in the flow. (3) For small velocity gradients we expect viscous forces, hence σ_{ij}, to depend only linearly on space derivatives of the velocity. A fluid that obeys this restriction is called a *Newtonian fluid.* [A discussion of *Stokesian fluids*, for which σ_{ij} depends quadratically on the velocity gradients, may be found in (**A1**, §§5.21 and 5.22). Non-Newtonian fluids of a very general nature are discussed in (**A2**).] (4) We expect viscous forces to be zero within an element of fluid in rigid rotation (because there is no slippage then). On these grounds, we expect no contribution to σ_{ij} from the vorticity tensor $\Omega_{ji} = (v_{i,j} - v_{j,i})$. We can exclude such a contribution for mathematical reasons as well because we have already shown that T_{ij}, hence σ_{ij}, must be symmetric [cf. (23.13)]. We therefore expect σ_{ij} to contain terms of the form $E_{ij} = \frac{1}{2}(v_{i,j} + v_{j,i})$, which, as we saw in §21, describe the rate of strain in the fluid.

The most general symmetrical tensor of rank two satisfying the above requirements is

$$\sigma_{ij} = \mu(v_{i,j} + v_{j,i}) + \lambda v^k_{,k}\,\delta_{ij} = 2\mu E_{ij} + \lambda(\nabla \cdot \mathbf{v})\,\delta_{ij}. \tag{25.2}$$

If we assume that the fluid is *isotropic*, so that there are no preferred directions, then λ and μ must be scalars; μ is called the *coefficient of shear viscosity* or the *coefficient of dynamical viscosity*, and λ is called the *dilatational coefficient of viscosity*, or the *second coefficient of viscosity*. For the present we regard both λ and μ as purely macroscopic coefficients that can be determined from experiment.

It is convenient to cast (25.2) into the slightly different form

$$\sigma_{ij} = \mu(v_{i,j} + v_{j,i} - \tfrac{2}{3}v^k_{,k}\,\delta_{ij}) + \zeta v^k_{,k}\,\delta_{ij}, \tag{25.3}$$

where

$$\zeta \equiv \lambda + \tfrac{2}{3}\mu \tag{25.4}$$

is known as the *coefficient of bulk viscosity*. The expression in parentheses in (25.3) has the mathematical property of being *traceless*, that is, it sums to zero when we contract on i and j. It also has the property of vanishing identically for a fluid that dilates symmetrically, that is, such that $(\partial v_1/\partial x^{(1)}) = (\partial v_2/\partial x^{(2)}) = (\partial v_3/\partial x^{(3)})$, and $(\partial v_i/\partial x^j) = 0$ for $i \neq j$. One can argue on intuitive physical grounds (**E2**, 19) that no frictional forces should be present in this case because there is no slipping of one part of the fluid relative to another; this will actually be true if and only if the coefficient of bulk viscosity ζ is identically zero.

Using (25.1) and (25.2) we see that the mean of the principal stresses is

$$\tfrac{1}{3}T_{ii} = -p + \zeta v^k_{,k}. \tag{25.5}$$

For an incompressible fluid $v^k_{,k} \equiv 0$, hence $-\frac{1}{3}T_{ii}$ equals the hydrostatic pressure p. For a compressible fluid we must identify p with the thermodynamic pressure given by the equation of state in order to be consistent with the requirements of hydrostatic and thermodynamic equilibrium.

If we call the mean of the principal stresses $-\bar{p}$, then

$$\bar{p}-p=-\zeta(\nabla\cdot\mathbf{v})=\zeta(D\ln\rho/Dt), \tag{25.6}$$

which shows that unless $\zeta\equiv 0$, there is in general a discrepancy between the scalar $\bar{p}$, which measures the isotropic part of the internal forces influencing the flow dynamics of a *moving* gas, and the thermostatic pressure p for the same gas *at rest* under identical thermodynamic conditions (i.e., same composition, density, temperature, ionization, etc.). This discrepancy can be quite significant for a gas undergoing very rapid (perhaps explosive) expansion or compression.

From kinetic theory one finds that for a perfect monatomic gas ζ is identically zero (cf. §32), as was first shown by Maxwell. The assumption that $\zeta\equiv 0$ for fluids in general was advanced by Stokes and is referred to as the *Stokes hypothesis*; the relation $\lambda=-\frac{2}{3}\mu$ (which implies $\zeta\equiv 0$) is called the *Stokes relation.* Fluids for which $\zeta=0$ are called *Maxwellian fluids.* In much of the classical work on the dynamics of viscous fluids the Stokes hypothesis is invoked from the outset. More recent work has been directed towards identifying the origins of bulk viscosity and its significance for fluid flow [e.g., see (**H2,** 521 and 644), (**O1,** 540–541), (**T1**), (**T3**), (**V1,** §10.8), and (**Z1,** 469)]. One finds that the bulk viscosity is nonzero (and indeed is of the same order of magnitude as the shear viscosity) when the gas undergoes a *relaxation process* on a time scale comparable to, or slower than, a characteristic fluid-flow time. Examples of such processes are the exchange of energy between translational motions and vibrational and rotational motions in polyatomic molecules, or between translational energy and ionization energy in an ionizing gas. When such processes occur, internal equilibration can lag flow-induced changes in the state of the gas, and this lag may give rise to irreversible processes that can, for example, cause the absorption and attenuation of sound waves (**L1,** §78) and affect the thermodynamic structure of shock waves (**Z1,** Chapter 7). In the derivations that follow we will not assume that $\zeta=0$, although this simplification will usually be made in later work. It should be noted that for incompressible flow $\nabla\cdot\mathbf{v}\equiv 0$, and the question of the correct value for ζ becomes irrelevant.

The covariant generalization of (25.1) valid in curvilinear coordinates is

$$T_{ij}=-pg_{ij}+\sigma_{ij}, \tag{25.7}$$

where from (25.3)

$$\sigma_{ij}=\mu(v_{i;j}+v_{j;i})+(\zeta-\tfrac{2}{3}\mu)v^k_{;k}g_{ij}=2\mu E_{ij}+(\zeta-\tfrac{2}{3}\mu)v^k_{;k}g_{ij}. \tag{25.8}$$

Here g_{ij} is the metric tensor of the coordinate system. Consider, for example, spherical coordinates. In what follows we shall need expressions for the rate of strain tensor E_{ij}, so we compute these first and then assemble T_{ij}. Thus using equation (A3.75), and retaining only the nonzero

Christoffel symbols given by equation (A3.63), we obtain

$$E_{11}=(\partial v_1/\partial r), \tag{25.9a}$$

$$E_{12}=\tfrac{1}{2}[(\partial v_1/\partial \theta)+(\partial v_2/\partial r)]-(v_2/r), \tag{25.9b}$$

$$E_{13}=\tfrac{1}{2}[(\partial v_1/\partial \phi)+(\partial v_3/\partial r)]-(v_3/r), \tag{25.9c}$$

$$E_{22}=(\partial v_2/\partial \theta)+rv_1, \tag{25.9d}$$

$$E_{23}=\tfrac{1}{2}[(\partial v_2/\partial \phi)+(\partial v_3/\partial \theta)]-v_3\cot\theta, \tag{25.9e}$$

$$E_{33}=(\partial v_3/\partial \phi)+v_1 r\sin^2\theta+v_2\sin\theta\cos\theta. \tag{25.9f}$$

Now converting to physical components via equations (A3.46) and the analogues of (A3.47) appropriate to covariant components we have

$$E_{rr}=\frac{\partial v_r}{\partial r}, \tag{25.10a}$$

$$E_{r\theta}=\frac{1}{2}\left(\frac{1}{r}\frac{\partial v_r}{\partial \theta}+\frac{\partial v_\theta}{\partial r}-\frac{v_\theta}{r}\right), \tag{25.10b}$$

$$E_{r\phi}=\frac{1}{2}\left(\frac{1}{r\sin\theta}\frac{\partial v_r}{\partial \phi}+\frac{\partial v_\phi}{\partial r}-\frac{v_\phi}{r}\right), \tag{25.10c}$$

$$E_{\theta\theta}=\frac{1}{r}\frac{\partial v_\theta}{\partial \theta}+\frac{v_r}{r}, \tag{25.10d}$$

$$E_{\theta\phi}=\frac{1}{2}\left(\frac{1}{r\sin\theta}\frac{\partial v_\theta}{\partial \phi}+\frac{1}{r}\frac{\partial v_\phi}{\partial \theta}-\frac{v_\phi\cot\theta}{r}\right), \tag{25.10e}$$

$$E_{\phi\phi}=\frac{1}{r\sin\theta}\frac{\partial v_\phi}{\partial \phi}+\frac{v_r}{r}+\frac{v_\theta\cot\theta}{r}. \tag{25.10f}$$

Finally, assembling T_{ij} in (25.7) and (25.8) we obtain

$$T_{rr}=-p+2\mu\left(\frac{\partial v_r}{\partial r}\right)+(\zeta-\tfrac{2}{3}\mu)\boldsymbol{\nabla}\cdot\mathbf{v}, \tag{25.11a}$$

$$T_{\theta\theta}=-p+2\mu\left(\frac{1}{r}\frac{\partial v_\theta}{\partial \theta}+\frac{v_r}{r}\right)+(\zeta-\tfrac{2}{3}\mu)\boldsymbol{\nabla}\cdot\mathbf{v}, \tag{25.11b}$$

$$T_{\phi\phi}=-p+2\mu\left(\frac{1}{r\sin\theta}\frac{\partial v_\phi}{\partial \phi}+\frac{v_r}{r}+\frac{v_\theta\cot\theta}{r}\right)+(\zeta-\tfrac{2}{3}\mu)\boldsymbol{\nabla}\cdot\mathbf{v}, \tag{25.11c}$$

$$T_{r\theta}=\mu\left[r\frac{\partial}{\partial r}\left(\frac{v_\theta}{r}\right)+\frac{1}{r}\frac{\partial v_r}{\partial \theta}\right], \tag{25.11d}$$

$$T_{r\phi}=\mu\left[r\frac{\partial}{\partial r}\left(\frac{v_\phi}{r}\right)+\frac{1}{r\sin\theta}\frac{\partial v_r}{\partial \phi}\right], \tag{25.11e}$$

$$T_{\theta\phi}=\mu\left[\frac{\sin\theta}{r}\frac{\partial}{\partial \theta}\left(\frac{v_\phi}{\sin\theta}\right)+\frac{1}{r\sin\theta}\frac{\partial v_\theta}{\partial \phi}\right], \tag{25.11f}$$

where $(\boldsymbol{\nabla}\cdot\mathbf{v})$ is given by equation (A3.88).

26. The Navier-Stokes Equations

CARTESIAN COORDINATES

Cauchy's equation of motion (23.4) or (23.5) is extremely general because it makes no particular assumptions about the form of the stress tensor. If we specialize Cauchy's equation to the case of a Newtonian fluid by using equations (25.1) and (25.3), we obtain the *Navier-Stokes equations*, which are the equations most commonly employed to describe the flow of a viscous fluid. For Cartesian coordinates, we find, by direct substitution,

$$\rho(Dv_i/Dt) = \rho(v_{i,t} + v_j v_{i,j}) = f_i - p_{,i} + [\mu(v_{i,j} + v_{j,i})]_{,j} + [(\zeta - \tfrac{2}{3}\mu)v_{k,k}]_{,i}. \tag{26.1}$$

(Notice that this equation is not in covariant form, and applies only in Cartesian coordinates. By introducing the metric tensor we could write it in a form that is covariant when ordinary derivatives are changed to covariant derivatives, but this would require too cumbersome a notation for our present purposes.)

There are several somewhat simplified forms of (26.1) that are of interest. For example, for one-dimensional flows in planar geometry, (26.1) reduces to

$$\rho \frac{Dv_z}{Dt} = f_z - \frac{\partial p}{\partial z} + \frac{\partial}{\partial z}\left[(\tfrac{4}{3}\mu + \zeta)\frac{\partial v_z}{\partial z}\right]. \tag{26.2}$$

We notice that for this class of flows the shear viscosity and the bulk viscosity have the same dynamical effect, and that their combined influence on the momentum equation can be accounted for by using an effective viscosity coefficient $\mu' = (\mu + \frac{3}{4}\zeta)$.

Viscous forces in the equations of motion for one-dimensional flows are often described in terms of a fictitious *viscous pressure*

$$Q \equiv -\tfrac{4}{3}\mu'(\partial v_z/\partial z) = \tfrac{4}{3}\mu'(D \ln \rho/Dt), \tag{26.3}$$

which can make a significant contribution to the dynamical pressure during a rapid compression of the gas. The momentum equation can then be written

$$\rho(Dv_z/Dt) = f_z - [\partial(p+Q)/\partial z]. \tag{26.4}$$

We will see in §27 that μ' and Q as defined here appear again in the same forms in the energy equation, so that they provide a completely consistent formalism for treating viscous processes in one-dimensional planar flow problems. Furthermore, we will find in §59 that the concept of viscous pressure plays an important role in developing a technique for stabilizing the solution of flow problems with shocks by means of an *artificial viscosity* (or *pseudoviscosity*).

Another simple form of (26.1) results for flows in which the thermodynamic properties of the fluid do not change much from point to point so that the coefficients of viscosity can be taken to be constant. Then, in

Cartesian coordinates, we can write

$$\sigma_{ij,j}=\mu(v_{i,j}+v_{j,i}-\tfrac{2}{3}v_{k,k}\,\delta_{ij})_{,j}+\zeta v_{k,ki}=\mu v_{i,jj}+(\zeta+\tfrac{1}{3}\mu)v_{k,ki}$$
$$=\mu\,\nabla^2 v_i+(\zeta+\tfrac{1}{3}\mu)(\nabla\cdot\mathbf{v})_{,i}, \tag{26.5}$$

from which we can see that the equations of motion simplify to

$$\rho(D\mathbf{v}/Dt)=\mathbf{f}-\nabla p+\mu\,\nabla^2\mathbf{v}+(\zeta+\tfrac{1}{3}\mu)\,\nabla(\nabla\cdot\mathbf{v}). \tag{26.6}$$

If the fluid is incompressible (a good approximation for liquids, and for gases at velocities well below the sound speed and over distances small compared to a scale height), then $\nabla\cdot\mathbf{v}\equiv 0$ and the equations of motions reduce to

$$(D\mathbf{v}/Dt)=(\mathbf{f}-\nabla p)/\rho+\nu\,\nabla^2\mathbf{v} \tag{26.7}$$

where $\nu\equiv\mu/\rho$ is called the *kinematic viscosity coefficient.*

CURVILINEAR COORDINATES

As mentioned in §23, the covariant generalization of Cauchy's equation is

$$\rho a^i=f^i+T^{ij}_{\ ;j}, \tag{26.8}$$

where a^i is the ith contravariant component of the acceleration as computed from (15.9). To evaluate (26.8) in any particular coordinate system, one can proceed as follows. (1) Calculate T_{ij} from (25.7) and (25.8). (2) Raise indices to obtain T^{ij}. (3) Use equation (A3.89) to evaluate the divergence $T^{ij}_{\ ;j}$. (4) Convert to physical components using the expressions given in §A3.7. The calculations are straightforward but usually lengthy.

For spherical coordinates we can shorten this process by using the expressions for the physical components of T given by equation (25.11) directly in equations (A3.91) for the divergence. For a Newtonian fluid one finds, with a bit of patience,

$$\begin{aligned}\rho a_r=f_r&-\frac{\partial p}{\partial r}+\frac{\partial}{\partial r}\left[2\mu\frac{\partial v_r}{\partial r}+(\zeta-\tfrac{2}{3}\mu)(\nabla\cdot\mathbf{v})\right]+\frac{1}{r}\frac{\partial}{\partial\theta}\left\{\mu\left[r\frac{\partial}{\partial r}\left(\frac{v_\theta}{r}\right)+\frac{1}{r}\frac{\partial v_r}{\partial\theta}\right]\right\}\\&+\frac{1}{r\sin\theta}\frac{\partial}{\partial\phi}\left\{\mu\left[\frac{1}{r\sin\theta}\frac{\partial v_r}{\partial\phi}+r\frac{\partial}{\partial r}\left(\frac{v_\phi}{r}\right)\right]\right\}\\&+\frac{\mu}{r}\left[4r\frac{\partial}{\partial r}\left(\frac{v_r}{r}\right)-\frac{2}{r\sin\theta}\frac{\partial}{\partial\theta}(v_\theta\sin\theta)-\frac{2}{r\sin\theta}\frac{\partial v_\phi}{\partial\phi}\right.\\&\left.+r\cot\theta\frac{\partial}{\partial r}\left(\frac{v_\theta}{r}\right)+\frac{\cot\theta}{r}\frac{\partial v_r}{\partial\theta}\right],\end{aligned} \tag{26.9a}$$

$$\begin{aligned}\rho a_\theta=f_\theta&-\frac{1}{r}\frac{\partial p}{\partial\theta}+\frac{\partial}{\partial r}\left\{\mu\left[r\frac{\partial}{\partial r}\left(\frac{v_\theta}{r}\right)+\frac{1}{r}\frac{\partial v_r}{\partial\theta}\right]\right\}\\&+\frac{1}{r}\frac{\partial}{\partial\theta}\left[\frac{2\mu}{r}\left(\frac{\partial v_\theta}{\partial\theta}+v_r\right)+(\zeta-\tfrac{2}{3}\mu)(\nabla\cdot\mathbf{v})\right]\\&+\frac{1}{r\sin\theta}\frac{\partial}{\partial\phi}\left\{\mu\left[\frac{\sin\theta}{r}\frac{\partial}{\partial\theta}\left(\frac{v_\phi}{\sin\theta}\right)+\frac{1}{r\sin\theta}\frac{\partial v_\theta}{\partial\phi}\right]\right\}\\&+\frac{\mu}{r}\left\{\frac{2\cot\theta}{r}\left[\sin\theta\frac{\partial}{\partial\theta}\left(\frac{v_\theta}{\sin\theta}\right)-\frac{1}{\sin\theta}\frac{\partial v_\phi}{\partial\phi}\right]+3r\frac{\partial}{\partial r}\left(\frac{v_\theta}{r}\right)+\frac{3}{r}\frac{\partial v_r}{\partial\theta}\right\}.\end{aligned} \tag{26.9b}$$

and

$$\rho a_\phi = f_\phi - \frac{1}{r \sin\theta}\frac{\partial p}{\partial \phi} + \frac{\partial}{\partial r}\left\{\mu\left[\frac{1}{r\sin\theta}\frac{\partial v_r}{\partial \phi} + r\frac{\partial}{\partial r}\left(\frac{v_\phi}{r}\right)\right]\right\}$$
$$+\frac{1}{r}\frac{\partial}{\partial\theta}\left\{\mu\left[\frac{\sin\theta}{r}\frac{\partial}{\partial\theta}\left(\frac{v_\phi}{\sin\theta}\right)+\frac{1}{r\sin\theta}\frac{\partial v_\theta}{\partial\phi}\right]\right\}$$
$$+\frac{1}{r\sin\theta}\frac{\partial}{\partial\phi}\left[\frac{2\mu}{r}\left(\frac{1}{\sin\theta}\frac{\partial v_\phi}{\partial\phi}+v_r+v_\theta\cot\theta\right)+(\zeta-\tfrac{2}{3}\mu)(\nabla\cdot\mathbf{v})\right]$$
$$+\frac{\mu}{r}\left\{\frac{3}{r\sin\theta}\frac{\partial v_r}{\partial\phi}+3r\frac{\partial}{\partial r}\left(\frac{v_\phi}{r}\right)\right. \tag{26.9c}$$
$$\left.+2\cot\theta\left[\frac{\sin\theta}{r}\frac{\partial}{\partial\theta}\left(\frac{v_\phi}{\sin\theta}\right)+\frac{1}{r\sin\theta}\frac{\partial v_\theta}{\partial\phi}\right]\right\},$$

where (a_r, a_θ, a_ϕ) are given by (15.11) and $\nabla\cdot\mathbf{v}$ is given by (A3.88).

Equations (26.9) are obviously very complicated, and, although they can actually be solved numerically with large high-speed computers, it is helpful to have simplified versions to work with. For example, expressions for zero bulk viscosity and constant dynamical viscosity are given in (**Y1**, 132), and for incompressible flow with constant viscosity in (**A1**, 183), (**L1**, 52), and (**Y1**, 132). A more useful simplification for our work is to consider a one-dimensional, spherically symmetric flow of a fluid with zero bulk viscosity. We then have

$$\rho\left(\frac{\partial v_r}{\partial t}+v_r\frac{\partial v_r}{\partial r}\right)=f_r-\frac{\partial p}{\partial r}+\frac{1}{r^3}\frac{\partial}{\partial r}\left[\tfrac{4}{3}\mu r^4\frac{\partial}{\partial r}\left(\frac{v_r}{r}\right)\right]. \tag{26.10}$$

As we will see in §27, in spherical geometry it is not possible to account completely for viscous effects by means of a scalar viscous pressure, as can be done for planar flows.

27. The Energy Equation

TOTAL, MECHANICAL, AND GAS ENERGY EQUATIONS

Conservation of total energy for a material volume $\mathcal{V}$ in a viscous fluid implies that

$$\frac{D}{Dt}\int_{\mathcal{V}}\rho(e+\tfrac{1}{2}v^2)\,dV=\int_{\mathcal{V}}\mathbf{f}\cdot\mathbf{v}\,dV+\int_S\mathbf{t}\cdot\mathbf{v}\,dS-\int_S\mathbf{q}\cdot d\mathbf{S}. \tag{27.1}$$

Here the left-hand side gives the time rate of change of the internal plus kinetic energy contained within $\mathcal{V}$, while the three integrals on the right-hand side can be interpreted as (1) the rate at which work is done on the fluid by external volume forces, (2) the rate at which work is done by surface forces arising from fluid stresses, and (3) the rate of energy loss out of the fluid element by means of a direct transport mechanism having an *energy flux* $\mathbf{q}$. The sign of the last integral is taken to be negative because when $\mathbf{q}$ is directed along the outward normal $\mathbf{n}$ of S, heat is lost from $\mathcal{V}$. For the present we assume that $\mathbf{q}$ results from *thermal conduction*; in Chapter 7 we include radiative effects.

We can transform (27.1) into a more useful form by writing

$$\mathbf{t}\cdot\mathbf{v} = v_i t^i = v_i T^{ij} n_j \tag{27.2}$$

and using the divergence theorem to convert the surface integrals to volume integrals while transforming the left-hand side by use of (19.12). We then have

$$\int_{\mathcal{V}} \{\rho[D(e+\tfrac{1}{2}v^2)/Dt] - v_i f^i - (v_i T^{ij})_{,j} + q^i_{,i}\}\, dV = 0. \tag{27.3}$$

Because $\mathcal{V}$ is arbitrary, the integrand must vanish if the integral is to vanish, hence

$$\rho[D(e+\tfrac{1}{2}v^2)/Dt] = v_i f^i + (v_i T^{ij} - q^j)_{,j} \tag{27.4}$$

and therefore

$$(\rho e + \tfrac{1}{2}\rho v^2)_{,t} + [\rho(e+\tfrac{1}{2}v^2)v^j - v_i T^{ij} + q^j]_{,j} = v_i f^i, \tag{27.5a}$$

or, equivalently,

$$(\rho e + \tfrac{1}{2}\rho v^2)_{,t} + [\rho(h+\tfrac{1}{2}v^2)v^j - v_i \sigma^{ij} + q^j]_{,j} = v_i f^i. \tag{27.5b}$$

Equation (27.5) is the generalization of the *total energy equation* (24.6) to the case of a viscous fluid, and has a completely analogous interpretation term by term.

By forming the dot product of the velocity with Cauchy's equation of motion (23.4) we obtain the *mechanical energy equation* for a viscous fluid:

$$\rho v_i (Dv^i/Dt) = \rho[D(\tfrac{1}{2}v^2)/Dt] = v_i f^i + v_i T^{ij}_{,j}. \tag{27.6}$$

This is the generalization of (24.8) to the case of a viscous fluid.

Subtracting (27.6) from (27.4) we obtain the *gas energy equation*

$$\rho(De/Dt) = v_{i,j} T^{ij} - q^j_{,j}, \tag{27.7}$$

which is the generalization of (24.9) to the case of a viscous fluid. Using (25.7) and (25.8) we have

$$\begin{aligned} v_{i,j} T^{ij} &= v_{i,j}[-p\,\delta^{ij} + 2\mu E^{ij} + (\zeta - \tfrac{2}{3}\mu) v^k_{,k}\,\delta^{ij}] \\ &= -p v^i_{,i} + 2\mu v_{i,j} E^{ij} + (\zeta - \tfrac{2}{3}\mu)(v^i_{,i})^2. \end{aligned} \tag{27.8}$$

Recalling the definition of E^{ij} from (21.6). and in particular that it is symmetric in i and j, we see that the second term on the right-hand side reduces to $2\mu E_{ij}E^{ij}$. Thus we can write

$$v_{i,j} T^{ij} = -p(\nabla\cdot\mathbf{v}) + \Phi \tag{27.9}$$

where the *dissipation function*

$$\Phi \equiv 2\mu E_{ij} E^{ij} + (\zeta - \tfrac{2}{3}\mu)(\nabla\cdot\mathbf{v})^2 \tag{27.10}$$

accounts for *viscous energy dissipation* in the gas. Hence the gas energy equation becomes

$$\rho\left[\frac{De}{Dt}+p\frac{D}{Dt}\left(\frac{1}{\rho}\right)\right]=\Phi-\nabla\cdot\mathbf{q}. \tag{27.11}$$

In Cartesian coordinates

$$\begin{aligned}\Phi=2\mu&\left[\left(\frac{\partial v_x}{\partial x}\right)^2+\left(\frac{\partial v_y}{\partial y}\right)^2+\left(\frac{\partial v_z}{\partial z}\right)^2\right]\\ &+\mu\left[\left(\frac{\partial v_x}{\partial y}+\frac{\partial v_y}{\partial x}\right)^2+\left(\frac{\partial v_y}{\partial z}+\frac{\partial v_z}{\partial y}\right)^2+\left(\frac{\partial v_z}{\partial x}+\frac{\partial v_x}{\partial z}\right)^2\right]+(\zeta-\tfrac{2}{3}\mu)(\nabla\cdot\mathbf{v})^2.\end{aligned} \tag{27.12}$$

For one-dimensional planar flow this expression simplifies to

$$\Phi=(\tfrac{4}{3}\mu+\zeta)(\partial v_z/\partial z)^2, \tag{27.13}$$

or, in terms of the viscous pressure Q defined in (26.3),

$$\Phi=-Q(\partial v_z/\partial z)=-Q(\nabla\cdot\mathbf{v})=-\rho Q[D(1/\rho)/Dt]. \tag{27.14}$$

Thus (27.11) can be rewritten as

$$\rho\left[\frac{De}{Dt}+(p+Q)\frac{D}{Dt}\left(\frac{1}{\rho}\right)\right]=-\frac{\partial q_z}{\partial z}=\frac{\partial}{\partial z}\left(K\frac{\partial T}{\partial z}\right) \tag{27.15}$$

which substantiates the claim made in §26 that the effects of viscosity in a one-dimensional planar flow can be accounted for completely and consistently through use of a viscous pressure Q. In writing the second equality in (27.15) we have anticipated equation (27.18).

ENTROPY GENERATION

By virtue of the first law of thermodynamics,

$$T(Ds/Dt)=(De/Dt)+p[D(1/\rho)/Dt], \tag{27.16}$$

equation (27.12) can be transformed into the *entropy generation equation*

$$T(Ds/Dt)=\rho^{-1}(\Phi-\nabla\cdot\mathbf{q}). \tag{27.17}$$

We will prove below that the dissipation function is always greater than (or equal to) zero; we thus see from (27.17) that viscous dissipation leads to an irreversible increase in the entropy of the gas.

We can strengthen this statement by appealing to *Fourier's law* of heat conduction, which, on the basis of experimental evidence, states that the heat flux in a substance is proportional to the temperature gradient in the material, and that heat flows from hotter to cooler regions. That is,

$$\mathbf{q}=-K\,\nabla T, \tag{27.18}$$

where K is the *coefficient of thermal conductivity*. Using (27.18) in (27.17), and integrating over a material volume $\mathscr{V}$, we have

$$\begin{aligned}\int_{\mathscr{V}} \rho \frac{Ds}{Dt}\,dV &= \int_{\mathscr{V}} \frac{1}{T}[\nabla\cdot(K\nabla T)+\Phi]\,dV \\ &= \int_{\mathscr{V}} \left[\nabla\cdot\left(\frac{K}{T}\nabla T\right)+\frac{K}{T^2}(\nabla T)^2+\frac{\Phi}{T}\right]dV \\ &= \int_S \frac{K}{T}(\nabla T)\cdot d\mathbf{S}+\int_{\mathscr{V}}\left[\frac{K}{T^2}(\nabla T)^2+\frac{\Phi}{T}\right]dV.\end{aligned} \tag{27.19}$$

The first term on the right-hand side of (27.19) is just

$$-\int_S T^{-1}\mathbf{q}\cdot\mathbf{n}\,dS = \int_{\mathscr{V}} T^{-1}(Dq/Dt)\,dV, \tag{27.20}$$

that is, the rate at which heat is delivered into $\mathscr{V}$ divided by the temperature at which the delivery is made; in writing (27.20) we have assumed that $\mathscr{V}$ is infinitesimal so that the variation of T over S and within $\mathscr{V}$ can be neglected. The second term on the right-hand side of (27.19) is manifestly positive. Thus (27.19) is consistent with the second law of thermodynamics (24.15), and shows that *both* viscous dissipation and heat conduction within the fluid lead to an irreversible entropy increase.

THE DISSIPATION FUNCTION

We must now prove that Φ as given by (27.10) is, in fact, positive. We may omit the term $\zeta(\nabla\cdot\mathbf{v})^2$ from further consideration because it is obviously positive. We need consider, therefore, only

$$\Psi \equiv 2E_{ij}E^{ij}-\tfrac{2}{3}(\nabla\cdot\mathbf{v})^2. \tag{27.21}$$

It is easy to show by direct substitution that

$$E_{ij}E^{ij}=A^2-2B, \tag{27.22}$$

where A and B are the invariants defined by (21.10) and (21.11). Recalling that $A=\nabla\cdot\mathbf{v}$, we have

$$\Psi=\tfrac{4}{3}A^2-4B. \tag{27.23}$$

Now that Ψ has been expressed in terms of invariants we may evaluate it in any coordinate system. Obviously it is most convenient to align the coordinate system along the principal axes of the rate-of-strain ellipsoid, in which case

$$\begin{aligned}\Psi &= \tfrac{4}{3}(E_{11}+E_{22}+E_{33})^2-4(E_{11}E_{22}+E_{22}E_{33}+E_{33}E_{11}) \\ &= \tfrac{4}{3}(E_{11}^2+E_{22}^2+E_{33}^2-E_{11}E_{22}-E_{22}E_{33}-E_{33}E_{11}).\end{aligned} \tag{27.24}$$

Choose the largest element $E_{(i)(i)}=E_{\max}$ as a normalization. Then we can write

$$\Psi=\tfrac{4}{3}E_{\max}^2(1+\alpha^2+\beta^2-\alpha-\beta-\alpha\beta) \tag{27.25}$$

where $\alpha \leq 1$ and $\beta \leq 1$. If $\alpha = \beta$, then

$$\Psi = \tfrac{4}{3}E_{\max}^2(1-2\alpha+\alpha^2) = \tfrac{4}{3}E_{\max}^2(1-\alpha)^2 \geq 0. \tag{27.26}$$

If $\alpha \neq \beta$, choose labels so that $\alpha > \beta$. Then

$$\Psi = \tfrac{4}{3}E_{\max}^2[(1-\alpha)^2 + (\alpha-\beta)(1-\beta)] \geq 0. \tag{27.27}$$

Thus in all cases $\Psi \geq 0$, which implies that $\Phi \geq 0$, which was to be proved.

Finally, it is easy to show by expanding the square that an alternative expression for Φ is

$$\Phi = 2\mu(E^{ij} - \tfrac{1}{3}\delta^{ij}\,\nabla\cdot\mathbf{v})(E_{ij} - \tfrac{1}{3}\delta_{ij}\,\nabla\cdot\mathbf{v}) + \zeta(\nabla\cdot\mathbf{v})^2. \tag{27.28}$$

One can see by inspection from this equation that $\Phi \geq 0$.

CURVILINEAR COORDINATES

The gas energy equation (27.11) is already in covariant form because the terms on the left-hand side are scalars, the divergence of the energy flux can be written as $\nabla\cdot\mathbf{q} = q^k_{;k}$, and Φ as defined by (27.10) is manifestly invariant under coordinate transformation. It is straightforward to evaluate these expressions in any coordinate system. Thus, in spherical coordinates,

$$\begin{aligned}-\nabla\cdot\mathbf{q} = \nabla\cdot(K\,\nabla T) = {}& \frac{1}{r^2}\frac{\partial}{\partial r}\left(r^2 K\frac{\partial T}{\partial r}\right) + \frac{1}{r^2\sin\theta}\frac{\partial}{\partial\theta}\left(K\sin\theta\frac{\partial T}{\partial\theta}\right)\\ &+\frac{1}{r^2\sin^2\theta}\frac{\partial}{\partial\phi}\left(K\frac{\partial T}{\partial\phi}\right).\end{aligned} \tag{27.29}$$

Next, using equations (25.9) or (25.10) to evaluate $E_{ij}E^{ij}$, we easily find

$$\begin{aligned}\Phi = 2\mu\Bigg\{&\left(\frac{\partial v_r}{\partial r}\right)^2 + \left(\frac{1}{r}\frac{\partial v_\theta}{\partial\theta} + \frac{v_r}{r}\right)^2 + \left(\frac{1}{r\sin\theta}\frac{\partial v_\phi}{\partial\phi} + \frac{v_r}{r} + \frac{v_\theta\cot\theta}{r}\right)^2\\ &+\frac{1}{2}\left[r\frac{\partial}{\partial r}\left(\frac{v_\theta}{r}\right) + \frac{1}{r}\frac{\partial v_r}{\partial\theta}\right]^2 + \frac{1}{2}\left[r\frac{\partial}{\partial\phi}\left(\frac{v_\phi}{r}\right) + \frac{1}{r\sin\theta}\frac{\partial v_r}{\partial\phi}\right]^2\\ &+\frac{1}{2}\left[\frac{\sin\theta}{r}\frac{\partial}{\partial\theta}\left(\frac{v_\phi}{\sin\theta}\right) + \frac{1}{r\sin\theta}\frac{\partial v_\theta}{\partial\phi}\right]^2\Bigg\}\\ &+(\zeta - \tfrac{2}{3}\mu)\left[\frac{1}{r^2}\frac{\partial}{\partial r}(r^2 v_r) + \frac{1}{r\sin\theta}\frac{\partial}{\partial\theta}(v_\theta\sin\theta) + \frac{1}{r\sin\theta}\frac{\partial v_\phi}{\partial\phi}\right]^2.\end{aligned} \tag{27.30}$$

For the interesting case of one-dimensional, spherically symmetric flow with zero bulk viscosity, (27.30) simplifies to

$$\Phi = \tfrac{4}{3}\mu\left[r\frac{\partial}{\partial r}\left(\frac{v_r}{r}\right)\right]^2, \tag{27.31}$$

hence the gas energy equation becomes

$$\rho\left[\frac{De}{Dt} + p\frac{D}{Dt}\left(\frac{1}{\rho}\right)\right] = \frac{1}{r^2}\frac{\partial}{\partial r}\left(Kr^2\frac{\partial T}{\partial r}\right) + \tfrac{4}{3}\mu\left[r\frac{\partial}{\partial r}\left(\frac{v_r}{r}\right)\right]^2. \tag{27.32}$$

From a comparison of (26.10) and (27.32) one easily sees that in spherical geometry it is not possible to choose a scalar viscous pressure Q such that the viscous force can be written $(\partial Q/\partial r)$ and the viscous dissipation as $Q[D(1/\rho)/Dt]$, as was possible in planar geometry. The reason is that viscous effects originate from a *tensor*, which is not, in general, isotropic even for one-dimensional flow. Thus in §59 we find that the artificial viscosity terms used to achieve numerical stability in computations of spherical flows with shocks are best derived from a tensor artificial viscosity.

The covariant generalization of the total energy equation is easily obtained from (27.5) by replacing the ordinary partial derivative in the divergence with the covariant derivative. In general, the resulting expression is lengthy, but it becomes relatively simple for one-dimensional flow:

$$\frac{\partial}{\partial t}[\rho(e+\tfrac{1}{2}v_r^2)]+\frac{1}{r^2}\frac{\partial}{\partial r}\left\{r^2\rho v_r(e+\tfrac{1}{2}v^2)+r^2v_r\left[p-\tfrac{4}{3}\mu r\frac{\partial}{\partial r}\left(\frac{v_r}{r}\right)\right]-r^2K\frac{\partial T}{\partial r}\right\}=f_rv_r. \tag{27.33}$$

In (27.33) we have taken $\zeta\equiv 0$.

STEADY FLOW

For one-dimensional steady flow we can again write an explicit integral of the energy equation that now accounts for the effects of viscous dissipation and thermal conduction. In this case (27.5), for planar geometry and a constant gravitational acceleration, g reduces to

$$\frac{d}{dz}\left[\rho v_z\left(h+\tfrac{1}{2}v_z^2-\tfrac{4}{3}\nu\frac{dv_z}{dz}\right)-K\frac{dT}{dz}\right]=-\rho v_z g, \tag{27.34}$$

where h is the specific enthalpy, ν is the kinematic viscosity, and we have set $\zeta=0$. Then recalling that $\dot{m}=\rho v_z=\text{constant}$, we can integrate (27.34) to obtain

$$\dot{m}[h+\tfrac{1}{2}v_z^2+gz-\tfrac{4}{3}\nu(dv_z/dz)]-K(dT/dz)=\text{constant}. \tag{27.35}$$

Similarly, for a one-dimensional, spherically symmetric steady flow under the action of an inverse-square gravitational force, one finds from (27.33),

$$\dot{\mathcal{M}}\{h+\tfrac{1}{2}v_r^2-(G\mathcal{M}/r)-\tfrac{4}{3}\nu r[d(v_r/r)/dr]\}-4\pi r^2K(dT/dr)=\text{constant}, \tag{27.36}$$

where $\dot{\mathcal{M}}=4\pi r^2\rho v_r$; again we have set $\zeta=0$.

28. *Similarity Parameters*

As is clear from the material presented above and in Chapter 2, the equations of fluid dynamics are, in general, rather complicated. It is therefore useful to have a simple means for judging both the relative importance of various phenomena that occur in a flow, and the flow's qualitative nature. This is most easily done in terms of a set of dimensionless numbers which provide a convenient characterization of the dominant physical processes in the flow. These numbers are called *similarity parameters* because flows whose physical properties are such that they produce, in combination, the same values of those numbers can be expected to be qualitatively similar, even though the value of any one quantity—say velocity or a characteristic length—may be substantially different from one flow to another.

THE REYNOLDS NUMBER

A very basic and important flow parameter is the *Reynolds number*

$$\mathrm{Re} \equiv vl/\nu, \tag{28.1}$$

where v is a characteristic velocity in the flow, l is a characteristic length, and ν is the kinematic viscosity. Rewriting the term on the right-hand side as $(\rho v)v/(\mu v/l)$ we see that the Reynolds number gives a measure of the ratio of inertial forces (momentum flux per unit area) to viscous forces (viscosity times velocity gradient) in the flow. Limiting flow types characterized by the Reynolds number are inviscid flow—which occurs at an infinitely large Reynolds number—and inertialess viscous flow (Stokes flow)—which occurs at a vanishingly small Reynolds number.

The Reynolds number also determines whether a flow is *laminar* (i.e., smooth and orderly) or *turbulent* (i.e., disorderly and randomly fluctuating). Experiments show that the transition from laminar to turbulent flow occurs when the Reynolds number exceeds some critical value (which depends on the presence or absence, and the nature, of boundary surfaces or of solid bodies immersed in the flow). When a flow becomes turbulent, its kinematic and dynamical properties change radically and are extremely complicated to describe. We shall not deal with turbulence here but refer the reader to several excellent books on the subject [e.g., (**B1**), (**H1**), (**T2**)].

Both experiment and theoretical analysis show that when the Reynolds number is large, the bulk of a flow, except near physical boundary surfaces, is essentially inviscid and nonconducting. Major modifications of the nature of the flow by viscosity and thermal conductivity are generally confined to thin *boundary layers*. The ratio of the thickness δ of the boundary layer to a typical flow length l is given by

$$\delta/l \sim \mathrm{Re}^{-1/2} \tag{28.2}$$

[see e.g., (**A1,** 129), (**O1,** 548)]. The flows with which we shall deal have no association with solid boundary surfaces, hence we shall have no occasion to deal with boundary layer theory [see (**S1**)].

THE PRANDTL NUMBER

Another important similarity parameter for fluid flow is the *Prandtl number*

$$\mathrm{Pr} \equiv c_p \mu / K \tag{28.3}$$

where c_p is the specific heat at constant pressure, μ is the shear viscosity, and K is the thermal conductivity. Rewriting the right-hand side as $(\mu/\rho)/(K/\rho c_p)$ we see that Pr is the ratio of the kinematic viscosity to the thermal diffusivity. If Λ is a characteristic thickness for the thermal boundary layer, then we expect $(\rho c_p T) \sim (KT/\Lambda^2)$. From (28.2) we know that the thickness δ of the velocity boundary layer is $\delta \sim (\nu/vl)^{1/2}$. Hence $\mathrm{Pr} \propto \delta^2/\Lambda^2$. For gases, the Prandtl number is found to be of order unity so that the thicknesses of the thermal and velocity boundary layers are comparable; in §§29 and 33 we derive from kinetic theory an expression for Pr in monatomic perfect gases. For other fluids (e.g., oil), the Prandtl number is very large, which implies that the thermal boundary layer is very thin and that viscous effects predominate over thermal conduction in the flow.

THE PECLET NUMBER

The *Peclet number*

$$\mathrm{Pe} \equiv \rho c_p v l / K \tag{28.4}$$

is related to the Reynolds and Prandtl numbers by

$$\mathrm{Pe} = \mathrm{Re} \cdot \mathrm{Pr}. \tag{28.5}$$

Rewriting the right-hand side of (28.4) as $(\rho c_p T v)/(KT/l)$ we see that the Peclet number is essentially the ratio of heat transported by the flow to heat transported by thermal conduction. In this sense the Peclet number is to heat transport what the Reynolds number is to momentum transport, and indeed in nondimensionalized forms of the Navier-Stokes equations the Peclet number appears explicitly in the heat transfer terms in the same way the Reynolds number appears in the momentum transfer terms.

THE NUSSELT NUMBER

The *Nusselt number* is defined as

$$\mathrm{Nu} \equiv \alpha l / K, \tag{28.6}$$

where α is the *heat transfer coefficient*

$$\alpha \equiv q/\Delta T, \tag{28.7}$$

where in turn q is the heat flux per unit area and time through some surface area in the fluid, and $(\Delta T/l)$ measures the temperature gradient near that surface. Thus the Nusselt number gives a measure of the ratio of the total heat flux (by all mechanisms including, for example, radiation) to the conductive heat flux through a surface.

THE MACH NUMBER

A final extremely important parameter for gas dynamics is the *Mach number*

$$\mathrm{M} \equiv v/a, \tag{28.8}$$

where v is a characteristic flow speed and a is the adiabatic speed of sound (cf. §48). Flows in which $\mathrm{M}<1$ are *subsonic* and flows in which $\mathrm{M}>1$ are *supersonic.* Flows that make a transition from subsonic to supersonic, or vice versa, are called *transonic.* Flows at small Mach numbers can be treated as being locally (i.e., on scales much smaller than a gravitation-induced scale height) incompressible, because density variations in such flows are usually small. In flows where M approaches or exceeds unity, the nonlinearity of the fluid-dynamical equations induces large density variations, and the gas must be treated as compressible from the outset.

3.2 Equations of Motion and Energy: The Kinetic Theory View

The continuum view of fluids used above is quite satisfactory for the derivation of the basic equations of fluid dynamics, provided that we accept on empirical grounds the existence of viscous forces and conductive energy transport. But for gases the need to invoke empirical results can be avoided completely by adopting a kinetic theory view, which allows one to derive the mathematical forms of the viscous and thermal conduction terms from first principles. Furthermore, this approach allows one to evaluate the transport coefficients (e.g., μ and K) by direct calculation.

In what follows we confine attention entirely to dilute gases, as defined in §6 and §7. We first show qualitatively how transport phenomena arise in a gas and give semiquantitative estimates of their effects. We then derive the equations of gas dynamics, and give a rigorous first-order solution for the particle distribution function.

29. The Mean Free Path and Transport Phenomena

We showed in §6 that the motion of rigid-sphere particles in a neutral gas can be considered to be a sequence of straight-line paths terminated abruptly by collisions from which particles emerge on new paths. Each of these flights between collisions is called a *free path* of a particle. Because of the random nature of the collision process not all free paths have equal lengths. They will, nevertheless, have a well-defined average value, the *mean free path.*

If a gas is homogeneous and in equilibrium, we know from the principle of detailed balancing that the net effect of exchanges of particles from one element of the gas to another is exactly zero. If, however, the gas is out of equilibrium because of inhomogeneities in temperature, density, or fluid velocity (i.e., the average particle velocity at each point), then the particles originating from one fluid element may have a higher energy or momentum than those returning from an adjacent fluid element. In this event, there will be a net transfer of energy or momentum from one fluid element to another, which is interpreted on the macroscopic level as a viscous force or as thermal conduction.

In this section we make a rough calculation of these *transport phenomena* in terms of the mean free path. This approach yields considerable physical insight and also gives numerical results of the right order of magnitude. We give a more rigorous discussion of these processes in §§32 and 33.

THE MEAN FREE PATH

Consider two distinct species of rigid-sphere particles, A and B, interacting through collisions. The *bimolecular collision frequency* C_{AB} is the average total rate, per unit volume of gas, of collisions of all kinds between these species. The average *collision time* for either species is the reciprocal of the collision frequency per particle of that species; thus

$$\tau_A = n_A / C_{AB} \tag{29.1}$$

where n_A is the number density of particles of species A. The mean free path for particles of species A is the average distance they can travel in a collision time with an average speed $\langle U_A \rangle$, that is,

$$\lambda_A = \langle U_A \rangle \tau_A = n_A \langle U_A \rangle / C_{AB}. \tag{29.2}$$

Our first task is to calculate C_{AB}. Let the particles have diameters d_A and d_B so that their interaction diameter is $d_{AB} = \frac{1}{2}(d_A + d_B)$. From (7.21) we know that the cross section for rigid spheres of this effective diameter is $\sigma = \frac{1}{4} d_{AB}^2$. From exactly the same argument that leads to (7.32) we can write the total collision frequency as

$$C_{AB} = \tfrac{1}{4} \iiiint f_A f_B g d_{AB}^2 \sin \chi \, d\chi \, d\varepsilon \, d^3u_A \, d^3u_B, \tag{29.3}$$

where the symbols have the same meaning as in §§6 and 7. The integrals over χ and ε are trivial, yielding [cf. (7.23)]

$$C_{AB} = \pi d_{AB}^2 \iint f_A f_B g \, d^3u_A \, d^3u_B. \tag{29.4}$$

To compute C_{AB} we obviously must know the distribution functions f_A and f_B; we can obtain a reasonable estimate by using the equilibrium distribution function f_0 given by (8.9) so that

$$C_{AB} = n_A n_B (m_A m_B)^{3/2} (2\pi kT)^{-3} \pi d_{AB}^2 \times \iint \exp[-(m_A u_A^2 + m_B u_B^2)/2kT] g \, d^3u_A \, d^3u_B. \tag{29.5}$$

From (7.6) we see that the argument in the exponential can be rewritten in terms of g, the relative speed, and G, the speed of the center of mass, of the two particles. Furthermore, we have shown earlier [cf. (7.33) and

(7.34)] that $d^3u_A\, d^3u_B = d^3g\, d^3G$. Thus (29.5) becomes

$$C_{AB} = 2n_A n_B (m_A m_B)^{3/2}(kT)^{-3} d_{AB}^2 \times \int_0^\infty e^{-(M_{AB}G^2/2kT)} G^2\, dG \int_0^\infty e^{-(\tilde{m}_{AB}g^2/2kT)} g^3\, dg. \quad (29.6)$$

The integrals are given in standard tables, and we obtain, finally,

$$C_{AB} = n_A n_B d_{AB}^2 (8\pi kT/\tilde{m}_{AB})^{1/2}. \quad (29.7)$$

Equation (29.7) gives the total collision rate per unit volume for two *distinct* species of particles A and B. If the gas consists of a single species so that all the particles are identical, then this rate should be divided by 2 because the integration counts a given pair of particles twice, once regarding one of them as having a velocity $\mathbf{u}_A$ and the other a velocity $\mathbf{u}_B$, and again with the first now having a velocity $\mathbf{u}_B$ and the second a velocity $\mathbf{u}_A$, when in fact both of these cases give the same collision. On the other hand, for the purpose of counting mean free paths, we recognize that each collision between two identical particles terminates *two* free paths of particles in the gas; this just cancels the factor of two mentioned above. Thus the collision time for species A is

$$\tau_A = (\tilde{m}_{AB}/8\pi kT)^{1/2}/(n_B d_{AB}^2) \quad (29.8)$$

whether species B is identical to species A or not. In the former case $\tilde{m}_{AB} = \frac{1}{2}m$, $n_B = N$, and $d_{AB} = d$, so that

$$\tau = (m/16\pi kT)^{1/2}(Nd^2). \quad (29.9)$$

Now using (29.9) in (29.2), along with (8.12) for $\langle U \rangle$, we find

$$\lambda = (2^{1/2} N\pi d^2)^{-1}. \quad (29.10)$$

From purely dimensional arguments, we would estimate that for a particle moving with a speed $\langle U \rangle$ through a field of N stationary particles per cm^3, the collision time is $\tau \sim 1/N\langle U \rangle \sigma$, and hence $\lambda \sim 1/N\sigma$ where σ is the interaction cross section πd^2. The additional factor of $\sqrt{2}$ appearing in (29.10) simply accounts for the correct average relative speeds of the test and field particles when the field particles are also allowed to move.

THE DISTRIBUTION OF FREE PATHS

Let us now calculate the *distribution of free paths*. Consider a large number N_0 of particles at some instant. Let $N(x)$ be the number of particles that have not yet suffered collisions after each particle in the group has traveled a distance x along its free path. Then if P_c is the *average collision probability* per unit length along the path, we must have

$$dN(x) = -N(x)P_c\, dx, \quad (29.11)$$

whence

$$N(x) = N_0 e^{-P_c x}. \quad (29.12)$$

But, by definition, the mean free path is the average distance a particle can travel between collisions; therefore the average collision probability per unit length must be $P_c = 1/\lambda$. Thus the number of particles that are able to travel a distance x without suffering a collision is

$$N(x) = N_0 \exp(-x/\lambda). \tag{29.13}$$

THE KNUDSEN NUMBER

When the mean free path λ is very small compared to a characteristic macroscopic length l in the flow, the gas can be considered to be a continuum and treated by the equations of continuum hydrodynamics. In the opposite extreme where $\lambda \gg l$, we are in the regime of *rarefied gas dynamics*, and must treat the problem of *free-molecule flow* in which the gas molecules act as if they were a large set of noninteracting particles. To categorize flows into a scheme between the extreme cases just described we introduce another dimensionless similarity parameter, the *Knudsen number*

$$\text{Kn} \equiv \lambda/l. \tag{29.14}$$

As we shall see in §32, for small Knudsen numbers we can derive the equations of gas dynamics from a first-order solution of the Boltzmann transport equation. This solution yields analytical expressions for the energy flux vector and the fluid stress tensor in terms of local properties and their gradients. The resulting conservation equations contain the same macroscopic terms as the equations for ideal fluids, plus additional terms describing momentum and energy transport by microscopic processes in the gas.

On the other hand, for large Knudsen numbers the local description mentioned above no longer suffices, because we must account for efficient exchange of particles over mean free paths that are of the same size as, or larger than, characteristic flow lengths. We must then employ a rather elaborate transport theory formulated in terms of integrations over finite (in some cases quite large) volumes of the gas.

The Knudsen number is small, and hence the continuum treatment is valid, for the material component of the radiating fluid in all astrophysical flows we consider in this book, even in such rarefied flows as stellar winds (cf. §62). For example, in the solar atmosphere we have a neutral atomic hydrogen density $N \sim 10^{15}\ \text{cm}^{-3}$, and can assign a typical cross section of $\sigma \sim \pi a_0^2 \sim 10^{-16}\ \text{cm}^2$ to these particles, from which we estimate $\lambda \sim 10$ cm; this may be compared to a pressure scale height H, which is of the order of 10^7 cm. Hence $\text{Kn} \sim 10^{-6}$, and a purely local, continuum treatment of the gas dynamics suffices (except for the effects of the interaction of radiation with the material). For the parameters just quoted and a temperature of the order of 10^4 K the collision time is $\tau \sim 10^{-5}$ sec. Thus, local nonuniformities in temperature, density, or velocity on scales of a few cm are substantially reduced by molecular exchange in times of the order of 10^{-5} sec.

TRANSPORT PROCESSES

Let us now evaluate the coefficients of viscosity and thermal conduction in terms of the mean free path. The basic assumption made here is that particles typically transport a momentum (or energy) characteristic of their point of origin through a mean free path, and then deposit it into the gas at the end of their flight. Consider a uniform gas in which the macroscopic flow velocity is $\mathbf{v} = (v_x, 0, 0)$, where $v_x = v_x(z)$. Then from (25.3) the viscous stress tensor is

$$\boldsymbol{\sigma} = \mu \begin{pmatrix} 0 & 0 & (\partial v_x/\partial z) \\ 0 & 0 & 0 \\ (\partial v_x/\partial z) & 0 & 0 \end{pmatrix}. \tag{29.15}$$

Hence from (22.3) the viscous force exerted on a unit area oriented parallel to the (x, y) plane, with normal $\mathbf{n} = (0, 0, 1)$, at a height $z = z_0$ is

$$\mathbf{t}_{\text{visc}} = [\mu(\partial v_x/\partial z)_0, 0, 0]. \tag{29.16}$$

This result is compatible with our qualitative expectation that each layer tends to drag the adjacent layer (lying below it, for the choice of $\mathbf{n}$ we have made) in the direction of its motion relative to that layer.

From the kinetic theory viewpoint, the viscous force $\mathbf{t}_{\text{visc}}$ must equal the net rate of transport in the z direction of momentum in the x direction across a unit area parallel to the (x, y) plane, per unit time. The number of particles crossing this plane from below in a unit time is

$$N_+ = \int_0^\infty \Phi(U_z) U_z \, dU_z \tag{29.17}$$

where Φ is the one-dimensional distribution of particle velocities along the z axis. We assume, for purposes of calculating a first approximation for N_+, that we can use the equilibrium distribution given by (8.15), despite the fact that the gas is not really in equilibrium. The integral in (29.17) is then elementary and one finds

$$N_+ = \tfrac{1}{4}N(8kT/\pi m)^{1/2} = \tfrac{1}{4}N\langle U \rangle. \tag{29.18}$$

The number N_- crossing in the downward direction equals N_+ by symmetry.

Now the average x velocity of particles crossing the unit plane at $z = z_0$ is not $v_x(z_0)$, but rather corresponds to the average x velocity in the layer whence these particles started their free paths. On the average we expect this layer to be located at $z = z_0 \pm \zeta_\mu \lambda$, where ζ_μ is an unknown numerical factor of order unity. Then to first order the net transport of x momentum across the unit area in a unit time from the positive to the negative side is

$$\begin{aligned} \tfrac{1}{4}N\langle U \rangle \{m[v_x(z_0) + \zeta_\mu \lambda (\partial v_x/\partial z)_0] - m[v_x(z_0) - \zeta_\mu \lambda (\partial v_x/\partial z)_0]\} \\ = \tfrac{1}{2}Nm\langle U \rangle \zeta_\mu \lambda (\partial v_x/\partial z)_0, \end{aligned} \tag{29.19}$$

which must equal the macroscopic force given by (29.16). Therefore we identify

$$\mu = \tfrac{1}{2}\zeta_\mu Nm\lambda\langle U\rangle = \tfrac{1}{2}\zeta_\mu \rho\lambda\langle U\rangle = (\zeta_\mu/\pi d^2)(mkT/\pi)^{1/2}. \qquad (29.20)$$

As we will see in §33, this result is in fortuitously good agreement with the exact result for rigid-sphere particles, for which one finds $\zeta_\mu = (1.016)\times(5\pi/16) = 0.9975$. Equation (29.20) shows explicitly the remarkable result, first noted by Maxwell, that for a dilute gas the coefficient of shear viscosity is independent of density (which Maxwell verified by experiment). For a rigid-sphere gas, μ varies as $T^{1/2}$; we will see in §33 that the temperature dependence of μ for an ionized gas is quite different.

By a similar analysis we can evaluate K, the coefficient of thermal conductivity. We now suppose the gas is at rest and is uniform except for a temperature gradient in the z direction [i.e., $T = T(z)$]. From the macroscopic viewpoint, the heat flux will be

$$q_z = -K(\partial T/\partial z)_0. \qquad (29.21)$$

From the kinetic theory viewpoint, the heat flux equals the net rate of thermal energy transport in the $+z$ direction by individual particles. The thermal energy per particle is $\tilde{e} = \tfrac{3}{2}kT$. We assume that, on the average, particles originate in layers at $z = z_0 \pm \zeta_K\lambda$, where again ζ_K is an unknown numerical factor of order unity. The net thermal energy transport is, therefore,

$$\begin{aligned}\tfrac{1}{4}N\langle U\rangle\tfrac{3}{2}k\{[T_0 - \zeta_K\lambda(\partial T/\partial z)_0] - [T_0 + \zeta_K\lambda(\partial T/\partial z)_0]\}\\ = -\tfrac{3}{4}kN\langle U\rangle\zeta_K\lambda(\partial T/\partial z)_0,\end{aligned} \qquad (29.22)$$

whence we identify

$$K = \tfrac{3}{4}\zeta_K kN\lambda\langle U\rangle, \qquad (29.23)$$

or, in view of (8.20),

$$K = \tfrac{1}{2}\zeta_K\rho\lambda\langle U\rangle c_v. \qquad (29.24)$$

Thus, from (29.20), we have

$$K = (\zeta_K/\zeta_\mu)c_v\mu. \qquad (29.25)$$

The results obtained in §33 show that for a rigid-sphere gas $\zeta_K = (1.025)\times\tfrac{5}{2}\times(5\pi/16) = 2.5160$, and hence $(\zeta_K/\zeta_\mu) = 2.5225$.

The Prandtl number for a rigid-sphere gas follows immediately from (29.25):

$$\mathrm{Pr} = (c_p\mu/K) = (\zeta_\mu/\zeta_K)(c_p/c_v) \approx \tfrac{2}{5}\gamma, \qquad (29.26)$$

where we have used the values of ζ_μ and ζ_K just quoted. For a perfect monoatomic gas we should thus have $\mathrm{Pr} \approx \tfrac{2}{3}$, which is in good agreement with experiment for noble gases (**H2**, 16). Equation (29.25) with $(\zeta_K/\zeta_\mu) \approx 2.5$ remains valid for power-law interactions in general and for Coulomb interactions in particular.

Although we have been able to obtain physical insight and even semi-quantitative results from mean-free-path calculations, the approach is obviously only heuristic. The calculation can be refined by allowing for differences in the lengths of free paths of different particles, effects of previous collisions on particle speeds ("persistence of velocity"), and correlation between the energy of a particle and the distance over which it can penetrate to the test area [see for example (**J1,** Chaps. 11–13), (**J2,** Chaps. 6 and 7)]. However, even in its most refined forms this approach remains ad hoc, applies only to rigid-sphere particles, and yields results of unknown reliability (their accuracy being determinable only when the calculation can be done by more precise methods). The basic flaw in this approach is that it does not provide a means of calculating the actual velocity distribution function of the particles, nor of specifying how the real distribution departs from the equilibrium distribution as a result of spatial variations in the macroscopic physical properties of the gas. Reliable results are obtained only when a direct calculation of the nonequilibrium velocity-distribution function is made along the lines discussed in §§32 and 33.

30. Moments of the Boltzmann Equation

The equations of gas dynamics can be deduced directly by calculating moments of the Boltzmann equation for quantities that are conserved in collisions of the particles composing the gas. This approach provides an independent derivation of the equations obtained earlier from macroscopic arguments, and deepens our understanding of the physical meaning of the terms that appear in these equations.

THE CONSERVATION THEOREM

We form a *moment* of the Boltzmann equation

$$(\partial f/\partial t)+u^i(\partial f/\partial x^i)+a^i(\partial f/\partial u^i)=(Df/Dt)_{\text{coll}} \tag{30.1}$$

by multiplying through by any physical quantity $Q(\mathbf{x}, \mathbf{u})$, and then integrating over velocity. Thus

$$\int Q\cdot\left(\frac{\partial f}{\partial t}+u^i\frac{\partial f}{\partial x^i}+a^i\frac{\partial f}{\partial u^i}\right)d^3u=\int Q\cdot\left(\frac{Df}{Dt}\right)_{\text{coll}}d^3u=I(Q), \tag{30.2}$$

where $I(Q)$ is defined by (7.39). If Q is *conserved* during the collision $(\mathbf{u}, \mathbf{u}_1)\rightarrow(\mathbf{u}', \mathbf{u}_1')$, in the sense that it is a summational invariant of the form

$$Q(\mathbf{x}, \mathbf{u})+Q(\mathbf{x}, \mathbf{u}_1)=Q(\mathbf{x}, \mathbf{u}')+Q(\mathbf{x}, \mathbf{u}_1'), \tag{30.3}$$

then it follows immediately from (7.39) and (7.42) that $I(Q)\equiv 0$, and (30.2) reduces to a *conservation law.* Henceforth, we assume that Q is conserved.

We can cast the left-hand side of (30.2) into a more useful form. Define the average value $\langle A\rangle$ of any quantity A as

$$\langle A\rangle=\left(\int Af\,d^3u\right)\Big/\left(\int f\,d^3u\right)=N^{-1}\int Af\,d^3u, \tag{30.4}$$

where the second equality follows from (6.1); here N is the number of particles per cm^3. Then, bearing in mind that t, x^j and u^j are all independent variables we have

$$\int Q\frac{\partial f}{\partial t}\,d^3u=\frac{\partial}{\partial t}\int Qf\,d^3u-\int\frac{\partial Q}{\partial t}f\,d^3u=\frac{\partial}{\partial t}(N\langle Q\rangle)-N\left\langle\frac{\partial Q}{\partial t}\right\rangle, \tag{30.5}$$

$$\begin{aligned}\int Qu^j\frac{\partial f}{\partial x^j}\,d^3u&=\frac{\partial}{\partial x^j}\int Qu^jf\,d^3u-\int u^j\frac{\partial Q}{\partial x^j}f\,d^3u\\&=\frac{\partial}{\partial x^j}(N\langle Qu^j\rangle)-N\left\langle u^j\frac{\partial Q}{\partial x^j}\right\rangle,\end{aligned} \tag{30.6}$$

and

$$\begin{aligned}\int Qa^j\frac{\partial f}{\partial u^j}\,d^3u&=\int\frac{\partial}{\partial u^j}(Qa^jf)\,d^3u-\int\frac{\partial}{\partial u^j}(Qa^j)f\,d^3u\\&=\sum_j\iint_{-\infty}^{\infty}du_l\,du_k\,\Big|_{-\infty}^{\infty}(Qa^jf)-N\left\langle\frac{\partial(Qa^j)}{\partial u^j}\right\rangle=-N\left\langle\frac{\partial(Qa^j)}{\partial u^j}\right\rangle.\end{aligned} \tag{30.7}$$

In (30.7) we have made use of the fact that as $u^j\to\pm\infty$, $f\to0$ so strongly that (Qa^jf) vanishes for all Q's of interest. We thus obtain the general *conservation theorem*, which states that

$$\frac{\partial}{\partial t}(N\langle Q\rangle)+\frac{\partial}{\partial x^j}(N\langle Qu^j\rangle)-N\left[\left\langle\frac{\partial Q}{\partial t}\right\rangle+\left\langle u^j\frac{\partial Q}{\partial x^j}\right\rangle+\left\langle\frac{\partial(Qa^j)}{\partial u^j}\right\rangle\right]=0, \tag{30.8}$$

for any conserved quantity Q. If we now restrict attention to velocity-independent external forces, and assume that Q is a function of $\mathbf{u}$, but not an explicit function of $\mathbf{x}$ or t, then (30.8) simplifies to

$$\frac{\partial}{\partial t}(N\langle Q\rangle)+\frac{\partial}{\partial x^j}(N\langle Qu^j\rangle)-Na^j\left\langle\frac{\partial Q}{\partial u^j}\right\rangle=0. \tag{30.9}$$

Consider now a gas of particles having no internal structure. For such a gas we can immediately write down five conserved quantities:

$$Q_1=m,\qquad Q_2=mu_1,\qquad Q_3=mu_2,\qquad Q_4=mu_3,\qquad\text{and}\qquad Q_5=\tfrac{1}{2}mu^2. \tag{30.10}$$

These are all obviously collisional invariants in view of (7.39) and (7.42), which show that, for purposes of calculating $I(Q)$, they are equivalent to the summational invariants (7.6) and the statement (7.7) that the total mass of the collision partners is conserved. We shall now show that if we use these five collisional invariants in (30.9), we recover the equation of continuity, the three equations of motion, and the energy equation for the gas.

THE EQUATION OF CONTINUITY

Choose $Q=m$; then (30.9) is a statement of mass conservation, which asserts that

$$(Nm)_{,t}+(Nm\langle u^j\rangle)_{,j}=0. \tag{30.11}$$

But $Nm=\rho$, the gas density, and $\langle u^j\rangle=v^j$, the jth component of the macroscopic flow velocity. Thus we recognize that (30.11) is simply the continuity equation

$$(\partial\rho/\partial t)+\nabla\cdot(\rho\mathbf{v})=0. \tag{30.12}$$

The vector $Nm\mathbf{v}=\rho\mathbf{v}$ can be interpreted physically either as the *momentum density vector* or as the *mass flux vector.*

THE MOMENTUM EQUATIONS

Now take $Q=mu^i$, the ith component of particle momentum; then (30.9) is an expression of momentum conservation, stating that

$$(Nm\langle u^i\rangle)_{,t}+(Nm\langle u^iu^j\rangle)_{,j}-Nma^j\,\delta^i_j=0. \tag{30.13}$$

But $Nm\langle u^i\rangle=\rho v^i$, and $Nma^i=\rho a^i=f^i$, the ith component of the external force per unit volume. Furthermore, using (6.4) and the fact that $\langle\mathbf{U}\rangle\equiv 0$, the tensor $Nm\langle u^iu^j\rangle$ can be rewritten

$$\begin{aligned}Nm\langle u^iu^j\rangle&=\rho\langle(v^i+U^i)(v^j+U^j)\rangle=\rho(v^iv^j+v^i\langle U^j\rangle+v^j\langle U^i\rangle+\langle U^iU^j\rangle)\\&=\rho v^iv^j+\rho\langle U^iU^j\rangle.\end{aligned} \tag{30.14}$$

This tensor can be identified as the *momentum flux density tensor* because it gives the product of Nu^i, the number flux of particles crossing a unit area oriented perpendicular to the ith coordinate axis, times mu^j, the jth component of their momentum, averaged over all particles. Thus $Nm\langle u^iu^j\rangle=\Pi^{ij}$, and by comparing (30.14) with (23.17) we see that we can now identify the stress tensor T^{ij} as

$$T^{ij}\equiv-\rho\langle U^iU^j\rangle. \tag{30.15}$$

Equation (30.15) shows explicitly that fluid stress results from momentum exchange on a microscopic level within the gas. Collecting results, we see that (30.13) reduces to

$$(\rho v^i)_{,t}+(\rho v^iv^j-T^{ij})_{,j}=f^i, \tag{30.16}$$

which is just the ith component of Cauchy's equation of motion (23.16) as derived earlier from macroscopic considerations.

As is done in macroscopic fluid dynamics, we define the negative of the pressure in the gas to be the average of the normal stresses. Thus

$$p = -\tfrac{1}{3}T_{ii} = \tfrac{1}{3}\rho\langle U_x^2 + U_y^2 + U_z^2\rangle = \tfrac{1}{3}\rho\langle U^2\rangle. \tag{30.17}$$

We can also define a translational kinetic temperature for particles in the gas via (8.13):

$$\langle U^2\rangle = 3kT/m. \tag{30.18}$$

Therefore we must have

$$p = \rho kT/m = NkT; \tag{30.19}$$

that is, the pressure is given by the perfect gas law, as expected for a gas of point particles. If we again write $T_{ij} = -p\,\delta_{ij} + \sigma_{ij}$ where σ_{ij} is the viscous stress tensor, we easily see from (30.15) and (30.18) that

$$\sigma_{ij} = -\rho(\langle U_iU_j\rangle - \tfrac{1}{3}\langle U^2\rangle\,\delta_{ij}), \tag{30.20}$$

which gives an explicit expression from which the viscous stresses can be calculated. It is clear that the results we obtain must necessarily depend on the functional form of the distribution function, which enters directly into the computation of the required averages of the velocity components. Using this decomposition of T_{ij} into a hydrostatic pressure and viscous stress, and recalling the equation of continuity, we can rewrite (30.16) into the standard form

$$\rho(Dv_i/Dt) = \rho(v_{i,t} + v_jv_{i,j}) = -p_{,i} + \sigma_{ij,j} + f_i \tag{30.21}$$

(for Cartesian coordinates only), which differs from its macroscopic counterpart (26.1) only in that we have not yet specified an analytic form for σ_{ij}.

THE ENERGY EQUATION

To obtain a conservation law for translational energy we set $Q = \tfrac{1}{2}mu^2$ in (30.9), which then becomes

$$(\tfrac{1}{2}Nm\langle u^2\rangle)_{,t} + (\tfrac{1}{2}Nm\langle u^2u^j\rangle)_{,j} = Nma^jv_j. \tag{30.22}$$

Now $Nm = \rho$; $Nma^jv_j = \rho a^jv_j = f^jv_j$;

$$\begin{aligned}\langle u^2\rangle &= \langle u_iu^i\rangle = \langle (v_i + U_i)(v^i + U^i)\rangle \\ &= v_iv^i + 2v_i\langle U^i\rangle + \langle U_iU^i\rangle = v^2 + \langle U^2\rangle;\end{aligned} \tag{30.23}$$

and

$$\begin{aligned}\langle u^2u^j\rangle &= \langle (v_i + U_i)(v^i + U^i)(v^j + U^j)\rangle \\ &= v_iv^i(v^j + \langle U^j\rangle) + v^j\langle U_iU^i\rangle + 2v^jv_i\langle U^i\rangle + 2v_i\langle U^iU^j\rangle + \langle U_iU^iU^j\rangle \\ &= v^j(v^2 + \langle U^2\rangle) + 2v_i\langle U^iU^j\rangle + \langle U^2U^j\rangle.\end{aligned} \tag{30.24}$$

Therefore (30.22) can be written

$$[\rho(\tfrac{1}{2}\langle U^2\rangle + \tfrac{1}{2}v^2)]_{,t} + [\rho(\tfrac{1}{2}\langle U^2\rangle + \tfrac{1}{2}v^2)v^j + \rho v_i\langle U^iU^j\rangle + \rho\langle \tfrac{1}{2}U^2U^j\rangle]_{,j} = v_jf^j. \tag{30.25}$$

Each of the terms in (30.25) admits of a simple physical interpretation. First, the translational energy associated with the random motion of a particle is $\tilde{e}_{\text{trans}} = \frac{1}{2}m\langle U^2\rangle$, which yields a specific internal energy (per gram) of $e = \frac{1}{2}\langle U^2\rangle$ for the gas. Next, from (30.15) we recall that $\rho\langle U^i U^j\rangle = -T^{ij} = p\,\delta^{ij} - \sigma^{ij}$. Finally, the last term on the left-hand side is the energy per particle, $\frac{1}{2}mU^2$, times the flux of particles NU^j along the jth axis in a frame moving with the fluid, averaged over all particles; this is the energy flux in the gas resulting from microscopic motions, hence we can identify the heat flux vector $\mathbf{q}$ as

$$\mathbf{q} = \rho\langle \tfrac{1}{2}U^2\mathbf{U}\rangle. \tag{30.26}$$

We then see that (30.25) can be rewritten as

$$(\rho e + \tfrac{1}{2}\rho v^2)_{,t} + [\rho(h + \tfrac{1}{2}v^2)v^j - v_i\sigma^{ij} + q^j]_{,j} = v_i f^i, \tag{30.27}$$

which is identical to the macroscopic total energy equation (27.5), and has the same physical interpretation as (27.5) and (24.5). Equation (30.27) can also be reduced, using the dot product of (30.21) with $\mathbf{v}$, to a gas energy equation identical to the macroscopic equation (27.11).

The conservation equations derived in this section are exact (for the adopted model of the gas), but have no practical value until we can evaluate σ^{ij} and q^i. In the macroscopic theory, this is done by an appeal to empirical results. Using kinetic theory we can evaluate these terms from first principles (see §§32 and 33).

31. *Conservation Equations for Equilibrium Flow*

In all flows of interest in astrophysical applications, both the velocity of the flow and the thermodynamic properties of the gas vary from point to point; indeed, gradients in the material properties such as pressure are often responsible (along with gravity) for driving the flow. Such flows are obviously not in equilibrium in the strict sense that the material is homogeneous and gradients of all physical quantities are zero. Nevertheless, if particle mean free paths are very small compared to characteristic flow lengths, and if gradients are sufficiently small, then the gas can be considered to be in *local equilibrium*. In this limit we assume that the translational energy-exchange time is negligible in comparison to characteristic flow times, and that all particles have a single Maxwellian velocity distribution at the local value of the temperature and density.

If we assume that $f(\mathbf{x}, \mathbf{u}, t) \equiv f_0(\mathbf{U})$, as given by (8.9) with $T = T(\mathbf{x})$ and $N = N(\mathbf{x})$, then we can evaluate q_i and σ_{ij} immediately from (30.26) and (30.20). Thus, for the heat flux

$$(q_i)_0 = \tfrac{1}{2}\rho(m/2\pi kT)^{3/2}\int_{-\infty}^{\infty} dU_k \int_{-\infty}^{\infty} dU_j \int_{-\infty}^{\infty} U_i U^2 \exp[-(mU^2/2kT)]\, dU_i, \tag{31.1}$$

(i, j, and k distinct; no sum on i). Now U^2 is an even function of all velocity components, while U_i is an odd function; therefore the integral over dU_i vanishes identically. Hence, in local equilibrium, $\mathbf{q}_0 \equiv 0$. Similarly, for an element of the viscous stress tensor

$$(\sigma_{ij})_0 = -\rho\left(\frac{m}{2\pi kT}\right)^{3/2} \int_{-\infty}^{\infty} e^{-(mU_k^2/2kT)}\, dU_k \times \int_{-\infty}^{\infty} U_j e^{-(mU_j^2/2kT)}\, dU_j \int_{-\infty}^{\infty} U_i e^{-(mU_i^2/2kT)}\, dU_i \tag{31.2}$$

(no sum on i or j). Here, the integrals over both dU_i and dU_j vanish identically because U_i and U_j are odd functions and the exponential is even. Therefore, in local equilibrium $(\sigma_{ij})_0 \equiv 0$, and $(T_{ij})_0 = -p\,\delta_{ij}$.

Using these results in the general conservation equations derived in §30, we obtain the gas dynamical equations for equilibrium flow:

$$\rho_{,t} + (\rho v^i)_{,i} = 0, \tag{31.3}$$

$$\rho(Dv_i/Dt) = \rho(v_{i,t} + v^j v_{i,j}) = f_i - p_{,i} \tag{31.4}$$

and

$$(\rho e + \tfrac{1}{2}\rho v^2)_{,t} + [\rho(h + \tfrac{1}{2}v^2)v^j]_{,j} = v_i f^i, \tag{31.5}$$

or, combining (31.4) and (31.5),

$$De/Dt + p[D(1/\rho)/Dt] = 0, \tag{31.6}$$

which are simply the hydrodynamical equations for an ideal fluid derived in Chapter 2. We thus see that these equations provide only a high-order idealization of the flow of a real gas, applying in the limit that gradients of physical properties in the flow field are vanishingly small, and that the velocity distribution departs negligibly from the local Maxwellian distribution. We further see that it is logically inconsistent to use the equilibrium distribution function f_0 to calculate the effects of transport phenomena and to evaluate transport coefficients as was done in §29, inasmuch as we have just seen that these phenomena are absent if the distribution function f actually equals f_0, and arise only *because* f differs from f_0.

32. The Chapman–Enskog Solution for Nonequilibrium Flow

To calculate the distribution function f, we must solve the Boltzmann equation. As we saw in §§8 and 9, it is possible to do this exactly for strict equilibrium, in which case we obtain the Maxwellian distribution f_0. The solution for nonequilibrium regimes is difficult even in simple physical situations, and in general we can obtain only approximate results. There are many ways in which approximate solutions of the Boltzmann equation can be constructed, but the most useful for our present purposes is the *Chapman–Enskog method*, which provides a self-consistent expansion scheme that yields a series of successive approximations for the deviations of the true distribution function f from the equilibrium distribution f_0.

It should be stressed from the outset that the Chapman–Enskog method does not provide a completely general solution of the Boltzmann equation, but rather only a particular class of solutions that depend on local values and local gradients of the temperature, density, and fluid velocity. The importance of this class of solutions is that a distribution function initially not of this type tends to become one of the class in a time of the order of a collision time (**C2,** §7.2).

The Chapman–Enskog theory is moderately complicated, so we confine our attention only to the lowest level of approximation that yields results of interest. Thus we evaluate only the first-order term in the expansion, which leads to the Navier-Stokes equations, and, in doing so, we solve only to lowest order certain integral equations that appear in the development of the theory. Furthermore, we confine attention to a single-species monatomic gas; other limitations of the method will be discussed at the end of the section. Very complete treatments of the Chapman–Enskog method in higher orders of approximation and for multicomponent gases can be found in (**B2,** Chaps. 7 and 9), (**C2,** Chaps. 7–10, 15, and 18), (**H2,** Chaps. 7 and 9), and (**V1,** Chap. 10).

EXPANSION PROCEDURE

To economize the notation, let us write

$$\mathcal{D}f \equiv (\partial f/\partial t) + u^i(\partial f/\partial x^i) + a^i(\partial f/\partial u^i) \tag{32.1}$$

for the differential operator that appears in Boltzmann's equation, and

$$J(f_i, f_j) \equiv \iint (f_i'f_j' - f_if_j)g\sigma(\mathbf{\Omega})\, d\Omega\, d^3u_j \tag{32.2}$$

for the collision integral of any two functions f_i and f_j.

To see how to develop an expansion procedure for solving Boltzmann's equation it is useful to transform the equation to a nondimensional form. Thus, express velocities in units of a reference speed v_0, lengths in units of a characteristic length l, time in units of (l/v_0), accelerations in units of (v_0^2/l), and densities in terms of a reference density n_0. Furthermore, express the differential collision rate, $ng\sigma\, d\Omega$, in units of a characteristic collision frequency ν_0. One then finds

$$\xi\hat{\mathcal{D}}\hat{f} = J(\hat{f}, \hat{f}). \tag{32.3}$$

Here $\hat{\mathcal{D}}$ is the differential operator in dimensionless form, $\hat{f}$ denotes the non-dimensionalized distribution function, and

$$\xi \equiv (v_0/l\nu_0) = (\lambda/l), \tag{32.4}$$

where λ is the mean free path.

The parameter ξ, which is inversely proportional to the density, provides a measure of departures from local translational equilibrium. Clearly as

$\xi \to 0$ the collision term dominates, and we recover the Maxwellian velocity distribution. Furthermore, when $\xi \ll 1$, we may expect f to depart only slightly from f_0. This suggests that we use ξ as an expansion parameter and assume that f can be written in the form

$$f = f_0 + f_1 + f_2 + \ldots = f_0(1 + \Phi_1 + \Phi_2 + \ldots) \tag{32.5}$$

where each successive term is of progressively higher order in ξ so that

$$\hat{f} = \hat{f}_0 + \xi \hat{f}_1 + \xi^2 \hat{f}_2 + \ldots = \hat{f}_0(1 + \xi\hat{\Phi}_1 + \xi^2\hat{\Phi}_2 + \ldots). \tag{32.6}$$

In principle, this procedure provides a systematic expansion of f in powers of (λ/l). In practice, because we know that $\xi \ll 1$ for flows of interest (cf. §29), the first-order solution is a very good approximation.

We now substitute (32.6) into (32.3) and sort terms into groups of equal order in the parameter ξ. Reverting to dimensional variables we obtain a hierarchy of equations of the form

$$J(f_0, f_0) = 0, \tag{32.7}$$

$$\mathscr{D} f_0 = J(f_0, f_1) + J(f_1, f_0), \tag{32.8}$$

$$\mathscr{D} f_1 = J(f_0, f_2) + J(f_1, f_1) + J(f_2, f_0), \tag{32.9}$$

and so on.

The zero-order equation (32.7) is the familiar one that yields the Maxwellian velocity distribution f_0. Given f_0, the first-order equation (32.8) provides an integral equation for f_1; it is this equation that we solve in what follows. Likewise, given f_0 and f_1, (32.9) provides an integral equation for f_2, and so on.

THE ZERO-ORDER EQUATION

We have already seen in §§8 and 9 that the solution of

$$J(f_0, f_0) = \iint [f_0(\mathbf{u}')f_0(\mathbf{u}_1') - f_0(\mathbf{u})f_0(\mathbf{u}_1)] g\sigma(\mathbf{\Omega})\, d\Omega\, d^3u_1 = 0 \tag{32.10}$$

is

$$f_0(\mathbf{u}) = N(m/2\pi kT)^{3/2} \exp[-m(\mathbf{u}-\mathbf{v})^2/2kT] \tag{32.11}$$

where $\mathbf{v} = \langle \mathbf{u} \rangle$. In the present context, we demand only that f_0 satisfy (32.10) locally, and do not impose the further requirement that the material be in strict equilibrium. This means that, because (32.10) by itself places no restrictions on the dependence of N, T, and $\mathbf{v}$ on $\mathbf{x}$ and t, we may regard

$$N = N(\mathbf{x}, t); \qquad T = T(\mathbf{x}, t); \qquad \text{and} \qquad \mathbf{v} = \mathbf{v}(\mathbf{x}, t) \tag{32.12}$$

as arbitrary functions of $\mathbf{x}$ and t.

From the functional form of f_0 one finds the purely mathematical results that

$$\int f_0 \, d^3u = N, \tag{32.13}$$

$$\int \mathbf{u} f_0 \, d^3u = N\mathbf{v}, \tag{32.14}$$

and

$$\int \tfrac{1}{2}m(\mathbf{u}-\mathbf{v})^2 f_0 \, d^3u = \tfrac{3}{2}NkT. \tag{32.15}$$

But, at the same time, we wish to interpret N, T, and $\mathbf{v}$ physically as the density, temperature, and local flow velocity of the gas, which from (6.1), (6.2), and (30.18) are

$$\int f \, d^3u = N, \tag{32.16}$$

$$\int \mathbf{u} f \, d^3u = N\mathbf{v}, \tag{32.17}$$

and

$$\int \tfrac{1}{2}m(\mathbf{u}-\mathbf{v})^2 f \, d^3u = \tfrac{3}{2}NkT, \tag{32.18}$$

where f is now the actual distribution function. For these two sets of equations to be compatible, we evidently must require that, for all terms f_i with $i>0$ in (32.5),

$$\int f_i \, d^3u = 0, \tag{32.19}$$

$$\int \mathbf{u} f_i \, d^3u = 0, \tag{32.20}$$

and

$$\int U^2 f_i \, d^3u = 0, \tag{32.21}$$

where, as before, $\mathbf{U} \equiv \mathbf{u}-\mathbf{v}$. Equations (32.19) to (32.21) place important constraints on the allowable functional form of f.

THE FIRST-ORDER SOLUTION

In (32.5) we wrote the first-order term f_1 in the expansion of f as $f_1 = f_0\Phi_1$, where according to (32.8), Φ_1 is a scalar function that satisfies the integral equation

$$-N^2 \mathcal{I}(\Phi_1) \equiv \iint [\Phi_1(\mathbf{u}') + \Phi_1(\mathbf{u}_1') - \Phi_1(\mathbf{u}) - \Phi_1(\mathbf{u}_1)] \times f_0(\mathbf{u}) f_0(\mathbf{u}_1) g\sigma(\mathbf{\Omega}) \, d\Omega \, d^3u_1 = \mathcal{D} f_0. \tag{32.22}$$

Here we have used $f_0(\mathbf{u}')f_0(\mathbf{u}_1') \equiv f_0(\mathbf{u})f_0(\mathbf{u}_1)$ from (32.10).

To evaluate the right-hand side of (32.22), we recall that f_0 depends on $\mathbf{x}$ and t through the functions (32.12). We therefore expand $\mathscr{D}f_0$ in terms of derivatives with respect to N, T, and $\mathbf{v}$, and derivatives of these quantities with respect to $\mathbf{x}$ and t. We then eliminate time derivatives of N, T, and $\mathbf{v}$ by means of the conservation equations for equilibrium flow (given in §31), which are the relations appropriate to $f=f_0$.

We notice first that

$$\begin{aligned}\mathscr{D}f_0 &= (\partial f_0/\partial t)+u^i(\partial f_0/\partial x^i)+a^i(\partial f_0/\partial u^i)\\ &= (Df_0/Dt)+U^i(\partial f_0/\partial x^i)+a^i(\partial f_0/\partial u^i).\end{aligned} \tag{32.23}$$

Then, taking $f_0=f_0[N(\mathbf{x}, t), T(\mathbf{x}, t), \mathbf{v}(\mathbf{x}, t)]$ we have

$$\begin{aligned}\mathscr{D}f_0 = &\frac{\partial f_0}{\partial N}\frac{DN}{Dt}+\frac{\partial f_0}{\partial v^i}\frac{Dv^i}{Dt}+\frac{\partial f_0}{\partial T}\frac{DT}{Dt}\\ &+U^i\left(\frac{\partial f_0}{\partial N}\frac{\partial N}{\partial x^i}+\frac{\partial f_0}{\partial v^j}\frac{\partial v^j}{\partial x^i}+\frac{\partial f_0}{\partial T}\frac{\partial T}{\partial x^i}\right)+a^i\frac{\partial f_0}{\partial u^i}.\end{aligned} \tag{32.24}$$

But, from (32.11), one immediately finds

$$(\partial f_0/\partial N)=f_0/N, \tag{32.25}$$

$$(\partial f_0/\partial v^i)=(m/kT)(u^i-v^i)f_0=(mU^i/kT)f_0, \tag{32.26}$$

$$(\partial f_0/\partial u^i)=-(mU^i/kT)f_0 \tag{32.27}$$

and

$$(\partial f_0/\partial T)=[(mU^2/2kT)-\tfrac{3}{2}](f_0/T). \tag{32.28}$$

Furthermore, from (31.3) to (31.6),

$$(DN/Dt)=-Nv^i_{,i} \tag{32.29}$$

$$(Dv_i/Dt)=a_i-(p_{,i}/\rho), \tag{32.30}$$

and

$$(De/Dt)=(3k/2m)(DT/Dt)=-(p/\rho)v^i_{,i}. \tag{32.31}$$

On substituting (32.25) to (32.31) into (32.24), and using the equation of state $p=NkT$, we obtain, after some reduction,

$$\mathscr{D}f_0=\{U_i[(mU^2/2kT)-\tfrac{5}{2}](\ln T)_{,i}+(m/kT)(U_iU_j-\tfrac{1}{3}U^2\,\delta_{ij})v_{i,j}\}f_0. \tag{32.32}$$

We can cast (32.32) into a more useful form by noticing that, because the last term is symmetric in i and j, we can replace $v_{i,j}$ by $\tfrac{1}{2}(v_{i,j}+v_{j,i})$, and furthermore, because $(U_iU_j-\tfrac{1}{3}U^2\,\delta_{ij})$ is obviously traceless, we can add to $\tfrac{1}{2}(v_{i,j}+v_{j,i})$ a scalar times δ_{ij}; in particular we can add $-(\tfrac{1}{3}v_{k,k})\,\delta_{ij}$. Thus we can also write

$$\begin{aligned}\mathscr{D}f_0 &= \{U_i[(mU^2/2kT)-\tfrac{5}{2}](\ln T)_{,i}+(m/kT)(U_iU_j-\tfrac{1}{3}U^2\,\delta_{ij})D_{ij}\}f_0\\ &= -N^2\mathscr{I}(\Phi_1),\end{aligned} \tag{32.33}$$

where

$$D_{ij}\equiv\tfrac{1}{2}(v_{i,j}+v_{j,i})-(\tfrac{1}{3}v_{k,k})\,\delta_{ij}. \tag{32.34}$$

The reason for introducing the traceless tensor D_{ij} here will become clear shortly.

We are now in a position to deduce the form of Φ_1. Because Φ_1 is a scalar, and because $\mathscr{I}(\Phi_1)$ is linear in Φ_1, we conclude that there must be a particular solution for Φ_1 which is a linear combination of the components of ∇T and of D_{ij}. Moreover, because f_0 is proportional to N, it follows from (32.22) and (32.33) that Φ_1 must be inversely proportional to N. Finally, we notice that in (32.33) the coefficients of $T_{,i}$ and D_{ij} depend only on the temperature T and the particle random velocities U_i.

We therefore can conclude that there must be a particular solution for Φ_1 of the form

$$\Phi_1 = -N^{-1}[(2kT/m)^{1/2}A_i(\ln T)_{,i} + B_{ij}D_{ij} + \Psi] \tag{32.35}$$

where A_i, B_{ij}, and Ψ are unknown functions of T and U_i. The minus sign and the factor multiplying A_i have been chosen for later convenience. When (32.35) is substituted into (32.33) we find that A_i, B_{ij}, and Ψ must satisfy the integral equations

$$N\mathscr{I}(A_i) = \mathscr{U}_i(\mathscr{U}^2 - \tfrac{5}{2})f_0, \tag{32.36}$$

$$N\mathscr{I}(B_{ij}) = 2(\mathscr{U}_i\mathscr{U}_j - \tfrac{1}{3}\mathscr{U}^2\,\delta_{ij})f_0, \tag{32.37}$$

and

$$\mathscr{I}(\Psi) = 0, \tag{32.38}$$

where, for brevity, we have introduced the dimensionless velocity variable

$$\mathscr{U} = (m/2kT)^{1/2}\mathbf{U}. \tag{32.39}$$

The central problem of the Chapman–Enskog method is to find solutions for equations (32.36) to (32.38). We will carry out this solution in §33, but we can make considerable further progress if we deduce a partial solution for A_i and B_{ij} from the following simple observations. First, $\mathbf{A}$ is a vector which is a function of T and of the components of $\mathscr{U}$. The only vector that can be constructed from these quantities is $\mathscr{U}$ itself, multiplied by some scalar function of the scalars $\mathscr{U}$ and T. Hence we must have

$$A_i = A(\mathscr{U}, T)\mathscr{U}_i. \tag{32.40}$$

Next, (32.37) shows that $\mathscr{I}(B_{ij} - B_{ji}) = \mathscr{I}(B_{ij}) - \mathscr{I}(B_{ji}) \equiv 0$; hence B_{ij} must be symmetric. Furthermore, (32.37) implies that $\mathscr{I}(B_{xx} + B_{yy} + B_{zz}) \equiv 0$; hence B_{ij} must be traceless. In short, B_{ij} is a traceless tensor function of T and of the components of $\mathscr{U}$. But the only second-rank, symmetric, traceless tensor that can be constructed from these quantities is $(\mathscr{U}_i\mathscr{U}_j - \tfrac{1}{3}\mathscr{U}^2\,\delta_{ij})$ times a scalar function of the scalars $\mathscr{U}$ and T. Hence we can write

$$B_{ij} = B(\mathscr{U}, T)(\mathscr{U}_i\mathscr{U}_j - \tfrac{1}{3}\mathscr{U}^2\,\delta_{ij}). \tag{32.41}$$

Equations (32.40) and (32.41) provide only partial solutions for A_i and B_{ij}. In contrast, for the function Ψ we can obtain a complete solution as follows. Equation (32.28) is satisfied if Ψ is chosen to be a linear combination of the summational invariants:

$$\Psi = \alpha_1 + \boldsymbol{\alpha}_2 \cdot \mathbf{U} + \alpha_3 U^2, \tag{32.42}$$

where α_1 and α_3 are scalars and $\boldsymbol{\alpha}_2$ is a vector. In fact, it can be shown that this is the only possible form for Ψ (**C2,** §4.41 and p. 50). Now, from (32.19) to (32.21), Φ_1 must satisfy the three conditions

$$\int f_0 \Phi_1 \, d^3U = 0, \tag{32.43}$$

$$\int U_i f_0 \Phi_1 \, d^3U = 0, \tag{32.44}$$

and

$$\int U^2 f_0 \Phi_1 \, d^3U = 0. \tag{32.45}$$

Using (32.35) and (32.40) to (32.42) in (32.43) to (32.45), and retaining only nonzero integrals, we have

$$\int (\alpha_1 + \alpha_3 U^2) f_0 \, d^3U = 0, \tag{32.46}$$

$$\int [\alpha_{2i} + A(\mathcal{U}, T)(\ln T)_{,i}] U^2 f_0 \, d^3U = 0, \tag{32.47}$$

and

$$\int (\alpha_1 + \alpha_3 U^2) U^2 f_0 \, d^3U = 0. \tag{32.48}$$

Then, using the explicit form for f_0, one finds that (32.46) and (32.48) imply that $\alpha_1 + \alpha_3 \langle U^2 \rangle = 0$ and $\alpha_1 + 3\alpha_3 \langle U^2 \rangle = 0$, which can be satisfied only if $\alpha_1 = \alpha_3 = 0$. Equation (32.47) implies that the vector $\boldsymbol{\alpha}_2$ is parallel to $\boldsymbol{\nabla} T$. Hence, in view of (32.40), the term $\boldsymbol{\alpha}_2 \cdot \mathbf{U}$ in Ψ can be absorbed into the term $\mathbf{A} \cdot \boldsymbol{\nabla} T$ in Φ_1 merely by modifying the value of $A(\mathcal{U}, T)$, which is as yet undetermined. Therefore in (32.35) we may simply set $\Psi \equiv 0$.

Finally, note in passing that (32.47) implies that

$$\int A(\mathcal{U}, T) \mathcal{U}^2 f_0 \, d^3\mathcal{U} = 0, \tag{32.49}$$

and

$$\int A(\mathcal{U}, T) \mathcal{U}_i^2 f_0 \, d^3\mathcal{U} = 0, \tag{32.50}$$

results that will prove useful later.

THE HEAT FLUX VECTOR

Now that we have an analytical form for f_1, we can evaluate the dominant term in the heat flux vector $\mathbf{q}$. If we use $f = f_0 + f_1$ in (30.26), and recall from §31 that $\mathbf{q}_0 \equiv 0$ when $f = f_0$, we find that

$$q_k = \tfrac{1}{2}m \int U_k U^2 f_0 \Phi_1 \, d^3U. \tag{32.51}$$

If we now substitute (32.35) into (32.51), we find that only a few terms survive the integration. In particular, the coefficient of D_{ij} contains integrals of the form $\int U_i U_j U_k U^2 B f_0 \, d^3U$ and $\int U_k U^4 B f_0 \, d^3U$. When we recall that B is a function only of the magnitude of U, and hence is an even function of the velocity components, it is clear that these integrals both vanish because the integrands are odd functions of the velocity components.

We therefore need consider only the term in ∇T. First, convert to dimensionless velocities via (32.39), and define

$$\mathscr{f}_0(\mathscr{U}) \equiv \pi^{-3/2} e^{-\mathscr{U}^2}, \tag{32.52}$$

while noting that $(m/2kT)^{3/2} \, d^3U = d^3\mathscr{U}$. Then

$$\begin{aligned} q_k &= \tfrac{1}{2}Nm[-N^{-1}(\ln T)_{,j}](2kT/m)^2 \int \mathscr{U}_k \mathscr{U}_j \mathscr{U}^2 A(\mathscr{U}, T) \mathscr{f}_0(\mathscr{U}) \, d^3\mathscr{U} \\ &= -(2k^2T/m)T_{,j}\,\delta_{jk} \int \mathscr{U}_l^2 \mathscr{U}^2 A(\mathscr{U}, T) \mathscr{f}_0(\mathscr{U}) \, d^3\mathscr{U} \\ &= -\left[(2k^2T/m) \int \mathscr{U}_l^2 \mathscr{U}^2 A(\mathscr{U}, T) \mathscr{f}_0(\mathscr{U}) \, d^3\mathscr{U}\right] T_{,k}. \end{aligned} \tag{32.53}$$

Comparing (32.53) with Fourier's law (27.18), we see that the factor in the square brackets is to be identified with the coefficient of thermal conductivity K.

The integral in (32.53) can be partly evaluated by integrating over angles. In this equation $\mathscr{U}_l$ denotes any single component of $\mathscr{U}$. For convenience, choose coordinate axes such that we can write $\mathscr{U}_l = \mathscr{U} \cos\theta$, and express $d^3\mathscr{U}$ as $d^3\mathscr{U} = \mathscr{U}^2 \, d\mathscr{U} \, d\omega = \mathscr{U}^2 \, d\mathscr{U} \, (\sin\theta \, d\theta \, d\phi)$, where $d\omega$ is an element of solid angle. The integrals over angles are trivial and we find that

$$K = (8\pi k^2 T/3m) \int A(\mathscr{U}, T) \mathscr{U}^6 \mathscr{f}_0(\mathscr{U}) \, d\mathscr{U}. \tag{32.54}$$

To complete the calculation of K we must determine the function $A(\mathscr{U}, T)$; we return to that problem in §33.

THE VISCOUS STRESS TENSOR

To evaluate the viscous stress tensor σ_{ij} we use $f = f_0 + f_1$ in (30.20). We already know from §31 that f_0 contributes nothing to σ_{ij}, and furthermore, from (32.21) we see that the second term in (30.20) vanishes identically for

$f=f_1$. Thus the only nonzero term in σ_{ij} is

$$\sigma_{ij}=-m\int U_iU_jf_0\Phi_1\,d^3U. \tag{32.55}$$

If we substitute (32.35) into (32.55), we again find that only a few terms survive the integration. In particular, the coefficient of $(\ln T)_{,k}$ contains an integral of the form $\int A(\mathcal{U},T)U_iU_jU_kf_0\,d^3U$, which is always zero because the integrand is an odd function of the velocity components. Thus only the terms in D_{ij} contribute to σ_{ij}, and we find

$$\begin{aligned}\sigma_{ij}&=(-Nm)(-N^{-1}D_{kl})(2kT/m)\int \mathcal{U}_i\mathcal{U}_jB_{kl}f_0(\mathcal{U})\,d^3\mathcal{U}\\ &=2kTD_{kl}\int \mathcal{U}_i\mathcal{U}_j(\mathcal{U}_k\mathcal{U}_l-\tfrac{1}{3}\mathcal{U}^2\,\delta_{kl})B(\mathcal{U},T)f_0(\mathcal{U})\,d^3\mathcal{U}.\end{aligned} \tag{32.56}$$

Only the first of the two integrals on the right-hand side of (32.56) survives because D_{ij} is traceless, which implies that $D_{kl}\,\delta_{kl}\equiv 0$. Suppose now that $i\neq j$. Then the only integrals that do not vanish are those in which $k=i$ and $l=j$, or $k=j$ and $l=i$. Recalling that D_{ij} is symmetric, we therefore have

$$\begin{aligned}\sigma_{ij}&=\left[2kT\int \mathcal{U}_i^2\mathcal{U}_j^2B(\mathcal{U},T)f_0(\mathcal{U})\,d^3\mathcal{U}\right](2D_{ij})\\ &=\left[2kT\int \mathcal{U}_i^2\mathcal{U}_j^2B(\mathcal{U},T)f_0(\mathcal{U})\,d^3\mathcal{U}\right](v_{i,j}+v_{j,i}-\tfrac{2}{3}v_{k,k}\,\delta_{ij}).\end{aligned} \tag{32.57}$$

By a slightly more complicated calculation, one can show that (32.57) remains valid when $i=j$. Comparing (32.57) with the viscous stress tensor for a Newtonian fluid as given by (25.3), we see that (1) the factor in the square brackets is to be identified with the coefficient of shear viscosity μ, and (2) kinetic theory predicts that the bulk viscosity is identically zero for a gas of monatomic structureless particles, a result mentioned earlier.

We can partially evaluate the integral in (32.57) by integrating over angles. Choose coordinates such that $\mathcal{U}_i=\mathcal{U}\cos\theta$ and $\mathcal{U}_j=\mathcal{U}\sin\theta\cos\phi$, and again write $d^3\mathcal{U}=\mathcal{U}^2\,d\mathcal{U}\,(\sin\theta\,d\theta\,d\phi)$. The angular integrals are straightforward and we obtain finally

$$\mu=(8\pi kT/15)\int B(\mathcal{U},T)\mathcal{U}^6f_0(\mathcal{U})\,d\mathcal{U}. \tag{32.58}$$

As was true for the thermal conductivity K, we must determine $B(\mathcal{U},T)$ in order to complete the calculation of μ; again we defer this problem until §33.

LIMITATIONS OF THE CHAPMAN–ENSKOG METHOD

While the Chapman–Enskog method is very powerful and provides transport coefficients that are in good agreement with experiment for a wide variety of gases and gas mixtures [cf. (**C2,** Chaps. 12–14)], it does nevertheless have limitations that should be borne in mind (**H2,** 18–21).

First, it is a classical theory, and hence fails in regimes where quantum effects dominate. For example, at low temperatures the de Broglie wavelength becomes large, and we can no longer consider the particles to be classical point particles moving on well-defined orbits. In addition, at high densities, quantum-statistical effects such as the Pauli exclusion principle come into operation, and the theory must be modified. Neither of these problems arises in the astrophysical media of interest to us, in which densities are relatively low and temperatures are high.

Second, the theory accounts only for binary collisions. Its validity can therefore be questioned in the important case of Coulomb collisions in an ionized gas for the reasons given in §10, where we saw that in a plasma, multiple, overlapping, weak collisions play the dominant role. However, as we will see in §33, it turns out that if the calculation is carried out to a sufficiently high degree of approximation, the Chapman–Enskog theory yields accurate results even in a plasma.

Third, the Chapman–Enskog method provides only a series solution to the Boltzmann equation, and there are questions about the convergence of the series. It is clear that the first-order approximation, the only one we discuss, can be expected to be valid only when $(\lambda/l) \ll 1$ and when gradients of physical properties are small. While these restrictions are usually met in the flows of interest to us, there are some situations (e.g., shocks) where gradients are so steep that the validity of both the Navier-Stokes equations and the Chapman–Enskog method becomes doubtful. In practice, this problem will not be a serious drawback for our work because we can rarely (if ever) observationally resolve shock structures in astrophysical flows; moreover, in shock computations the details of the shock structure are often purposely smeared out by use of an artificial viscosity in order to gain numerical stability. In contrast, for radiation, which can be considered to be a gas composed of photons, the condition $(\lambda/l) \ll 1$ will be met only in the interior of a star, whereas in the atmosphere $(\lambda/l) \gtrsim 1$, and photons escape freely into space. In the former case, one can write analogues to the Chapman–Enskog solution (cf. §§80 and 97); but in the latter case such an approach is worthless even in principle, and one must confront the full equation of transfer.

Fourth, the theory discussed above applies only to monatomic gases with no internal degrees of freedom. The effects of internal excitation and of ionization are thus ignored; we discuss these effects briefly below.

EFFECTS OF EXCITATION AND IONIZATION

The atoms in a gas have internal excitation states and can also ionize; molecules have, in addition, rotational and vibrational energy states. The existence of these extra energy states changes not only local thermodynamic properties (e.g., pressure or specific internal energy) of a gas, but also affects the transport of momentum and energy in nonuniform media. Thus in analyzing the flow of an excited and ionized fluid we must be careful to use the correct constitutive relations for quantities such as p and e, and also the correct transport coefficients. In what follows, we assume the material is in LTE (local thermodynamic equilibrium).

The pressure in a gas results from exchange of momentum associated with translational motions and hence is unaffected by the existence of internal excitation states. Ionization, in contrast, contributes extra particles and therefore alters the equation of state $p(\rho, T)$ (cf. §14).

The specific internal energy of a gas is affected to first order by the availability of nontranslational energy states. Thus, an excited and ionizing monatomic gas will have $e = e_{\text{trans}} + e_{\text{exc}} + e_{\text{ion}}$ [see equation (14.11)], and in a gas that includes molecules one must allow for yet other terms in e_{vib}, e_{rot}, and e_{dissoc}. Similar modifications can be made for other thermodynamic properties such as specific heats.

In the calculation of transport coefficients, internal excitation and ionization can contribute significantly to the transport of energy. For example when we calculate the heat flux vector in a gas we should account for both the translational kinetic energy and the internal excitation energy transported by the translational motions of the particles. Thus, in place of (30.26), we should write

$$\mathbf{q} = \rho\langle \tfrac{1}{2}U^2\mathbf{U}\rangle + \rho\langle e_{\text{exc}}\mathbf{U}\rangle. \tag{32.59}$$

The second term should, in principle, include all forms of nontranslational energy available to the particles (electronic, rotational, vibrational, ionization, dissociation). The evaluation of this term is complicated by the need to specify how rapidly the internal energy transported by a particle is communicated to other particles in its surroundings, which requires large amounts of atomic cross-section data to describe the various kinds of interactions that can occur. In the absence of such data, resort has been made to some rather artificial models [cf. (**C2,** Chap. 11)], and an approximate result known as the *Eucken correction* has often been used (**H2,** §7.66). We cite more accurate detailed work for hydrogen-helium mixtures in §33. Another major effect of ionization in a strongly ionized gas is to change the interparticle interaction from a short-range potential, appropriate for neutral particles, to the long-range Coulomb interaction; as we shall see in §33, this transition radically alters the temperature dependence of both K and μ. As was mentioned earlier in §25, detailed analysis also shows that the finite time required to exchange energy between translational and internal excitation modes results in a nonzero bulk viscosity (**C2,** §11.51), (**H2,** §7.6c), (**V1,** §10.8).

33. Evaluation of the Transport Coefficients

In order to calculate the thermal conductivity and the shear viscosity from equations (32.54) and (32.58), we must solve the integral equations (32.36) and (32.37) for the scalar functions $A(\mathcal{U}, T)$ and $B(\mathcal{U}, T)$. The standard approach used to obtain these solutions is to expand A and B in a series of special polynomials, the *Sonine polynomials*, which are orthogonal for weighting functions of the form $x^n e^{-x^2}$. Chapman and Cowling use an infinite series of these polynomials and express the transport coefficients as ratios of infinite determinants (**C2**, §§7.51 and 7.52). Their results have practical value because one finds that the ratio of the determinants converges very rapidly as rows and columns are added, so that it is actually necessary to consider only a few terms in the expansion. Alternatively, A and B can be expanded in Sonine polynomials and the integral equations solved by application of a variational principle (**H2**, §§7.3 and 7.4). In their full generality, both methods are quite complicated, and it would take us too far afield to discuss them here. Instead we apply the variational principle to the simplest possible trial functions for A and B, which, despite their simplicity, yield precisely the same first-order results as are obtained from the other methods [cf. (**B2**, §§7.9 and 7.10)].

THE VARIATIONAL PRINCIPLE

We show below that the expressions for both K and μ can be cast into the general form

$$H = C\left[2\int FG\, d^3\mathcal{U} - \int F\mathcal{J}(F)\, d^3\mathcal{U}\right]. \tag{33.1}$$

Here H denotes either K or μ, and the function F is the solution of the integral equation

$$\mathcal{J}(F) = G, \tag{33.2}$$

where $\mathcal{J}$ is the Boltzmann collision operator (32.22).

Consider now how the value of H will change for small variations δF of F around its correct value as given by the exact solution of (33.2). Writing out $K \equiv \int F\mathcal{J}(F)\, d^3\mathcal{U}$ in a form similar to equation (7.42), it is easy to show that $\delta K = 2\int \delta F\mathcal{J}(F)\, d^3\mathcal{U}$, hence

$$\delta H = 2C\int \delta F\,[G - \mathcal{J}(F)]\, d^3\mathcal{U}. \tag{33.3}$$

If F is, in fact, the solution of (33.2), then $\delta H = 0$. That is, H is *stationary* with respect to small variations of F around its true form; therefore it is reasonable to expect that even an approximate expression for F will yield a good estimate of H.

In particular, suppose we choose a trial function F_0, and write $F = \alpha F_0$, where α is a variational parameter to be determined in such a way as to optimize the value of H. The optimum choice of α is that for which H is stationary with respect to variations in α. Thus, starting with

$$H(\alpha) = C\left[2\alpha\int F_0 G\, d^3\mathcal{U} - \alpha^2\int F_0\mathcal{J}(F_0)\, d^3\mathcal{U}\right], \tag{33.4}$$

we demand that

$$\delta H(\alpha)=2C\left[\int F_0 G\, d^3\mathcal{U}-\alpha\int F_0\mathcal{I}(F_0)\, d^3\mathcal{U}\right]\delta\alpha\equiv 0. \tag{33.5}$$

which yields

$$\alpha=\left(\int F_0 G\, d^3\mathcal{U}\right)\Big/\left(\int F_0\mathcal{I}(F_0)\, d^3\mathcal{U}\right), \tag{33.6}$$

whence the optimum estimate of H is

$$H=\frac{C\left(\int F_0 G\, d^3\mathcal{U}\right)^2}{\int F_0\mathcal{I}(F_0)\, d^3\mathcal{U}}. \tag{33.7}$$

We must now show that the expressions for K and μ given in §32 can actually be cast into the form of (33.1). Consider first K. From (32.49) we have

$$-\tfrac{5}{2}\int A(\mathcal{U},T)\mathcal{U}^2 f_0(\mathcal{U})\, d^3\mathcal{U}\equiv 0, \tag{33.8}$$

thus (32.54) can be rewritten as

$$K=(2k^2T/3m)\int A(\mathcal{U},T)\mathcal{U}^2(\mathcal{U}^2-\tfrac{5}{2})f_0(\mathcal{U})\, d^3\mathcal{U}. \tag{33.9}$$

Furthermore, by virtue of (32.40), $A(\mathcal{U},T)\mathcal{U}^2=A(\mathcal{U},T)\mathcal{U}_i\mathcal{U}_i=A_i\mathcal{U}_i$; therefore

$$K=(2k^2T/3m)\int A_i\mathcal{U}_i(\mathcal{U}^2-\tfrac{5}{2})f_0(\mathcal{U})\, d^3\mathcal{U}. \tag{33.10}$$

But recall that A_i also satisfies (32.36); hence (33.10) is equivalent to

$$K=(2k^2T/3m)\int A_i\mathcal{I}(A_i)\, d^3\mathcal{U}. \tag{33.11}$$

Thus, combining (33.10) and (33.11), we have

$$K=(2k^2T/3m)\left[2\int A_i\mathcal{U}_i(\mathcal{U}^2-\tfrac{5}{2})f_0(\mathcal{U})\, d^3\mathcal{U}-\int A_i\mathcal{I}(A_i)\, d^3\mathcal{U}\right] \tag{33.12}$$

which is clearly of the form (33.1) with $F=A_i$ and $G=\mathcal{U}_i(\mathcal{U}^2-\tfrac{5}{2})f_0(\mathcal{U})$.

Consider now μ. We can cast (32.58) into a more useful form by defining the traceless tensor

$$\mathcal{U}_l^\circ\mathcal{U}_m\equiv\mathcal{U}_l\mathcal{U}_m-\tfrac{1}{3}\mathcal{U}^2\,\delta_{lm}, \tag{33.13}$$

and by noting the identity

$$(\mathcal{U}_l^\circ\mathcal{U}_m)(\mathcal{U}_l^\circ\mathcal{U}_m)=\tfrac{2}{3}\mathcal{U}^4, \tag{33.14}$$

which follows directly by substitution from (33.13) when we recall that $\delta_{lm}\,\delta_{lm}=3$. Then (32.58) becomes

$$\mu=\tfrac{1}{5}kT\int B(\mathcal{U},T)\mathcal{U}_l^\circ\mathcal{U}_m\mathcal{U}_l^\circ\mathcal{U}_m f_0(\mathcal{U})\,d^3\mathcal{U} \tag{33.15a}$$

$$=\tfrac{1}{10}kT\int B_{lm}2\mathcal{U}_l^\circ\mathcal{U}_m f_0(\mathcal{U})\,d^3\mathcal{U} \tag{33.15b}$$

$$=\tfrac{1}{10}kT\int B_{lm}\mathcal{I}(B_{lm})\,d^3\mathcal{U}, \tag{33.15c}$$

where we have used (32.41) and (32.37). Combining (33.15b) and (33.15c) we have

$$\mu=\tfrac{1}{10}kT\left[2\int B_{lm}2\mathcal{U}_l^\circ\mathcal{U}_m f_0(\mathcal{U})\,d^3\mathcal{U}-\int B_{lm}\mathcal{I}(B_{lm})\,d^3\mathcal{U}\right], \tag{33.16}$$

which again is clearly of the form (33.1) with $F=B_{lm}$ and $G=2\mathcal{U}_l^\circ\mathcal{U}_m f_0(\mathcal{U})$.

CALCULATION OF K AND μ

In order to apply (33.7) to (33.12) and (33.16) we must make an appropriate choice of the trial function F_0 in each case. In view of (32.40), to calculate K we write

$$F=\alpha F_0=\alpha A_i^0=\alpha A^0\mathcal{U}_i \tag{33.17}$$

where α is the variational parameter and A^0 is a trial function for $A(\mathcal{U},T)$. Similarly, to calculate μ we choose, bearing in mind (32.41),

$$F=\alpha B_{lm}^0=\alpha B^0\mathcal{U}_l^\circ\mathcal{U}_m, \tag{33.18}$$

where B^0 is a trial function for $B(\mathcal{U},T)$.

Both A^0 and B^0 must be chosen such that Φ_1 will still satisfy the restrictions imposed by (32.43) to (32.45) if we use our guesses for A_i and B_{ij} in (32.35); in fact we can impose these requirements on each trial function separately. If we do this, it is easy to see that the simplest choice of B^0 that meets the requirements is $B^0=\text{constant}$. In this case we can absorb the constant into α, and set $B^0\equiv 1$ in (33.18). The situation for A^0 is more complicated. The choice $A^0=\text{constant}$ is unsuccessful because it fails to satisfy (32.49), which is a consequence of (32.43) to (32.45). Comparing (32.36) and (32.40) leads one to suspect that $A(\mathcal{U},T)$ should be expanded in even powers of $\mathcal{U}$, a conjecture shown to be valid by the Sonine polynomial analysis [cf. (**C2**, §7.51)]. Hence we try $A^0=a+b\mathcal{U}^2$, noting that one of the constants can be absorbed into the variational parameter, so that only their ratio matters. Substituting this trial solution into (32.49), one finds that $(a/b)=-\tfrac{5}{2}$; hence in (33.17) we set $A^0=(\mathcal{U}^2-\tfrac{5}{2})$.

Using these trial functions in (33.7) we find

$$[K]_1 = \left(\frac{2k^2T}{3m}\right) \frac{\left[\int (\mathcal{U}^2 - \frac{5}{2})^2 \mathcal{U}^2 f_0(\mathcal{U})\, d^3\mathcal{U}\right]^2}{\left[\int (\mathcal{U}^2 - \frac{5}{2}) \mathcal{U}_i \mathcal{I}[\mathcal{U}_i(\mathcal{U}^2 - \frac{5}{2})\right] d^3\mathcal{U}} \equiv \left(\frac{2k^2T}{3m}\right) \frac{I_1^2}{I_2}, \tag{33.19}$$

and, recalling (33.14),

$$[\mu]_1 = \left(\frac{kT}{10}\right) \frac{\left[\frac{4}{3}\int \mathcal{U}^4 f_0(\mathcal{U})\, d^3\mathcal{U}\right]^2}{\int \mathcal{U}_l^\circ \mathcal{U}_m \mathcal{I}(\mathcal{U}_l^\circ \mathcal{U}_m)\, d^3\mathcal{U}} \equiv \left(\frac{kT}{10}\right) \frac{I_3^2}{I_4}. \tag{33.20}$$

The notation emphasizes that these expressions are only first approximations to K and μ.

The integrals I_1 and I_3 are straightforward, and with the help of standard tables, we find $I_1 = \frac{15}{4}$ and $I_3 = 5$. The integrals I_2 and I_4 are more complicated. Consider first I_2. Introducing the dimensionless relative velocity $\boldsymbol{\gamma} \equiv (m/2kT)^{1/2}\mathbf{g}$, and using the definition of $\mathcal{I}$, we have

$$I_2 = (2kT/m)^{1/2} \int d^3\mathcal{U} \int d^3\mathcal{U}_1 \oint \sigma(\boldsymbol{\Omega})\, d\Omega [\![f_0(\mathcal{U}) f_0(\mathcal{U}_1) \gamma \times \{(\mathcal{U}^2 - \tfrac{5}{2})\boldsymbol{\mathcal{U}} \cdot [\boldsymbol{\mathcal{U}}(\mathcal{U}^2 - \tfrac{5}{2}) + \boldsymbol{\mathcal{U}}_1(\mathcal{U}_1^2 - \tfrac{5}{2}) - \boldsymbol{\mathcal{U}}'(\mathcal{U}'^2 - \tfrac{5}{2}) - \boldsymbol{\mathcal{U}}_1'(\mathcal{U}_1'^2 - \tfrac{5}{2})]\}]\!] \tag{33.21}$$

From the results of §7 in dimensionless form for particles of equal masses, we can write

$$\boldsymbol{\mathcal{U}} + \boldsymbol{\mathcal{U}}_1 = \boldsymbol{\mathcal{U}}' + \boldsymbol{\mathcal{U}}_1' = 2\boldsymbol{\Gamma}, \tag{33.22}$$

$$\mathcal{U}^2 + \mathcal{U}_1^2 = \mathcal{U}'^2 + \mathcal{U}_1'^2 = 2\Gamma^2 + \tfrac{1}{2}\gamma^2, \tag{33.23}$$

$$\boldsymbol{\mathcal{U}} = \boldsymbol{\Gamma} + \tfrac{1}{2}\boldsymbol{\gamma} \tag{33.24a}$$

$$\boldsymbol{\mathcal{U}}' = \boldsymbol{\Gamma} + \tfrac{1}{2}\boldsymbol{\gamma}', \tag{33.24b}$$

and

$$\boldsymbol{\mathcal{U}}_1 = \boldsymbol{\Gamma} - \tfrac{1}{2}\boldsymbol{\gamma} \tag{33.24c}$$

$$\boldsymbol{\mathcal{U}}_1' = \boldsymbol{\Gamma} - \tfrac{1}{2}\boldsymbol{\gamma}', \tag{33.24d}$$

where $\boldsymbol{\Gamma}$ is the dimensionless center of mass velocity. Then from (7.35) we have

$$d^3\mathcal{U}\, d^3\mathcal{U}_1 = (4\pi\gamma^2\, d\gamma)(4\pi\Gamma^2\, d\Gamma)(d\omega_\gamma/4\pi)(d\omega_\Gamma/4\pi), \tag{33.25}$$

where $d\omega_\gamma$ and $d\omega_\Gamma$ are the solid angles describing the orientations of $\boldsymbol{\gamma}$ and $\boldsymbol{\Gamma}$. Furthermore,

$$f_0(\mathcal{U}) f_0(\mathcal{U}_1) = \pi^{-3} \exp[-(\mathcal{U}^2 + \mathcal{U}_1^2)] = \pi^{-3} \exp[-(2\Gamma^2 + \tfrac{1}{2}\gamma^2)], \tag{33.26}$$

and

$$\mathcal{U}^2 - \tfrac{5}{2} = (\Gamma^2 + \tfrac{1}{4}\gamma^2 - \tfrac{5}{2}) + \boldsymbol{\gamma} \cdot \boldsymbol{\Gamma}. \tag{33.27}$$

From (33.22) one immediately sees that all terms containing the numerical factor $\frac{5}{2}$ in the square bracket of (33.21) cancel one another and can be dropped henceforth. Next, if we expand the remaining terms in the square bracket, we find

$$\mathcal{U}\cdot[\mathcal{U}^2\mathcal{U}+\mathcal{U}_1^2\mathcal{U}_1-\mathcal{U}'^2\mathcal{U}'-\mathcal{U}_1'^2\mathcal{U}_1'] \\ =(\boldsymbol{\gamma}\cdot\boldsymbol{\Gamma})^2-(\boldsymbol{\gamma}'\cdot\boldsymbol{\Gamma})^2+\tfrac{1}{2}\gamma^2[\boldsymbol{\gamma}\cdot\boldsymbol{\Gamma}-(\boldsymbol{\gamma}'\cdot\boldsymbol{\Gamma})\cos\chi] \tag{33.28}$$

where χ is the angle between $\boldsymbol{\gamma}$ and $\boldsymbol{\gamma}'$, as usual. The product of (33.26), (33.27), and (33.28) must be integrated over all $d\omega_\Gamma$ for fixed $\boldsymbol{\gamma}$ and $\boldsymbol{\gamma}'$. Notice that all the terms in the parenthesis of (33.27) are independent of the orientation of $\boldsymbol{\Gamma}$. When integrated over $(d\omega_\Gamma/4\pi)$, the product of these terms with those in (33.28) must average to zero, because if ψ and ψ' are the angles between $\boldsymbol{\Gamma}$ and $\boldsymbol{\gamma}$, and between $\boldsymbol{\Gamma}$ and $\boldsymbol{\gamma}'$, respectively, then the angle average of the first two terms is of the form $\gamma^2\Gamma(\langle\cos^2\psi\rangle-\langle\cos^2\psi'\rangle)\equiv 0$, and the averages of the last two terms contain the factors $\langle\cos\psi\rangle=\langle\cos\psi'\rangle\equiv 0$. Here we use the result from (7.8) that $\gamma'=\gamma$. Therefore, only the term in $\boldsymbol{\gamma}\cdot\boldsymbol{\Gamma}$ from (33.27) survives, and it, in turn, clearly can yield a nonzero average only with the two terms in the square bracket of (33.28).

Let us now decompose $\boldsymbol{\gamma}'$ into components along $\boldsymbol{\gamma}$ and along a unit vector $\mathbf{n}$ perpendicular to $\boldsymbol{\gamma}$; then

$$\boldsymbol{\gamma}'=(\cos\chi)\boldsymbol{\gamma}+\gamma(\sin\chi)\mathbf{n}, \tag{33.29}$$

and the terms to be integrated over $d\omega_\Gamma$ are $\frac{1}{2}\gamma^2[(\boldsymbol{\gamma}\cdot\boldsymbol{\Gamma})^2(1-\cos^2\chi)-\gamma\sin\chi\cos\chi(\boldsymbol{\gamma}\cdot\boldsymbol{\Gamma})(\mathbf{n}\cdot\boldsymbol{\Gamma})]$. Because $\boldsymbol{\gamma}$ and $\mathbf{n}$ are orthogonal, the last term must average to zero, and we thus need compute only the integral of $\frac{1}{2}\gamma^4\Gamma^2\cos^2\psi(1-\cos^2\chi)$. Hence (33.21) has been reduced to

$$I_2=\frac{1}{2\pi^3}\left(\frac{2kT}{m}\right)^{1/2}\int_0^\infty d\gamma(4\pi\gamma^7e^{-\gamma^2/2})\oint d\Omega\,\sigma(\gamma,\chi)\sin^2\chi \\ \times\int_0^\infty d\Gamma(4\pi\Gamma^4e^{-2\Gamma^2})\oint\frac{\cos^2\psi}{4\pi}\,d\omega_\Gamma, \tag{33.30}$$

where we have omitted the trivial integral over $d\omega_\gamma$.

In (33.30), the fourth integral is trivial; the third integral is given in standard tables; and the second integral is $\sigma_{(2)}(\gamma)$ as defined by equation (7.24). Thus, if we rewrite the first integral in terms of a new variable $y^2\equiv\frac{1}{2}\gamma^2$, we obtain, finally,

$$I_2=4(kT/\pi m)^{1/2}\int_0^\infty y^7\sigma_{(2)}(y)e^{-y^2}\,dy \tag{33.31}$$

and hence

$$[K]_1=\tfrac{75}{32}(\pi k^3T/m)^{1/2}\Big/\int_0^\infty y^7\sigma_{(2)}(y)e^{-y^2}\,dy. \tag{33.32}$$

By an analysis similar to that outlined above it is not difficult to show that

$I_4 \equiv I_2$, and therefore that

$$[\mu]_1 = \tfrac{5}{8}(\pi mkT)^{1/2} \Big/ \int_0^\infty y^7 \sigma_{(2)}(y) e^{-y^2}\,dy. \tag{33.33}$$

From (33.32) and (33.33) we have

$$[K]_1 = \tfrac{15}{4}(k/m)[\mu]_1 = \tfrac{5}{2}c_v[\mu]_1 = \tfrac{3}{2}c_p[\mu]_1, \tag{33.34}$$

which shows that the Prandtl number for a monatomic gas is

$$\text{Pr} = c_p\mu/K = \tfrac{2}{3} \tag{33.35}$$

as stated earlier in §29.

RESULTS FOR SIMPLE MOLECULAR MODELS

For a given choice of an interparticle potential we can determine $\sigma_{(2)}(\gamma)$ as described in §7, and hence evaluate $[K]_1$ and $[\mu]_1$ explicitly. The simplest case is for rigid spheres, for which $\sigma_{(2)}$ as given by (7.25) is a constant. Substituting this value into (33.33) we obtain

$$[\mu]_1 = (5/16d^2)(mkT/\pi)^{1/2}, \tag{33.36}$$

which differs from the result given by a mean-free-path analysis only by a factor of order unity ($5\pi/16 = 0.982$). Given $[\mu]_1$, $[K]_1$ can be computed from (33.34). Chapman and Cowling calculate the correction from the first to the fourth degree of approximation (which is essentially exact) and show that $[\mu]_4 = 1.01600[\mu]_1$, $[K]_4 = 1.02513[K]_1$, and $[K]_4 = 2.5225[\mu]_4 c_v$.

The next simplest model is for an inverse-power potential. Using (7.26) in (33.33), one finds

$$[\mu]_1 = \frac{5}{8}\left(\frac{mkT}{\pi}\right)^{1/2}\left(\frac{2kT}{\alpha C_\alpha}\right)^{2/\alpha} \Big/ \left[A_2(\alpha)\Gamma\left(4 - \frac{2}{\alpha}\right)\right], \tag{33.37}$$

where Γ denotes the standard gamma function. The thermal conductivity $[K]_1$ is again given in terms of $[\mu]_1$ by (33.34). Both $[\mu]_1$ and $[K]_1$ are proportional to T^s where

$$s = \tfrac{1}{2} + (2/\alpha). \tag{33.38}$$

Rigid-sphere molecules represent the limit $\alpha \to \infty$, in which case $s \to \frac{1}{2}$, in agreement with (33.36). Chapman and Cowling (**C2,** 173) give formulae that correct (33.37) to the second approximation.

The models discussed above are manifestly crude (although they agree remarkably well with experiment for many gases). More realistic interaction potentials allow for a strong repulsive term at small particle separations and a weak attractive term at large separations. The calculation of transport coefficients for such potentials is rather more complicated than for the models considered above; the details are discussed extensively in (**C2,** §§10.4 and 10.5), (**H2,** §8.4), and the references cited therein.

IONIZED GASES

The results derived above apply to neutral gases with short-range potentials. For ionized gases, the interaction is described by the Coulomb potential, which has $\alpha = 1$. For this value of α we expect from (33.38) that both K and μ will vary as $T^{5/2}$, which is markedly different from the temperature dependence found in a neutral gas. This result can be understood physically by the following dimensional argument. From (29.23) we see that the thermal conductivity is proportional to $(\tilde{e}/T)N\lambda\langle U\rangle$ where $\tilde{e}$ is the energy per particle, N is the number density of particles, λ is the mean free path, and $\langle U\rangle$ is the average transport velocity. But $(\tilde{e}/T)$ is merely a numerical factor, while $N\lambda \sim N/(N\sigma) \sim 1/\sigma$ where σ is the collision cross section. Writing $\sigma \approx \pi d_c^2$ where d_c is a typical collision radius, we estimate d_c by equating a particle's kinetic energy to the potential energy of the interaction: $\frac{1}{2}mU^2 \approx e^2/d_c$, which implies that $d_c \propto U^{-2}$ and thus $\sigma \propto U^{-4}$. Hence $K \sim \langle U\rangle/\sigma \sim U^5$; but $U \sim T^{1/2}$, hence $K \sim T^{5/2}$. From (29.25), one sees that the viscosity coefficient will have the same temperature dependence.

Although (33.38) gives the correct temperature dependence of K and μ, (33.37) cannot be applied immediately because $A_2(1)$ calculated from (7.27) diverges. To prevent this divergence, one must introduce a cutoff $b_{\max}$ in (7.16) and (7.27), for the same reasons a cutoff was needed in the calculations of relaxation times in §10. Once again we set $b_{\max} = D$, the Debye length [cf. equation (10.15)]; one then finds

$$A_2(1) = 2[\ln(1+x^2) - x^2/(1+x^2)] \tag{33.39}$$

where $x \equiv 4kTD/e^2$. With this value for A_2 we can then write

$$[\mu]_1 = \tfrac{5}{8}(mkT/\pi)^{1/2}(2kT/e^2)^2/A_2(1) \tag{33.40}$$

and

$$[K]_1 = (75/32)(k^3T/\pi m)^{1/2}(2kT/e^2)^2/A_2(1). \tag{33.41}$$

If we set $m = m_e$ or $m = m_p$ in (33.40) and (33.41), we obtain transport coefficients for a gas of pure electrons or protons respectively (even though such gases of course do not exist in nature). We then notice that $K_e = (m_p/m_e)^{1/2}K_p \approx 43K_p$, and $\mu_e = (m_e/m_p)^{1/2}\mu_p \approx \mu_p/43$, so that thermal conduction is dominantly by electrons, and viscous forces are dominated by protons. This is what one would expect intuitively because in equilibrium both electrons and protons have the same thermal kinetic energies, but the electron random velocities, and hence the flux of electrons, are larger than the proton velocities by a factor of $(m_p/m_e)^{1/2}$, which implies that the electron thermal energy flux is larger by that factor. In contrast, for viscosity it is momentum transport that matters, and here the protons dominate because their momenta in the macroscopic flow are a factor of (m_p/m_e) larger than the electrons' momenta. Thus even though the proton flux across a surface is a factor of $(m_e/m_p)^{1/2}$ smaller than the electron flux, protons still transport more momentum by a net factor of $(m_p/m_e)^{1/2}$.

To obtain useful results for a real hydrogen plasma one must consider a *binary gas mixture* of protons and electrons. The theory then becomes much more complicated, and lies too far afield to discuss here, so we will merely quote results; the interested reader can refer to (**C2**, Chaps. 8 and 9) or (**H2**, §8.2) for details. Chapman (**C1**) has given a concise discussion of transport coefficients for completely ionized gases. He derives expressions that give the ionic and electronic contributions to both μ and K in a form that displays explicitly how each contribution is affected by collisions of the ions and electrons with other ions and electrons. The final results for completely ionized pure hydrogen are that $[\mu/\mu_p]_1 = 0.96$ and $[K/K_e]_1 = 0.324$, where μ_p and K_e are given by (33.40) and (33.41). These results apply only in the first approximation; in higher approximations both the basic computation of μ_p, μ_e, K_p, and K_e, and the formulae for combining these into transport coefficients for the mixed gas become more complicated. However, because μ is dominated by μ_p, and K by K_e, it is reasonable to suppose that the next approximation to K and μ will behave in much the same way as the next approximation for K_e and μ_p. From the formulae given by Chapman and Cowling (**C2**, 173) one finds $[\mu_p]_2 = 1.15[\mu_p]_1$ and $[K_e]_2 = 1.25[K_e]_1$. Work by Landshoff (**L2**), (**L3**) shows that changes produced by yet higher approximations are quite small.

A rather different approach was taken by Spitzer and Härm (**S3**), who solved numerically the Fokker-Planck equation for a completely ionized hydrogen plasma; this approach accounts fully for all mutual interactions among the ions and electrons and provides an accurate nonequilibrium distribution function. Results are presented for the electrical and thermal conductivity of the plasma (but not its viscosity); these results are conveniently expressed in terms of the conductivity of a *Lorentz gas* (an hypothetical fully ionized gas in which the electrons do not interact with one another and the protons are at rest), which is given by

$$K_L = 40(2k^7T^5/\pi^3 m)^{1/2}/e^4 \ln \Lambda \qquad (33.42)$$

where Λ is defined by (10.18). In an actual plasma $K = \delta_T K_L$; for hydrogen, Spitzer and Härm calculate $\delta_T = 0.225$. The results of the Fokker-Planck computation are found to be in good agreement with those from the Chapman–Enskog method provided that the latter are carried out to a sufficiently high degree of approximation.

A temperature gradient in a plasma alters the velocity distribution in such a way as to produce a heat flux transported by electrons; this net flow of electrons also implies that an electric current flows. But in a steady-state plasma, no current must flow in the direction of the temperature gradient because it would rapidly rise without limit. This paradox is resolved by recognizing that a secondary electric field is produced, which exactly cancels the current induced by the temperature gradient; at the same time it necessarily reduces the heat flow. To account for these *thermoelectric effects* the conductivity $K = \delta_T K_L$ calculated thus far must be reduced to an effective conductivity εK; for a pure hydrogen plasma Spitzer and Härm compute $\varepsilon = 0.419$. [See (**S2,** §5.5) or (**S3**) for further discussion.]

All of the results described above apply to completely ionized plasmas. For partially ionized gases, the case of interest in most stellar envelopes, the situation is yet more complex, and we can only cite results. Extensive calculations for μ and K for a partially ionized plasma of hydrogen and helium over a wide range of temperatures and pressures have been tabulated and discussed in detail by Edmonds (**E1**). Comprehensive calculations of the thermal conductivity in hydrogen-helium plasmas, using improved interaction cross sections and fourth-order Chapman–Enskog theory have been made by Ulmschneider (**U1**). The results are given in graphical form for a wide range of temperatures and pressures.

In conclusion it is worth remarking that energy transport by thermal conduction is usually negligible in comparison with radiative transport in most stellar atmospheres and envelopes [cf. Tables 2 and 3 of (**E1**)]. The important exception arises in stellar coronae where million-degree temperatures imply a high thermal conductivity, and these temperatures and low densities imply a high degree of ionization and hence low radiative efficiency. In the same vein, viscosity plays an important role in determining shock structure, but as was mentioned earlier, these structures cannot normally be resolved observationally or numerically. Even so, accurate viscosity coefficients prove very useful in a variety of applications, for example, estimates of wave damping.

References

(A1) Aris, R. (1962) *Vectors, Tensors, and the Basic Equations of Fluid Mechanics.* Englewood Cliffs: Prentice-Hall.

(A2) Astarita, G. and Marrucci, G. (1974) *Principles of Non-Newtonian Fluid Mechanics.* New York: McGraw-Hill.

(B1) Batchelor, G. K. (1953) *The Theory of Homogeneous Turbulence.* Cambridge: Cambridge University Press.

(B2) Bond, J. W., Watson, K. M., and Welch, J. A. (1965) *Atomic Theory of Gas Dynamics.* Reading: Addison-Wesley.

(C1) Chapman, S. (1954) *Astrophys. J.*, **120,** 151.

(C2) Chapman, S. and Cowling, T. G. (1970) *The Mathematical Theory of Non-Uniform Gases.* (3rd ed.) Cambridge: Cambridge University Press.

(E1) Edmonds, F. N. (1957) *Astrophys. J.*, **125,** 535.

(E2) Einstein, A. (1955) *The Meaning of Relativity.* Princeton: Princeton University Press.

(H1) Hinze, J. O. (1959) *Turbulence.* New York: McGraw-Hill.

(H2) Hirschfelder, J. O., Curtiss, C. F., and Bird, R. B. (1954) *Molecular Theory of Gases and Liquids.* New York: Wiley.

(J1) Jeans, J. H. (1925) *Dynamical Theory of Gases.* (4th ed.) Cambridge: Cambridge University Press.

(J2) Jeans, J. H. (1940) *An Introduction to the Kinetic Theory of Gases.* Cambridge: Cambridge University Press.

(L1) Landau, L. D. and Lifshitz, E. M. (1959) *Fluid Mechanics.* Reading: Addison-Wesley.
(L2) Landshoff, R. (1949) *Phys. Rev.*, **76**, 904.
(L3) Landshoff, R. (1951) *Phys. Rev.*, **82**, 442.
(O1) Owczarek, J. (1964) *Fundamentals of Gas Dynamics.* Scranton: International Textbook Co.
(S1) Schlichting, H. (1960) *Boundary Layer Theory.* New York: McGraw-Hill.
(S2) Spitzer, L. (1956) *Physics of Fully Ionized Gases.* New York: Interscience.
(S3) Spitzer, L. and Härm, R. (1953) *Phys. Rev.*, **89**, 977.
(T1) Tisza, L. (1942) *Phys. Rev.*, **61**, 531.
(T2) Townsend, A. A. (1976) *The Structure of Turbulent Shear Flow.* Cambridge: Cambridge University Press.
(T3) Truesdell, C. (1952) *Physik*, **131**, 273.
(U1) Ulmschneider, P. (1970) *Astron. and Astrophys.*, **4**, 144.
(V1) Vincenti, W. G. and Kruger, C. H. (1965) *Introduction to Physical Gas Dynamics.* New York: Wiley.
(Y1) Yuan, S. W. (1967) *Foundations of Fluid Mechanics.* Englewood Cliffs: Prentice-Hall.
(Z1) Zeldovich, Ya. B. and Raizer, Yu. P. (1967) *Physics of Shock Waves and High-Temperature Hydrodynamic Phenomena.* New York: Academic Press.

4

Relativistic Fluid Flow

In many radiation hydrodynamics problems of astrophysical interest, the fluid moves at extremely high velocities, and relativistic effects become important. Examples of such flows are supernova explosions, the cosmic expansion, and solar flares. To account for relativistic effects on a macroscopic level, it is usually adequate to adopt a continuum view, without inquiring in detail into the nature of the fluid itself; such an approach is obviously appropriate for a high-velocity flow of moderate-temperature, low-density gas. In some cases, however, the fluid exhibits relativistic effects on a microscopic level. These situations require a kinetic theory approach, which, in addition, has the advantage of providing precise definitions of, and relations among, the thermodynamic properties of the material.

In what follows we shall develop both the continuum and kinetic theory views of the dynamics of relativistic ideal fluids, thereby retaining parallelism with our earlier work in the nonrelativistic limit, while at the same time laying a thorough groundwork for the treatment of radiation in Chapters 6 and 7. For relativistic nonideal fluids, we consider the continuum view only, obtaining covariant generalizations of the results in Chapter 3.

The flows that are of primary importance to us in this book can be treated entirely within the framework of special relativity. Nevertheless, many of the equations derived in this chapter are completely covariant and apply in general relativity. Only in §§95 and 96 will we need to forsake inertial frames and work in a Riemannian spacetime. Excellent accounts of the theory of general relativistic flows are given in (**L4**); Chapters 5, 22, and 26 of (**M3**); (**T1**); and Chapter 11 of (**W2**). Numerical methods for solving general-relativistic flow problems are discussed in (**M1**), (**M2**), and (**W3**).

4.1 Basic Concepts of Special Relativity

In this section we summarize the ideas from special relativity needed to obtain the equations of hydrodynamics in covariant form. For a more complete discussion the reader can consult the many texts available [e.g., (**A1**), (**A2**), (**B1**), (**L1**), (**L2**), (**L3**), (**M4**), (**R1**), (**S1**), and (**W2**)].

34. The Relativity Principle

In Newtonian mechanics, one presupposes the existence of an *absolute space* of three dimensions in which one can choose rigid *reference frames*; furthermore, one assumes that time is a *universal* independent variable. Among all such frames the preferred frames are the nonaccelerating *inertial frames.* From an operational point of view one attempts to define a reference frame fixed in absolute space through observations of extremely distant objects (e.g., distant galaxies and quasars). We thus obtain a *fundamental reference system* that (1) does not rotate with respect to the large-scale distribution of matter in the Universe and (2) is symmetrical in the sense that when we account for the expansion of the Universe, all material appears to recede isotropically from the observer.

All frames moving uniformly with respect to this fundamental system are inertial frames. Consider a system S' moving uniformly with velocity $\mathbf{v}$ relative to a system S. Accepting the Newtonian view of space and time, one would transform coordinates between S and S' by means of a *Galilean transformation*:

$$x' = x - v_x t, \tag{34.1a}$$

$$y' = y - v_y t, \tag{34.1b}$$

$$z' = z - v_z t, \tag{34.1c}$$

and

$$t' = t. \tag{34.1d}$$

Because S' does not accelerate with respect to S, an isolated body moving with constant velocity $\mathbf{v}_0$ in S will appear to move with a (different) constant velocity $\mathbf{v}_0'$ in S'. Furthermore, because (34.1) implies that $\ddot{\mathbf{x}}' = (d^2\mathbf{x}'/dt^2) \equiv (d^2\mathbf{x}/dt^2) = \ddot{\mathbf{x}}$, if we assume with Newton that a mechanical force $\mathbf{f}$ is the same in all inertial frames, then we conclude that $m\ddot{\mathbf{x}} = \mathbf{f} = m\ddot{\mathbf{x}}'$. Thus Newton's laws of motion are *invariant* under a Galilean transformation. Therefore the dynamical behavior of all mechanical systems governed by Newton's laws is identical in all inertial frames; that is, in Newtonian mechanics all inertial frames are dynamically equivalent, a property referred to as *Newtonian* (or *Galilean*) *relativity.*

With the development of Maxwell's theory of electromagnetism, it was realized that light travels with a unique speed c in vacuum; it was hypothesized that this propagation occurs in a "luminiferous ether," which was assumed to be at rest relative to Newtonian absolute space. A consequence of applying the Galilean transformation to Maxwell's equations is that the velocity of light measured by an observer should depend on his motion relative to the ether. In particular, because laboratories on Earth share its distinctive motion through absolute space, one should be able to detect the drift of the ether past the lab. As is well known, sensitive experiments reveal no such effect, and one is forced to conclude that the Newtonian concepts of space and time are faulty.

The difficulties just described were completely overcome by Einstein's theory of *special relativity*, which is based on two fundamental principles.

First, Einstein asserted his *relativity principle*, which states that all inertial frames are completely equivalent for performing all physical experiments. This is a sweeping generalization, because it implies that *all* laws of physics, not just Newton's laws, must have an invariant form (i.e., must be *covariant*) when we change from one inertial frame to another. In particular, a correct formulation of a physical law must not contain any reference, explicit or implicit, to the velocity, relative to some "absolute space", of the coordinate system in which the phenomenon is described. The demand for covariance of valid physical laws immediately suggests that one must attempt to formulate them as *tensor equations*, which, as we already know, have precisely this property.

Second, Einstein asserted (in agreement with experiment but contrary to "common sense") that the speed of light is the same in all inertial frames, independent of the motion of the source. This postulate has profound implications because it is incompatible with Galilean transformations, in which space and time are independent (cf. §35). Instead, a new transformation, the *Lorentz transformation*, is required, which couples space and time into a single entity, *spacetime*, defined in such a way that Einstein's second postulate is satisfied. We then choose as valid physical laws those formulations that are covariant under Lorentz transformation.

35. The Lorentz Transformation

THE SPECIAL LORENTZ TRANSFORMATION

Let system S' move with uniform speed v along the z axis of system S, and at the instant the origins of S and S' coincide, set $t = t' = 0$. At that instant, suppose that a light pulse is emitted from the origin. According to the two postulates of special relativity, observers in both frames must observe a spherical wavefront, centered on the origin of their system, propagating with velocity c. That is, an observer in S will describe the wavefront by the equation

$$x^2 + y^2 + z^2 - c^2t^2 = 0, \tag{35.1}$$

and an observer in S' by the equation

$$x'^2 + y'^2 + z'^2 - c^2t'^2 = 0. \tag{35.2}$$

Direct substitution from (34.1) shows that with the Galilean transformation we cannot satisfy both of these equations simultaneously.

We seek to modify the transformation law so that it guarantees compatibility of (35.1) and (35.2), while reducing to the Galilean transformation in the limit of very low velocities, where our everyday experience applies. For $v_x = v_y = 0$, the terms in x and y offer no trouble, but we must derive new transformations for z and t. If we assume that the transformation preserves the *homogeneity* of space, so that all points in space and time have equivalent transformation properties, then we conclude that the transformation equations must be *linear.* Thus we hypothesize a transformation of the form

$$z' = \gamma(z - vt), \tag{35.3}$$

where γ is a constant to be determined, which reduces to unity for vanishingly small velocities. We must also modify (34.1d) because no transformation of space coordinates alone can yield wavefronts that are simultaneously spheres in both systems. We thus try the linear transformation

$$t' = At + Bz \tag{35.4}$$

where A and B must also be determined.

Substituting (35.3) and (35.4) into (35.2) and comparing with (35.1) we find

$$\gamma^2 - B^2c^2 = 1, \tag{35.5}$$

$$A^2c^2 - \gamma^2 v^2 = c^2, \tag{35.6}$$

and

$$ABc^2 + \gamma^2 v = 0. \tag{35.7}$$

From these three equations we find that A, B, and γ are

$$\gamma = A = (1 - v^2/c^2)^{-1/2} \tag{35.8}$$

and

$$B = -\gamma v/c^2. \tag{35.9}$$

Using the customary notation $\beta \equiv v/c$, the Lorentz transformation equations are

$$x' = x, \tag{35.10a}$$

$$y' = y, \tag{35.10b}$$

$$z' = \gamma(z - vt), \tag{35.10c}$$

and

$$t' = \gamma(t - \beta z/c), \tag{35.10d}$$

For $v/c \ll 1$, (35.10a) to (35.10c) reduce to the Galilean transformation; however, the transformation from t to t' still depends on v to first order, a fact ignored in the Galilean transformation through the assumption of the existence of an absolute universal time.

Thus the Lorentz transformation turns out to be a *four-dimensional* transformation in spacetime. In spacetime, let us now choose coordinates $x^{(0)} \equiv ct$, $x^{(1)} \equiv x$, $x^{(2)} \equiv y$, and $x^{(3)} \equiv z$. Then if $\mathbf{x}$ is a four-component column vector, (35.10) can be written in matrix notation as

$$\mathbf{x}' = \mathbf{L}\mathbf{x} \tag{35.11}$$

where **L** is the matrix

$$\mathbf{L}=\begin{pmatrix} \gamma & 0 & 0 & -\beta\gamma \\ 0 & 1 & 0 & 0 \\ 0 & 0 & 1 & 0 \\ -\beta\gamma & 0 & 0 & \gamma \end{pmatrix}. \tag{35.12}$$

In component form, (35.11) can be written

$$x'^{\alpha} = L^{\alpha'}_{\beta} x^{\beta} \tag{35.13}$$

where $L^{\alpha'}_{\beta}$ is the element in the α'th row and βth column of **L**; from (35.13) it is obvious that

$$(\partial x^{\alpha'}/\partial x^{\beta}) = L^{\alpha'}_{\beta}. \tag{35.14}$$

Notice that **L** is symmetric, so that $\mathbf{L}^t \equiv \mathbf{L}$. Furthermore, it is easy to show by direct calculation that $|\mathbf{L}|=1$. We demand that all valid Lorentz transformations have unit determinant; the significance of this requirement will become clearer below when we discuss Minkowski coordinates.

The matrix **L** transforms quantities from S to S'; clearly there must be an *inverse transformation* $\mathbf{L}^{-1}$ such that

$$\mathbf{x}=\mathbf{L}^{-1}\mathbf{x}'. \tag{35.15}$$

By direct calculation of the inverse of **L** one readily finds

$$\mathbf{L}^{-1}=\begin{pmatrix} \gamma & 0 & 0 & \beta\gamma \\ 0 & 1 & 0 & 0 \\ 0 & 0 & 1 & 0 \\ \beta\gamma & 0 & 0 & \gamma \end{pmatrix}. \tag{35.16}$$

This result is precisely what we would expect on physical grounds because if observer O in S sees S' moving with velocity v along the z axis, then from the principle of relativity O′ in S' must see S moving with velocity $-v$ along his z' axis. Both points of view are equally valid, hence the transformation from S' to S must be the same as that from S to S', but with the sign of v reversed; equation (35.16) has precisely this property. In component form

$$x^{\alpha} = L^{\alpha}_{\beta'} x^{\beta'}, \tag{35.17}$$

where $L^{\alpha}_{\beta'}$ is the element in the αth row and β'th column of $\mathbf{L}^{-1}$. Hence

$$(\partial x^{\alpha}/\partial x'^{\beta}) = L^{\alpha}_{\beta'}. \tag{35.18}$$

Finally, note in passing that the inverse relationship between **L** and $\mathbf{L}^{-1}$ implies that

$$L^{\alpha}_{\gamma'} L^{\gamma'}_{\beta} = \delta^{\alpha}_{\beta} \tag{35.19a}$$

and

$$L^{\alpha'}_{\gamma} L^{\gamma}_{\beta'} = \delta^{\alpha'}_{\beta'}. \tag{35.19b}$$

LORENTZ-FITZGERALD CONTRACTION AND TIME DILATION

Let us now examine some physical consequences of the Lorentz transformation. One is that a standard measuring rod will be found to have a different length by observers in the two frames S and S'. From (35.10) we see that $\Delta x' = \Delta x$ and $\Delta y' = \Delta y$, hence the lengths of intervals perpendicular to $\mathbf{v}$ are the same in both frames. But suppose that we have a rod of length Δz at rest along the z axis in S, and an observer in S' measures its length at a given instant t'. Then, using $z = \gamma(z' + \beta ct')$ from (35.15), we see that

$$\Delta z' = \Delta z/\gamma; \tag{35.20}$$

that is, the rod appears to have contracted by a factor $(1 - v^2/c^2)^{1/2}$ when it moves relative to the observer making the measurement. This is known as the *Lorentz–Fitzgerald contraction effect.* The length of a rod is greatest when it is at rest relative to an observer; this is its *proper length.*

Similarly, suppose we have a clock at rest in S which is observed by O′ in S'; then from (35.10d), we find

$$\Delta t' = \gamma\, \Delta t; \tag{35.21}$$

that is, the clock appears to run more slowly when it is in motion relative to the observer. This is the *time-dilation effect.* When a clock is at rest relative to an observer it keeps *proper time,* and appears to go at its fastest rate.

Equations (35.20) and (35.21) imply that the *spacetime volume element* is invariant, that is,

$$dV\, dt = dx\, dy\, dz\, dt \equiv dx'\, dy'\, dz'\, dt' = dV'\, dt', \tag{35.22}$$

where dV and dV' denote ordinary three-dimensional volume elements in S and S'. We shall use this result repeatedly.

THE SPACETIME INTERVAL

An *event* (or *world point*) in spacetime is specified by its four coordinates $(x^{(0)}, x^{(1)}, x^{(2)}, x^{(3)}) = (ct, x, y, z)$ which tell when and where the event occurs. The sequence of world points belonging to a real particle is called its *world line,* which describes the particle's motion in spacetime. The *spacetime interval ds* between two events is defined to be

$$ds^2 = -c^2\, dt^2 + dx^2 + dy^2 + dz^2 \equiv -c^2\, d\tau^2; \tag{35.23}$$

(35.23) also defines the interval of *proper time,* $d\tau$.

The arrangement of signs $(-;\ +,\ +,\ +)$ in (35.23) is known as the *signature* of the metric; the choice made here is the "spacelike" convention. Some authors use the "timelike" convention $(+; -, -, -)$. Both choices are equally valid, but formulae using the two conventions can differ in the signs given to various terms; it is essential to check the signatures of the metrics when comparing formulae from different sources.

The importance of the expression (35.23) for the spacetime interval is that it is a *world scalar*, that is, it is invariant under Lorentz transformation, and hence provides a coordinate-independent measure of the "distance" between events in spacetime. The invariance of both the form and the value of ds^2 under the special Lorentz transformation derived above can be easily verified by direct substitution from (35.10). We will demand that general Lorentz transformations (derived below), which describe the motion of S' relative to S along an arbitrary direction, must also preserve invariance of the spacetime interval; this requirement helps to determine the form of the general transformation laws.

We can write the interval, which is the *line element* of spacetime, in standard metric form (cf. §A3.4) as

$$ds^2 = \eta_{\alpha\beta}\, dx^\alpha\, dx^\beta, \tag{35.24}$$

where Greek indices run from 0 to 3. In this tensor form it is obvious that the interval has an invariant value under coordinate transformation. The tensor

$$\boldsymbol{\eta} = \begin{pmatrix} -1 & 0 & 0 & 0 \\ 0 & 1 & 0 & 0 \\ 0 & 0 & 1 & 0 \\ 0 & 0 & 0 & 1 \end{pmatrix} \tag{35.25}$$

is called the *Lorentz metric*. While the metric in ordinary three-dimensional space is positive definite (being the sum of squares), the Lorentz metric is *indefinite*, and ds^2 may be positive, negative, or zero.

Transformations of coordinate systems imply transformations of the metric tensor according to the standard rule [cf. equation (A3.17)]

$$\eta'_{\alpha\beta} = \frac{\partial x^\varepsilon}{\partial x'^\alpha} \frac{\partial x^\zeta}{\partial x'^\beta}\, \eta_{\varepsilon\zeta} = L^\varepsilon_{\ \alpha'} L^\zeta_{\ \beta'} \eta_{\varepsilon\zeta} \tag{35.26}$$

where we have used (35.18). In view of the invariance of the *form* of ds^2 under Lorentz transformation we know that $\eta'_{\alpha\beta} = \eta_{\alpha\beta}$; hence we accept as valid Lorentz transformations only those transformations for which

$$L^\varepsilon_{\ \alpha'} L^\zeta_{\ \beta'} \eta_{\varepsilon\zeta} \equiv \eta_{\alpha\beta}. \tag{35.27}$$

It is easily verified that (35.27) is satisfied by the special Lorentz transformation derived above. Finally, it is worth noting that the reciprocal tensor satisfies $\eta^{\alpha\beta} \equiv \eta_{\alpha\beta}$, and that $\eta_{\alpha\beta}\eta^{\beta\gamma} = \delta^\gamma_\alpha$.

Intervals can be classified into three categories: they are called *spacelike* if $ds^2 > 0$, *timelike* if $ds^2 < 0$, and *null* if $ds^2 = 0$. Because the interval is invariant this categorization is unique, and an interval which is, say, spacelike in one frame will be spacelike in all frames. For two events separated by a spacelike interval there always exists a Lorentz transformation to a particular frame in which the two events occur simultaneously at two different locations. Similarly, for two events separated by a timelike interval there always exists a Lorentz transformation such that in some particular frame the two events occur at the same location at two successive times. From this fact one sees that the world lines of physical particles must be timelike.

Null intervals describe photon paths. From any given event in spacetime the ensemble of all photon paths generates a *null cone* along which light signals propagate. The volume around the x^0 axis contained within the cone comprises all timelike intervals, and can be separated into an absolute future and an absolute past relative to the event at the origin. The volume of spacetime exterior to the null cone comprises all spacelike intervals and represents a conditional "present" in which events are absolutely separated in space.

Finally, let us reconsider the invariance of the spacetime volume element in the light of the metric form for the interval. We know from general considerations [cf. equation (A3.21)] that

$$\sqrt{-\eta}\, dx^{(0)}\, dx^{(1)}\, dx^{(2)}\, dx^{(3)} = \sqrt{-\eta'}\, dx'^{(0)}\, dx'^{(1)}\, dx'^{(2)}\, dx'^{(3)}, \quad (35.28)$$

where $\eta \equiv |\eta_{\alpha\beta}|$ and $\eta' \equiv |\eta'_{\alpha\beta}|$; minus signs appear in (35.28) because the determinants are negative. But $\eta'_{\alpha\beta} \equiv \eta_{\alpha\beta}$, hence $\eta' \equiv \eta$. Thus (35.28) immediately leads back to (35.22).

THE GENERAL LORENTZ TRANSFORMATION

The special Lorentz transformation derived above applies when S' moves along the z axis of S. Suppose now that S' moves with a velocity $\mathbf{v}$ in an arbitrary direction relative to S (again assuming that the origins coincide at $t = t' = 0$ and that the axes in the two systems are parallel). We can calculate the corresponding general Lorentz transformation by introducing two additional frames S_0 and S_0' rotated with respect to S and S', respectively, such that $\mathbf{v}$ lies along the z_0 and z_0' axes. We first apply the special Lorentz transformation between S_0 and S_0', and then undo the rotations to recover the transformation between S and S'. The calculation is straightforward but tedious, and we can infer the form of the general Lorentz transformation more easily by arguing along a somewhat different line.

Adopting for the moment the three-vector notation $\mathbf{x} = (x, y, z)$, $\mathbf{v} = (v_x, v_y, v_z)$, and $\boldsymbol{\beta} = \mathbf{v}/c$, we notice that for the particular choice $\mathbf{v} = (0, 0, v)$ we can write

$$\beta^{-2}\boldsymbol{\beta\beta} \cdot \mathbf{x} = (0, 0, z), \quad (35.29)$$

hence

$$\mathbf{x} - \beta^{-2}\boldsymbol{\beta\beta} \cdot \mathbf{x} = (x, y, 0), \quad (35.30)$$

and therefore

$$[\mathsf{I} + (\gamma - 1)\beta^{-2}\boldsymbol{\beta\beta}] \cdot \mathbf{x} = (x, y, \gamma z). \quad (35.31)$$

Thus the lower right-hand (3×3) matrix in the special Lorentz transformation (35.12) can be regarded as the limiting form of the dyadic $[\mathsf{I} + (\gamma - 1)\beta^{-2}\boldsymbol{\beta\beta}]$. Similarly the row and column three-vectors flanking the

(0, 0) element in (35.12) are limiting forms of the vector $-\boldsymbol{\beta}\gamma$. We therefore propose that the general Lorentz transformation can be written in the form

$$\begin{pmatrix} ct' \\ \mathbf{x}' \end{pmatrix} = \begin{pmatrix} \gamma & -\gamma\boldsymbol{\beta} \\ -\gamma\boldsymbol{\beta} & \mathsf{I}+(\gamma-1)\beta^{-2}\boldsymbol{\beta}\boldsymbol{\beta} \end{pmatrix} \begin{pmatrix} ct \\ \mathbf{x} \end{pmatrix} = \mathbf{L}\begin{pmatrix} ct \\ \mathbf{x} \end{pmatrix}, \tag{35.32}$$

or, equivalently,

$$t' = \gamma(t - \mathbf{v}\cdot\mathbf{x}/c^2) \tag{35.33}$$

and

$$\mathbf{x}' = \mathbf{x} + \{[(\gamma-1)(\mathbf{v}\cdot\mathbf{x})/v^2] - \gamma t\}\mathbf{v}. \tag{35.34}$$

We shall find these results useful later.

Equation (35.32) can be verified by the direct calculation mentioned above. It is also easy to show by direct calculation that this transformation satisfies the basic requirement (35.27). Finally, by a lengthy and tedious calculation, one can show that $|\mathbf{L}| = 1$, as required. Thus (35.32) does indeed give the correct Lorentz transformation. The matrix $\mathbf{L}$ is often called the *boost matrix.*

FOUR-VECTORS

We have thus far considered only transformations of the coordinates or of coordinate increments. But clearly we can adopt the Lorentz transformation as a general transformation for any four-vector in spacetime. Thus let A^α be a general contravariant four-vector; often it is convenient to represent it as

$$A^\alpha = (A^{(0)}, \mathbf{a}), \tag{35.35}$$

where $\mathbf{a}$ is an ordinary three-vector composed of the three space components $(A^{(1)}, A^{(2)}, A^{(3)})$ of $\mathbf{A}$. Using the standard transformation law for contravariant vectors, we find that under Lorentz transformation $\mathbf{A}$ becomes $\mathbf{A}'$ where

$$A'^\alpha = (\partial x'^\alpha/\partial x^\beta)A^\beta = L_\beta^{\alpha'} A^\beta. \tag{35.36}$$

In matrix notation, which is convenient for calculation, $\mathbf{A}' = \mathbf{L}\mathbf{A}$ and $\mathbf{A} = \mathbf{L}^{-1}\mathbf{A}'$ where $\mathbf{A}$ and $\mathbf{A}'$ now denote four-element column vectors.

We can define the general covariant vector B_α as

$$B_\alpha = \eta_{\alpha\beta}B^\beta. \tag{35.37}$$

We then have $B_0 = -B^{(0)}$, $B_1 = B^{(1)}$, $B_2 = B^{(2)}$, and $B_3 = B^{(3)}$ or

$$B_\alpha = (-B^{(0)}, \mathbf{b}). \tag{35.38}$$

Using the standard transformation law for covariant vectors, we find that under Lorentz transformation, $\mathbf{B}$ becomes $\mathbf{B}'$ where

$$B'_\alpha = (\partial x^\beta/\partial x'^\alpha)B_\beta = L_{\alpha'}^\beta B_\beta. \tag{35.39}$$

In matrix notation $\mathbf{B}'^t = \mathbf{B}^t\mathbf{L}^{-1}$, or transposing and recalling that $\mathbf{L}^{-1}$ is symmetric, $\mathbf{B}' = \mathbf{L}^{-1}\mathbf{B}$.

We can assign a length l to a four-vector through the relation

$$l^2 = \eta_{\alpha\beta}A^\alpha A^\beta = \eta^{\alpha\beta}A_\alpha A_\beta = A^\alpha A_\alpha; \tag{35.40}$$

the result is manifestly an invariant, hence we can uniquely classify general four-vectors as spacelike, timelike, or null.

Second-order contravariant four-tensors transform as

$$T'^{\alpha\beta} = (\partial x'^\alpha/\partial x^\mu)(\partial x'^\beta/\partial x^\nu)T^{\mu\nu} = L^{\alpha'}_{\mu}L^{\beta'}_{\nu}T^{\mu\nu}, \tag{35.41}$$

which in matrix notation is $\mathsf{T}' = \mathbf{L}\mathsf{T}\mathbf{L}^t = \mathbf{L}\mathsf{T}\mathbf{L}$ because $\mathbf{L}$ is symmetric. Similarly

$$T^{\alpha\beta} = (\partial x^\alpha/\partial x'^\mu)(\partial x^\beta/\partial x'^\nu)T'^{\mu\nu} = L^{\alpha}_{\mu'}L^{\beta}_{\nu'}T'^{\mu\nu}, \tag{35.42}$$

or $\mathsf{T} = \mathbf{L}^{-1}\mathsf{T}'(\mathbf{L}^{-1})^t = \mathbf{L}^{-1}\mathsf{T}\mathbf{L}^{-1}$ because $\mathbf{L}^{-1}$ is symmetric. Similar relations can be written for second-order covariant tensors.

From the above considerations we see that Lorentz transformations of four-vectors and four-tensors are merely special examples of the standard tensor formalism outlined in the appendix. It follows that physical laws expressed as tensor equations containing four-vectors and four-tensors will automatically be covariant under Lorentz transformation, and hence will satisfy Einstein's principle of relativity. We shall therefore seek relativistic generalizations of familiar nonrelativistic relations by attempting to restate them in four-tensor form.

MINKOWSKI COORDINATES

An interesting perspective on the geometrical nature of Lorentz transformation can be obtained by considering the properties of spacetime in a coordinate system introduced by H. Minkowski. Minkowski's formalism was of great importance in the development of relativity theory, and often provides an effective tool for deepening one's understanding of physical problems, particularly in relativistic kinematics.

Minkowski chose the coordinate system $(x^{(0)}, x^{(1)}, x^{(2)}, x^{(3)}) = (ict, x, y, z)$, for which $ds^2 = \mu_{\alpha\beta}\,dx^\alpha\,dx^\beta$; the *Minkowski metric* is $\mu_{\alpha\beta} = \delta_{\alpha\beta}$. Thus, in this coordinate system, the metric is positive definite, providing an obvious similarity between the spacetime interval and the line element of three-space. In Minkowski coordinates we have $\mathbf{x}' = \mathbf{\Lambda x}$, where the Lorentz transformation is now given by

$$\mathbf{\Lambda} = \begin{pmatrix} \gamma & 0 & 0 & -i\beta\gamma \\ 0 & 1 & 0 & 0 \\ 0 & 0 & 1 & 0 \\ i\beta\gamma & 0 & 0 & \gamma \end{pmatrix}. \tag{35.43}$$

We see that Λ is *Hermitian*, that is, $\Lambda = \Lambda^\dagger$ where "$\dagger$" denotes the *adjoint* (i.e., conjugate transpose) *matrix* [see, e.g., (**M5**, §§49–51)]. More important, it is easily shown that $\mathbf{\Lambda}^{-1} = \mathbf{\Lambda}^{t}$, so that in Minkowski coordinates the Lorentz transformation is *orthogonal* (more precisely it is *unitary*). Moreover one easily sees that $|\mathbf{\Lambda}| = 1$. Hence we conclude that in Minkowski coordinates a Lorentz transformation is merely a *rotation* in spacetime. This result applies not only to four-vectors but also to four-tensors; for example it is easy to show that the transformation law for a second-order contravariant tensor becomes $\mathsf{T}' = \mathbf{\Lambda}\mathsf{T}\mathbf{\Lambda}^{-1}$, that is, it is a *similarity transformation*, as expected for a rotation.

Given Minkowski's interpretation of the geometrical significance of the Lorentz transformation as a pure rotation, it becomes intuitively obvious that the form (and value) of the spacetime interval must remain invariant. Even though Minkowski's approach can sometimes provide clear and beautiful insights, in practice it is a nuisance to work with complex vectors and transformation matrices; therefore, having considered the conceptually important results just discussed, we will not use these coordinates further but will work exclusively with the Lorentz metric.

36. Relativistic Kinematics of Point Particles

FOUR-VELOCITY

To treat the relativistic kinematics of particles we must develop four-dimensional generalizations of the concepts of velocity and acceleration. If a particle moves on some path $[x(t), y(t), z(t)]$ in three-space, then its three-velocity has components $\mathbf{v} = (dx/dt, dy/dt, dz/dt)$. We cannot use this definition in four-space because t is not invariant under Lorentz transformation; hence the vector $\mathbf{v}$ will not have the transformation properties appropriate to a four-vector. We overcome this difficulty by using the proper time τ as the independent variable because $d\tau$ is a world scalar.

From (35.23) we see that the relationship between proper time and lab-frame time for a particle moving with an instantaneous three-velocity $\mathbf{v}$ is

$$d\tau = \left\{1 - \frac{1}{c^2}\left[\left(\frac{dx}{dt}\right)^2 + \left(\frac{dy}{dt}\right)^2 + \left(\frac{dz}{dt}\right)^2\right]\right\}^{1/2} dt = (1 - v^2/c^2)^{1/2}\, dt. \quad (36.1)$$

Clearly proper time is the time measured by a clock in the frame in which the particle is always at rest, the *proper frame* (or *comoving* frame). We will usually distinguish *proper quantities* (i.e., those measured in the proper frame) with a subscript zero, but in the case of proper time it is convenient to use a special symbol. It should be noted that a particle's comoving frame is not generally an inertial frame, a point to which we will return in our discussion of fluid flow.

Given that proper time is a world scalar, we define the *four-velocity* as

$$V^\alpha \equiv (dx^\alpha/d\tau); \quad (36.2)$$

it is obvious that V^α is a genuine contravariant four-vector whose space components reduce to the ordinary three-velocity $\mathbf{v}$ in the limit that $v/c \ll 1$. Hence V^α is indeed the correct four-dimensional generalization of $\mathbf{v}$. In components,

$$V^{(0)} = c(dt/d\tau) = \gamma c, \tag{36.3a}$$

and

$$V^i = (dt/d\tau)(dx^i/dt) = \gamma v^i, \qquad (i = 1, 2, 3), \tag{36.3b}$$

hence

$$V^\alpha = \gamma(c, \mathbf{v}). \tag{36.4}$$

By a trivial calculation one sees that the four-velocity has a constant magnitude:

$$V_\alpha V^\alpha = -c^2. \tag{36.5}$$

Equation (36.5) shows that V^α is timelike. Indeed, the four-velocity of a particle evaluated in its proper frame is $(V^\alpha)_0 = (c, 0, 0, 0)$, hence at any chosen location, $(V^\alpha)_0$ defines the direction of the local proper-time axis in four-dimensional spacetime.

FOUR-ACCELERATION

In ordinary three-space the acceleration is $\mathbf{a} = (d\mathbf{v}/dt)$. Again this expression is not Lorentz covariant, but if we define

$$A^\alpha \equiv (dV^\alpha/d\tau), \tag{36.6}$$

then A^α is a contravariant four-vector whose space components reduce to the ordinary three-acceleration $\mathbf{a}$ in the limit $v/c \ll 1$. Differentiating (36.5) with respect to proper time we find

$$[d(V_\alpha V^\alpha)/d\tau] = 2V_\alpha(dV^\alpha/d\tau) \equiv 0, \tag{36.7}$$

hence

$$V_\alpha A^\alpha \equiv 0, \tag{36.8}$$

which shows that the four-acceleration is orthogonal to the four-velocity in spacetime. Using the easily derived expression

$$(d\gamma/dt) = \gamma^3 \mathbf{v} \cdot \mathbf{a}/c^2, \tag{36.9}$$

we can obtain an explicit expression for A^α in terms of the three-acceleration $\mathbf{a}$, namely

$$A^\alpha = \gamma \frac{d}{dt}[\gamma(c, \mathbf{v})] = \gamma^2 \left[\frac{\gamma^2 \mathbf{v} \cdot \mathbf{a}}{c}, \mathbf{a} + \left(\frac{\gamma^2 \mathbf{v} \cdot \mathbf{a}}{c^2}\right)\mathbf{v}\right], \tag{36.10}$$

and thus the magnitude of A^α is given by

$$A^2 = A_\alpha A^\alpha = \gamma^4[a^2 + (\gamma \mathbf{v} \cdot \mathbf{a}/c)^2], \tag{36.11}$$

which shows that the four-acceleration is spacelike.

37. *Relativistic Dynamics of Point Particles*

FOUR-MOMENTUM

The Newtonian momentum of a particle is $\mathbf{p} = m\mathbf{v}$, where $\mathbf{v}$ is the particle's velocity, and m is its mass, which is assumed to be a unique constant that depends only on the internal constitution of the particle. A relativistic generalization of this expression must replace $\mathbf{v}$ with the four-velocity V^α, and, if the *four-momentum* P^α is to be a genuine four-vector, we can at most multiply V^α by a world scalar. We thus adopt the definition

$$P^\alpha = m_0 V^\alpha, \tag{37.1}$$

where m_0 is the *proper mass* (or *rest mass*) of the particle, that is, its mass in a frame in which it is at rest.

In view of (36.3), equation (37.1) can be written as

$$P^{(0)} = \gamma m_0 c \tag{37.2a}$$

and

$$P^i = \gamma m_0 v^i, \qquad (i = 1, 2, 3), \tag{37.2b}$$

or, more compactly,

$$P^\alpha = \gamma m_0(c, \mathbf{v}). \tag{37.3}$$

We can recover the usual Newtonian definition of momentum from the space components of P^α if we define the *relativistic mass* (or *relative mass*) to be

$$m \equiv \gamma m_0 = m_0/(1 - v^2/c^2)^{1/2} \tag{37.4}$$

and write

$$P^\alpha = m(c, \mathbf{v}) = (mc, \mathbf{p}). \tag{37.5}$$

Although it would take us too far afield to discuss this point in detail, one should realize that the concept of rest mass is nontrivial. It implies that we can resolve a real particle into basic constituents (e.g., electrons, protons, . . . , quarks, . . .) that we can *count*, and to each of which we can assign an intrinsic property called "mass" on the basis of *ab initio* calculation using a fundamental theory (e.g., quantum electrodynamics, quantum chromodynamics, . . .) and/or by astute experimentation. This view is quite different from the macroscopic approach in which mass is typically defined operationally by means of experiments (e.g., collisions) that actually deal with particle *momenta* [see e.g., (**L2,** §16)]. Thus while it is often stated that (37.4) implies that the mass of a particle varies with its velocity, we consider this view to be misleading at best (and perhaps flatly wrong), arising from a prerelativistic lack of appreciation for the intrinsically four-dimensional nature of spacetime. We adopt the position that the "real" mass of a particle is its rest mass, and that the variation predicted by (37.4) reflects not a property of matter itself, but rather the variation of its dynamical *effect* [as judged by measurements of particle momentum (and energy—see below)] produced by the relativistic relationship between space and time that results when they are unified into spacetime.

In Newtonian mechanics we demand that the sum $\sum_i \mathbf{p}_i$ of the momenta of all particles be conserved in a system of particles not subject to external forces. The relativistic generalization of this conservation law is that the sum of the particles' four-momenta be conserved. We then have four conservation equations instead of three; as we will soon see, the fourth is a statement of conservation of energy, because the time-component $P^{(0)}$ of the four-momentum is proportional to a particle's energy.

FOUR-FORCE

From Newton's second law we have the prerelativistic relation

$$\boldsymbol{\phi} = (d\mathbf{p}/dt) = d(m\mathbf{v})/dt, \tag{37.6}$$

where $\boldsymbol{\phi}$ is the ordinary three-force acting on a particle. The natural covariant generalization of (37.6) is

$$\Phi^\alpha = (dP^\alpha/d\tau) = m_0(dV^\alpha/d\tau) = m_0 A^\alpha. \tag{37.7}$$

Equivalently,

$$\Phi^\alpha = \left(\frac{dt}{d\tau}\right)\left(\frac{dP^\alpha}{dt}\right) = \gamma \frac{d}{dt}(mc, \mathbf{p}) = \gamma(\dot{m}c, \boldsymbol{\phi}). \tag{37.8}$$

From (37.7) and (36.8) we see that

$$V_\alpha \Phi^\alpha \equiv 0, \tag{37.9}$$

which implies that

$$c^2 \dot{m} = \boldsymbol{\phi} \cdot \mathbf{v} \tag{37.10}$$

and hence

$$\Phi^\alpha = \gamma(\boldsymbol{\phi} \cdot \mathbf{v}/c, \boldsymbol{\phi}). \tag{37.11}$$

Thus the space components of the four-force are proportional to the ordinary (Newtonian) force acting on the particle, while the time component is proportional to $1/c$ times the work done by that force. All four components reduce to their Newtonian values in the limit of small particle velocities. Comparing (37.11) with (36.10) we see that

$$\boldsymbol{\phi} = m[\mathbf{a} + (\gamma^2 \mathbf{v} \cdot \mathbf{a}/c^2)\mathbf{v}], \tag{37.12}$$

which shows that because of relativistic effects the acceleration of a particle is generally not in the same direction as the applied force.

If we evaluate Φ^α in a particle's comoving frame, denoting the value in that frame as Φ_0^α, then we see that $\Phi_0^0 \equiv 0$. This result, which we use again in Chapter 7, is true for all ordinary body forces (e.g., gravity, ignoring general relativistic effects) that act on point particles without changing their internal state. We shall see below how this result must be modified if the internal structure (e.g., chemical or nuclear composition, or internal excitation state) of the particle is affected by the forces acting on it.

ENERGY

Classically, the rate at which the force $\boldsymbol{\phi}$ does work on a particle equals the rate of increase of its kinetic energy $T = \frac{1}{2}mv^2$. But from (37.10) we see that if the rate of work done is to be identified with the rate of increase of a particle's energy, its total energy must be defined to be $\tilde{e} = mc^2 + \text{constant}$. The zero point cannot be determined by experiment, so we adopt as the relativistic expression of a particle's energy the relation

$$\tilde{e} = mc^2 = \gamma m_0 c^2 = m_0 c^2 / (1 - v^2/c^2)^{1/2}, \tag{37.13}$$

For $v/c \ll 1$ we can expand (37.13) as

$$\tilde{e} = m_0 c^2 + \tfrac{1}{2} m_0 v^2 [1 + \tfrac{3}{4}(v^2/c^2) + \tfrac{5}{8}(v^4/c^4) + \ldots]. \tag{37.14}$$

The first component of the second term on the right-hand side of (37.14) is the classical Newtonian formula for kinetic energy, and subsequent terms are relativistic corrections to this formula.

The first term on the right-hand side of (37.14) is nonclassical and implies that matter has a *rest energy* $m_0 c^2$ associated with its rest mass. The rest energy of matter is enormous (9×10^{20} ergs/gram), dwarfing ordinary chemical energies (i.e., excitation, ionization) by many orders of magnitude; for example, the ionization energy of hydrogen corresponds to about 10^{13} ergs/gram. It is this equivalence of mass and energy that results in the release of vast amounts of electromagnetic and thermal energy in stellar interiors, in nuclear explosions, and in power reactors of various kinds, via nuclear reactions that yield products having slightly lower rest masses than the original input nuclei. More to the point, Einstein emphasized that *all* forms of energy (mechanical, thermal, electromagnetic, nuclear, etc.) affect the mass, and hence inertia, of a particle, and that energy itself has inertia.

Using (37.13) we can rewrite the four-momentum given by (37.5) as

$$P^\alpha = (\tilde{e}/c, \mathbf{p}), \tag{37.15}$$

which shows explicitly that conservation of four-momentum in a system of particles implies both energy and momentum conservation, as stated earlier. From (36.5) and (37.1) it follows that

$$P_\alpha P^\alpha = -m_0^2 c^2, \tag{37.16}$$

hence from (37.15) we obtain the important formula

$$\tilde{e}^2 = p^2 c^2 + m_0^2 c^4. \tag{37.17}$$

Similarly, using (37.13) in (37.8), we can write the four-force as

$$\Phi^\alpha = \gamma(\dot{\tilde{e}}/c, \boldsymbol{\phi}), \tag{37.18}$$

a form that will prove useful below.

We have thus far tacitly assumed that the rest mass of a particle is a strict invariant, and insofar as the particle is truly elementary we can defend this assumption along the lines advanced in our discussion of four-momentum. Now suppose that the "particle" is not elementary, but has an internal structure into which energy can be fed. For example, suppose the particle is an atom, which can absorb or emit energy; in this case one may take the view that the proper mass of our nonelementary particle changes by the mass-equivalent of the energy absorbed or emitted. We must then modify equations (37.7) to (37.11).

For a variable rest mass the four-force becomes

$$\Phi^\alpha = m_0(dV^\alpha/d\tau) + (dm_0/d\tau)V^\alpha. \tag{37.19}$$

Equation (36.8) remains valid, hence we find

$$V_\alpha \Phi^\alpha = -c^2(dm_0/d\tau), \tag{37.20}$$

which shows that V^α and Φ^α are no longer orthogonal. Substituting from (36.4) and (37.18) we find the modified rate-of-work equation

$$\dot{\tilde{e}} = \boldsymbol{\phi} \cdot \mathbf{v} + (c^2 - v^2)(dm_0/d\tau). \tag{37.21}$$

We interpret this equation physically as (rate of increase of particle energy) = (rate of work done by the applied force) + (rate of energy exchange through other, say radiative, mechanisms). We thus deduce that the rate of energy input from nonmechanical sources is

$$W = (c^2 - v^2)(dm_0/d\tau), \tag{37.22}$$

which, evaluated in the comoving frame of the particle, is $W_0 = c^2(dm_0/d\tau)$. We must therefore modify (37.11) to read

$$\Phi^\alpha = \gamma[(\boldsymbol{\phi} \cdot \mathbf{v} + W)/c, \boldsymbol{\phi}], \tag{37.23}$$

which shows that in the presence of nonmechanical effects that modify the internal state of the "particle", we can no longer assume that $\Phi_0^0 \equiv 0$; rather, these effects perform "work" as if some additional "force" were acting. We will find terms of precisely this kind in the equations of radiation hydrodynamics discussed in §§93 and 96.

PHOTONS

Suppose we choose photons as the particles to be considered. From quantum mechanics we know that a photon's energy is $\tilde{e} = h\nu$ and its momentum is $p = h\nu/c$, where ν is its frequency. From (37.17) we see that $(m_0)_{\text{photon}} \equiv 0$, that is, *photons have zero rest mass*. Because a photon's rest mass vanishes the definition of four-momentum given by (37.1) is no longer useful.

Nonetheless equation (37.15) remains valid, hence we can write the photon four-momentum as

$$M^\alpha = (P^\alpha)_{\text{photon}} = (h\nu/c)(1, \mathbf{n}) = \hbar(k, \mathbf{k}) \equiv \hbar K^\alpha, \tag{37.24}$$

where $\mathbf{n}$ is the unit vector in the photon's direction of propagation, and

$k=(2\pi\nu/c)=(2\pi/\Lambda)$ is the photon's *wavenumber*. In (37.24), K^α is the *photon-propagation four-vector*, which, as one expects, is a null vector,

$$K_\alpha K^\alpha \equiv 0, \tag{37.25}$$

as is the photon four-momentum

$$M_\alpha M^\alpha \equiv 0. \tag{37.26}$$

4.2 Relativistic Dynamics of Ideal Fluids

Let us now consider the relativistic dynamics of a compressible ideal fluid, that is, we ignore, for the present, viscous and conduction effects. In §§38–42 we view the gas as a continuum; even in this case it will sometimes be convenient to use particle-counting arguments. We adopt a kinetic-theory view in §43.

38. Kinematics

In seeking relativistic generalizations of the usual classical expressions that describe the kinematics of a fluid, let us first consider the question of reference frames. In general, the velocity of a fluid as measured in a fixed laboratory frame is a function of both space and time: $\mathbf{v}=\mathbf{v}(\mathbf{x}, t)$. Therefore when we speak of the *comoving* (or *proper*) *frame* of a fluid parcel, we are in general dealing with a non-inertial frame, because the fluid can accelerate as it moves. In what follows, we need to apply Lorentz transformations to relate quantities measured in the comoving frame to those measured in the laboratory frame. Here we encounter a problem because, strictly speaking, the Lorentz transformation applies only between inertial frames, which have a constant velocity with respect to one another. To deal with this problem, special relativity hypothesizes that we can consider the comoving frame for any particular fluid parcel to comprise a *sequence* of inertial frames, each of which has a velocity instantaneously coinciding with that of the fluid parcel; it is then assumed that a Lorentz transformation applies between each of these inertial frames and the lab frame. When this is done, the resulting formulation is internally consistent and yields results in agreement with experiment. Further discussion of this point can be found in, for example, (**S1,** Chapters 4 and 6), and the references cited therein.

How do we now describe the motion of a fluid and its time evolution? The covariant generalizations to be used for the velocity and acceleration of a fluid element are, of course, its four-velocity and four-acceleration. To describe time evolution, we must develop a covariant form for the Lagrangean time derivative (D/Dt). From a Newtonian view, (D/Dt) is the time derivative evaluated following the motion of the fluid; put another way, it is $(D/Dt)_0$, the time derivative evaluated in the comoving frame, which, in relativistic terms, is just the derivative with respect to proper time

$(D/D\tau)$. We thus generalize the Lagrangean derivative to mean $(D/D\tau)$, which is a scalar operator in spacetime because proper time is a world scalar.

It is easy to write an expression for $(D/D\tau)$, namely

$$\frac{D}{D\tau}=\left(\frac{dt}{d\tau}\right)\frac{\partial}{\partial t}+\left(\frac{dx^i}{d\tau}\right)\frac{\partial}{\partial x^i}=V^0\frac{1}{c}\frac{\partial}{\partial t}+V^i\frac{\partial}{\partial x^i}, \tag{38.1}$$

or, more compactly,

$$(D/D\tau)=V^\alpha(\partial/\partial x^\alpha), \tag{38.2}$$

which is manifestly Lorentz covariant. In the limit $v/c \ll 1$, $(D/D\tau)$ clearly reduces to the Lagrangean (D/Dt).

As written, (38.2) applies only in Cartesian coordinates in a flat spacetime. We can easily generalize to curvilinear coordinates in flat spacetime ["flat" implying that all components of the Riemann curvature tensor are zero; see, for example, (**A1,** 149) or (**M3,** 283)] by replacing the ordinary derivatives with covariant derivatives to obtain

$$(Df/D\tau)=V^\alpha f_{;\alpha}. \tag{38.3}$$

Here f is any differentiable function. From (38.3) we recognize that $(D/D\tau)$ is the *intrinsic derivative* with respect to proper time in a four-dimensional spacetime (cf. §15 and §A3.10). If spacetime is indeed flat then the covariant derivative in (38.3) merely accounts for curvature of the three-space coordinate mesh (say spherical coordinates) in the "ordinary space" part of the Lorentz metric. But it is worth mentioning that (38.3) is also valid in curved spacetime for general line elements of the form $ds^2 = g_{\mu\nu}\,dx^\mu\,dx^\nu$, where $g_{\mu\nu}=g_{\mu\nu}(x^{(0)}, x^{(1)}, x^{(2)}, x^{(3)})$; hence (38.3) also applies in general relativity.

39. The Equation of Continuity

In Newtonian hydrodynamics the density is the mass per unit volume, or the number of particles per unit volume times the mass of each particle; because both of these quantities are presumed to be invariants, the Newtonian density is considered to be the same in all frames (e.g., in the laboratory and comoving frames). When relativistic effects become important, however, the situation is more complicated, and several definitions of density, each useful in certain contexts, can be made.

First, suppose that in the comoving frame we have N_0 particles per unit proper volume, each of proper mass m_0; then the *density of proper mass in the comoving (proper) frame* is

$$\rho_0=N_0 m_0. \tag{39.1}$$

As measured in the laboratory frame, the density of proper mass will be different. If we choose a comoving volume element δV_0, then the *number* of particles in it will be $N_0\,\delta V_0$; if we count the same particles in the lab frame we will, of course, get the same number, hence

$$N\,\delta V = N_0\,\delta V_0, \tag{39.2}$$

where N denotes the lab-frame particle density and δV is the lab-frame volume element corresponding to δV_0. But owing to Lorentz contraction [cf. (35.20)],

$$\delta V = \delta V_0/\gamma \tag{39.3}$$

hence

$$N = \gamma N_0. \tag{39.4}$$

Therefore the *density of proper mass in the laboratory frame* is

$$\rho = N m_0 = \gamma \rho_0. \tag{39.5}$$

From the Newtonian view this quantity can be considered, as our choice of notation implies, to be "the" density. It is also sometimes useful to define the *density of relative mass* as measured in the lab frame to be

$$\rho' = Nm = \gamma^2 N_0 m_0 = \gamma\rho = \gamma^2 \rho_0, \tag{39.6}$$

in terms of which we can write the momentum density as $\rho'\mathbf{v}$, where $\mathbf{v}$ is the ordinary three-velocity of the fluid.

To derive a relativistic version of the equation of continuity we start from the standard Newtonian equation

$$\rho_{,t} + (\rho v^i)_{,i} = 0, \tag{39.7}$$

which, using (39.5), can be rewritten as

$$(\gamma\rho_0)_{,t} + (\gamma\rho_0 v^i)_{,i} = 0. \tag{39.8}$$

Now recalling that $V^\alpha = \gamma(c, \mathbf{v})$, we see that (39.8) is simply

$$(\rho_0 V^\alpha)_{,\alpha} = 0, \tag{39.9}$$

which is manifestly covariant under Lorentz transformation. We thus accept (39.9) as the correct covariant generalization of the equation of continuity in a flat spacetime.

If we choose curvilinear coordinates or have a curved spacetime, the further generalization of (39.9) is

$$(\rho_0 V^\alpha)_{;\alpha} = 0. \tag{39.10}$$

For example, see (**P1,** 230) for (39.10) written in spherical coordinates.

The vector $N_0 V^\alpha$ is the four-dimensional *particle flux density vector.* The equation of continuity is thus merely a statement that the particle flux density is conserved in spacetime, that is, that particles are neither created nor destroyed. If particles are *not* conserved (e.g., nuclear reactions occur), then a source-sink term must appear on the right-hand sides of (39.9) and (39.10).

40. The Material Stress-Energy Tensor

In §23 we saw that in Newtonian fluid dynamics the three equations of momentum conservation can be formulated as

$$(\rho v^i)_{,t} + \Pi^{ij}_{,j} = f^i, \tag{40.1}$$

where $\mathbf{\Pi}$ is the momentum flux density tensor

$$\Pi^{ij} \equiv \rho v^i v^j + p\,\delta^{ij}, \tag{40.2}$$

and $\mathbf{f}^i$ is the externally applied force density. We wish to construct a similar formulation in spacetime, and we therefore seek four-dimensional generalizations of $\mathbf{\Pi}$ and $\mathbf{f}$. We defer the question of the force density to §41 and concentrate here on obtaining an appropriate expression for the *material stress-energy tensor* M, which is the Lorentz-covariant generalization of the Newtonian momentum flux density tensor. Notice also that the left-hand side of each equation in (40.1) can be expressed as the four-divergence of a suitable four-vector, hence we expect to be able to cast the dynamical equations into the form of a four-divergence of M.

It is clear from the outset that we will arrive at four conservation relations rather than three. Of these, three will be momentum conservation equations. From (40.1) we see that the space components M^{ij} of the stress-energy tensor should be generalizations of Π^{ij}, and hence account for the momentum flux density, both macroscopic and microscopic (i.e., pressure), in the fluid, while the zeroth column M^{i0} must reduce to c times a component of the momentum density. [The factor of c is needed because in the four-divergence operator, $(\partial/\partial x^0)$ is $c^{-1}(\partial/\partial t)$.] Furthermore, we recall from §37 that when we generalized the conservation law for the total Newtonian momentum of a group of particles to conservation of their total four-momentum, the fourth equation turned out to be an expression of energy conservation. The same is true for a fluid, hence we expect the stress-energy tensor to contain an element M^{00} representing the total energy density of the fluid and a vector M^{0j} representing $(1/c)$ times the energy flux in the jth direction of the flow. (Again the factor of $1/c$ is needed to balance the same factor in $\partial/\partial x^0$.)

Thus on the basis of the Newtonian equations alone we expect M to be of the general form

$$\mathsf{m} = \begin{pmatrix} \rho_{00}c^2 & c\rho\mathbf{v} \\ c\rho\mathbf{v} & \rho v^i v^j + p\,\delta^{ij} \end{pmatrix} \tag{40.3}$$

where we have written m instead of M to emphasize that (40.3) is not yet the relativistic stress-energy tensor, inasmuch as it is not covariant. Following L. H. Thomas (**T2**), we have written

$$\rho_{00} \equiv \rho_0(1 + e/c^2) \tag{40.4}$$

where e is the specific internal energy of the gas produced by the microscopic motions of its constituent particles, ρ_{00} is the total mass density of the fluid, including the mass equivalent of its thermal energy, and $\rho_{00}c^2$ is the total energy density of the fluid. Both ρ_0 and e are defined in the proper frame (see §43) and are world scalars, as is ρ_{00}.

In constructing an expression for M we are guided by three general principles: (1) it must be covariant, and hence must contain only world scalars, four-vectors, and four-tensors; (2) it must give the correct fluid energy density and hydrostatic pressure in the comoving frame; (3) it must yield the correct nonrelativistic equations in the laboratory frame when $v/c \ll 1$.

From (40.3) with $\mathbf{v}=0$ we see that in the comoving frame we must have

$$\mathsf{M}_0 = \begin{pmatrix} \rho_{00}c^2 & 0 & 0 & 0 \\ 0 & p & 0 & 0 \\ 0 & 0 & p & 0 \\ 0 & 0 & 0 & p \end{pmatrix}. \tag{40.5}$$

We could now obtain the components of M in the lab frame by applying a Lorentz transformation to (40.5) as in equation (35.42) [see e.g., (**W2,** 48)]. But a simpler approach is to notice that in the comoving frame $V_0^\alpha = (c, 0, 0, 0)$, hence in this frame

$$(V^\alpha V^\beta)_0 = \begin{pmatrix} c^2 & 0 & 0 & 0 \\ 0 & 0 & 0 & 0 \\ 0 & 0 & 0 & 0 \\ 0 & 0 & 0 & 0 \end{pmatrix}, \tag{40.6}$$

while in this same frame the *projection tensor* is

$$(P^{\alpha\beta})_0 \equiv \eta^{\alpha\beta} + c^{-2}(V^\alpha V^\beta)_0 = \begin{pmatrix} 0 & 0 & 0 & 0 \\ 0 & 1 & 0 & 0 \\ 0 & 0 & 1 & 0 \\ 0 & 0 & 0 & 1 \end{pmatrix}. \tag{40.7}$$

We use the projection tensor extensively in §4.3. Substituting (40.6) and (40.7) into (40.5) we conclude that

$$(M^{\alpha\beta})_0 = (\rho_{00} + p/c^2)(V^\alpha V^\beta)_0 + p\eta^{\alpha\beta}. \tag{40.8}$$

Again we follow Thomas (**T2**) and define

$$\rho_{000} \equiv \rho_{00} + p/c^2 = \rho_0(1 + h/c^2), \tag{40.9}$$

where h is the specific enthalpy of the fluid. If we can assume that p is a world scalar (see §43) then we see that

$$M^{\alpha\beta} = \rho_{000} V^\alpha V^\beta + p\eta^{\alpha\beta} \tag{40.10}$$

is a fully covariant expression for M, which reduces to (40.8) in the comoving frame. Notice that M is symmetric.

It is obvious by inspection that (40.10) satisfies requirements (1) and (2) stated above. To check whether it also satisfies requirement (3), we write $\rho=\gamma\rho_0$ as in (39.5), note that in a nonrelativistic fluid $e/c^2 \ll 1$ and $p/\rho_0 c^2 \ll 1$, and expand in powers of v/c. We find

$$\begin{aligned} M^{ij} &= \gamma^2\rho_0 v^i v^j[1+(\rho_0 e+p)/\rho_0 c^2]+p\,\delta^{ij} \rightarrow \rho v^i v^j + p\,\delta^{ij} + O(v^2/c^2) \\ &= \Pi^{ij} + O(v^2/c^2), \end{aligned} \tag{40.11}$$

which is the correct nonrelativistic expression, while

$$M^{0i} = M^{i0} = \gamma^2\rho_0 c v^i[1+(\rho_0 e+p)/\rho_0 c^2] \rightarrow c\rho v^i + O(v^2/c^2), \tag{40.12}$$

which is c times the momentum density, as expected. Furthermore,

$$\begin{aligned} M^{00} &= \gamma^2\rho_0(c^2+e+pv^2/\rho_0 c^2) \\ &= \gamma\rho(c^2+e)+O(v^2/c^2) \rightarrow \rho c^2 + \tfrac{1}{2}\rho v^2 + \rho e + O(v^2/c^2), \end{aligned} \tag{40.13}$$

which is the correct nonrelativistic energy density (including rest energy) of the fluid. Hence (40.10) does in fact provide a fully satisfactory expression for M.

The stress-energy tensor does not in itself provide a complete description of the fluid. To obtain a complete system of equations we also require constitutive relations that describe the microphysics of the gas. We need at least a caloric equation of state relating e to p and ρ_0, and perhaps also an equation of state of the form $p=p(\rho_0, T)$ where T is the *proper temperature* of the fluid. In practice these relations must be obtained from microscopic kinetic-theoretic considerations, and are operationally definable only in the comoving frame of the fluid (see §43).

Finally, it is worth noting that (40.10) also applies in general relativity if we replace the Lorentz metric $\eta^{\alpha\beta}$ with a general metric $g^{\alpha\beta}$.

41. The Four-Force Density

To obtain a Lorentz-covariant generalization of the right-hand side of (40.1) we use the *four-force density*

$$F^\mu \equiv \Phi^\mu/\delta V_0 = (\gamma/\delta V_0)(\boldsymbol{\phi}\cdot\mathbf{v}/c, \boldsymbol{\phi}), \tag{41.1}$$

where Φ^μ and $\boldsymbol{\phi}$ are, respectively, the four-force and the Newtonian force acting on a finite element of material contained in the proper volume δV_0. F^μ is a four-vector because Φ^μ is a four-vector and δV_0 is a world scalar. In any arbitrary frame we define the *ordinary force density* to be $\mathbf{f}=\boldsymbol{\phi}/\delta V$. Therefore

$$F^\mu = (\gamma\,\delta V/\delta V_0)(\mathbf{f}\cdot\mathbf{v}/c, \mathbf{f}). \tag{41.2}$$

But from (39.3) $\delta V_0 = \gamma\,\delta V$, hence

$$F^\mu = (\mathbf{f}\cdot\mathbf{v}/c, \mathbf{f}). \tag{41.3}$$

Thus in the relativistic equations of hydrodynamics the four-force density in any frame has space components equal to the Newtonian force density in that frame, and has a time component equal to the rate of work, per unit volume, done by the Newtonian force density. Put another way, the four-force density has space components equal to the rate of increase (per unit volume) of the momentum, and a time component equal to $1/c$ times the rate of increase of the energy, of the material, as measured in the frame of reference adopted.

42. *The Dynamical Equations*

GENERAL FORM

Arguing by analogy to the Newtonian equations (40.1), we expect the relativistic fluid-dynamical equations to have the general form

$$M^{\alpha\beta}_{;\beta} = F^{\alpha}, \tag{42.1}$$

or

$$(\rho_{000} V^{\alpha} V^{\beta} + p g^{\alpha\beta})_{;\beta} = F^{\alpha}. \tag{42.2}$$

Here $g^{\alpha\beta}$ may be the Lorentz metric $\eta^{\alpha\beta}$ if we use Cartesian coordinates in a flat spacetime, may have space components appropriate to curvilinear coordinates imbedded in a flat spacetime, or may be a general metric in curved spacetime.

Writing (42.1) and (42.2) out in Cartesian coordinates, using (40.10) and (41.3), and defining $\rho_1 \equiv \gamma^2 \rho_{000}$, we obtain

$$(\rho_1 - p/c^2)_{,t} + (\rho_1 v^j)_{,j} = f^0/c = (\mathbf{f} \cdot \mathbf{v}/c^2), \tag{42.3}$$

and

$$(\rho_1 v^i)_{,t} + (\rho_1 v^i v^j + p\,\delta^{ij})_{,j} = f^i, \qquad (i = 1, 2, 3), \tag{42.4a}$$

or

$$(\rho_1 v_i)_{,t} + (\rho_1 v_i v^j)_{,j} + p_{,i} = f_i, \qquad (i = 1, 2, 3). \tag{42.4b}$$

In this form the equations bear a close resemblance to their Newtonian counterparts. Equations (42.4) are the momentum equations. Equation (42.3), while bearing a superficial resemblance to the Newtonian continuity equation (to which it reduces in the limit $c \to \infty$), is actually the energy equation; this becomes more apparent if we write out ρ_1 to display all physical variables explicitly:

$$[\gamma^2(\rho_0 c^2 + \rho_0 e + p) - p]_{,t} + [\gamma^2(\rho_0 c^2 + \rho_0 e + p) v^j]_{,j} = \mathbf{f} \cdot \mathbf{v}, \tag{42.5}$$

and

$$[\gamma^2(\rho_0 c^2 + \rho_0 e + p) v_i]_{,t} + [\gamma^2(\rho_0 c^2 + \rho_0 e + p) v_i v^j]_{,j} + c^2 p_{,i} = c^2 f_i \qquad (i = 1, 2, 3). \tag{42.6}$$

If in (42.5) one again defines $\rho = \gamma\rho_0$, expands the other factor γ, drops terms of $O(v^2/c^2)$ and higher, and, finally, subtracts c^2 times the continuity equation (39.9), one recovers the nonrelativistic energy equation (24.6). We will discuss a similar reduction of the momentum equation (42.6) shortly.

Detailed expressions for (42.5) and (42.6) in spherical coordinates are given in (**P1,** 230–231).

THE GAS ENERGY EQUATION

We can cast the energy equation (42.3) into a much more revealing form by reducing it to a gas energy equation following L. H. Thomas (**T2**). Form the inner product of (42.2) with V_α to obtain

$$V_\alpha V^\alpha(\rho_{000}V^\beta)_{;\beta} + \rho_{000}V^\beta(V_\alpha V^\alpha_{;\beta}) + g^{\alpha\beta}V_\alpha p_{,\beta} = V_\alpha F^\alpha. \qquad (42.7)$$

But $V_\alpha V^\alpha = -c^2$, which implies that $V_\alpha V^\alpha_{;\beta} \equiv 0$; hence (42.7) reduces to

$$c^2(\rho_{000}V^\alpha)_{;\alpha} - V^\alpha p_{,\alpha} = -V_\alpha F^\alpha. \qquad (42.8)$$

Now subtracting $(c^2 + e + p/\rho_0)$ times the continuity equation (39.10) from (42.8) we find

$$\rho_0 V^\alpha[(\partial e/\partial x^\alpha) - (p/\rho_0^2)(\partial\rho_0/\partial x^\alpha)] = -V_\alpha F^\alpha, \qquad (42.9)$$

or, recalling that $V^\alpha(\partial/\partial x^\alpha) \equiv (D/D\tau)$,

$$\rho_0\left[\frac{De}{D\tau} + p\frac{D}{D\tau}\left(\frac{1}{\rho_0}\right)\right] = -V_\alpha F^\alpha. \qquad (42.10)$$

From (37.23) and (41.1) it follows that if we are dealing with ordinary body forces, which conserve particle numbers and rest masses, $V_\alpha F^\alpha = 0$. The flow is then adiabatic and we obtain

$$(De/D\tau) + p[D(1/\rho_0)/D\tau] = 0. \qquad (42.11)$$

as the relativistic generalization of the Newtonian gas-energy equation for an ideal gas. Equation (42.11) is the one that most simply relates e, p, and ρ_0, which are all defined in the comoving frame. Moreover, it obviously reduces to the Newtonian gas-energy equation as $c \to \infty$, and differs from it in general by terms of $O(v^2/c^2)$.

Of course, if physical processes operate in which the numbers or rest masses of the particles in the fluid are not conserved (nuclear reactions), or through which the fluid can dissipate or transport energy internally (viscosity and conductivity), or exchange energy with an external source (radiation), then $V_\alpha F^\alpha \neq 0$, and the flow is no longer adiabatic. We consider such cases in §4.3 and Chapter 7.

THE MOMENTUM EQUATION

The momentum equations (42.4) can also be manipulated into a much more useful form. Multiplying (42.3) by v_i and subtracting the result from (42.4b) we obtain

$$\rho_1(v_{i,t} + v^j v_{i,j}) + p_{,i} + (v_i/c^2)p_{,t} = f_i - v_i(\mathbf{f}\cdot\mathbf{v}/c^2). \qquad (42.12)$$

Then using (38.2) for $(D/D\tau)$ and defining $\rho_* \equiv \gamma\rho_{000}$, which is the relative density corresponding to the proper mass density plus the mass equivalent of the fluid enthalpy, we have

$$\rho_*(D\mathbf{v}/D\tau) = \mathbf{f} - \nabla p - c^{-2}\mathbf{v}(p_{,t} + \mathbf{f}\cdot\mathbf{v}), \tag{42.13}$$

which is the relativistic generalization of Euler's equation of motion (23.6) for an ideal fluid. These equations assume their simplest form in the comoving frame where $\mathbf{v} = 0$; we then have

$$\rho_{000}(D\mathbf{v}/D\tau) = \rho_{000}(D\mathbf{v}/Dt)_0 = \mathbf{f}_0 - \nabla p. \tag{42.14}$$

Equation (42.14) now appears almost identical to Euler's equation.

For problems of fluid flow, a characteristic time-increment is $\Delta t \sim \Delta x/v$, where Δx is a characteristic length and v is a typical velocity in the flow. Therefore the term in $(\partial p/\partial t)$ in (42.13) is $O(v^2/c^2)$ compared to ∇p; it is thus apparent that (42.13) differs from its Newtonian counterpart only by terms that are $O(v^2/c^2)$. In Chapter 6 we find that when radiative effects are taken into account the situation is quite different, because frame-dependent terms that are $O(v/c)$ appear.

GENERAL RELATIVISTIC EQUATIONS

In general relativity the hydrodynamic equations can be written in the very compact form

$$M^{\alpha\beta}_{;\beta} = 0. \tag{42.15}$$

Here one assumes that spacetime is Riemannian, with an *intrinsic curvature*, and is described by a general metric $g_{\alpha\beta}$ of which each element can be a function of $x^{(0)}, \ldots, x^{(3)}$. In such a formulation, the quantities interpreted as forces in special relativity are found not to be independent physical entities, but instead are results of spacetime curvature, which, when one computes covariant derivatives in the curved manifold, leads to additional terms interpretable as forces. For a detailed discussion of this view the reader should consult one of the texts on general relativity cited at the end of this chapter.

43. *The Kinetic Theory View*

Instead of considering the fluid to be a continuum, let us now suppose it to be composed of a large number of paricles, each having a rest mass m_0. We can use kinetic theory to calculate particle-energy and particle-momentum densities and fluxes; in doing so, we will find that we have constructed the fluid stress-energy tensor directly from microscopic considerations. Our treatment parallels that given in (**L1,** §10) and (**P1,** Chap. 9) to which the reader can refer for further details. We do not attempt to treat particle collisions or general relativistic effects; these topics are discussed in (**D1**), (**E2**), and (**S2**), and the references cited therein.

THE DISTRIBUTION FUNCTION

Let $f(\mathbf{x}, \mathbf{p}, t)$ be the *particle distribution function* defined such that at time t the number of particles in a volume element $d\mathbf{x}$ centered on $\mathbf{x}$, and with momenta in a momentum-space element $d\mathbf{p}$ centered on $\mathbf{p}$, is $f(\mathbf{x}, \mathbf{p}, t)\,d\mathbf{x}\,d\mathbf{p}$; here all quantities are measured in the laboratory frame. In what follows we will focus on a single point in spacetime; to economize the notation we suppress reference to $\mathbf{x}$ and t.

If a particle's velocity in the lab frame is $\mathbf{u}$, decompose it into

$$\mathbf{u}=\mathbf{v}+\mathbf{U} \tag{43.1}$$

where

$$\mathbf{v}\equiv\int \mathbf{u}f(\mathbf{p})\,d\mathbf{p}\Big/\int f(\mathbf{p})\,d\mathbf{p} \tag{43.2}$$

is the *flow velocity* of the fluid, that is, the average velocity of the particles in a small neighborhood of $\mathbf{x}$, and $\mathbf{U}$ is a particle's *random velocity* (measured in the lab frame) relative to the flow velocity.

The frame moving with the flow velocity is the comoving frame; quantities measured in this frame are denoted with subscript zero. For example, the random velocity of a particle in this frame is $\mathbf{U}_0$, its momentum is $\mathbf{p}_0$. and the distribution function is $f_0(\mathbf{p}_0)$. The third set of frames of interest comprises the rest frames of the particles, each of which moves with one of the particles. Quantities measured in one of these frames will be denoted with a prime, except for a particle's rest mass, which we will still call m_0.

INVARIANCE OF THE DISTRIBUTION FUNCTION

Suppose we choose a definite group of particles. Observers in both the lab and comoving frames will count the same *number* of particles in the group, even though the particles will be observed to be in different phase-space volume elements. We therefore have

$$f(\mathbf{p})\,d\mathbf{x}\,d\mathbf{p}=f_0(\mathbf{p}_0)\,d\mathbf{x}_0\,d\mathbf{p}_0. \tag{43.3}$$

Consider a proper volume element $d\mathbf{x}'$ in the rest frame of some particle. According to (39.3) an observer in the comoving frame will measure its volume to be

$$d\mathbf{x}_0=(1-U_0^2/c^2)^{1/2}\,d\mathbf{x}', \tag{43.4}$$

while an observer in the lab frame will measure its volume as

$$d\mathbf{x}=(1-u^2/c^2)^{1/2}\,d\mathbf{x}'. \tag{43.5}$$

Hence

$$d\mathbf{x}_0=\left(\frac{1-U_0^2/c^2}{1-u^2/c^2}\right)^{1/2}d\mathbf{x}. \tag{43.6}$$

On the other hand, if the particle has rest mass m_0, (37.13) implies that its energy in the comoving frame is

$$\tilde{e}_0=m_0c^2/(1-U_0^2/c^2)^{1/2}, \tag{43.7}$$

and in the lab frame it is

$$\tilde{e} = m_0 c^2/(1-u^2/c^2)^{1/2}; \tag{43.8}$$

comparing with (43.6) we conclude that

$$d\mathbf{x}_0 = (\tilde{e}/\tilde{e}_0)\, d\mathbf{x}. \tag{43.9}$$

We can relate $d\mathbf{p}$ to $d\mathbf{p}_0$ by applying the general Lorentz transformation (35.32) with velocity $-\mathbf{v}$ (the velocity of the lab frame as seen from the comoving frame) to a particle's four-momentum measured in the comoving frame. We obtain

$$\tilde{e} = \gamma(\tilde{e}_0 + \mathbf{v} \cdot \mathbf{p}_0), \tag{43.10}$$

and

$$\mathbf{p} = \mathbf{p}_0 + [\gamma(\tilde{e}_0/c^2) + (\gamma-1)\mathbf{v} \cdot \mathbf{p}_0/v^2]\mathbf{v}. \tag{43.11}$$

In general we can write

$$d\mathbf{p} = J(p^{(1)}, p^{(2)}, p^{(3)}/p_0^{(1)}, p_0^{(2)}, p_0^{(3)})\, d\mathbf{p}_0, \tag{43.12}$$

where J is the Jacobian of the transformation from the comoving system to the lab system. To simplify the calculation we do not use (43.11) directly, but instead rotate the coordinate axes such that in the new coordinate system the comoving frame moves with velocity v along $x^{(3)}$. In this new system we have from (35.15).

$$(\tilde{e}/c, p^{(1)}, p^{(2)}, p^{(3)}) = [\gamma(\tilde{e}_0/c + \beta p_0^{(3)}), p_0^{(1)}, p_0^{(2)}, \gamma(p_0^{(3)} + \beta \tilde{e}_0/c)], \tag{43.13}$$

whence

$$J = \begin{vmatrix} 1 & 0 & 0 \\ 0 & 1 & 0 \\ 0 & 0 & \gamma\left(1 + \dfrac{\beta}{c}\dfrac{\partial \tilde{e}_0}{\partial p_0^{(3)}}\right) \end{vmatrix} = \gamma\left(1 + \frac{\beta}{c}\frac{\partial \tilde{e}_0}{\partial p_0^{(3)}}\right). \tag{43.14}$$

But from (37.17) we know that $\tilde{e}_0^2 = p_0^2 c^2 + m_0^2 c^4$, hence

$$(\partial \tilde{e}_0/\partial p_0^{(3)}) = p_0^{(3)} c^2/\tilde{e}_0, \tag{43.15}$$

and therefore

$$J = \gamma(1 + v p_0^{(3)}/\tilde{e}_0) = \tilde{e}/\tilde{e}_0, \tag{43.16}$$

where the second equality follows from (43.13). The latter expression for J contains no reference to the orientation of the coordinate axes, and hence applies in the original coordinate system as well. Combining (43.12) with (43.16) we have

$$d\mathbf{p} = (\tilde{e}/\tilde{e}_0)\, d\mathbf{p}_0, \tag{43.17}$$

hence from (43.9) we find

$$d\mathbf{x}\, d\mathbf{p} = d\mathbf{x}_0\, d\mathbf{p}_0. \tag{43.18}$$

Using (43.18) in (43.3) we see that the distribution function f is *invariant* under Lorentz transformation, that is,

$$f(\mathbf{p}) = f_0(\mathbf{p}_0). \tag{43.19}$$

THE COMOVING-FRAME NUMBER DENSITY, ENERGY DENSITY, AND PRESSURE

From the point of view of thermodynamics, the fluid energy density and the pressure are best defined in the comoving frame of the fluid. First, we express the comoving-frame *particle number density* as the integral of the distribution function over all momenta:

$$N_0 \equiv \int f_0(\mathbf{p}_0)\, d\mathbf{p}_0. \tag{43.20}$$

Given N_0, the density of proper mass in the comoving frame is $\rho_0 = N_0 m_0$.

The *total energy density* in the comoving frame is obtained by taking the sum over all momenta of the product of the number of particles at a given momentum times the energy of those particles, that is,

$$\rho_{00}c^2 = \int \tilde{e}_0 f_0(\mathbf{p}_0)\, d\mathbf{p}_0. \tag{43.21}$$

Here ρ_{00} has the same meaning as in §40. Combined with (40.4) and (43.20), equation (43.21) provides an operational definition of the fluid's specific internal energy e.

The stress tensor in the comoving frame is obtained by calculating the rate of momentum transfer across a unit area, as was done to derive equation (30.15). We find

$$-T^{ij} = \int U_0^i p_0^j f_0(\mathbf{p}_0)\, d\mathbf{p}_0. \tag{43.22}$$

To recover the correct equations of hydrodynamics for an ideal fluid we must now assume that the distribution function $f_0(\mathbf{p}_0)$ is isotropic in the comoving frame. The stress tensor is then diagonal, $T^{ij} = -p_m\, \delta^{ij}$, where the *pressure* is given by

$$p_m = \int U_0^i p_0^i f_0(\mathbf{p}_0)\, d\mathbf{p}_0. \tag{43.23}$$

In (43.23) there is no sum on i, and the subscript "m" for "material" has been used to avoid confusion with the magnitude p of the momentum vector $\mathbf{p}$.

Because f_0 is isotropic, we can evaluate (43.23) along any axis. Thus we can equally well write

$$p_m = \int (\mathbf{U}_0 \cdot \mathbf{l})(\mathbf{p}_0 \cdot \mathbf{l}) f_0(\mathbf{p}_0)\, d\mathbf{p}_0 \tag{43.24}$$

where $\mathbf{l}$ is an arbitrarily oriented unit vector; this form will prove useful in what follows.

THE LABORATORY-FRAME PARTICLE DENSITY, ENERGY DENSITY, AND MOMENTUM DENSITY

Let us now calculate the particle density, energy density, and momentum density of the fluid as measured in the laboratory frame. The *particle density* is

$$N = \int f(\mathbf{p})\, d\mathbf{p}. \tag{43.25}$$

Using (43.10), (43.17), and (43.19) we can rewrite (43.25) as

$$N = \gamma \int [1 + (\mathbf{v} \cdot \mathbf{p}_0)/\tilde{e}_0] f_0(\mathbf{p}_0)\, d\mathbf{p}_0. \tag{43.26}$$

Because f_0 is isotropic the integral containing $(\mathbf{v} \cdot \mathbf{p}_0)$ vanishes, and we have

$$N = \gamma N_0, \tag{43.27}$$

in agreement with (39.4).

The *energy density* (*ED*) in the laboratory frame is

$$ED = \int \tilde{e} f(\mathbf{p})\, d\mathbf{p} = \gamma^2 \int (1/\tilde{e}_0)(\tilde{e}_0 + \mathbf{v} \cdot \mathbf{p}_0)^2 f_0(\mathbf{p}_0)\, d\mathbf{p}_0, \tag{43.28}$$

where we again used (43.10), (43.17), and (43.19). Expanding the square, and again noting that the integral containing $(\mathbf{v} \cdot \mathbf{p}_0)$ vanishes because f_0 is isotropic, we have

$$ED = \gamma^2 \int [\tilde{e}_0 + (\mathbf{v} \cdot \mathbf{p}_0)^2/\tilde{e}_0] f_0(\mathbf{p}_0)\, d\mathbf{p}_0. \tag{43.29}$$

From (37.3) and (37.13) we have

$$\mathbf{p}_0 = \tilde{e}_0 \mathbf{U}_0/c^2. \tag{43.30}$$

Hence (43.29) can be rewritten as

$$ED = \gamma^2 \int [\tilde{e}_0 + (\mathbf{U}_0 \cdot \mathbf{v})(\mathbf{p}_0 \cdot \mathbf{v})/c^2] f_0(\mathbf{p}_0)\, d\mathbf{p}_0, \tag{43.31}$$

and if we choose $\mathbf{l} = (\mathbf{v}/v)$, we see from (43.21) and (43.24) that

$$ED = \gamma^2 [\rho_{00} c^2 + (v^2/c^2) p_m]. \tag{43.32}$$

Using the identity $(v^2/c^2) = 1 - \gamma^{-2}$ we obtain finally

$$ED = \gamma^2 (\rho_{00} c^2 + p_m) - p_m = \gamma^2 \rho_{000} c^2 - p_m, \tag{43.33}$$

which is identical to M^{00} in the stress-energy tensor, as expected.

The *momentum density* (*MD*) in the laboratory frame is

$$MD = \int \mathbf{p} f(\mathbf{p})\, d\mathbf{p} = \int (\tilde{e}/\tilde{e}_0) \mathbf{p} f_0(\mathbf{p}_0)\, d\mathbf{p}_0. \tag{43.34}$$

Using (43.10) and (43.11) we obtain

$$MD = \gamma\int [1+(\mathbf{v}\cdot\mathbf{p}_0)/\tilde{e}_0][\mathbf{p}_0+(\gamma-1)v^{-2}(\mathbf{v}\cdot\mathbf{p}_0)\mathbf{v}+(\gamma\tilde{e}_0/c^2)\mathbf{v}]f_0(\mathbf{p}_0)\,d\mathbf{p}_0. \tag{43.35}$$

Expanding the integrand, discarding those terms that integrate to zero for isotropic f_0, and using (43.30) we find

$$MD = (\gamma/c^2)\int [\gamma\tilde{e}_0\mathbf{v}+(\gamma-1)v^{-2}(\mathbf{U}_0\cdot\mathbf{v})(\mathbf{p}_0\cdot\mathbf{v})\mathbf{v}+(\mathbf{U}_0\cdot\mathbf{v})\mathbf{p}_0]f_0(\mathbf{p}_0)\,d\mathbf{p}_0. \tag{43.36}$$

Again using (43.21) and (43.24) with $\mathbf{l}=(\mathbf{v}/v)$, we obtain

$$MD = \gamma^2\rho_{00}\mathbf{v}+\gamma(\gamma-1)(p_m/c^2)\mathbf{v}+(\gamma/c^2)\int(\mathbf{U}_0\cdot\mathbf{v})\mathbf{p}_0 f_0(\mathbf{p}_0)\,d\mathbf{p}_0. \tag{43.37}$$

By resolving $\mathbf{p}_0$ along $\mathbf{l}=(\mathbf{v}/v)$ and a unit vector $\mathbf{m}$ orthogonal to $\mathbf{l}$, it is easy to show that, for an isotropic f_0,

$$\int(\mathbf{U}_0\cdot\mathbf{v})\mathbf{p}_0 f_0(\mathbf{p}_0)\,d\mathbf{p}_0 \equiv \mathbf{v}\int(\mathbf{U}_0\cdot\mathbf{l})(\mathbf{p}_0\cdot\mathbf{l})f_0(\mathbf{p}_0)\,d\mathbf{p}_0, \tag{43.38}$$

which is just $p_m\mathbf{v}$. Thus (43.37) reduces to

$$MD = \gamma^2(\rho_{00}+p_m/c^2)\mathbf{v} = \gamma^2\rho_{000}\mathbf{v}, \tag{43.39}$$

which is $(1/c)$ times M^{i0} in the stress-energy tensor, consistent with our earlier interpretation of those components.

THE LABORATORY-FRAME PARTICLE FLUX, ENERGY FLUX, AND MOMENTUM FLUX

The *particle flux density vector* (*PF*) in the laboratory frame is

$$PF = \int \mathbf{v}f(\mathbf{p})\,d\mathbf{p}. \tag{43.40}$$

Again using (43.17), (43.19), and (43.30) restated in the lab frame (i.e., $\tilde{e}\mathbf{v}=c^2\mathbf{p}$), we find

$$PF = c^2\int(\mathbf{p}/\tilde{e}_0)f_0(\mathbf{p}_0)\,d\mathbf{p}_0. \tag{43.41}$$

Using (43.11) for $\mathbf{p}$, and retaining only the terms that survive the integration we have

$$PF = \gamma\mathbf{v}\int f_0(\mathbf{p}_0)\,d\mathbf{p}_0 = \gamma N_0\mathbf{v} = N\mathbf{v}, \tag{43.42}$$

which is identical to the particle flux density vector N_0V^α introduced in §39.

The *particle energy flux vector* (EF) in the laboratory frame is

$$EF = \int \tilde{e}\mathbf{v}f(\mathbf{p})\,d\mathbf{p}. \tag{43.43}$$

Proceeding as above, and using (43.34) and (43.39), this expression reduces to

$$EF = \int (\tilde{e}^2/\tilde{e}_0)\mathbf{v}f_0(\mathbf{p}_0)\,d\mathbf{p}_0 = c^2\int (\tilde{e}/\tilde{e}_0)\mathbf{p}f_0(\mathbf{p}_0)\,d\mathbf{p}_0 = \gamma^2\rho_{000}c^2\mathbf{v}. \tag{43.44}$$

Thus the energy flux is c times the components M^{0i} in the stress-energy tensor, as expected from the interpretation given earlier for those components. It is also equal to c^2 times the momentum density (43.39), consistent with the relationship between M^{0i} and M^{i0}.

Finally, the *particle momentum flux tensor* in the laboratory frame is

$$\Pi^{ij} = \int v^i p^j f(\mathbf{p})\,d\mathbf{p}. \tag{43.45}$$

Again proceeding as above we find

$$\begin{aligned}
\Pi^{ij} &= \int v^i p^j\left(\frac{\tilde{e}}{\tilde{e}_0}\right)f_0(\mathbf{p}_0)\,d\mathbf{p}_0 = c^2\int \frac{p^i p^j}{\tilde{e}_0}f_0(\mathbf{p}_0)\,d\mathbf{p}_0\\
&= c^2\int \frac{1}{\tilde{e}_0}\left[p_0^i + \frac{(\gamma-1)(\mathbf{v}\cdot\mathbf{p}_0)}{v^2}v^i + \frac{\gamma\tilde{e}_0}{c^2}v^i\right]\\
&\quad\times\left[p_0^j + \frac{(\gamma-1)(\mathbf{v}\cdot\mathbf{p}_0)}{v^2}v^j + \frac{\gamma\tilde{e}_0}{c^2}v^j\right]f_0(\mathbf{p}_0)\,d\mathbf{p}_0\\
&= c^2\int \frac{1}{\tilde{e}_0}\left[p_0^i p_0^j + \frac{(\gamma-1)^2(\mathbf{v}\cdot\mathbf{p}_0)^2}{v^4}v^i v^j + \frac{\gamma^2\tilde{e}_0^2}{c^4}v^i v^j\right.\\
&\quad\left.+\frac{(\gamma-1)(\mathbf{v}\cdot\mathbf{p}_0)}{v^2}(p_0^i v^j + p_0^j v^i)\right]f_0(\mathbf{p}_0)\,d\mathbf{p}_0\\
&= \int\left[U_0^i p_0^j + \frac{(\gamma-1)^2(\mathbf{v}\cdot\mathbf{p}_0)(\mathbf{v}\cdot\mathbf{U}_0)}{v^4}v^i v^j + \frac{\gamma^2\tilde{e}_0}{c^2}v^i v^j\right.\\
&\quad\left.+\frac{(\gamma-1)(\mathbf{v}\cdot\mathbf{U}_0)}{v^2}(p_0^i v^j + p_0^j v^i)\right]f_0(\mathbf{p}_0)\,d\mathbf{p}_0\\
&= p_m\left\{\delta^{ij} + [(\gamma-1)^2 + 2(\gamma-1)]\frac{v^i v^j}{v^2}\right\} + \gamma^2\rho_{00}v^i v^j,
\end{aligned} \tag{43.46}$$

where we have used (43.11), (43.17), (43.21), (43.24), (43.30), (43.38), and the isotropy of f_0. Using the definition of γ to collapse the term in square brackets we obtain finally

$$\Pi^{ij} = p_m\,\delta^{ij} + \gamma^2(\rho_{00} + p_m/c^2)v^i v^j = p_m\,\delta^{ij} + \gamma^2\rho_{000}v^i v^j, \tag{43.47}$$

which is identical to the space-components M^{ij} in the stress-energy tensor, as expected.

We have thus been able to derive the entire stress-energy tensor from a microscopic point of view.

THE EQUATION OF STATE

Thus far, the only restriction we have placed on $f_0(\mathbf{p}_0)$ is that it be isotropic in the comoving frame. But if the particles move with only nonrelativistic velocities, then it is reasonable to assume that f_0 is the Maxwellian distribution. In the nonrelativistic limit $\tilde{e}_0 = m_0(c^2 + \frac{1}{2}U_0^2)$, hence from (43.21) one easily finds

$$\rho_{00}c^2 = \rho_0 c^2 + \tfrac{3}{2}N_0 kT \tag{43.48}$$

where T is the *material temperature* measured in the comoving frame. Hence the specific internal energy is $e = \frac{3}{2}(N_0 kT)/m_0$, and the energy per unit volume is $\hat{e} = \frac{3}{2}N_0 kT$. Similarly, from (43.22) one finds

$$-T^{ij} = \rho_0\langle U_0^i U_0^j\rangle = N_0 kT\,\delta^{ij}, \tag{43.49}$$

whence $p_m = N_0 kT$, and therefore

$$p_m = \tfrac{2}{3}\hat{e} \qquad \text{(N.R.)}. \tag{43.50}$$

All other thermodynamic properties in the comoving frame are the same as those derived in Chapter 1 for a perfect gas.

In the extreme relativistic limit, we see from (37.17) that $\tilde{e}_0 \to p_0 c$, and from (40.4) that $\rho_{00}c^2 \to \rho_0 e = \hat{e}$. Thus from (43.21)

$$\hat{e} = c\int p_0 f_0(\mathbf{p}_0)\, d\mathbf{p}_0 = 4\pi c\int p_0 f_0(p_0) p_0^2\, dp_0. \tag{43.51}$$

In this limit we can also write $\mathbf{U}_0 \approx c\mathbf{n}$ where $\mathbf{n} \equiv (\mathbf{p}_0/p_0)$ is the unit vector along $\mathbf{p}_0$. Then, from (43.24), we have

$$p_m = c\int (\mathbf{n}\cdot\mathbf{l})^2 p_0 f_0(\mathbf{p}_0)\, d\mathbf{p}_0 = \tfrac{1}{3}(4\pi c)\int p_0 f_0(p_0) p_0^2\, dp_0 \tag{43.52}$$

because $\langle \mathbf{n}\cdot\mathbf{l}\rangle^2 = \frac{1}{3}$. Thus we obtain the important relation

$$p_m = \tfrac{1}{3}\hat{e}, \qquad \text{(E.R.)} \tag{43.53}$$

which may be contrasted with (43.50) for a nonrelativistic gas. To obtain (43.53) we invoked only the isotropy of f_0. To derive the other thermodynamic quantities for the gas we need an explicit expression for f_0. If the gas is nondegenerate we can use the relativistic generalization of the Maxwellian distribution obtained by eliminating the velocity $\mathbf{v}$ in favor of the momentum $\mathbf{p}$. One then easily can show [cf. (**C1**, Chap. 10)] that $\hat{e} = 3N_0 kT$, $p_m = N_0 kT$, $c_v = 3R$, $c_p = 4R$, and $\Gamma = \frac{4}{3}$ (here Γ is the adiabatic exponent). Following a different route we shall derive, in §69, the same value for Γ and the same relation between p and $\hat{e}$ for equilibrium radiation (which, of course, is a gas composed of ultrarelativistic particles: photons). Relativistic effects can also be important when the gas is degenerate; see (**C1,** Chap. 24) for details. An in-depth general discussion of relativistic gases is given in (**D1**) and (**S2**).

4.3 Relativistic Dynamics of Nonideal Fluids

Let us now consider a relativistic viscous and heat-conducting fluid viewed as a continuum. We will essentially follow Eckart's ground-breaking analysis (**E1**); for more detailed treatments and discussions of related topics the reader should consult (**L5,** Chap. 5), (**M3,** Chap. 22), (**T1**), (**W1**), and (**W2,** 53–57). Because it causes no additional complication to do so, we assume a general metric tensor $g_{\alpha\beta}$ (with spacelike signature); most of the equations written below will then hold in general relativity as well as special relativity.

44. Kinematics

THE ECKART DECOMPOSITION THEOREM

We define the *projection tensor* to be

$$P^{\alpha}_{\beta} \equiv \delta^{\alpha}_{\beta} + c^{-2} V^{\alpha} V_{\beta}, \tag{44.1}$$

or, in covariant components,

$$P_{\alpha\beta} = g_{\alpha\gamma} P^{\gamma}_{\beta} = g_{\alpha\beta} + c^{-2} V_{\alpha} V_{\beta}, \tag{44.2}$$

with a similar expression for $P^{\alpha\beta}$. It is easy to prove by direct calculation the following useful relations:

$$P^{\alpha}_{\beta} P^{\beta}_{\gamma} \equiv P^{\alpha}_{\gamma}, \tag{44.3}$$

$$P^{\alpha\beta} \equiv P^{\alpha\gamma} P^{\beta}_{\gamma}, \tag{44.4}$$

and

$$P^{\alpha\beta} P_{\alpha\beta} = 3. \tag{44.5}$$

More important, using (36.5), one easily finds that

$$V_{\alpha} P^{\alpha}_{\beta} = V^{\beta} P^{\alpha}_{\beta} \equiv 0, \tag{44.6}$$

which shows that P^{α}_{β} produces a projection orthogonal to V^{α} in spacetime. Recalling that at each point in spacetime V^{α} lies along the local axis of proper time, we see that P^{α}_{β} selects three directions along the local axes of proper space, and thus is the local spatial projection operator in the comoving frame. Indeed, one sees by direct evaluation that

$$(P^{\alpha}_{\beta})_0 = \begin{pmatrix} 0 & 0 & 0 & 0 \\ 0 & 1 & 0 & 0 \\ 0 & 0 & 1 & 0 \\ 0 & 0 & 0 & 1 \end{pmatrix}. \tag{44.7}$$

Using P^{α}_{β} we can decompose any vector A^{α} into a scalar a that gives its projection onto the proper time axis, and a vector a^{α} which is the projection of A^{α} into proper space. Thus if we define

$$a \equiv -c^{-2}V_{\alpha}A^{\alpha}, \tag{44.8}$$

and

$$a^{\alpha} \equiv P^{\alpha}_{\beta}A^{\beta}, \tag{44.9}$$

then we can write

$$A^{\alpha} = aV^{\alpha} + a^{\alpha}. \tag{44.10}$$

It is easy to show by direct calculation that (44.8) to (44.10) are mutually consistent.

Similarly, we can decompose any tensor $W^{\alpha\beta}$ into its proper components by defining the scalar

$$w \equiv c^{-4}V_{\alpha}V_{\beta}W^{\alpha\beta}, \tag{44.11}$$

the vector

$$w^{\alpha} \equiv -c^{-2}P^{\alpha}_{\beta}W^{\beta\gamma}V_{\gamma}, \tag{44.12}$$

and the tensor

$$w^{\alpha\beta} \equiv P^{\alpha}_{\gamma}P^{\beta}_{\delta}W^{\gamma\delta}. \tag{44.13}$$

We can then write

$$W^{\alpha\beta} = wV^{\alpha}V^{\beta} + w^{\alpha}V^{\beta} + w^{\beta}V^{\alpha} + w^{\alpha\beta}, \tag{44.14}$$

which is known as the *Eckart decomposition theorem*. Again, it is easy to verify by direct calculation that (44.11) to (44.14) are mutually consistent.

THE VELOCITY-GRADIENT TENSOR

Consider the fluid velocity-gradient tensor $V_{\alpha;\beta}$. Notice first that

$$V_{\alpha;\beta}V^{\beta} \equiv (\delta V_{\alpha}/\delta\tau), \tag{44.15}$$

the intrinsic derivative of V_{α} with respect to proper time; this quantity is the fluid *four-acceleration* A_{α}. We therefore write $V_{\alpha;\beta}$ in terms of an acceleration component along the local time axis, and a set of spatial components, by means of the decomposition

$$V_{\alpha;\beta} = V_{\alpha;\gamma}P^{\gamma}_{\beta} - c^{-2}A_{\alpha}V_{\beta}. \tag{44.16}$$

To verify the correctness of this decomposition we note that

$$V^{\beta}V_{\alpha;\beta} = (V^{\beta}P^{\gamma}_{\beta})V_{\alpha;\gamma} - c^{-2}(V^{\beta}V_{\beta})A_{\alpha} = 0 + A_{\alpha}, \tag{44.17}$$

and

$$P^{\gamma}_{\beta}V_{\alpha;\gamma} = (P^{\gamma}_{\beta}P^{\delta}_{\gamma})V_{\alpha;\delta} - c^{-2}(P^{\gamma}_{\beta}V_{\gamma})A_{\alpha} = P^{\delta}_{\beta}V_{\alpha;\delta} + 0. \tag{44.18}$$

We can decompose the right-hand side of (44.16) further by defining the antisymmetric *rotation tensor*

$$\Omega_{\alpha\beta} \equiv \tfrac{1}{2}(V_{\alpha;\gamma}P^{\gamma}_{\beta} - V_{\beta;\gamma}P^{\gamma}_{\alpha}), \tag{44.19}$$

and the symmetric *shear tensor* (or *rate of strain tensor*)

$$E_{\alpha\beta} \equiv \tfrac{1}{2}(V_{\alpha;\gamma}P^{\gamma}_{\beta} + V_{\beta;\gamma}P^{\gamma}_{\alpha}), \tag{44.20}$$

which are given these names because in the comoving frame they reduce to the covariant generalizations of the Newtonian expressions for these quantities (cf. §21). Clearly $E_{\alpha\beta} + \Omega_{\alpha\beta} \equiv V_{\alpha;\gamma}P^{\gamma}_{\beta}$, hence we can rewrite (44.16) as

$$V_{\alpha;\beta} = E_{\alpha\beta} + \Omega_{\alpha\beta} - c^{-2}A_{\alpha}V_{\beta}. \tag{44.21}$$

It is actually more convenient to choose the shear tensor to be

$$D_{\alpha\beta} \equiv E_{\alpha\beta} - \tfrac{1}{3}\theta P_{\alpha\beta}, \tag{44.22}$$

where

$$\theta \equiv V^{\alpha}_{;\alpha} \tag{44.23}$$

is the *expansion* of the fluid. One sees that $D_{\alpha\beta}$ is the covariant generalization of the Newtonian traceless shear tensor [cf. (25.3) and (32.34)].

Thus we can write, finally,

$$V_{\alpha;\beta} = D_{\alpha\beta} + \Omega_{\alpha\beta} + \tfrac{1}{3}\theta P_{\alpha\beta} - c^{-2}A_{\alpha}V_{\beta}, \tag{44.24}$$

which is the *relativistic generalization of the Cauchy-Stokes theorem* discussed in §21. It shows that in spacetime a fluid is accelerated along its proper time axis, and experiences shear, rotation, and expansion along its local space axes. Explicit expressions for P^{α}_{β}, A_{α}, θ, $\Omega_{\alpha\beta}$, and $D_{\alpha\beta}$ in terms of the ordinary velocity $\mathbf{v}$ and lab-frame space and time derivatives are easily derived. These expressions are too lengthy to reproduce here, but are given by Greenberg in (**G1,** 764–765); the reader should note that the signs in some of Greenberg's formulae conflict with those in our formulae because Greenberg uses a metric with a timelike signature.

Finally, note in passing that from (44.6),

$$V^{\alpha}D_{\alpha\beta} = V_{\alpha}D^{\alpha\beta} \equiv 0, \tag{44.25}$$

and

$$V^{\alpha}\Omega_{\alpha\beta} = V_{\alpha}\Omega^{\alpha\beta} \equiv 0. \tag{44.26}$$

Furthermore, by virtue of (44.4) and (44.5) we have

$$\begin{aligned} P_{\alpha\beta}D^{\alpha\beta} &= P^{\alpha\beta}D_{\alpha\beta} = \tfrac{1}{2}(V_{\alpha;\gamma}P^{\alpha\gamma} + V_{\beta;\gamma}P^{\beta\gamma}) - \theta \\ &= V_{\alpha;\gamma}(g^{\alpha\gamma} + c^{-2}V^{\alpha}V^{\gamma}) - \theta = V^{\gamma}_{;\gamma} - \theta \equiv 0, \end{aligned} \tag{44.27}$$

where we used (36.8) and (44.23). We will find all of these results useful in §§46 and 47.

45. The Stress-Energy Tensor

Given the results of §44 it is fairly straightforward to deduce the form of the stress-energy tensor for viscous and heat-conducting fluids by seeking suitable covariant generalizations of the usual Newtonian expressions. Thus the viscous terms in the stress-energy tensor must be

$$-\mathsf{M}_{\text{viscous}} = 2\mu\mathsf{D} + \zeta\theta\mathsf{P}, \tag{45.1}$$

where D is the shear tensor (44.22), P is the projection tensor (44.2), and μ and ζ are, respectively, the coefficients of shear and bulk viscosity. One can see by inspection that, in the comoving frame, (45.1) reduces to the covariant generalization of the Newtonian expression (25.3). (The minus sign appears because of the sign convention for the Cauchy stress tensor; cf. §22.)

Consider now the contribution from heat flow. Classically we describe heat flow in the comoving frame by a vector $\mathbf{q}$, whose components give the rate of energy flow per unit area along each coordinate axis. We saw in §§40 and 43 that the energy flux is c times the elements M^{0i} of the stress-energy tensor. Therefore if we take $\mathbf{Q}$ to be the four-vector generalization of $\mathbf{q}$, then in the comoving frame we must have $(M^{0i})_0 = Q_0^i/c$; we can obtain precisely such a contribution to M_0 from a term of the form $c^{-2}V^\alpha Q^\beta$. But if we introduce such a term, then because M must be symmetric, and from the Eckart decomposition theorem, we know that M must also contain a term of the form $c^{-2}Q^\alpha V^\beta$. We thus conclude that

$$M_{\text{heat}}^{\alpha\beta} = c^{-2}(V^\alpha Q^\beta + Q^\alpha V^\beta). \tag{45.2}$$

As before, the $(i, 0)$ elements of M_{heat} can be interpreted as c times a momentum density. That such terms should be present is reasonable because heat is energy and, by Einstein's mass-energy equivalence, has inertia; hence in the comoving frame a heat flux q^i gives rise to an equivalent mass flux or a momentum density equal to $c^{-2}q^i$, which is, in fact, identical to $(M_{\text{heat}}^{i0})_0/c$.

Adding the viscous and heat-flow contributions to the stress-energy tensor (40.10) for an ideal gas, we conclude that the complete material stress-energy tensor for a viscous, heat-conducting fluid is

$$M_{\alpha\beta} = \rho_{000}V_\alpha V_\beta + pg_{\alpha\beta} - 2\mu D_{\alpha\beta} - \zeta\theta P_{\alpha\beta} + c^{-2}(Q_\alpha V_\beta + V_\alpha Q_\beta), \tag{45.3}$$

or

$$M_{\alpha\beta} = \rho_{00}V_\alpha V_\beta + pP_{\alpha\beta} - 2\mu D_{\alpha\beta} - \zeta\theta P_{\alpha\beta} + c^{-2}(Q_\alpha V_\beta + V_\alpha Q_\beta). \tag{45.4}$$

Notice that a one-to-one correspondence can be made between the terms in (45.4) and those appearing in the Eckart decomposition theorem (44.14). In particular, we can identify Q^α with w^α as given by (44.12); then in view of (44.6) we see that

$$V_\alpha Q^\alpha \equiv 0, \tag{45.5}$$

a result we will find useful below.

Explicit expressions for $M^{\alpha\beta}$ in terms of the ordinary velocity, and lab-frame space and time derivatives, are given in (**G1,** 769–770). To translate the symbols used in (**G1**) to those used here, make the substitutions $\nu \to 2\mu$, $\beta \to 3\zeta$, and $\lambda \to K$ (the thermal conductivity, see §46).

46. The Energy Equation

We are now in a position to derive the equations of hydrodynamics for a relativistic nonideal fluid. As for an ideal fluid, the general equations of motion follow from

$$M^{\alpha\beta}_{;\beta} = F^\alpha. \tag{46.1}$$

For simplicity we assume that both μ and ζ are constants. Differentiating each term in (45.4) we have

$$(\rho_{00}V^\alpha V^\beta)_{;\beta} = [(D\rho_{00}/D\tau) + \rho_{00}\theta]V^\alpha + \rho_{00}A^\alpha, \tag{46.2}$$

$$(pP^{\alpha\beta})_{;\beta} = p_{,\beta}P^{\alpha\beta} + pP^{\alpha\beta}_{;\beta}, \tag{46.3}$$

$$\zeta(\theta P^{\alpha\beta})_{;\beta} = \zeta\theta_{,\beta}P^{\alpha\beta} + \zeta\theta P^{\alpha\beta}_{;\beta}, \tag{46.4}$$

$$(Q^\alpha V^\beta)_{;\beta} = \theta Q^\alpha + (DQ^\alpha/D\tau), \tag{46.5}$$

and

$$(V^\alpha Q^\beta)_{;\beta} = V^\alpha Q^\beta_{;\beta} + Q^\beta g^{\alpha\gamma}V_{\gamma;\beta} = V^\alpha Q^\beta_{;\beta} + Q_\beta(D^{\alpha\beta} + \Omega^{\alpha\beta} + \tfrac{1}{3}\theta P^{\alpha\beta}). \tag{46.6}$$

Then, collecting terms and using the relation

$$P^{\alpha\beta}_{;\beta} = (g^{\alpha\beta} + c^{-2}V^\alpha V^\beta)_{;\beta} = c^{-2}(A^\alpha + \theta V^\alpha), \tag{46.7}$$

we have

$$\begin{aligned} M^{\alpha\beta}_{;\beta} = {} & \{(D\rho_{00}/D\tau) + \theta[\rho_{00} + c^{-2}(p - \zeta\theta)]\}V^\alpha \\ & + [\rho_{00} + c^{-2}(p - \zeta\theta)]A^\alpha + P^{\alpha\beta}(p - \zeta\theta)_{,\beta} - 2\mu D^{\alpha\beta}_{;\beta} \\ & + c^{-2}[(DQ^\alpha/D\tau) + \tfrac{4}{3}\theta Q^\alpha + V^\alpha Q^\beta_{;\beta} + Q_\beta(D^{\alpha\beta} + \Omega^{\alpha\beta})] = F^\alpha. \end{aligned} \tag{46.8}$$

We obtain the energy equation by taking the "time" component of (46.8), that is, by projecting it onto V_α. Thus forming $V_\alpha M^{\alpha\beta}_{;\beta}$ and using (36.5), (36.8), (44.6), (44.25), (44.26), and (45.5) we find

$$c^2(D\rho_{00}/D\tau) + (\rho_{00}c^2 + p)\theta = -2\mu V_\alpha D^{\alpha\beta}_{;\beta} + \zeta\theta^2 - [Q^\beta_{;\beta} - c^{-2}V_\alpha(DQ^\alpha/D\tau)], \tag{46.9}$$

where, as in §42, we have assumed that the only forces acting are such that $V_\alpha F^\alpha \equiv 0$.

We can rewrite the left-hand side of (46.9) in a more useful form by noticing that

$$\begin{aligned} V^\alpha(\rho_{00})_{,\alpha} + \rho_{000}V^\alpha_{;\alpha} &= V^\alpha(\rho_{000})_{,\alpha} + \rho_{000}V^\alpha_{;\alpha} - c^{-2}V^\alpha p_{,\alpha} \\ &= (\rho_{000}V^\alpha)_{;\alpha} - c^{-2}V^\alpha p_{,\alpha}. \end{aligned} \tag{46.10}$$

Similarly, on the right-hand side, using (45.5) we have

$$V_\alpha(DQ^\alpha/D\tau)=[D(V_\alpha Q^\alpha)/D\tau]-Q^\alpha A_\alpha\equiv -Q^\alpha A_\alpha, \tag{46.11}$$

and using (44.25) we have

$$V_\alpha D^{\alpha\beta}_{;\beta}=(V_\alpha D^{\alpha\beta})_{;\beta}-D^{\alpha\beta}V_{\alpha;\beta}\equiv -D^{\alpha\beta}V_{\alpha;\beta}. \tag{46.12}$$

Then substituting for $V_{\alpha;\beta}$ from (44.24), and using (44.25) and (44.27) we see that

$$D^{\alpha\beta}V_{\alpha;\beta}\equiv D^{\alpha\beta}D_{\alpha\beta}. \tag{46.13}$$

Using (46.10) to (46.13) in (46.9) we find

$$(\rho_{000}c^2V^\alpha)_{;\alpha}-V^\alpha p_{,\alpha}=2\mu D^{\alpha\beta}D_{\alpha\beta}+\zeta\theta^2-(Q^\beta_{;\beta}+c^{-2}Q^\alpha A_\alpha), \tag{46.14}$$

Then by the same steps that lead from (42.8) to (42.10) we can reduce (46.14) to

$$\rho_0 T\frac{Ds}{D\tau}\equiv\rho_0\left[\frac{De}{D\tau}+p\frac{D}{D\tau}\left(\frac{1}{\rho_0}\right)\right]=2\mu D^{\alpha\beta}D_{\alpha\beta}+\zeta\theta^2-\left(Q^\beta_{;\beta}+\frac{Q^\alpha}{c^2}A_\alpha\right), \tag{46.15}$$

where s is the specific entropy.

Equation (46.15) is the relativistic generalization of the entropy generation equation (27.11) when we use (27.28) for the dissipation function Φ. As in the nonrelativistic limit, the first two terms on the right-hand side correspond to irreversible heat generation by viscous dissipation. The third term gives the rate of heat flow into the material from its surroundings. The fourth term is purely relativistic in origin and predicts an additional deposition of heat when the material accelerates into the heat flow ($\mathbf{a}$ and $\mathbf{q}$ antiparallel), which is reasonable because if we suppose that the heat flux arises from radiation, then we see that more heat can be delivered to material accelerating into the radiation flow because photons will be blueshifted to higher energies as they enter the fluid element and redshifted to lower energies as they leave it.

Thus far we have left the form of Q^α unspecified, although we expect it to reduce to Fourier's law $\mathbf{q}=-K\nabla T$ in the nonrelativistic limit. We can deduce an expression for Q^α as follows. From (27.19), we know that classically we must have

$$\int_{\mathcal{V}}\rho\frac{Ds}{Dt}\,dV+\int_S\frac{\mathbf{q}\cdot\mathbf{n}}{T}\,dS=\int_{\mathcal{V}}\left[\frac{\partial}{\partial t}(\rho s)+\nabla\cdot\left(\rho s\mathbf{v}+\frac{\mathbf{q}}{T}\right)\right]dV\geq 0, \tag{46.16}$$

where we used (19.3) and the divergence theorem. Because $\mathcal{V}$ is arbitrary,

$$(\rho s)_{,t}+\nabla\cdot[\rho s\mathbf{v}+(\mathbf{q}/T)]\geq 0. \tag{46.17}$$

From (46.17) we see that the vector $\mathbf{s}=\rho s\mathbf{v}+(\mathbf{q}/T)$ can be interpreted classically as the entropy flux density in a heat-conducting fluid [in an ideal fluid the term $\mathbf{q}/T$ is, of course, absent; cf. (27.14)]. The covariant generalization of $\mathbf{s}$ is

$$S^{\alpha} = \rho_0 s V^{\alpha} + (Q^{\alpha}/T), \tag{46.18}$$

which we take to be the *entropy flux density four-vector.* As the covariant generalization of (46.17) we therefore take

$$S^{\alpha}_{;\alpha} \geq 0, \tag{46.19}$$

which is, in essence, a relativistic statement of the second law of thermodynamics.

Substituting (46.18) into (46.19), and using (39.10), we have

$$\rho_0(Ds/D\tau) + (Q^{\alpha}/T)_{;\alpha} \geq 0. \tag{46.20}$$

Combining (46.20) and (46.15) we thus find

$$\rho_0(Ds/D\tau) + (Q^{\alpha}/T)_{;\alpha} = (2\mu D^{\alpha\beta} D_{\alpha\beta} + \zeta\theta^2)/T - (Q^{\alpha}/T^2)(T_{,\alpha} + c^{-2} T A_{\alpha}) \geq 0. \tag{46.21}$$

The first term on the right-hand side is obviously positive, so we must merely choose Q^{α} in such a way as to guarantee that the second term will be positive. Eckart (**E1**) noted that the simplest way to do this is to take

$$Q^{\alpha} = -KP^{\alpha\beta}(T_{,\beta} + c^{-2} T A_{\beta}), \tag{46.22}$$

which (1) is consistent with Fourier's law in the classical limit, (2) is consistent with the requirements of the Eckart decomposition theorem [cf. (44.12) and (45.5)], and (3) makes the second term on the right-hand side a positive perfect square, guaranteeing positivity, as desired. The term TA_{β} is relativistic in origin and implies a flow of heat in accelerated matter even if the material is isothermal; the flow is in the direction opposite to the acceleration, and can be ascribed to the inertia of the heat energy [see (**E1**) for further discussion and interpretation]. Finally, using (46.22) in (46.21), and evaluating the result in the comoving frame we find

$$\rho_0\left(\frac{Ds}{D\tau}\right) = \frac{K}{T} \nabla \cdot \left(\nabla T + \frac{T\mathbf{a}_0}{c^2}\right) + \frac{K}{T^2}\left(\nabla T + \frac{T\mathbf{a}_0}{c^2}\right)^2 + \frac{\Phi_0}{T}, \tag{46.23}$$

which is a direct analogue of the classical result (27.19) when the terms containing $\mathbf{a}_0$ are suppressed.

Explicit expressions for Q^{α} and for the energy equation (46.15) in terms of the ordinary velocity, and lab-frame space and time derivatives, are given in (**G1,** 765–766); again conflicts of signs arise in the formulae because of Greenberg's choice of a timelike signature.

47. *The Equations of Motion*

To obtain the equations of motion for a relativistic nonideal fluid we take the "space" components of (46.8) by calculating $P_{\alpha\gamma} M^{\gamma\beta}_{;\beta}$. Noting that

$$P_{\alpha\gamma} A^{\gamma} = A_{\alpha} \tag{47.1}$$

and

$$P_{\alpha\gamma}F^{\gamma} = F_{\alpha}, \tag{47.2}$$

we easily find

$$\begin{aligned}[\rho_{00} + c^{-2}(p - \zeta\theta)](DV_{\alpha}/D\tau) = F_{\alpha} - P^{\beta}_{\alpha}(p - \zeta\theta)_{,\beta} + 2\mu P_{\alpha\gamma}D^{\gamma\beta}_{;\beta} \\ - c^{-2}[P_{\alpha\gamma}(DQ^{\gamma}/D\tau) + \tfrac{4}{3}\theta Q_{\alpha} + P_{\alpha\gamma}Q_{\beta}(D^{\gamma\beta} + \Omega^{\gamma\beta})].\end{aligned} \tag{47.3}$$

To obtain the nonrelativistic limit we let $c \to \infty$. Then in Cartesian coordinates $V_i \to v_i$, $(D/D\tau) \to (D/Dt)$, $P_{ij} \to \delta_{ij}$, $P^i_j \to \delta^i_j$, and we recover the usual Navier-Stokes equation (26.1), as expected.

Explicit expressions for the momentum equations in terms of the ordinary velocity, and lab-frame space and time derivatives, are given in (**G1**, 767–768). As before, it is necessary to translate some of the notation.

References

(A1) Adler, R., Bazin, M., and Schiffer, M. (1975) *Introduction to General Relativity*. (2nd ed.) New York: McGraw-Hill.

(A2) Anderson, J. L. (1967) *Principles of Relativity Physics*. New York: Academic Press.

(B1) Bergmann, P. G. (1976) *Introduction to the Theory of Relativity*. New York: Dover.

(C1) Cox, J. P. and Giuli, T. R. (1968) *Principles of Stellar Structure*. New York: Gordon and Breach.

(D1) de Groot, S. R., van Leeuwen, W. A., and van Weert, Ch. G. (1981) *Relativistic Kinetic Theory*. Amsterdam: North-Holland.

(E1) Eckart, C. (1940) *Phys. Rev.*, **58,** 919.

(E2) Ehlers, J. (1971) in *General Relativity and Cosmology*, ed. R. K. Sachs, pp. 1–70. New York: Academic Press.

(G1) Greenberg, P. J. (1975) *Astrophys. J.*, **195,** 761.

(L1) Landau, L. D. and Lifshitz, E. M. (1975) *The Classical Theory of Fields*. (4th ed.) Oxford: Pergamon.

(L2) Lawden, D. F. (1968) *An Introduction to Tensor Calculus and Relativity*. London: Chapman and Hall.

(L3) Leighton, R. B. (1959) *Principles of Modern Physics*. New York: McGraw-Hill.

(L4) Lichnerowicz, A. (1967) *Relativistic Hydrodynamics and Magnetohydrodynamics*. New York: Benjamin.

(L5) Lightman, A. P., Press, W. H., Price, R. H., and Teukolsky, S. A. (1975) *Problem Book in Relativity and Gravitation*. Princeton: Princeton University Press.

(L6) Lord, E. A. (1976) *Tensors, Relativity, and Cosmology*. New Delhi: Tata McGraw-Hill.

(M1) May, M. M. and White, R. H. (1966) *Phys. Rev.*, **141,** 1232.

(M2) May, M. M. and White, R. H. (1967) in *Methods in Computational Physics*, Vol. **7.** New York: Academic Press.

(M3) Misner, C. W., Thorne, K. S., and Wheeler, J. A. (1973) *Gravitation*. San Francisco: Freeman.

(M4) Møller, C. (1972) *The Theory of Relativity*. London: Oxford University Press.
(M5) Murdoch, D. C. (1957) *Linear Algebra for Undergraduates*. New York: Wiley.
(P1) Pomraning, G. C. (1973) *The Equations of Radiation Hydrodynamics*. Oxford: Pergamon.
(R1) Rindler, W. (1969) *Essential Relativity*. New York: Van Nostrand Reinhold.
(S1) Schwartz, H. M. (1968) *Introduction to Special Relativity*. New York: McGraw-Hill.
(S2) Synge, J. L. (1957) *The Relativistic Gas*. Amsterdam: North-Holland.
(T1) Taub, A. H. (1978) *Ann. Rev. Fluid Mech.*, **10,** 301.
(T2) Thomas, L. H. (1930) *Quart. J. of Math.*, **1,** 239.
(W1) Weinberg, S. (1971) *Astrophys. J.*, **168,** 175.
(W2) Weinberg, S. (1972) *Gravitation and Cosmology: Principles and Applications of the General Theory of Relativity*. New York: Wiley.
(W3) Wilson, J. R. (1979) in *Sources of Gravitational Radiation*, ed. L. Smarr, p. 423. Cambridge: Cambridge University Press.

5

Waves, Shocks, and Winds

Having derived the equations of fluid dynamics, we are now in a position to analyze some flows of interest. We shall confine attention to only a few illustrative examples of astrophysical importance. We consider first the propagation of small-amplitude disturbances in both homogeneous and stratified media, that is, *acoustic waves* and *acoustic-gravity waves* such as are observed in the solar atmosphere. Here it is adequate to use linearized equations of hydrodynamics. We then consider the nonlinear equations in the context of the generation, structure, propagation, and dissipation of *shocks*, which are important in a wide variety of astrophysical problems. Finally, we examine a nonlinear, radial, steady-flow problem that provides a first rough approximation to the physics of *stellar winds*, which are responsible for stellar mass-loss via supersonic flow into interstellar space.

5.1 Acoustic Waves

Acoustic waves are small-amplitude disturbances that propagate in a compressible medium through the interplay between fluid inertia and the restoring force of pressure. In order to isolate distinctly the characteristic properties of pure acoustic waves we assume that the medium is homogeneous, isotropic, and of infinite extent, and that no externally imposed forces act.

48. The Wave Equation

Take the ambient medium to be a perfect gas at rest with constant density ρ_0 and pressure p_0. Impose a small disturbance that perturbs these quantities locally to $\rho = \rho_0 + \rho_1$ and $p = p_0 + p_1$, where $|\rho_1/\rho_0| \ll 1$ and $|p_1/p_0| \ll 1$. The fluid acquires a small fluctuating velocity $\mathbf{v}_1$ such that $|\mathbf{v}_1|/a \ll 1$, where a is the speed of sound (see below). Velocity gradients are assumed to be so small that viscous effects are negligible, and the time scale for conductive heat transport is assumed to be so large compared to a characteristic fluctuation time that energy exchange by conduction can be ignored. In the absence of these dissipative processes, the wave-induced changes in gas properties are adiabatic, and because the undisturbed medium is homogeneous, the resulting flow is isentropic.

LINEARIZED FLUID EQUATIONS

Because all the perturbations ρ_1, p_1, and $\mathbf{v}_1$ are small, we can linearize the fluid equations, discarding all terms of second or higher order in these quantities. The equation of continuity (19.4) then becomes

$$(\partial \rho_1/\partial t) + \rho_0 \nabla \cdot \mathbf{v}_1 = 0 \tag{48.1}$$

and Euler's equation (23.6) becomes

$$\rho_0(\partial \mathbf{v}_1/\partial t) + \nabla p_1 = 0. \tag{48.2}$$

We can dispense with the energy equation because the disturbance is adiabatic, which implies variations in material properties can be related through derivatives taken at constant entropy. In particular,

$$p_1 = (\partial p/\partial \rho)_s \rho_1. \tag{48.3}$$

Taking the curl of (48.2) we find

$$\partial(\nabla \times \mathbf{v}_1)/\partial t \equiv 0, \tag{48.4}$$

hence the vorticity $\boldsymbol{\omega}$ of the disturbed fluid remains constant in time. As the undisturbed fluid was initially at rest it follows that $\boldsymbol{\omega} \equiv 0$ at all times, and the disturbance is a potential flow with

$$\mathbf{v}_1 = \nabla \phi_1, \tag{48.5}$$

where ϕ_1 is the velocity potential.

THE WAVE EQUATION

Taking $(\partial/\partial t)$ of (48.1) and subtracting the divergence of (48.2) we find

$$(\partial^2 \rho_1/\partial t^2) - \nabla^2 p_1 = 0, \tag{48.6}$$

which, by virtue of (48.3), implies that

$$(\partial^2 \rho_1/\partial t^2) - a^2 \nabla^2 \rho_1 = 0 \tag{48.7}$$

and

$$(\partial^2 p_1/\partial t^2) - a^2 \nabla^2 p_1 = 0, \tag{48.8}$$

where

$$a^2 \equiv (\partial p/\partial \rho)_s. \tag{48.9}$$

Furthermore, using (48.5) and (48.9) in (48.2) we find

$$\rho_0[\partial(\nabla \phi_1)/\partial t] + a^2 \nabla \rho_1 = \nabla[\rho_0(\partial \phi_1/\partial t) + a^2 \rho_1] = 0, \tag{48.10}$$

which implies

$$\rho_0(\partial \phi_1/\partial t) + a^2 \rho_1 = C \tag{48.11}$$

where C is a constant over all space. But both ϕ_1 and ρ_1 vanish in the undisturbed fluid (i.e., at infinite distance from the wave), hence we can set

$C \equiv 0$. In addition, (48.5) and (48.1) imply

$$(\partial \rho_1/\partial t) + \rho_0 \nabla^2 \phi_1 = 0. \tag{48.12}$$

Combining (48.11) and (48.12) we obtain

$$(\partial^2 \phi_1/\partial t^2) - a^2 \nabla^2 \phi_1 = 0. \tag{48.13}$$

Equations (48.7), (48.8), and (48.13) are all *wave equations*, and show that acoustic disturbances propagate as waves.

SOLUTION OF THE WAVE EQUATION

Consider the special case in which all perturbations are functions of one coordinate only (z). The wave equation then reduces to

$$(\partial^2 \phi_1/\partial t^2) - a^2(\partial^2 \phi_1/\partial z^2) = 0 \tag{48.14}$$

which has the general solution

$$\phi_1 = f_1(z - at) + f_2(z + at), \tag{48.15}$$

where f_1 and f_2 are arbitrary functions of their arguments. This solution shows that an initial disturbance $f_1(t=0) = f_{10}(z)$ propagates with unaltered shape along the positive z axis, while an initial disturbance $f_2(t=0) = f_{20}(z)$ propagates along the negative z axis, both with speed a. Thus (48.15) represents a superposition of two *traveling plane waves*, and a as defined by (48.9) is the *adiabatic speed of sound*. Moreover the only nonzero component of the wave velocity $\mathbf{v}_1 = \nabla\phi_1$ lies along the propagation axis, hence acoustic waves are *longitudinal* waves.

More generally, in Cartesian coordinates (48.13) admits solutions of the form

$$\phi_1 = f_1(\mathbf{x} \cdot \mathbf{n} - at) + f_2(\mathbf{x} \cdot \mathbf{n} + at), \tag{48.16}$$

which are plane waves traveling with speed a along $\pm\mathbf{n}$, the unit vector defining the direction of wave propagation. As implied by (48.16), the propagation of acoustic waves is isotropic because the ambient medium is homogeneous and isotropic. From (48.16) one finds

$$\mathbf{v}_1 = (\partial f_1/\partial \xi)\mathbf{n} + (\partial f_2/\partial \eta)\mathbf{n}, \tag{48.17}$$

where ξ and η denote the arguments $\mathbf{x} \cdot \mathbf{n} \mp at$; again, the waves are longitudinal, with nonzero velocity components only along $\mathbf{n}$.

For plane waves the perturbation amplitudes ρ_1, p_1, and v_1 can all be related simply. Thus choosing $\phi_1 = f(z - at)$ we have $v_1 = (\partial\phi_1/\partial z) = f'$, while (48.11) implies that $\rho_1 = -(\rho_0/a^2)(\partial\phi_1/\partial t) = (\rho_0/a)f'$. Therefore

$$\rho_1 = (v_1/a)\rho_0, \tag{48.18}$$

hence from (48.3) and (48.9)

$$p_1 = a^2 \rho_1 = a\rho_0 v_1. \tag{48.19}$$

Furthermore, using the general thermodynamic relation

$$(\partial T/\partial p)_s = \beta T/\rho c_p, \tag{48.20}$$

which follows from (2.27) and (5.7), we have

$$T_1/T_0 = (\beta/\rho_0 c_p)p_1 = (a\beta/c_p)v_1. \tag{48.21}$$

Here β is the coefficient of thermal expansion as defined by (2.14). Equations (48.18) to (48.21) show that in acoustic waves ρ_1, p_1, T_1, and v_1 are all in phase.

THE SPEED OF SOUND

Let us now derive explicit formulae for the speed of sound. For an adiabatic perfect gas, $p = p_0(\rho/\rho_0)^\gamma$ where γ is constant, hence

$$a^2 = \gamma p_0/\rho_0 = \gamma RT, \tag{48.22}$$

which shows that the speed of sound in a perfect gas is a function of the temperature only. To account for ionization effects, we merely replace γ with $\Gamma_1 \equiv (\partial \ln p/\partial \ln \rho)_s$ as given by (14.29), obtaining

$$a^2 = \Gamma_1 p_0/\rho_0. \tag{48.23}$$

For a perfect gas, (48.19) and (48.21) reduce to

$$p_1/p_0 = \gamma\rho_1/\rho_0 = \gamma v_1/a \tag{48.24a}$$

and

$$T_1/T_0 = (\gamma - 1)v_1/a. \tag{48.24b}$$

To account for ionization effects we replace γ in (48.24a) with Γ_1 and $(\gamma - 1)$ in (48.24b) with $\Gamma_3 - 1 \equiv (\partial \ln T/\partial \ln \rho)_s$ as given by (14.30).

Equation (48.23) yields the correct sound speed only for a nonrelativistic fluid; hence it fails at very high temperatures, in degenerate material, or if the fluid comprises both matter and radiation, and the latter contributes significantly to the pressure and energy density of the composite fluid. To obtain a relativistically correct expression for the sound speed we linearize the relativistic dynamical equations (42.3) and (42.4), obtaining

$$(\partial \hat{e}_1/\partial t) + (\hat{e} + p)_0 \nabla \cdot \mathbf{v}_1 = 0 \tag{48.25}$$

and

$$(\hat{e} + p)_0(\partial \mathbf{v}_1/\partial t) + c^2 \nabla p_1 = 0 \tag{48.26}$$

where $\hat{e} \equiv \rho_{00}c^2 = \rho_0(c^2 + e)$ is the total energy density of the fluid (including rest energy). Combining (48.25) and (48.26) we obtain

$$(\partial^2 \hat{e}_1/\partial t^2) - c^2 \nabla^2 p_1 = 0, \tag{48.27}$$

which implies that an acoustic wave propagates with a speed

$$a = c[(\partial p/\partial \hat{e})_s]^{1/2} = c[(\partial p/\partial \rho_0)_s(\partial \rho_0/\partial \hat{e})_s]^{1/2}. \tag{48.28}$$

But

$$(\partial\hat{e}/\partial\rho_0)_s = (\hat{e}/\rho_0) + \rho_0(\partial e/\partial\rho_0)_s, \tag{48.29}$$

and $T\,ds \equiv 0 = de - (p/\rho_0^2)\,d\rho_0$ implies that

$$(\partial e/\partial\rho_0)_s = p/\rho_0^2, \tag{48.30}$$

whence

$$(\partial\hat{e}/\partial\rho_0)_s = (\hat{e}+p)/\rho_0. \tag{48.31}$$

Therefore (48.28) becomes

$$a = c[\Gamma_1 p/(\hat{e}+p)]^{1/2}. \tag{48.32}$$

See also (**L7**), (**T4**), and (**W2**).

For a nonrelativistic gas, $\hat{e} \to \rho_0 c^2 \gg p$, and (48.32) gives $a(\text{N.R.}) = (\Gamma_1 p/\rho)^{1/2}$, in agreement with (48.23). For an extremely relativistic gas we recall from §43 that $\hat{e} \to \rho_0 e = 3p$ [cf. (43.53)] and that $\Gamma_1 \to \frac{4}{3}$, hence (48.32) gives $a(\text{E.R.}) = c/\sqrt{3}$. As we will see in §69, this result also holds for a gas composed of pure thermal radiation.

49. Propagation of Acoustic Waves

MONOCHROMATIC PLANE WAVES

Let us now consider the propagation of a *monochromatic wave* (i.e., a wave having a sinusoidal time variation at a definite frequency ω). This special case is important because an arbitrary wave packet can be synthesized from a linear combination of monochromatic waves (*Fourier components*) whose relative amplitudes and phases are determined by a Fourier analysis of the packet.

Thus consider a wave of the form

$$p_1 = P \exp[i(\omega t - \mathbf{k}\cdot\mathbf{x})], \tag{49.1}$$

$$\rho_1 = R \exp[i(\omega t - \mathbf{k}\cdot\mathbf{x})], \tag{49.2}$$

$$T_1 = \Theta \exp[i(\omega t - \mathbf{k}\cdot\mathbf{x})], \tag{49.3}$$

and

$$\phi_1 = \Phi \exp[i(\omega t - \mathbf{k}\cdot\mathbf{x})], \tag{49.4}$$

where P, R, Θ, and Φ are complex constants that are interrelated by (48.18) to (48.24). The use of complex quantities is convenient mathematically because it facilitates computation of the relative phases of different variables, but for the purposes of physical interpretation we use the *real parts* of (49.1) to (49.4).

According to (49.1) to (49.4), at a given instant the wave comprises a regular (sinusoidal) spatial sequence of compressions and rarefactions coupled to a sinusoidal velocity field. Likewise, at a given spatial point the fluid density, pressure, and temperature fluctuate sinusoidally in time around their equilibrium values, and the fluid particles oscillate sinusoidally around their equilibrium positions.

In (49.1) to (49.4) $\mathbf{k}$ is the *wave vector* or *propagation vector*, which determines the direction $\mathbf{n}$ of wave propagation via

$$\mathbf{k} = k\mathbf{n}. \tag{49.5}$$

The *wavenumber* k is related to the *wavelength* Λ of the wave by

$$k = 2\pi/\Lambda. \tag{49.6}$$

Substituting into (48.7), (48.8), and (48.13) we find that (49.1) to (49.4) are valid solutions of the wave equation provided that

$$\omega^2 = a^2k^2. \tag{49.7}$$

From (49.7) it follows that the planes of constant phase perpendicular to $\mathbf{n}$ propagate along $\mathbf{n}$ with speed a; thus for pure acoustic waves the *phase speed* v_p equals the speed of sound. Simple geometric considerations show that two planes of constant phase separated by one wavelength Λ along $\mathbf{n}$ are separated by a distance

$$\Lambda/n_i = 2\pi/kn_i = 2\pi/k_i \tag{49.8}$$

along the ith coordinate axis. Because the constant-phase surfaces succeed one another in a *period*

$$\tau = 2\pi/\omega, \tag{49.9}$$

one sees from (49.8) and (49.9) that the *phase trace speed* along the ith axis is

$$(v_t)_i = \omega/k_i = a/n_i. \tag{49.10}$$

Notice that the trace speed is infinite in the planes perpendicular to $\mathbf{n}$, which is expected because these are planes of constant phase, hence a local change in phase at any point in a plane must "propagate" instantaneously over the entire plane (i.e., over the entire wave front). Infinite phase or trace speeds are not in violation of relativistic causality because they represent only the behavior of a mathematical "marker", not a physically significant quantity like momentum or energy.

Equations (49.5) and (49.7) show that pure acoustic waves propagate isotropically with a unique speed a, and have a simple proportionality between wavenumber and frequency. In §§52–54 we will see that the behavior of acoustic-gravity waves in a stratified medium is markedly different.

MONOCHROMATIC SPHERICAL WAVES

Thus far we have discussed only plane waves, but from symmetry considerations one expects that an isotropic medium will also support one-dimensional *spherical waves* emanating from a point source. Thus specialize (48.8) to

$$\left(\frac{\partial^2 p_1}{\partial t^2}\right)-\frac{a^2}{r^2}\frac{\partial}{\partial r}\left(r^2\frac{\partial p_1}{\partial r}\right)=0. \tag{49.11}$$

The substitution $p_1=f/r$ reduces (49.11) to $(\partial^2 f/\partial t^2)-a^2(\partial^2 f/\partial r^2)$, which leads to a general solution of the form

$$p_1=[f_1(r-at)/r]+[f_2(r+at)/r] \tag{49.12}$$

where f_1 and f_2 are arbitrary functions. The two terms in (49.12) represent spherical disturbances diverging from, and converging on, the origin. The wave amplitude falls off as r^{-1}, hence the wave intensity (proportional to the square of the amplitude) varies as r^{-2}.

Specializing now to monochromatic waves we choose a solution of the form

$$p_1=Pe^{i(\omega t-kr)}/r, \tag{49.13}$$

which satisfies (49.11) only if $\omega=ak$. The thermodynamic relations (48.3) and (48.19) are independent of geometry, and show that in a spherical acoustic wave both ρ_1 and T_1 are proportional to, and in phase with, p_1. From (48.11) we obtain

$$\phi_1=(i/\rho_0\omega)p_1, \tag{49.14}$$

which shows that ϕ_1 leads p_1 in phase by 90°. Calculating $\mathbf{v}_1$ from (48.5) we have

$$\mathbf{v}_1=[k-(i/r)](p_1/\rho_0\omega)\hat{\mathbf{r}}. \tag{49.15}$$

As $r\to\infty$, (49.15) reduces to (48.19), as it should because a wave is locally planar when its radius of curvature approaches infinity. However, near the origin v_1 lags p_1 by 90°, and grows in amplitude as r^{-2}.

GROUP VELOCITY

To determine the speed of energy propagation in waves we study the behavior of *wave packets*. A packet is a localized disturbance that can be described mathematically as a superposition of a large (perhaps infinite) number of Fourier components that interfere in such a way as to produce a finite wave amplitude only in a strongly localized region in space and time. Intuitively one expects that such a localized disturbance must result from a concentration of wave energy, and that the motion of the packet tracks the flow of wave energy in the medium.

Consider a packet composed of plane waves $\exp[i(\omega t-\mathbf{k}\cdot\mathbf{x})]$ whose wave vectors $\mathbf{k}$ lie in the volume $\mathbf{k}_0\pm\Delta\mathbf{k}$ of $\mathbf{k}$ space; we assume that all waves having their wave vectors within this volume have unit amplitude, and that all others have zero amplitude. We further assume that ω is a general function of (k_x, k_y, k_z). The amplitude of the wave packet is then

$$A(\mathbf{x}, t) = \int_{k_{0x}-\Delta k_x}^{k_{0x}+\Delta k_x} dk_x \int_{k_{0y}-\Delta k_y}^{k_{0y}+\Delta k_y} dk_y \int_{k_{0z}-\Delta k_z}^{k_{0z}+\Delta k_z} dk_z e^{i[\omega(k_x,k_y,k_z)t - k_x x - k_y y - k_z z]}. \tag{49.16}$$

For a small enough volume $\Delta\mathbf{k}$, we can use the linear expansion

$$\omega(\mathbf{k}) = \omega(\mathbf{k}_0) + (\partial\omega/\partial k_x)_0\, \delta k_x + (\partial\omega/\partial k_y)_0\, \delta k_y + (\partial\omega/\partial k_z)_0\, \delta k_z, \tag{49.17}$$

whence

$$\begin{aligned} A(\mathbf{x}, t) &\approx e^{i(\omega_0 t - \mathbf{k}_0\cdot\mathbf{x})} \int_{-\Delta k_x}^{\Delta k_x} d(\delta k_x) \int_{-\Delta k_y}^{\Delta k_y} d(\delta k_y) \int_{-\Delta k_z}^{\Delta k_z} d(\delta k_z) \\ &\times \exp\left[i\left\{ \left[\left(\frac{\partial\omega}{\partial k_x}\right)_0 t - x \right] \delta k_x + \left[\left(\frac{\partial\omega}{\partial k_y}\right)_0 t - y \right] \delta k_y + \left[\left(\frac{\partial\omega}{\partial k_z}\right)_0 t - z \right] \delta k_z \right\} \right] \\ &\approx 8 e^{i(\omega_0 t - \mathbf{k}_0\cdot\mathbf{x})} \frac{\sin\{[(\partial\omega/\partial k_x)_0 t - x]\,\Delta k_x\}}{(\partial\omega/\partial k_x)_0 t - x} \\ &\times \frac{\sin\{[(\partial\omega/\partial k_y)_0 t - y]\,\Delta k_y\}}{(\partial\omega/\partial k_y)_0 t - y} \\ &\times \frac{\sin\{[(\partial\omega/\partial k_z)_0 t - z]\,\Delta k_z\}}{(\partial\omega/\partial k_z)_0 t - z}. \end{aligned} \tag{49.18}$$

Equation (49.18) shows that the packet is an *amplitude-modulated plane wave*. The maximum of each of the factors $(\sin \xi)/\xi$ is attained as $\xi \to 0$, hence the packet has maximum amplitude at

$$x = (\partial\omega/\partial k_x)_0 t, \tag{49.19a}$$

$$y = (\partial\omega/\partial k_y)_0 t, \tag{49.19b}$$

and

$$z = (\partial\omega/\partial k_z)_0 t. \tag{49.19c}$$

Thus the wave packet has a *group velocity* (or *packet velocity*)

$$\mathbf{v}_g = \nabla_k \omega \tag{49.20}$$

where

$$\nabla_k \equiv (\partial/\partial k_x)\hat{\mathbf{i}} + (\partial/\partial k_y)\hat{\mathbf{j}} + (\partial/\partial k_z)\hat{\mathbf{k}}. \tag{49.21}$$

These results hold for an arbitrary dependence of ω on $\mathbf{k}$.

In the particular case of a packet of pure acoustic waves, for which $\omega = ak$, one readily finds

$$\mathbf{v}_g = a[(k_x/k)_0\hat{\mathbf{i}} + (k_y/k)_0\hat{\mathbf{j}} + (k_z/k)_0\hat{\mathbf{k}}] = a\mathbf{n}_0. \tag{49.22}$$

Thus, for pure acoustic waves, the group velocity equals the phase velocity, and an acoustic wave packet propagates with the speed of sound along $\mathbf{n}_0$ (or $\mathbf{k}_0$); as we will see in §53, the behavior of gravity-modified acoustic waves in a stratified medium is more complicated (and interesting).

50. Wave Energy and Momentum

We obtained the mechanical energy equation for a fluid by taking the dot product of the flow velocity with the momentum equation (cf. §§24 and 27). Similarly we can derive a mechanical energy equation for an acoustic wave by taking the dot product of the velocity fluctuation $\mathbf{v}_1$ with the linearized momentum equation (48.2), obtaining

$$\rho_0\mathbf{v}_1\cdot(\partial\mathbf{v}_1/\partial t)=(\tfrac{1}{2}\rho_0 v_1^2)_{,t}=-\mathbf{v}_1\cdot\nabla p_1, \qquad (50.1)$$

which states that the rate of change of the kinetic energy density in the wave equals the rate of work done on the fluid by the wave-induced pressure gradient. Notice that all quantities in (50.1) are *second order.*

In place of the gas energy equation, we have the adiabatic relation (48.3), which can be combined with the continuity equation (48.1), to give

$$(p_1/a^2\rho_0)(\partial p_1/\partial t)=(\tfrac{1}{2}p_1^2/a^2\rho_0)_{,t}=-p_1\nabla\cdot\mathbf{v}_1. \qquad (50.2)$$

In order to interpret (50.2) physically, consider the energy density $\hat{e}=\rho e$ of the disturbed fluid. If we take $e=e(\rho, s)$, then because the wave is adiabatic

$$e\rho=e_0\rho_0+[\partial(\rho e)/\partial\rho]_s\rho_1+\tfrac{1}{2}[\partial^2(\rho e)/\partial\rho^2]_s\rho_1^2. \qquad (50.3)$$

But from (3.12), $(\partial e/\partial\rho)_s=p/\rho^2$, hence $[\partial(\rho e)/\partial\rho]_s=h$. Therefore

$$[\partial^2(\rho e)/\partial\rho^2]_s=(\partial h/\partial\rho)_s=(\partial h/\partial p)_s(\partial p/\partial\rho)_s=a^2/\rho, \qquad (50.4)$$

where we used (2.33) to obtain $(\partial h/\partial p)_s$. Thus

$$\rho e=\rho_0 e_0+h_0\rho_1+\tfrac{1}{2}(a^2\rho_1^2/\rho_0)=\rho_0 e_0+h_0\rho_1+\tfrac{1}{2}(p_1^2/a^2\rho_0). \qquad (50.5)$$

The first term on the right-hand side of (50.5) is the energy density of the unperturbed fluid and is unrelated to the presence of the wave. The second term averages to zero over a sufficiently large volume because the total mass of the fluid cannot be changed by a wave, hence $\int\rho_1\,dV=0$. (Likewise, this term averages to zero over time for harmonic disturbances.) Hence the third term represents the nonvanishing net change in the fluid's internal energy density resulting from the presence of a wave; this energy (a second-order quantity) is called the *compressional energy* of the wave. Thus (50.2) is analogous to the first law of thermodynamics, stating that the rate of change of the compressional energy stored in the wave equals the negative of the rate of work done by the wave's pressure perturbation on the wave-induced expansion and compression of the fluid.

Taking the sum of (50.1) and (50.2) we obtain a *wave energy equation* in conservation form, namely

$$(\partial \varepsilon_w/\partial t) = -\nabla \cdot \boldsymbol{\phi}_w \tag{50.6}$$

where the *wave energy density* is

$$\varepsilon_w = \tfrac{1}{2}\rho_0 v_1^2 + \tfrac{1}{2}(p_1^2/a^2\rho_0) \tag{50.7}$$

and the *wave energy flux* is

$$\boldsymbol{\phi}_w = p_1\mathbf{v}_1. \tag{50.8}$$

The momentum density in the perturbed fluid is $\boldsymbol{\mu} = \rho\mathbf{v} = \rho_0\mathbf{v}_1 + \rho_1\mathbf{v}_1$. Using the fact that $\mathbf{v}_1 = \nabla\phi_1$, one sees from equation (A2.64) that

$$\rho_0 \int_V \mathbf{v}_1 \, dV = \rho_0 \int_S \phi_1 \mathbf{n} \, dS. \tag{50.9}$$

The latter integral is identically zero if S lies in the unperturbed fluid where $\phi_1 \equiv 0$, so in a sufficiently large volume the net *wave momentum density* is

$$\boldsymbol{\mu}_w = \rho_1\mathbf{v}_1 = p_1\mathbf{v}_1/a^2 = \boldsymbol{\phi}_w/a^2. \tag{50.10}$$

The above formulae simplify for plane waves. Thus from (48.19) we find

$$\varepsilon_w = \rho_0 v_1^2 \tag{50.11}$$

and

$$\boldsymbol{\phi}_w = a\rho_0 v_1^2 \mathbf{n} = a\varepsilon_w \mathbf{n}. \tag{50.12}$$

The *instantaneous* values of ε_w and ϕ_w are of little interest; rather we usually wish to know the *time averages* $\langle\varepsilon_w\rangle$ and $\langle\phi_w\rangle$. In particular, for monochromatic waves, we calculate averages over a cycle (one wave period). When complex representations like (49.1) to (49.4) are used it is important to remember that in computing, say, $\langle p_1\mathbf{v}_1\rangle$ we must take the time average of the *real parts* of p_1 and $\mathbf{v}_1$. This is most easily done by noting that for any time-harmonic complex quantities $\alpha = \alpha_0 e^{i\omega t} = \alpha_R + i\alpha_I$ and $\beta = \beta_0 e^{i(\omega t + \psi)} = \beta_R + i\beta_I$ we have the general identities

$$\tfrac{1}{2}(\alpha^*\beta + \alpha\beta^*) \equiv \alpha_0\beta_0 \cos\psi \tag{50.13}$$

and

$$\begin{aligned}\langle\alpha_R\beta_R\rangle &= \alpha_0\beta_0\langle\cos(\omega t)\cos(\omega t + \psi)\rangle \equiv \tfrac{1}{2}\alpha_0\beta_0\cos\psi \\ &= \alpha_0\beta_0\langle\sin(\omega t)\sin(\omega t + \psi)\rangle = \langle\alpha_I\beta_I\rangle.\end{aligned} \tag{50.14}$$

Hence for monochromatic plane waves we have the useful result

$$\langle\alpha_R\beta_R\rangle \equiv \tfrac{1}{4}(\alpha^*\beta + \alpha\beta^*). \tag{50.15}$$

From (50.15) we see that for a monochromatic wave the average kinetic and potential energy densities each equal

$$\tfrac{1}{2}\rho_0\langle v_{1R}^2\rangle = \tfrac{1}{4}\rho_0 v_1 v_1^* = \tfrac{1}{4}\rho_0 \mathbf{v}_1 \cdot \mathbf{v}_1^*, \tag{50.16}$$

hence the average wave energy density is

$$\langle \varepsilon_w \rangle = \tfrac{1}{2}\rho_0 \mathbf{v}_1 \cdot \mathbf{v}_1^*, \tag{50.17}$$

and the average wave energy flux is

$$\langle \boldsymbol{\Phi}_w \rangle = \langle p_{1R} \mathbf{v}_{1R} \rangle = \tfrac{1}{4}(p_1 \mathbf{v}_1^* + p_1^* \mathbf{v}_1). \tag{50.18}$$

For a wave packet whose velocity amplitude can be represented by a sum over monochromatic components, that is,

$$\mathbf{V}(t) = \sum_j \mathbf{V}_j e^{i\omega_j t}, \tag{50.19}$$

one easily sees that the only terms surviving in a time average yield

$$\langle V_R^2 \rangle = \tfrac{1}{2}\langle \mathbf{V} \cdot \mathbf{V}^* \rangle = \tfrac{1}{2}\sum_j \mathbf{V}_j \cdot \mathbf{V}_j^*. \tag{50.20}$$

Hence the average energy density in the wave equals the sum of the average energy densities in the monochromatic components. Likewise

$$\langle \boldsymbol{\Phi}_w \rangle = \sum_j \langle \boldsymbol{\Phi}_{wj} \rangle. \tag{50.21}$$

51. Damping of Acoustic Waves by Conduction and Viscosity

Thus far we have ignored viscosity and thermal conduction and have supposed that acoustic waves propagate adiabatically. We now inquire what happens to a wave when these dissipative processes are operative. We assume the fluid is a perfect gas, and consider the propagation of a plane wave along the x axis.

The linearized continuity equation (48.1) is unaffected by viscosity or conductivity, while the linearized momentum equation (26.2) is

$$\rho_0(\partial v_1/\partial t) = -(\partial p_1/\partial x) + \mu'(\partial^2 v_1/\partial x^2) \tag{51.1}$$

and the linearized energy equation (27.11) is

$$\rho_0(\partial e_1/\partial t) = -p_0(\partial v_1/\partial x) + K(\partial^2 T_1/\partial x^2). \tag{51.2}$$

Here $\mu' \equiv \frac{4}{3}\mu + \zeta$ is the effective one-dimensional viscosity. Notice that no viscous term appears in (51.2) because the dissipation function $\mu'(\partial v_1/\partial x)^2$ is a second-order quantity.

For a perfect gas $p = \rho RT = (\gamma - 1)\rho e$, hence

$$T_1 = [p_1 - (p_0/\rho_0)\rho_1]/R\rho_0 \tag{51.3}$$

and

$$e_1 = [p_1 - (p_0/\rho_0)\rho_1]/(\gamma - 1)\rho_0. \tag{51.4}$$

Using (51.3) and (51.4) and eliminating $(\partial v_1/\partial x)$ via (48.1), we can rewrite (51.2) as

$$\frac{\partial}{\partial t}(p_1 - a^2\rho_1) = \gamma\chi \frac{\partial^2}{\partial x^2}(p_1 - a^2\rho_1), \tag{51.5}$$

where χ is the *thermal diffusivity* [cf. (28.3)]

$$\chi \equiv K/\rho_0 c_p = (\gamma-1)K/\gamma R\rho_0. \tag{51.6}$$

Now take a plane-wave solution

$$p_1 = P \exp[i(\omega t - kx)] \tag{51.7}$$

$$\rho_1 = R \exp[i(\omega t - kx)] \tag{51.8}$$

and

$$\phi_1 = \Phi \exp[i(\omega t - kx)], \tag{51.9}$$

where $v_1 = (\partial\phi_1/\partial x)$. Substituting (51.7) to (51.9) in (48.1), (51.1), and (51.5) we find

$$i\omega R - \rho_0 k^2 \Phi = 0, \tag{51.10}$$

$$-ikP + (\rho_0 k\omega - i\mu' k^3)\Phi = 0, \tag{51.11}$$

and

$$(i\omega + \gamma\chi k^2)P - a^2(i\omega + \chi k^2)R = 0. \tag{51.12}$$

We obtain a nontrivial solution of (51.10) to (51.12) only if the determinant of the coefficients vanishes. Enforcing this condition, we obtain the *dispersion relation*

$$\omega^2 = a^2 k^2 \left[\frac{1 - i(\chi k^2/\omega)}{1 - i(\gamma\chi k^2/\omega)}\right] + \frac{i\mu' k^2 \omega}{\rho_0}. \tag{51.13}$$

Notice that when $\chi = \mu' = 0$ we recover our previous result $\omega^2 = a^2 k^2$.

Suppose first that both μ' and χ are very small. Then set $k = k_0 + \delta k = (\omega/a) + \delta k$, expand to first order in small quantities, and solve for δk to obtain

$$\delta k = -(i\omega^2/2a^3\rho_0)[\mu' + (\gamma-1)\rho_0\chi]. \tag{51.14}$$

The solution for, say, p_1 is then of the form

$$p_1 = Pe^{i(\omega t - k_0 x)} e^{-x/L} \tag{51.15}$$

where $L \equiv -i/\delta k$, and similarly for ρ_1, T_1, ϕ_1, and v_1.

Equation (51.15) shows that the wave still propagates with the sound speed a, but its amplitude steadily diminishes with a characteristic decay-length L. Thus small amounts of viscosity and conductivity *damp* acoustic waves by an irreversible conversion of wave energy into entropy. Recalling from §29 that $(\mu/\rho) \sim \chi \sim a\lambda$, where λ is a particle mean free path, we see that $(L/\Lambda) \sim (\Lambda/\lambda)$, hence the decay length is of the order of (Λ/λ) wavelengths. The assumptions under which we solved (51.13) imply $(\Lambda/\lambda) \gg 1$, hence the decay is slow. Equation (51.14) shows that high-frequency waves are more heavily damped than low-frequency waves, which is not surprising because for a given amplitude they will have steeper gradients of velocity and temperature over smaller physical distances (a wavelength).

In reality the physics of the problem is more complicated, and more interesting, than indicated above. To simplify the analysis suppose the gas has zero viscosity but a finite thermal conductivity. (Recalling from §§29 and 33 that in a real gas $K \propto \mu$ this assumption may, at first sight, seem hypothetical. But as we will see in §101 it is actually realistic for a radiating gas where radiation provides an efficient energy transport mechanism while viscosity and thermal conduction—by particles—are both negligible). The dispersion relation then becomes

$$k^4 - [(\gamma\omega^2/a^2) - i(\omega/\chi)]k^2 - i(\omega^3/a^2\chi) = 0, \qquad (51.16)$$

which yields immediately

$$2k^2 = \left(\frac{\gamma\omega^2}{a^2} - \frac{i\omega}{\chi}\right) \pm \left[\frac{\gamma^2\omega^4}{a^4} - \frac{\omega^2}{\chi^2} - \frac{2i(\gamma-2)\omega^3}{a^2\chi}\right]^{1/2}. \qquad (51.17)$$

In general, one must calculate $k(\omega)$ from (51.17) numerically. But we can obtain analytical expressions in two limiting regimes. First, suppose that the dimensionless ratio $\varepsilon \equiv (\omega\chi/a^2) \ll 1$ because ω and/or χ is very small; from the scaling relation $\chi \sim a\lambda$ one recognizes that in this regime $\lambda/\Lambda \ll 1$. Expanding (51.17) to second order in ε we have

$$2k^2 \approx (\omega/\chi)(\gamma\varepsilon - i) \pm (i\omega/\chi)[1 + i(\gamma-2)\varepsilon - 2(\gamma-1)\varepsilon^2]. \qquad (51.18)$$

Taking the positive root we have

$$k_+^2 \approx (\omega/a)^2[1 - i(\gamma-1)(\chi\omega/a^2)] \qquad (51.19)$$

whence we obtain

$$k_1 \approx \pm[(\omega/a) - i(\gamma-1)(\chi\omega^2/2a^3)]. \qquad (51.20)$$

This root corresponds to the same mode obtained in (51.14), that is, a slowly damped acoustic wave propagating with the adiabatic sound speed.

Taking the negative root in (51.18) and retaining only leading terms we find

$$k_-^2 \approx -i\omega/\chi \qquad (51.21)$$

whence we obtain

$$k_2 \approx \pm(\omega/2\chi)^{1/2}(1-i). \qquad (51.22)$$

This root corresponds to a new mode, a propagating *thermal wave* [cf. (**L2**, §77) and §§101 and 103]. This particular mode propagates very slowly, with phase speed $v_p = (2\varepsilon)^{1/2}\, a \ll a$; because the real and imaginary parts of k are of the same magnitude, the wave is very heavily damped, decaying away in a few wavelengths. Moreover $\Lambda_2/\Lambda_0 = k_0/k_{2R} = (2\varepsilon)^{1/2} \ll 1$, hence the physical distance over which this mode can propagate is less than a wavelength of an undamped acoustic wave of the same frequency.

Suppose now that the dimensionless ratio $\varepsilon' \equiv (a^2/\omega\chi) \ll 1$ because ω and/or χ is very large; estimating χ as before we see that in this regime

$\Lambda/\lambda \ll 1$. We now expand (51.17) as

$$2k^2 \approx \left(\frac{\gamma\omega^2}{a^2}\right)\left(1-\frac{i\varepsilon'}{\gamma}\right) \pm \left(\frac{\gamma\omega^2}{a^2}\right)\left[1+\frac{i(2-\gamma)\varepsilon'}{\gamma^2}\right]. \tag{51.23}$$

Taking the positive root we have

$$k_+^2 \approx (\omega^2/a_T^2)[1-i(\gamma-1)(\varepsilon'/\gamma^2)], \tag{51.24}$$

whence

$$k_3 \approx \pm[(\omega/a_T)-i(\gamma-1)(a_T/2\gamma\chi)]. \tag{51.25}$$

Here a_T is the *isothermal sound speed*

$$a_T^2 \equiv (\partial p/\partial\rho)_T = p/\rho = a^2/\gamma. \tag{51.26}$$

This root again corresponds to a damped acoustic wave, but with the qualitatively important difference that now the wave propagates at *constant temperature* because wave-induced temperature fluctuations are efficiently

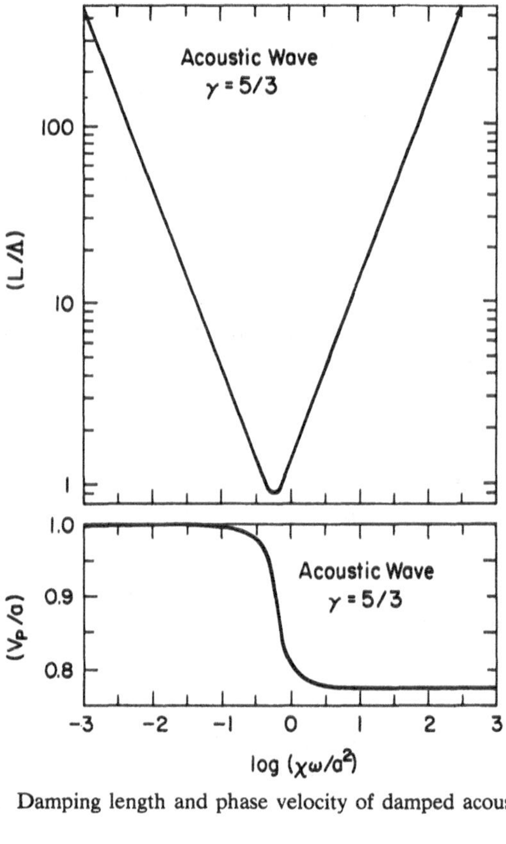

Fig. 51.1 Damping length and phase velocity of damped acoustic mode.

eradicated by the high thermal conductivity and/or steep temperature gradients. Isothermal acoustic waves propagate a factor of $\gamma^{-1/2}$ more slowly than adiabatic acoustic waves; the decay length is now independent of frequency and is proportional to the thermal conductivity. Because $|k_{3I}/k_{3R}| \ll 1$ the wave survives over many wavelengths before it decays. This is not to say that it is slowly damped, however, because by scaling arguments one finds that $L \approx \lambda$, hence the wave decays significantly over a particle mean free path!

Taking the negative root in (51.23) we have

$$k_-^2 \approx -i\omega/\gamma\chi, \tag{51.27}$$

whence we obtain

$$k_4 \approx \pm(\omega/2\gamma\chi)^{1/2}(1-i). \tag{51.28}$$

This root corresponds to another thermal wave, which is heavily damped in the sense that it decays away over a few wavelengths. However the phase

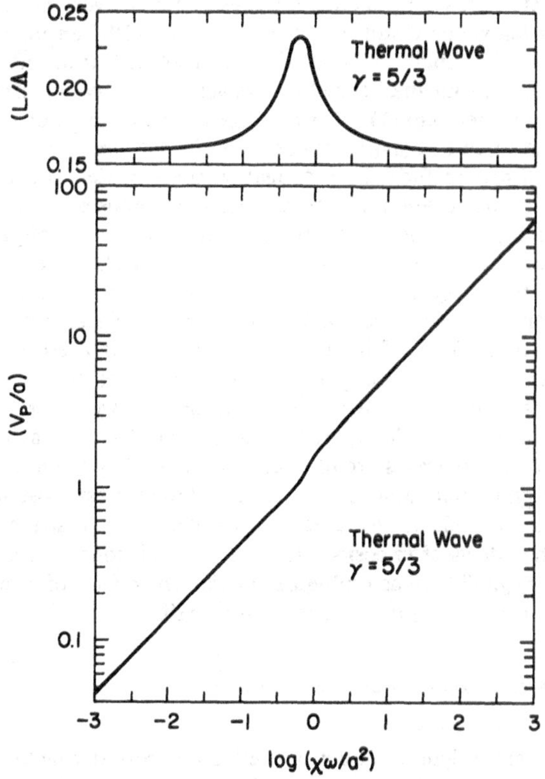

Fig. 51.2 Damping length and phase velocity of thermal mode.

speed of this mode is very large, $v_p = (2\gamma\chi\omega)^{1/2} = (2\gamma/\varepsilon')^{1/2}\ a \gg a$, as is its wavelength compared to that of an undamped acoustic wave of the same frequency: $\Lambda_4/\Lambda_0 = (2\gamma/\varepsilon')^{1/2} \gg 1$. Hence this mode propagates rapidly over a physical distance of the order of $(a\lambda/\omega)^{1/2}$ before it is damped.

Plots of v_p/a and L/Λ obtained from a numerical solution of (51.17) for $\gamma = \frac{5}{3}$ are shown in Figures 51.1 and 51.2. Here one sees that the acoustic mode is most heavily damped when $(\chi\omega/a^2) \sim 1$, which is also where v_p drops from a to a_T for that mode. In contrast, the thermal wave mode is least damped when $(\chi\omega/a^2) \sim 1$.

The damping of acoustic waves in relativistic fluids is discussed by Weinberg (**W2**), who gives an expression for δk [cf. his equations (2.55) to (2.57)] which reduces to (51.14) in the nonrelativistic limit.

5.2 Acoustic-Gravity Waves

Let us now consider wave propagation in a compressible medium stably stratified in a gravitational field, such as the atmosphere of the Earth, the Sun, or a star. In such a medium, waves can be driven by both compressional and *buoyancy* restoring forces, each of which can induce harmonic oscillations of a fluid element slightly displaced from its equilibrium position. As a result the behavior of waves is more complicated than in a homogeneous medium: (1) Their propagation characteristics are *anisotropic* because the force of gravity imposes a preferred direction in the fluid. (2) They are *dispersive* (i.e., the propagation speed varies as k and/or ω are varied). (3) Stratification of the atmosphere imposes a *cutoff frequency* below which *gravity-modified acoustic waves* cannot propagate, and thus restricts the region of $(k_x - \omega)$ space in which such waves can exist. (4) Buoyancy forces give rise to a new class of propagating waves: the low-frequency pressure-modified *internal gravity waves*, which are inherently *two-dimensional*. (The adjective "internal" is used to discriminate these waves from surface gravity waves found at fluid interfaces, for example, the surface of the ocean; for brevity we refer to these two different wave modes simply as "acoustic waves" and "gravity waves".)

Acoustic-gravity waves are of interest because they are ubiquitous in the terrestrial and solar atmospheres, and certainly must exist in stellar atmospheres as well. In §§52 to 54 we consider adiabatic fluctuations because the effects of viscosity and thermal conduction on acoustic-gravity waves are negligible in astrophysical media. The effects of radiative damping, which can be major, are discussed in §102.

52. The Wave Equation and Wave Energy

BUOYANCY OSCILLATIONS

We can obtain insight into buoyancy effects by considering the motion of a small fluid parcel rising and falling adiabatically in a stratified medium. We assume strictly vertical motion and write the vertical displacement as ζ_1. When the parcel is displaced from its equilibrium position it experiences a net force $-g(\rho-\rho_0)$ where ρ is the density inside the parcel and ρ_0 is the density in the ambient atmosphere; hence its motion is governed by

$$\rho_1(\partial^2\zeta_1/\partial t^2) = -g(\rho-\rho_0). \tag{52.1}$$

For small displacements

$$\rho_0(\zeta_1) = \rho_0(0) + (d\rho/dz)_{at}\zeta_1 \tag{52.2a}$$

and

$$\rho(\zeta_1) = \rho_0(0) + (d\rho/dz)_{ad}\zeta_1, \tag{52.2b}$$

where the subscripts denote "atmosphere" and "adiabatic". Thus

$$(\partial^2\zeta_1/\partial t^2) = -\omega_{BV}^2\zeta_1 \tag{52.3}$$

where

$$\omega_{BV}^2 \equiv (g/\rho)[(d\rho/dz)_{ad} - (d\rho/dz)_{at}] \tag{52.4}$$

is known as the *Brunt-Väisälä frequency.*

Equation (52.3) admits two distinct classes of solutions:

$$\zeta_1(t) = \zeta_1(0)\exp(\pm i\,|\omega_{BV}|\,t) \tag{52.5a}$$

for $\omega_{BV}^2 > 0$, and

$$\zeta_1(t) = \zeta_1(0)\exp(\pm\,|\omega_{BV}|\,t) \tag{52.5b}$$

for $\omega_{BV}^2 < 0$.

When $\omega_{BV}^2 > 0$ the atmosphere is *stably stratified* and fluid parcels undergo harmonic oscillations of bounded amplitude. But when $\omega_{BV}^2 < 0$, a displaced fluid parcel experiences further force in the direction of its displacement, and the perturbation grows exponentially. In the latter case the atmosphere is *convectively unstable*; indeed the inequality $(d\rho/dz)_{ad} < (d\rho/dz)_{at}$ is just the standard *Schwarzschild criterion* for instability against convection (**C9,** 264). When $\omega_{BV}^2 = 0$ the atmosphere is in adiabatic equilibrium, and is *neutrally stable*; neither buoyancy oscillations nor convection can then occur.

In a simple buoyancy oscillation the displaced fluid expands and cools adiabatically as it rises, and slows in response to gravitational braking as the density in the parcel exceeds that in the surrounding medium. At the top of the cycle $T_1 < 0$, $\rho_1 > 0$, and $v_1 = 0$. The denser fluid element is then accelerated downward, and passes through its equilibrium position, where $\rho_1 = T_1 = 0$, with maximum downward velocity. Thus ρ_1, T_1, and v_1 are out of phase, T_1 leading v_1 by 90° and ρ_1 lagging v_1 by 90°, in strong contrast with a pure acoustic wave where ρ_1, T_1, and v_1 are all exactly in phase (cf. §48).

Because buoyancy oscillations are slow (see §§53 and 54), sound waves have time to run back and forth within a displaced fluid parcel and to establish pressure equilibrium between it and the surrounding atmosphere.

We can therefore take the pressure gradient in the parcel to be the same as in the unperturbed atmosphere, and using the relation $\delta\rho = (\partial\rho/\partial p)_{ad}\,\delta p = \delta p/a^2$ inside the parcel we can rewrite (52.4) in the useful form

$$\omega_{BV}^2 = (g/a^2\rho)[(dp/dz)_{at} - a^2(d\rho/dz)_{at}]. \tag{52.6}$$

FLUID EQUATIONS

The dynamical behavior of acoustic-gravity waves is determined by the equation of continuity (19.4), Euler's equation (23.6) with $\mathbf{f} \equiv \rho\mathbf{g}$, and the gas energy equation

$$\rho\{(De/Dt) + p[D(1/\rho)/Dt]\} = (Dq/Dt), \tag{52.7}$$

where (Dq/Dt) is the net rate of energy input, per unit volume, to the gas from external sources. For adiabatic fluctuations $(Dq/Dt) \equiv 0$.

Before writing linearized fluid equations, it is convenient to derive some alternative forms of (52.7), which will prove useful later. We first develop some necessary thermodynamic relations. Thus expanding dp as a function of (ρ, T) we have the general expression

$$(d \ln p/d \ln \rho) = (\partial \ln p/\partial \ln T)_\rho (d \ln T/d \ln \rho) + (d \ln p/d \ln \rho)_T. \tag{52.8}$$

Using the cyclic relation (2.7) one finds $(\partial p/\partial T)_\rho = \beta/\kappa_T$ where β and κ_T are defined by (2.14) and (2.15). We can thus rewrite (52.8) as

$$(d \ln p/d \ln \rho) = (T\beta/p\kappa_T)(d \ln T/d \ln \rho) + (1/p\kappa_T). \tag{52.9}$$

This relation is general, hence holds for adiabatic changes in particular. Thus using (14.19) and (14.21) we obtain the important identity

$$\Gamma_1 = (T\beta/p\kappa_T)(\Gamma_3 - 1) + (1/p\kappa_T). \tag{52.10}$$

Next, we obtain a general expression for the ratio c_p/c_v by applying (5.17) to an adiabatic process, which yields

$$c_p/c_v = \rho\kappa_T(\partial p/\partial\rho)_s = p\kappa_T\Gamma_1. \tag{52.11}$$

Substituting (52.10) and (52.11) into (5.7) we then find

$$c_v = \beta/\kappa_T\rho(\Gamma_3 - 1) \tag{52.12}$$

whence, using (14.22) and (52.11) we obtain

$$c_p = (\beta p/\rho)\Gamma_2/(\Gamma_2 - 1). \tag{52.13}$$

Suppose now we choose T and p as fundamental variables in (52.7). Then using (2.15), (2.22), and (5.11) in the expansion

$$\rho\left[\left(\frac{\partial e}{\partial T}\right)_p - \frac{p}{\rho^2}\left(\frac{\partial\rho}{\partial T}\right)_p\right] dT + \rho\left[\left(\frac{\partial e}{\partial p}\right)_T - \frac{p}{\rho^2}\left(\frac{\partial\rho}{\partial p}\right)_T\right] dp = dq \tag{52.14}$$

we have

$$c_p\rho\, dT - \beta T\, dp = dq \tag{52.15}$$

which, with the aid of (52.13), can be rewritten as

$$c_p\rho T\left[\frac{dT}{T} - \frac{(\Gamma_2 - 1)}{\Gamma_2}\frac{dp}{p}\right] = dq. \tag{52.16}$$

Alternatively, suppose we choose p and ρ as fundamental variables. Then using (2.28), (2.29), and (52.11) in the expansion

$$\rho\left(\frac{\partial e}{\partial p}\right)_\rho dp + \rho\left[\left(\frac{\partial e}{\partial \rho}\right)_p - \frac{p}{\rho^2}\right] d\rho = dq \tag{52.17}$$

we get

$$(\kappa_T c_v \rho/\beta)\, dp - (c_p/\beta)\, d\rho = (\kappa_T c_v \rho/\beta)[dp - (\Gamma_1 p/\rho)\, d\rho] = dq. \tag{52.18}$$

Then using (52.12) and (48.23) we find

$$(\Gamma_3 - 1)^{-1}(dp - a^2\, d\rho) = dq. \tag{52.19}$$

Finally, suppose we choose ρ and T as fundamental variables. Then using (2.11), (2.17), and (5.7) in the expansion

$$\rho\left(\frac{\partial e}{\partial T}\right)_\rho dT + \rho\left[\left(\frac{\partial e}{\partial \rho}\right)_T - \frac{p}{\rho^2}\right] d\rho = dq \tag{52.20}$$

we find

$$\rho c_v\, dT - (\beta T/\kappa_T \rho)\, d\rho = dq. \tag{52.21}$$

Hence from (52.12)

$$\rho c_v T[(dT/T) - (\Gamma_3 - 1)(d\rho/\rho)] = dq. \tag{52.22}$$

LINEARIZED FLUID EQUATIONS

Assume that the ambient atmosphere is static and in hydrostatic equilibrium so that

$$\nabla p_0 = \rho_0 \mathbf{g}, \tag{52.23}$$

where $\mathbf{g} = -g\mathbf{k}$ is constant. Then for small-amplitude adiabatic disturbances the linearized fluid equations are

$$(\partial \rho_1/\partial t) + \nabla \cdot (\rho_0 \mathbf{v}_1) = 0, \tag{52.24}$$

$$\rho_0(\partial \mathbf{v}_1/\partial t) = \rho_1 \mathbf{g} - \nabla p_1, \tag{52.25}$$

and, from (52.19),

$$(Dp_1/Dt) - a^2(D\rho_1/Dt) = 0, \tag{52.26}$$

which can be rewritten as

$$(p_1 - a^2\rho_1)_{,t} + \mathbf{v}_1 \cdot (\nabla p_0 - a^2 \nabla \rho_0) = 0 \tag{52.27}$$

or

$$\begin{aligned}(\partial p_1/\partial t) &= -\mathbf{v}_1 \cdot \nabla p_0 - a^2 \rho_0 \nabla \cdot \mathbf{v}_1 \\ &= -\mathbf{v}_1 \cdot \nabla p_0 - \Gamma_1 p_0 \nabla \cdot \mathbf{v}_1.\end{aligned} \tag{52.28}$$

WAVE EQUATION

To derive a wave equation we first differentiate (52.25) with respect to time, obtaining

$$\rho_0(\partial^2 \mathbf{v}_1/\partial t^2) = (\partial \rho_1/\partial t)\mathbf{g} - \nabla(\partial p_1/\partial t). \tag{52.29}$$

We then eliminate $(\partial \rho_1/\partial t)$ and $(\partial p_1/\partial t)$ via (52.24) and (52.28); after some simple reductions one finds

$$\begin{aligned}(\partial^2 \mathbf{v}_1/\partial t^2) = {} & a^2\, \nabla(\nabla \cdot \mathbf{v}_1) + (a^2\, \nabla \cdot \mathbf{v}_1)\nabla \ln \Gamma_1 + (\Gamma_1 - 1)(\nabla \cdot \mathbf{v}_1)\mathbf{g} \\ & + \rho_0^{-1}[\nabla(\mathbf{v}_1 \cdot \nabla p_0) - \mathbf{g}(\mathbf{v}_1 \cdot \nabla \rho_0)].\end{aligned} \tag{52.30}$$

In the ambient atmosphere there is a unique relation between p_0 and ρ_0, that is, we can write $\rho_0 = f(p_0)$. Hence $\nabla \rho_0 = (df/dp_0)\nabla p_0 = (df/dp_0)\rho_0 \mathbf{g}$; therefore (52.30) reduces to

$$(\partial^2 \mathbf{v}_1/\partial t^2) = a^2\, \nabla(\nabla \cdot \mathbf{v}_1) + (a^2\, \nabla \cdot \mathbf{v}_1)\nabla \ln \Gamma_1 + (\Gamma_1 - 1)(\nabla \cdot \mathbf{v}_1)\mathbf{g} + \nabla(\mathbf{g} \cdot \mathbf{v}_1), \tag{52.31}$$

which is the fundamental equation governing the propagation of acoustic-gravity waves. For a gas with constant ratio of specific heats, $\Gamma_1 \equiv \gamma$, and (52.31) simplifies to

$$(\partial^2 \mathbf{v}_1/\partial t^2) = a^2\, \nabla(\nabla \cdot \mathbf{v}_1) + (\gamma - 1)(\nabla \cdot \mathbf{v}_1)\mathbf{g} + \nabla \cdot (\mathbf{g} \cdot \mathbf{v}_1), \tag{52.32}$$

an expression first derived by Lamb (**L1,** 555).

In some applications it is more convenient to work with $\mathbf{x}_1$, the displacement of a fluid element from its equilibrium position, instead of its velocity $\mathbf{v}_1$. To first order in small quantities

$$\mathbf{v}_1 = (\partial \mathbf{x}_1/\partial t) \tag{52.33}$$

hence

$$\mathbf{x}_1(\mathbf{x}_0, t) = \int_0^t \mathbf{v}_1(\mathbf{x}_0, t')\, dt'. \tag{52.34}$$

Thus integrating (52.31) with respect to time we find

$$(\partial^2 \mathbf{x}_1/\partial t^2) = a^2\, \nabla(\nabla \cdot \mathbf{x}_1) + (a^2\, \nabla \cdot \mathbf{x}_1)\nabla \ln \Gamma_1 + (\Gamma_1 - 1)(\nabla \cdot \mathbf{x}_1)\mathbf{g} + \nabla \cdot (\mathbf{g} \cdot \mathbf{x}_1). \tag{52.35}$$

The same approach may be used to rewrite (52.24), (52.25), (52.28), and (52.32) in terms of $\mathbf{x}_1$.

WAVE ENERGY DENSITY AND FLUX

To obtain a wave energy equation we multiply (52.54) by p_1/ρ_0, take the dot product of (52.25) with $\mathbf{v}_1$, multiply (52.27) by $p_1/a^2\rho_0$, and add, obtaining

$$\frac{\partial}{\partial t}\left(\frac{1}{2}\rho_0 v_1^2+\frac{1}{2}\frac{p_1^2}{a^2\rho_0}\right)=-\nabla\cdot(p_1\mathbf{v}_1)+\rho_1\mathbf{g}\cdot\mathbf{v}_1-\left(\frac{p_1}{a^2\rho_0}\right)\mathbf{v}_1\cdot\nabla p_0$$
$$=-\nabla\cdot(p_1\mathbf{v}_1)+(w_1 g/a^2)(p_1-a^2\rho_1). \qquad (52.36)$$

In (52.36), w_1 denotes the vertical component of $\mathbf{v}_1$.

Now from (52.6) and (52.27) we have

$$(p_1-a^2\rho_1)_{,t}=-w_1[(dp_0/dz)-a^2(d\rho_0/dz)]=-a^2\rho_0\omega_{BV}^2 w_1/g, \qquad (52.37)$$

which, when integrated with respect to time, yields

$$p_1-a^2\rho_1=-a^2\rho_0\omega_{BV}^2\zeta_1/g, \qquad (52.38)$$

where ζ_1 is the vertical component of the displacement $\mathbf{x}_1$. Hence the last term in (52.36) equals $-\rho_0\omega_{BV}^2\zeta_1(\partial\zeta_1/\partial t)$, whence we see that (52.36) can be rewritten as a conservation law

$$(\partial\varepsilon_w/\partial t)+\nabla\cdot\boldsymbol{\phi}_w=0, \qquad (52.39)$$

where the wave energy density is

$$\varepsilon_w=\tfrac{1}{2}\rho_0 v_1^2+\tfrac{1}{2}(p_1^2/a^2\rho_0)+\tfrac{1}{2}\rho_0\omega_{BV}^2\zeta_1^2, \qquad (52.40)$$

and the wave energy flux is

$$\boldsymbol{\phi}_w=p_1\mathbf{v}_1. \qquad (52.41)$$

We thus obtain the same expression for $\boldsymbol{\phi}_w$ in an acoustic-gravity wave as in a pure acoustic wave [cf. (50.8)]. In contrast, ε_w for acoustic-gravity waves contains a *buoyancy energy density* (or *gravitational energy density*) $\frac{1}{2}\rho_0\omega_{BV}^2\zeta_1^2$ in addition to the kinetic energy and compressional energy densities found previously for pure acoustic waves. The compressional and buoyancy terms both are potential energies for the oscillating fluid parcel. Generalizations of these results to include the effects of magnetic fields for magnetoacoustic-gravity waves are given in (**A2,** 458) and (**B10,** 252).

Although (52.39) is a genuine conservation relation connecting ε_w and $\boldsymbol{\phi}_w$, it is not a complete energy equation because a consistent second-order expansion of the nonlinear energy equation contains, in general, nonzero contributions from other second-order quantities such as ρ_2, p_2, etc. For this reason Eckart (**E1,** 53) called ε_w the *external energy density* of the wave, remarking that it "has little relation to *e*, the 'internal' energy of thermodynamics". While this terminology is often adopted in the literature, we will not use it because the compressional energy term in ε_w does, in fact, originate from the internal energy of the gas [cf. (50.3) to (50.5)]. We merely caution the reader to remember that ε_w and $\boldsymbol{\phi}_w$ do not represent the *total* second-order energy density and flux associated with a wave.

53. *Propagation of Acoustic-Gravity Waves in an Isothermal Medium*

For the special case of a static *isothermal atmosphere* it is possible to describe the properties of acoustic-gravity waves analytically in some detail. While this model of the atmosphere is restrictive, it yields considerable physical insight; moreover it is actually not a bad approximation to the temperature-minimum region joining the upper photosphere and the lower chromosphere in the solar atmosphere (see §54).

For simplicity, assume that the material is a perfect gas with constant specific heats. In the ambient atmosphere we then have

$$p_0(z) = p_0(0)e^{-z/H} \tag{53.1a}$$

and

$$\rho_0(z) = \rho_0(0)e^{-z/H} \tag{53.1b}$$

where the *scale height* is

$$H \equiv RT/g = a^2/\gamma g. \tag{53.2}$$

For the special case we are considering, the pressure and density scale heights are equal, which is not true in general (cf. §54).

THE DISPERSION RELATION

Consider monochromatic plane waves of the general form

$$\mathbf{x}_1 = \mathbf{X} \exp [i(\omega t - \mathbf{k} \cdot \mathbf{x})]. \tag{53.3}$$

We confine attention to steady-state oscillations and therefore take ω to be real. Because the atmosphere is homogeneous in layers perpendicular to the direction of gravity, there is no preferred direction of propagation in the (x, y) plane, hence it suffices to consider propagation in the (x, z) plane only. Thus we take $\mathbf{x}_1 = (\xi_1, 0, \zeta_1)$ or $\mathbf{X} = (X, 0, Z)$ where X and Z are complex constants. Likewise we set $\mathbf{k} = (K_x, 0, K_z)$. In general K_z will be complex because a wave can grow or decay in amplitude as it propagates vertically in a stratified medium; in contrast, the horizontal homogeneity of the ambient atmosphere implies that waves should neither grow nor decay horizontally, hence we can take $K_x \equiv k_x$, a real number.

Substituting (53.3) into the wave equation (52.32), we obtain two homogeneous equations for X and Z, which yield a nontrivial solution only if the determinant of coefficients

$$\begin{vmatrix} a^2k_x^2 - \omega^2 & a^2k_xK_z - igk_x \\ a^2k_xK_z - i(\gamma-1)gk_x & a^2K_z^2 - igK_z - \omega^2 \end{vmatrix} \equiv 0. \tag{53.4}$$

Evaluating (53.4) we obtain the dispersion relation

$$\omega^4 - [a^2(k_x^2 + K_z^2) - i\gamma g K_z]\omega^2 + (\gamma-1)g^2k_x^2 = 0. \tag{53.5}$$

Demanding that the real and imaginary parts of (53.5) both be zero, we find

$$K_z = k_z + (i/2H) \tag{53.6}$$

where k_z is real. From (53.6) and (53.3) it follows that the amplitudes of $\mathbf{x}_1$ and $\mathbf{v}_1$ both grow as $e^{z/2H}$ with increasing height in the atmosphere. Note in passing that this result and (53.1b) imply that the kinetic energy density $\frac{1}{2}\rho_0 v_1^2$ in the wave is constant with height, a point to which we return later.

Using (53.6) we can rewrite (53.5) as

$$\omega^4-[\omega_a^2+a^2(k_x^2+k_z^2)]\omega^2+\omega_g^2 a^2 k_x^2=0 \tag{53.7}$$

where

$$\omega_a \equiv \gamma g/2a = a/2H \tag{53.8}$$

and

$$\omega_g \equiv (\gamma-1)^{1/2} g/a = (\gamma-1)^{1/2} a/\gamma H. \tag{53.9}$$

From (52.6) one sees that ω_g is just the Brunt-Väisälä frequency for an isothermal medium, and is thus relevant to buoyancy oscillations and gravity waves. We will see shortly that ω_a is the minimum frequency for acoustic-wave propagation. Note that

$$\omega_a/\omega_g = \gamma/2(\gamma-1)^{1/2}, \tag{53.10}$$

hence ω_a is always larger than ω_g for the physically relevant range $1 \leq \gamma \leq \frac{5}{3}$.

THE DIAGNOSTIC DIAGRAM

We can determine a great deal about the behavior of acoustic-gravity waves from an analysis of the dispersion relation. Solving (53.7) for k_z we find

$$k_z^2 = a^{-2}(\omega^2-\omega_a^2)-(\omega^2-\omega_g^2)(k_x^2/\omega^2). \tag{53.11}$$

One sees immediately that $k_z^2<0$ if $\omega_g \leq \omega \leq \omega_a$, hence no progressive wave can exist in this frequency range, a result that contrasts strongly with that for a homogeneous medium, where there is no restriction on the frequency of propagating waves. Furthermore, we see that as $\omega \to \infty$, $k^2=(k_x^2+k_z^2) \to \omega^2/a^2$, as in (49.7) for pure acoustic waves; thus waves with $\omega \geq \omega_a$ can be regarded as acoustic waves modified by gravity.

To clarify the picture further, it is helpful to study the domains of wave propagation in the *diagnostic diagram*, a plot of ω versus k_x. We can delineate three distinct domains in the (k_x, ω) plane by finding the *propagation boundary curves* along which $k_z^2=0$; these separate regions of real and imaginary vertical wavenumber. Thus setting $k_z^2=0$ in (53.11) we have

$$\omega^2 = \tfrac{1}{2}\{\omega_a^2+a^2k_x^2 \pm [(\omega_a^2+a^2k_x^2)^2-4a^2k_x^2\omega_g^2]^{1/2}\}, \tag{53.12}$$

which has two branches, as shown in Figure 53.1 for $\gamma=1.4$. Along the propagation boundary curves we also have

$$k_x^2=\omega^2(\omega^2-\omega_a^2)/a^2(\omega^2-\omega_g^2) \tag{53.13}$$

and

$$v_p^2/a^2=(\omega^2-\omega_g^2)/(\omega^2-\omega_a^2), \tag{53.14}$$

where v_p is the phase speed of the wave.

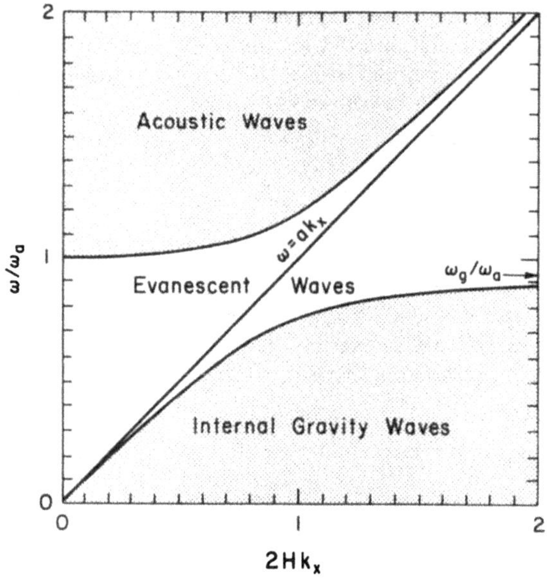

Fig. 53.1 Diagnostic diagram for acoustic-gravity waves in an isothermal atmosphere; $\gamma = 1.4$.

From the remark made above we know that acoustic waves lie in the region above the upper branch of (53.12), and we expect the region below the lower branch to contain gravity waves. Consider first the properties of waves on the upper propagation boundary curve ($\omega \geq \omega_a$). As $k_x \to \infty$, $\omega \to ak_x$, and $v_p \to a$; thus in the high-frequency limit we recover essentially pure acoustic waves, as one would expect because for wavelengths small compared to a scale height the medium is essentially homogeneous (unstratified) over a wavelength. As $k_x \to 0$, $\omega \to \omega_a$, and $v_p \to \infty$. The significance of this result can be appreciated more fully by considering vertically propagating waves.

For vertically propagating waves ($k_x \equiv 0$) we have from (53.11)

$$k_z^2 = (\omega^2 - \omega_a^2)/a^2 \tag{53.15}$$

and

$$v_p^2/a^2 = \omega^2/(\omega^2 - \omega_a^2), \tag{53.16}$$

which show that vertical propagation is possible only if $\omega > \omega_a$; that is, *only* acoustic waves can propagate vertically in a stratified medium. Furthermore, we see that as $\omega \to \omega_a$ there is an *atmospheric resonance* phenomenon, first recognized by Lamb. Because of this resonance, an imposed vertical disturbance at $\omega = \omega_a$ cannot propagate, but rather in effect lifts (or drops) the whole atmosphere *coherently* (which implies that $k_z = 0$ and also

$v_p = \infty$). For vertically propagating acoustic waves, we find from, (53.15) and (53.16), that

$$v_g = (d\omega/dk) = k_z a^2/\omega = a^2/v_p \tag{53.17}$$

or

$$v_p v_g = a^2. \tag{53.18}$$

Notice that because $v_p \to \infty$ as $\omega \to \omega_a$, $v_g \to 0$, hence no energy is propagated by an acoustic disturbance at the atmospheric resonance. Thus ω_a is the *acoustic cutoff frequency, below* which acoustic waves cannot propagate in a stratified atmosphere.

Now consider the lower propagation boundary curve ($\omega \leq \omega_g$). From (53.13) and (53.14) we see that as $k_x \to 0$, $\omega \to (\omega_g/\omega_a)ak_x \to 0$, and $v_p \to (\omega_g/\omega_a)a$. As $k_x \to \infty$, $\omega \to \omega_g$ and $v_p \to 0$. Hence ω_g is a cutoff frequency *above* which gravity waves cannot propagate.

An important property of gravity waves is that they propagate in only a limited range of angles above and below the horizontal. Thus writing $k_x = k \cos \alpha$ we can rearrange (53.7) as

$$\omega^2/a^2k^2 = v_p^2/a^2 = (\omega^2 - \omega_g^2 \cos^2 \alpha)/(\omega^2 - \omega_a^2). \tag{53.19}$$

For an acoustic wave $\omega > \omega_a$, and (53.19) places no restriction on α. But for a gravity wave, $\omega < \omega_g$, the phase velocity is real only if $\omega^2 < \omega_g^2 \cos^2 \alpha$, or if

$$|\alpha| \leq \cos^{-1}(\omega/\omega_g). \tag{53.20}$$

For a fixed k_x, gravity waves can propagate within an ever-increasing range of angles as $\omega \to 0$; but as ω increases toward the lower propagation boundary curve, $k_z \to 0$ hence only horizontal propagation is possible. We can also rewrite (53.7) as

$$\cos^2 \alpha = k_x^2/k^2 = (\omega^2/\omega_g^2) + [\omega^2(\omega_a^2 - \omega^2)/\omega_g^2 a^2 k_x^2], \tag{53.21}$$

which shows that at fixed ω the range of α within which gravity waves can propagate opens from zero on the lower propagation boundary curve to the limiting value $\pm\alpha_{max} = \cos^{-1}(\omega/\omega_g)$ as $k_x \to \infty$.

The region between the two domains of propagation contains *evanescent waves*. Here k_z is imaginary, hence the wave amplitudes grow or decay exponentially with height. For these waves the pressure perturbation p_1 and the vertical component of $\mathbf{v}_1$ are 90° out of phase, hence they have zero vertical energy flux (though they may transport energy horizontally). Evanescent waves have infinite vertical phase velocity, and therefore represent *standing waves*.

The domain into which a wave specified by a particular (k_x, ω) falls depends on γ, g, and T for the atmosphere. For a given γ and g, (53.8) and (53.9) show that both ω_a and ω_g vary as $T^{-1/2}$, hence both the acoustic-wave and gravity-wave cutoffs decrease with increasing temperature. Propagation boundary curves for $\gamma = \frac{5}{3}$, $g = 3 \times 10^4$ (appropriate to the solar

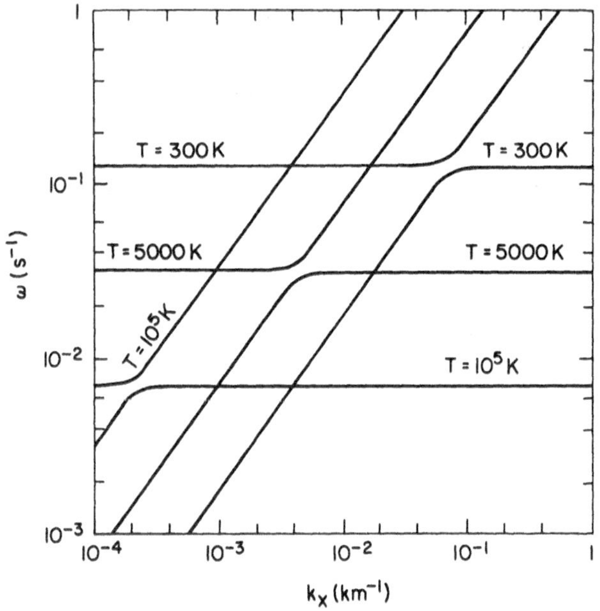

Fig. 53.2 Propagation boundary curves for acoustic-gravity waves in an isothermal atmosphere for several temperatures; $\gamma = \frac{5}{3}$, $g = 3 \times 10^4$.

atmosphere), and several values of T are shown in Figure 53.2. Notice that a wave that can propagate at one temperature may be evanescent at another temperature.

GROUP VELOCITY

Rearranging the dispersion relation as

$$(\omega^2 - \omega_g^2)k_x^2 + \omega^2 k_z^2 = \omega^2(\omega^2 - \omega_a^2)/a^2 \tag{53.22}$$

we see that in wavenumber space the contours of constant ω are conic sections, known as *slowness surfaces.* For acoustic waves, $\omega > \omega_a$, all of the coefficients in (53.22) are positive, hence the contours are ellipses as shown in Figure 53.3a; for gravity waves, $\omega < \omega_g$, the coefficients in (53.22) alternate in sign, hence the contours are hyperbolae, as shown in Figure 53.3b.

The curves shown in Figure 53.3 reveal a great deal about the propagation characteristics of acoustic-gravity waves. When $\omega \gg \omega_a$, the constant-ω ellipses approach the unit circle; as $\omega \to \omega_a$ they shrink to a point at the origin. For all the ellipses $a^2(k_x^2 + k_z^2)/\omega^2 \leq 1$, hence for acoustic waves the phase speed always exceeds the sound speed. As $\omega \to \infty$, $v_p \to a$, and as $\omega \to \omega_a$, $v_p \to \infty$. When $\omega \ll \omega_g$ the constant-ω hyperbolae collapse to the asymptotes $ak_x/\omega = \omega_a/\omega_g$. As $\omega \to \omega_g$ from below, we must have $k_x \gg 1$ to

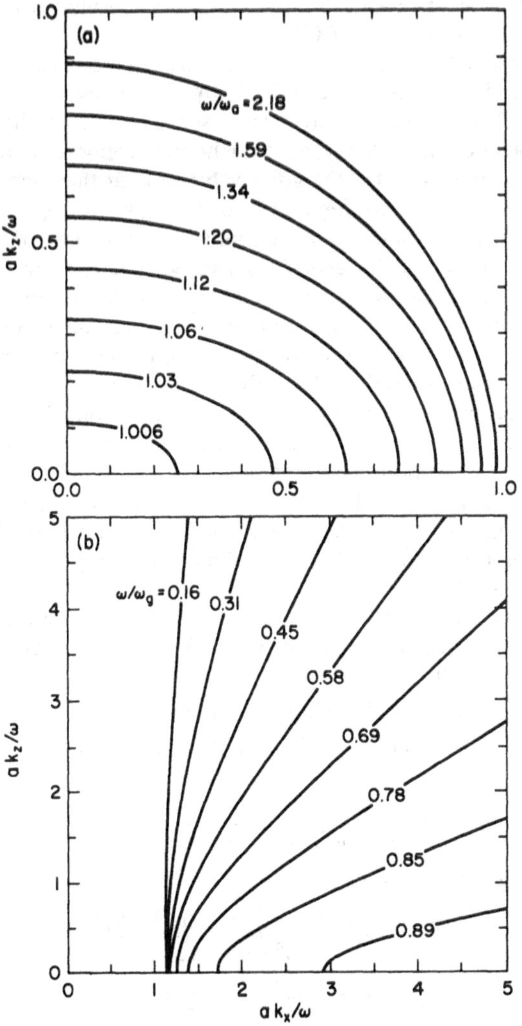

Fig. 53.3 Slowness surfaces in an isothermal atmosphere for (a) acoustic waves, (b) gravity waves.

assure that k_z remains real, hence the vertices of the hyperbolae move toward infinity, and the asymptotes collapse onto the k_x axes. It is evident from Figure 53.3b that $a^2(k_x^2+k_z^2)/\omega^2>1$ for all the hyperbolae, hence for gravity waves the phase speed is always less than the sound speed. Indeed, as $\omega \to 0$, $v_p \to (\omega_g/\omega_a)a$, and as $\omega \to \omega_g$, $v_p \to 0$. Physically the case $\omega = \omega_g$ corresponds to the stationary buoyancy oscillation described in §52, which does not propagate, hence has $v_p = 0$. All of these results for v_p also follow, of course, from an analysis of (53.19).

As we saw in §49, the direction of phase propagation lies along the wave vector $\mathbf{k}$, and is thus characterized by the angle $\alpha = \cos^{-1}(k_x/k) = \tan^{-1}(k_z/k_x)$. Therefore for a wave with a specified (k_x, ω), the direction of phase propagation lies along the straight line connecting the origin in Figure 53.3 to the curve for the given value of ω at the appropriate value of k_x. On the other hand, the group velocity is given by $\mathbf{v}_g = \nabla_k \omega$, hence it is perpendicular to contours of constant ω. Inspection of Figure 53.3 immediately shows that, for acoustic waves, $\mathbf{v}_p$ and $\mathbf{v}_g$ point almost in the same direction, becoming coincident as $\omega \rightarrow \infty$. In contrast, for gravity waves $\mathbf{v}_g$ is usually nearly *perpendicular* to $\mathbf{v}_p$ (becoming exactly perpendicular in the limit $\omega \rightarrow 0$); because horizontal phase and energy propagation are in the same direction, this orthogonality implies that vertical phase and energy propagation are oppositely directed. From Figure 53.3b one sees that as $\omega \rightarrow 0$, phase propagation in a gravity wave is nearly vertical while energy propagation is essentially horizontal; as $\omega \rightarrow \omega_g$, phase propagation becomes more nearly horizontal and energy propagation essentially vertical. Along the $k_z = 0$ axis, corresponding to the lower propagation boundary curve in Figure 53.1, both phase and energy propagation in a gravity wave are horizontal.

We can make the above geometric considerations more quantitative by calculating $(v_g)_i = (\partial\omega/\partial k_i)$ directly from the dispersion relation. Writing $\mathbf{v}_g = (u_g, 0, w_g)$ we find

$$\frac{u_g}{a} = \frac{\omega(\omega^2 - \omega_g^2)ak_x}{\omega^4 - \omega_g^2 a^2 k_x^2} = \frac{(\omega^2 - \omega_g^2)ak_x}{\omega(2\omega^2 - \omega_a^2 - a^2k^2)} \tag{53.23a}$$

and

$$\frac{w_g}{a} = \frac{\omega^3 ak_z}{\omega^4 - \omega_g^2 a^2 k_x^2} = \frac{\omega ak_z}{2\omega^2 - \omega_a^2 - a^2k^2}, \tag{53.23b}$$

which imply a ratio of group speed to sound speed of

$$\frac{v_g}{a} = \frac{\omega[\omega^4 a^2 k^2 + \omega_g^2(\omega_g^2 - 2\omega^2)a^2k_x^2]^{1/2}}{\omega^4 - \omega_g^2 a^2 k_x^2} = \frac{[\omega^4 a^2 k^2 + \omega_g^2(\omega_g^2 - 2\omega^2)a^2k_x^2]^{1/2}}{\omega(2\omega^2 - \omega_a^2 - a^2k^2)}. \tag{53.24}$$

For high-frequency acoustic waves with $\omega \gg \omega_a$ and $\omega^4 \gg \omega_g^2 a^2 k_x^2$ we find

$$u_g/a \approx ak_x/\omega \approx \cos\alpha \tag{53.25a}$$

and

$$w_g/a \approx ak_z/\omega \approx \sin\alpha \tag{53.25b}$$

where we noted that for such waves $\omega \approx ak$. For low-frequency gravity waves with $\omega \ll \omega_g$ and $\omega^4 \ll \omega_g^2 a^2 k_x^2$ we find

$$u_g \approx \omega/k_x \tag{53.26a}$$

and

$$w_g \approx -\omega^3 k_z/\omega_g^2 k_x^2 \approx -\text{sgn}\,(k_z)(\omega/\omega_g)(\omega/k_x) \approx -\text{sgn}\,(k_z)(\omega/\omega_g)u_g \tag{53.26b}$$

because for such waves $|k_z| \approx k \approx (\omega_g/\omega)k_x$. Here $\text{sgn}\,(k_z)$ denotes the algebraic sign of k_z. Note that for $\omega \ll \omega_g$, $|w_g| \ll |u_g|$.

Using the expression for k^2 obtained from (53.19) to eliminate k and k_x from (53.24), we can express v_g in terms of ω, ω_a, ω_g, and α, the angle of phase propagation, as

$$\frac{v_g}{a} = \frac{\{(\omega^2-\omega_a^2)(\omega^2-\omega_g^2\cos^2\alpha)[\omega^4+\omega_g^2(\omega_g^2-2\omega^2)\cos^2\alpha]\}^{1/2}}{\omega^4+\omega_g^2(\omega_a^2-2\omega^2)\cos^2\alpha}. \tag{53.27}$$

Notice that $v_g = 0$ for gravity waves propagating at the limiting angle $\alpha_{\max} = \cos^{-1}(\omega/\omega_g)$.

Alternatively, if β is the angle between $\mathbf{v}_g$ and the horizontal, then $u_g = v_g \cos\alpha$ and $w_g = v_g \sin\alpha$, hence from (53.23)

$$\tan\beta = [\omega^2/(\omega^2-\omega_g^2)]\tan\alpha. \tag{53.28}$$

Using (53.28) in (53.27) one finds

$$\frac{v_g}{a} = \frac{[(\omega^2-\omega_g^2)^3(\omega^2-\omega_a^2)(\omega^2-\omega_g^2\sin^2\beta)]^{1/2}}{\omega[(\omega^2-\omega_g^2)^2+\omega_g^2(\omega_a^2-\omega_g^2)\cos^2\beta]}, \tag{53.29}$$

which shows that energy propagation for gravity waves can occur only for angles less than $\pm\beta_{\max} = \sin^{-1}(\omega/\omega_g)$ away from the horizontal.

POLARIZATION RELATIONS

To obtain a more complete description of an acoustic-gravity wave write the fluctuations in (ρ, p, T) and $\mathbf{v}_1 \equiv (u_1, 0, w_1)$ as

$$\frac{\rho_1}{\rho_0 R} = \frac{p_1}{p_0 P} = \frac{T_1}{T_0 \Theta} = \frac{u_1}{U} = \frac{w_1}{W} = \frac{\xi_1}{X} = \frac{\zeta_1}{Z} = e^{z/2H} e^{i(\omega t - k_x x - k_z z)} \tag{53.30}$$

where the amplitudes R, P, Θ, U, W, X, and Z are complex constants. Substituting these representations into the linearized fluid equations (52.24), (52.25), and (52.27) we obtain

$$i\omega R - ik_x U - [(1/2H) + ik_z]W = 0, \tag{53.31a}$$

$$-ik_x P + i\omega U = 0, \tag{53.31b}$$

$$gR - [(1/2H) + ik_z]P + i\omega W = 0, \tag{53.31c}$$

and

$$-ia^2\omega R + i\omega P + [(\gamma-1)a^2/\gamma H]W = 0. \tag{53.31d}$$

For these equations to have a nontrivial solution, the determinant of the coefficients must be zero; enforcing this requirement we recover the dispersion relation (53.7).

The system (53.31) also yields *polarization relations* that specify the relative amplitudes and phases among the perturbed variables. One finds

$$R = \frac{\omega}{(\omega^2 - a^2k_x^2)}\left\{k_z + \frac{i}{H}\left[\frac{(\gamma-1)a^2k_x^2}{\gamma\omega^2} - \frac{1}{2}\right]\right\}W, \tag{53.32a}$$

$$P = \frac{a^2\omega}{(\omega^2 - a^2k_x^2)}\left[k_z + \frac{i}{H}\left(\frac{\gamma-2}{2\gamma}\right)\right]W, \tag{53.32b}$$

and

$$U = \frac{a^2k_x}{(\omega^2 - a^2k_x^2)}\left[k_z + \frac{i}{H}\left(\frac{\gamma-2}{2\gamma}\right)\right]W; \tag{53.32c}$$

furthermore, from the perfect gas law $(T_1/T_0) = (p_1/p_0) - (\rho_1/\rho_0)$, hence

$$\Theta = \frac{(\gamma-1)\omega}{(\omega^2 - a^2k_x^2)}\left[k_z + \frac{i}{H}\left(\frac{1}{2} - \frac{a^2k_x^2}{\gamma\omega^2}\right)\right]W. \tag{53.32d}$$

Finally, by integrating with respect to time we find $X = U/i\omega$ and $Z = W/i\omega$.

The relative amplitudes of the wave perturbations are thus

$$\left|\frac{\rho_1}{\rho_0}\right| = \frac{\omega a k_z}{|\omega^2 - a^2k_x^2|}\left\{1 + \left(\frac{1}{2k_zH}\right)^2\left[\frac{2(\gamma-1)a^2k_x^2}{\gamma\omega^2} - 1\right]^2\right\}^{1/2}\left|\frac{w_1}{a}\right|, \tag{53.33a}$$

$$\left|\frac{p_1}{p_0}\right| = \frac{\gamma\omega a k_z}{|\omega^2 - a^2k_x^2|}\left[1 + \left(\frac{\gamma-2}{2\gamma}\right)^2\left(\frac{1}{2k_zH}\right)^2\right]^{1/2}\left|\frac{w_1}{a}\right|, \tag{53.33b}$$

$$\left|\frac{u_1}{a}\right| = \frac{a^2k_xk_z}{|\omega^2 - a^2k_x^2|}\left[1 + \left(\frac{\gamma-2}{2\gamma}\right)^2\left(\frac{1}{2k_zH}\right)^2\right]^{1/2}\left|\frac{w_1}{a}\right|, \tag{53.33c}$$

and

$$\left|\frac{T_1}{T_0}\right| = \frac{(\gamma-1)\omega a k_z}{|\omega^2 - a^2k_x^2|}\left[1 + \left(\frac{1}{2k_zH}\right)^2\left(1 - \frac{2a^2k_x^2}{\gamma\omega^2}\right)^2\right]^{1/2}\left|\frac{w_1}{a}\right|. \tag{53.33d}$$

In the high-frequency limit, $\omega \gg \omega_a$, and for nearly vertical phase propagation (so that $\omega \gg ak_x$), $(2k_zH)^{-1} \approx \omega_a/\omega \ll 1$, and the perturbation amplitudes limit to

$$|\rho_1/\rho_0| : |p_1/p_0| : |T_1/T_0| : |u_1/a| : |w_1/a| = \sin\alpha : \gamma\sin\alpha : (\gamma-1)\sin\alpha : \sin\alpha\cos\alpha : 1. \tag{53.34}$$

Here $\sin\alpha = k_z/k$. Notice that the relative amplitudes are the same as for a pure acoustic wave, and do not depend on k or ω. The angular factors merely describe the orientation of $\mathbf{k}$ and have no further significance.

In the low-frequency limit, $\omega \ll \omega_g$ and $\omega^2 \ll a^2k_x^2$, we have $k_z \approx (\omega_g/\omega)k_x$, and the perturbation amplitudes limit to

$$|\rho_1/\rho_0| : |p_1/p_0| : |T_1/T_0| : |u_1/a| : |w_1/a| = (\gamma-1)^{1/2}(\omega_g/\omega) : \gamma\omega_g/ak_x : (\gamma-1)^{1/2}(\omega_g/\omega) : \omega_g/\omega : 1. \tag{53.35}$$

Thus for low-frequency gravity waves $|\rho_1/\rho_0| = |T_1/T_0|$, and we see that $|\rho_1/\rho_0|$, $|T_1/T_0|$, and $|u_1/a|$ can become arbitrarily large relative to $|w_1/a|$ as $\omega \to 0$; in contrast, the fractional pressure fluctuation can be either large or small, depending on the relative size of ak_x and ω_g. Essentially the pressure perturbation drives the horizontal flow and is large for gravity waves with large horizontal wavelengths.

To determine relative phases of the perturbations, note that the phase shift δ_{AB} between any two complex quantities A and B is given by

$$\tan \delta_{AB} = \text{Im}\,(A/B)/\text{Re}\,(A/B). \tag{53.36}$$

In using (53.36) one must be careful about quadrants. A positive (negative) phase shift implies that A leads (lags) B in time by $\delta_{AB}/2\pi$ periods. From (53.36) and (53.32) we find

$$\tan \delta_{PW} = (2k_zH)^{-1}[(\gamma-2)/\gamma], \tag{53.37a}$$

$$\tan \delta_{RW} = (2k_zH)^{-1}\{[2(\gamma-1)a^2k_x^2/\gamma\omega^2]-1\}, \tag{53.37b}$$

and

$$\tan \delta_{\Theta W} = (2k_zH)^{-1}[1-(2a^2k_x^2/\gamma\omega^2)]. \tag{53.37c}$$

Note that for an adiabatic wave $\delta_{PU} \equiv 0$, that is, the pressure and the horizontal velocity fluctuations are always exactly in phase; therefore the horizontal wave energy flux propagates in the same direction as the horizontal phase velocity and is nonzero unless $k_x \equiv 0$.

Now consider the high- and low-frequency limits of (53.37); for definiteness, assume upward propagating waves ($k_z > 0$). For high-frequency acoustic waves in which $\omega^2 \gg a^2k_x^2$ we see from (53.32) that in general $-90° \leq \delta_{RW} \leq 0°$, $-90° \leq \delta_{PW} \leq 0°$, and $0° \leq \delta_{\Theta W} \leq 90°$. For waves with $\omega \gg \omega_a$ and $ak_x \ll \omega$, $(2k_zH)^{-1} \approx \omega_a/\omega \ll 1$, and all the phase shifts are small, of order ω_a/ω. T_1 leads, and ρ_1 and p_1 lag, w_1 slightly. The vertical energy flux propagates in the same direction as the vertical phase velocity. For waves near the propagation boundary curve, where $k_z \to 0$, $\delta_{PW} \to -90°$, and therefore the vertical energy flux vanishes.

For low-frequency gravity waves in which $\omega^2 \ll a^2k_x^2$ and $\omega^2 \ll \omega_g^2$ we see from (53.32) that in general $-180° \leq \delta_{RW} \leq -90°$, $90° \leq \delta_{PW} \leq 180°$, and $90° \leq \delta_{\Theta W} \leq 180°$. When $\omega \ll \omega_g$ the ω^{-2} terms in (53.37) dominate, and we see that $\delta_{RW} \to -90°$ and $\delta_{\Theta W} \to 90°$, in agreement with physical descriptions of buoyancy oscillations given at the beginning of this section. Furthermore, for such waves $2k_zH \approx \omega\omega_a/\omega_g ak_x$, hence $\delta_{PW} \to 180°$, which implies downward energy propagation when there is upward phase propagation, and vice versa.

WAVE ENERGY DENSITY AND FLUX

Using (50.15) we find the time-averaged kinetic energy density in an acoustic-gravity wave is

$$\tfrac{1}{4}\rho_0 e^{z/H}(UU^* + WW^*) = \rho_0(0)(\omega^4 - \omega_g^2a^2k_x^2)WW^*/4\omega^2(\omega^2 - a^2k_x^2), \tag{53.38}$$

the time-averaged compressional energy density is

$$(\rho_0/4a^2)e^{z/H}PP^* = \rho_0(0)(\omega^2-\omega_g^2)WW^*/4(\omega^2-a^2k_x^2), \qquad (53.39)$$

and the time-averaged buoyancy energy density is

$$\tfrac{1}{4}\rho_0\omega_g^2\zeta_1\zeta_1^* = \tfrac{1}{4}\rho_0(0)(\omega_g/\omega)^2WW^*, \qquad (53.40)$$

hence the time-averaged wave energy density is

$$\varepsilon_w = \rho_0(0)(2\omega^2-\omega_a^2-a^2k^2)WW^*/2(\omega^2-a^2k_x^2). \qquad (53.41)$$

Note that the sum of the compressional and buoyancy energies exactly equals the kinetic energy, as one would expect from the virial theorem.

The time-averaged energy flux is

$$(\phi_w)_x = \tfrac{1}{4}(p_1u_1^*+p_1^*u_1) = \rho_0(0)a^2(\omega^2-\omega_g^2)k_xWW^*/2\omega(\omega^2-a^2k_x^2) \qquad (53.42a)$$

and

$$(\phi_w)_z = \tfrac{1}{4}(p_1w_1^*+p_1^*w_1) = \rho_0(0)a^2\omega k_zWW^*/2(\omega^2-a^2k_x^2). \qquad (53.42b)$$

Comparing (53.41) and (53.23) with (53.42) we see that $\boldsymbol{\phi}_w = \varepsilon_w\mathbf{v}_g$, as expected. Notice that all components of ε_w and $\boldsymbol{\phi}_w$ are independent of z, hence the wave energy density and flux of adiabatic waves are constant with height in an isothermal atmosphere.

From (53.38) and (53.39) one sees that

$$(\varepsilon_w)_b = [\omega_g^2(\omega^2-a^2k_x^2)/(\omega^4-\omega_g^2a^2k_x^2)](\varepsilon_w)_k \qquad (53.43a)$$

and

$$(\varepsilon_w)_c = [\omega^2(\omega^2-\omega_g^2)/(\omega^4-\omega_g^2a^2k_x^2)](\varepsilon_w)_k \qquad (53.43b)$$

where the subscripts denote "buoyancy", "compressional", and "kinetic". Thus for acoustic waves, $\omega^2 \gg \omega_g^2$,

$$(\varepsilon_w)_b \to (\omega_g^2/\omega^2)(\varepsilon_w)_k \qquad (53.44a)$$

and

$$(\varepsilon_w)_c \to [1-(\omega_g^2/\omega^2)](\varepsilon_w)_k \qquad (53.44b)$$

for small k_x, where $\omega^2 \gg a^2k_x^2$; whereas for large k_x, where $a^2k_x^2 \approx \omega^2-\omega_a^2$,

$$(\varepsilon_w)_b \to (\omega_a^2\omega_g^2/\omega^4)(\varepsilon_w)_k \qquad (53.45a)$$

and

$$(\varepsilon_w)_c \to [1-(\omega_a^2\omega_g^2/\omega^4)](\varepsilon_w)_k. \qquad (53.45b)$$

For gravity waves, $\omega^2 \ll \omega_g^2$, we find

$$(\varepsilon_w)_b \to [(a^2k_x^2-\omega^2)/a^2k_x^2](\varepsilon_w)_k \qquad (53.46a)$$

and

$$(\varepsilon_w)_c \to (\omega^2/a^2k_x^2)(\varepsilon_w)_k. \qquad (53.46b)$$

Thus for acoustic waves nearly all the potential energy is compressional, whereas for gravity waves it is nearly all gravitational.

54. *Propagation of Acoustic-Gravity Waves in a Stellar Atmosphere*

We now consider the propagation of acoustic-gravity waves in a stratified medium in which the temperature and ionization state of the medium vary with height, for example, the atmospheres of the Sun and other stars. Strong impetus to the development of acoustic-gravity wave theory in astrophysics was given by the discovery of prominent oscillatory motions in the solar atmosphere (**L6**), (**N1**). Later research has shown that the observed motions are the evanescent tails of standing, gravity modified, acoustic eigenmodes, trapped in a subphotospheric resonant cavity. Strictly speaking the theory developed here is not applicable to these trapped modes because we omit discussion of the effects of boundary conditions; even so, it is often instructive to consider them simply as evanescent acoustic-gravity waves that happen to exist for a discrete set of (ω, k_x) combinations. Further motivation for studying acoustic-gravity waves is provided by observations of both propagating and trapped acoustic waves in the solar atmosphere, and by the existence of a substantial nonthermal broadening of solar spectrum lines, much of which is thought to be caused by unresolved, small-amplitude wave motions.

To study acoustic-gravity wave propagation in realistic models of stellar atmospheres we must understand the phenomena of wave refraction and reflection and their implications for wave tunneling and trapping. Furthermore, we must account for ionization effects and temperature gradients in (1) calculation of the sound speed, scale height, and Brunt-Väisälä frequency; (2) formulation of the wave equation; and (3) the expression relating the temperature perturbation to the vertical velocity.

REFLECTION, REFRACTION, TUNNELING, AND TRAPPING

In a nonisothermal atmosphere, propagating waves experience *refraction* and *partial reflection*. If they encounter a semi-infinite region in which they are evanescent, they can be *totally reflected* with no transmission of energy beyond the point of reflection. If, however, the evanescent layer is finite, some fraction of the incident wave energy may leak through the region by *tunneling*, and the wave continues to propagate, with reduced amplitude, on the other side of the barrier. We discuss these phenomena first for discontinuous changes in atmospheric properties at definite boundaries, and then for a slowly varying atmosphere, where we can apply the WKB approximation.

(*a*) *Reflection and Refraction at an Interface* When a wave encounters a discontinuity in material properties, (1) the propagation vector **k** changes in both direction and magnitude; (2) the absolute and relative amplitudes of the wave perturbations change; and (3) phase differences between perturbations may change.

To illustrate the basic properties of refraction and partial reflection we consider a plane, monochromatic, acoustic wave propagating from one homogeneous medium, A, with sound speed a_A, into a second homogeneous medium, B, with sound speed a_B. The interface between the two media is the plane $z=z_1$; A is the region $z\leq z_1$ and B is $z\geq z_1$. Choose the x axis such that $\mathbf{k}=(k_x, 0, k_z)$. Continuity at the interface requires that k_x and ω be the same in both media, whereas k_z, given by the dispersion relation $k_z^2=a^2\omega^2-k_x^2$, changes. Therefore the angle of propagation $\theta=\tan^{-1}(k_x/k_z)$ between $\mathbf{k}$ and the z axis changes. We readily find that

$$\sin\theta_A/\sin\theta_B=(k_x/k_A)/(k_x/k_B)=a_A/a_B. \tag{54.1}$$

Thus if $a_B>a_A$ (i.e., $T_B>T_A$) the wave refracts away from the vertical, and refracts towards it if $a_B<a_A$.

The behavior of a wave incident on a discontinuity follows from continuity conditions at the interface. First, pressure equilibrium dictates that $p_{0A}(z_1)=p_{0B}(z_1)$ and $p_{1A}(z_1)=p_{1B}(z_1)$. The former condition implies a density discontinuity given by $(\rho_{0A}/\rho_{0B})=(T_{0B}\mu_A/T_{0A}\mu_B)$ where μ is the mean molecular weight. Second, w_1 must be continuous to avoid a vacuum or interpenetration of material elements. Third, the energy flux must be continuous because there can be no sources or sinks in an interface of zero thickness. Continuity of p_1, w_1, and $\boldsymbol{\phi}_w$ provide three independent equations that determine the amplitudes of the reflected and refracted waves, and the phase shift of the reflected wave, in terms of the amplitude of the incident wave.

Following (53.30) we write w_1 in medium A as

$$w_{1A}=W_A^+e^{i(\omega t-k_xx)}e^{-ik_{zA}(z-z_0)}+W_A^-e^{i(\omega t-k_xx)}e^{ik_{zA}(z-z_0)}, \tag{54.2}$$

where W_A^+ and W_A^- are complex amplitude functions for the incident and reflected waves, respectively. We set $\exp(z/2H)\equiv 1$ because H is infinite (each medium is homogeneous). Similarly

$$w_{1B}=W_B^+e^{i(\omega t-k_xx)}e^{-ik_{zB}(z-z_1)}. \tag{54.3}$$

The pressure perturbation follows from (53.30) and (53.31c) with $g=0$ and $H=\infty$, namely $p_1^\pm=[\rho_0\omega/(\pm k_z)]W^\pm$. Hence

$$p_{1A}=(\rho_{0A}\omega/k_{zA})(W_A^+-W_A^-) \tag{54.4a}$$

and

$$p_{1B}=(\rho_{0B}\omega/k_{zB})W_B^+. \tag{54.4b}$$

To simplify the notation, write the amplitude functions $W^\pm$ in terms of real (positive) amplitudes and real phases: $W_A^+\equiv A_1e^{i\delta_A^+}$, $W_A^-\equiv A_2e^{i\delta_A^-}$, and $W_B^+\equiv Be^{i\delta_B^+}$. Then factoring W_A^+ out of both w_{1A} and w_{1B}, and defining $A\equiv A_2/A_1$, $B\equiv B_1/A_1$, $\delta_A\equiv\delta_A^-+2k_{zA}(z_1-z_0)$, and $\delta_B\equiv\delta_B^+-\delta_A^-$ we can write

$$w_{1A}(z_1)=A_1e^{i\delta_A^+}e^{i[\omega t-k_xx-k_z(z_1-z_0)]}(1+Ae^{i\delta_A}) \tag{54.5a}$$

and

$$w_{1B}(z_1)=A_1e^{i\delta_A^+}e^{i(\omega t-k_xx)}Be^{i\delta_B}. \tag{54.5b}$$

In terms of the amplitudes and phases just defined and the parameter

$$r\equiv \rho_{0B}k_{zA}/\rho_{0A}k_{zB} \tag{54.6}$$

the three continuity conditions

$$w_{1A}(z_1)=w_{1B}(z_1), \tag{54.7a}$$

$$p_{1A}(z_1)=p_{1B}(z_1), \tag{54.7b}$$

and

$$\tfrac{1}{4}(p_{1A}^*w_{1A}+p_{1A}w_{1A}^*)=\tfrac{1}{4}(p_{1B}^*w_{1B}+p_{1B}w_{1B}^*) \tag{54.7c}$$

can be written

$$e^{-ik_{zA}(z_1-z_0)}(1+Ae^{i\delta_A})=Be^{i\delta_B}, \tag{54.8a}$$

$$e^{-ik_{zA}(z_1-z_0)}(1-Ae^{i\delta_A})=rBe^{i\delta_B}, \tag{54.8b}$$

and

$$1-A^2=rB^2. \tag{54.8c}$$

Multiplying (54.8a) and (54.8b) by their complex conjugates we have

$$1+A^2+2A\cos\delta_A=B^2, \tag{54.9a}$$

$$1+A^2-2A\cos\delta_A=r^2B^2, \tag{54.9b}$$

and

$$1-A^2=rB^2, \tag{54.9c}$$

Solving (54.14) we find that the square of the ratio of the reflected wave amplitude to the incident wave amplitude is

$$A^2=|W_A^-|^2/|W_A^+|^2=(1-r)^2/(1+r)^2<1, \tag{54.10}$$

which is sometimes called the *reflectance* (**L2**). When $r\ll 1$ or $r\gg 1$, $A^2\approx 1$ and the wave is almost totally reflected; as the discontinuity becomes small, $r\to 1$, $A^2\to 0$, and the wave is almost totally transmitted. Using (54.1) and (54.6) we can write the reflectance in terms of material properties and the angle of incidence as

$$A^2=\left[\frac{\rho_{0B}a_A\cos\theta_A-\rho_{0A}(a_A^2-a_B^2\sin^2\theta_A)^{1/2}}{\rho_{0B}a_B\cos\theta_A-\rho_{0A}(a_A^2-a_B^2\sin^2\theta_A)^{1/2}}\right]^2. \tag{54.11}$$

From (54.9), the ratio of the amplitude of the refracted (transmitted) wave to the amplitude of the incident wave is

$$B=|W_B^+|/|W_A^+|=2/(1+r), \tag{54.12}$$

which is greater than or less than unity depending on whether r is less than or greater than unity. This result is not, as it might seem, paradoxical; when the amplitude of w_1 in the transmitted wave exceeds that in the incident wave ($r<1$, $B>1$), there is a compensating decrease in the amplitude of p_1 such that the wave energy fluxes in A and B are exactly equal. In fact, (54.11) and (54.12) imply

$$\phi_{wA} = \tfrac{1}{2}(\omega\rho_{0A}/k_{zA})A_1^2(1-A^2) = (\omega\rho_{0A}/k_{zA})A_1^2[2r/(1+r)^2] \quad (54.13)$$

and

$$\phi_{wB} = \tfrac{1}{2}(\omega\rho_{0B}/k_{zB})A_1^2B^2 = (\omega\rho_{0B}/k_{zB})A_1^2[2/(1+r)^2], \quad (54.14)$$

which are seen to be equal by recalling the definition of r. Note that the flux is very small whenever the discontinuity is large: when $r \gg 1$, $2r/(1+r)^2 \to 2/r \ll 1$, and when $r \ll 1$, $2r/(1+r)^2 \to 2r \ll 1$. Thus little energy is transmitted across the boundary when $a_A \gg a_B$ or $a_B \gg a_A$.

(*b*) *An Interface Between Evanescent and Propagation Regions* The foregoing analysis assumes that the wave is propagating ($k_z^2>0$) in both regions. Consider now the case where the wave is evanescent either in A ($k_{zA}^2<0$) or in B ($k_{zB}^2<0$); the case of evanescence in both is uninteresting. In an evanescent wave, the energy density decreases exponentially with distance into the region of evanescence. The wavenumber k_z is a pure imaginary, hence ik_z is real, and the z-momentum equation shows that p_1 is 90° out of phase with w_1; therefore the wave carries no energy flux.

Suppose first that $k_{zB}^2<0$, and write $\kappa_B = ik_{zB}$. Assume that medium B is semi-infinite so that we have an upward-decaying solution $w_{1B} = W_B^+ \exp[i(\omega t - k_x x)] \exp[-\kappa_B(z-z_1)]$. The equations corresponding to (54.9) then become

$$1 + A^2 + 2A\cos\delta_A = B^2, \quad (54.15a)$$

$$1 + A^2 - 2A\cos\delta_A = r^2B^2, \quad (54.15b)$$

and

$$(\omega\rho_{0A}/k_{zA})A_1^2(1-A^2) = 0, \quad (54.15c)$$

where now

$$r \equiv \rho_{0B}k_{zA}/\rho_{0A}\kappa_B. \quad (54.16)$$

These yield

$$A = 1, \quad (54.17a)$$

$$B^2 = 4/(1+r^2), \quad (54.17b)$$

$$\cos\delta_A = (1-r^2)/(1+r^2), \quad (54.17c)$$

and $\phi_{wA} = \phi_{wB} = 0$. We thus have total reflection in medium A, and excite an evanescent disturbance in medium B.

Suppose now that the wave is evanescent in a finite region A ($k_{zA}^2<0$), but can propagate in medium B, which is semi-infinite. In medium A we can now admit both exponentially decaying and growing solutions. Defining $\kappa_A \equiv ik_{zA}$ we can write w_{1A} with respect to some reference level z_0 as

$$w_{1A} = A_1' e^{i\delta}[e^{-\kappa_A(z-z_0)} + A' e^{\kappa_A(z-z_0)} e^{i\delta_A}] \tag{54.18}$$

or

$$w_{1A}(z_1) = A_1' e^{i\delta}(e^{-\Delta} + A' e^{\Delta} e^{i\delta_A}) \tag{54.19}$$

where $\Delta \equiv \kappa_A(z_1 - z_0)$. To write the continuity conditions at z_1 we define $A_1 \equiv A_1' e^{-\Delta}$, $A \equiv A' e^{2\Delta}$, and $B \equiv B_1/A_1 = B_1/(A_1' e^{-\Delta})$. Then at z_1, $w_{1A}(z_1) = A_1 e^{i\delta}(1 + Ae^{i\delta_A})$ and $w_{1B}(z_1) = A_1 B e^{i\delta_B}$, and continuity of w_1, p_1, and $\boldsymbol{\phi}_w$ imply

$$1 + A^2 + 2A\cos\delta_A = B^2, \tag{54.20a}$$

$$1 + A^2 - 2A\cos\delta_A = r^2 B^2, \tag{54.20b}$$

and

$$2A\sin\delta_A = rB^2 \tag{54.20c}$$

where now

$$r \equiv \rho_{0B}\kappa_A/\rho_{0A}k_{zB}. \tag{54.21}$$

Solving these equations we again find

$$A = 1 \tag{54.22a}$$

and

$$B^2 = 4/(1 + r^2). \tag{54.22b}$$

The phase shifts are now

$$\tan\delta_A = 2r/(1 - r^2) \tag{54.23a}$$

and

$$\tan\delta_B = r. \tag{54.23b}$$

Interference between the growing and decaying evanescent waves in medium A produces a phase lag between w_{1A} and p_{1A} that is not exactly 90°. One finds that the energy flux in region A is

$$\phi_{wA} = (2\omega\rho_{0B}/k_{zB})A_1' e^{-\Delta}/(1 + r^2), \tag{54.24}$$

which decreases exponentially with increasing Δ.

(*c*) *Tunneling* Wave tunneling, which is related to the second case just described, occurs when a wave that is propagating in medium A encounters a finite layer B in which it is evanescent (with growing and decaying solutions), and then emerges into a layer C (perhaps semi-infinite) in which it can again freely propagate ($k_{zc}^2 > 0$). In layer A we have both an incident and reflected wave; in layer C we have only an outward-propagating transmitted wave.

If the interface between A and B is at z_1, and that between B and C is at z_2, the thickness of layer B is $z_2 - z_1$. The vertical velocities in the three regions are

$$w_{1A} = A_1 e^{i(\omega t - k_x x)} e^{ik_{zA}(z-z_0)}(1+Ae^{i\delta_A}) \tag{54.25a}$$

$$w_{1B} = A_1 e^{i(\omega t - k_x x)}[B_1 e^{-\kappa_B(z-z_1)} e^{i\delta_B^+} + B^- e^{\kappa_B(z-z_1)} e^{i\delta_B^-}], \tag{54.25b}$$

and

$$w_{1C} = A_1 e^{i(\omega t - k_x x)} C e^{ik_{zC}(z-z_2)} e^{i\delta_C^+}. \tag{54.25c}$$

The matching conditions at z_1 and z_2 yield six equations in A_1, B_1, B_2, C, δ_A, and $\delta_B \equiv \delta_B^- - \delta_B^+$. At z_1 we have

$$1 + A^2 + 2A \cos \delta_A = B_1^2 + B_2^2 + 2B_1 B_2 \cos \delta_B, \tag{54.26a}$$

$$1 + A^2 - 2A \cos \delta_A = r_1^2(B_1^2 + B_2^2 - B_1 B_2 \cos \delta_B), \tag{54.26b}$$

and

$$1 - A^2 = 2r_1 B_1 B_2 \sin \delta_B. \tag{54.26c}$$

At z_2, defining $b_1 \equiv B_1 e^{-\Delta}$, $b_2 \equiv B_2 e^{\Delta}$, and $\Delta \equiv \kappa_B(z_2 - z_1)$ we have

$$b_1^2 + b_2^2 + 2b_1 b_2 \cos \delta_B = C^2, \tag{54.27a}$$

$$b_1^2 + b_2^2 - 2b_1 b_2 \cos \delta_B = r_2^2 C^2, \tag{54.27b}$$

and

$$2b_1 b_2 \sin \delta_B = r_2 C^2. \tag{54.27c}$$

Here

$$r_1 \equiv \rho_{0B} k_{zA} / \rho_{0A} \kappa_B \tag{54.28a}$$

and

$$r_2 \equiv \rho_{0C} \kappa_B / \rho_{0B} k_{zC}. \tag{54.28b}$$

The solution at z_2 is $b_1 = b_2$, $C = 4b_1/(1+r_2^2)$, $\tan \delta_B = 2r_2/(1-r_2^2)$, and $\tan \delta_C = r_2$, which then gives at z_1:

$$1 + A^2 + 2A \cos \delta_A = b_1^2 K_1, \tag{54.29a}$$

$$1 + A^2 - 2A \cos \delta_A = r_1^2 b_1^2 K_2, \tag{54.29b}$$

and

$$1 - A^2 = 2b_1^2 K_3, \tag{54.29c}$$

where

$$K_1 \equiv 2\{\cosh 2\Delta + [(1-r_2^2)/(1+r_2^2)]\}, \tag{54.30a}$$

$$K_2 \equiv 2\{\cosh 2\Delta - [(1-r_2^2)/(1+r_2^2)]\}, \tag{54.30b}$$

and

$$K_3 \equiv 2r_2/(1+r_2^2). \tag{54.30c}$$

Solving (54.29) we find

$$b_1^2 = 4(K_1 + r_1^2 K_2 + 4r_1 K_3)^{-1} \tag{54.31}$$

and

$$A^2 = \frac{(K_1 + r_1^2 K_2 - 4r_1 K_3)}{(K_1 + r_1^2 K_2 + 4r_1 K_3)} \tag{54.32}$$

or

$$A^2 = \frac{(1+r_1^2)(1+r_2^2)\cosh 2\Delta + (1-r_1^2)(1-r_2^2) - 4r_1r_2}{(1+r_1^2)(1+r_2^2)\cosh 2\Delta + (1-r_1^2)(1-r_2^2) + 4r_1r_2}. \tag{54.33}$$

Note that when $\Delta \to 0$, so that the evanescent region vanishes, $A^2 = (1-r_1r_2)^2/(1+r_1r_2)^2$, which from the definitions of r_1 and r_2 is (54.10) with $r \equiv \rho_{0C}k_{zA}/\rho_{0A}k_{zC}$. As Δ increases, $\cosh 2\Delta$ increases exponentially and A^2 rapidly approaches unity, that is, the total reflection limit.

The energy flux is given by

$$\begin{aligned}\phi_w &= \frac{A_1^2}{2}\left(\frac{\omega\rho_{0A}}{k_{zA}}\right)(1-A^2)\\ &= \frac{\omega\rho_{0A}A_1^2}{k_{zA}}\left[\frac{4r_1r_2}{(1+r_1^2)(1+r_2^2)\cosh 2\Delta + (1-r_1^2)(1-r_2^2) + 4r_1r_2}\right]\end{aligned} \tag{54.34}$$

which diminishes rapidly with increasing Δ because of the factor $\cosh 2\Delta$ in the denominator. For fairly small values of $\Delta = \kappa_B(z_2 - z_1)$, which occurs when the wavelength of the disturbance is large compared to the thickness of the evanescent zone, a nonnegligible fraction of the energy flux can leak through layer B, appearing in C as a propagating wave. This process is closely analogous to quantum mechanical tunneling, and is observed to occur for various types of waves in both the Earth's and the Sun's atmospheres.

The mathematical description of refraction and partial reflection of acoustic-gravity waves incident on a discontinuity is considerably more complicated, and will not be included here as it adds little physical insight.

(*d*) *Trapping* Another interesting effect of atmospheric structure is wave trapping. Here one has a region in which waves can freely propagate, bounded on both top and bottom by layers in which the waves are nonpropagating. The waves are thus totally reflected at the boundaries of the propagating layer, hence two waves with the same ω and k_x having equal but opposite k_z's will interfere destructively unless their phases are such that they form a *standing wave*.

Standing wave conditions are easiest to describe for waves confined in a *cavity* between two rigid boundaries. Consider pure acoustic waves with $k^2 = k_x^2 + k_z^2 = \omega^2/a^2$, and let the distance between the boundaries of the cavity be $D = z_2 - z_1$. At z_1 and z_2 the vertical velocity must vanish, hence we have

$$w_1(z_1) = A_1 e^{i(\omega t - k_x x)} + A_2 e^{i(\omega t - k_x x)} = 0 \tag{54.35a}$$

and

$$w_1(z_2) = A_1 e^{i(\omega t - k_x x)} e^{-ik_z(z_2 - z_1)} + A_2 e^{i(\omega t - k_x x)} e^{ik_z(z_2 - z_1)} = 0. \tag{54.35b}$$

The first condition gives $A_2 = -A_1$. Thus $|A_1| = |A_2|$ and the phase shift between w^+ and w^- is 180°. The second condition gives

$$e^{-ik_zD} - e^{ik_zD} = -2i \sin(k_zD) = 0, \tag{54.36}$$

which implies that $k_z = n\pi/D$, $(n = 0, 1, 2, \ldots)$.

Because k_z is fixed, once ω and k_x are chosen (for a given a), (54.36) shows that standing waves exist only for certain combinations of k_x and ω, namely

$$k_x^2 = (\omega/a)^2 - (n\pi/D)^2. \tag{54.37}$$

Equation (54.37) shows that $n = 0$ corresponds to a horizontally propagating wave. For $n = 1$, $k_z = \pi/D$ and $\Lambda_z = 2D$, hence there is half a wavelength between z_1 and z_2. For $n = 2$ there is exactly one wavelength between z_1 and z_2. In general $\Lambda_z = 2D/n$ or $D = n\Lambda_z/2$, that is, the cavity is spanned by an integral number of half wavelengths.

If we allow the upper boundary to be open, so that $p_1(z_2) = 0$ while the lower boundary remains rigid, then $w_1(z_1) = 0$ again implies $A_1 = -A_2$, while at z_2

$$p_1(z_2) = -ik_zA_1e^{-ik_zD} + ik_zA_2e^{ik_zD} = 0 \tag{54.38}$$

implies

$$e^{-ik_zD} + e^{ik_zD} = 2\cos(k_zD) = 0. \tag{54.39}$$

Therefore

$$k_z = (n + \tfrac{1}{2})\pi/D, \qquad (n = 0, 1, 2, \ldots), \tag{54.40}$$

which implies

$$\Lambda_z = 4D/(2n+1) \tag{54.41}$$

or

$$D = (2n+1)\Lambda_z/4. \tag{54.42}$$

That is, the cavity is spanned by an odd number of quarter wavelengths.

(*e*) *Wave Refraction in a Continuously Varying Medium* Now consider the propagation of an acoustic disturbance driven at frequency ω through a static medium in which the sound speed is a slowly varying function of spatial position, but is constant in time. The frequency of the wave remains unchanged as it propagates (because the oscillation is driven), but in general its amplitude and its wave vector will vary with spatial position.

In the limit that the wavelength of the disturbance is much smaller than the characteristic length over which the medium varies, we can use the language of *geometrical acoustics* (in analogy with geometrical optics) to describe the disturbance as a wave packet moving along a *ray*. The ray is the curve tangent to the propagation vector at each point in the medium, and is thus generated by the differential equation

$$(d\mathbf{x}/ds) = \mathbf{k}/k \equiv \mathbf{n}, \tag{54.43}$$

starting from initial conditions $\mathbf{k}=\mathbf{k}_0$ and $\mathbf{x}=\mathbf{x}_0$ at $s=s_0$. We clearly get a different ray, hence a different $\mathbf{x}(s)$ and $\mathbf{k}(s)$, for each choice of $\mathbf{k}_0$ at a fixed $\mathbf{x}_0$. Thus on a ray we must regard $\mathbf{x}$ and $\mathbf{k}$ as independent (canonical) variables.

Adopting this formalism we write $\omega=\omega(\mathbf{x},\mathbf{k},t)$. But ω is a constant of motion along the ray; therefore

$$\dot{\omega}\equiv(D\omega/Dt)_{\rm ray}=(\partial\omega/\partial t)+\dot{\mathbf{x}}\cdot\nabla\omega+\dot{\mathbf{k}}\cdot\nabla_k\omega\equiv 0, \tag{54.44}$$

where ∇ denotes the gradient with respect to spatial coordinates holding $\mathbf{k}$ fixed, and ∇_k denotes the gradient with respect to wave-vector coordinates holding $\mathbf{x}$ fixed. For a driven wave in a static medium $(\partial\omega/\partial t)\equiv 0$. Furthermore, the velocity of the packet along the ray, $\dot{\mathbf{x}}$, is just the group velocity

$$\dot{\mathbf{x}}=\mathbf{v}_g=\nabla_k\omega=a\mathbf{n}. \tag{54.45}$$

Here we noted that for an acoustic wave $\omega=ak$, and that a depends on $\mathbf{x}$ but not $\mathbf{k}$. From (54.45) and (54.44) we conclude that

$$\dot{\mathbf{k}}=-\nabla\omega, \tag{54.46}$$

which for an acoustic wave yields

$$\dot{\mathbf{k}}=-k\nabla a. \tag{54.47}$$

Thus in a homogeneous medium $\mathbf{k}$ is constant along a ray, as expected. Alternatively

$$\dot{\mathbf{k}}\equiv D(k\mathbf{n})/Dt=k\dot{\mathbf{n}}+\dot{k}\mathbf{n}=k\dot{\mathbf{n}}-k(\dot{a}/a)\mathbf{n} \tag{54.48}$$

because $k=\omega/a$ and $\dot{\omega}\equiv 0$. But

$$\dot{a}=(\partial a/\partial t)+\dot{\mathbf{x}}\cdot\nabla a=a\mathbf{n}\cdot\nabla a \tag{54.49}$$

hence

$$\dot{\mathbf{n}}=-\nabla a+(\mathbf{n}\cdot\nabla a)\mathbf{n} \tag{54.50a}$$

or

$$(d\mathbf{n}/ds)=[-\nabla a+(\mathbf{n}\cdot\nabla a)\mathbf{n}]/a, \tag{54.50b}$$

and

$$\dot{k}=-k\mathbf{n}\cdot\nabla a. \tag{54.51}$$

Equations (54.50) and (54.51) show that if $\mathbf{n}$ lies along ∇a the direction of propagation $\mathbf{n}$ remains unchanged but the magnitude of $\mathbf{k}$ varies. On the other hand, if $\mathbf{n}$ is perpendicular to ∇a, then k is constant but $\mathbf{n}$ rotates away from the direction of ∇a. More generally, writing $G\equiv|\nabla a|$ and G_n for the component of ∇a along $\mathbf{n}$, we have

$$(\nabla a)\cdot d\mathbf{n}=-(ds/a)(G^2-G_n^2)\leq 0. \tag{54.52}$$

Equation (54.52) shows that the change in the direction of $\mathbf{k}$ is always *away* from ∇a; that is, rays always refract away from regions of higher sound speed (i.e., higher temperature) toward regions of lower sound speed (i.e., lower temperature).

The change in wave amplitude can be found from a WKB analysis. For definiteness assume the properties of the medium to vary in z only, and write w_1 in terms of a constant amplitude and the phase functions $\phi(x, t)$ and $\psi(z)$:

$$w_1 = Ae^{i\phi(x,t)}e^{i\psi(z)}. \tag{54.53}$$

For steady-state wave motion in a static medium homogeneous in x, $\phi = \omega t - k_x x$. Pure acoustic waves satisfy the differential equation $(d^2w_1/dz^2) + k^2(z)w_1 = 0$. Acoustic-gravity waves satisfy

$$(d^2W/dz^2) + k^2(z)W = 0, \tag{54.54}$$

where $w_1 = W \exp(\int dz/2H)$. Using (54.53) in (54.54) we find

$$i\psi''(z) - [\psi'(z)]^2 + k^2(z) = 0, \tag{54.55}$$

where primes denote differentiation with respect to z.

A first approximation invoking the assumption of slow variation is obtained by setting $\psi'' = 0$, whence $\psi(z) \approx \pm\int k(z')\,dz'$. Using this value to estimate ψ'' in (54.55) we then have

$$[\psi'(z)]^2 \approx k^2(z) \pm ik'(z) \tag{54.56}$$

and thus

$$\psi(z) \approx \pm\int k(z')\{1 \pm i[k'(z')/k^2(z')]\}^{1/2}\,dz'. \tag{54.57}$$

Because we assume slow variations, $|k'(z)/k^2(z)| \ll 1$, hence $[1 \pm i(k'/k^2)]^{1/2} \approx 1 \pm i(k'/2k^2)$, thus

$$\psi(z) \approx \pm\int k(z')\,dz' + \tfrac{1}{2}i \ln k(z). \tag{54.58}$$

Using (54.58) in (54.53) we find

$$w_1(x, z, t) = \frac{A}{[k(z)]^{1/2}}\,e^{i(\omega t - k_x x)}e^{\pm i\int k(z')dz'} = \frac{A[a(z)]^{1/2}}{\omega^{1/2}}\,e^{i\phi}e^{\pm i\int k(z')dz'} \tag{54.59}$$

for acoustic waves, and

$$w_1 = W \exp\left(\int \frac{dz'}{2H}\right) = e^{\int dz'/2H}\frac{A}{[k(z)]^{1/2}}\,e^{i\phi}e^{\pm i\int k(z')dz'} \tag{54.60}$$

for acoustic-gravity waves.

(f) Wave Reflection in a Continuously Varying Medium We saw above that a pure acoustic wave is totally reflected at a discontinuity if the sound speed (i.e., temperature) in the second medium is high enough that $k_{zB}^2 = (\omega/a)^2 - k_x^2 \leq 0$. In a continuously varying medium, the ray path of a pure acoustic wave propagating into regions of ever-increasing temperature

continually bends away from the direction of ∇T until it turns around, that is, until the wave reflects. For a one-dimensional temperature variation this process is symmetric about ∇T (assumed to lie along the z axis), so if at some height z_0 the incoming wave has wave-vector components $[k_x, k_z(z_0)]$, the reflected wave at z_0 has components $[k_x, -k_z(z_0)]$.

In a stratified medium, wave reflection is governed by several effects. First consider gravity-modified acoustic waves for which the dominant terms in the dispersion relation (53.7) are obtained by ignoring buoyancy effects (equivalent to setting $\omega_g = 0$), which yields

$$k_z^2 = (\omega/a)^2 - (1/4H^2) - k_x^2 \tag{54.61}$$

where H is the density scale height. Suppose first the wave is propagating into regions of increasing temperature. With increasing temperature, $(\omega/a)^2$ decreases as T^{-1} but $1/4H^2$ decreases as T^{-2}; thus at sufficiently high temperatures, we tend to recover the pure acoustic limit $k_z^2 = (\omega/a)^2 - k_x^2$ and reflection occurs as described in the preceding paragraph.

On the other hand, suppose a wave at height z_1, whose frequency is not much larger than the local value of the acoustic cutoff frequency $\omega_a(z_1)$, is propagating into regions of decreasing temperature. If, for the moment, we assume that $k_x^2 \ll (1/4H^2)$, then (54.61) simplifies to

$$k_z^2 = (\omega^2 - \omega_a^2)/a^2. \tag{54.62}$$

Now ω_a rises as T^{-1}, hence it is clear that $k_z \to 0$, hence the wave is reflected at some height z_r where $\omega \approx \omega_a(z_r)$. Including the k_x^2 dependence from (54.61), we find that reflection occurs where

$$\omega = [\omega_a^2(z_r) + a^2 k_x^2]^{1/2}, \tag{54.63}$$

that is, at a slightly smaller value of ω_a than given by (54.62). Reflection occurs when $\omega \to \omega_a$ because, as discussed in §53, the wave runs into an atmospheric resonance where it in effect tries to move the whole atmosphere simultaneously ($k_z = 0$), but is unable to overcome the inertia of all that material. More detailed analysis shows that as a wave propagates into regions of decreasing density scale height, p_1 lags farther and farther behind w_1 until they become 90° out of phase and the wave ceases to transport energy; beyond that point the wave appears only as an evanescent disturbance.

Reflection of internal gravity waves ($\omega \lesssim \omega_{BV}$) occurs for different reasons. Writing the dispersion relation as

$$k_z^2 = [(\omega_{BV}/\omega)^2 - 1]k_x^2 + (\omega^2 - \omega_a^2)/a^2 \tag{54.64}$$

we see that the first term on the right-hand side is positive (and dominates for low-frequency waves with $\omega \ll \omega_{BV}$) while the second term is negative because $\omega_a > \omega_{BV}$ [cf. (53.10)]. Thus, as discussed in §53, for a fixed k_x, $k_z^2 \to 0$ as ω increases from very small values to the value set by the lower propagation boundary curve given by (53.12). The maximum frequency attained on this curve is ω_{BV}, which is reached asymptotically as $k_x \to \infty$. Thus in a varying medium, a gravity wave that can propagate with frequency ω at a given height is surely reflected at any height where the local value of $\omega_{BV}(z)$ decreases to ω. Physically this occurs because, as described in §52, ω_{BV} is the natural frequency for pure buoyancy oscillations in which buoyant fluid bobs vertically up and down at the effective free-fall rate. If one attempts to drive the oscillation faster, the fluid cannot fall back to its equilibrium position at the rate it is being driven, and the motion becomes purely evanescent in the vertical direction.

For small k_x, the lower propagation boundary is given by

$$\omega_0 = (\omega_{BV}/\omega_a)ak_x, \tag{54.65}$$

hence gravity waves can be reflected even when $\omega \ll \omega_{BV}$ if they propagate from a region where $\omega < \omega_0$ to one where $\omega = \omega_0$. Because the ratio (ω_{BV}/ω_a) is a slowly varying function (except in an ionizing medium, see below), this type of reflection tends to occur when gravity waves with small k_x propagate into regions of decreasing a.

We noted in §52 that $\omega_{BV}^2 > 0$ only in a stably stratified medium, and that gravity waves cannot exist in a convectively unstable region, where $\omega_{BV}^2 < 0$. It follows that near an interface between stably and unstably stratified regions, ω_{BV}^2 will fall to zero from the value characteristic of the stable region, hence *all* gravity waves will be reflected back into the stable layer, with only evanescent disturbances penetrating into the convective layer.

TEMPERATURE-GRADIENT AND IONIZATION EFFECTS ON THE BRUNT-VÄISÄIÄ FREQUENCY

For a perfect gas in hydrostatic equilibrium, $(d \ln \rho_0/dz) = (d \ln p_0/dz) - (d \ln T_0/dz)$, and $(dp_0/dz) = -\rho_0 g$; using these expressions in (52.6) we find

$$\omega_{BV}^2 = (\gamma - 1)(g/a)^2 + g(d \ln T_0/dz) = \omega_g^2(z) + g(d \ln T_0/dz). \tag{54.66}$$

In (54.66), $\omega_g(z)$ denotes the local value of the buoyancy frequency in the absence of a temperature gradient. We see that if the temperature increases upward ω_{BV} is increased over ω_g and the possibility of gravity-wave propagation is enhanced. If the temperature declines upward, ω_{BV} is smaller than ω_g, and gravity-wave propagation occurs only for a more restricted range of frequencies. Indeed, if (dT/dz) is sufficiently negative, $\omega_{BV}^2 < 0$ and gravity waves are completely suppressed, the atmosphere becoming convectively unstable.

For an ionizing gas we can rewrite (52.6) as

$$\omega_{BV}^2 = g[(1/H) - (1/\Gamma_1 H_p)] \tag{54.67}$$

where Γ_1 is given by (14.19) and H and H_p are, respectively, the density and pressure scale heights:

$$H^{-1} = -(d \ln \rho_0/dz) = H_p^{-1} + (d \ln T_0/dz) - (d \ln \mu/dz) \quad (54.68)$$

and

$$H_p = -(d \ln p_0/dz)^{-1} = p_0/\rho_0 g = \mathcal{R}T/\mu g. \quad (54.69)$$

Ionization effects must be accounted for in the mean molecular weight μ and in $(d \ln \mu/dz)$; for example, for pure hydrogen $\mu = 1/(1+x)$ [cf. (14.32)] where x is the degree of ionization [cf. (14.6)]. Note that the approach of Γ_1 toward unity in an ionization zone can cause the difference between $(1/H)$ and $(1/\Gamma_1 H_p)$ to become very small, thus sharply diminishing ω_{BV}^2 in that region.

An alternative expression for ω_{BV} can be obtained (**T3**) by expanding the density derivatives in (52.4) as $(d\rho/dz) = (\partial\rho/\partial p)_T(dp/dz) + (\partial\rho/\partial T)_p(dT/dz)$ and demanding that $(dp/dz)_{ad}$ inside the element equal (dp/dz) in the ambient atmosphere. One then finds

$$\omega_{BV}^2 = g\beta[(dT/dz)_{ad} - (dT/dz)_{at}], \quad (54.70)$$

where, from (14.24) and (14.27),

$$\beta = -(\partial \ln \rho/\partial T)_p = T^{-1}\{1 + \tfrac{1}{2}x(1-x)[\tfrac{5}{2} + (\varepsilon_H/kT)]\} \quad (54.71)$$

for ionizing hydrogen. Furthermore, we can evaluate $(dT/dz)_{ad}$ as

$$(dT/dz)_{ad} = (\partial T/\partial p)_s(dp/dz)_{ad} = (-\rho g)(T/p)(\Gamma_2 - 1)/\Gamma_2 = -(\Gamma_3 - 1)gT/a^2, \quad (54.72)$$

where a^2 is given by (48.23). All thermodynamic quantities in (54.70) to (54.72) are to be evaluated allowing for ionization effects; for example Γ_1 and $(\Gamma_3 - 1)$ for ionizing hydrogen are given by (14.29) and (14.30).

In an ionizing medium, the acoustic cutoff frequency is again given by $\omega_a = a/2H$ as in (53.3), but now H is defined by (54.68) and a^2 by (48.23). If the gradients in T and μ are small enough to be neglected, then in an ionizing medium $H \approx H_p = a^2/\Gamma_1 g$, whence

$$\omega_{BV}^2 \approx (\Gamma_1 - 1)g^2/a^2. \quad (54.73)$$

The ratio (ω_{BV}/ω_a) in (54.65) is then

$$(\omega_{BV}/\omega_a) \approx 2(\Gamma_1 - 1)^{1/2}/\Gamma_1, \quad (54.74)$$

which varies with height only if Γ_1 varies.

FORMULATION OF THE WAVE EQUATION

To derive a wave equation for acoustic-gravity waves in a general stratified medium, it is convenient to work with scaled variables, as in §53. When the temperature, and therefore the density scale height H, varies with height, the ambient density is given by

$$\rho_0(z) = \rho_0(z_1) \exp\left[-\int_{z_1}^{z} dz'/H(z')\right] \tag{54.75}$$

where z_1 is an arbitrary reference height chosen at a convenient location. Thus defining

$$E(z) \equiv \exp\left[\int_{z_1}^{z} dz'/2H(z')\right] \tag{54.76}$$

we can write $\rho_0(z) = \rho_0(z_1)/E^2(z)$, which is the generalization of (53.16).

In scaling the perturbation variables, we note that k_z is no longer constant with height. We therefore absorb the dependence on k_z into *depth-dependent amplitude functions* defined by

$$\frac{\rho_1}{\rho_0 R(z)} = \frac{p_1}{\rho_0 P(z)} = \frac{T_1}{T_0 \Theta(z)} = \frac{u_1}{U(z)} = \frac{w_1}{W(z)} = e^{i(\omega t - k_x x)} E(z). \tag{54.77}$$

Using $i\omega U(z) = ik_x P(z)$ to eliminate $U(z)$ in favor of $P(z)$, the linearized fluid equations (52.24), (52.25), and (52.27) become

$$i\omega R(z) - (ik_x^2/\omega)P(z) - [(1/2H) - (d/dz)]W(z) = 0, \tag{54.78a}$$

$$gR(z) - [(1/2H) - (d/dz)]P(z) + i\omega W(z) = 0, \tag{54.78b}$$

and

$$i\omega R(z) - (i\omega/a^2)P(z) + \rho_0^{-1}[(d\rho_0/dz) - a^{-2}(dp_0/dz)]W(z) = 0. \tag{54.78c}$$

The coefficient of W in the energy equation (54.78c) reduces to

$$\begin{aligned} \rho_0^{-1}[(d\rho_0/dz) - a^{-2}(dp_0/dz)] &= -(1/H) + a^{-2}(p_0/\rho_0 H_p) \\ &= -[(1/H) - (1/\Gamma_1 H_p)] = -\omega_{BV}^2/g, \end{aligned} \tag{54.79}$$

hence (54.78c) can be rewritten as

$$i\omega R(z) - (i\omega/a^2)P(z) - (\omega_{BV}^2/g)W(z) = 0. \tag{54.78d}$$

The derivatives (dW/dz) and (dP/dz), which are $(-ik_z W)$ and $(-ik_z P)$ in an isothermal medium, now depend on gradients of T_0 and μ, and cannot be written in simpler form.

We eliminate R, first between (54.78a) and (54.78d), and then between (54.78b) and (54.78d) to obtain two equations relating P and W:

$$P(z) = [i\omega a^2/(\omega^2 - a^2 k_x^2)][(\omega_{BV}^2/g) - (1/2H) + (d/dz)]W(z) \tag{54.80a}$$

and

$$W(z) = [i\omega/(\omega^2 - \omega_{BV}^2)][(g/a^2) - (1/2H) + (d/dz)]P(z). \tag{54.80b}$$

From (54.78d) the density amplitude function is then

$$R(z)=[i\omega/(\omega^2-a^2k_x^2)][(\omega_{BV}^2/g)(a^2k_x^2/\omega^2)-(1/2H)+(d/dz)]W(z). \quad (54.81)$$

The temperature perturbation for an ionizing gas can be obtained by first writing $(p_1/p_0)=(\partial \ln p_0/\partial \ln T_0)_\rho(T_1/T_0)+(\partial \ln p_0/\partial \ln \rho_0)_T(\rho_1/\rho_0)$, which is general, and then applying the cyclic relation $(\partial \ln p_0/\partial \ln \rho_0)_T/(\partial \ln p_0/\partial \ln T_0)_\rho=(\partial \ln T_0/\partial \ln \rho_0)_p$ to obtain

$$(T_1/T_0)=(\partial \ln T_0/\partial \ln \rho_0)_p[(\partial \ln \rho_0/\partial \ln p_0)_T(p_1/p_0)-(\rho_1/\rho_0)]. \quad (54.82)$$

Alternatively we can write $p=\rho kT/\mu m_H$, where μ is variable, which implies

$$(\partial \ln \rho_0/\partial \ln T_0)_p \equiv T\beta=1-(\partial \ln \mu/\partial \ln T_0)_p \equiv Q \quad (54.83)$$

[cf. (2.14) and (14.33)], and then using $(\partial \ln \rho_0/\partial \ln p_0)_T=p_0\kappa_T$ [cf. (2.15)] we have

$$T_1/T_0=[\kappa_T p_1-(\rho_1/\rho_0)]/Q \quad (54.84a)$$

or

$$\Theta(z)=[\kappa_T p_0 P(z)-R(z)]/Q. \quad (54.84b)$$

Substituting (54.80a) for P and (54.81) for R, and using (52.10) with T from (54.83) in the form

$$p_0\kappa_T\Gamma_1=Q(\Gamma_3-1)+1 \quad (54.85)$$

we obtain, after some reduction,

$$\Theta(z)=\frac{i\omega_{BV}^2}{gQ\omega}W(z)+\frac{i\omega(\Gamma_3-1)}{(\omega^2-a^2k_x^2)}\left(\frac{\omega_{BV}^2}{g}-\frac{1}{2H}+\frac{d}{dz}\right)W(z). \quad (54.86)$$

Recalling from (54.73) that ω_{BV}^2 is proportional to (Γ_1-1) when gradient terms can be neglected, we see that $\Theta(z)$ becomes small relative to $W(z)$ in ionization zones, where both Γ_1 and Γ_3 approach unity.

Because H and ω_{BV}^2 contain gradients of T and μ, we cannot simplify (54.80) and (54.81) as we did (53.32). Nevertheless (54.80a) and (54.80b) can be combined into a single wave equation for W (or P), namely

$$\begin{aligned}\frac{d^2W}{dz^2}&-\left[\frac{d}{dz}\ln\left(k_x^2-\frac{\omega^2}{a^2}\right)\right]\frac{dW}{dz}\\&+\left[\frac{\omega_{BV}^2k_x^2}{\omega^2}+\frac{\omega^2}{a^2}-k_x^2-\frac{1}{4H^2}+\left(\frac{\omega_{BV}^2}{g}-\frac{1}{2H}\right)\frac{d}{dz}\ln\left(k_x^2-\frac{\omega^2}{a^2}\right)\right.\\&\left.+\frac{d}{dz}\left(\frac{\omega_{BV}^2}{g}-\frac{1}{2H}\right)\right]W=0.\end{aligned} \quad (54.87)$$

In an isothermal medium, (54.87) reduces to $(d^2W/dz^2)+k_z^2W=0$, where k_z^2 is given by (53.11) with ω_g^2 replaced by ω_{BV}^2 and a suitably general expression for ω_a. For a nonisothermal medium we can still obtain the same simple form by defining a new variable ψ as

$$\psi(z)=W(z)\{k_x^2-[\omega/a(z)]^2\}/\{k_x^2-[\omega/a(z_1)]^2\}, \tag{54.88}$$

where z_1 is a convenient reference height. Let us also define

$$h_0(z)=[(\omega^2-\omega_a^2)/a^2]+(\omega_{BV}^2-\omega^2)(k_x^2/\omega^2); \tag{54.89}$$

in the isothermal limit $h_0=k_z^2$ and is constant. Then we find

$$[d^2\psi(z)/dz^2]+h(z)\psi(z)=0 \tag{54.90}$$

where

$$h(z)\equiv h_0(z)+\frac{d}{dz}\left(\frac{\omega_{BV}^2}{g}-\frac{1}{2H}\right)-\left(\frac{\omega_{BV}^2}{g}-\frac{1}{2H}\right)\frac{\omega^2}{a^2K^2}\frac{d\ln a^2}{dz}$$
$$-\left[\frac{3}{4}\left(\frac{\omega^2}{a^2K^2}\right)^2+\frac{\omega^2}{2a^2K^2}\right]\left(\frac{d^2\ln a^2}{dz^2}\right)^2+\frac{\omega^2}{a^2K^2}\frac{d^2\ln a^2}{dz^2}, \tag{54.91}$$

with $K^2\equiv k_x^2-a^2\omega^2$.

SOLUTION OF THE WAVE EQUATION

We have reduced the linearized fluid equations (52.24) to (52.26) to a single, second-order, ordinary differential equation (54.90). This equation can be represented by a difference equation and solved as a two-point, boundary-value problem along lines discussed in §59. In practice, it is important to allow a variable step size Δz in the difference representation of $(d^2\psi/dz^2)$ because the vertical wavelength $2\pi/[h(z)]^{1/2}$ varies substantially over distances comparable to a wavelength, especially for internal gravity waves.

At the lower boundary, we specify a velocity or pressure perturbation that drives the wave or else specify the net upward energy flux; because the problem is linear these conditions are all essentially equivalent. At the upper boundary, the easiest condition to impose is to allow only an outgoing wave. The justification for this condition is that in a stratified atmosphere the velocity amplitudes increase exponentially with height, and we can always place the upper boundary above the height range where we would expect the waves (in a nonlinear treatment) to become highly nonlinear and dissipate, and thus be unable to reflect back into the region of interest.

From the (complex) solution for $\psi(z)$, $W(z)$ can be determined from (54.88) and (dW/dz) can then be obtained from (54.87). The scaled amplitude functions $P(z)$, $R(z)$, and $\Theta(z)$ follow from (54.80a), (54.81), and (54.86). The linear wave perturbations ρ_1, p_1, T_1, u_1, and w_1 as functions of (x, z, t) are then determined from (54.77). All the perturbations are complex variables; we take the physical perturbation to be the real part of the corresponding perturbation variable. Thus at point (x_i, z_i, t_i) in the fluid, the velocity is

$$\mathbf{v}_1(x_i, z_i, t_i) = \{\mathrm{Re}\,[u_1(x_i, z_i, t_i)], 0, \mathrm{Re}\,[w_1(x_i, z_i, t_i)]\}, \tag{54.92}$$

the gas pressure is

$$p(x_i, z_i, t_i) = p_0(z_i) + \mathrm{Re}\,[p_1(x_i, z_i, t_i)], \tag{54.93}$$

and similarly for the other variables.

The magnitudes and phases of the wave perturbations are obtained from the standard formulae for complex variables. From (54.77) one sees that the phase lag between any two variables (say p_1 and w_1) is the same as that between their amplitude functions [i.e., $P(z)$ and $W(z)$]. Moreover one sees that although the phase of each variable changes with x and t, the phase *lag* between two variables is a function of z only. Thus

$$\delta_{PW}(z_i) = \delta_P(x_i, z_i, t_i) - \delta_W(x_i, z_i, t_i) \equiv \delta_P(0, z_i, 0) - \delta_W(0, z_i, 0). \tag{54.94}$$

Because the derivative (dW/dz) in (54.80), (54.81), and (54.86) cannot in general be replaced by $-ik_zW$ we cannot write analytic expressions like (53.32) for the ratios (P/W), (R/W), and (Θ/W) or like (53.37) for phase differences. Instead, these must be computed from the numerical solutions for the amplitude functions. However, in the special case that k_z and H are almost constant, so that we can take $(dW/dz) \approx -ik_zW$ and $H \approx H_p$, we find the phase differences for an ionizing gas are

$$\tan\delta_{PW} \approx (2k_zH)^{-1}[(\Gamma_1-2)/\Gamma_1], \tag{54.95a}$$

$$\tan\delta_{RW} \approx (2k_zH)^{-1}\{[2(\Gamma_1-1)a^2k_x^2/\Gamma_1\omega^2]-1\}, \tag{54.95b}$$

and

$$\tan\delta_{\Theta W} \approx (2k_zH)^{-1}[(\Gamma_1-2)/\Gamma_1] + (k_zH)^{-1}[1-(ak_x/\omega)^2][(\Gamma_1-1)/\Gamma_1(\Gamma_3-1)Q]. \tag{54.95c}$$

STRUCTURE OF THE SOLAR ATMOSPHERE

To illustrate the theory developed above, we discuss the propagation of linear acoustic-gravity waves in the solar atmosphere using a semiempirical model derived from an analysis of spectral data (**V2**), (**V3**). The model is approximately in hydrostatic equilibrium, but small adjustments are necessary to match observed scale heights. The required nongravitational forces are usually *parameterized* in terms of a "turbulent pressure" gradient; the ultimate origin of these forces is presently unknown but presumably they result from small-scale fluid flow and magnetic fields.

The temperature structure, shown in Figure 54.1, exhibits an initial decline as implied by radiative equilibrium at an open boundary (cf. §82). This region, known as the *photosphere*, contains the surface (about one photon mean free path into the Sun) from which most of the visible light is emitted. Moreover, this is the region where radiation interacts strongly

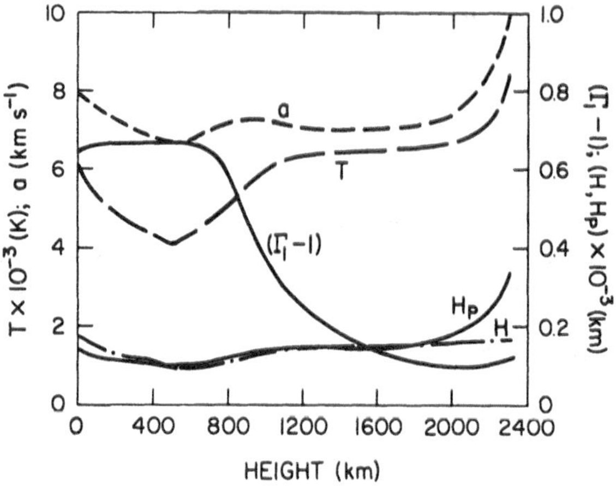

Fig. 54.1 Atmospheric structural parameters in a model solar atmosphere.

with wave-induced local temperature perturbations (cf. §102). At about 500 km above the visible surface, the temperature passes through the *temperature minimum region* and then rises outward in the *chromosphere* where the cores of strong spectral lines in the solar spectrum are formed. The temperature rise is thought to result from dissipation of mechanical (i.e., wave) and magnetic energy. The initial rise is followed by a plateau from about 1000 to 2000 km; here nonradiative energy input continues to increase the internal energy of the gas, but nearly all this energy is consumed in ionizing hydrogen, and the temperature rises only slightly. Above the plateau, the temperature rises abruptly through the *transition region* (whose thermal structure is determined by a balance between radiative losses, nonradiative energy dissipation, and thermal conduction) into the *corona*, a tenuous envelope at about 1.5×10^6 K, which is the seat of the *solar wind* (cf. §§61 and 62).

The propagation of acoustic-gravity waves, including their refraction and reflection properties, is governed by the average values and gradients of the temperature, mean molecular weight, and ionization fraction, which determine Γ_1, Γ_2, and Γ_3, H and H_p, and the parameters a^2, ω_a^2, and ω_{BV}^2 that appear in the wave equation and dispersion relation.

As shown in Figure 54.1, the adiabatic exponents are all near $\frac{5}{3}$ in the photosphere and begin to decrease a short distance above the temperature minimum, as hydrogen begins to ionize. They continue to decrease until the top of the temperature plateau near 2200 km, at which point they rise sharply back toward $\frac{5}{3}$ as hydrogen ionization becomes essentially complete.

The sound speed, also shown in Figure 54.1, exhibits relatively little variation over the height range 0 to 2000 km; a varies only as $T^{1/2}$ and

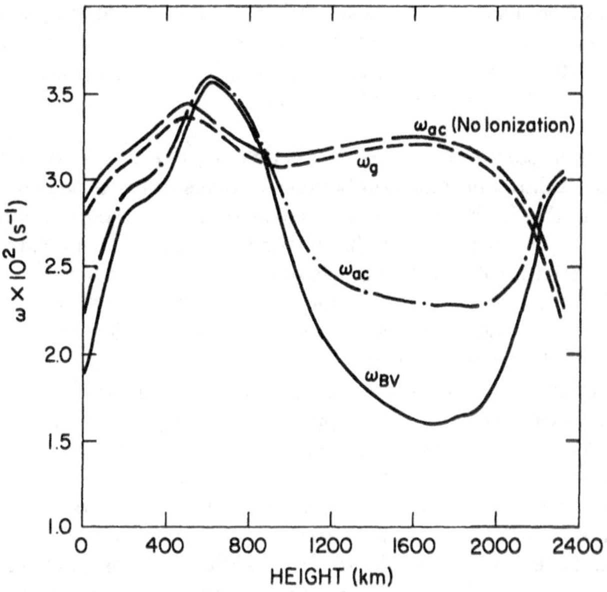

Fig. 54.2 *Upper curves:* acoustic cutoff frequency and buoyancy frequency in a model solar atmosphere, assuming $\Gamma_1 \equiv \frac{5}{3}$. *Lower curves:* acoustic cutoff frequency and Brunt–Väisälä frequency in a model solar atmosphere, allowing for ionization and temperature gradient effects.

even this variation is largely offset by the decrease of Γ_1 in the chromosphere. In the transition region a rises sharply as T increases to coronal values. The acoustic cutoff frequency $\omega_a = a/2H$, shown in Figure 54.2, has a distinct maximum near 750 km where H has a definite minimum and a has a weaker maximum. Above that height ω_a decreases in the chromosphere because H increases as T increases and μ decreases [cf. (54.68)] while a decreases slightly owing to the decrease in Γ_1. In the transition region ω_a decreases sharply $\propto T^{-1/2}$ as T rises to coronal values.

The Brunt-Väisälä frequency exhibits much more dramatic changes, primarily as the result of changes in Γ_1. From Figure 54.1 one sees that H and H_p are nearly equal in most of the chromosphere, hence in this region $\omega_{BV}^2 \approx (\Gamma_1 - 1)g/\Gamma_1 H$ [cf. (54.67)]. Noting that H varies relatively slowly, one infers that ω_{BV} in the chromosphere should respond mainly to changes in $(\Gamma_1 - 1)/\Gamma_1$. The correctness of this inference is seen from Figure 54.2, which shows ω_{BV} calculated with realistic values of Γ_1 and with $\Gamma_1 \equiv \frac{5}{3}$; almost all of the chromospheric drop in ω_{BV} is caused by ionization effects. As hydrogen becomes fully ionized in the upper chromosphere, ω_{BV} rises slightly again and then decreases sharply to very small values in the corona in response to extremely high coronal temperatures. In addition, ω_{BV}

decreases from its maximum near 750 km downward into the photosphere. Indeed, just below the photosphere there is a convection zone that extends deep into the solar envelope; here $\omega_{BV}^2 < 0$, hence ω_{BV} must pass through zero near the bottom of the photosphere.

TEMPERATURE-GRADIENT AND IONIZATION EFFECTS ON THE DIAGNOSTIC DIAGRAM

Figure 54.3 shows propagation boundary curves in the diagnostic diagram for four heights in the model solar atmosphere: (1) in the low photosphere; (2) at 750 km, where ω_a has its maximum value of about $3.4 \times 10^{-2}\ \mathrm{s}^{-1}$ ($\sim$185 s period); (3) at 1700 km where ω_{BV} has a pronounced minimum at about $1.2 \times 10^{-2}\ \mathrm{s}^{-1}$ ($\sim$525 s period); and (4) in a 1.5×10^6 K corona of fully ionized hydrogen and helium. Full allowance is made for temperature gradients and ionization effects.

The two shaded areas on the diagram indicate ranges of (k_x, ω) for which acoustic-gravity waves can be trapped in some region of the solar atmosphere. Region I corresponds to acoustic waves trapped in the *chromospheric cavity* that extends from about 750 km height to the transition region or corona. Acoustic waves with $\omega > 1.6 \times 10^{-2}\ \mathrm{s}^{-1}$ can propagate in the middle to upper chromosphere; those propagating downward are reflected back up if $\omega \lesssim 3.4 \times 10^{-2}\ \mathrm{s}^{-1}$, while those propagating upward refract away from the steep temperature rise to the corona, and totally reflect if they have horizontal wavenumbers $k_x > 10^{-4}\ \mathrm{km}^{-1}$ ($\Lambda_x \lesssim 63{,}000$ km). Note also that acoustic waves with ω slightly less than

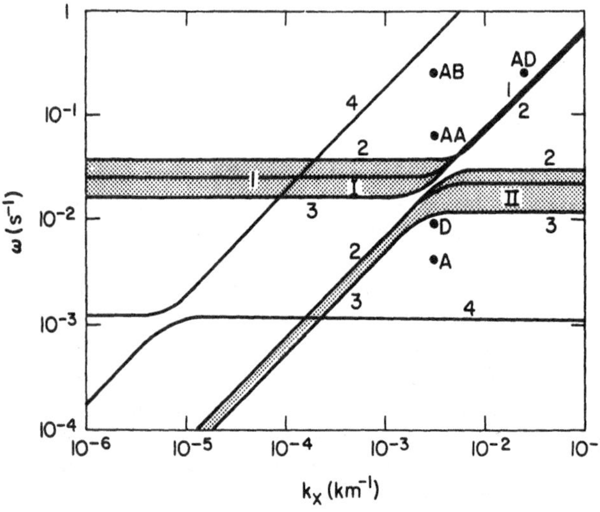

Fig. 54.3 Propagation boundary curves at four heights in a model solar atmosphere. Shaded areas indicate ranges of (k_x, ω) in which waves may be trapped in a cavity. Lettered dots mark (k_x, ω) values for representative waves discussed in text.

$\omega_{a,\max}$ can propagate freely both just above and just below the layer at 750 km where $\omega_a = \omega_{a,\max}$. Thus acoustic waves at these frequencies propagating upward from the subphotospheric convection zone tunnel through a thin layer around 750 km as evanescent waves and penetrate into the chromospheric cavity, where they are trapped, with most of their energy remaining.

Internal gravity waves can be trapped similarly in the photosphere and low-chromosphere region for the (k_x, ω) values shown as region II in Figure 54.3. As mentioned above, ω_{BV} rapidly decreases toward zero near the bottom of the photosphere. Hence, downward propagating gravity waves will be reflected upward at the interface between the stably stratified photosphere and the convection zone. Gravity waves propagating upward in the photosphere having frequencies greater than about $1.2 \times 10^{-2}\,s^{-1}$ will be reflected downward from the middle chromosphere where ω_{BV} drops to a local minimum. Gravity waves with $\omega < 1.2 \times 10^{-2}\,s^{-1}$ and wavenumbers in the small-k_x end of the propagation domain are likewise reflected downward from the chromosphere. Gravity waves with ω only slightly above $1.2 \times 10^{-2}\,s^{-1}$ can tunnel into the upper chromosphere, where they propagate until reflected by the coronal temperature rise. In the corona only gravity waves with periods greater than about two hours can propagate.

ADIABATIC ACOUSTIC-GRAVITY WAVES IN THE SOLAR ATMOSPHERE

Results obtained from numerical solutions of (54.90) in the model solar atmosphere just described exhibit the refraction, partial reflection, and tunneling effects described earlier in this section. They also demonstrate the general trends for phase differences and amplitude ratios of high- and low-frequency wave perturbations as discussed in §53 [cf. (53.33) and (53.37)]. The waves chosen as representative examples are shown as lettered dots in Figure 54.3. They comprise sets of freely propagating gravity waves and acoustic waves, some of which have very small k_z somewhere in the computational domain, plus one case of a fully reflected gravity wave that tunnels through a thin evanescent layer. More extensive results are given in (**M3**) and (**M4**).

One readily sees a strong response to the dominant features of the solar model in the computed height dependence of the eigenfunctions, phase differences, and relative amplitudes of the waves. Phase lags are shown in Figure 54.4 for two of the freely propagating gravity waves, for the reflected gravity wave, and for an acoustic wave with relatively small k_z (long period). The phase lags discriminate readily between the acoustic and gravity-wave regimes and exhibit effects of partial reflection more strikingly than do the other wave properties. From the discussion following (53.73), we expect all the perturbations to be approximately in phase for acoustic waves, with the largest phase differences occurring where ω_a/ω is largest; the magnitudes of the phase differences in Figure 54.4d do in fact

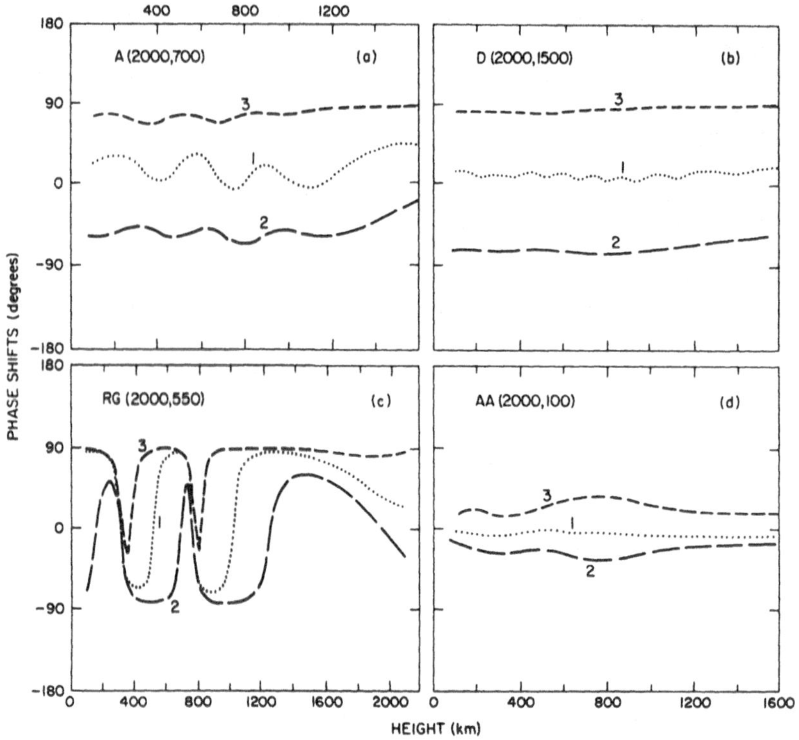

Fig. 54.4 Phase shifts for acoustic-gravity waves in a model solar atmosphere. (1) δ_{PW}. (2) δ_{RW}. (3) δ_{TW}.

follow closely the variations in ω_a shown in Figure 54.2. Also, as expected from (53.37), p_1 and ρ_1 lag w_1 for the acoustic wave, and T_1 leads. The example shown in Figure 54.4d is a relatively low-frequency acoustic wave; at higher frequencies the phase differences are much smaller, as can be seen in the eigenfunctions for a high-frequency acoustic wave shown in Figure 54.5d.

Gravity waves, which experience substantial partial reflection from the large variation of ω_{BV} in the chromosphere, show oscillations in the phase differences on a scale of half the vertical wavelength. These oscillations are an interference pattern that results from the superposition of upward and downward propagating waves. Figures 54.4a,b,c show phase differences for gravity waves that propagate energy upward, hence have negative k_z and propagate phase downward. In this case p_1 slightly leads w_1, while T_1 leads and ρ_1 lags by amounts that approach 90° as ω/ω_{BV} and ω/ak_x become very small. The phase oscillations approach ±90° as reflection becomes nearly complete, which accounts for the extreme behavior seen in Figure 54.4c for a gravity wave that is evanescent between about 1525 and 1750 km in height.

The eigenfunctions for the same three gravity waves and for a high-frequency acoustic wave are shown in Figure 54.5; the relative perturbation amplitudes for this set of waves are shown in Figure 54.6. Here we see other effects of the variation of temperature and ionization in the solar model. The eigenfunctions again show the marked difference in phase behavior between gravity (a, b, c) and acoustic (d) waves.

Figure 54.6d shows that acoustic waves in the solar atmosphere behave very nearly in accordance with the asymptotic (isothermal) relations (53.34). Here $k_z \approx k$, hence $\sin \alpha \approx 1$ and $\cos \alpha$ is very small. One sees that $|u_1|/a \approx (|w_1|/a) \cos \alpha$ is indeed small, and that $|p_1|/p_0 \approx \Gamma\,|w_1|/a$ and $|T_1|/T_0 \approx (\Gamma - 1)^{1/2}\,|w_1|/a$ both appear to have the perfect-gas value of Γ at

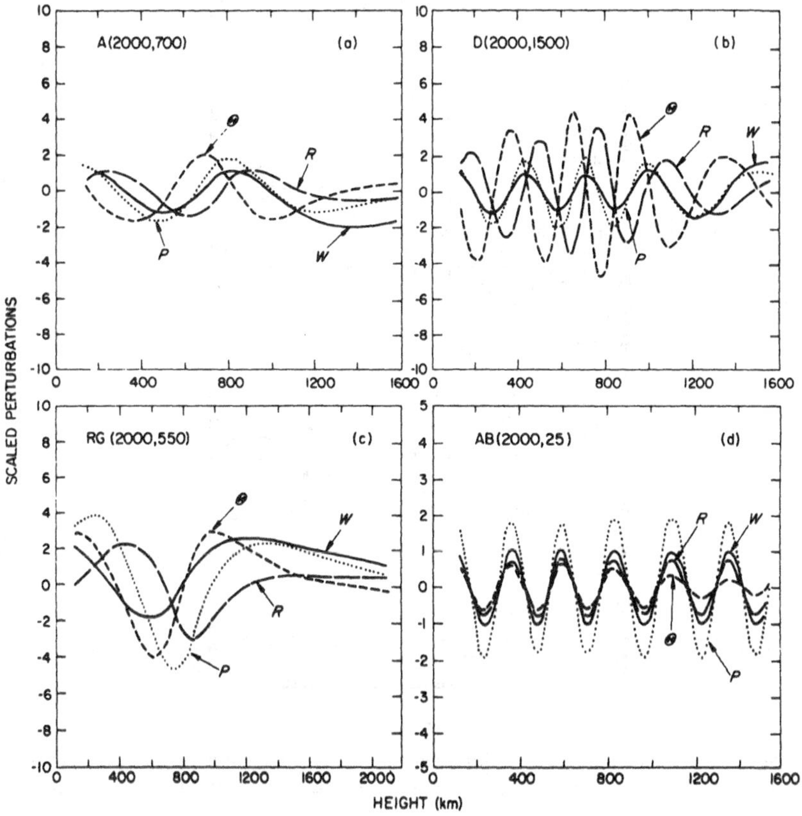

Fig. 54.5 Scaled eigenfunctions for acoustic-gravity waves in a model solar atmosphere.

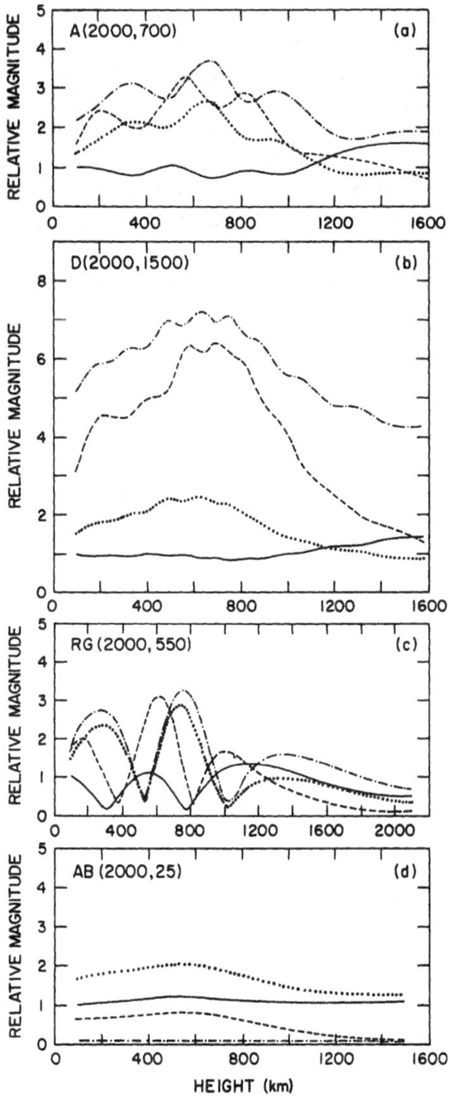

Fig. 54.6 Relative amplitudes for acoustic-gravity waves in a model solar atmosphere. *Solid curves: W. Dot-dash curves: U. Dotted curves: P. Dashed curves:* Θ.

low heights, and to vary in the manner predicted by (53.34) as Γ decreases nearly to unity with increasing height in the chromosphere.

Similarly the amplitude ratios for the gravity waves in Figure 54.6a and 54.6b follow closely the relations (53.35). The ratios $|T_1|/T_0$ and $|u_1|/a$, both of which are proportional to (ω_{BV}/ω), are much larger for the lower-frequency wave D than for the higher-frequency wave A, and both follow the increase in ω_{BV} in the low chromosphere. The ratio $|T_1|/T_0$ drops sharply below $|u_1|/a$ in the middle chromosphere because Γ approaches unity. The ratio $|\rho_1|/\rho_0$, not shown in Figure 54.6, follows $|T_1|/T_0$ very closely for gravity waves; the physical reason is that in gravity waves the density perturbation is produced mainly by the difference in temperature between the adiabatically oscillating fluid and its surroundings. Figure 54.6 contains only gravity waves of a single horizontal wavelength, hence it does not show the dependence of $|p_1|/p_0$ on k_x^{-1} which results from the fact that in gravity waves the pressure perturbation acts mainly to drive horizontal flow; hence as the horizontal extent of the flow increases (Λ_x becomes larger) the pressure perturbation needed to drive it increases.

Figure 54.6c illustrates in another way the characteristics of standing waves: for a perfect standing wave the nodes of $|p_1|$, $|\rho_1|$, and $|T_1|$ would all

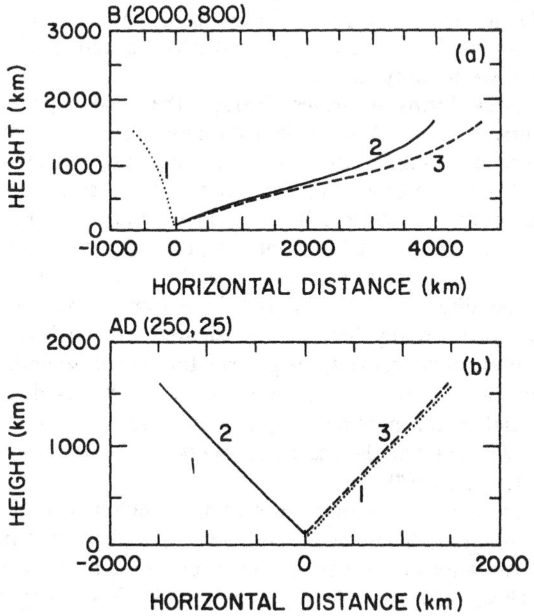

Fig. 54.7 Constant phase path (curve 1), phase-velocity path (curve 2), and group-velocity path (curve 3) for acoustic-gravity waves in a model solar atmosphere.

fall midway between the nodes of $|w_1|$, with two nodes per vertical wavelength, and the amplitude would be exactly zero at each node.

Figure 54.7 illustrates refraction of acoustic-gravity waves in response to the continuous variation of properties of the solar atmosphere. Because the sound speed changes little with height, high-frequency waves show little bending of the direction of the phase velocity $\mathbf{v}_p$ or group velocity $\mathbf{v}_g$. Gravity waves, in contrast, show strong refraction from the large changes in ω_{BV}, with $\mathbf{v}_p$ bending away from the vertical as ω_{BV} decreases, and $\mathbf{v}_g$ bending toward the vertical. Though the direction of $\mathbf{v}_g$ tends toward the vertical, the magnitudes of both u_g and w_g decrease to zero as ω_{BV} decreases to ω.

5.3 Shock Waves

The theory developed in §§5.1 and 5.2 applies only to small-amplitude disturbances, which propagate essentially adiabatically and are damped only slowly by dissipative processes. As the wave amplitude increases, this simple picture breaks down because of the effects of the *nonlinear* terms in the equations of hydrodynamics. When nonlinear phenomena become important, the character of the flow alters markedly. In particular, in an acoustic disturbance a region of compression tends to overrun a rarefaction that precedes it; thus as an acoustic wave propagates, the leading part of the profile progressively steepens, eventually becoming a near discontinuity, which we identify as a *shock*.

Once a shock forms it moves through the fluid supersonically and therefore outruns preshock acoustic disturbances by which adjustments in local fluid properties might otherwise take place; it can therefore persist as a distinct entity in the flow until it is damped by dissipative mechanisms. The material behind a shock is hotter, denser, and has a higher pressure and entropy than the material in front of it; the *stronger* the shock (i.e., the higher its velocity) the more pronounced is the change in material properties across the discontinuity. The rise in entropy across a shock front implies that wave energy has been dissipated irreversibly; this process damps, and ultimately destroys, the propagating shock (sometimes rapidly).

In contrast to acoustic waves, internal gravity waves do not develop shocks. Instead in the nonlinear regime they *break* and degenerate into turbulence. We will not discuss these phenomena in this book; see for example, (**M3**) and (**M4**).

Shock phenomena are of tremendous importance in astrophysics. As we saw in §5.2, the growth of waves to finite amplitude occurs naturally and inevitably in an atmosphere having an exponential density falloff. Thus, as Biermann (**B3**), (**B4**) and Schwarzschild (**S8**) first recognized, small-amplitude acoustic disturbances generated by turbulence in a stellar convection zone can propagate outward with ever-increasing amplitude until they steepen into shocks that dissipate their energy, thus heating the ambient atmosphere. Indeed, this mechanism is thought to provide part of the heating responsible for the outward temperature rise in stellar chromospheres (**A2**, Chaps. 9 and 10), (**B10**, Chap. 7), (**K3**), (**S5**), (**S6**), (**S7**), (**U4**), (**U6**).

Most of the shocks formed by spontaneous growth of randomly generated acoustic waves are rather weak. Much more impressive phenomena are produced in pulsating stars (e.g., Cepheids and RR Lyraes) in which a coherent velocity pulse generated by a radial motion of the entire stellar envelope propagates outward and drives a shock strong enough (1) to alter radically both the thermodynamic properties (e.g., degree of ionization) and the dynamical state (e.g., some layers are lofted outward and subsequently free-fall back) of the atmosphere and (2) to produce interesting spectroscopic phenomena (e.g., emission lines). Even more dramatic are the exceedingly strong shocks, essentially *blast waves*, generated in supernova explosions, which blow away the entire outer envelope of a star.

Similar phenomena also occur in laboratory situations, for example: when a projectile or aircraft moves supersonically through the atmosphere, when a piston is driven rapidly into a tube of gas (a shock tube), in the blast wave produced by a strong explosion, or when rapidly flowing gas encounters a constriction in a flow channel or runs into a wall.

55. The Development of Shocks

Let us now construct a solution of the full nonlinear hydrodynamical equation for a pulse propagating into an infinite homogeneous medium. We assume the flow is one dimensional (along the x axis) and is adiabatic, thus neglecting, for the moment, dissipation. The density, pressure, and velocity of the flow are then completely determined by the continuity and momentum equations and an equation of state $p=p(\rho, s)$ [or $p=p(\rho)$ because s is constant].

In §§48 and 49 we saw that all the physical quantities (ρ, p, T, etc.) and the fluid velocity u in a small-amplitude traveling wave are functions of a single argument $x \pm at$; this implies that any quantity can be expressed as a function of any other [e.g., $p=p(u)$, $\rho=\rho(u)$, etc.] independent of position and time. For finite-amplitude waves, the simple relationships obtained earlier no longer apply. But, as Riemann (**R5**, 157) first showed, it is possible to obtain a general solution of the full nonlinear equations for a traveling wave, in which all physical properties and the fluid velocity are again functions of a single argument $x \pm vt$; but now the propagation speed v of each point on the wave profile is a function of the fluid velocity u at that point in the disturbance. Hence, even in the nonlinear case it is possible to express any physical property of the wave as a function of any other; in particular we can regard the density as a unique function of the fluid velocity.

Thus assuming $\rho = \rho(u)$, we can write the continuity equation as

$$(d\rho/du)(\partial u/\partial t) + [u(d\rho/du) + \rho](\partial u/\partial x) = 0 \tag{55.1a}$$

or

$$(\partial u/\partial t) + [u + \rho(du/d\rho)](\partial u/\partial x) = 0, \tag{55.1b}$$

and the momentum equation as

$$\frac{\partial u}{\partial t} + \left[u + \frac{1}{\rho}\left(\frac{\partial p}{\partial \rho}\right)_s \frac{d\rho}{du}\right]\frac{\partial u}{\partial x} = 0. \tag{55.2}$$

Comparing (55.1b) with (55.2) we see that

$$(du/d\rho) = \pm[(\partial p/\partial \rho)_s]^{1/2}/\rho = \pm a/\rho \tag{55.3}$$

where a is the adiabatic sound speed, regarded here as a function of ρ [recall that $p = p(\rho)$]. Hence, the general relation between the fluid velocity and the density or pressure in the wave is

$$u = \pm \int_{\rho_0}^{\rho} (a/\rho)\, d\rho = \pm \int_{p_0}^{p} dp/\rho a, \tag{55.4}$$

where ρ_0 and p_0 are ambient values in the undisturbed fluid. Note that for a small-amplitude disturbance with $\rho = \rho_0 + \rho_1$, where $|\rho_1|/\rho_0 \ll 1$, (55.4) reduces to $u = \pm a_0 \rho_1/\rho_0$, where a_0 is the sound speed in the undisturbed medium, in agreement with (48.18).

Using (55.3) in (55.1b) or (55.2), we obtain

$$(\partial u/\partial t) + (u \pm a)(\partial u/\partial x) = 0. \tag{55.5}$$

Similarly, by inverting the function $\rho(u)$, we can write the continuity equation as

$$(\partial \rho/\partial t) + [\rho(du/d\rho) + u](\partial \rho/\partial x) = 0, \tag{55.6}$$

which, from (55.3), implies

$$(\partial \rho/\partial t) + (u \pm a)(\partial \rho/\partial x) = 0. \tag{55.7}$$

Equations (55.5) and (55.7) yield general solutions of the form

$$u = F_1[x - (u \pm a)t] \tag{55.8}$$

and

$$\rho = F_2[x - (u \pm a)t] \tag{55.9}$$

where F_1 and F_2 are arbitrary functions that fix the run of u and ρ at $t = 0$.

Equations (55.8) and (55.9) represent traveling waves known as *simple waves*. One sees that a particular value of, say, ρ or u propagates through the ambient medium with phase speed

$$v_p(u) = u \pm a(u) \tag{55.10}$$

where $a(u)$ is given by (55.3) and (55.4). In (55.8) to (55.10), we choose the positive (negative) sign for waves traveling in the positive (negative) x

direction. Because $\rho=\rho(u)$, $p=p[\rho(u)]$, etc., all physical variables in the wave propagate in the same manner as u.

To make the results derived above more concrete, consider a simple wave in a perfect gas. Then $a^2 \propto p/\rho \propto \rho^{\gamma-1}$ implies that $(\gamma-1)(d\rho/\rho)=2(da/a)$, hence (55.4) yields

$$u=\pm 2(a-a_0)/(\gamma-1) \tag{55.11}$$

or

$$a=a_0\pm\tfrac{1}{2}(\gamma-1)u, \tag{55.12}$$

which implies

$$v(u)=\tfrac{1}{2}(\gamma+1)u\pm a_0, \tag{55.13}$$

Using the polytropic gas laws we readily find from (55.12) that

$$\rho=\rho_0[1\pm\tfrac{1}{2}(\gamma-1)(u/a_0)]^{2/(\gamma-1)}, \tag{55.14}$$

$$p=p_0[1\pm\tfrac{1}{2}(\gamma-1)(u/a_0)]^{2\gamma/(\gamma-1)}, \tag{55.15}$$

and

$$T=T_0[1\pm\tfrac{1}{2}(\gamma-1)(u/a_0)]^2. \tag{55.16}$$

Consider a finite pulse having an initial sinusoidal shape as sketched in Figure 55.1, moving to the right. In the small-amplitude limit, we recover the acoustic equations (48.18), (48.24a), and (48.24b) from (55.14) to (55.16), and the disturbance propagates to the right with unchanged shape at speed a_0. But in the finite-amplitude regime, (55.10) to (55.16) plainly show that the more compressed parts of the pulse have a larger fluid velocity u, are hotter, have a higher sound speed $a(u)$, and move to the right with a higher velocity than the less compressed regions. Thus the crest of the pulse continuously gains on the pulse front and, as shown in Figure 55.1, the wave front progressively steepens.

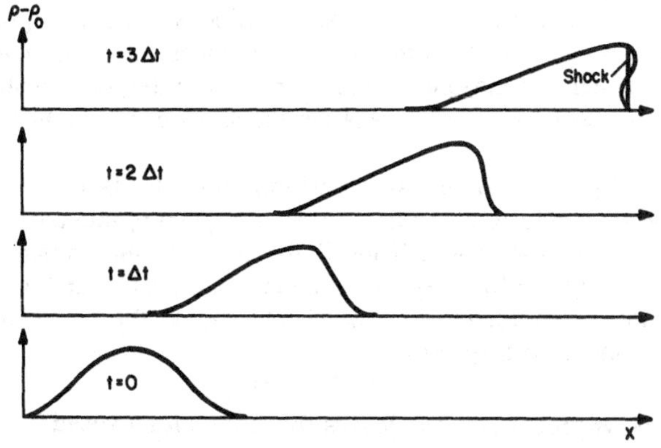

Fig. 55.1 Nonlinear steepening of a simple wave into a shock.

According to (55.12) to (55.16), the wave crest eventually overtakes the pulse front, and at later times the solution becomes multiple valued as sketched in Figure 55.1 for $t = 3\,\Delta t$. This result is unphysical and indicates a break-down of the theory. In reality, the front steepens into a shock, in which all variables change abruptly through a very thin layer, within which, owing to steep gradients, viscosity and thermal conductivity come into play to determine the detailed structure of the front. A continuation of the construction illustrated in Figure 55.1 shows that, as time progresses, the pulse becomes more and more triangular in shape and the velocity amplitude of the shock front decreases monotonically, implying that once a shock forms the wave continuously dissipates energy and is damped. By a similar construction, one can see that in a periodic acoustic wave the crests overrun the troughs, and the wave changes shape from a sinusoid to a train of shocks separated by the wavelength of the original wave. Such a shock train is called a *sawtooth wave* or N *wave*. We discuss the propagation and damping of pulses and sawtooth waves in an exponential atmosphere in §58.

56. Steady Shocks

In laboratory experiments (e.g., flow in a nozzle), it is possible to achieve steady flow, so that if shocks form they are fixed in space and have upstream and downstream properties that are time independent. For such *steady shocks* one can derive analytical expressions relating the values of the physical variables on the upstream and downstream sides of the front. More important, it is usually possible to consider propagating shocks as instantaneously steady because, as shown in §57, the shock thickness is only of the order of a particle mean free path λ, whereas the distance over which the properties of the upstream material can change significantly is some characteristic structural length in the fluid, say a scale height H. Consequently, the ratio of the time required for material to cross the shock front to the time needed for upstream conditions to change appreciably is very small, roughly equal to the Knudsen number $\mathrm{Kn} = \lambda/H$, which is only $\sim 10^{-6}$ in a stellar atmosphere, and even smaller deeper in a star. Therefore, at any instant even a propagating shock is steady to a high degree of approximation.

To simplify the analysis, we transform from the laboratory frame in which the shock moves along the x axis with speed v_s into a frame moving with the shock front itself. Thus if the material into which the shock propagates has a lab-frame speed v_1, and the material behind the shock has a lab-frame speed v_2, then in the shock's frame the upstream material enters the shock with speed

$$u_1 = v_1 - v_s \tag{56.1}$$

and the downstream material leaves the front with a speed

$$u_2 = v_2 - v_s. \tag{56.2}$$

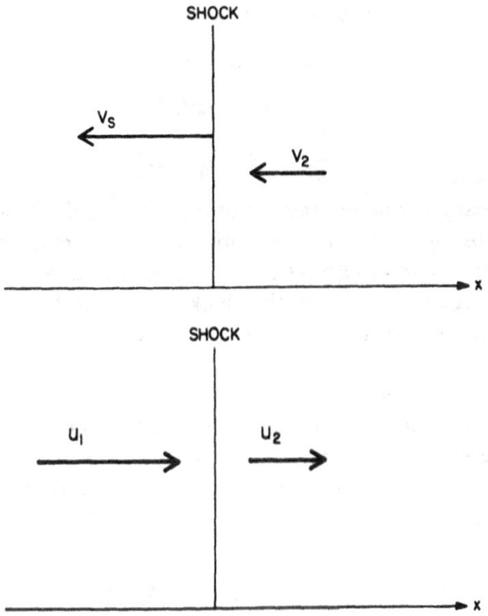

Fig. 56.1 Fluid velocities near shock, measured in lab frame (top), and shock's frame (bottom).

In the case illustrated in Figure 56.1, $v_s < 0$ and $v_1 = 0$, hence $u_1 > 0$, as is u_2.

THE CONSERVATION LAWS

In the frame of the shock the flow is steady, hence the equations of continuity, momentum, and energy (again ignoring viscosity and conduction) reduce to

$$\frac{d(\rho u)}{dx} = 0, \tag{56.3}$$

$$\frac{d}{dx}(\rho u^2 + p) = 0, \tag{56.4}$$

and

$$\frac{d}{dx}[\rho u(h + \tfrac{1}{2}u^2)] = 0, \tag{56.5}$$

which are conservation relations stating that the mass, momentum, and energy fluxes per unit area are constant throughout the flow, and in particular must be constant across the shock front. If we integrate (56.3) to (56.5) across the shock thickness, say from $-\frac{1}{2}\delta$ to $+\frac{1}{2}\delta$, and formally take the limit as $\delta \to 0$ (because $\mathrm{Kn} \ll 1$), we obtain

$$\rho_1 u_1 = \rho_2 u_2 \equiv \dot{m}, \tag{56.6}$$

$$\rho_1 u_1^2 + p_1 = \rho_2 u_2^2 + p_2, \tag{56.7}$$

and

$$h_1 + \tfrac{1}{2} u_1^2 = h_2 + \tfrac{1}{2} u_2^2; \tag{56.8}$$

here all upstream variables have subscript "1" and all downstream variables have subscript "2". The quantity $\dot{m}$ is the mass flux through the shock. Note that these equations remain valid for curved shock fronts (e.g., in a spherical medium) because the thickness of the front is almost always negligible compared to its radius of curvature.

GENERAL JUMP RELATIONS

The conservation relations can be manipulated into other useful forms. Let $V \equiv 1/\rho$ be the specific volume of the material (this notation is inconsistent with that used in Chapter 1, but is adopted here to avoid confusion between volume and velocity). Then $u_1 = \dot{m} V_1$ and $u_2 = \dot{m} V_2$, hence (56.7) gives

$$p_2 - p_1 = \dot{m}^2 (V_1 - V_2), \tag{56.9}$$

which shows that in a (p, V) diagram the initial and final states of the material are connected by a straight line with slope $-\dot{m}^2$ (see Figure 56.2).

Alternatively, using $u_1/u_2 = V_1/V_2$ in (56.7) to eliminate u_1 or u_2 we find

$$u_1^2 = V_1^2 (p_2 - p_1)/(V_1 - V_2) \tag{56.10}$$

and

$$u_2^2 = V_2^2 (p_2 - p_1)/(V_1 - V_2), \tag{56.11}$$

hence

$$u_1^2 - u_2^2 = (p_2 - p_1)(V_1 + V_2). \tag{56.12}$$

Substituting (56.10) and (56.11) into (56.8) we obtain

$$h_2 - h_1 = \tfrac{1}{2}(V_1 + V_2)(p_2 - p_1) \tag{56.13}$$

or

$$e_2 - e_1 = \tfrac{1}{2}(V_1 - V_2)(p_1 + p_2), \tag{56.14}$$

which are known as the *Rankine-Hugoniot* relations (**R2**), (**H17**).

Suppose we are given the upstream conditions ρ_1, p_1, and u_1, and that the material obeys a caloric equation of state $e(\rho, p)$ or $h(\rho, p)$, which may be quite general, including, for example, excitation and ionization effects. Then, in the (p, V) plane, either (56.13) or (56.14) defines a unique curve $p(V) = f(p_1, V_1, V)$, called the *Hugoniot curve*, passing through (p_1, V_1) and having the general shape sketched in Figure 56.2. Equation (56.9) combined with either (56.13) or (56.14) determines p_2 and V_2, hence u_2; the solution is shown graphically in Figure 56.2 as the intersection of the Hugoniot curve with the straight line given by (56.9). In general one finds that (p_2, V_2) differs substantially from (p_1, V_1), the discrepancy between the

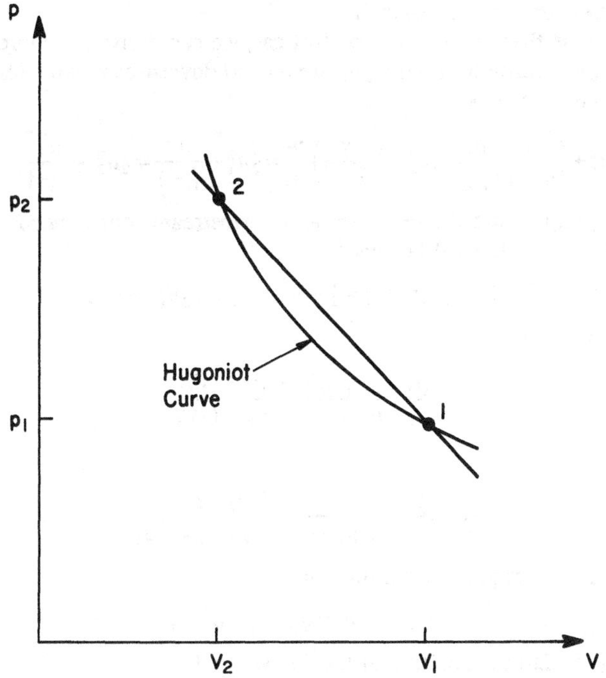

Fig. 56.2 Shock Hugoniot curve in (p, V) diagram.

two growing larger with increasing Mach number. We thus conclude that, to the level of approximation at which we are now working, the equations of hydrodynamics admit a *discontinuous jump* in the physical variables across a shock; such jumps are called *weak solutions* of the fluid equations.

In principle, (56.9) to (56.14) permit two types of solutions: (1) those in which $p_2 > p_1$, $V_2 < V_1$ (or $\rho_2 > \rho_1$), and $u_2 < u_1$, or (2) those in which the inequalities are all reversed. The former are *compression shocks* and the latter are *rarefaction discontinuities*; we see below that while both solutions are permitted mathematically, only compression shocks can exist physically. Rarefactions, when they occur, are always continuous (**C6,** Chap. 3).

If we choose solutions with $p_2 > p_1$, then it follows from (56.13) and (56.14) that $h_2 > h_1$ and $e_2 > e_1$. Further, using the fact that $(u_1 - u_2) = \dot{m}(V_1 - V_2)$ in (56.9), we find

$$u_1 - u_2 = [(p_2 - p_1)(V_1 - V_2)]^{1/2} \tag{56.15}$$

where we chose the positive root on physical grounds. Alternatively from (56.6) and (56.7), we can write

$$u_1 - u_2 = (p_2/\rho_2 u_2) - (p_1/\rho_1 u_1); \tag{56.16}$$

these results will prove useful shortly.

JUMP RELATIONS FOR A PERFECT GAS

If we assume that the fluid is a perfect gas, we can derive a comprehensive set of explicit formulae relating upstream and downstream variables. In this case, (56.8) becomes

$$\tfrac{1}{2}u_1^2+\left(\frac{\gamma}{\gamma-1}\right)\frac{p_1}{\rho_1}=\tfrac{1}{2}u_2^2+\left(\frac{\gamma}{\gamma-1}\right)\frac{p_2}{\rho_2}=\tfrac{1}{2}u_1^2+\frac{a_1^2}{\gamma-1}=\tfrac{1}{2}u_2^2+\frac{a_2^2}{\gamma-1}, \quad (56.17)$$

where a_1 and a_2 are the upstream and downstream sound speeds, respectively. Likewise, (56.13) becomes

$$\gamma(p_2V_2-p_1V_1)=\tfrac{1}{2}(\gamma-1)(V_1+V_2)(p_2-p_1), \quad (56.18)$$

whence we obtain

$$\frac{p_2}{p_1}=\frac{(\gamma+1)V_1-(\gamma-1)V_2}{(\gamma+1)V_2-(\gamma-1)V_1} \quad (56.19)$$

or

$$\frac{V_2}{V_1}=\frac{(\gamma+1)p_1+(\gamma-1)p_2}{(\gamma-1)p_1+(\gamma+1)p_2}=\frac{\rho_1}{\rho_2}=\frac{u_2}{u_1}. \quad (56.20)$$

From the perfect gas law we then have

$$T_2/T_1=p_2V_2/p_1V_1=a_2^2/a_1^2. \quad (56.21)$$

Using (56.20) in (56.10) and (56.11) we find

$$u_1^2=\tfrac{1}{2}V_1[(\gamma-1)p_1+(\gamma+1)p_2] \quad (56.22)$$

and

$$u_2^2=\tfrac{1}{2}V_1[(\gamma+1)p_1+(\gamma-1)p_2]^2/[(\gamma-1)p_1+(\gamma+1)p_2], \quad (56.23)$$

and from (56.22) and $\dot{m}=u_1/V_1$ we obtain

$$\dot{m}^2=[(\gamma-1)p_1+(\gamma+1)p_2]/2V_1. \quad (56.24)$$

In one-dimensional steady flows, it is sometimes convenient to introduce the *critical velocity* u_c at which the flow speed equals the local sound speed. Then, from (56.17),

$$\tfrac{1}{2}u_1^2+\left(\frac{\gamma}{\gamma-1}\right)\frac{p_1}{\rho_1}=\tfrac{1}{2}u_2^2+\left(\frac{\gamma}{\gamma-1}\right)\frac{p_2}{\rho_2}=\frac{(\gamma+1)}{2(\gamma-1)}u_c^2, \quad (56.25)$$

hence

$$u_1+\left(\frac{2\gamma}{\gamma-1}\right)\frac{p_1}{\rho_1u_1}=\left(\frac{\gamma+1}{\gamma-1}\right)\frac{u_c^2}{u_1} \quad (56.26)$$

and

$$u_2+\left(\frac{2\gamma}{\gamma-1}\right)\frac{p_2}{\rho_2u_2}=\left(\frac{\gamma+1}{\gamma-1}\right)\frac{u_c^2}{u_2}, \quad (56.27)$$

whence

$$u_1-u_2+\left(\frac{2\gamma}{\gamma-1}\right)\left(\frac{p_1}{\rho_1 u_1}-\frac{p_2}{\rho_2 u_2}\right)=\left(\frac{\gamma+1}{\gamma-1}\right)\frac{(u_2-u_1)u_c^2}{u_1 u_2}. \tag{56.28}$$

In view of (56.16), (56.28) reduces to the *Prandtl relation*

$$u_1 u_2 = u_c^2. \tag{56.29}$$

The relationship between upstream and downstream flow quantities can be expressed concisely in terms of $\mathscr{p}\equiv\Delta p/p_1=(p_2-p_1)/p_1$, the *fractional pressure jump* across the shock. Thus from (56.20) we find the *compression ratio*

$$\rho_2/\rho_1 = V_1/V_2 = [2\gamma+(\gamma+1)\mathscr{p}]/[2\gamma+(\gamma-1)\mathscr{p}], \tag{56.30}$$

and from (56.21) we have

$$T_2/T_1=(1+\mathscr{p})[2\gamma+(\gamma-1)\mathscr{p}]/[2\gamma+(\gamma+1)\mathscr{p}]. \tag{56.31}$$

Furthermore, from (56.22) we find

$$M_1^2-1=\tfrac{1}{2}(\gamma+1)\mathscr{p}/\gamma \tag{56.32}$$

and from (56.23)

$$M_2^2-1=-(\gamma+1)\mathscr{p}/2\gamma(1+\mathscr{p}), \tag{56.33}$$

whence we see that because $\mathscr{p}\geq 0$, $M_1^2\geq 1$ while $M_2^2\leq 1$. That is, the upstream flow is always supersonic relative to a shock front, and the downstream flow is always subsonic. Note that if $M_1=1$, then $\mathscr{p}=0$ and the jump in all physical quantities vanishes, that is, there is no shock.

For very strong shocks $\mathscr{p}\to\infty$, which implies that

$$\rho_2/\rho_1 \to (\gamma+1)/(\gamma-1) \tag{56.34}$$

and

$$M_2^2\to(\gamma-1)/2\gamma, \tag{56.35}$$

while $M_1^2\to(\gamma+1)\mathscr{p}/2\gamma\to\infty$ and $T_2/T_1\to(\gamma-1)\mathscr{p}/(\gamma+1)\to\infty$. Hence for a monatomic gas, $\gamma=\frac{5}{3}$, the limiting compression ratio in an extremely strong shock is $(\rho_2/\rho_1)_{max}=4$, and the limiting value of the downstream Mach number is $(M_2)_{min}=1/\sqrt{5}$.

For weak shocks (i.e., $\mathscr{p}\ll 1$), we have

$$(\rho_2/\rho_1)-1\approx\mathscr{p}/\gamma, \tag{56.36}$$

$$(T_2/T_1)-1\approx(\gamma-1)\mathscr{p}/\gamma, \tag{56.37}$$

$$M_1^2-1\approx\tfrac{1}{2}(\gamma+1)\mathscr{p}/\gamma, \tag{56.38}$$

and

$$M_2^2-1\approx-\tfrac{1}{2}(\gamma+1)\mathscr{p}/\gamma. \tag{56.39}$$

Equations (56.36) and (56.37) are merely linear expansions of the polytropic relations between ρ, p, and T, and show that to first order in $\mathscr{p}$ weak shocks are essentially adiabatic [but see (56.51) and (56.56)].

Upstream and downstream flow properties can also be related in terms of the upstream Mach number M_1. Thus from (56.32) we find

$$p_2/p_1 = [2\gamma M_1^2 - (\gamma - 1)]/(\gamma + 1), \tag{56.40}$$

hence from (56.20)

$$\rho_2/\rho_1 = (\gamma + 1)M_1^2/[(\gamma - 1)M_1^2 + 2] = u_1/u_2, \tag{56.41}$$

and from (56.21)

$$T_2/T_1 = [2\gamma M_1^2 - (\gamma - 1)][(\gamma - 1)M_1^2 + 2]/(\gamma + 1)^2 M_1^2. \tag{56.42}$$

Then using (56.41) and (56.42) in $M_2^2 = (u_2/a_2)^2 = M_1^2(u_2/u_1)^2(a_1/a_2)^2$ we find

$$M_2^2 = [(\gamma - 1)M_1^2 + 2]/[2\gamma M_1^2 - (\gamma - 1)]. \tag{56.43}$$

For strong shocks $M_1 \to \infty$ and we obtain the same limiting values for ρ_2/ρ_1 and M_2 stated above, while $p_2/p_1 \to 2\gamma M_1^2/(\gamma+1) \to \infty$ and $T_2/T_1 \to 2\gamma(\gamma-1)M_1^2/(\gamma+1)^2 \to \infty$.

For weak shocks with $M_1^2 = 1 + m$, $m \ll 1$, we find

$$(p_2/p_1) - 1 = 2\gamma m/(\gamma + 1) \tag{56.44}$$

$$(\rho_2/\rho_1) - 1 = 2m/(\gamma + 1) \tag{56.45}$$

$$(T_2/T_1) - 1 = 2(\gamma - 1)m/(\gamma + 1) \tag{56.46}$$

and

$$M_2^2 = 1 - m. \tag{56.47}$$

THE ENTROPY JUMP

In the (p, V) diagram, the adiabats form a one-parameter family of curves $p = p(V, s)$ where the specific entropy s is fixed along each curve. In contrast, the Hugoniot curves form a *two*-parameter family, with the curve passing through (p_1, V_1) having the form $p(V, p_1, V_1)$. In general, the Hugoniot through (p_1, V_1) is not identical with the adiabat through (p_1, V_1)—a fact demonstrated below for weak shocks in general materials, and for shocks of arbitrary strengths in a perfect gas. Thus in general Hugoniots *cross* adiabats, which implies that the entropy of the material changes as it passes through a shock; therefore the entropy experiences a discrete jump across the front by an amount determined by the shock strength.

According to the second law of thermodynamics, the entropy of a substance cannot be decreased by internal processes alone (cf. §3); thus the downstream specific entropy in a shock must equal or exceed its upstream value, $s_2 \geq s_1$. This entropy increase, predicted by the mass, momentum, and energy conservation relations alone, implies an irreversible dissipation of energy, even for an ideal fluid, entirely independently of the existence of a dissipation mechanism, which at first sight seems paradoxical. This apparent paradox is easily resolved by studying shock structure for a real gas (§57). We then find that a shock is not a mathematical discontinuity, but is actually a thin transition layer, a few particle mean free paths λ thick, where dissipative mechanisms (which generate entropy) are strongly operative in response to steep gradients. The ideal fluid is merely a degenerate case obtained when we suppress the internal transport properties of a fluid, which is equivalent to letting $\lambda \rightarrow 0$, which in turn implies that the transition layer collapses to a discontinuity. We get the same total entropy jump for given upstream conditions in both cases because the entropy, like any other thermodynamic variable, can be regarded as a function of any two other variables, say (p, V). As we have seen, the downstream values of these variables are uniquely fixed by the hydrodynamical equations alone, regardless of the detailed physical properties of the fluid.

Consider first a weak shock, and examine the implications of (56.13). Take $h = h(p, s)$ and expand in powers of $\Delta p \equiv p_2 - p_1$ and $\Delta s \equiv s_2 - s_1$. Anticipating the result that Δs is $O(\Delta p^3)$, we retain only first-order terms in Δs and terms up to third order in Δp, obtaining

$$h_2 - h_1 = (\partial h/\partial s)_p\, \Delta s + (\partial h/\partial p)_s\, \Delta p + \tfrac{1}{2}(\partial^2 h/\partial p^2)_s\, \Delta p^2 + \tfrac{1}{6}(\partial^3 h/\partial p^3)_s\, \Delta p^3. \tag{56.48}$$

From (2.33) we have $(\partial h/\partial s)_p = T$ and $(\partial h/\partial p)_s = 1/\rho = V$, hence

$$h_2 - h_1 = T_1\, \Delta s + V_1\, \Delta p + \tfrac{1}{2}(\partial V/\partial p)_s\, \Delta p^2 + \tfrac{1}{6}(\partial^2 V/\partial p^2)_s\, \Delta p^3. \tag{56.49}$$

Similarly, take $V = V(p, s)$; inasmuch as $(V_1 + V_2)$ in (56.13) is already multiplied by Δp we can expand V to only second order in Δp, and omit the term in Δs, obtaining

$$V_2 = V_1 + (\partial V/\partial p)_s\, \Delta p + \tfrac{1}{2}(\partial^2 V/\partial p^2)_s\, \Delta p^2. \tag{56.50}$$

Substituting (56.49) and (56.50) into (56.13) we find

$$s_2 - s_1 = \tfrac{1}{12}(\partial^2 V/\partial p^2)_s (p_2 - p_1)^3/T_1. \tag{56.51}$$

From (56.51) we see that Δs is nonzero (unless $\Delta p \equiv 0$), and the requirement that $\Delta s > 0$ fixes the sign of Δp once the sign of $(\partial^2 V/\partial p^2)_s$ is known. For "normal" substances both experiment and theory show that $(\partial^2 V/\partial p^2)_s > 0$, that is, adiabats are concave upward in the (p, V) diagram; for example it follows from (4.16) that for a perfect gas $(\partial^2 V/\partial p^2)_s = (\gamma + 1) V/\gamma^2 p^2$. Thus for normal substances we conclude that $\Delta p > 0$ across a shock front, as asserted earlier. Given that $\Delta p > 0$, (56.9) implies $\Delta V < 0$ $(\Delta \rho > 0)$, (56.12) implies $\Delta u < 0$, (56.13) implies $\Delta h > 0$, and (56.14) implies $\Delta e > 0$.

For a weak shock, (56.9) yields

$$\dot{m} \approx [-(\partial p/\partial V)_s]^{1/2}, \tag{56.52}$$

and to the same order, (56.10) and (56.11) yield

$$u_1 \approx u_2 \approx \dot{m} V = [(\partial p/\partial \rho)_s]^{1/2} = a. \tag{56.53}$$

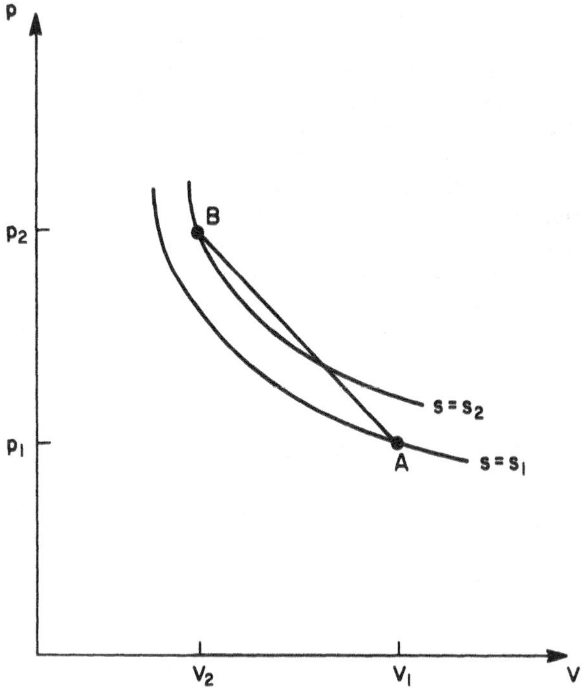

Fig. 56.3 Shock transition connecting two adiabats in (p, V) diagram.

These results can be refined by reference to Figure 56.3. There we see that because the initial and final states are joined by the chord AB whose slope is $-\dot{m}^2$ [cf. (56.9)], we must have $\dot{m}^2 > [-(\partial p/\partial V)_s]_1$. Therefore at point A

$$u_1^2 = V_1^2\dot{m}^2 > -V_1^2[(\partial p/\partial V)_s]_1 = [(\partial p/\partial \rho)_s]_1 = a_1^2, \tag{56.54}$$

that is, $u_1 > a_1$. By a similar analysis at point B we find $u_2 < a_2$. These general results are consistent with (56.38), (56.39), and (56.47), which, however, apply only for a perfect gas. Indeed it can be shown that all of the inequalities stated above are true for shocks of arbitrary strength provided only that $(\partial^2 V/\partial p^2)_s > 0$; see (**L2,** §84).

For the special case of a perfect gas, we can write explicit formulae for Δs in shocks of arbitrary strength. Thus from (4.10) we have

$$\begin{aligned}\Delta s &= c_v \ln (p_2\rho_1^\gamma/p_1\rho_2^\gamma)\\ &= c_v\left\{\ln (1+\not{p}) + \gamma \ln \left[\frac{2\gamma+(\gamma-1)\not{p}}{2\gamma+(\gamma+1)\not{p}}\right]\right\}\\ &= c_v\left\{\ln \left[\frac{2M_1^2-(\gamma-1)}{\gamma+1}\right] + \gamma \ln \left[\frac{(\gamma-1)M_1^2+2}{(\gamma+1)M_1^2}\right]\right\}.\end{aligned} \tag{56.55}$$

From (56.55) it is straightforward to show that Δs is a monotone increasing function of $\not{p}$ and of M_1, and furthermore that for $\not{p} \ll 1$ and $m \ll 1$

$$\Delta s = (\gamma - 1)c_v \not{p}^3/12\gamma^2 = 2\gamma(\gamma - 1)c_v m^3/3(\gamma + 1)^2 \geq 0; \qquad (56.56)$$

thus Δs is, in fact, always greater than or equal to zero. Equation (56.56) is, of course, consistent with (56.51) and (56.44).

STABILITY

We have seen that the requirement that entropy not decrease across shock fronts implies that $\Delta p > 0$, hence that rarefaction discontinuities do not exist. There are additional reasons why rarefaction discontinuities cannot exist. If such a discontinuity did exist, it would have $u_1 < a_1$ and $u_2 > a_2$, and would therefore propagate subsonically through the undisturbed medium. But then any small disturbance, which would travel as an acoustic wave at the speed of sound, produced in the flow at the jump could outrun the discontinuity. Therefore the rarefaction region behind the discontinuity would tend to spread into the gas in front of the discontinuity faster than the discontinuity itself could propagate, and in doing so would erode away any initial jump in material properties. That is, a rarefaction discontinuity is immediately smoothed into a continuous transition. Furthermore, because a rarefaction discontinuity would move supersonically with respect to downstream material, it could not be influenced by any process or change in conditions occurring behind the jump. That is, no feedback on the wave is possible, and in that sense the wave is uncontrolled. Both of these properties imply that rarefaction discontinuities are *mechanically unstable*, and disintegrate immediately.

In contrast, in a compression shock the entropy increases. The front outruns acoustic waves that might tend to smear it out, and the upstream material remains "unaware" of the shock until it slams into it; hence the shock can propagate as a sharp discontinuity. Furthermore, the shock propagates subsonically with respect to downstream material, hence the material behind the front can influence the front's behavior; if the downstream gas is strongly compressed and heated, it tends to strengthen the shock; if the downstream material cools rapidly (e.g., by radiation losses) the driving force behind the shock front weakens and eventually the shock dissipates. Thus compression shocks not only satisfy entropy constraints but are, in addition, mechanically stable. Further discussion of these issues can be found in (**C6,** Chap. 3), (**L2,** §84), and (**Z1,** §1.17).

RELATIVISTIC SHOCKS

We can obtain jump conditions across a shock in a relativistic flow by expressing the continuity equation (39.9) and the dynamical equations (42.2) to (42.6) in the frame in which the shock is stationary, and then subjecting them to the same analysis as led to (56.6) to (56.8). Thus writing N for the number of particles per unit proper volume, and $U_x = \gamma u_x$ for the

x component of the four-velocity of the material relative to the front, particle conservation implies continuity of the particle flux:

$$N_1 U_{x1} = N_2 U_{x2} \equiv j. \tag{56.57}$$

Similarly, energy and momentum conservation imply continuity of the energy flux

$$[(\hat{e}+p)U_0 U_x]_1 = [(\hat{e}+p)U_0 U_x]_2 \tag{56.58a}$$

or

$$(\hat{e}+p)_1 \gamma_1 U_{x1} = (\hat{e}+p)_2 \gamma_2 U_{x2}, \tag{56.58b}$$

and of the momentum flux

$$(\hat{e}+p)_1 U_{x1}^2 + p_1 c^2 = (\hat{e}+p)_2 U_{x2}^2 + p_2 c^2. \tag{56.59}$$

Here $\hat{e} = \rho_0(c^2+e) = Nm_0(c^2+e)$ is the total proper energy density of the fluid.

Rewrite (56.57) as

$$U_{x1} = j\tilde{V}_1 \tag{56.60a}$$

and

$$U_{x2} = j\tilde{V}_2, \tag{56.60b}$$

where $\tilde{V} \equiv 1/N$ is the volume per particle. Then (56.59) becomes

$$p_2 - p_1 = j^2(\tilde{h}_1 \tilde{V}_1 - \tilde{h}_2 \tilde{V}_2)/c^2, \tag{56.61}$$

while (56.58) reduces to

$$\gamma_1 \tilde{h}_1 = \gamma_2 \tilde{h}_2. \tag{56.62}$$

Here

$$\tilde{h} \equiv m_0(c^2+e) + (p/N) \tag{56.63}$$

is the total enthalpy per particle. In the nonrelativistic limit, (56.61) reduces to (56.9). Multiplying (56.61) by $(\tilde{h}_1 \tilde{V}_1 + \tilde{h}_2 \tilde{V}_2)$ and using (56.60) we find

$$(\tilde{h}_1 U_{x1}/c)^2 - (\tilde{h}_2 U_{x2}/c)^2 = (p_2 - p_1)(\tilde{h}_1 \tilde{V}_1 + \tilde{h}_2 \tilde{V}_2). \tag{56.64}$$

Then adding the square of (56.62) we obtain, finally,

$$\tilde{h}_2^2 - \tilde{h}_1^2 = (\tilde{h}_1 \tilde{V}_1 + \tilde{h}_2 \tilde{V}_2)(p_2 - p_1), \tag{56.65}$$

which reduces to (56.13) in the nonrelativistic limit. Equations (56.61) and (56.65), first derived by Taub (**T1**), are the relativistic generalizations of the Rankine-Hugoniot jump relations.

It is possible to obtain relativistic generalizations for essentially all of the results derived above for nonrelativistic shocks. Thus one can show (**T2**) that the generalization of (56.51) for the entropy jump across a weak shock is

$$s_2 - s_1 = \tfrac{1}{12}[\partial^2(\tilde{h}\tilde{V})/\partial p^2]_s (p_2 - p_1)^3/\tilde{h}T, \tag{56.66}$$

which, as long as $[\partial^2(\tilde{h}\tilde{V})/\partial p^2]_s > 0$, implies that $p_2 > p_1$, $\tilde{h}_2 > \tilde{h}_1$, $N_2 > N_1$,

$\tilde{V}_2 < \tilde{V}_1$, $U_{x2} < U_{x1}$, $U_{x1}/a_1 > 1$, $U_{x2}/a_2 < 1$ in a shock in which entropy increases. By straightforward manipulation of (56.58) and (56.59), one can also show that

$$\frac{u_{x1}}{c} = \left[\frac{(p_2 - p_1)(\hat{e}_2 + p_1)}{(\hat{e}_2 - \hat{e}_1)(\hat{e}_1 + p_2)}\right]^{1/2} \tag{56.67}$$

and

$$\frac{u_{x2}}{c} = \left[\frac{(p_2 - p_1)(\hat{e}_1 + p_2)}{(\hat{e}_2 - \hat{e}_1)(\hat{e}_2 + p_1)}\right]^{1/2}, \tag{56.68}$$

where u_x is the ordinary velocity (i.e., the three-velocity) of the material relative to the shock. In the nonrelativistic limit, $\hat{e} \to \rho c^2 \gg p$, and (56.67) and (56.68) reduce to (56.10) and (56.11). In the extreme relativistic limit, $p \to \frac{1}{3}\hat{e}$, hence

$$\frac{u_{x1}}{c} \to \left[\frac{3\hat{e}_2 + \hat{e}_1}{3(3\hat{e}_1 + \hat{e}_2)}\right]^{1/2} \tag{56.69}$$

and

$$\frac{u_{x2}}{c} \to \left[\frac{3\hat{e}_1 + \hat{e}_2}{3(3\hat{e}_2 + \hat{e}_1)}\right]^{1/2}. \tag{56.70}$$

For weak shocks, $\hat{e}_2 \approx \hat{e}_1$ and $u_{x1} \approx u_{x2} \approx c/\sqrt{3}$; for strong shocks $\hat{e}_2 \gg \hat{e}_1$ and $u_{x1} \to c$ while $u_{x2} \to c/3$. Using the relativistic law for the addition of velocities, that is,

$$\Delta u = (u_{x1} - u_{x2})/[1 - (u_{x1}u_{x2}/c^2)], \tag{56.71}$$

(which follows from boosting the four-velocity of a particle moving with velocity u_{x1} in a frame S into a frame S' moving with velocity u_{x2} relative to S), we find that the relative velocity of the gas on the two sides of the shock is

$$\frac{\Delta u}{c} = \left[\frac{(p_2 - p_1)(\hat{e}_2 - \hat{e}_1)}{(\hat{e}_1 + p_2)(\hat{e}_2 + p_1)}\right]^{1/2}. \tag{56.72}$$

In the nonrelativistic limit, (56.72) reduces to (56.15).

More complete discussions of relativistic shocks are given in (**I2**), (**L7**), (**L8**), (**M2**), (**T2**), and (**T4**).

57. Shock Structure

Let us now investigate how dissipative processes—viscosity and thermal conduction—determine the structure and thickness of shock fronts. We expect these processes to play a key role within the front because gradients are very steep there (indeed, infinitely steep according to the idealized analysis of §56).

The conservation relations for the steady flow in the shock's frame now are [cf. (26.2) and (27.34)]

$$\rho u = \rho_1 u_1 = \dot{m}, \tag{57.1}$$

$$\rho u^2 + p - \mu'(du/dx) = \rho_1 u_1^2 + p_1, \tag{57.2}$$

and

$$\rho u(h + \tfrac{1}{2}u^2) - \mu' u(du/dx) - K(dT/dx) = \rho_1 u_1(h_1 + \tfrac{1}{2}u_1^2), \tag{57.3}$$

and the entropy-generation equation (27.17) in the shock's frame is

$$\rho u T\left(\frac{ds}{dx}\right) = \mu'\left(\frac{du}{dx}\right)^2 + \frac{d}{dx}\left(K\frac{dT}{dx}\right). \tag{57.4}$$

Here $\mu' \equiv \frac{4}{3}\mu + \zeta$ denotes the effective one-dimensional viscosity.

Equations (57.1) to (57.4) apply at all points in the flow and determine the physical properties (ρ, p, u, etc.) as functions of x across the shock front. The constants on the right-hand sides of (57.2) and (57.3) are evaluated in the upstream flow far from the shock, where $(du/dx) = (dT/dx) \equiv 0$. If we evaluate the left-hand sides of (57.2) and (57.3) in the downstream flow far from the shock where (du/dx) and (dT/dx) again vanish, we recover the ideal-fluid jump relations (56.7) and (56.8).

VISCOUS SHOCKS

Consider first a hypothetical fluid having a finite viscosity but zero thermal conductivity. In order to simplify the discussion we assume that μ' is constant and that the fluid is a perfect gas. We can then rewrite (57.2) and (57.3) as

$$p + \dot{m}u - \mu'(du/dx) = p_1 + \dot{m}u_1 \tag{57.5}$$

and

$$u\{\tfrac{1}{2}\dot{m}u + [\gamma p/(\gamma-1)]\} - \mu' u(du/dx) = \dot{m}\{\tfrac{1}{2}u_1^2 + [\gamma p_1/(\gamma-1)\rho_1]\}. \tag{57.6}$$

Multiplying (57.5) by $\gamma u/(\gamma-1)$ and subtracting (57.6) we obtain

$$-\nu u(du/dx) = a_1^2(u-u_1) + u_1^2[\gamma u - \tfrac{1}{2}(\gamma-1)u_1] - \tfrac{1}{2}(\gamma+1)u_1 u^2, \tag{57.7}$$

where $\nu = \mu'/\rho$ is the effective kinematic viscosity. Let $w \equiv u_1 - u$; then (57.7) can be rewritten

$$\nu(dw/dx) = w[u_1^2 - a_1^2 - \tfrac{1}{2}(\gamma+1)u_1 w]/(u_1 - w). \tag{57.8}$$

The *velocity drop* w varies from $w = 0$ far upstream to

$$w_{\text{max}} = u_1 - u_2 = 2(u_1^2 - a_1^2)/(\gamma+1)u_1 \tag{57.9}$$

far downstream; the second equality in (57.9) follows from the Prandtl relation (56.29).

From (57.8) and (57.9), one finds that $(dw/dx) \geq 0$, with $(dw/dx) = 0$ at $w = 0$ and $w = w_{\text{max}}$; therefore $w(x)$ is monotone increasing and $u(x)$ is

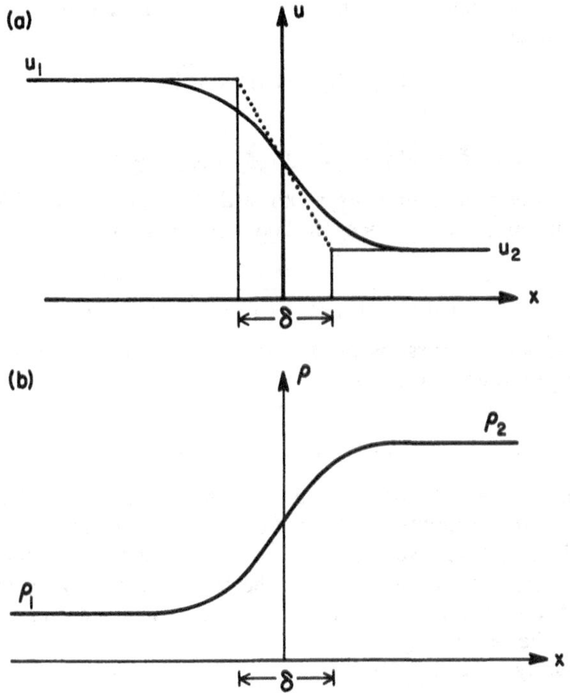

Fig. 57.1 Velocity and density variation in a viscous shock.

monotone decreasing. Furthermore, w has an inflection point because

$$\nu(d^2w/dx^2) = u_1[u_1^2 - a_1^2 - (\gamma+1)w(u_1 - \tfrac{1}{2}w)]/(u_1 - w)^2, \quad (57.10)$$

which shows that $(d^2w/dx^2) > 0$ at $w = 0$, $(d^2w/dx^2) < 0$ at $w = w_{\max}$, and $(d^2w/dx^2) = 0$ at $w = u_1 - \sqrt{u_1 u_2}$. Thus $u(x)$ varies as sketched in Figure 57.1a. From continuity it follows that $\rho(x)$ is monotone increasing, as sketched in Figure 57.1b.

Using (57.8) in (57.5) we find

$$p = p_1 + \rho_1 w[a_1^2 + \tfrac{1}{2}(\gamma - 1)u_1 w]/(u_1 - w), \quad (57.11)$$

which shows that $p(x)$ is monotone increasing, like the sketch in Figure 57.1b. Combining (57.1) and (57.11) we find

$$T/T_1 = (p/p_1)(\rho_1/\rho) = 1 + (\gamma - 1)(w/u_1) + \tfrac{1}{2}\gamma(\gamma - 1)(w/a_1)^2, \quad (57.12)$$

which shows that $T(x)$ increases monotonically, like the sketch in Figure 57.1b.

As an estimate of the shock width δ we take

$$\delta \approx [w/(dw/dx)]_0, \quad (57.13)$$

where x_0 is the point at which $w = \frac{1}{2}w_{max}$. Substituting from (57.9) into (57.8) one finds

$$(dw/dx)_0 = \tfrac{1}{2}(u_1^2 - a_1^2)^2/\nu(\gamma u_1^2 + a_1^2), \tag{57.14}$$

whence

$$\delta = 2\nu(\gamma u_1^2 + a_1^2)/[(\gamma+1)u_1(u_1^2 - a_1^2)]. \tag{57.15}$$

From mean free path arguments we know that $\nu = \mu_1/\rho_1 \sim a_1\lambda$. Thus in the weak-shock limit, where $u_1 \approx a$, we find, using (56.38),

$$\delta = \frac{2\nu}{a_1^2(M_1^2-1)} = \frac{4\gamma\nu}{(\gamma+1)a_1\not{p}} \sim \left(\frac{4\gamma}{\gamma+1}\right)\left(\frac{\lambda}{\not{p}}\right). \tag{57.16}$$

Thus the shock thickness is of the order of a particle mean free path divided by the fractional pressure jump.

For a strong shock ($M_1 \gg 1$), (57.15) yields

$$\delta \sim [2\gamma/(\gamma+1)](\lambda/M_1), \tag{57.17}$$

which formally predicts that δ becomes much smaller than λ when $M_1 \gg 1$. This result is incompatible with a fluid description, and comes from taking $\nu \sim a_1\lambda$. If instead we assume that the viscous dissipation occurs mainly in the hot material at the back edge of the transition zone and take $\nu \sim \bar{u}\lambda \approx a_2\lambda \sim M_1 a_1 \lambda$, then $\delta \sim C\lambda$ where C is a number of order unity; hence δ remains of the order of λ.

For a purely viscous shock

$$\rho u T(ds/dx) = \dot{m}T(ds/dx) = \mu'(du/dx)^2; \tag{57.18}$$

hence s increases monotonically through the shock like the sketch in Figure 57.1b. We can use (57.18) to estimate the entropy jump across the shock by replacing derivatives with finite differences, writing

$$\dot{m}T_1(\Delta s/\Delta x) \approx \mu'(u_2 - u_1)^2/\Delta x^2. \tag{57.19}$$

Using (56.16) to write $\Delta u = \Delta p/\dot{m} \approx \Delta p/\rho a$ for a weak shock, and adopting $\Delta x \approx \delta$ as given by (57.16), we find that (57.19) gives the same result as (56.56) to within a numerical factor of order unity.

The analysis presented above shows the fundamental role played by viscosity in determining shock structure: it leads to an irreversible conversion of kinetic energy of the inflowing material into heat. Put differently, it transforms *ordered* flow motion of the particles in the gas into *random* motions via the mechanism of dissipation of particle momentum.

CONDUCTING SHOCKS

Now consider a fluid with zero viscosity ($\mu' \equiv 0$), but finite thermal conductivity. These assumptions are of more than hypothetical interest because, as we mentioned in §51, they are realistic for a radiating gas in which radiative energy transport can strongly influence shock structure even when viscous and thermal-conduction effects are negligible (cf. §104). As we will see, the structure of an inviscid conducting shock can be qualitatively different from that of a pure viscous shock.

When $\mu' \equiv 0$, the momentum and energy conservation relations are

$$p + \dot{m}u = p_1 + \dot{m}u_1 \tag{57.20}$$

and

$$u\{\tfrac{1}{2}\dot{m}u + [\gamma p/(\gamma-1)]\} + q = \dot{m}\{\tfrac{1}{2}u_1^2 + [\gamma p_1/(\gamma-1)\rho_1]\}, \tag{57.21}$$

where $q = -K(dT/dx)$. The entropy-generation equation reduces to

$$(ds/dx) = (K/\dot{m}T)(d^2T/dx^2), \tag{57.22}$$

where K has been assumed to be constant.

Rewrite (57.20) as

$$p = p_1 + \dot{m}u_1(1-\eta) \tag{57.23}$$

where η is the *volume ratio*

$$\eta \equiv V/V_1 = \rho_1/\rho = u/u_1. \tag{57.24}$$

Clearly the pressure is a monotone increasing function through the shock front, rising to p_2 when η equals

$$\eta_2 = V_2/V_1 = [(\gamma-1)M_1^2 + 2]/(\gamma+1)M_1^2, \tag{57.25}$$

as given by (56.41). Using the perfect gas law we then find

$$T/T_1 = \eta[\gamma M_1^2(1-\eta) + 1] \tag{57.26}$$

which shows that $T(\eta)$ is a quadratic function of η, as sketched in Figure 57.2. $T(\eta)$ reaches its maximum value at

$$\eta_{\max} = (\gamma M_1^2 + 1)/2\gamma M_1^2. \tag{57.27}$$

Note in passing that as $M_1 \to \infty$, $\eta_{\max} \to \frac{1}{2}$. Finally, from (57.21) and (57.23) to (57.25), we obtain

$$q = -\dot{m}u_1^2(\gamma+1)(1-\eta)(\eta-\eta_2)/2(\gamma-1), \tag{57.28}$$

which shows that $q \le 0$ for $\eta_2 \le \eta \le 1$. More precisely, $q = 0$ at $\eta = 1$ and at $\eta = \eta_2$, and q reaches an absolute minimum at the midpoint of the compression, that is, at

$$\eta_0 = \tfrac{1}{2}(1+\eta_2). \tag{57.29}$$

Consider first a weak shock for which $\eta_2 > \eta_{\max}$, for example, the transition to point A sketched in Figure 57.2. From (57.26) we see that in this case $T(x)$ increases monotonically from T_1 to T_2, which implies $(dT/dx) \ge 0$, which is consistent with the conclusion that $q \le 0$. Furthermore, the fact that q achieves a minimum at $\eta = \eta_0$ implies that $(dq/dx)_0 = -K(d^2T/dx^2)_0 = 0$; therefore $T(x)$ has an inflection point at η_0 and varies as sketched in Figure 57.3a. These conclusions further imply that the

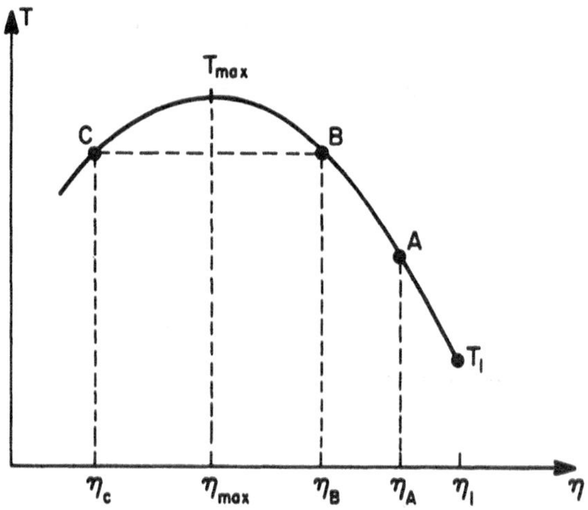

Fig. 57.2 Temperature variation as a function of volume ratio in a conducting shock.

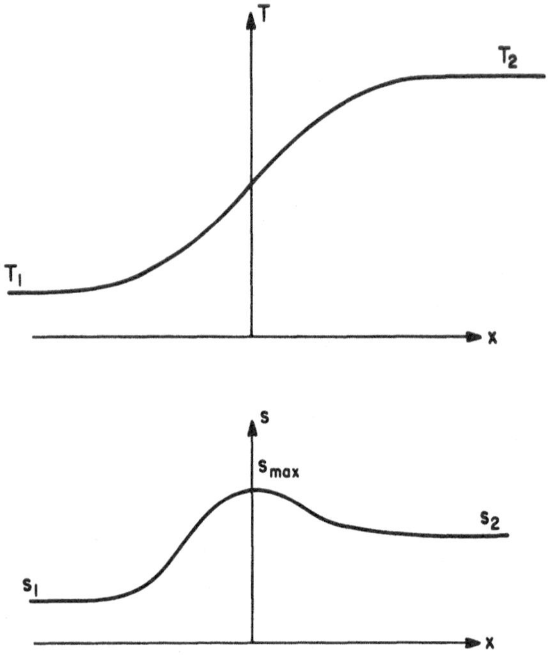

Fig. 57.3 Temperature and entropy variation as a function of spatial position in a weak conducting shock.

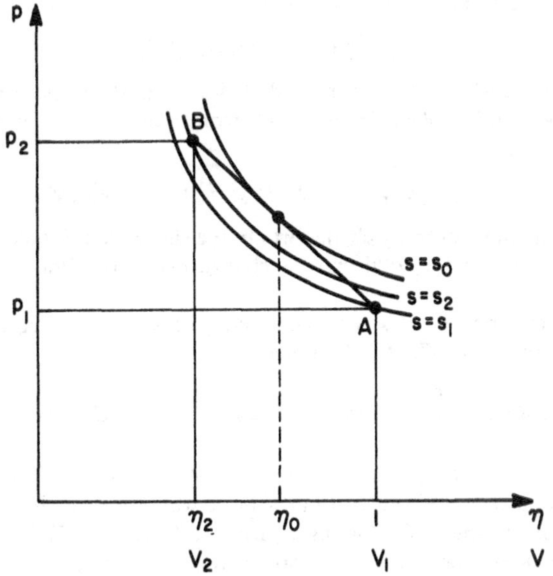

Fig. 57.4 Shock transition and initial, final, and maximum entropy adiabats in (p, V) diagram for a conducting shock.

entropy must achieve a local maximum at $\eta = \eta_0$ because $(d^2T/dx^2) = 0$ implies $(ds/dx) = 0$; thus $s(x)$ varies as sketched in Figure 57.3b.

We can calculate the maximum entropy increase in the shock as follows. According to (57.23), in the (p, V) plane the gas follows the straight line AB shown in Figure 57.4, and the material reaches its maximum entropy where this line is just tangent to an isentrope, say at $s = s_0$. For a weak shock the equation for the straight line is

$$p - p_1 = \frac{(p_2 - p_1)(V - V_1)}{(V_2 - V_1)} \approx \left(\frac{\partial p}{\partial V}\right)_{s_1}(V - V_1) + \frac{1}{2}\left(\frac{\partial^2 p}{\partial V^2}\right)_{s_1}(V_2 - V_1)(V - V_1), \tag{57.30}$$

where we have ignored a term in $(\partial p/\partial s)_V$ because $(s_2 - s_1)$ is third order in Δp or ΔV. Similarly, the equation for the isentrope $s = s_0$ is

$$p - p_1 \approx \left(\frac{\partial p}{\partial V}\right)_{s_1}(V - V_1) + \frac{1}{2}\left(\frac{\partial^2 p}{\partial V^2}\right)_{s_1}(V - V_1)^2 + \left(\frac{\partial p}{\partial s}\right)_{V_1}(s_0 - s_1), \tag{57.31}$$

where again we neglect third-order terms.

To enforce tangency of the two curves, we demand that $[(\partial p/\partial V)_{\text{line}}]_0 = [(\partial p/\partial V)_{\text{isentr}}]_0$ at the point where $(\partial s/\partial V) = 0$ (hence s is a maximum), whence we find that $V_0 = \frac{1}{2}(V_1 + V_2)$, in agreement with (57.29). Then, demanding $(p_{\text{line}})_0 = (p_{\text{isentr}})_0$, we equate the right-hand sides of (57.30) and

(57.31) evaluated at $V = V_0$, obtaining

$$s_0 - s_1 = \tfrac{1}{8}[(\partial^2 p/\partial V^2)_s/(\partial p/\partial s)_V]_1 (V_1 - V_2)^2. \tag{57.32}$$

From (4.16) and (5.15) we find that for a perfect gas $(\partial^2 p/\partial V^2)_s = \gamma(\gamma+1)p/V^2$ and $(\partial p/\partial s)_V = p/c_v$, and from (56.36) we have $\Delta V/V = -\Delta p/\gamma p$, hence

$$s_0 - s_1 = \tfrac{1}{8}\gamma(\gamma+1)c_v(\Delta V/V)^2 = \tfrac{1}{8}[(\gamma+1)/\gamma]c_v \not{h}^2. \tag{57.33}$$

Thus the maximum entropy change *within* the shock front is second order in Δp or ΔV, whereas the total entropy change *across* the front is only third order.

To estimate the thickness of a conducting shock, we calculate the total entropy jump from (57.22), obtaining

$$(\dot{m}/K)\,\Delta s = \int_{-\infty}^{\infty} T^{-1}(d^2T/dx^2)\,dx = \int_{-\infty}^{\infty} [T^{-1}(dT/dx)]^2\,dx, \tag{57.34}$$

where we integrated by parts and noted that $(dT/dx) = 0$ at $x = \pm\infty$. The integral is approximately equal to $(q/KT)_0^2\,\delta$, where q is the heat flux and δ is the shock thickness. Evaluating q_0 from (57.28) and (57.29), using the scaling rule $K \sim a\lambda\rho c_v$, and using (56.56) for Δs, we find

$$\delta \sim \tfrac{16}{3}[(\gamma-1)/(\gamma+1)](\lambda/\not{h}), \tag{57.35}$$

which agrees with (57.15) for a weak viscous shock to within a numerical factor of order unity.

We have thus shown that for shocks below a certain critical strength [i.e., for which the downstream volume ratio η_2 is greater than η_{max} defined by (57.27)] in an inviscid, conducting fluid all physical properties vary continuously through the shock front over a distance of the order of a few particle mean free paths. The shock structure is qualitatively similar to that of a viscous shock, with the velocity decreasing monotonically, while ρ, p, and T rise monotonically; it differs only in that the entropy passes through a local maximum instead of increasing monotonically.

The situation for strong shocks in an inviscid, conducting fluid is quite different; here, as Rayleigh (**R3**) first noted, only the temperature varies continuously, while the other variables experience a discontinuous jump within the shock front [see also (**B2**)]. Thus suppose the material is compressed to some $\eta_2 < \eta_{max}$, for example, to point C in Figure 57.2; this case arises naturally because, as noted above, as $M_1 \to \infty$, $\eta_{max} \to \frac{1}{2}$ while $\eta_2 \to (\gamma-1)/(\gamma+1) \le \frac{1}{4}$. Equation (57.28) shows that $q = -K(dT/dx) \le 0$ for $\eta_2 \le \eta \le 1$, regardless of whether η_2 is less than η_{max} or not. We therefore must guarantee that

$$(dT/dx) = (dT/d\eta)(d\eta/dx) \ge 0 \tag{57.36}$$

throughout the shock front. Now $(d\eta/dx)$ is always ≤ 0, hence only those portions of the curve $T(\eta)$ for which $(dT/d\eta) \le 0$ are physically accessible.

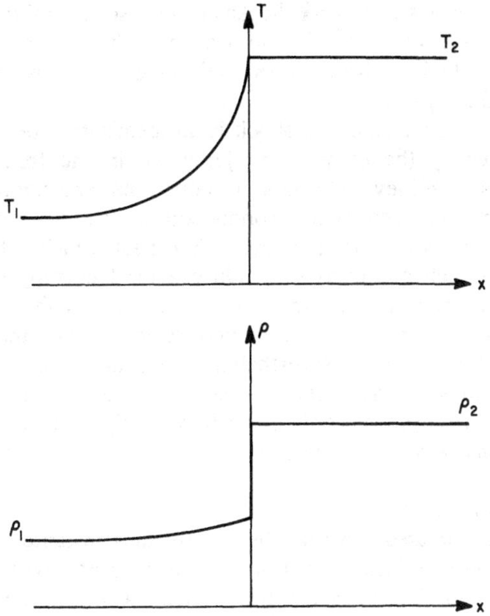

Fig. 57.5 Temperature and density variation as a function of spatial position in an isothermal shock.

Hence as the material is compressed from $\eta = 1$ to $\eta = \eta_C$, it is not possible for the temperature to rise to $T_{\max}$ and then track down the descending branch to $T = T_C$. The only way the material can actually make the transition to point C is for the temperature to rise continuously from T_1 to $T_B = T_C$ at $\eta = \eta_B > \eta_{\max}$, and then remain constant while the relative volume collapses *discontinuously* from η_B to η_C, giving rise to a density jump like that sketched in Figure 57.5. Because this density discontinuity (which accounts for most of the total density jump across the shock) occurs at a single temperature, this type of solution is called an *isothermal shock.*

Thus in an inviscid, conducting fluid, all shocks above some critical strength will be isothermal. Combining (57.27) with (56.41) we see that the critical Mach number at which η_2 equals $\eta_{\max}$ is

$$(M_1^2)_{\text{crit}} = (3\gamma - 1)/\gamma(3-\gamma), \tag{57.37}$$

which, from (56.40), implies that

$$(p_2/p_1)_{\text{crit}} = (\gamma + 1)/(3-\gamma), \tag{57.38}$$

and, from (57.27),

$$(\eta_2)_{\text{crit}} = (\gamma + 1)/(3\gamma - 1). \tag{57.39}$$

For a monatomic gas with $\gamma = \frac{5}{3}$ shocks become isothermal for $M_1 \geq (\frac{9}{5})^{1/2} \approx 1.18$, or $p_2/p_1 \geq 2$, or $\eta_2 \leq \frac{2}{3}$. Conditions for an isothermal shock are met even more easily in polyatomic or ionizing gases, or when radiation is present, for then $\gamma < \frac{5}{3}$.

The notion of an isothermal shock is an idealization because in reality the strong density (hence velocity) jump within the front implies that viscous effects must inevitably come into operation and smooth the discontinuity. When *both* viscosity and conduction act (as they must in any real gas), viscosity converts flow momentum into heat, which is transported by conduction in such a way as to produce a local entropy maximum. All properties vary continuously through the front, though the density and pressure may rise rapidly in a limited region where the temperature changes less swiftly (**B2**). Nevertheless, the analysis presented above is instructive because it shows that we can *guarantee* a continuous solution for arbitrarily strong shocks only through the dissipative effects of viscosity, a point of considerable significance for numerical calculations (cf. §59).

THE RELAXATION LAYER

We have thus far assumed that the gas remains instantaneously in local thermodynamic equilibrium as it flows through a narrow transition zone, the *dissipation zone*, at the shock front, having a thickness δ of only a few particle mean free paths. In reality, the material may not be able to remain in equilibrium because the characteristic flow time $t_f \sim \lambda/u_1$ through the front may be much shorter than the time required for some thermodynamically important process (e.g., ionization of the material) to occur. Thus while some degrees of freedom may equilibrate within the dissipation layer (always true for the translational degrees of freedom of each particle species), others may be far from equilibrium when the material emerges from that layer.

In this event, the dissipation zone is followed downstream by a *relaxation layer* within which internal relaxation processes operate to bring the material to its final equilibrium state. If the characteristic relaxation time for some process is t_{relax}, the thickness of the relaxation layer associated with that process is $\Delta \sim u_2 t_{\text{relax}}$. Clearly $\Delta \gg \delta$ whenever $t_{\text{relax}} \gg t_f$. Several relaxation processes may occur simultaneously (or even sequentially), and the full thickness of the layer (i.e., the distance required to reach the point where the downstream conditions predicted by the Rankine-Hugoniot relations are achieved) is determined by the slowest process.

Relaxation processes can sometimes be described by phenomenological equations of the form

$$(dn/dt) = (n_{\text{equib}} - n)/t_{\text{relax}}, \tag{57.40}$$

where n represents the number of particles in the desired state (e.g., ionized as a result of passing through the shock) and n_{equib} is the number that would be in that state if the material were in equilibrium at the downstream values of the material properties (i.e., ρ_2, p_2, T_2, etc.). Equation (57.40) implies an exponential relaxation of the form

$$n = n_0 \exp(-t/t_{\text{relax}}) + n_{\text{equib}}[1 - \exp(-t/t_{\text{relax}})] \tag{57.41}$$

where $t \sim x/u_2$, x being the distance downstream from the shock front and u_2 the downstream flow velocity. To obtain a more accurate picture we must specify the *rates* of the relevant relaxation processes, write *kinetic equations* that describe how these processes determine the distribution of particles over various states, and solve these equations (usually numerically) simultaneously with the equations of hydrodynamics.

In general, the problem can be quite complicated because on the one hand the relaxation rates depend on the thermodynamic state of the material, hence the dynamics of the flow, but on the other hand the relaxation processes determine the thermodynamic state of the material (hence the flow dynamics), for example by setting the rate of thermal energy loss into ionization (or the rate of energy gain by recombinations). We will discuss rate coefficients and kinetic equations in §85, and give examples of solutions of the set of coupled equations in §105. For the present, it suffices merely to describe qualitatively some of the basic processes that occur in the absence of radiation in order to get a physical feeling for their relative importance in different regimes.

(*a*) *Molecular Gas* The extent to which any particular process plays a significant role in determining the structure of the relaxation layer depends strongly on the degree of ionization of the gas. Consider first a neutral gas composed of atoms and diatomic molecules. The most rapid of all relaxation processes is the establishment of equilibrium among the translational degrees of freedom (i.e., of a Maxwellian velocity distribution). Typically only a few collisions are required to effect a complete randomization of particle motions and kinetic energy, hence a Maxwellian is usually established within a few mean free paths. Indeed, to a good approximation, the translational relaxation layer is coincident with the dissipation layer, and we can assign a unique kinetic temperature to each particle species at every point in the flow. Similarly, molecular rotation is typically quite easily excited in only a few collisions, and this degree of freedom usually remains in equilibrium with translational motions.

In contrast, molecular vibrational modes, which first become excited at temperatures of the order of 10^3 K, may require hundreds to thousands of collisions to come into equilibrium, and the vibrational relaxation layer in cool material and/or weak shocks may be much thicker than the dissipation layer. However, as temperatures rise to a few thousand kelvins, either because the upstream material is hot or because the shock is strong, vibrational relaxation proceeds much more rapidly, and is displaced from its role as the slowest process by molecular dissociation. When temperatures reach about 10^4 K in the downstream material, molecular dissociation proceeds very rapidly and the limiting process becomes ionization, which we discuss further below.

To gain insight into the effects of relaxation processes on shock structure in a neutral gas we make the idealization that the shock is composed of two distinct regions: (1) a very thin dissipation zone (also called the *external relaxation zone*) in which viscosity and conduction effects are large, and within which equilibrium of the translational (and perhaps other) degrees of freedom is achieved, followed by (2) a relaxation zone (also called the *internal relaxation zone*) in which viscosity and conduction are unimportant, but some hitherto incompletely excited degree of freedom comes into equilibrium. These zones are assumed to be separated by a definite interface. As before upstream and downstream quantities are denoted by subscripts "1" and "2" respectively; properties at the interface are denoted by a subscript "i". Then the conservation relations are

$$\rho u = \rho_1 u_1 = \rho_i u_i = \rho_2 u_2 \equiv \dot{m}, \tag{57.42}$$

$$p + \rho u^2 = p_1 + \rho_1 u_1^2 = p_i + \rho_i u_i^2 = p_2 + \rho_2 u_2^2, \tag{57.43}$$

and

$$h + \tfrac{1}{2}u^2 = h_1 + \tfrac{1}{2}u_1^2 = h_i + \tfrac{1}{2}u_i^2 = h_2 + \tfrac{1}{2}u_2^2. \tag{57.44}$$

where unsubscripted variables denote quantities measured downstream from the interface. The enthalpy h_i includes only contributions from the translational and other rapidly excited degrees of freedom, other degrees of freedom still being frozen at their upstream values.

In this idealized description the material undergoes the transition sketched in Figure 57.6. Joining the initial state A to the final state C is the straight line (56.9). If all degrees of freedom were excited as rapidly as translational motions, in the dissipation zone the material would jump essentially discontinuously from A to C as defined by the intersection of the equilibrium Hugoniot with the straight line. But if some degrees of freedom are frozen during passage through the dissipation zone, the material has, in effect, a larger γ than it would in equilibrium [recall from kinetic theory that $\gamma = (n+2)/n$ where n = number of available degrees of freedom]. Therefore in the dissipation zone the material jumps essentially discontinuously from A only to point B, defined by the intersection of the straight line with a nonequilibrium Hugoniot which has a steeper slope than the equilibrium curve. Point B corresponds to (p_i, V_i) at the interface. The material then slowly relaxes along the straight line to its downstream equilibrium state C.

From (57.43) one has

$$(p_2 - p_i)/(p_2 - p_1) = (\eta_i - \eta_2)/(1 - \eta_2). \tag{57.45}$$

In a strong shock $\eta_i \approx \frac{1}{4}$ even if only translational motions are excited, and given that $\eta_2 \geq 0$, we see that the fractional pressure rise in the relaxation zone is always small, less than 25 percent of the total pressure jump in the

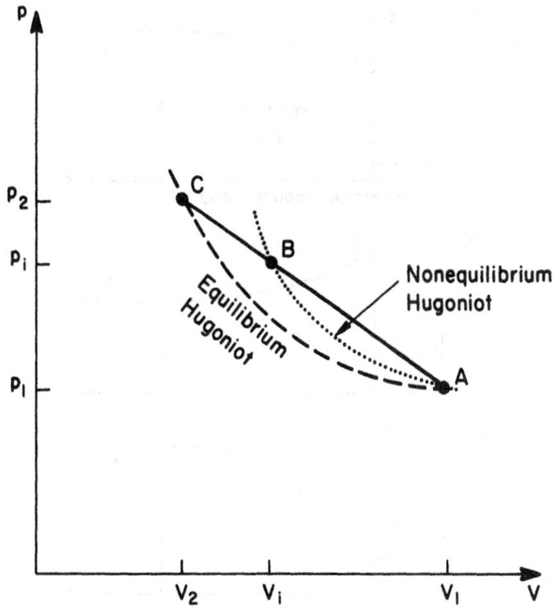

Fig. 57.6 Shock transition in material with a nonequilibrium relaxation layer.

shock. The pressure variation in the relaxation zone is sketched qualitatively in Figure 57.7a. The enthalpy increase behind the interface is even smaller. From (57.44) we have

$$(h_2 - h_i)/(h_2 - h_1) = (\eta_i^2 - \eta_2^2)/(1 - \eta_2^2) \tag{57.46}$$

and by the same reasoning we see that the fractional enthalpy increase in the relaxation zone is less than 6 per cent of the total.

Exploiting the result that h is nearly constant in the relaxation zone we can write

$$T_i/T_2 \approx \gamma_2(\gamma_i - 1)/\gamma_i(\gamma_2 - 1), \tag{57.47}$$

where γ_i and γ_2 are, respectively, the effective adiabatic exponents at the interface and far downstream. Because $\gamma_i \gtrsim \gamma_2$, (57.47) implies that the shock has a significant *temperature overshoot* immediately behind the dissipation zone, as sketched in Figure 57.7b, followed by a long, inviscid, nonconducting *tail* in which the temperature decreases to its equilibrium value. Such shocks are called *partly dispersed*, with part of the total shock dissipation occurring in classical dissipation mechanisms (viscosity, conductivity) and part in a lagging relaxation process. If the shock is sufficiently weak it can become *fully dispersed* by relaxation processes alone, and the solution is continuous even in the absence of viscosity and thermal conductivity.

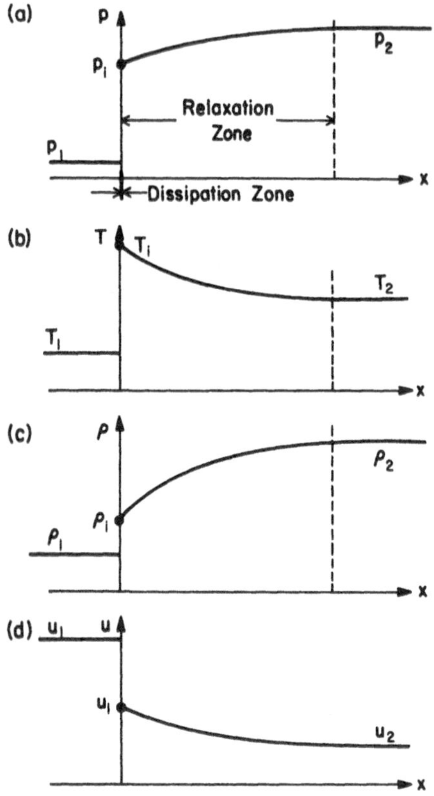

Fig. 57.7 Pressure, temperature, density, and velocity variation as a function of spatial position in a shock with a nonequilibrium relaxation zone.

The postshock temperature overshoot can be quite substantial. For example, if $\gamma_i = \frac{5}{3}$ (only translational motions) and $\gamma_2 = \frac{7}{5}$ (translation plus molecular rotation), $T_i/T_2 = 1.4$; if $\gamma_2 = \frac{9}{7}$ (translation, rotation, and vibration), $T_i/T_2 = 1.8$. Furthermore, noting that $\rho \sim p/T$, we see that the modest rise in p coupled with a significant drop in T leads to a fairly large rise in ρ in the relaxation zone, as sketched in Figure 57.7c. By continuity, u varies inversely as ρ, as sketched in Figure 57.7d.

(*b*) *Fully Ionized Plasma* Suppose now that the gas is sufficiently hot that all molecules have been dissociated, and it is composed of atoms, ions, and electrons. Indeed, consider first the extreme case of a completely ionized hydrogen plasma containing only electrons and protons (**I1**), (**S10**). As a first step in describing the shock structure we suppose that thermal conduction in the plasma can be neglected. Then the only relaxation phenomenon that occurs is the equilibration of the postshock kinetic temperatures of the two species of particles.

Initially, in the upstream material, $T_e = T_p = T_1$. As the material passes through the shock front, Coulomb interactions among the protons produce viscous forces that dissipate a large fraction of the protons' directed kinetic energy into thermal motions, producing a large proton temperature rise $\Delta T_p \sim m_p u_1^2/k \sim (m_p a_1^2/k)M_1^2$ within a layer of thickness $\delta \sim u_2 t_{pp}$, where t_{pp} is the proton self-collision time [cf. (10.26)]. Because the electron self-collision time $t_{ee} = (m_e/m_p)^{1/2} t_{pp} \approx \frac{1}{43} t_{pp}$, the electrons also have adequate time to convert their own directed motions into thermal energy within the dissipation zone. However, this mechanism leads to an electron temperature increase of only $\Delta T_e \sim m_e u_1^2/k = (m_e/m_p)\Delta T_p \sim \frac{1}{1800}\Delta T_p$, which is clearly negligible. Furthermore, the electron-proton energy-exchange time is much too long $[t_{ep} = (m_p/m_e)^{1/2} t_{pp} = (m_p/m_e)t_{ee}]$ to permit significant energy transfer from the protons to the electrons within the dissipation zone, hence an opportunity for a large discrepancy between T_e and T_p arises.

The strong Coulomb forces coupling the electrons and protons assure that there can be no charge separation over distances much larger than a Debye length. Therefore, as the protons are compressed in the shock, the electrons are also compressed by the same amount, and because the electrons cannot exchange energy with the protons in the time available (and, for the present, we are ignoring thermal conduction), this compression occurs essentially adiabatically. Hence the electron temperature just downstream from the front is $T_{e,i} \approx (\rho_2/\rho_1)^{\gamma-1} T_1$, or, for a strong shock $(\rho_2/\rho_1 = 4)$ in a monatomic gas, $T_{e,i} \approx 2.5 T_1$. This is a large rise, but still much smaller than that experienced by the protons for large Mach numbers.

Within the framework of assumptions made above, we can derive a simple quantitative relation between T_e and T_p in the downstream flow. Behind the dissipation zone the pressure is nearly constant, hence

$$p = n_e k T_e + n_p k T_p \approx p_2 = 2 n_{e2} k T_2, \tag{57.48}$$

which, because the plasma is fully ionized $(n_e = n_p)$ and the postshock density is nearly constant $(n_e \approx n_{e2})$, implies that

$$T_e + T_p \approx 2T_2 \approx \tfrac{5}{8} M_1^2 T_1. \tag{57.49}$$

For a completely ionized plasma of electrons and ions of charge Z, (57.49) generalizes to

$$Z T_e + T_{\text{ion}} = (Z+1) T_{\text{ion},2} = \tfrac{5}{16}(Z+1) M_1^2 T_1. \tag{57.50}$$

For example, if $M_1 = 4$ in hydrogen, $T_2 = 5T_1$, and given that $T_{e,i} \approx 2.5 T_1$ (adiabatic compression of the electrons), (57.49) shows that $T_{p,i} \approx 7.5 T_1$. As the plasma flows downstream, energy exchange between the electrons and protons finally occurs, producing a large temperature-relaxation layer

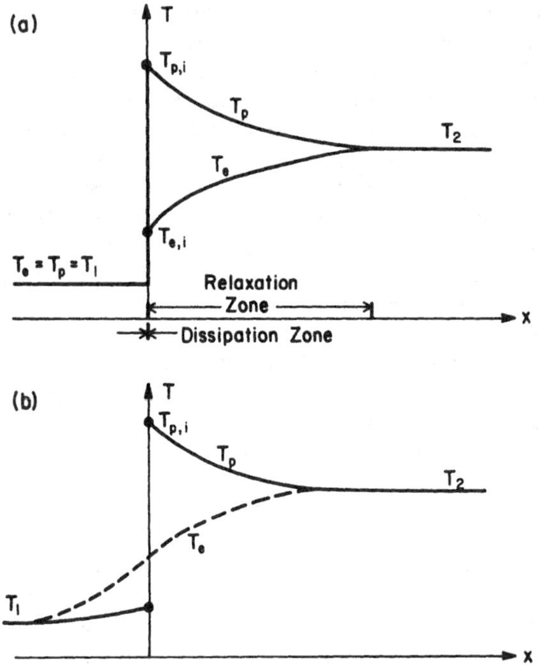

Fig. 57.8 Temperature variation as a function of spatial position in a shock in a two-fluid fully ionized hydrogen plasma. (a) Electron conduction omitted. (b) Electron conduction included.

of thickness

$$\Delta \sim u_2 t_{ep} = (m_p/m_e)^{1/2} u_2 t_{pp} \sim (m_p/m_e)^{1/2} \lambda_{\text{proton}} \sim (m_p/m_e)^{1/2}\, \delta \quad (57.51)$$

in which T_p decreases, and T_e increases, towards T_2, in accordance with (57.49). The resulting temperature profile is sketched in Figure 57.8a.

The picture described above is seriously inadequate, however, because we have ignored conduction effects. This omission is appropriate for protons because the characteristic length scale of the temperature gradient in the relaxation zone greatly exceeds λ_{proton}, the length scale over which viscous and/or conduction effects by protons are important. However the situation is quite different for electrons. From (10.23) or (10.26) we see that the mean free path of particles in a plasma is independent of a particle's mass, hence in ionized hydrogen ($Z=1$), $\lambda_e = \lambda_{\text{proton}}$, which implies that the thermal diffusivity for electrons $\chi_e \propto \bar{v}_e \lambda_e$ is $\chi_e = (m_p/m_e)^{1/2} \chi_p$ (as noted also in §33). Thus the characteristic length scale over which electron conduction is important is

$$l_e \sim (\chi_e/u_2) \sim (m_p/m_e)^{1/2} (\chi_p/u_2) \sim (m_p/m_e)^{1/2} \lambda_{\text{proton}}, \quad (57.52)$$

that is, l_e is of the same order of thickness as the relaxation zone.

Therefore, thermal conduction by electrons transports energy very efficiently throughout the entire relaxation zone, and thus strongly heats the electrons immediately behind the dissipation zone, while simultaneously promoting a more rapid equilibration of T_2 with T_p because heat is transferred to the postshock electrons by heat conduction at about the same rate as energy is transferred from the protons. Much more important, because the electron velocities behind the shock are roughly a factor of $(m_p/m_e)^{1/2}$ larger than the downstream flow speed, the electrons can overtake the shock front and conduct heat into the upstream material before the shock front arrives. This *conduction precursor* efficiently preheats the upstream electrons, which then transfer some of their excess energy to the upstream ions. The result is a temperature structure like that sketched in Figure 57.8b. (We will find a similar phenomenon, the *radiation precursor*, for shocks in radiating fluids, see §§104 and 105).

Shafranov (**S10**) made detailed numerical calculations of shock structure in ionized hydrogen, including the effects of electron thermal conduction, for a wide range of upstream Mach numbers. In the limit of very strong shocks, $M_1 \gg 1$, he obtains the results listed in Table 57.1, which apply immediately in front of and immediately behind the viscous dissipation zone. Note that the electron temperature is now continuous across the dissipation zone (for which reason such shocks are sometimes called *electron-isothermal* shocks), and has almost achieved its final downstream value already at the shock front. Similarly the protons are preheated to about 15 percent of the downstream temperature, and the postshock proton temperature overshoot is now only about 25 percent of the downstream temperature, the rest of the excess proton energy predicted by (57.49) having been consumed in heating the electrons.

In his numerical work Shafranov found that below a certain critical Mach number electron conduction is sufficient to produce fully dispersed shocks (i.e., all variables continuous across the front). For a plasma of electrons and positive ions of charge Z, Imshennik (**I1**) derived analytical formulae for the critical Mach number

$$(M_1^2)_{\text{crit}} = [\gamma^2 + (3Z+1)\gamma - Z]/\gamma[(3Z+1) - \gamma(Z-1)], \qquad (57.53)$$

Table 57.1. Physical Properties at Shock Front in a Hydrogen Plasma for $M_1 \gg 1$

Quantity	Preshock	Postshock
ρ/ρ_1	1.131	3.526
T_e/T_1	$0.29M_1^2$	$0.29M_1^2$
T_e/T_2	0.928	0.928
T_p/T_1	$1.2 + 0.05M_1^2$	$0.387M_1^2$
T_p/T_2	$0.16 + (3.84/M_1^2)$	1.238

the critical pressure jump

$$(p_2/p_1)_{\text{crit}} = (Z+1)(\gamma+1)/[(Z+1)(\gamma+1)-2Z(\gamma-1)], \quad (57.54)$$

and the critical volume ratio

$$(\eta_2)_{\text{crit}} = [(Z+1)\gamma(\gamma+1)-Z(\gamma^2-1)]/[(Z+1)\gamma(\gamma+1)-Z(\gamma-1)^2], \quad (57.55)$$

separating fully dispersed and discontinuous shocks. As $Z\rightarrow\infty$, (57.53) to (57.55) reduce to the single-fluid Rayleigh formulae (57.37) to (57.39). For $Z=1$ shocks are fully dispersed when $M_1^2 \leq \frac{19}{15} \approx 1.125$, $p_2/p_1 \leq \frac{4}{3}$, and $\eta_2 \geq \frac{16}{19} \approx 0.842$. For any $\gamma > 1$ the critical Mach number, or pressure jump, above which we get discontinuous electron-isothermal shocks is a monotonc increasing function of Z.

(c) *Weakly Ionized Monatomic Gas* If instead of a fully ionized plasma we start with a weakly ionized monatomic gas and generate downstream temperatures of about 10^4 K to 2×10^4 K in the shock, the slowest post-shock relaxation process is ionization of the gas. If the material is originally completely neutral, the first few ionizations behind the shock are produced by atom-atom collisons; this mechanism is slow, and if it were the only ionization mechanism the material would remain neutral for very large distances downstream. However, once a few *seed electrons* have been produced, subsequent ionizations occur efficiently via electron-atom collisions, which are far more effective than atom-atom collisions because the electrons: (1) move much faster than the atoms, hence collide with many more particles per unit time and (2) are charged, hence interact strongly with atoms via the long-range Coulomb potential. (In a multicomponent gas, e.g., a stellar atmosphere, the seed electrons may come from trace elements such as Na, Mg, Al, K, and Ca, which have low ionization potentials and are thus easily ionized while the dominant constituents H and He, which have much higher ionization potentials, remain completely neutral.) Thus the seed electrons rapidly produce yet more electrons and generate an *electron avalanche*, which runs away exponentially until ultimately quenched when the ion density becomes large enough that recombinations can equilibrate against the rate of ionization.

Because the amount of energy required to ionize each atom is typically much larger than the average thermal energy of an electron, only the electrons far out on the Maxwellian tail are effective. After they collisionally ionize an atom these electrons end up with much lower velocities, hence the electron gas is cooled. The tail of the distribution can be replenished in a few electron self-collision times, but the factor limiting the replenishment (hence the growth rate of the electron avalanche) is the rate of energy transfer from the shock-heated atoms to the electrons. Initially this energy exchange is very slow because electron-atom collisions transfer

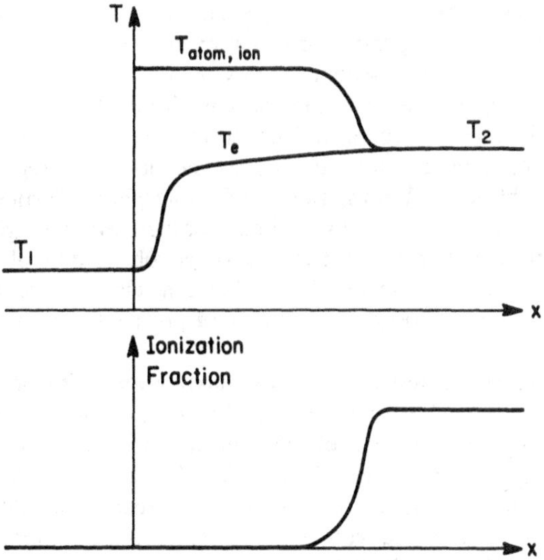

Fig. 57.9 Temperature and ionization-fraction variation as a function of spatial position in a shock in ionizing material.

energy inefficiently. But as the plasma becomes ionized, energy is transferred more rapidly in a two-step process: (1) atoms transfer energy to ions efficiently (because they have equal masses) and (2) the ions transfer energy to the electrons via Coulomb collisions (which are effective because of the long-range potential). Eventually the drain of energy to the electrons cools the atoms and ions, and the temperatures of all three particle species equalize at T_2, as sketched in Figure 57.9. Similarly, the ionization fraction saturates to its equilibrium value.

In most situations of astrophysical interest, radiation plays a more dominant role in determining excitation and ionization within the shock front and relaxation zone than the mechanisms just described. The radiating shock problem is more complex because photon mean free paths usually greatly exceed particle mean free paths. Hence radiation can force a nonlocal coupling of conditions at one point to those at widely separated points and can drive substantial departures from local thermodynamic equilibrium. Examples of such phenomena are described in (**K1**) and (**K2**); we discuss radiating shocks in Chapter 8.

58. Propagation of Weak Shocks

Having considered steady shocks in some detail we turn to the propagation and dissipation of shocks (both single pulses and trains of shocks) in a stratified medium such as a stellar atmosphere. For the present we confine attention to weak shocks ($m \lesssim 1$ or $\not{p} \lesssim 1$), which can be treated analytically, returning to strong-shock propagation in §60.

A major goal of weak-shock theory is to account for shock-wave dissipation. Dissipation is important because it bleeds energy from shocks and ultimately quenches them. Indeed, from the outset dissipation retards the growth of acoustic disturbances into the nonlinear regime and thus raises the height of shock formation in the atmosphere. Furthermore, once a shock forms, dissipation reduces (or at least retards the growth of) the shock's amplitude, hence extends the range of validity of weak-shock theory. Finally, dissipation provides a basic mechanism for nonradiative heat input into the atmosphere, a matter of great interest in astrophysical calculations.

In constructing the theory we must make several simplifying assumptions. (1) The material is a perfect gas with constant ratio of specific heats γ. We thus neglect ionization effects, which can be an important sink of thermal energy in shock-heated gas. (2) The shocks propagate strictly vertically in an isothermal atmosphere in hydrostatic equilibrium. (This model provides a rough caricature of the temperature-minimum region of the solar atmosphere.) We thus suppress refraction and reflection effects. (3) We ignore the back reaction of the shocks on the ambient medium. This is an important omission because shock heating may significantly alter the thermodynamic state of the atmosphere, and deposition of shock momentum may extend the atmosphere (i.e., increase its scale height). (4) We ignore the gravitational potential energy (buoyancy energy) in, and transported by, the wave. Therefore the theory can be accurate only for waves with frequencies much higher than the acoustic cutoff frequency. (5) Finally, we ignore radiative energy exchange, which is important in astrophysical applications; we return to this aspect of the problem in Chapter 8.

Despite its obvious limitations, weak-shock theory provides useful insight into the physics of shock propagation and has been popular in a wide variety of applications. The theory developed here follows the approach in (**S13**), to which the reader is referred for further details; see also (**B10**, Chaps. 6 and 7), (**B5**), and (**B6**).

PROPAGATION OF N WAVES

Consider the propagation of a small-amplitude, periodic acoustic wave. Because the phase velocity

$$v_p(u) = a_0 + \tfrac{1}{2}(\gamma+1)u \tag{58.1}$$

is largest at wave crests and smallest in the troughs, the wave steepens, the crests eventually overtake the troughs, shock, and produce a propagating N wave.

If the velocity profile is initially sinusoidal, $u = \frac{1}{2}u_0 \sin(2\pi z/\Lambda)$, then the crest of the wave overtakes the wave front with a speed $\frac{1}{4}(\gamma+1)u_0$, which is also the speed with which the wave front overtakes the preceding trough. Therefore the peak and trough coalesce into a vertical front when

$$\tfrac{1}{4}(\gamma+1)\int_0^t u_0(t')\,dt'=\tfrac{1}{4}(\gamma+1)a_0^{-1}\int_0^z u_0(z)\,dz=\tfrac{1}{4}\Lambda. \tag{58.2}$$

For a uniform medium $u_0(z)\equiv u_0$, hence shocks form after the wave propagates a distance

$$Z=a_0\Lambda/(\gamma+1)u_0. \tag{58.3}$$

In an isothermal atmosphere the velocity amplitude of an acoustic wave scales as $u_0(z)\propto\rho_0^{-1/2}\propto\exp(z/2H)$ where H is the scale height. In this case, (58.2) yields a shock-formation distance

$$Z=2H\ln\left[\frac{a_0\Lambda}{2(\gamma+1)u_0H}+1\right]. \tag{58.4}$$

Notice that in both cases the distance for shock formation increases with increasing Λ, hence short-period waves steepen into shocks sooner than long-period waves (see Figure 58.1).

After the wave travels a distance Z it has become an **N** wave with velocity profile

$$u(z)=\tfrac{1}{2}u_0[1-(2z/\Lambda)],\qquad(0\leq z\leq\Lambda), \tag{58.5}$$

which implies that at a fixed position in the medium the velocity varies as

$$u(t)=\tfrac{1}{2}u_0[1-(2t/\tau)],\qquad(0\leq t\leq\tau). \tag{58.6}$$

The shock travels with the velocity of the wave crest, namely

$$v_{\text{shock}}=a_0+\tfrac{1}{4}(\gamma+1)u_0 \tag{58.7}$$

hence

$$m=M^2-1\approx\tfrac{1}{2}(\gamma+1)u_0/a_0. \tag{58.8}$$

The energy transported by the wave in one period is

$$E_w=\tau\phi_w=\int_0^\tau(p-p_0)u\,dt\qquad(\text{ergs cm}^{-2}). \tag{58.9}$$

But for a weak shock $p-p_0\approx\gamma p_0u/a_0$ [cf. (48.24a) or (56.44)], hence

$$E_w=(\gamma p_0/a_0)\int_0^\tau u^2\,dt. \tag{58.10}$$

In particular, for an **N** wave

$$E_w=\tfrac{1}{12}\gamma p_0u_0^2\tau/a_0=\tfrac{1}{12}\rho_0u_0^2\Lambda=\gamma p_0m^2\Lambda/3(\gamma+1)^2. \tag{58.11}$$

Taking the logarithmic derivative of (58.11) and rearranging, we find an

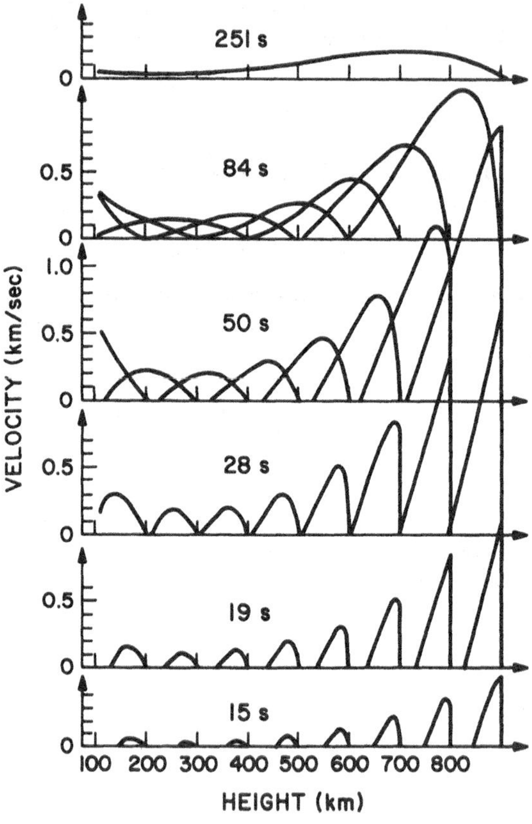

Fig. 58.1 Development of monochromatic acoustic waves into shocks in a model solar atmosphere. From (**U3**), by permission.

equation for the variation of the Mach number with height:

$$\frac{1}{m}\frac{dm}{dz}=\frac{1}{2}\left(\frac{1}{H}+\frac{1}{E_w}\frac{dE_w}{dz}-\frac{1}{\Lambda}\frac{d\Lambda}{dz}\right), \tag{58.12}$$

where H is the pressure scale height of the ambient atmosphere.

We can evaluate (dE_w/dz) by noting, from (56.56), that the amount of energy dissipated, per gram, by the shock is

$$\Delta q = T\,\Delta s = 2\gamma p_0 m^3/3(\gamma+1)^2\rho_0 \qquad (\text{ergs g}^{-1}) \tag{58.13}$$

while the mass flux into the shock front is $\rho_0 v_{\text{shock}}$ (g cm^{-2} s^{-1}), hence the average rate of energy loss by the shock is $(dE_w/dt)=-\rho_0 v_{\text{shock}}\Delta q$ (ergs cm^{-2} s^{-1}). Dividing by the shock velocity we obtain

$$(dE_w/dz) = -2\gamma p_0 m^3/3(\gamma+1)^2. \tag{58.14}$$

Combining (58.11) and (58.14) we see that for an N wave

$$(dE_w/dz) = -2mE_w/\Lambda. \tag{58.15}$$

In a weak N wave, each pulse peak moves as fast as the one that precedes it, hence the spacing between pulses remains constant, $\Lambda \equiv \Lambda_0$. Using this result and (58.15) in (58.12) we obtain the propagation equation

$$\frac{d}{dz}\left(\frac{1}{m}\right) + \frac{1}{2Hm} = \frac{1}{\Lambda_0}, \tag{58.16}$$

which has the solution

$$\frac{1}{m} = \frac{2H}{\Lambda_0} + \left(\frac{1}{m_0} - \frac{2H}{\Lambda_0}\right)e^{-z/2H}, \tag{58.17}$$

where m_0 measures the Mach number at a convenient reference height $z = 0$. From (58.17) we see that for $z \gg H$

$$m \rightarrow \Lambda_0/2H, \tag{58.18}$$

which leads to the remarkable conclusion that in an isothermal atmosphere an N wave propagates asymptotically with *constant shape* (i.e., constant wavelength Λ_0 and amplitude m). The tendency for the wave amplitude to grow exponentially as the density decreases is exactly balanced by the increased rate of dissipation, hence damping, resulting from a larger amplitude. This result is fortuitous; we will now show that a single pulse behaves quite differently.

PROPAGATION OF A PULSE

Let us now consider the propagation of a single pulse of total width Λ. Suppose the initial velocity profile is the sinusoid $u = u_0 \sin(\pi z/\Lambda)$. A shock forms when some part of the profile becomes vertical. One can easily show that this condition is first met right at the front of the wave, in a time T given, for a uniform medium, by

$$x = \tfrac{1}{2}(\gamma+1)u_0 \sin(\pi x/\Lambda)T. \tag{58.19}$$

For $x \ll \Lambda$, (58.19) implies that the shock forms in a distance

$$Z = a_0 T = 2a_0\Lambda/\pi(\gamma+1)u_0, \tag{58.20}$$

which is a factor of $(2/\pi)$ smaller than for an N wave. We can apply the same factor to (58.4) to estimate the distance for shock formation in a stratified atmosphere.

Once the pulse has steepened into a shock the velocity profile becomes

$$u(z) = u_0[1-(z/\Lambda)], \qquad (0 \leq z \leq \Lambda), \tag{58.21}$$

so that at a fixed location the velocity varies in time as

$$u(t) = u_0[1-(t/\tau)], \qquad (0 \leq t \leq \tau). \tag{58.22}$$

Using (58.22) in (58.10) we find that the energy transported by a pulse is

$$E_w = \tfrac{1}{3}\rho_0 u_0^2 \Lambda = 4\gamma p_0 m^2 \Lambda / 3(\gamma+1)^2, \tag{58.23}$$

that is, exactly four times the energy in a single period of an N wave of the same wavelength and total velocity jump at the front. The N wave transports less energy because the downward motion in the tail of the wave partially cancels the effect of the upward motion at the head of the wave.

Unlike an N wave, a pulse changes shape as it propagates because the head of the pulse, traveling with speed Ma_0, always outruns the tail of the pulse, traveling with speed a_0. Thus

$$(d\Lambda/dt) = (M-1)a_0 \approx \tfrac{1}{2}ma_0, \tag{58.24}$$

or

$$(d\Lambda/dz) = \tfrac{1}{2}m. \tag{58.25}$$

Using (58.23) in (58.14) we have

$$(dE_w/dz) = -mE_w/2\Lambda, \tag{58.26}$$

and using (58.26) and (58.25) in (58.12) we have

$$m^{-1}(dm/dz) = (1/2H) - (m/2\Lambda). \tag{58.27}$$

Equations (58.26) and (58.27) completely describe the propagation of the pulse.

Using (58.25) to eliminate z from (58.27) and solving we find

$$m = (\Lambda/2H) - (K/\Lambda), \tag{58.28}$$

where K is a constant. Substituting (58.28) into (58.25) we then have

$$(d\Lambda/dz) = (\Lambda/4H) - (K/2\Lambda) \tag{58.29}$$

which has the solution

$$\Lambda^2 = K' e^{z/2H} + 2HK. \tag{58.30}$$

Evaluating the constants so that $\Lambda = \Lambda_0$ and $m = m_0$ at $z = 0$, we find

$$\Lambda = \Lambda_0[(2Hm_0/\Lambda_0)(e^{z/2H} - 1) + 1]^{1/2} \tag{58.31}$$

and

$$m = m_0 e^{z/2H}[(2Hm_0/\Lambda_0)(e^{z/2H} - 1) + 1]^{-1/2}. \tag{58.32}$$

From (58.32) we see that for $(z/H) \gg 1$,

$$m \propto e^{z/4H} \propto \rho_0^{-1/4}, \tag{58.33}$$

in contrast to small-amplitude waves, for which the velocity amplitude increases as $\rho_0^{-1/2}$; this slower growth is attributable to a loss of wave energy via dissipation as the wave propagates.

APPLICATIONS TO THE SOLAR CHROMOSPHERE

Following the early work of Schatzman (**S3**), various forms of weak shock theory have been applied to the propagation and dissipation of shocks in the solar chromosphere by several authors [see e.g., (**O1**), (**J1**), (**U1**), (**U2**), (**U3**), and the summary in (**B10,** Chaps. 6 and 7)]. Most of this work makes allowance for radiative energy losses in addition to viscous dissipation. Some typical results showing the steepening of monochromatic acoustic waves are displayed in Figure 58.1. The waves are drawn whenever the wavefront reaches a multiple of 100 km. One sees clearly that short-period waves steepen into shocks sooner than long-period waves, as predicted by (58.4). The waves are heavily damped by radiative losses at heights below about 500 km, but develop into shocks quickly thereafter. The calculations cited suggest that the observed radiative energy loss by the chromosphere can be sustained by the dissipation of weak, short-period shocks with $m \leq 0.4$ to 0.5, for which weak-shock theory should be valid.

A critical assessment of the accuracy of weak-shock theory can be made by comparing its predictions with the results of full nonlinear calculations, as was done by Stein and Schwartz (**S13**), (**S14**). They find that the theory gives reasonable results so long as the wave period τ is much shorter than the acoustic cutoff period $\tau_a \approx 200$ s; as $\tau \rightarrow \tau_a$ the contribution of gravitational terms omitted from the theory described above become increasingly important, and the quality of the results deteriorates rapidly.

Stein and Schwartz also found that weak-shock theory always tends to overestimate the rate of growth of m with height; the effect is small for $\tau \leq 25$ s, but is major for $\tau \geq 50$ s. A consequence of this too-rapid growth of m is that the dissipation rate predicted by weak-shock theory is too large; for $\tau \sim 100$ s it is in error by almost a factor of 10 (**S14**), and for $\tau \sim 400$ s the error is several orders of magnitude. Indeed, even for the same Mach number, weak-shock theory predicts a larger dissipation rate than the nonlinear theory, by about 10 percent for $m = 0.28$ and about 50 percent for $m = 3$ ($M = 2$); the latter value should not be surprising, however, because weak-shock theory explicitly assumes that $m \ll 1$. The calculations show that almost 90 percent of the shock energy is deposited of heights less than 2000 km, and the damping length for short-period waves is only 500 km. Furthermore, it is found that it is essential to account for ionization effects and radiative losses in calculating the shock-induced temperature rise of the material. The temperature increase calculated by assuming that the gas is adiabatic is a factor of three too large.

CRITIQUE

Despite its frequent application in astrophysics, it is clear that weak-shock theory has only limited validity; an interesting critique of this approach is given in (**C10**). Besides being only linear, the theory contains numerous other approximations, which must be invoked in order to obtain analytical results. Ultimately it is based on Brinkley-Kirkwood theory (**B11**), which makes very simplified assumptions about the thermodynamic path followed by the material through and behind the shock, and further assumes that the postshock flow can be described by a similarity solution (cf. §60).

Most astrophysical calculations using weak-shock theory have been made for one-dimensional, infinite, monochromatic wave trains. This picture is a gross oversimplification for flows (e.g., in the solar chromosphere) containing a field of large-amplitude waves having different periods and directions of propagation, and neglects completely the possibility of wave-wave interactions (e.g., when a shock overruns another shock or a rarefaction). Moreover the theory formulated above omits gravitational energy terms, which are very important for waves with $\tau \approx \tau_{ac}$; allowance for such terms has been made in a theory developed by Saito (**S1**), which is discussed briefly in (**B10,** p. 296 and p. 343). The most serious flaw in the theory discussed thus far is the omission of radiation losses; some 60 to 80 percent of the energy in short-period waves is lost to radiative damping. We remedy this defect in Chapter 8.

The difficulties described above demonstrate the need for a more powerful method. We therefore turn to numerical techniques, which not only can handle the full nonlinear equations, but are also versatile and flexible enough to (1) permit a detailed description of the microphysics of the gas, (2) allow for structural complexities in the ambient medium (e.g., temperature and ionization gradients), (3) allow for wave-wave interactions and the back reaction of the waves on the background atmosphere, (4) be generalized easily to include the transport of energy and momentum by radiation, and (5) account for radiation-induced departures from local thermodynamic equilibrium.

59. Numerical Methods

One of the most effective methods for solving the equations of hydrodynamics is to replace the original differential equations by a set of *finite difference equations* that determine the physical properties of the fluid on discrete space and time meshes. Given suitable initial and boundary conditions we follow the evolution of the fluid by solving this discrete algebraic system at successive timesteps. Two problems to be faced are that (1) we must assure that the finite difference equations are numerically *stable* and (2) an efficient scheme must be found for handling shocks, which can produce discontinuities in the solution at or between mesh points.

In this section we do not attempt to discuss state-of-the-art methods, but confine our attention to one basic technique that has been successful in many applications; this example provides a good introduction to the vast literature on the subject. A fundamental reference on the numerical solution of fluid-flow problems is the classic book by Richtmyer and Morton (**R4**). More powerful modern techniques are discussed in (**W7**).

NUMERICAL SIMULATION OF ACOUSTIC WAVES

To obtain insight we start with a simple physical problem: the propagation of adiabatic acoustic waves in a perfect gas with no external forces. In planar geometry the Lagrangean dynamical equations are

$$(Dv/Dt) = -(\partial p/\partial m), \tag{59.1}$$

$$(Dx/Dt) = v, \tag{59.2}$$

$$V \equiv 1/\rho = (\partial x/\partial m), \tag{59.3}$$

and

$$(De/Dt) = -p(DV/Dt). \tag{59.4}$$

The system is closed by adjoining an equation of state of the form $p = p(\rho, e)$.

We now discretize the system in both space (i.e., mass) and time. First we divide the medium into a set of *mass cells* by choosing a set of values $\{m_j\}$ giving the Lagrangean coordinate of *cell surfaces*, as sketched in Figure 59.1a. The cell surfaces are located at spatial positions $\{x_j\}$, which

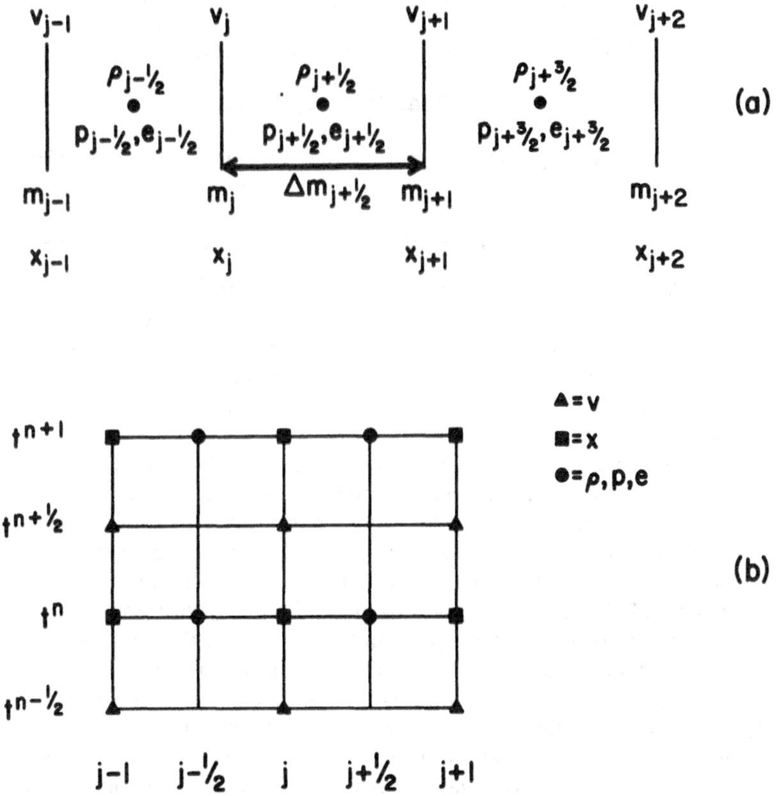

Fig. 59.1 (a) Centering of physical variables on Lagrangean mesh. (b) Spacetime centering of physical variables on Lagrangean mesh.

vary in time as the fluid moves; in contrast the Lagrangean coordinates $\{m_j\}$ remain fixed in time. We assign all thermodynamic properties to *cell centers*, and denote them by half-integer subscripts: $\rho_{j+(1/2)}$, $p_{j+(1/2)}$, $e_{j+(1/2)}$. Similarly, the (fixed) mass contained in a cell is $\Delta m_{j+(1/2)} = m_{j+1} - m_j$. In order to track the motion of the Lagrangean grid $\{m_j\}$ through physical space (x), velocities are assigned to cell surfaces (e.g., v_j).

Next we choose a discrete set of times $\{t^n\}$ at which the thermodynamic properties of the fluid are to be determined. We label variables with superscripts corresponding to their location in time [e.g., $\rho^n_{j+(1/2)}$ for ρ at the center of cell $(j, j+1)$ at time t^n]. Noting that we wish to know interface positions at the same time levels as the thermodynamic variables (because the density of a cell of fixed mass is determined by its volume), it is clear that velocities should be time centered midway between these levels [i.e., at $t^{n+(1/2)} \equiv \frac{1}{2}(t^n + t^{n+1})$]. The location of the variables in spacetime is sketched in Figure 59.1b.

The differential equations (59.1) to (59.4) are then replaced by

$$[v_j^{n+(1/2)} - v_j^{n-(1/2)}]/\Delta t^n = -[p^n_{j+(1/2)} - p^n_{j-(1/2)}]/\Delta m_j, \tag{59.5}$$

$$(x_j^{n+1} - x_j^n)/\Delta t^{n+(1/2)} = v_j^{n+(1/2)}, \tag{59.6}$$

$$V^{n+1}_{j+(1/2)} = 1/\rho^{n+1}_{j+(1/2)} = (x^{n+1}_{j+1} - x^{n+1}_j)/\Delta m_{j+(1/2)}, \tag{59.7}$$

and

$$e^{n+1}_{j+(1/2)} - e^n_{j+(1/2)} = -p^n_{j+(1/2)}[V^{n+1}_{j+(1/2)} - V^n_{j+(1/2)}]. \tag{59.8}$$

Here $\Delta t^{n+(1/2)} \equiv t^{n+1} - t^n$, $\Delta t^n \equiv \frac{1}{2}[\Delta t^{n-(1/2)} + \Delta t^{n+(1/2)}]$, and $\Delta m_j \equiv \frac{1}{2}[\Delta m_{j-(1/2)} + \Delta m_{j+(1/2)}]$. Given $\{v_j\}^{n-(1/2)}$, $\{x_j\}^n$, $\{\rho_{j+(1/2)}\}^n$, $\{p_{j+(1/2)}\}^n$ and $\{e_{j+(1/2)}\}^n$, either from initial conditions or from a preceding integration step, we solve (59.5) to (59.8) in the order indicated to obtain $\{v_j\}^{n+(1/2)}$, $\{x_j\}^{n+1}$, $\{\rho_{j+(1/2)}\}^{n+1}$, $\{e_{j+(1/2)}\}^{n+1}$, and finally $\{p_{j+(1/2)}\}^{n+1} = \{p[V^{n+1}_{j+(1/2)}, e^{n+1}_{j+(1/2)}]\}$.

As written, the system is *explicit*, that is, each variable is determined by direct evaluation using information already available. But it should be noted that the accuracy of (59.8) is impaired because p in (59.2) should really be evaluated at the *midpoint* of the integration interval, say as $\frac{1}{2}[p^n_{j+(1/2)} + p^{n+1}_{j+(1/2)}]$, instead of at one end (t^n). If this is done, then *two* unknowns, $p^{n+1}_{j+(1/2)}$ and $e^{n+1}_{j+(1/2)}$, appear in (59.8), which must now be solved simultaneously with the equation of state. In the special case of a perfect gas with constant specific heats, we can use the relation $p = (\gamma - 1)\rho e$ to eliminate either p or e and thus recover an explicit equation for either $p^{n+1}_{j+(1/2)}$ or $e^{n+1}_{j+(1/2)}$. But in general, when the physics of the gas is more complicated (e.g., by ionization), some kind of iteration procedure must be used to solve the coupled equations. As we will see below, in practice it is almost always necessary to make this additional effort anyway to assure numerical stability and energy conservation.

THE PROBLEM OF NUMERICAL STABILITY

If one attempts to solve (59.5) to (59.8) numerically, starting from smooth initial conditions, one finds that for sufficiently small timesteps Δt the solution remains smooth, and provides a good approximation to the analytical solution of (59.1) to (59.4). If, however, Δt is greater than some critical value, then starting from the same initial data the numerical solution quickly develops unphysical oscillations that rapidly grow and eventually swamp the true solution. This behavior is the result of a *numerical instability* of the finite difference equations, which, under certain circumstances, can allow small errors (e.g., roundoff) in the calculation to become progressively amplified until they dominate the computation. Experience has shown that in solving complex physical problems it is essential to perform a *stability analysis* of the difference equations used to model the system.

One of the basic tools of stability theory for initial-value problems is the *von Neumann local stability analysis*, which exploits the fact that a difference equation

$$y_j^n = L(x, t, y), \tag{59.9}$$

where L is a *linear difference operator* on a *uniform* spacetime mesh ($x_j = j\,\Delta x$, $t^n = n\,\Delta t$) can be solved exactly by a Fourier series of the general form

$$y_j^n = \sum_k A_k e^{ikj\,\Delta x} \xi_k^n. \tag{59.10}$$

In (59.10) the coefficients are determined by the imposed initial and boundary conditions. Each harmonic grows or decays independently of the others; ξ_k is the *amplification factor* (or *growth factor*) for the kth Fourier component over the time interval Δt. If the solution is to be stable, we must guarantee that no harmonic can become unbounded. Thus a *necessary* condition for stability is that the modulus $\|\xi_k\| \leq 1$ for all k. If this condition is met, no harmonic will be amplified; if it is violated, some harmonic can grow without limit and the solution becomes unstable.

As an example, consider the equation

$$(\partial\rho/\partial t) + v_0(\partial\rho/\partial x) = 0, \tag{59.11}$$

which describes the advection of material by a constant velocity field; we assume that $v_0 > 0$. The general solution of (59.11) is $\rho = f(x - v_0 t)$ where f is an arbitrary function fixed by initial conditions; this solution is a traveling wave in which the original distribution of the material is displaced to the right without an alteration in form. Suppose we represent (59.11) by the difference equation

$$(\rho_j^{n+1} - \rho_j^n)/\Delta t = -v_0(\rho_{j+1}^n - \rho_{j-1}^n)/2\,\Delta x, \tag{59.12}$$

which is centered in space and explicit in time. Taking a trial solution of the form

$$\rho_j^n(k) = \xi_k^n e^{ikj\,\Delta x}, \tag{59.13}$$

we find that (59.12) implies that

$$\xi_k = 1 - i\alpha \sin k\Delta x, \tag{59.14}$$

where $\alpha \equiv v_0\,\Delta t/\Delta x$. The modulus of ξ_k is thus

$$\|\xi_k\| = (1+\alpha^2 \sin^2 k\Delta x)^{1/2} \geq 1. \tag{59.15}$$

Hence the difference scheme is *unconditionally unstable*, that is, it is unstable for *any* finite Δt, no matter how small!

Alternatively, suppose we represent (59.11) by

$$(\rho_j^{n+1} - \rho_j^{n-1})/2\,\Delta t = -v_0(\rho_{j+1}^n - \rho_{j-1}^n)/2\,\Delta x, \tag{59.16}$$

which is centered in space and a *leapfrog* in time. Substituting (59.13) we now find

$$\xi_k^2 + 2i\beta\xi_k - 1 = 0, \tag{59.17}$$

where $\beta \equiv \alpha \sin k\Delta x$. Equation (59.17) has the solution

$$\xi_k = -i\beta \pm (1-\beta^2)^{1/2}. \tag{59.18}$$

When $\beta^2 > 1$, $\|\xi_k\| > 1$; when $\beta^2 \leq 1$, $\|\xi_k\| = 1$. Thus (59.16) is *conditionally stable*, that is, it is stable provided that $|(v_0\,\Delta t/\Delta x) \sin k\Delta x| \leq 1$ for all k, which implies that we must choose Δt such that

$$v_0\,\Delta t/\Delta x \leq 1. \tag{59.19}$$

Equation (59.19) is an example of the famous *Courant condition* (**C7**) which, in physical terms, states that the timestep must be sufficiently small that the wave cannot propagate over more than one spatial cell Δx in the interval Δt. We will encounter the Courant condition again in many other contexts.

Another stable difference-equation representation of (59.11) is obtained by using *upstream* (or *upwind*) *differencing*, writing

$$(\rho_j^{n+1} - \rho_j^n)/\Delta t = -v_0(\rho_j^n - \rho_{j-1}^n)/\Delta x, \tag{59.20}$$

which is explicit. Physically, (59.20) recognizes that material flows into cell j from cell $j-1$. Substituting (59.13) and calculating the amplification factor, we find

$$\|\xi_k\| = 1 + 2\alpha(1-\alpha)(\cos k\Delta x - 1). \tag{59.21}$$

For $0 \leq \alpha \leq 1$, $\|\xi_k\|$ reaches its maximum when $\cos k\Delta x = 1$, in which case $\|\xi_k\| = 1$, and the difference equation is stable. For $\alpha > 1$, $\|\xi_k\|$ is maximized when $\cos k\Delta x = -1$, in which case $\|\xi_k\| = (2\alpha - 1)^2 > 1$. Hence (59.20) is stable only if the Courant condition is satisfied. We thus see that a given differential equation may have more than one stable difference-equation representation, as well as unstable representations.

The above analysis applies to a single difference equation. In practice, we are more often interested in the stability of a system of M equations of the form

$$\mathbf{y}^n = \mathbf{L}(x, t, \mathbf{y}), \tag{59.22}$$

where $\mathbf{L}$ is a linear difference operator coupling $\mathbf{y}^n$ and $\mathbf{y}^{n-1}$, and $\mathbf{y}^n(x)$ denotes the numerical approximation, obtained from the difference equations, to the true solution $\mathbf{y}(x, n\Delta t) = [y_1(x, n\Delta t), \ldots, y_M(x, n\Delta t)]$ of the differential equations. Representing the solution by a Fourier series with amplitudes $\mathbf{a}_k$, one can show that (59.22) is equivalent to

$$\mathbf{a}_k^n = \mathbf{G}(\Delta t, k)\mathbf{a}_k^{n-1} \tag{59.23}$$

where $\mathbf{G}(\Delta t, k)$ is the $M \times M$ *amplification matrix* for the kth harmonic. The stability of the system of difference equations clearly depends on the behavior of $[\mathbf{G}(\Delta t, k)]^n$ for $0 \le n\Delta t \le T$, and intuitively it is obvious that we can achieve stability only if there exists some number τ such that for $(0 \le \Delta t \le \tau)$ and $(0 \le n\Delta t \le T)$ the matrices $[\mathbf{G}(\Delta t, k)]^n$ are *uniformly bounded* for all k, in the sense defined below.

We define the *bound* of a matrix $\mathbf{F}$ to be

$$\|\mathbf{F}\| \equiv \max_{|\mathbf{v}|=1} |\mathbf{F}\mathbf{v}| = \max_{|\mathbf{v}|\neq 0} (|\mathbf{F}\mathbf{v}|/|\mathbf{v}|); \tag{59.24}$$

here $|\mathbf{v}|$ is the usual Euclidian norm of $\mathbf{v}$, that is, $|\mathbf{v}| = (v_1^2 + \ldots + v_M^2)^{1/2}$. Thus the bound of $\mathbf{F}$ is the maximum norm of the vectors resulting from the operation of a transformation $\mathbf{F}$ on the M-dimensional vector space $(\mathbf{v}, \mathbf{w}, \ldots \text{etc.})$. If the eigenvalues of $\mathbf{F}$ are $\lambda_1, \ldots, \lambda_M$, we define the *spectral radius* R of $\mathbf{F}$ to be $\max |\lambda_i|$, $(i = 1, \ldots, M)$. It is easy to see that $\|\mathbf{F}\| \ge R$ because the maximum value of $|\mathbf{F}\mathbf{v}|/|\mathbf{v}|$ cannot be less than the value obtained when $\mathbf{v}$ is the eigenvector corresponding to R, and may exceed this value if we can choose $\mathbf{v}$ astutely (e.g., as a linear combination of eigenvectors) such that $|\mathbf{v}|$ becomes very small while $|\mathbf{F}\mathbf{v}|$ remains of order R.

The spectral radius of $\mathbf{F}^n$ is R^n because the eigenvalues of $\mathbf{F}^n$ are $\lambda_1^n, \ldots, \lambda_M^n$. Moreover

$$\|\mathbf{F}^2\| = \max_{|\mathbf{v}|\neq 0} \frac{|\mathbf{F}(\mathbf{F}\mathbf{v})|}{|\mathbf{v}|} = \max_{|\mathbf{v}|\neq 0} \frac{|\mathbf{F}(\mathbf{F}\mathbf{v})|}{|\mathbf{F}\mathbf{v}|}\frac{|\mathbf{F}\mathbf{v}|}{|\mathbf{v}|} \le \max_{\substack{|\mathbf{v}|\neq 0\\|\mathbf{w}|\neq 0}} \frac{|\mathbf{F}\mathbf{w}|}{|\mathbf{w}|}\frac{|\mathbf{F}\mathbf{v}|}{|\mathbf{v}|} = \|\mathbf{F}\|^2, \tag{59.25}$$

therefore $R^2 \le \|\mathbf{F}^2\| \le \|\mathbf{F}\|^2$, hence by induction $R^n \le \|\mathbf{F}^n\| \le \|\mathbf{F}\|^n$. In particular if $R(\Delta t, k)$ is the spectral radius of $\mathbf{G}(\Delta t, k)$, it follows from (59.25) that

$$[R(\Delta t, k)]^n \le \|[\mathbf{G}(\Delta t, k)]^n\| \le \|\mathbf{G}(\Delta t, k)\|^n, \tag{59.26}$$

hence a *necessary* condition for stability is that there exist numbers τ and C_1 such that for $(0 \le \tau \le \Delta t)$, $(0 \le n\Delta t \le T)$,

$$[R(\Delta t, k)]^n \le C_1 \tag{59.27}$$

for all k.

Without loss of generality we can assume $C_1 \geq 1$, hence $R(\Delta t, k) \leq C_1^{1/n}$, and in particular

$$R(\Delta t, k) \leq C_1^{\Delta t/T} \tag{59.28}$$

because $n \leq T/\Delta t$. For $0 \leq \Delta t \leq \tau$, the exponential $C_1^{\Delta t/T}$ can always be bounded from above by an expression of the form $(1+C_2\Delta t)$. Hence, we find the *von Neumann necessary condition for stability* is that if $\lambda_1, \ldots, \lambda_M$ are the eigenvalues of the amplification matrix $\mathbf{G}(\Delta t, x)$, then

$$|\lambda_i| \leq 1 + O(\Delta t) \tag{59.29}$$

for all $i = 1, \ldots, M$, for $0 \leq \Delta t \leq \tau$, and for all k.

Equation (59.29) merits two comments. (1) In our earlier analysis we required that the modulus of the amplification factors $\|\xi_k\|$ all be less than unity. The present result is less restrictive, and allows for the possibility of a legitimate exponential growth of the solution. (2) Although (59.29) is a *necessary* condition for stability, it is not, in general, *sufficient*. Furthermore, it is a purely *local* criterion and does not account for boundary conditions. In practice (59.29) is sometimes found to be both necessary *and* sufficient, but in general the derivation of sufficient stability criteria for a given system requires a much more extensive (and difficult) analysis; see (**R4**).

To illustrate the von Neumann analysis, consider the system (59.1) to (59.4). In order to simplify the algebra, rewrite (59.1) as

$$(Dv/Dt) = -(\partial p/\partial \rho)_s(\partial \rho/\partial m) = (a^2/V^2)(\partial V/\partial m), \tag{59.30}$$

where a is the adiabatic sound speed. Difference (59.30) as

$$v_j^{n+(1/2)} - v_j^{n-(1/2)} = (\Delta t/\Delta m)(a^2/V^2)_j^n[V_{j+(1/2)}^n - V_{j-(1/2)}^n], \tag{59.31}$$

and couple this equation to the time difference of (59.7), namely

$$V_{j+(1/2)}^{n+1} - V_{j+(1/2)}^n = (\Delta t/\Delta m)[v_{j+1}^{n+(1/2)} - v_j^{n+(1/2)}]. \tag{59.32}$$

Now examine the growth of the kth Fourier component, taking trial solutions

$$v_j^{n+(1/2)}(k) = (A_k)^{n+(1/2)}e^{ikj\,\Delta m} \tag{59.33}$$

and

$$V_{j+(1/2)}^n(k) = (B_k)^n e^{ik[j+(1/2)]\Delta m}. \tag{59.34}$$

One finds that the amplification matrix is

$$\mathbf{G}(\Delta t, k) = \begin{pmatrix} 1 & i\beta^2/\alpha \\ i\alpha & 1-\beta^2 \end{pmatrix}, \tag{59.35}$$

where $\alpha \equiv 2(\Delta t/\Delta m)\sin(\frac{1}{2}k\Delta m)$ and $\beta \equiv (a/V)\alpha$. The eigenvalues of $\mathbf{G}$ are

$$\lambda = \tfrac{1}{2}\{(2-\beta^2) \pm [(2-\beta^2)^2 - 4]^{1/2}\}. \tag{59.36}$$

If $\beta^2 \leq 4$, λ lies on the unit circle; if $\beta^2 > 4$, $\|\lambda\| > 1$. Thus for stability we must have $-1 \leq (\rho a\, \Delta t/\Delta m) \sin(\frac{1}{2}k\Delta m) \leq 1$, which is guaranteed if

$$\Delta t \leq (\Delta m/\rho)/a = \Delta x/a, \tag{59.37}$$

that is, if the Courant condition is satisfied. In practice we choose the smallest value of Δt found from (59.37) for the entire mesh.

IMPLICATIONS OF SHOCK DEVELOPMENT

As we saw in §55, nonlinear effects in wave propagation inevitably lead to shock formation. We further saw in §57 that the thickness of a shock is of the order of a few particle mean free paths, and indeed that in an ideal fluid the shock is a mathematical discontinuity. Once shocks form, the differential equations governing the flow must be supplemented by jump conditions, in effect internal boundary conditions within the flow, in order to obtain a unique solution.

If we were to attempt to simulate a flow containing shocks using the Lagrangean difference equations written above, we would immediately encounter severe difficulties. First, these equations contain no dissipative terms, and therefore cannot account for the entropy increase produced by a shock; consequently they will not yield even approximately the right answers behind a shock. Second, even if we were to include dissipative terms using realistic values of the molecular viscosity and thermal conductivity of the gas, the shock thickness would generally be several orders of magnitude smaller than the spacing of grid points on which the difference equations are solved. A brute-force attempt to reduce the grid size to a mean free path is doomed from the outset because if we reduce $\Delta x \to \Delta x/k$, it follows from the Courant condition that we must also reduce the timestep $\Delta t \to \Delta t/k$, hence the total computing effort needed to follow the flow for a definite time rises as k^2, which rapidly becomes prohibitive.

One option is to use the *method of characteristics* to follow discontinuities, and then impose the Rankine-Hugoniot relations to do *shock fitting*. Although this method has been highly developed [see e.g., (**H16**)], it is relatively complex and cumbersome because generally the shock is not at a gridpoint, and an iterative solution of a moderately complicated set of nonlinear equations is required to locate it before the jump conditions can be applied [see (**H16**), (**R4,** 304–311)]. While shock fitting is relatively easy to apply when shocks propagate into undisturbed fluid or at least occur regularly in a train (**U5**), (**U6**), it becomes harder to use when, say, shocks overrun one another, or one shock collides with another propagating in the opposite direction. Moreover, because shocks can develop spontaneously anywhere in a flow, one must also develop strategies for *detecting* when shocks have formed.

To overcome these difficulties, von Neumann and Richtmyer (**V4**) devised a scheme that handles shocks *automatically*, wherever and whenever they arise. The essence of the method is to introduce into the difference equations an artificial dissipative process that models the real dissipation mechanisms in a gas and that gives the correct entropy jump (hence the correct physical properties in the postshock flow), but which smears the shock over a few mesh points in the difference-equation grid, instead of leaving it unresolved on a subgrid level. The shock discontinuity is thus replaced by a transition layer within which the fluid properties vary rapidly but not discontinuously, across which the basic conservation relations are satisfied, and whose thickness can be adjusted to match the grid size, which is chosen according to the physical requirements of the problem. The difference equations then apply everywhere, and the computations proceed completely automatically, without shock fitting.

THE VON NEUMANN–RICHTMYER ARTIFICIAL-VISCOSITY METHOD

We showed in §57 that both viscosity and thermal conduction produce entropy in a shock. We found that viscosity yields a smooth variation of all quantities in shocks of all strengths, whereas thermal conduction yields a smooth transition only for shocks below a certain strength, while stronger inviscid conducting shocks sustain discontinuities in density and pressure. As smoothness of the numerical solution is of paramount importance, we choose viscosity as our dissipation mechanism.

In §§26 and 27 we saw that the momentum and energy equations for one-dimensional planar viscous flow can be written

$$\rho(Dv/Dt) = f - [\partial(P+Q)/\partial z] \tag{59.38}$$

and

$$(De/Dt) + (p+Q)[D(1/\rho)/Dt] = \dot{q}, \tag{59.39}$$

where $\dot{q}$ is the energy input per unit mass from "external" sources (e.g., radiation or thermonuclear reactions), and Q is an equivalent viscous pressure

$$Q = -\tfrac{4}{3}\mu'(\partial v/\partial z). \tag{59.40}$$

Here $\mu' = \mu + \tfrac{3}{4}\zeta$ is the effective viscosity. To handle shocks in the difference equations we propose to use an *artificial viscosity* (or *pseudoviscosity*), choosing a suitable form for Q. One option would be to use (59.40) with a large constant value for μ', chosen such that the implied molecular mean free path would be of the same order as the grid spacing Δz. This approach is not satisfactory, however, for as we saw in §57 the thickness of a shock, for a given λ, is inversely proportional to its strength, hence we would obtain sharp strong shocks, but weak shocks would be spread over many gridpoints. Moreover such a large viscosity would spuriously reduce the Reynolds number of the flow in regions devoid of shocks, and would therefore seriously degrade the accuracy of the overall solution.

Von Neumann and Richtmyer realized that these problems could be overcome by using a *nonlinear* artificial viscosity that is large in shocks but very small elsewhere. In particular, they suggested using a Q that is *quadratic* in the rate of shear, and adopted

$$Q = \begin{cases} \tfrac{4}{3}\rho l^2(\partial v/\partial z)^2 & \text{for} \quad (\partial v/\partial z) < 0, \qquad (59.41a) \\ 0 & \text{for} \quad (\partial v/\partial z) \geq 0. \qquad (59.41b) \end{cases}$$

Recalling that μ has dimensions (dyne s cm^{-2}) = (g cm^{-3})(cm^2 s^{-1}) one sees that l has the dimensions of length. Typically l is chosen to be some small multiple of the grid spacing Δz.

We can also view (59.41) as a viscous pressure that is linear in the rate of shear

$$Q = -\tfrac{4}{3}\mu_Q(\partial v/\partial z), \qquad (59.42)$$

with an artificial viscosity coefficient that is proportional to the rate of compression of the fluid:

$$\mu_Q = \begin{cases} l^2(D\rho/Dt) & \text{for} \quad (D\rho/Dt) > 0, \qquad (59.43a) \\ 0 & \text{for} \quad (D\rho/Dt) \leq 0. \qquad (59.43b) \end{cases}$$

The relevance of this interpretation will emerge below.

The artificial viscosity given by (59.41) to (59.43): (1) comes into action only when the gas is compressed (both a prerequisite and a characteristic property of shock formation) and (2) is very small or zero in regions away from shocks. Both of these properties are highly desirable.

From the analysis of §57 it follows that (59.38) to (59.40) will lead to the Rankine-Hugoniot relations regardless of the precise analytical form of Q, provided only that $Q \to 0$ where velocity gradients vanish in the upstream and downstream flow far from a shock front. As the von Neumann-Richtmyer pseudoviscosity manifestly meets this requirement, we are assured that it will produce the correct jump in entropy (and all other variables) across the shock, as well as the correct shock propagation speed. [These attributes can also be verified by direct analysis (**V4**), (**R4,** 314–316).] A large body of computational work has demonstrated that the von Neumann-Richtmyer method gives good results as long as the resulting shock thickness, typically 3 to 4 Δz, is not too coarse to permit an accurate representation of other physical processes of interest (e.g., radiation transport).

THE IMPLICATIONS OF DIFFUSIVE ENERGY TRANSPORT

We have thus far assumed that the gas is essentially adiabatic and have ignored energy transport by diffusion processes such as thermal conduction (or radiation diffusion). However, we know that these processes do occur in a real fluid, and we should inquire into their consequences for the numerical stability of finite-difference representations of the energy equation.

To gain insight, consider a simple linear heat-conduction problem as posed by the parabolic equation

$$(\partial T/\partial t) = \sigma(\partial^2 T/\partial x^2) \qquad (59.44)$$

along with suitable initial and boundary conditions. Suppose we represent (59.44) by the *explicit formula*

$$(T_j^{n+1} - T_j^n)/\Delta t = \sigma(T_{j+1}^n - 2T_j^n + T_{j-1}^n)/\Delta x^2. \tag{59.45}$$

Using (59.13) as a trial solution for the kth Fourier component we find

$$\xi_k = 1 + \alpha(\cos k\Delta x - 1) \tag{59.46}$$

where $\alpha \equiv 2\sigma\,\Delta t/\Delta x^2$. Clearly ξ_k is always ≤ 1. To bound ξ_k away from -1 we must have $\alpha \leq 1$; hence the stability criterion is

$$\Delta t = \Delta x^2/2\sigma, \tag{59.47}$$

which is quite restrictive because if we refine the spatial grid by a factor $1/k$, we must decrease Δt by a factor $1/k^2$, so the computing effort to span a given time interval increases as k^3. The timestep set by (59.47) may be much smaller than the natural hydrodynamic timestep, particularly in regions of high conductivity and/or low-heat capacity (i.e., large σ). Thus use of an explicit formula for diffusion processes may seriously impair one's ability to follow a flow numerically. We must therefore find a more stable difference scheme.

The difficulty just described is overcome by using an *implicit* difference equation, that is, one that contains information about T^{n+1} in both the time and space operators. In particular, consider formulae of the general form

$$T_j^{n+1} - T_j^n = \tfrac{1}{2}\alpha[\theta(\delta^2 T)_j^{n+1} + (1-\theta)(\delta^2 T)_j^n], \tag{59.48}$$

with $0 \leq \theta \leq 1$. Here $(\delta^2 T)_j$ denotes the centered second difference $(T_{j+1} - 2T_j + T_{j-1})$ at the time levels indicated. The von Neumann local stability analysis now leads to

$$\xi_k = [1 - (1-\theta)\alpha(1 - \cos k\Delta x)]/[1 + \theta\alpha(1 - \cos k\Delta x)]. \tag{59.49}$$

As before, ξ_k is always ≤ 1. Furthermore, ξ_k is a monotone decreasing function of $\gamma \equiv \alpha(1 - \cos k\Delta x)$, hence for a given α and θ, ξ_k is minimized when $\cos k\Delta x = -1$. To bound ξ_k away from -1 we therefore demand that

$$-1 \leq [1 - 2(1-\theta)\alpha]/(1 + 2\theta), \tag{59.50}$$

which implies that

$$(1 - 2\theta)\alpha \leq 1. \tag{59.51}$$

For $0 \leq \theta \leq \frac{1}{2}$, Δt is restricted by

$$\Delta t \leq \Delta x^2/[2\sigma(1 - 2\theta)], \tag{59.52}$$

which reduces to (59.47) when $\theta = 0$. For $\frac{1}{2} \leq \theta \leq 1$, (59.51) is satisfied for *all* α; the difference equation (59.48) is then *unconditionally stable*, and can be solved using arbitrarily large values of Δt, thus surmounting any incompatibility with the hydrodynamically determined timestep. For $\theta = \frac{1}{2}$

we have the *Crank-Nicholson scheme*, which is time centered and has a truncation error of $O(\Delta t^2)$. For $\theta = 1$ we have the *backward Euler* (or *fully implicit*) *scheme*, which has a worse truncation error, $0(\Delta t)$, but is very stable.

The advantages of an implicit scheme are manifest; but to enjoy them we must pay a price. Unlike the explicit scheme, where T_j^{n+1} is calculated directly from preexisting information, in an implicit scheme, T_j^{n+1} is coupled to T^{n+1} at adjacent space points, hence we must solve a linear system of the form

$$-a_j T_{j-1}^{n+1} + b_j T_j^{n+1} - c_j T_{j+1}^{n+1} = r_j, \qquad (j = 1, \ldots, J). \tag{59.53}$$

Boundary conditions guarantee that $a_1 = c_J \equiv 0$. Equation (59.53) is solved by *Gaussian elimination*; we first perform a forward elimination to compute

$$d_j \equiv c_j/(b_j - a_j d_{j-1}) \tag{59.54}$$

and

$$e_j \equiv (r_j + a_j e_{j-1})/(b_j - a_j d_{j-1}) \tag{59.55}$$

for $j = 1, \ldots, J$, and then calculate T_j^{n+1} by back substitution

$$T_j^{n+1} = d_j T_{j+1}^{n+1} + e_j \tag{59.56}$$

for $j = J, J-1, \ldots, 1$. The computational effort required to solve (59.53) scales only linearly with the number of mesh points [as does an update of T^{n+1} via the explicit scheme (59.45)]. In any realistic problem, solving either (59.45) or (59.53) is typically only a small fraction of the total effort required to advance the dynamics a timestep, and the additional cost of using an implicit system is usually greatly outweighed by the ability to take large timesteps.

In this book we are not concerned with solving the heat-conduction equation per se, but rather with solving the equations of hydrodynamics when the energy equation contains diffusive terms. The main lesson we learn from the above analysis is that even if we can use an explicit scheme for the continuity and momentum equations, we must generally use an implicit energy equation in order to avoid unacceptable timestep restrictions; therefore in what follows we will always write the energy equation in implicit form.

ILLUSTRATIVE DIFFERENCE EQUATIONS

Let us now examine some illustrative *examples* of difference equations for solving one-dimensional Lagrangean flow problems in a gravitationally stratified medium.

(*a*) *Explicit Hydrodynamics*; *Planar Geometry* Consider first explicit hydrodynamics in planar geometry. Choose the independent variable to be the column mass measured inward into the medium from an upper boundary (as would be appropriate for a stellar atmosphere viewed from the outside). Then

$$m(z) \equiv \int_z^\infty \rho(z')\,dz'; \tag{59.57}$$

note that now $dm = -\rho\,dz$, which is opposite to the sign convention used in (59.1) to (59.8). The equations to be solved are

$$(Dv/Dt) = [\partial(p+Q)/\partial m] - g, \tag{59.58}$$

$$(Dz/Dt) = v, \tag{59.59}$$

$$V = 1/\rho = -(\partial z/\partial m), \tag{59.60}$$

and

$$(De/Dt) + (p+Q)[D(1/\rho)/Dt] = \dot{q}. \tag{59.61}$$

Choose a mass grid $\{m_d\}$, $d = 1, \ldots, D+1$ marking the surfaces of D discrete slabs. At the upper boundary, m_1 will be nonzero if we assume there is material lying above that surface. The momentum equation is represented by the explicit formula

$$[v_d^{n+(1/2)} - v_d^{n-(1/2)}]/\Delta t^n = -g + [p_{d+(1/2)}^{n+\lambda} - p_{d-(1/2)}^{n+\lambda} + Q_{d+(1/2)}^{n-(1/2)} - Q_{d-(1/2)}^{n-(1/2)}]/\Delta m_d. \tag{59.62}$$

Here Δm_d, $\Delta m_{d+(1/2)}$, Δt^n, and $\Delta t^{n+(1/2)}$ are defined as in (59.5) to (59.8). In general, the timesteps $\Delta t^{n-(1/2)}$ and $\Delta t^{n+(1/2)}$ are unequal; to improve the accuracy of (59.62) we can center the pressure gradient by defining

$$t^{n+\lambda} \equiv \tfrac{1}{2}\{[t^n - \tfrac{1}{2}\Delta t^{n-(1/2)}] + [t^n + \tfrac{1}{2}\Delta t^{n+(1/2)}]\} = t^n + \tfrac{1}{4}[\Delta t^{n+(1/2)} - \Delta t^{n-(1/2)}], \tag{59.63}$$

and using the approximate extrapolation

$$p_{d+(1/2)}^{n+\lambda} \equiv p_{d+(1/2)}^n + \tfrac{1}{4}[\Delta t^{n+(1/2)} - \Delta t^{n-(1/2)}][p_{d+(1/2)}^n - p_{d+(1/2)}^{n-1}]/\Delta t^{n-(1/2)}. \tag{59.64}$$

The artificial viscosity is computed from a difference representation of (59.41), that is,

$$Q_{d+(1/2)}^{n-(1/2)} = k_Q^2 \tfrac{1}{2}[\rho_{d+(1/2)}^{n-1} + \rho_{d+(1/2)}^n][v_{d+1}^{n-(1/2)} - v_d^{n-(1/2)}]^2 \tag{59.65}$$

if $\rho_{d+(1/2)}^n > \rho_{d+(1/2)}^{n-1}$, and $Q_{d+(1/2)}^{n-(1/2)} = 0$ otherwise. Here $k_Q \equiv l/\Delta z$ is a pure number, typically 1.5 to 2. In astrophysical applications, where zone thicknesses may range over several orders of magnitude in a single flow, it is often more satisfactory to use a fixed length l in (59.41) than to choose a constant k_Q (**W7**). Notice that the pseudoviscous pressure is *lagged* at $t^{n-(1/2)}$ in (59.62). In general this lack of centering does not produce large errors; it is possible to improve the centering, but the result is inconvenient and may even be unstable (**R4,** 319).

To find $v_1^{n+(1/2)}$ and $v_{D+1}^{n+(1/2)}$ we apply boundary conditions. At the lower boundary we assume that v is a known function of time:

$$v_{D+1}^{n+(1/2)} = f[t^{n+(1/2)}]. \tag{59.66}$$

Typical choices are $v_{D+1} \equiv 0$, or v_{D+1} equals the velocity of a driving "piston." At the upper boundary there are several commonly used choices. For a *free boundary* (no net force across the first cell) we have

$$v_1^{n+(1/2)} \equiv v_2^{n+(1/2)}. \tag{59.67}$$

For a *transmitting boundary* we demand

$$(\partial v/\partial t) = -a(\partial v/\partial z) = a\rho(\partial v/\partial m), \tag{59.68}$$

where a is the local sound speed, which implies that an incident wave is propagated through the boundary without alteration [cf. (59.11)]. In finite form,

$$v_1^{n+(1/2)} - v_1^{n-(1/2)} = \Delta t^n a_{3/2}^n \rho_{3/2}^n [v_2^{n-(1/2)} - v_1^{n-(1/2)}]/\Delta m_{3/2}. \tag{59.69}$$

For *zero surface pressure*, that is $(p+Q)_1 \equiv 0$, we have

$$v_1^{n+(1/2)} - v_1^{n-(1/2)} = -g\, \Delta t^n + [(\Delta t^n/\Delta m_1)(p_{3/2}^{n+\lambda} + Q_{3/2}^{n-(1/2)})], \tag{59.70}$$

where

$$\Delta m_1 \equiv \tfrac{1}{2}\Delta m_{3/2} + m_1 \tag{59.71}$$

includes any atmospheric mass (m_1) assumed to lie above the first cell.

Having updated velocities we calculate

$$z_d^{n+1} = z_d^n + v_d^{n+(1/2)}\, \Delta t^{n+(1/2)} \tag{59.72}$$

and

$$V_{d+(1/2)}^{n+1} = 1/\rho_{d+(1/2)}^{n+1} = (z_d^{n+1} - z_{d+1}^{n+1})/\Delta m_{d+(1/2)}, \tag{59.73}$$

and can then calculate the artificial viscosity $Q_{d+(1/2)}^{n+(1/2)}$ at $t^{n+(1/2)}$.

Finally we solve the implicit energy equation

$$\begin{aligned} e_{d+(1/2)}^{n+1} - e_{d+(1/2)}^n + \{\tfrac{1}{2}[p_{d+(1/2)}^n + p_{d+(1/2)}^{n+1}] + Q_{d+(1/2)}^{n+(1/2)}\}[V_{d+(1/2)}^{n+1} - V_{d+(1/2)}^n] \\ = \Delta t^{n+(1/2)}[(1-\theta)\dot{q}_{d+(1/2)}^n + \theta\dot{q}_{d+(1/2)}^{n+1}] \equiv \Delta t^{n+(1/2)}\langle \dot{q}_{d+(1/2)}\rangle^{n+(1/2)} \end{aligned} \tag{59.74}$$

given constitutive equations $e = e(\rho, T)$ and $p = p(\rho, T)$ or $e = e(\rho, p)$. Notice that in (59.74) the pseudoviscous pressure is time centered so that the correct total work is computed. In calculating $\langle \dot{q} \rangle$ one would choose $\theta = \frac{1}{2}$ for accuracy, and $\theta = 1$ to enhance stability or for quasistatic energy transport (e.g., by radiation—see introduction to §6.5 and also §98).

To solve (59.74) we start from an estimate of T^{n+1}, say T^*, and then *linearize*:

$$p_{d+(1/2)}^{n+1} \approx p[\rho_{d+(1/2)}^{n+1}, T_{d+(1/2)}^*] + [(\partial p/\partial T)_\rho]_{d+(1/2)}\, \delta T_{d+(1/2)} \tag{59.75}$$

and

$$e_{d+(1/2)}^{n+1} \approx e[\rho_{d+(1/2)}^{n+1}, T_{d+(1/2)}^*] + [(\partial e/\partial T)_\rho]_{d+(1/2)}\, \delta T_{d+(1/2)}. \tag{59.76}$$

We likewise linearize the external source $\dot{q}$. If $\dot{q}$ depends only on local values of (ρ, T) we can solve (59.74) pointwise, iterating to consistency (i.e., $\delta T \to 0$). If $\dot{q}$ contains diffusive terms, the linearization process yields a banded system like (59.53), which is solved by Gaussian elimination and iterated to convergence.

A stability analysis of the full system (59.62) plus (59.72) to (59.74) is complicated, so we will only quote results (**R4,** 12.12). Outside a shock, where the pseudoviscosity vanishes, the usual Courant condition must be satisfied. In a very strong shock, the analysis implies that the timestep must be restricted further by a factor $f_Q = \gamma^{1/2}/2k_Q$, which is about $\frac{1}{3}$ in typical problems. Trial calculations show that this theoretical restriction is too strict, and that choosing $f_Q \approx \frac{1}{2}$ is usually sufficient to assure stability. One must also impose timestep restrictions to assure *accuracy* of the solution as well as stability; thus one may restrict the fractional change in any variable to be less than some prechosen value, whose size is set from experience.

The efficacy of the artificial viscosity method is illustrated in Figure 59.2, which shows two test calculations for a propagating shock with a pressure ratio $p_2/p_1 = 5$ in a gas with $\gamma = 2$. The results in part (a) were obtained using $k_Q = 2$, and in part (b) using $k_Q = 0$. Without pseudoviscosity, there are large oscillations in the postshock fluid, and the shock speed is 10 per cent too low. These oscillations do not grow in time, and are not numerical instabilities (Δt was chosen to be 0.22 times the Courant limit). They are motions of the mass cells reminiscent of random motions of molecules in the shock-heated gas; the effect of artificial viscosity is to damp these motions and to convert their kinetic energy into internal energy of the gas.

The differential equations (59.58) to (59.61) can be combined into a total energy equation

$$\frac{D}{Dt}(e+\tfrac{1}{2}v^2) - \frac{\partial}{\partial m}[(p+Q)v] = -gv + \dot{q}, \tag{59.77}$$

which has the integral

$$\frac{D}{Dt}\int (e+\tfrac{1}{2}v^2+gz)\,dm = \int \dot{q}\,dm + v_{D+1}(p+Q)_{D+1} - v_1(p+Q)_1. \tag{59.78}$$

Physically, (59.78) states that the change in the total (internal plus kinetic plus potential) energy of the fluid over some time interval equals the time-integrated energy input from external sources minus the work done by the fluid against its boundaries. A discretized energy conservation relation is obtained by replacing the integrals over mass and time by sums over mass cells and timesteps, starting from some initial time t^0 when the

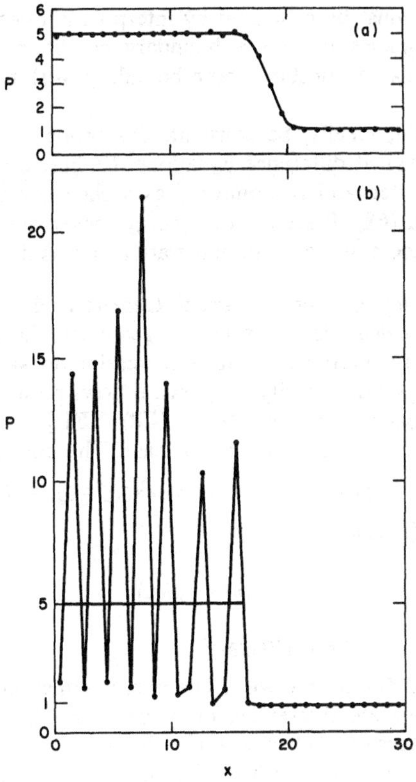

Fig. 59.2 Shock computed (a) with and (b) without artificial viscosity. From (**R4**) by permission.

state of the flow is known. Thus at t^{n+1}

$$\begin{aligned}\mathscr{E}^{n+1} &= \sum_d \{e^{n+1}_{d+(1/2)}\,\Delta m_{d+(1/2)} + [g(z_d^{n+1}-z_d^0)+\tfrac{1}{2}(v_d^{n+1})^2]\,\Delta m_d\} \\ &\quad + \mathscr{W}_1^{n+1} - \mathscr{W}_{D+1}^{n+1} - \mathscr{q}^{n+1} \\ &= \sum_d [e^0_{d+(1/2)}\,\Delta m_{d+(1/2)} + \tfrac{1}{2}(v_d^0)^2\,\Delta m_d] = \text{Constant},\end{aligned} \tag{59.79}$$

where

$$\mathscr{W}_i^{n+1} \equiv \sum_{k=0}^{n} [\tfrac{1}{2}(p_i^k + p_i^{k+1}) + Q_i^{k+(1/2)}](z_i^{k+1} - z_i^k) \tag{59.80}$$

and

$$\mathscr{q}^{n+1} \equiv \sum_{k=0}^{n} \Delta t^{k+(1/2)} \sum_d \langle \dot{q}_{d+(1/2)} \rangle^{k+(1/2)}\,\Delta m_{d+(1/2)}. \tag{59.81}$$

In (59.79), v^{n+1} must be estimated by interpolation between $v^{n+(1/2)}$ and $v^{n+(3/2)}$. For problems using the boundary conditions $(p+Q)_1 \equiv 0$ and $v_{D+1} \equiv 0$ (hence $z_{D+1} \equiv \text{constant}$), both boundary work terms $\mathscr{W}^{n+1}$ vanish identically.

While $\mathscr{E}$ should, ideally, be constant, this property is not guaranteed exactly by the explicit difference equations. Rather, $\mathscr{E}$ is calculated after each integration step and is monitored as a check on the quality of the solution, cf. (**C8**), (**F1**). If satisfactory energy conservation is not obtained, the timestep is decreased and the integration step is done over.

(*b*) *Explicit Hydrodynamics*; *Spherical Geometry* For one-dimensional Lagrangean flow calculations in spherical geometry, the independent variable is M_r, the mass interior to radius r, increasing outward in the medium. The choice of a pseudoviscosity in spherical flows requires some care; the customary approach [see e.g., (**C5**), (**C8**), (**F1**)] is to use a scalar pseudoviscous pressure as in the planar case. The equations to be solved are then

$$(Dv/Dt) = -(GM_r/r^2) - 4\pi r^2[\partial(p+Q)/\partial M_r], \tag{59.82}$$

$$(Dr/Dt) = v, \tag{59.83}$$

and

$$V = 1/\rho = \tfrac{4}{3}\pi(\partial r^3/\partial M_r), \tag{59.84}$$

with

$$Q \equiv \tfrac{4}{3}\rho l^2(\partial v/\partial r)^2 \tag{59.85}$$

if $(\partial v/\partial r) < 0$, and $Q = 0$ otherwise. The energy equation is again (59.61).

These equations are discretized on a mass grid $\{M_i\}$, $i = 1, \ldots, I+1$, defining the surfaces of I spherical shells; M_i is the mass interior to the ith surface, and the mass of the ith shell is $\Delta M_{i+(1/2)} \equiv M_{i+1} - M_i$. An explicit difference representation of the momentum equation is

$$\begin{aligned}[v_i^{n+(1/2)} - v_i^{n-(1/2)}]/\Delta t^n &= -[GM_i/(r_i^{n+\lambda})^2] \\ &\quad - 4\pi(r_i^{n+\lambda})^2[p_{i+(1/2)}^{n+\lambda} - p_{i-(1/2)}^{n+\lambda} + Q_{i+(1/2)}^{n-(1/2)} - Q_{i-(1/2)}^{n-(1/2)}]/\Delta M_i,\end{aligned} \tag{59.86}$$

where $Q_{i+(1/2)}^{n-(1/2)}$ is obtained from a discretized version of (59.85), $p^{n+\lambda}$ is defined by (59.64),

$$r_i^{n+\lambda} \equiv r_i^n + \tfrac{1}{4}[\Delta t^{n+(1/2)} - \Delta t^{n-(1/2)}]v_i^{n-(1/2)} \tag{59.87}$$

and

$$\Delta M_i \equiv \tfrac{1}{2}[\Delta M_{i-(1/2)} + \Delta M_{i+(1/2)}]. \tag{59.88}$$

At the inner boundary we impose (59.66). At the outer boundary we impose (59.67) to (59.71), modified for spherical geometry.

After updating velocities we compute

$$r_i^{n+1} = r_i^n + v_i^{n+(1/2)}\,\Delta t^{n+(1/2)} \tag{59.89}$$

and

$$V_{i+(1/2)}^{n+1} = 1/\rho_{i+(1/2)}^{n+1} = \tfrac{4}{3}\pi[(r_{i+1}^{n+1})^3 - (r_i^{n+1})^3]/\Delta M_{i+(1/2)}, \tag{59.90}$$

and solve the same energy equation as in the planar case.

Although (59.86) to (59.90) have been applied widely, using (59.85) for the pseudoviscosity can cause serious numerical difficulties in certain problems, for example, accretion flows in star formation (**A1**), (**T5**). In particular, because radii converge as $r \to 0$, inflowing material ($v<0$) can experience compression even when $(\partial v/\partial r)>0$; this material should be subject to a pseudoviscous pressure, yet (59.85) predicts $Q=0$ in this case. Furthermore, in some problems (59.85) produces a spurious diffusion of radial momentum.

These difficulties are overcome by use of a *tensor artificial viscosity* (**T5**). Writing $\mathsf{T}=-p\mathsf{I}+\mathsf{Q}$, we demand that Q have the same analytical form as the molecular viscosity $\boldsymbol{\sigma}$, but permit a different viscosity coefficients. Thus we write

$$Q_{ij} = \mu_Q(v_{i;j} + v_{j;i} - \tfrac{2}{3}v^k_{;k}\,\delta_{ij}) \tag{59.91}$$

where, by analogy with the planar case, we set

$$\mu_Q = -\rho l^2 v^k_{;k} = l^2(D\rho/Dt) \tag{59.92}$$

for $v^k_{;k}<0$ or $(D\rho/Dt)>0$, and $\mu_Q=0$ otherwise. Equations (59.91) and (59.92) have the following desirable properties: (1) μ_Q is positive for compression and zero for expansion regardless of the direction of the flow. (2) Trace $\mathsf{Q}=0$, hence the pseudoviscosity is zero for homologous contraction ($\mathbf{v}=-k\mathbf{r}$), as is also true for molecular viscosity (the no-slip condition). (3) They reduce to the previous formulae in the planar limit, while discriminating correctly between the velocity divergence (scalar) and velocity gradient (tensor), which are fortuitously identical in one-dimensional planar flows. In practice (59.91) and (59.92) give very satisfactory results.

For one-dimensional spherical flow (59.91) is

$$\mathsf{Q}=2\mu_Q\begin{pmatrix}(\partial v/\partial r)-\frac{1}{3}\nabla\cdot\mathbf{v} & 0 & 0\\ 0 & (v/r)-\frac{1}{3}\nabla\cdot\mathbf{v} & 0\\ 0 & 0 & (v/r)-\frac{1}{3}\nabla\cdot\mathbf{v}\end{pmatrix}. \tag{59.93}$$

From (A3.91) and the fact that Q is traceless we find the radial component of the pseudoviscous force to be

$$(\nabla\cdot\mathsf{Q})_r = r^{-3}[\partial(r^3 Q_{rr})/\partial r] \tag{59.94}$$

where

$$Q_{rr} = 2\mu_Q[(\partial v/\partial r)+\tfrac{1}{3}(D\ln\rho/Dt)] \equiv -Q. \tag{59.95}$$

The momentum equation is

$$(Dv/Dt) = -(GM_r/r^2) - 4\pi r^2(\partial p/\partial M_r) - (4\pi/r)[\partial(r^3 Q)/\partial M_r], \tag{59.96}$$

and the energy equation is

$$(De/Dt)+p[D(1/\rho)/Dt]=\dot{q}+\Phi, \tag{59.97}$$

where, from (27.31), the dissipation function is

$$\Phi = 3\mu_Q[(\partial v/\partial r) + \tfrac{1}{3}(D \ln \rho/Dt)]^2. \tag{59.98}$$

The momentum equation has the discrete representation

$$\frac{v_i^{n+(1/2)} - v_i^{n-(1/2)}}{\Delta t^n} = \frac{-GM_i}{(r_i^{n+\lambda})^2} - \frac{4\pi}{\Delta M_i}\left((r_i^{n+\lambda})^2[p_{i+(1/2)}^{n+\lambda} - p_{i-(1/2)}^{n+\lambda}] + \frac{1}{r_i^n}\{[r_{i+(1/2)}^n]^3 Q_{i+(1/2)}^{n-(1/2)} - [r_{i-(1/2)}^n]^3 Q_{i-(1/2)}^{n-(1/2)}\}\right) \tag{59.99}$$

where the cell-center radius $r_{i+(1/2)}$ is chosen so as to contain half the volume (or mass) of the shell (r_i, r_{i+1}),

$$r_{i+(1/2)}^3 \equiv \tfrac{1}{2}(r_i^3 + r_{i+1}^3), \tag{59.100}$$

and where $r^{n+\lambda}$ and $p^{n+\lambda}$ are defined by (59.87) and (59.64). Equations (59.89) and (59.90) remain unchanged, while the energy equation becomes

$$e_{i+(1/2)}^{n+1} - e_{i+(1/2)}^n + \tfrac{1}{2}[p_{i+(1/2)}^n + p_{i+(1/2)}^{n+1}][V_{i+(1/2)}^{n+1} - V_{i+(1/2)}^n] = \Delta t^n[\langle \dot{q}_{i+(1/2)}\rangle^{n+(1/2)} + \Phi_{i+(1/2)}^{n+(1/2)}]. \tag{59.101}$$

In (59.99) and (59.101)

$$Q_{i+(1/2)}^{n-(1/2)} = -2(\mu_Q)_{i+(1/2)}^{n-(1/2)}\left[\frac{v_{i+1}^{n-(1/2)} - v_i^{n-(1/2)}}{r_{i+1}^{n-(1/2)} - r_i^{n-(1/2)}} + \frac{1}{3}\frac{\ln \rho_{i+(1/2)}^n - \ln \rho_{i+(1/2)}^{n-1}}{\Delta t^{n-(1/2)}}\right] \tag{59.102}$$

and

$$\Phi_{i+(1/2)}^{n+(1/2)} = 3(\mu_Q)_{i+(1/2)}^{n+(1/2)}\left[\frac{v_{i+1}^{n+(1/2)} - v_i^{n+(1/2)}}{r_{i+1}^{n+(1/2)} - r_i^{n+(1/2)}} + \frac{1}{3}\frac{\ln \rho_{i+(1/2)}^{n+1} - \ln \rho_{i+(1/2)}^n}{\Delta t^{n+(1/2)}}\right]^2, \tag{59.103}$$

where

$$(\mu_Q)_{i+(1/2)}^{n-(1/2)} = l^2[\rho_{i+(1/2)}^n - \rho_{i+(1/2)}^{n-1}]/\Delta t^{n-(1/2)} \tag{59.104}$$

if $\rho_{i+(1/2)}^n > \rho_{i+(1/2)}^{n-1}$, and is zero otherwise. In (59.104), either $l = k_Q \Delta r$ where k_Q is a number of order unity, or l is a prechosen, fixed length. The radii $r^{n\pm(1/2)}$ in (59.102) and (59.103) must be estimated by interpolation in time.

It follows from (27.4) [cf. also (27.33)] that for one-dimensional, spherically symmetric viscous flows we have a total energy equation of the form

$$\frac{D}{Dt}(e + \tfrac{1}{2}v^2) + \frac{\partial}{\partial M_r}[4\pi r^2 v(p+Q)] = \frac{-GM_r v}{r^2} + \dot{q}, \tag{59.105}$$

where Q is a scalar pressure as in (59.85) or the radial component of a tensor as in (59.95). Equation (59.105) yields a conservation relation like (59.78). As was true in planar geometry, this conservation relation follows from the *differential* equations governing the flow, but is not guaranteed by the explicit *difference* equations written above; rather, it is again monitored as a check on the quality of the solution. However, we will see below that it is possible to write a set of implicit difference equations that do yield an exact total energy conservation relation provided that the pseudoviscosity enters as a scalar pressure in both the energy and momentum equations. This is not the case when one uses the tensor formulation described above. A compromise is to use a pseudoviscosity that is mathematically equivalent to a scalar, but which is tailored to mimic the basic physical properties of the tensor pseudoviscosity. In particular, if we replace v/r in (59.93) by $(\partial v/\partial r)$ we obtain the isotropic tensor

$$\mathsf{Q} = 2\mu_Q[(\partial v/\partial r) - \tfrac{1}{3}\nabla \cdot \mathbf{v}]\mathsf{I} \equiv -Q\mathsf{I}, \tag{59.106}$$

where μ_Q is given by (59.92). This choice, while heuristic, has the following desirable properties. (1) Q is nonzero for compression and zero for expansion; (2) Q is zero for homologous contraction; (3) Q is isotropic, hence we may use Q as a scalar pressure. Note, however, that (59.106) does not yield trace $\mathsf{Q} = 0$, as did (59.93).

(*c*) *Implicit Hydrodynamics*; *Spherical Geometry* The schemes described above use explicit hydrodynamics. A rationale for that approach is that if we model wave phenomena and/or shocks, the physical system changes significantly in a flow time $t_f \sim l/a$, which is generally of the same order as the Courant time $\Delta x/a$. Because we want to follow these very changes there is no reason to take longer timesteps. A counterexample to this argument is provided by stellar evolution calculations, where we are interested in changes on a nuclear burning—rather than hydrodynamic—timescale, and we need to use very large timesteps to be able to follow the evolution of a star through a long sequence of near-equilibrium states. Furthermore, it is necessary to use an implicit scheme in order to avoid unnecessarily restrictive timestep limitations from thin zones and/or regions of high sound speed (e.g., in a stellar interior).

An implicit scheme (**K4**) suitable for calculations of both quasistatic stellar evolution and hydrodynamic events such as nova explosions is

$$(r_i^{n+1} - r_i^n)/\Delta t^{n+(1/2)} = \langle v \rangle^{n+(1/2)}, \tag{59.107}$$

$$(v_i^{n+1} - v_i^n)/\Delta t_{n+(1/2)} = \langle a_i \rangle^{n+(1/2)}, \tag{59.108}$$

$$V_{i+(1/2)}^{n+1} = 1/\rho_{i+(1/2)}^{n+1} = \tfrac{4}{3}\pi[(r_{i+1}^{n+1})^3 - (r_i^{n+1})^3]/\Delta M_{i+(1/2)}, \tag{59.109}$$

and

$$e_{i+(1/2)}^{n+1} - e_{i+(1/2)}^n + \langle (p+Q)_{i+(1/2)} \rangle^{n+(1/2)} [V_{i+(1/2)}^{n+1} - V_{i+(1/2)}^n] = \langle \dot{q}_{i+(1/2)} \rangle^{n+(1/2)} \Delta t^{n+(1/2)}. \tag{59.110}$$

Here the time averages $\langle\ \rangle^{n+(1/2)}$ are defined as

$$\langle x\rangle^{n+(1/2)} \equiv (1-\theta)x^n + \theta x^{n+1}, \tag{59.111}$$

and in (59.108) the acceleration is

$$a_i^m \equiv -GM_i/(r_i^m)^2 - 4\pi(r_i^m)^2[(p+Q)^m_{i+(1/2)} - (p+Q)^m_{i-(1/2)}]/\Delta M_i. \tag{59.112}$$

For Q we use a discrete representation of the scalar pressure defined by (59.106). Notice that there is no difficulty in time-centering variables in an implicit scheme, and special interpolation or extrapolation procedures such as are used in explicit schemes become unnecessary. In stellar evolution calculations, the equations are often rewritten in terms of the logarithms of physical variables, such as ρ and p, that run over several orders of magnitude (**K4**).

The system written above is unconditionally stable for $\frac{1}{2} \leq \theta \leq 1$. Nevertheless, (59.107) and (59.108) may be unsatisfactory for collapse problems on a Kelvin-Helmholtz timescale because a centered formula ($\theta = \frac{1}{2}$) may lead to growing oscillations. Equation (59.107) should then be made fully implicit ($\theta = 1$). But then (59.108) with $\theta = 1$ may artificially damp real oscillations; in such cases special formulae (e.g., containing three time levels) may be needed (**B7**).

Given constitutive relations for p and e, and a specification of $\dot{q}$, (59.107) to (59.112) are linearized around a trial solution at t^{n+1} and iterated to consistency. The linearized system is typically block tridiagonal, and is solved by Gaussian elimination (see §§83, 97, and 98).

As Fraley (**F2**) pointed out, the momentum equation can also be written

$$\begin{aligned}(v_i^{n+1} - v_i^n)/\Delta t^{n+(1/2)} &= -GM_i\langle 1/r^2\rangle_i \\ &\quad - 4\pi\langle r^2\rangle_i[\langle (p+Q)_{i+(1/2)}\rangle^{n+(1/2)} - \langle (p+Q)_{i-(1/2)}\rangle^{n+(1/2)}]/\Delta M_i\end{aligned} \tag{59.113}$$

where the angular brackets denote suitable time averages. In particular, if we choose the special averages

$$\langle r^2\rangle_i \equiv \tfrac{1}{3}[(r_i^n)^2 + r_i^n r_i^{n+1} + (r_i^{n+1})^2] \tag{59.114}$$

and

$$\langle 1/r^2\rangle_i \equiv 1/(r_i^n r_i^{n+1}), \tag{59.115}$$

then on multiplying (59.113) through by $v_i^{n+(1/2)} \equiv \frac{1}{2}(v_i^n + v_i^{n+1})$ and using the fact that $r_i^{n+1} - r_i^n = v_i^{n+(1/2)}\,\Delta t^{n+(1/2)}$ we obtain the *exact* conservation relation

$$\begin{aligned}&\sum_i \{[\tfrac{1}{2}(v_i^{n+1})^2 - (GM_i/r_i^{n+1})]\,\Delta M_i - \langle (p+Q)_{i+(1/2)}\rangle^{n+(1/2)} V_{i+(1/2)}^{n+1}\,\Delta M_{i+(1/2)}\} \\ &= \sum_i \{[\tfrac{1}{2}(v_i^n)^2 - (GM_i/r_i^n)]\,\Delta M_i - \langle (p+Q)_{i+(1/2)}\rangle^{n+(1/2)} V_{i+(1/2)}^{n}\,\Delta M_{i+(1/2)}\}.\end{aligned} \tag{59.116}$$

Here we assumed, for simplicity, zero contribution from the work terms at the boundary surfaces. Provided that we use precisely the same average for $\langle p+Q\rangle$ in both (59.113) and (59.110) we then obtain an exact total energy conservation relation

$$\sum_i \{e_{i+(1/2)}^{n+1}\,\Delta M_{i+(1/2)}+[\tfrac{1}{2}(v_i^{n+1})^2-(GM_i/r_i^{n+1})]\,\Delta M_i\} - \sum_{k=0}^{n} \Delta t^{k+(1/2)} \sum_i \langle \dot{q}_{i+(1/2)}\rangle^{k+1}\,\Delta M_{i+(1/2)} = \text{constant}. \quad (59.117)$$

Fraley's form of the momentum equation with $\theta=\frac{1}{2}$ in $\langle p+Q\rangle$ is often used in stellar pulsation calculations (**C1**), (**W9**) where it is important to obtain precise total energy conservation in order to avoid artificial damping of self-excited oscillations.

CRITIQUE

In numerical simulations of fluid flow, the choice of the best computational method requires good judgment because one's desire for accuracy and stability must be balanced against limitations in computer speed and capacity. In addition, one must sometimes face (perhaps unresolved) questions about the ability of the equations used to model faithfully the real physics of the flow. A short but illuminating discussion of these points is given in (**H1,** 86–90).

The worst problem inherent in one-dimensional Lagrangean schemes is their limited ability to resolve features with very steep gradients of material properties as they move through the fluid. Important examples are propagating shocks, and the cyclic motion of the hydrogen ionization zone in pulsating stars. In the case of shocks, artificial viscosity smears the front over a few zones. Although the Rankine-Hugoniot relations are still satisfied in the upstream and downstream flows, and the effect of the shock on the large-scale structure of the ambient medium is given correctly, each zone contains much more mass, and therefore is much more opaque, than the actual shock front itself; hence a calculation of radiative transport through the shock can be falsified badly. One approach to overcoming this difficulty is to use upstream and downstream conditions determined from coarse-zone calculations to do after-the-fact shock fitting with an extremely fine-zone model that resolves the shock structure and permits an accurate transfer calculation (**H7**).

The hydrogen-helium ionization zone in pulsating stars is even more troublesome because it contains the thermodynamic "engine" that drives the pulsation, and accurate modeling of its structure is therefore essential. As the star pulsates, a steep temperature and ionization front having a characteristic thickness of about a thousandth of a scale height sweeps back and forth over several pressure scale heights, hence through several Lagrangean mass zones. To avoid prohibitively small timesteps, relatively coarse zones are used; but coarse zoning produces unphysical bumps in the luminosity light curve and in the surface velocity. The only satisfactory solution is to use *rezoning* or *adaptive-mesh* schemes, in which the computational mesh is neither fixed in space (Eulerian), nor attached to definite material elements (Lagrangean). Instead, the mesh moves both in space and through the fluid in such a way as to track physically significant phenomena, such as shocks and ionization fronts, in the flow. Adaptive-mesh algorithms are discussed in (**C1**), (**T5**), and (**W7**).

60. Propagating Strong Shocks

LARGE-AMPLITUDE WAVES IN THE SOLAR CHROMOSPHERE

The propagation of strong shocks through a stratified atmosphere can be modeled using the methods discussed in §59. Instructive examples are provided by the work of Stein and Schwartz (**S13**), (**S14**), who solved (59.58) to (59.61) for periodic strong shocks in an isothermal atmosphere ($\gamma \equiv 1$). They chose $T = 5700$ K and $g \approx 3 \times 10^4$ cm s^{-2} (appropriate for the Sun), which imply a sound speed $a \approx 7$ km s^{-1} and an acoustic-cutoff period

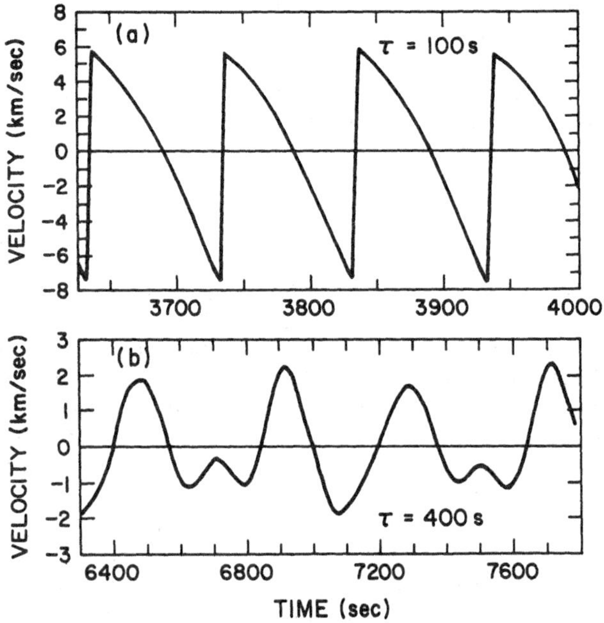

Fig. 60.1 Periodic wave trains in an isothermal stratified atmosphere. (a) Shock train generated by disturbance with period of 100 s. (b) Oscillation generated by disturbance with period of 400 s. From (**S14**) by permission.

$\tau_{ac} \approx 200$ s. The waves were excited by imposing a periodic velocity $v = 0.32 \sin(t/\tau)$ km s^{-1} at the lower boundary.

Their results reveal an important qualitative difference between waves with $\tau < \tau_{ac}$ and those with $\tau > \tau_{ac}$. The short-period waves steepen into shocks and form N waves, as expected from the discussion in §§58 and 59. The steady-state velocity variation of a mass element whose initial height in the atmosphere was 1000 km is shown in Figure 60.1a for a wave with period $\tau = 100$ s.

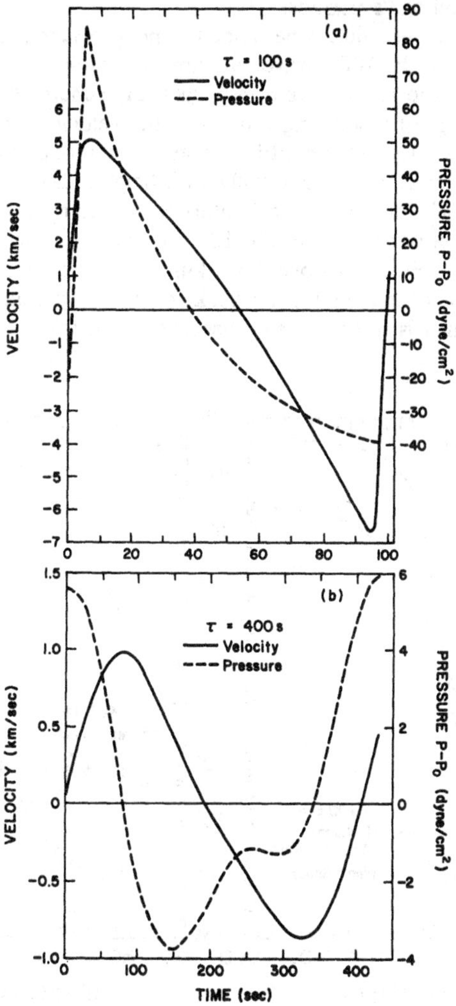

Fig. 60.2 Velocity and pressure perturbations for a single cycle of waves shown in Figure 60.1. From (**S14**) by permission.

In contrast, waves with $\tau > \tau_{ac}$ tend to lift the medium quasi-rigidly. The velocity variation of a mass element whose initial height was 1000 km is shown in Figure 60.1b for a wave with period $\tau = 400$ s. Note that the velocity varies nearly sinusoidally, shows no indication of shocks, and has only about one-third the amplitude of the short-period wave even though both are excited with the same driving term. The 400 s oscillation also shows beats with the 200 s natural period of the atmosphere, which indicates that a steady state has not yet been achieved in the calculation even after 19 full wave periods.

As shown in Figure 60.2, the velocity and pressure perturbations are nearly in phase for the 100 s wave, and approximately 90° out of phase for the 400 s wave. The 100 s wave transports a large energy flux, whereas the 400 s wave is almost a standing wave with the velocity nearly in phase at all heights, and transports very little energy. Even though the amplitude of the 400 s wave increases substantially with height, its small energy transport retards shock formation and inhibits wave-energy dissipation.

The dissipation per period in the 100 s wave as a function of height is shown in Figure 60.3a. Above 1000 km the dissipation rate is nearly constant, and the estimate given by weak shock theory is almost a factor of 10 too low, which is not surprising because the shock is no longer weak

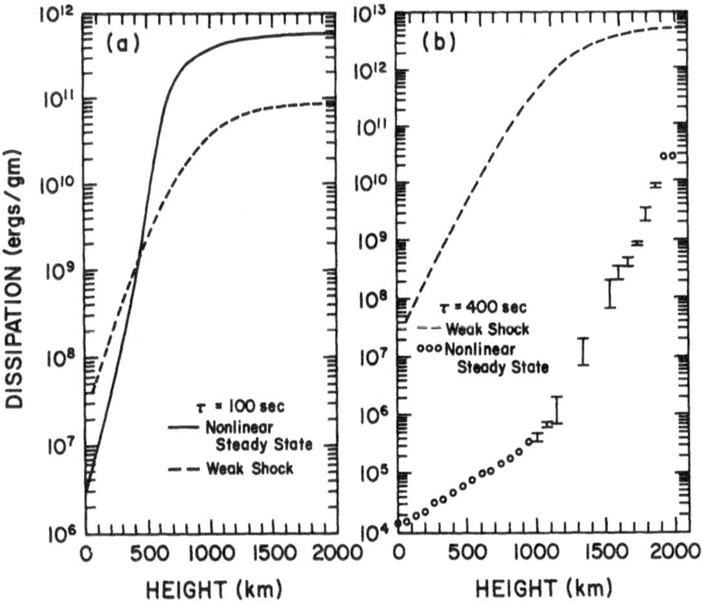

Fig. 60.3 Dissipation per period as a function of height for the waves shown in Figure 60.1. From (**S14**) by permission.

($M \approx 2$) at these heights. Below 500 km, weak shock theory predicts too large a dissipation rate because it assumes the existence of a shock before one has actually developed. As shown in Figure 60.3b the failure of weak shock theory for the 400-s wave is much more dramatic. The actual dissipation in this (nearly) standing wave is about 10^4 times smaller than predicted by weak shock theory. These results clearly show that weak shock theory must be used with great caution, and that the full nonlinear equations must be solved before meaningful estimates of chromospheric heating by shock dissipation can be made.

The results discussed above are only illustrative because radiative losses are omitted; more realistic calculations are discussed in §105.

SIMILARITY SOLUTIONS

An alternative to numerical modeling of shocks is to develop analytical solutions for idealized problems that are reasonable representations of situations of interest; this approach offers physical insight and provides benchmarks against which numerical calculations can be compared. An effective method of constructing such solutions is to carry out *similarity* (or *dimensional*) *analyses* of *self-similar flows.* In these flows the spatial *profiles* (i.e., *distributions*) of the physical variables are time-independent functions of an appropriate *similarity variable*; the time evolution of the flow is described fully by the time variations of the *scaling* of these profiles and of the similarity variable. Similarity methods have been highly developed by Sedov and his co-workers, and have been applied to a wide variety of problems (**S9**); here we consider only two examples of astrophysical interest.

Consider first the *blast wave* driven by a *point explosion.* Here we imagine the essentially instantaneous release of a large amount of energy $\mathscr{E}$ into a very small volume, which drives a spherical shock into a homogeneous medium of density ρ_1. We assume that the material is a perfect gas with a constant ratio of specific heats γ. We seek to describe the motion of the shock at a time when the mass of the material set in motion by the blast is large compared to that in which the initial energy release occurred, but when the shock strength is still so large that we can neglect the exterior gas pressure (*backpressure*) relative to the postshock pressure. We thus neglect the internal energy of the ambient gas compared to the explosion energy.

From the conservation laws (56.6) and (56.7) we have

$$\rho_1 v_s = \rho_2(v_s - v_2) \tag{60.1}$$

and

$$\rho_1 v_s^2 = \rho_2(v_s - v_2)^2 + p_2, \tag{60.2}$$

where v_s is the lab-frame shock velocity, and subscripts 1 and 2 denote pre- and post-shock quantities, respectively. Neglecting p_1 is equivalent to assuming an infinite shock strength, hence

$$\rho_2/\rho_1 = (\gamma+1)/(\gamma-1). \tag{60.3}$$

From (60.1) and (60.2) we obtain

$$p_2 = \rho_1 v_s v_2, \tag{60.4}$$

and from (60.1) and (60.3) we obtain

$$v_2 = 2v_s/(\gamma+1), \tag{60.5}$$

whence we find

$$p_2 = 2\rho_1 v_s^2/(\gamma+1). \tag{60.6}$$

For a self-similar expansion of the shock front, the pressure, density, and velocity distribution in the flow can be written

$$p(r, t) = p_2(t)\hat{p}(\xi), \tag{60.7a}$$

$$\rho(r, t) = \rho_2(t)\hat{\rho}(\xi), \tag{60.7b}$$

and

$$v(r, t) = v_2(t)\hat{v}(\xi). \tag{60.7c}$$

The profiles $\hat{p}$, $\hat{\rho}$, and $\hat{v}$ depend on the dimensionless variable ξ, which is related to physical distance by a transformation of the form

$$\xi = r/R(t). \tag{60.8}$$

We can determine $R(t)$ from dimensional arguments. The nature of the flow depends only on the two quantities $\mathscr{E}$ and ρ_1, having dimensions $[\mathscr{E}] = \text{ergs} = \text{g cm}^2\,\text{s}^{-2}$ and $[\rho_1] = \text{g cm}^{-3}$. The only combination of $\mathscr{E}$ and ρ_1 that contains only length and time is the ratio $\mathscr{E}/\rho_1$, which has dimensions $\text{cm}^5\,\text{s}^{-2}$. Hence self-similar motion of the flow can depend on length and time only through the dimensionless combination $\mathscr{E}t^2/\rho_1 r^5$, which implies that

$$\xi = (\rho_1/\mathscr{E})^{1/5}(r/t^{2/5}) \tag{60.9}$$

is an appropriate similarity variable.

For a given value of γ the shock front is characterized by a fixed value of ξ, say ξ_s; hence the position of the shock at time t is given by

$$r_s = \xi_s(\mathscr{E}/\rho_1)^{1/5}t^{2/5}, \tag{60.10}$$

and the shock speed is

$$v_s = (dr_s/dt) = \tfrac{2}{5}\xi_s(\mathscr{E}/\rho_1)^{1/5}t^{-3/5} = \tfrac{2}{5}\xi_s^{5/2}(\mathscr{E}/\rho_1)^{1/2}r_s^{-3/2}. \tag{60.11}$$

The speed of the gas behind the front, v_2, then follows from (60.5). One can understand the scaling in (60.11) intuitively by noting that the kinetic energy per unit volume in the blast wave as measured by either $\rho_2 v_s^2$ or $\rho_2 v_2^2$ must scale as $\mathscr{E}/r_s^3$.

According to (60.6) and (60.11) the pressure behind the shock is

$$p_2 = K\xi_s^2(\rho_1^3\mathscr{E}^2/t^6)^{1/5} = K\xi_s^5\mathscr{E}/r_s^3, \tag{60.12}$$

where $K \equiv 8/25(\gamma+1)$. Thus (for a given γ and ρ_1) the postshock pressure in explosions of different strengths reaches the same value at times and distances that are proportional to $\mathscr{E}^{1/3}$. We can understand this scaling intuitively by recalling that the pressure in a gas is proportional to the average energy per unit volume, hence to $\mathscr{E}/r_s^3$.

The density behind the shock remains fixed at its limiting value (60.3) as long as the postshock pressure p_2 is much greater than the back pressure p_1.

To complete the solution, one must determine the dimensionless shock-position parameter ξ_s and the profile functions $\hat{p}$, $\hat{\rho}$, and $\hat{v}$. The latter follow from solving three first-order ordinary differential equations obtained by transforming the equations of gasdynamics into dimensionless variables and converting derivatives with respect to r and t into derivatives with respect to ξ. The solution must satisfy the constraints $\hat{p}=\hat{\rho}=\hat{v}=1$ at $\xi=\xi_s$. The value of ξ_s is obtained by demanding total energy conservation, which implies

$$\int_0^{r_s} (e+\tfrac{1}{2}v^2)4\pi r^2\,dr = \mathscr{E}, \tag{60.13}$$

where e is the internal energy of the gas.

An exact solution of the point-explosion problem was obtained by Sedov (**S9**). It shows that the density drops very rapidly behind the shock front. In fact, nearly all the gas contained within r_s is concentrated into a very thin shell ($\Delta r/r_s \lesssim 0.08$) near the front because the shock slows as it sweeps up more and more material. Therefore, gas closer to the origin has a higher expansion speed than gas farther out in the flow, and tends to overrun the front. The pressure drops by a factor of 2 to 3 a short distance behind the shock and then remains roughly constant everywhere inside.

A rigorous derivation of the solution is moderately complicated. But by making the simplifying assumption that all the mass inside the blast wave is concentrated into the thin shell behind the shock, one can derive (**C4**), (**Z1,** 97–99) the approximate formula

$$\xi_s \approx [75(\gamma-1)(\gamma+1)^2/16\pi(3\gamma-1)]^{1/5}, \tag{60.14}$$

which is found to be in good agreement with the exact results.

Another astrophysically interesting problem that can be treated by similarity methods is the propagation of a strong shock in an exponentially stratified medium. Examples are a global shock resulting from a stellar pulsation or supernova explosion passing outward through a stellar envelope, or perhaps a shock emanating from a point source such as a man-made explosion in the Earth's atmosphere or an impulsive flare in the Sun's atmosphere. A shock from a point explosion may initially weaken as it sweeps up material (cf. discussion above), but eventually density-gradient effects dominate and the upward-propagating part of the shock strengthens monotonically as it passes outward into regions of ever-decreasing density. A global shock, of course, responds only to the density drop and strengthens continuously. Ultimately the shock's speed becomes so large that it escapes to infinity in a finite time; a blast wave can thus *break out* (or *vent*) from an atmosphere, perhaps even transporting material from the heart of the explosion into space if the explosion is sufficiently strong and occurs at a sufficiently high altitude.

Self-similar upward-propagating shocks are discussed in (**G1**), (**H5**), (**H6**), and (**R1**) [see also (**Z1,** Chap. 12)]. Consider a planar atmosphere of an ideal gas with constant γ, having a density stratification $\rho(z)=\rho_0 e^{-z}$, where z is the distance above some convenient reference level, in units of the scale height H. Then similarity analysis shows (**G1**) that the density, shock speed, postshock material speed, and pressure have the forms

$$\rho = A\psi(\zeta)e^{-Z}, \tag{60.15a}$$

$$v_s = v_{s0}e^{\alpha Z}, \tag{60.15b}$$

$$v_2 = v_s\theta(\zeta), \tag{60.15c}$$

and

$$p = Av_{s0}^2\phi(\zeta)e^{-\beta Z}, \tag{60.15d}$$

where $\zeta \equiv z - Z$ is the distance behind the shock (in scale heights). Note that $\zeta \leq 0$, and that velocities are measured in (scale heights s^{-1}). The parameters α and β depend only on the ratio of specific heats γ, and are tabulated in (**G1**) for a wide range of γ; for example, $(\alpha, \beta)=(0.176, 0.646)$ for $\gamma=\frac{4}{3}$ and $(0.204, 0.591)$ for $\gamma=\frac{5}{3}$. Near the shock front $(\zeta=0)$ one has

$$\rho(\zeta) \approx \rho(0)e^{-\zeta}, \tag{60.16a}$$

$$p(\zeta) \approx p(0)e^{-\beta\zeta}, \tag{60.16b}$$

$$v(\zeta) \approx v_s e^{\alpha\zeta}, \tag{60.16c}$$

and

$$\mu(\zeta) \approx e^{(1-\beta)\zeta} \tag{60.16d}$$

where $\mu \equiv \gamma p/\rho v_s^2 = a^2/v_s^2 = 1/M^2$. Equations (60.15) and (60.16) are found to be in excellent agreement with numerical computations.

By integrating (60.15b), we find that the shock position as a function of time is

$$Z = \frac{1}{\alpha}\ln\left(\frac{1}{1-\alpha v_{s0}t}\right) = \frac{1}{\alpha}\ln\left(\frac{t_\infty}{t_\infty - t}\right), \tag{60.17}$$

which shows that the shock approaches infinity in the finite time

$$t_\infty = 1/\alpha v_{s0}. \tag{60.18}$$

Because the shock accelerates rapidly, it can outrun even large perturbations in the flow if they are located beyond some (small) critical distance behind the front; such shocks are sometimes called *self-propagating*. As a result, the asymptotic behavior of the shock is very insensitive to details of the original explosion; in particular it is essentially identical for any energy-deposition time Δt on the range $0 \leq \Delta t \leq t_\infty$, and is modified significantly only if the deposition rate is strongly singular near $t = t_\infty$.

Moreover, for $t \gg t_\infty$ the fluid variables at points near the origin are independent of the shock position. Thus the ratios p_∞/p_1 and v_∞/v_1 are the same for all fluid elements in the flow, where v_1 and p_1 are an element's initial speed and pressure, and v_∞ and p_∞ are its speed and pressure for $t \gg t_\infty$. Similarly, the distance l_∞ through which a fluid element moves between the time the shock passes over it and t_∞ is the same for all elements. Numerical calculations (**G1**) yield $(l_\infty, v_\infty/v_1, p_\infty/p_1) = (4.60, 1.72, 0.099)$ for $\gamma = \frac{5}{3}$ and $(6.23, 1.75, 0.075)$ for $\gamma = \frac{4}{3}$.

5.4 Thermally Driven Winds

The Sun is surrounded by a tenuous, extremely hot envelope ($n_e \sim 4 \times 10^8 \text{ cm}^{-3}$, $T \sim 1.5 \times 10^6$ K) called the corona; other late-type stars have similar structures. As first recognized by Chapman (**C3**), at such high temperatures thermal conduction by electrons becomes an efficient energy transport mechanism, and an unavoidable consequence of this fact is that the corona must extend far out into interplanetary space, even enveloping the Earth in a low-density, high-temperature plasma. Subsequently, Parker showed (**P1**) that if the corona were hydrostatic, the gas pressure at infinite distance from the Sun would exceed the total pressure in the surrounding interstellar medium by orders of magnitude; therefore a static corona is actually impossible. Instead, the corona undergoes a continuous *dynamical expansion* and produces a transsonic flow known as the *solar wind*. The solar wind is driven, ultimately, by conversion of the high specific enthalpy of coronal gas into kinetic energy of fluid motion; such winds are known as *thermal winds* to distinguish them from winds accelerated by direct momentum input to the fluid by intense radiation fields (cf. §107).

Several excellent reviews of thermal winds are available [e.g., (**B9**), (**H13**), (**H14**), (**H15**), (**H18**), and (**P2**)]. We discuss here only the most elementary aspects of the theory.

61. *Basic Model*

To obtain insight into the dynamics of thermal winds we consider the highly idealized problem of a steady, spherically symmetric flow of a fully ionized pure hydrogen plasma in a gravitational field. We first briefly recapitulate Chapman's and Parker's arguments about the inevitability of coronal expansion.

CORONAL EXPANSION

In a fully ionized hydrogen plasma, the electron heat flux is $\mathbf{q} = -K\,\nabla T$, where the thermal conductivity $K = K_0 T^{5/2}$ with $K_0 \approx 8 \times 10^{-7}$ ergs $\text{cm}^{-1}\,\text{s}^{-1}\,\text{K}^{-7/2}$ (cf. §33). At coronal temperatures the conductivity of the plasma exceeds that of ordinary metallic laboratory conductors. Following Chapman, suppose that thermal conduction is the only energy transport mechanism operative; then $\nabla \cdot \mathbf{q} = 0$, which implies that

$$\frac{1}{r^2}\frac{d}{dr}\left(r^2 K_0 T^{5/2}\frac{dT}{dr}\right) = 0. \tag{61.1}$$

Integrating (61.1) and demanding that $T \to 0$ as $r \to \infty$ we find

$$T(r) = T_0(r_0/r)^{2/7}, \tag{61.2}$$

where r_0 is a suitable reference level (say $r_0 = R_\odot \approx 7\times10^{10}$ cm). The falloff predicted by (61.2) is very slow; for $T_0 \sim 1.5\times10^6$ K, $T \sim 3\times10^5$ K at the Earth's orbit ($r_\oplus \approx 1.5\times10^{13}$ cm).

Suppose further that the corona is in hydrostatic equilibrium so that

$$(dp/dr) = -G\mathcal{M}_\odot \rho/r^2. \tag{61.3}$$

For fully ionized hydrogen, $n_e = n_p \equiv n$, $p = 2nkT$, and $\rho = nm_\mathrm{H}$. If we assume that $T(r)$ is given by (61.2), then (61.3) becomes

$$\frac{d}{dx}(nx^{-2/7}) = -\left(\frac{r_0}{H}\right)\frac{n}{x^2}. \tag{61.4}$$

Here $x \equiv r/r_0$, and

$$H \equiv 2kT_0 r_0^2/G\mathcal{M}_\odot m_\mathrm{H} \tag{61.5}$$

is a scale height; for the solar corona $H \sim 10^5$ km. Integrating (61.4) we find

$$n(r) = n_0 x^{2/7}\exp\left[-\tfrac{7}{5}(r_0/H)(1-x^{-5/7})\right], \tag{61.6}$$

which implies that $n \sim 10^5$ near the Earth's orbit if $n_0 \sim 4\times10^8$ in the corona. We thus arrive at Chapman's conclusion that the Earth must be enveloped in hot, dense (compared to the interstellar medium) plasma extending from the solar corona.

Combining (61.2) and (61.6) we see that the pressure distribution in a static corona,

$$p(r) = p_0 \exp\left[-\tfrac{7}{5}(r_0/H)(1-x^{-5/7})\right], \tag{61.7}$$

implies that the coronal pressure does not vanish as $x \to \infty$, but instead approaches a finite value. For $p_0 \sim 0.2$ dynes cm^{-2} we find that $p_\infty \sim 10^{-5}$ dynes cm^{-2}, which is six to seven orders of magnitude larger than the total pressure in the interstellar medium. We thus arrive at Parker's conclusion that the corona *must* expand.

STEADY ONE-DIMENSIONAL FLOW

Consider now a steady, one-dimensional wind flow in spherical geometry. We must solve the continuity equation

$$r^{-2}[d(r^2\rho v)/dr] = 0, \tag{61.8}$$

the momentum equation

$$\rho v(dv/dr) = -(dp/dr) - G\mathcal{M}_{\odot}\rho/r^2, \tag{61.9}$$

and the energy equation

$$\frac{1}{r^2}\frac{d}{dr}[r^2\rho v(\tfrac{1}{2}v^2+h)] = -\rho v\left(\frac{G\mathcal{M}_{\odot}}{r^2}\right) + \frac{1}{r^2}\frac{d}{dr}\left(r^2 K\frac{dT}{dr}\right), \tag{61.10}$$

where $h = e + (p/\rho) = 5kT/m_H$ is the specific enthalpy of the plasma.

The continuity equation has the integral

$$4\pi r^2 n m_H v = \dot{\mathcal{M}} \equiv m_H \mathcal{F} \tag{61.11}$$

where $\mathcal{F}$ is the particle flux. The energy equation has the integral

$$\dot{\mathcal{M}}[\tfrac{1}{2}v^2 + h - (G\mathcal{M}_{\odot}/r)] - 4\pi r^2 K(dT/dr) = \mathcal{E} \tag{61.12}$$

where $\mathcal{E}$ is the total energy flux. The two first-order differential equations (61.9) and (61.12) determine the structure of a thermal wind. When this system is integrated, two more integration constants appear; hence a total of four conditions (boundary conditions, specifications of the behavior of the solution, or choices of free parameters) must be imposed in order to determine a unique solution.

ISOTHERMAL WINDS

Before discussing the general problem posed above, it is instructive to consider the case of an isothermal wind. In physical terms we, in effect, invoke some hypothetical heating mechanism to maintain a constant temperature in the face of the tendency of the gas to cool adiabatically as it expands. In practical terms we can then dispense with (61.12) and solve only (61.9) subject to (61.11).

Setting $T \equiv T_0$, and using (61.11) to eliminate n we can rewrite (61.9) as

$$\tfrac{1}{2}[1 - (2kT_0/m_H v^2)](dv^2/dr) = -(G\mathcal{M}_{\odot}/r^2)[1 - (4kT_0 r/G\mathcal{M}_{\odot} m_H)]. \tag{61.13}$$

Equation (61.13) admits several types of solution. Notice first that in the solar corona $(4kT_0 r_0/G\mathcal{M}_{\odot} m_H) \sim 0.3$, hence the right-hand side of (61.13) passes from negative to positive as r increases from r_0 to ∞, and vanishes at the *critical radius*

$$r_c \equiv G\mathcal{M}_{\odot} m_H/4kT_0. \tag{61.14}$$

At $r = r_c$ the left-hand side of (61.13) must therefore vanish. There are two options: either

$$(dv/dr)_{r_c} = 0, \tag{61.15}$$

or (dv/dr) is nonzero but v equals the *critical velocity*

$$v_c = (2kT_0/m_H)^{1/2}, \tag{61.16}$$

which is also the isothermal sound speed at $r = r_0$.

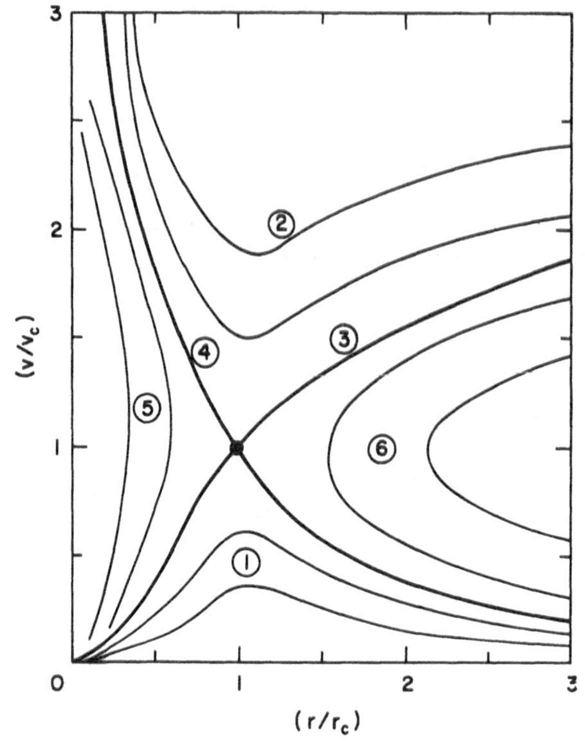

Fig. 61.1 Topology of thermal wind solutions.

If we demand that both v and (dv/dr) be single valued and continuous, we find four types of solutions as sketched in Figure 61.1. First, if we suppose $(dv/dr)_{r_c} = 0$ we can construct solutions in which $[1-(2kT_0/m_H v^2)]$ has the same sign for all r. If we choose $v(r_c) < v_c$, then $v(r)$ has an absolute maximum at r_c, and v is everywhere subsonic (type 1). If $v(r_c) > v_c$, then $v(r)$ has an absolute minimum at r_c, and v is everywhere supersonic (type 2). Alternatively, if we assume that $v(r_c) = v_c$, then we obtain two unique *critical solutions* that pass through (r_c, v_c) with finite slope. Both solutions are transonic, either with v increasing monotonically from subsonic $(v < v_c)$ for $r < r_c$ to supersonic for $r > r_c$ (type 3), or with v decreasing monotonically from supersonic to subsonic (type 4). If we drop the requirement that v be single valued we find two additional families of solutions, types 5 and 6 in Figure 61.1; their significance will emerge below.

To choose a model for the real solar wind, one appeals to observation. It is known that flow velocities at the base of the corona are much smaller than $v_c \approx 170 \text{ km s}^{-1}$, hence we can immediately exclude solutions of types 2 and 4. To choose between solutions of types 1 and 3 we integrate (61.13) to obtain

$$(v/v_c)^2 - \ln (v/v_c)^2 = 4[\ln (r/r_c) + (r_c r)] + C. \tag{61.17}$$

For solutions of type 1, $v < v_c$, and v decreases as $r \to \infty$; hence for $r \gg r_c$ the dominant term on the left-hand side is $\ln (v/v_c)^2$ and on the right-hand side it is $4 \ln (r/r_c)$. Thus for type-1 solutions $v \propto r^{-2}$ as $r \to \infty$, which implies [cf. (61.11)] that n, hence p, remains finite. In fact these solutions yield values of p_∞ that greatly exceed interstellar pressures, and can therefore be rejected on the same grounds as the hydrostatic solution was.

In contrast, for the critical solution (type 3), $v > v_c$ and increases with increasing r, hence $v(r) \sim 2v_c\ [\ln (r/r_c)]^{1/2}$ for large r, which implies that $n(r) \propto (1/r^2 v) \to 0$ as $r \to \infty$. The critical solution can thus match a zero-pressure boundary condition at infinity. This fact led Parker to conclude (**P1**) that the solar wind is an accelerating transonic flow, a conclusion verified by observations from space vehicles.

The unbounded velocity of the critical solution as $r \to \infty$ is unphysical. It is an artifact of insisting the flow be isothermal, for, as noted earlier, this assumption implies a continuous deposition of thermal energy into the gas, leading to an infinite reservoir of energy which can accelerate the flow without limit. Parker overcame this difficulty by demonstrating that one can obtain satisfactory wind models, in which $n \to 0$ and v approaches a finite value v_∞ as $r \to \infty$, by assuming that the corona either (1) is isothermal for $R_\odot \leq r \leq r_*$ and then expands adiabatically ($\gamma = \frac{5}{3}$) for $r \geq r_*$, or (2) is everywhere polytropic with $\gamma \leq \frac{3}{2}$. Although it is obviously highly oversimplified, model (1) provides at least a plausible caricature of a thermal wind.

HEAT-CONDUCTING WINDS

While instructive, the isothermal winds just discussed are too idealized; to model realistic thermal winds we must retain the energy equation (61.12) in order to determine $T(r)$ consistently with $n(r)$ and $v(r)$. For a systematic survey of solutions, it is convenient to transform to dimensionless variables (**C2**), writing

$$\tau \equiv T/T_0, \tag{61.18a}$$

$$\psi \equiv v^2 \mu m_H / kT_0, \tag{61.18b}$$

and

$$\lambda \equiv G\mathcal{M}_\odot \mu m_H / kT_0 r, \tag{61.18c}$$

where μ is the number of atomic mass units per particle ($\frac{1}{2}$ for ionized hydrogen). In these variables the dynamical equations become

$$n\psi^{1/2}/\lambda^2 = (kT_0/\mu m_H)^{3/2} \mathcal{F}/(4\pi G^2 \mathcal{M}_\odot) \equiv \mathcal{C}, \tag{61.19}$$

$$\tfrac{1}{2}[1 - (\tau/\psi)](d\psi/d\lambda) = 1 - 2(\tau/\lambda) - (d\tau/d\lambda), \tag{61.20}$$

and

$$\mathcal{A}\tau^{5/2}(d\tau/d\lambda) = \varepsilon_\infty - \tfrac{1}{2}\psi + \lambda - \tfrac{5}{2}\tau \tag{61.21}$$

where

$$\varepsilon_\infty \equiv \mu\mathscr{E}/kT_0\mathscr{F} \tag{61.22}$$

is the residual energy, per particle, at infinity, in units of kT_0, and

$$\mathscr{A} \equiv 4\pi K_0 G\mu^2 m_H \mathscr{M}_\odot T_0^{3/2}/k^2\mathscr{F}. \tag{61.23}$$

To integrate these equations we must specify ε_∞ and $\mathscr{A}$, and impose the requirements that $\tau \to 0$ as $\lambda \to 0$ and that the solution pass through the critical point if appropriate (see below).

The solutions of (61.19) to (61.23) fall into two basic classes: (1) transonic critical *winds*, resembling the isothermal solution of type 3 and (2) subsonic *breezes*, similar to the isothermal solutions of type 1. The winds all have $\mathscr{E} > 0$, whereas the breezes all have $\mathscr{E} = 0$. Breeze solutions played an important role in the development of stellar wind theory (**C2**), (**R7**), but will not be discussed further here.

The wind solutions display three distinct asymptotic behaviors of $T(r)$ as $r \to \infty$ [(**D1**), (**D2**), (**D3**), (**H18**, 47), (**P3**), (**R7**), (**W4**)]. The behavior of any particular solution is determined by the dominant heat-transport mechanism as $r \to \infty$. Suppose first that the heat-conduction flux at infinity, $\mathscr{E}_c(\infty)$, remains finite while the enthalpy flux goes to zero. Then $[r^2 T^{5/2}(dT/dr)]_\infty =$ constant, which implies $T(r) \propto r^{-2/7}$ asymptotically; this is the kind of solution discussed by Chapman and by Parker (**P3**). Next suppose that both the conduction and enthalpy fluxes go to zero in such a way that their ratio $(nvr^2 5kT)/[r^2 K_0 T^{5/2}(dT/dr)]$ remains finite as $r \to \infty$. This condition can be achieved if $T(r) \propto r^{-2/5}$ asymptotically, the solution first discovered by Whang and Chang (**W4**). Finally, suppose that the conduction flux vanishes more rapidly than the enthalpy flux as $r \to \infty$; in this case there is no energy exchange within the gas and it expands adiabatically. Thus as $r \to \infty$, $T \propto \rho^{(\gamma-1)} \propto r^{-2(\gamma-1)}$ (because $\rho \propto r^{-2}$); hence $T(r) \propto r^{-4/3}$ for $\gamma = \frac{5}{3}$, the solution first discovered by Durney (**D1**).

In fact, a continuous sequence of solutions exhibiting these different asymptotic behaviors can be obtained by fixing the coronal base temperature T_0 and choosing different values for the base density n_0. For small values of n_0, the critical solutions have large values of $e(\infty) \equiv \mathscr{E}/\mathscr{F}$, the energy flux at infinity per particle, and $T(r) \propto r^{-2/7}$. As n_0 is increased, an ever-larger fraction of $e_c(\infty)$, the conduction flux at infinity, is consumed in driving a more and more massive flow. Eventually at some critical value of n_0, say n_0^*, $e_c(\infty) \to 0$, and $T \propto r^{-2/5}$ for this particular solution. If we increase n_0 still further, $e_c(\infty)$ remains zero and at large r the gas expands adiabatically with $T \propto r^{-4/3}$. As more and more material is added to the flow, more and more thermal energy is consumed in the adiabatic expansion, and $e(\infty)$ decreases. Finally, for a sufficiently large n_0, $e(\infty)$ vanishes, and the solution abruptly changes from a transonic wind to a subsonic breeze.

For wind solutions (only) the further transformation $\tau_* \equiv \tau/e_\infty$, $\psi_* \equiv \psi/e_\infty$, and $\lambda_* \equiv \lambda/e_\infty$ reduces the number of arbitrary constants to one; equation (61.20) has the same form in the new variables, while (61.21) becomes

$$\mathcal{K}\tau_*^{5/2}(d\tau_*/d\lambda_*) = 1 - \tfrac{1}{2}\psi_* + \lambda_* - \tfrac{5}{2}\tau_* \tag{61.24}$$

where $\mathcal{K} \equiv e_\infty^{3/2}\mathcal{A}$. A large number of solutions for a wide range of $\mathcal{K}$ are given in (**D2**); each of these can be redimensionalized into an infinite number of physical solutions for differing choices of, say, n_0 and T_0. For a typical solar wind model, one finds that near the Earth's orbit the wind speed is $\sim 300 \text{ km s}^{-1}$ and the particle density is $\sim 10 \text{ cm}^{-3}$. The rate of mass loss in the wind is $\sim 10^{-14}\, \mathcal{M}_\odot$/year, which is negligible compared to the rate of mass loss via thermonuclear energy release.

TRANSITION TO THE INTERSTELLAR MEDIUM

The above discussion tacitly assumes that the flow expands into a vacuum. In reality, the wind ultimately stagnates against the ambient interstellar medium, forming a stationary shock across which the flow velocity drops suddenly from highly supersonic to subsonic, while both the temperature and density rise sharply. The wind solution thus jumps discontinuously from the critical solution to one of the solutions of type 6 in Figure 61.1; the correct subsonic solution is chosen by matching conditions in the interstellar medium as $r \to \infty$.

By imposing the Rankine-Hugoniot relations across the jump, one can determine the radius r_s at which the shock front is located. To obtain a rough estimate of r_s, we can equate the *impact pressure* $nm_H v_\infty^2$ of the flow to the interstellar gas pressure p_i. Noting that $r^2 n(r) = r_\oplus^2 n(r_\oplus)$ because the flow speed is already essentially v_∞ at $r = r_\oplus$, we find

$$r_s/r_\oplus \approx (n_\oplus m_H v_\infty^2/p_i)^{1/2}. \tag{61.25}$$

For typical values of v_∞, p_i, and $n_\oplus$ one obtains $r_s \gtrsim 30 r_\oplus$.

Accretion flows, in which material falls in from infinity onto a star, are also possible (**H15**); in this case the solution runs inward along the critical solution of type 4 and jumps discontinuously to a solution of type 5 through an *accretion shock* near the surface of the star.

62. Physical Complications

While the model described in §61 gives basic insight into the nature of thermal winds, it is a terribly oversimplified description of the real solar wind. It is therefore worthwhile to mention some of the physical complications that occur in the real solar wind as an introduction to more sophisticated treatments that include phenomena which change the picture substantially, sometimes even qualitatively. Good general reviews are given in (**H13**) and (**H14**).

FLUID PROPERTIES

In §16 we assumed that a thermal wind can be considered to be a steady flow of an equilibrium, inviscid (although heat-conducting) fluid. Each of these assumptions requires scrutiny. Dimensional arguments (**P2**) show that because the physical scale lengths in stellar winds are so large, the Knudsen number is usually (but not always) small enough that the material can be treated as a continuum rather than as individual particles. On the other hand, interparticle collision frequencies are so low (because of low densities) that collisional equilibrium between electrons and protons in the plasma cannot be maintained (**S15**), (**H2**). Instead, we must use a two-fluid model, allowing the electrons and protons to have different temperatures. Indeed, space observations show that near $r_\oplus$, $T_e \approx 1.5\times10^5$ K while $T_p \approx 4\times10^4$ K; the higher electron temperature is maintained by the larger heat flux transported by the electrons. The analytical properties of two-fluid polytropic flows are discussed in (**S17**) and a comprehensive review of two-fluid, solar-wind models is given in (**H11**). A problem with many two-fluid models of the solar wind is that the computed difference between T_e and T_p is much larger than is observed; a number of additional "noncollisional" energy-exchange mechanisms, including a variety of plasma instabilities, have been hypothesized to remedy this problem.

The inclusion of viscous terms has a large impact on the mathematical structure of the stellar wind problem because they eliminate the singularity at the critical point (**W5**), raise the order of the system of differential equations, and admit a richer variety of boundary conditions. Early calculations including viscous terms [e.g., (**S2**)] predicted major changes in the flow; these conclusions are now known to be incorrect because the integration scheme yielded spurious solutions of the Navier-Stokes equations (**S16**). More recent calculations [e.g., (**W8**)] show that viscous terms have relatively little effect on the dynamics of the flow but do contribute to heating the proton component of the plasma.

Thermal conduction by electrons is a major energy-transport mechanism in thermal winds. But the density in the flow is so low that the correctness of the standard Spitzer-Härm conduction coefficient, valid in a collision-dominated plasma, becomes questionable. In the outer parts of the flow one finds that the ratio of the collisional mean free path λ to the scale length l of the temperature gradient exceeds unity. In this regime, the thermal flux predicted by the standard formula $\mathbf{q} \propto -\lambda\,\nabla T$ may exceed the physical upper bound set by assuming the entire electron thermal energy $\frac{3}{2}n_e kT$ is transported at the mean thermal speed v_{th} of the electrons. We must therefore impose *flux limiting* on the thermal conduction (a problem discussed in the context of radiative energy diffusion at the end of §97). A variety of schemes [e.g., (**P4**), (**H10**), (**W8**)] that inhibit the heat flux when $\lambda \gtrsim l$ have been suggested. These modifications typically improve agreement between theory and observation; nevertheless they are ad hoc, and the difficult problem of calculating accurate transport coefficients in a collisionless plasma remains to be solved, although progress is being made (**S11**).

MAGNETIC FIELDS

The assumption of spherically symmetric steady flow is also inadequate for the real solar wind. It has been known for some time that the properties of the solar wind measured near the Earth vary over a wide range on time scales of hours to days. Prominent features include strong flare-induced shocks, and high-speed *plasma streams* often observed to recur with a solar rotation period. In the customary picture of the solar wind, these phenomena were viewed as distinct events superposed on a "quiet solar wind" presumably described by models like those discussed in §61. This picture had to be abandoned in the face of X-ray observations, which show the corona to be extremely inhomogeneous and highly structured by magnetic fields (**Z2**). These structures strongly affect the character of the wind emanating from any particular volume element. On one hand, little, if any, wind originates from the hot, dense, magnetically confined *coronal loops*; these structures are cooled mainly by thermal conduction downward into the chromosphere and by radiative losses. On the other hand, high-speed wind streams originate in the *coronal holes*, which are rarefied, relatively cool regions with rapidly diverging magnetic fields that open into interplanetary space; here the wind itself is an important cooling mechanism. Thus the solar wind is not a smooth, steady flow perturbed by occasional "atypical" events. Instead, the observed strongly fluctuating, complex flow is *representative* of the wind, indeed *is* the wind.

Magnetic fields may also strongly affect energy and momentum input into the wind. It is now believed that the dominant coronal heating mechanism is direct dissipation of magnetic energy into the plasma (**V1**), (**L4**). The earlier concept of shock heating seems inconsistent with current observations. Recent work has shown that the dissipation of Alfven waves can deposit substantial energy and momentum in a wind flow; models based on this mechanism seem to provide a satisfactory representation of the winds observed in many late-type stars (**H3**), (**M1**).

More realistic wind models attempt to allow for rapidly diverging flow geometries and nonthermal momentum and energy sources. These phenomena have dramatic effects on the flow. For example, the solutions may exhibit multiple critical points. A sketch of possible topologies for three-critical-point solutions (**H12**) is shown in Figure 62.1; one sees that the flow can become rather complicated. Momentum and energy inputs to the flow can also strongly affect the mass flux and terminal flow speed of a wind. Thus energy input to the supersonic part of the flow increases the terminal flow speed but has no effect on the mass flux (as expected because the mass flux is already fixed in the subsonic part of the flow because information about changes in the supersonic part of the flow cannot propagate upstream). In contrast, energy addition to the subsonic part of

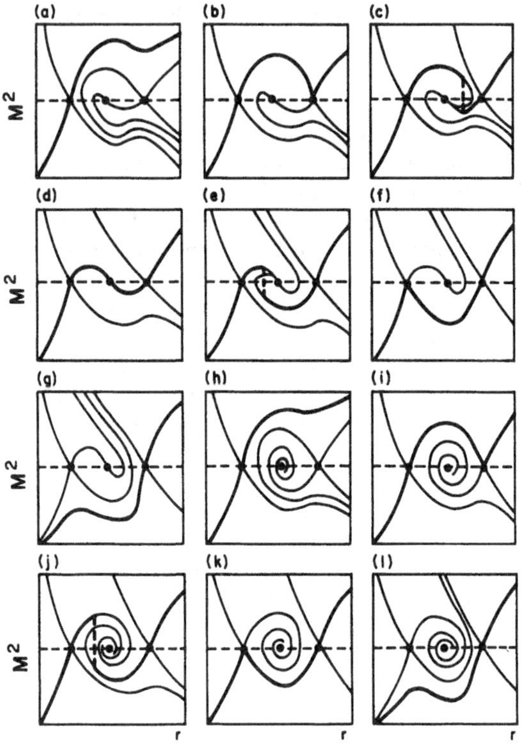

Fig. 62.1 Topology of thermal wind solutions with multiple critical points. From (**H12**) by permission.

the flow increases the mass flux but has little effect on the terminal wind speed. Momentum input in the subsonic part of the flow may actually reduce the terminal wind speed (**L3**).

Magnetic fields can also induce an azimuthal component in the flow in a stellar wind from a rotating star, thereby allowing the wind to exert a torque on the star and to act as a sink of stellar angular momentum (**W1**), (**B9,** §3.7). Whereas the rate of mass loss into the solar wind is negligible, the rate of angular-momentum loss is substantial and is responsible for the Sun's present low rotation rate.

Present research in this area attempts to address fully three-dimensional magnetohydrodynamic flow from a structured corona, a topic that lies far outside the scope of this book.

References

(A1) Appenzeller, I. (1970) *Astron. and Astrophys.*, **5,** 355.

(A2) Athay, R. G. (1976) *The Solar Chromosphere and Corona: Quiet Sun.* Dordrecht: Reidel.

(B1) Becker, E. (1968) *Gas Dynamics.* New York: Academic Press.
(B2) Becker, R. (1922) *Z. fur Physik,* **8,** 321.
(B3) Biermann, L. (1946) *Naturwiss.,* **33,** 118.
(B4) Biermann, L. (1948) *Z. fur Astrophys.,* **25,** 161.
(B5) Bird, G. A. (1964) *Astrophys. J.,* **139,** 684.
(B6) Bird, G. A. (1964) *Astrophys. J.,* **140,** 288.
(B7) Bodenheimer, P. (1968) *Astrophys. J.,* **153,** 483.
(B8) Bond, J. W., Watson, K. M., and Welch, J. A. (1965) *Atomic Theory of Gas Dynamics.* Reading: Addison-Wesley.
(B9) Brandt, J. C. (1970) *Introduction to the Solar Wind.* San Francisco: Freeman.
(B10) Bray, R. J. and Loughhead, R. E. (1974) *The Solar Chromosphere.* London: Chapman and Hall.
(B11) Brinkley, S. R. and Kirkwood, J. G. (1947) *Phys. Rev.,* **71,** 606.
(C1) Castor, J. I., Davis, C. G., and Davison, D. K. (1977) *Los Alamos Scientific Laboratory Report No. LA-6664.* Los Alamos: University of California.
(C2) Chamberlain, J. (1961) *Astrophys. J.,* **133,** 675.
(C3) Chapman, S. (1959) *Proc. Roy. Soc. (London),* **A253,** 450.
(C4) Chernyi, G. G. (1957) *Doklady Akad. Nauk SSR,* **112,** 213.
(C5) Christy, R. F. (1964) *Rev. Mod. Phys.,* **36,** 555.
(C6) Courant, R. and Friedrichs, K. O. (1948) *Supersonic Flow and Shock Waves.* New York: Interscience.
(C7) Courant, R., Friedrichs, K. O., and Lewy, H. (1928) *Math. Ann.,* **100,** 32.
(C8) Cox, A. N., Brownlee, R. R., and Eilers, D. D. (1966) *Astrophys. J.,* **144,** 1024.
(C9) Cox, J. P. and Giuli, R. T. (1968) *Principles of Stellar Structure.* New York: Gordon and Breach.
(C10) Cram, L. E. (1977) *Astron. Astrophys.,* **59,** 151.
(D1) Durney, B. R. (1971) *Astrophys. J.,* **166,** 669.
(D2) Durney, B. R. and Roberts, P. H. (1971) *Astrophys. J.,* **170,** 319.
(D3) Durney, B. R. and Werner, N. E. (1972) *Astrophys. J.,* **171,** 609.
(E1) Eckart, C. (1960) *Hydrodynamics of Oceans and Atmospheres.* Oxford: Pergamon.
(F1) Falk, S. W. and Arnett, W. D. (1977) *Astrophys. J. Suppl.,* **33,** 515.
(F2) Fraley, G. S. (1968) *Astrophys. Space Sci.,* **2,** 96.
(G1) Grover, R. and Hardy, J. W. (1966) *Astrophys. J.,* **143,** 48.
(H1) Harlow, F. H. and Amsden, A. A. (1971) *Los Alamos Scientific Laboratory Report No. LA-4700.* Los Alamos: University of California.
(H2) Hartle, R. E. and Sturrock, P. A. (1968) *Astrophys. J.,* **151,** 1155.
(H3) Hartmann, L. and MacGregor, K. B. (1980) *Astrophys. J.,* **242,** 260.
(H4) Hayes, W. D. (1960) *Gasdynamic Discontinuities.* Princeton: Princeton University Press.
(H5) Hayes, W. D. (1968) *J. Fluid Mech.,* **32,** 305.
(H6) Hayes, W. D. (1968) *J. Fluid Mech.,* **32,** 317.
(H7) Hill, S. J. (1972) *Astrophys. J.,* **178,** 793.
(H8) Hines, C. O. (1960) *Can. J. Phys.,* **38,** 1441.
(H9) Hines, C. O. (1968) Syllabus on *Internal Gravity and Acoustic Waves in Planetary and Solar Atmospheres.* Boulder: University of Colorado.
(H10) Hollweg, J. V. (1976) *J. Geophys. Res.,* **81,** 1649.
(H11) Hollweg, J. V. (1978) *Rev. Geophys. Space Phys.,* **16,** 689.
(H12) Holzer, T. E. (1977) *J. Geophys. Res.,* **82,** 23.

(H13) Holzer, T. E. (1979) in *Solar System Plasma Physics*. ed. C. F. Kennel, L. J. Lanzerotti, and E. N. Parker. p. 101. Amsterdam: North-Holland.

(H14) Holzer, T. E. (1980) in *Cool Stars, Stellar Systems, and the Sun*. SAO Special Report No. 389. ed. A. K. Dupree. p. 153. Cambridge: Smithsonian Astrophysical Observatory.

(H15) Holzer, T. E. and Axford, W. I. (1970) *Ann. Rev. Astron. Astrophys.*, **8**, 31.

(H16) Hoskin, N. E. (1964) in *Methods in Computational Physics*. ed. B. Alder, S. Fernbach, and M. Rotenberg. Vol. **3**, p. 265. New York: Academic Press.

(H17) Hugoniot, A. (1889) *J. de l'Ecole Polytech.*, **58**, 1.

(H18) Hundhausen, A. J. (1972) *Coronal Expansion and Solar Wind*. New York: Springer.

(I1) Imshinnik, V. S. (1962) *Soviet Phys.-JETP*, **15**, 167.

(I2) Israel, W. (1960) *Proc. Roy. Soc. London*, **A259**, 129.

(J1) Jordan, S. D. (1970) *Astrophys. J.*, **161**, 1189.

(K1) Klein, R. I., Stein, R. F., and Kalkofen, W. (1976) *Astrophys. J.*, **205**, 499.

(K2) Klein, R. I., Stein, R. F., and Kalkofen, W. (1978) *Astrophys. J.*, **220**, 1024.

(K3) Kuperus, M. (1969) *Space Sci. Rev.*, **9**, 713.

(K4) Kutter, G. S. and Sparks, W. M. (1972) *Astrophys. J.*, **175**, 407.

(L1) Lamb, H. (1945) *Hydrodynamics*. New York: Dover.

(L2) Landau, L. D. and Lifshitz, E. M. (1959) *Fluid Mechanics*. Reading: Addison-Wesley.

(L3) Leer, E. and Holzer, T. E. (1980) *J. Geophys. Res.*, **85**, 4681.

(L4) Leibacher, J. W. and Stein, R. F. (1981) in *Second Cambridge Workshop on Cool Stars, Stellar Systems, and the Sun*. SAO Special Report No. 392. ed. M. S. Gimpapa and L. Golub, p. 23. Cambridge: Smithsonian Astrophysical Observatory.

(L5) Leibacher, J. W. and Stein, R. F. (1981) in *The Sun as a Star*. ed. S. D. Jordan, p. 263. Washington: National Aeronautics and Space Administration.

(L6) Leighton, R. B., Noyes, R. W., and Simon, G. W. (1962) *Astrophys. J.*, **135**, 474.

(L7) Liang, E. P. (1977) *Astrophys. J.*, **211**, 361.

(L8) Lichnerowicz, A. (1967) *Relativistic Hydrodynamics and Magnetohydrodynamics*. New York: Benjamin.

(L9) Lighthill, J. (1978) *Waves in Fluids*. Cambridge: Cambridge University Press.

(M1) MacGregor, K. B. (1981) in *Second Cambridge Workshop on Cool Stars, Stellar Systems, and the Sun*. SAO Special Report No. 392. ed. M. S. Giampapa and L. Golub, p. 83. Cambridge: Smithsonian Astrophysical Observatory.

(M2) McKee, C. R. and Colgate, S. A. (1973) *Astrophys. J.*, **181**, 903.

(M3) Mihalas, B. R. W. (1979) *Ph.D. Thesis*, University of Colorado, Boulder.

(M4) Mihalas, B. W. and Toomre, J. (1981) *Astrophys. J.*, **249**, 349.

(M5) Mihalas, B. W. and Toomre, J. (1982) *Astrophys. J.*, **263**, 386.

(N1) Noyes, R. W. and Leighton, R. B. (1963) *Astrophys. J.*, **138**, 631.

(O1) Osterbrock, D. E. (1961) *Astrophys. J.*, **134**, 347.

(O2) Owczarek, J. A. (1964) *Fundamentals of Gas Dynamics*. Scranton: International Textbook Company.

(P1) Parker, E. N. (1958) *Astrophys. J.*, **128**, 664.

(P2) Parker, E. N. (1963) *Interplanetary Dynamical Processes.* New York: Interscience.

(P3) Parker, E. N. (1965) *Astrophys. J.*, **141,** 1463.

(P4) Perkins, F. W. (1973) *Astrophys. J.*, **179,** 637.

(R1) Raizer, Yu. P. (1964) *Zh. Prikl. Math. Tech. Fiz.*, **4,** 49.

(R2) Rankine, W. J. M. (1870) *Phil. Trans. Roy. Soc.*, **160,** 277.

(R3) Rayleigh, J. W. S. Lord (1910) *Proc. Roy. Soc.* (*London*), **A84,** 247.

(R4) Richtmyer, R. D. and Morton, K. W. (1967) *Difference Methods for Initial-Value Problems.* (2nd ed.) New York: Interscience.

(R5) Riemann, B. (1953) *Collected Works of Bernhard Riemann.* (2nd ed.) New York: Dover.

(R6) Roberts, P. H. (1971) *Astrophys. Letters*, **9,** 79.

(R7) Roberts, P. H. and Soward, A. M. (1972) *Proc. Roy. Soc.* (*London*), **A328,** 185.

(S1) Saito, M. (1964) *Pub. Astron. Soc. Japan*, **16,** 179.

(S2) Scarf, F. L. and Noble, L. M. (1965) *Astrophys. J.*, **141,** 1479.

(S3) Schatzman, E. (1949) *Ann. d. Astrophys.*, **12,** 203.

(S4) Schatzman, E. and Souffrin, P. (1967) *Ann. Rev. Astron. Astrophys.*, **5,** 67.

(S5) Schmitz, F. and Ulmschneider, P. (1980) *Astron. Astrophys.*, **84,** 93.

(S6) Schmitz, F. and Ulmschneider, P. (1980) *Astron. Astrophys.*, **84,** 191.

(S7) Schmitz, F. and Ulmschneider, P. (1981) *Astron. Astrophys.*, **93,** 178.

(S8) Schwarzschild, M. (1948) *Astrophys. J.*, **107,** 1.

(S9) Sedov, L. I. (1959) *Similarity and Dimensional Methods in Mechanics.* New York: Academic Press.

(S10) Shafranov, V. D. (1957) *Soviet Phys.-JETP*, **5,** 1183.

(S11) Shoub, E. C. (1977) *Astrophys. J. Supp.*, **34,** 259.

(S12) Stein, R. F. and Leibacher, J. (1974) *Ann. Rev. Astron. Astrophys*, **12,** 407.

(S13) Stein, R. F. and Schwartz, R. A. (1972) *Astrophys. J.*, **177,** 807.

(S14) Stein, R. F. and Schwartz, R. A. (1973) *Astrophys. J.*, **186,** 1083.

(S15) Sturrock, P. A. and Hartle, R. E. (1966) *Phys. Rev. Letters*, **16,** 628.

(S16) Summers, D. (1980) *Astrophys. J.*, **241,** 468.

(S17) Summers, D. (1982) *Astrophys. J.*, **257,** 881.

(T1) Taub, A. (1948) *Phys. Rev.*, **74,** 328.

(T2) Taub, A. (1978) *Ann. Rev. Fluid Mech.*, **10,** 301.

(T3) Thomas, J. H., Clark, P. A., and Clark, A. (1971) *Solar Phys.*, **16,** 51.

(T4) Thorne, K. S. (1973) *Astrophys. J.*, **179,** 897.

(T5) Tscharnuter, W. M. and Winkler, K.-H. (1979) *Comp. Phys. Comm.*, **18,** 171.

(U1) Ulmschneider, P. (1970) *Solar Phys.*, **12,** 403.

(U2) Ulmschneider, P. (1971) *Astron. Astrophys.*, **12,** 297.

(U3) Ulmschneider, P. (1971) *Astron. Astrophys.*, **14,** 275.

(U4) Ulmschneider, P. and Kalkofen, W. (1977) *Astron. Astrophys.*, **57,** 199.

(U5) Ulmschneider, P., Kalkofen, W., Nowak, T., and Bohn, H. U. (1977) *Astron. Astrophys.* **54,** 61.

(U6) Ulmschneider, P., Schmitz, F., Kalkofen, W., and Bohn, H. U. (1978) *Astron. Astrophys.*, **70,** 487.

(V1) Vaiana, G. S. (1980) in *Cool Stars, Stellar Systems, and the Sun.* SAO Special Report No. 389. ed. A. K. Dupree, p. 195. Cambridge: Smithsonian Astrophysical Observatory.

(V2) Vernazza, J. E., Avrett, E. H., and Loeser, R. (1973) *Astrophys. J.*, **184,** 605.

(V3) Vernazza, J. E., Avrett, E. H., and Loeser, R. (1976) *Astrophys. J. Supp.* **30,** 1.

(V4) Von Neumann, J. and Richtmyer, R. D. (1950) *J. Appl. Phys.*, **21,** 232.

(W1) Weber, E. J. and Davis, L. (1967) *Astrophys. J.*, **148,** 217.

(W2) Weinberg, S. (1971) *Astrophys. J.*, **168,** 175.

(W3) Weymann, R. (1960) *Astrophys. J.*, **132,** 452.

(W4) Whang, Y. C. and Chang, C. C. (1965) *J. Geophys. Res.*, **70,** 4175.

(W5) Whang, Y. C., Liu, C. K., and Chang, C. C. (1965) *Astrophys. J.*, **145,** 255.

(W6) Whitam, G. B. (1974) *Linear and Nonlinear Waves.* New York: Wiley.

(W7) Winkler, K.-H. (1984) in *Astrophysical Radiation Hydrodynamics.* ed. K.-H. Winkler and M. Norman. Dordrecht: Reidel.

(W8) Wolff, C. L., Brandt, J. C., and Southwick, R. G. (1971) *Astrophys. J.*, **165,** 181.

(W9) Wood, P. R. (1974) *Astrophys. J.*, **190,** 605.

(Y1) Yuan, S. W. (1967) *Foundations of Fluid Mechanics.* Englewood Cliffs: Prentice-Hall.

(Z1) Zel'dovich, Ya. B. and Raizer, Yu. P. (1966) *Physics of Shock Waves and High-Temperature Hydrodynamic Phenomena.* New York: Academic Press.

(Z2) Zirker, J. B. (1977) *Coronal Holes and High Speed Wind Streams.* Boulder: Colorado Associated University Press.

6

Radiation and Radiative Transfer

In the preceding chapters we treated the physics of nonradiating fluids; we now extend the analysis to *radiating fluids* comprising both material and radiation. Radiation adds to the total energy density, momentum density, stress, and energy flux in the fluid. We must therefore define these quantities for radiation and derive equations that describe the coupling among them and their coupling to the material.

Our first goal is to develop an understanding of the "microphysics" of the radiation field and of transport processes in the combined matter-radiation fluid along lines conceptually similar to our study of gases in Chapters 1 and 3. For this purpose it suffices to assume that the material is static, which is what we generally do in this chapter. Detailed discussion of how radiation transports energy and momentum through moving media, and couples to the dynamics of flows, is reserved for Chapter 7.

In §6.1 we derive expressions that specify the basic dynamical properties of the radiation field, in particular its energy density, energy flux, and stress tensor; we specialize these to the case of thermal equilibrium in §6.2. We then turn to the principal task of this chapter: the formulation and solution of the *transfer equation*, which determines how radiation is transported through the material. In §6.3 we describe the interaction of radiation with material in terms of macroscopic absorption and emission coefficients. Then in §6.4 we derive the transfer equation, which is the equivalent of the Boltzmann equation for photons (cf. §92), and discuss the significance of its moments.

In §6.5 we discuss methods for solving the transfer equation. In opaque material, such as the interior of a star, photons are trapped and the radiation field is nearly isotropic and approaches thermal equilibrium; the photon mean free path λ_p is much smaller than a characteristic structural length l in the material. In this limit, radiative energy transport can be described as a diffusion process, and we can derive an asymptotic solution of the transfer equation, which is similar to the Chapman–Enskog solution of the Boltzmann equation describing transport phenomena in gases (see also §97). Like the Chapman–Enskog solution, radiation diffusion theory is valid only for $\lambda_p \ll 1$.

But as we approach a boundary surface of a radiating medium (e.g., the atmosphere of a star), the material becomes transparent, and photon mean free paths can vastly exceed characteristic structural lengths. Here a *nonlocal* treatment is needed. We must solve the full transfer equation describing photon exchange within the fluid, thereby in effect constructing a nonlocal kinetic theory for photons. The nonlocal nature of the transfer problem is exacerbated by the *scattering* of photons by matter. After being thermally emitted, a photon may scatter many times, changing essentially only its direction of travel, before being absorbed and destroyed by reconversion into thermal energy. In doing so, a photon migrates a large distance called the *photon destruction length.* The radiation field is then no longer uniquely determined by local conditions but at any point may be determined by conditions within a large *interaction volume* whose size is set by a photon destruction length, not a mean free path. Hence *the radiation field is not, in general, a local variable.*

Because the transfer equation is (superficially) linear in the radiation field, it is possible to solve it for very general physical situations by powerful numerical techniques. Actually the transfer problem is linear only to the extent that we consider the material absorption and emission coefficients as *given.* But in reality these coefficients depend on the internal excitation and ionization state of the material, and, as we discuss in §6.6, this state is fixed in part by radiative processes that populate and depopulate atomic levels. We therefore find that in general *the radiation field and the internal state of the matter must be determined simultaneously and self-consistently.*

In the diffusion regime, the radiation field and level populations have their *thermal equilibrium* distributions and the coupling between radiation and matter presents no difficulty. Somewhat nearer to a radiating surface we reach a regime in which significant nonlocal radiation transport occurs, but collisional processes still dominate the state of the material, which can be calculated from the equations of statistical mechanics evaluated at local values of the temperature and density—the *local thermodynamic equilibrium* (LTE) regime. When the medium is very transparent, and photons escape freely from a boundary surface into space, the radiation field takes on a strongly *nonequilibrium* character. We must then reconsider the microphysics of the gas, allowing for a nonequilibrium interaction between the radiation and material: this poses a difficult problem both mathematically and conceptually because the local state of the material is then coupled by photon exchange to the state of the material within an entire interaction volume. The techniques required to solve this interlocked problem are discussed in §6.7.

This chapter forms essential background for the discussion of radiation hydrodynamics in Chapter 7. Conceptually the goal of these two chapters is to develop formalisms that describe accurately the strong interactions between radiation and matter in radiating fluids. It will pay the reader to reread this chapter after reading Chapter 7, having the benefits of insights gained there. Although it is our intent to give a self-contained account of the topics treated in this chapter, some of the material is technical, and we must sometimes omit details. We recommend that the reader consult references such as (**A2**), (**A3**), (**J1**), (**K1**), (**M2**), (**P1**), (**P3**), (**S1**), (**T2**), (**U1**) and (**W1**) for further background and amplification.

6.1 The Radiation Field

63. The Specific Intensity and Photon Distribution Function

The radiation field is, in general, a function of position and time, and at any given position has a distribution in both angle and frequency. We define the *specific intensity* $I(\mathbf{x}, t; \mathbf{n}, \nu)$ of radiation at position $\mathbf{x}$ and time t, traveling in direction $\mathbf{n}$ with frequency ν, to be such that the amount of energy transported by radiation of frequencies $(\nu, \nu+d\nu)$ across a surface element dS, in a time dt, into a solid angle $d\omega$ around $\mathbf{n}$, is

$$d\mathscr{E} = I(\mathbf{x}, t; \mathbf{n}, \nu)\, dS \cos\alpha\, d\omega\, d\nu\, dt, \tag{63.1}$$

where α is the angle between $\mathbf{n}$ and the normal to dS. In cgs units, I has dimensions ergs $\text{cm}^{-2}\,\text{s}^{-1}\,\text{Hz}^{-1}\,\text{sr}^{-1}$.

In most of what follows, we confine attention to one-dimensional structures and flows in planar or spherical geometry. In the planar case, we assume that the material is homogeneous in the horizontal direction, with properties varying only as a function of z and t. The intensity then has azimuthal symmetry around the unit vector $\mathbf{k}$; its angular distribution can be described completely in terms of the polar angle Θ or $\mu \equiv \cos\Theta = \mathbf{n}\cdot\mathbf{k}$. Hence $I = I(z, t; \mu, \nu)$. We assume that z is positive in the direction opposite to gravity, and explicit mention of z and t will normally be suppressed.

In spherical geometry, a position is specified by (r, θ, ϕ), and the direction of radiation at that position by polar and azimuthal angles (Θ, Φ) measured with respect to the radial unit vector $\hat{\mathbf{r}}$. For *spherical symmetry* I depends on r only, and is independent of Φ; therefore $I = I(r, t; \mu, \nu)$ where now $\mu = \mathbf{n}\cdot\hat{\mathbf{r}}$. Explicit mention of r and t will usually be suppressed.

The specific intensity provides a complete *macroscopic* description of the radiation field. From a *microscopic* view, the radiation field is composed of photons, and we define the *photon number density* ψ such that $\psi(\mathbf{x}, t; \mathbf{n}, \nu)\, d\omega\, d\nu$ is the number of photons per unit volume at $(\mathbf{x}, t)$ with frequencies $(\nu, \nu+d\nu)$, traveling with velocity c into a solid angle $d\omega$ around $\mathbf{n}$. The number of photons crossing a surface element $d\mathbf{S}$ in time dt is then $\psi\,(\mathbf{n}\cdot d\mathbf{S})(d\omega\, d\nu)(c\, dt)$. Each photon has energy $h\nu$, so the energy transported is

$$d\mathscr{E} = ch\nu\psi\, dS \cos\alpha\, d\omega\, d\nu\, dt. \tag{63.2}$$

Comparing (63.2) with (63.1) we find

$$I(\mathbf{x}, t; \mathbf{n}, \nu) = ch\nu\psi(\mathbf{x}, t; \mathbf{n}, \nu). \tag{63.3}$$

A relative of ψ is the *photon distribution function* f_R, defined such that $f_R(\mathbf{x}, t; \mathbf{n}, p)\, d^3p$ is the number of photons per unit volume at $(\mathbf{x}, t)$ with momenta $(\mathbf{p}, \mathbf{p}+d\mathbf{p})$, where $\mathbf{p}=(h\nu/c)\mathbf{n}$. Using $d^3p = p^2\, dp\, d\omega = (h/c)^3\nu^2\, d\nu\, d\omega$, we find $(h^3\nu^2/c^3)f_R\, d\nu\, d\omega \equiv \psi\, d\nu\, d\omega$, and therefore

$$I(\mathbf{x}, t; \mathbf{n}, \nu) = (h^4\nu^3/c^2)f_R(\mathbf{x}, t; \mathbf{n}, \nu). \tag{63.4}$$

The function f_R is a relativistically invariant distribution function (cf. §43) describing certain massless, extreme-relativistic particles (photons) in a six-dimensional phase space; it is completely analogous to the invariant particle distribution function used in §43 to construct a kinetic theory for a relativistic gas. In §§90 and 91 we use a kinetic theory approach to develop expressions for the radiative energy density, energy flux, and stress in terms of f_R and its moments. Furthermore, in §§92 and 95 we describe the interaction between radiation and matter by a Boltzmann equation for f_R. But for the present, we emphasize the continuum view, and most of the analysis in this chapter is done in terms of I.

64. The Mean Intensity and Radiation Energy Density

The *mean intensity* J is defined as the average of the specific intensity over all solid angles, that is,

$$J_\nu = J(\mathbf{x}, t; \nu) \equiv (4\pi)^{-1} \oint I(\mathbf{x}, t; \mathbf{n}, \nu)\, d\omega. \tag{64.1}$$

J_ν has dimensions ergs cm^{-2} s^{-1} Hz^{-1} sr^{-1}. The mean intensity is the *zeroth moment* of the radiation field over angles.

In a planar atmosphere I is independent of Φ. Thus, noting that $d\omega = \sin\Theta\, d\Theta\, d\Phi = -d\mu\, d\Phi$, we then have

$$J(z, t; \nu) = (4\pi)^{-1} \int_0^{2\pi} d\Phi \int_{-1}^{1} d\mu I(z, t; \mu, \nu) = \tfrac{1}{2} \int_{-1}^{1} I(z, t; \mu, \nu)\, d\mu. \tag{64.2}$$

This result also holds in spherical symmetry with z replaced by r.

The *monochromatic radiation energy density* at frequency ν is the number density of photons at that frequency, summed over all solid angles, times their energy $h\nu$. That is,

$$E_\nu = E(\mathbf{x}, t; \nu) = h\nu \oint \psi(\mathbf{x}, t; \mathbf{n}, \nu)\, d\omega. \tag{64.3}$$

Using (63.3) and (64.1) we see that

$$E_\nu = c^{-1} \oint I(\mathbf{x}, t; \mathbf{n}, \nu)\, d\omega = (4\pi/c)J_\nu. \tag{64.4}$$

E_ν has dimensions ergs cm^{-3} Hz^{-1}. The *total radiation energy density* is

$$E = E(\mathbf{x}, t) \equiv \int_0^\infty E(\mathbf{x}, t; \nu)\, d\nu = (4\pi/c) \int_0^\infty J(\mathbf{x}, t; \nu)\, d\nu = (4\pi/c)J(\mathbf{x}, t), \tag{64.5}$$

which has dimensions ergs cm^{-3}.

65. The Radiative Energy Flux and Momentum Density

We define the *monochromatic radiation flux* $\mathbf{F}(\mathbf{x}, t; \nu)$ to be a vector such that $\mathbf{F} \cdot d\mathbf{S}$ gives the net rate of radiant energy flow per unit frequency interval, at frequency ν, across $d\mathbf{S}$. The net number flux of photons crossing $d\mathbf{S}$ per unit time and frequency interval from all solid angles is

$$N = \left(\oint \psi(\mathbf{x}, t; \mathbf{n}, \nu) c\mathbf{n} \, d\omega \right) \cdot d\mathbf{S}, \tag{65.1}$$

which, multiplied by the energy per photon, $h\nu$, gives the energy flux. Recalling that $I = ch\nu\psi$ we see that

$$\mathbf{F}_\nu = \mathbf{F}(\mathbf{x}, t; \nu) = \oint I(\mathbf{x}, t; \mathbf{n}, \nu)\mathbf{n} \, d\omega, \tag{65.2}$$

or, in components

$$\mathbf{F}_\nu = (F_x, F_y, F_z)_\nu = \left(\oint I_\nu n_x \, d\omega, \oint I_\nu n_y \, d\omega, \oint I_\nu n_z \, d\omega \right), \tag{65.3}$$

where $n_x = (1 - \mu^2)^{1/2} \cos \Phi$, $n_y = (1 - \mu^2)^{1/2} \sin \Phi$, and $n_z = \mu$. In cgs units F_ν has dimensions ergs $\text{cm}^{-2}\,\text{s}^{-1}\,\text{Hz}^{-1}$. The flux is the *first moment* of the radiation field over angle. Summing over all frequencies we obtain the *integrated radiation flux*

$$\mathbf{F} = \mathbf{F}(\mathbf{x}, t) \equiv \int_0^\infty \mathbf{F}(\mathbf{x}, t; \nu) \, d\nu \tag{65.4}$$

which has dimensions ergs $\text{cm}^{-2}\,\text{s}^{-1}$.

For azimuthal symmetry around $\mathbf{k}$, F_x and F_y are identically zero; the remaining component F_z is therefore often called "the" flux

$$F_\nu = F(z, t; \nu) \equiv 2\pi \int_{-1}^{1} I(z, t; \mu, \nu)\mu \, d\mu, \tag{65.5}$$

as if it were a scalar. Following Eddington it is customary to define

$$H_\nu = H(z, t; \nu) \equiv (4\pi)^{-1} F(z, t; \nu) = \tfrac{1}{2} \int_{-1}^{1} I(z, t; \mu, \nu)\mu \, d\mu, \tag{65.6}$$

which is similar to (64.1) for J_ν. Equations (65.5) and (65.6) also apply in spherical symmetry with z replaced by r.

The momentum of a photon with energy $h\nu$ is $(h\nu/c)\mathbf{n}$; therefore the net rate of radiative momentum transport across $d\mathbf{S}$ at frequency ν is $c^{-1}\mathbf{F}_\nu \cdot d\mathbf{S}$. This transport is effected by particles moving with a speed c, hence the *monochromatic radiation momentum density* vector is

$$\mathscr{G}_\nu = c^{-2}\mathbf{F}_\nu. \tag{65.7}$$

Integrating over all frequencies we see that the *total radiation momentum density is*

$$\mathscr{G} = c^{-2}\mathbf{F} \tag{65.8}$$

where $\mathbf{F}$ is the total radiation flux. This result also follows from considerations of the form of the radiation stress-energy tensor (cf. §91).

66. The Radiation Pressure Tensor

As for material particles, we define the *radiation stress tensor*, or *pressure tensor*, P such that P^{ij} is the net rate of transport, per unit area of a surface oriented perpendicular to the jth coordinate axis, of the ith component of momentum. The number of photons of frequency ν, moving in direction n^j, crossing a unit area in a unit time, is $\psi_\nu c n^j$; each has momentum $(h\nu n^i/c)$ in the ith direction. Thus summing over all solid angles we obtain the *monochromatic radiation pressure tensor*

$$P^{ij}(\mathbf{x}, t; \nu) = \oint \psi(\mathbf{x}, t; \mathbf{n}, \nu)(h\nu n^i/c)(cn^j)\, d\omega, \tag{66.1}$$

or

$$P^{ij}(\mathbf{x}, t; \nu) = c^{-1} \oint I(\mathbf{x}, t; \mathbf{n}, \nu) n^i n^j\, d\omega. \tag{66.2}$$

In dyadic notation

$$\mathsf{P}_\nu = \mathsf{P}(\mathbf{x}, t; \nu) = c^{-1} \oint I(\mathbf{x}, t; \mathbf{n}, \nu)\mathbf{nn}\, d\omega. \tag{66.3}$$

P_ν is manifestly symmetric, and is clearly the *second moment* of the radiation field over angle; the components of P_ν have dimensions dynes $\text{cm}^{-2}\,\text{Hz}^{-1}$.

The rate of momentum transport across an oriented surface element $d\mathbf{S}$ with normal $\mathbf{l}$, by photons of frequency ν, is $P_\nu^{ij} l_j\, d\mathbf{S}$. Integrating over a closed surface S surrounding a volume V, we find that the total rate of flow of the ith component of radiation momentum at frequency ν out of V is

$$\int_S P_\nu^{ij} l_j\, dS = \int_V (P_\nu^{ij})_{,j}\, dV. \tag{66.4}$$

In the absence of momentum exchange between radiation and matter (and of body forces that affect photons, that is, general relativity effects), this flow decreases the momentum density $\mathcal{G}_\nu$ in V. We therefore must have

$$(\partial \mathcal{G}_\nu/\partial t) = c^{-2}(\partial \mathbf{F}_\nu/\partial t) = -\nabla \cdot \mathsf{P}_\nu, \tag{66.5}$$

which is identical to the momentum equation (23.5) for an ideal fluid in the absence of body forces. We emphasize that (66.5) applies only in the absence of material; interactions with matter are treated in §§78, 93, and 96.

Because I_ν is independent of Φ in a one-dimensional medium, direct calculation from (66.2) yields

$$\mathsf{P}_\nu = \begin{pmatrix} P_\nu & 0 & 0 \\ 0 & P_\nu & 0 \\ 0 & 0 & P_\nu \end{pmatrix} - \frac{1}{2}\begin{pmatrix} 3P_\nu - E_\nu & 0 & 0 \\ 0 & 3P_\nu - E_\nu & 0 \\ 0 & 0 & 0 \end{pmatrix}. \tag{66.6}$$

Here the scalar P_ν is defined as

$$P_\nu \equiv (4\pi/c)K_\nu, \tag{66.7}$$

where, in turn,

$$K_\nu \equiv \tfrac{1}{2}\int_{-1}^{1} I(z, t; \mu, \nu)\mu^2\, d\mu. \tag{66.8}$$

In planar geometry P_{11}, P_{22}, and P_{33} in (66.6) represent P_{xx}, P_{yy}, and P_{zz}; in spherical geometry they represent the physical components $P_{\theta\theta}$, $P_{\phi\phi}$, and P_{rr} relative to the orthonormal triad $(\hat{\boldsymbol{\theta}}, \hat{\boldsymbol{\phi}}, \hat{\mathbf{r}})$. Equations (66.6) show that in the special case of a one-dimensional medium two scalars, P_ν and E_ν, suffice to specify the full tensor P_ν. Moreover, derivatives with respect to (x, y) or (θ, ϕ) must be identically zero by symmetry. Therefore, in planar geometry the only nonvanishing component of $\nabla \cdot \mathsf{P}_\nu$ is

$$(\nabla \cdot \mathsf{P}_\nu)_z = \partial P_\nu/\partial z, \tag{66.9}$$

and, from (A3.91), the only nonvanishing component in spherical symmetry is

$$(\nabla \cdot \mathsf{P}_\nu)_r = (\partial P_\nu/\partial r) + (3P_\nu - E_\nu)/r. \tag{66.10}$$

From (66.9) and (66.10) one can understand why in one-dimensional problems it is customary to refer to the scalar P_ν as "the" radiation pressure. But it is important to bear in mind that because the second term in (66.6) is not necessarily zero, P_ν is not, in general, isotropic, and therefore does not reduce to a simple hydrostatic pressure. The anisotropy of P_ν reflects an anisotropic distribution of $I(\mu, \nu)$, which is induced by efficient photon exchange between regions with significantly different physical properties, particularly in the presence of strong gradients, and/or an open boundary.

Because the trace P_ν^{ii} of P_ν is an invariant, it is sometimes used to define a *mean radiation pressure*

$$\bar{P}_\nu = \bar{P}(\mathbf{x}, t; \nu) \equiv \tfrac{1}{3}P^{ii}(\mathbf{x}, t; \nu) = \tfrac{1}{3}E(\mathbf{x}, t; \nu), \tag{66.11}$$

the last equality following directly from (66.6). While (66.11) is true in general, note that $\bar{P}_\nu$ does *not*, in general, equal P_ν, nor does it have any particular dynamical significance.

The radiation pressure tensor will be isotropic for any distribution of the radiation field $I(\mu, \nu)$ that yields $P_\nu = \frac{1}{3}E_\nu$. A particular example is *isotropic radiation*, for which $K_\nu = \frac{1}{3}J_\nu$ from (64.2) and (66.8), hence $P_\nu = \frac{1}{3}E_\nu$ from (64.4) and (66.7). In this case

$$\mathsf{P}(\mathbf{x}, t; \nu) = \begin{pmatrix} P_\nu & 0 & 0 \\ 0 & P_\nu & 0 \\ 0 & 0 & P_\nu \end{pmatrix} \tag{66.12}$$

and, for computational purposes, the entire radiation pressure tensor can be replaced by a scalar hydrostatic pressure $P_\nu = \frac{1}{3}E_\nu$. This is a case of great practical importance because it holds true in the diffusion regime (cf. §80), which shows why one can always use scalar pressures in stellar interior calculations. Note also from (66.10) that in this case $(\nabla \cdot \mathsf{P})_r$ reduces to just $(\partial P/\partial r)$. The small departures of P_ν from isotropy in the diffusion limit in moving media are discussed in §97.

On the other hand, at the boundary of a medium with a positive temperature gradient inward, the radiation field is peaked (i.e., is largest) in the direction of outward flow (cf. §§79 and 82). Moreover, from (66.7) and (64.4) we see that such radiation, with $\mu \approx 1$, is more heavily weighted in P_ν than in E_ν. Hence near boundary surfaces the ratio P_ν/E_ν usually exceeds $\frac{1}{3}$. The extreme example is a *plane wave* traveling along the z axis, for which $I(\mu) = I_0\, \delta(\mu - 1)$. In this case, called the *streaming limit*, $J_\nu = H_\nu = K_\nu$, hence $P_\nu = E_\nu$, and P_ν has only one nonzero element, namely P_{zz}.

From the discussion above, we see that it is useful to define the dimensionless ratio

$$f_\nu = f(\mathbf{x}, t; \nu) \equiv P(\mathbf{x}, t; \nu)/E(\mathbf{x}, t; \nu) = K_\nu/J_\nu, \qquad (66.13)$$

which is known as the *variable Eddington factor*, to give a measure of the degree of anisotropy of the radiation field. For an opaque medium with a boundary, f_ν typically lies in the range $\frac{1}{3}$ to 1. We will see in §§78 and 83 that f_ν can be used to close the system of moments of the transfer equation; this important idea was first suggested by B. E. Freeman (**F2**), and cast into an easily applied form by G. R. Spillman (**S4**).

All quantities defined in this section have frequency-integrated counterparts. For example, the scalar describing the total radiation pressure in a one-dimensional medium is

$$P = P(\mathbf{x}, t) = \int_0^\infty P(\mathbf{x}, t; \nu)\, d\nu = (4\pi/c) \int_0^\infty K(\mathbf{x}, t; \nu)\, d\nu, \qquad (66.14)$$

which is also the total hydrostatic pressure for isotropic radiation; it has dimensions dynes cm^{-2}. Similarly, the *total radiation pressure tensor* is

$$\mathsf{P} = \mathsf{P}(\mathbf{x}, t) = c^{-1} \int_0^\infty d\nu \oint d\omega I(\mathbf{x}, t; \mathbf{n}, \nu)\mathbf{nn}. \qquad (66.15)$$

6.2 Thermal Radiation

In the important limiting case of thermodynamic equilibrium, the radiation field is described by a unique distribution function that depends on only one state variable, the absolute temperature T. Such *thermal radiation* exists in a *hohlraum*, an isolated enclosure in thermal equilibrium at a uniform temperature; in the laboratory, close approximations to a hohlraum are provided by carefully insulated ovens. Another close approximation is the deep interior of a star. For example, in the Sun the mean temperature gradient from center to surface is about $10^7\,\text{K}/10^{11}\,\text{cm} = 10^{-4}\,\text{K cm}^{-1}$, while photon mean free paths are a fraction of a centimeter; the radiation field is thus in equilibrium with material at a very uniform temperature, and approaches perfect thermal radiation very closely.

67. Planck's Law

Thermal radiation is described by the *Planck function* $B_\nu(T)$. The functional form of $B_\nu(T)$ follows immediately from Bose–Einstein quantum statistics [see, e.g., (**C5,** Chap. 10) or (**H2,** Chap. 12)]; but for our purposes a brief semiclassical derivation suffices [cf. (**S2,** Chap. 6)].

Planck hypothesized that radiation comes in discrete quanta of energy $h\nu$, and that the radiation field in a hohlraum is a superposition of quantized oscillations or *modes*. In a rectangular cavity of dimensions (X, Y, Z) each mode is characterized by positive integers (n_x, n_y, n_z) such that the propagation vector $\mathbf{k}$ has components $(n_x\pi/X,\ n_y\pi/Y,\ n_z\pi/Z)$, which guarantees that the modes are standing waves. For each $\mathbf{k}$ there are two senses of polarization of the field, with electric vectors orthogonal to each other and to $\mathbf{k}$, defining two modes.

Let us count the number of modes with frequencies on the range $(\nu, \nu+d\nu)$. The magnitude of $\mathbf{k}$ is $k = 2\pi/\Lambda = 2\pi\nu/c$; hence

$$\nu = (ck/2\pi) = \tfrac{1}{2}c[(n_x/X)^2 + (n_y/Y)^2 + (n_z/Z)^2]^{1/2}. \tag{67.1}$$

Surfaces of constant frequency are thus the ellipsoids

$$(n_x/a_x)^2 + (n_y/a_y)^2 + (n_z/a_z)^2 = 1, \tag{67.2}$$

where $(a_x, a_y, a_z) = (2\nu/c)(X, Y, Z)$. The number of normal modes with frequencies $\nu' \le \nu$ equals twice (for two polarizations) the number of points with integer coordinates within one octant (all n's≥ 0) of the ellipsoid (67.2). The volume of an ellipsoid is $(4\pi a_x a_y a_z/3)$, hence

$$N(\nu) = 2\times\tfrac{1}{8}\times(4\pi/3)(8XYZ\nu^3/c^3) = (8\pi\nu^3/3c^3)V, \tag{67.3}$$

where $V = XYZ$ is the volume of the hohlraum. Therefore the number of modes with frequencies on $(\nu, \nu+d\nu)$ is

$$dN = (8\pi\nu^2/c^3)V\,d\nu. \tag{67.4}$$

To calculate the average energy associated with these modes we assume, with Planck, that the energy of n active modes of frequency ν is $nh\nu$. In equilibrium at temperature T, the relative probability of a set of modes having total energy ε_n is

$$\pi_n = \exp(-\varepsilon_n/kT) \Big/ \sum_n \exp(-\varepsilon_n/kT), \tag{67.5}$$

whence the average energy of all modes at frequency ν is

$$\langle\varepsilon\rangle = \left[\sum_{n=0}^{\infty} nh\nu \exp(-nh\nu/kT)\right] \Big/ \left[\sum_{n=0}^{\infty} \exp(-nh\nu/kT)\right]. \tag{67.6}$$

Writing $x \equiv \exp(-h\nu/kT)$ we can rewrite the denominator of (67.6) as

$$1+x+x^2+\ldots=(1-x)^{-1}=(1-e^{-h\nu/kT})^{-1}, \tag{67.7}$$

while the numerator is

$$\begin{aligned} h\nu(x+2x^2+3x^3+\ldots) &= h\nu x\frac{d}{dx}(1+x+x^2+\ldots) \\ &= h\nu x\frac{d}{dx}\left(\frac{1}{1-x}\right)=\frac{h\nu x}{(1-x)^2}. \end{aligned} \tag{67.8}$$

Hence

$$\langle\varepsilon\rangle = h\nu x/(1-x) = h\nu/(e^{h\nu/kT}-1). \tag{67.9}$$

The energy density in the hohlraum is the number of modes per unit volume times the average energy per mode. Hence from (67.4) and (67.9) we have

$$E_\nu^* = E^*(\nu, T) = (8\pi h\nu^3/c^3)/(e^{h\nu/kT}-1), \tag{67.10}$$

where the asterisk indicates thermal equilibrium. From (64.4) we then see that $B_\nu(T)$, the (isotropic) specific intensity in thermal equilibrium, is

$$B_\nu(T) = (2h\nu^3/c^2)/(e^{h\nu/kT}-1); \tag{67.11}$$

this distribution characterizes the radiation, usually called *blackbody radiation*, emitted by a *perfect radiator* or *black body*.

68. Stefan's Law

The integrated energy density for thermal radiation is

$$E^*(T) = \int_0^\infty E^*(\nu, T)\,d\nu = (8\pi h/c^3)\int_0^\infty \nu^3(e^{h\nu/kT}-1)^{-1}\,d\nu. \tag{68.1}$$

Writing $x \equiv h\nu/kT$ we have

$$E^*(T) = (8\pi k^4T^4/c^3h^3)\int_0^\infty x^3e^{-x}(1+e^{-x}+e^{-2x}+\ldots)\,dx. \tag{68.2}$$

Integrating the series term by term we find

$$6(1+2^{-4}+3^{-4}+\ldots) = 6\zeta_4 = \pi^4/15, \tag{68.3}$$

where ζ_4 is the Riemann zeta function of order four (**A1**, 807). We thus obtain *Stefan's law*, which states that in thermal equilibrium the total radiation energy density is proportional to the fourth power of the absolute temperature, or

$$E^*(T) = a_R T^4, \tag{68.4}$$

where

$$a_R \equiv 8\pi^5k^4/15c^3h^3. \tag{68.5}$$

From (68.4) and (64.5) one sees that the integrated Planck function is

$$B(T) = (a_R c/4\pi)T^4. \tag{68.6}$$

It is customary to define the *Stefan-Boltzmann constant* σ_R such that

$$\pi B(T) = \sigma_R T^4, \tag{68.7}$$

whence $\sigma_R = \frac{1}{4}a_R c$. The rationale for this definition follows from calculating the radiation flux emergent from a black body, namely

$$F_{BB}(\nu) = 2\pi \int_0^1 B_\nu(T)\mu \, d\mu = \pi B_\nu(T), \tag{68.8}$$

which yields an integrated flux

$$F_{BB} = \pi B(T) = \sigma_R T^4. \tag{68.9}$$

Note that (68.8) and (68.9) apply to radiation *emerging* from a hohlraum; the isotropy of equilibrium radiation implies that the net flux *within* the hohlraum is identically zero.

69. Thermodynamics of Equilibrium Radiation

The radiation field within an equilibrium cavity has associated with it both an energy density and a stress. Energy can be fed into or withdrawn from the cavity, and the radiation field can do mechanical work. In §§69 to 71 we examine the thermodynamic properties of equilibrium radiation, both by itself and accompanied by material. Inasmuch as radiation has no mass, it is awkward to work with intensive variables defined per unit mass; we will therefore use extensive variables. Conversion of our results to quantities per unit volume or mass is straightforward.

Because thermal radiation is isotropic, the monochromatic thermal radiation pressure is, from (66.12)

$$P_\nu^* = \tfrac{1}{3}E_\nu^* = (4\pi/3c)B_\nu(T), \tag{69.1}$$

whence the total thermal radiation pressure is

$$P^* = \tfrac{1}{3}E^* = \tfrac{1}{3}a_R T^4. \tag{69.2}$$

To calculate the entropy of thermal radiation we apply the first law of thermodynamics to radiation in an enclosure. Thus

$$T\,dS_{\text{rad}} = d\mathscr{E} + P\,dV = d(E^*V) + P^*\,dV \tag{69.3}$$

implies that

$$dS_{\text{rad}} = (4a_R T^2 V)\,dT + (\tfrac{4}{3}a_R T^3)\,dV = d(\tfrac{4}{3}a_R T^3 V). \tag{69.4}$$

Hence the entropy of equilibrium radiation is

$$S_{\text{rad}} = \tfrac{4}{3}a_R T^3 V. \tag{69.5}$$

If the volume of the cavity is changed adiabatically, $dS \equiv 0$, and from (69.4) we have

$$TV^{1/3} = \text{constant}, \quad (69.6)$$

which from (69.2) implies that

$$P^* V^{4/3} = \text{constant} \quad (69.7)$$

and hence

$$P^* T^{-4} = \text{constant}. \quad (69.8)$$

Comparison of these results with (4.13) to (4.15) shows that the polytropic laws for thermal radiation are identical to those for a perfect gas with $\gamma = \frac{4}{3}$, as mentioned in §43. Furthermore, from (14.19) to (14.21), with ρ replaced by V^{-1}, we find for thermal radiation that $\Gamma_1 = \Gamma_2 = \Gamma_3 = \gamma = \frac{4}{3}$.

The heat capacity of radiation at constant volume is

$$C_v = (\partial \mathscr{E}/\partial T)_v = \partial(a_R T^4 V)/\partial T = 4a_R T^3 V. \quad (69.9)$$

However the heat capacity at constant pressure is *not* $C_p = \gamma C_v$, and in this sense the analogy between thermal radiation and a perfect gas with $\gamma = \frac{4}{3}$ fails. In fact, C_p is infinite. To understand this result physically, consider introducing heat into the enclosure while holding P^* constant. From (69.2), T remains constant, while V increases to accommodate the increase in energy of the system. Thus in (2.4), $dQ > 0$ while $dT \equiv 0$, hence $C_p = \infty$.

Finally, from (48.32) with $\Gamma_1 = \frac{4}{3}$ and $[p/(\hat{e}+p)]_{\text{rad}} = P^*/(E^* + P^*) = \frac{1}{4}$, we see that the speed of an "acoustic" disturbance in a gas of pure thermal radiation is $c/\sqrt{3}$.

70. *Thermodynamics of Equilibrium Radiation Plus a Perfect Gas*

Now consider a two-component gas comprising thermal radiation and a perfect gas of particles with mass $m = \mu_0 m_H$. The gas occupies a volume V and contains $\mathcal{N}$ particles. To simplify the notation in this section and in §71 we write

$$p_g = p_{\text{gas}} = \mathcal{N}kT/V, \quad (70.1)$$

$$p = p_{\text{total}} = p_g + P^* = (\mathcal{N}kT/V) + \tfrac{1}{3}a_R T^4, \quad (70.2)$$

and define

$$\alpha \equiv P^*/p_g. \quad (70.3)$$

The total internal energy in the volume is

$$\mathscr{E} = \tfrac{3}{2}\mathcal{N}kT + a_R T^4 V, \quad (70.4)$$

hence the specific internal energy per unit mass is

$$e = \tfrac{3}{2}RT + (a_R T^4/\rho). \quad (70.5)$$

To calculate the entropy of the system, we use the first law of thermodynamics

$$T\,dS = d\mathscr{E} + p\,dV. \quad (70.6)$$

Substituting (70.2) and (70.4) we find

$$dS = \mathcal{N}k\, d[\ln (T^{3/2}V)] + d(\tfrac{4}{3}a_R T^3 V), \tag{70.7}$$

which implies that

$$S = \mathcal{N}k \ln (T^{3/2}V) + \tfrac{4}{3}a_R T^3 V + \text{Constant}. \tag{70.8}$$

Equation (70.8) states that the total entropy of the composite gas equals the sum of the entropies of the radiation field and of the translational motion of the particles.

The heat capacity at constant volume is

$$\begin{aligned} C_v &= (\partial \mathscr{E}/\partial T)_v = \tfrac{3}{2}\mathcal{N}k + 4a_R T^3 V \\ &= \tfrac{3}{2}\mathcal{N}k(1+8\alpha) = 4a_R T^3 V[(1/8\alpha)+1], \end{aligned} \tag{70.9}$$

which clearly yields the correct limits as $\alpha \to 0$ and $\alpha \to \infty$. The specific heat at constant volume, per unit mass, is

$$c_v = (\tfrac{3}{2}k/m)(1+8\alpha). \tag{70.10}$$

The heat capacity at constant pressure follows from

$$C_p = (\partial \mathscr{H}/\partial T)_p, \tag{70.11}$$

where the total enthalpy is

$$\mathscr{H} = \mathscr{E} + pV = \tfrac{5}{2}\mathcal{N}kT + \tfrac{4}{3}a_R T^4 V. \tag{70.12}$$

Thus

$$C_p = \tfrac{5}{2}\mathcal{N}k + \tfrac{16}{3}a_R T^3 V + \tfrac{4}{3}a_R T^4 (\partial V/\partial T)_p. \tag{70.13}$$

From (70.3) one easily finds

$$(\partial V/\partial T)_p = (V/T)(1+4\alpha), \tag{70.14}$$

hence

$$C_p = \mathcal{N}k(\tfrac{5}{2} + 20\alpha + 16\alpha^2) \tag{70.15}$$

or

$$c_p = \tfrac{5}{2}(k/m)(1+8\alpha+\tfrac{32}{5}\alpha^2). \tag{70.16}$$

Equations (70.15) and (70.16) go to the correct limit as $\alpha \to 0$, and diverge as $\alpha \to \infty$, as expected from §69.

For an adiabatic change $dS \equiv 0$, and (70.7) implies that

$$\left(\frac{3}{2}\frac{\mathcal{N}kT}{V} + 4a_R T^4\right)\left(\frac{V}{T}\right) dT_s + \left(\frac{\mathcal{N}kT}{V} + \tfrac{4}{3}a_R T^4\right) dV_s = 0, \tag{70.17}$$

whence

$$\begin{aligned} \Gamma_3 - 1 \equiv -(\partial \ln T/\partial \ln V)_s &= (p_g + 4P^*)/(\tfrac{3}{2}p_g + 12P^*) \\ &= (1+4\alpha)/(\tfrac{3}{2} + 12\alpha). \end{aligned} \tag{70.18}$$

From (70.3) we have

$$dp = (p_g + 4P^*)(dT/T) - p_g(dV/V), \tag{70.19}$$

hence

$$\Gamma_2/(\Gamma_2-1) \equiv (\partial \ln p/\partial \ln T)_s = [1+4\alpha-(\partial \ln V/\partial \ln T)_s]/(1+\alpha). \tag{70.20}$$

Using (70.18) in (70.20) we have

$$(\Gamma_2-1)/\Gamma_2 = (1+5\alpha+4\alpha^2)/(\tfrac{5}{2}+20\alpha+16\alpha^2). \tag{70.21}$$

In the limit as $\alpha \to 0$, (70.21) yields the same result as (4.15) for $\gamma=\frac{5}{3}$; as $\alpha \to \infty$ we recover (69.8) for pure radiation. To calculate Γ_1 we use (14.22), (70.18), and (70.21), obtaining

$$\Gamma_1 = (\tfrac{5}{2}+20\alpha+16\alpha^2)/[(\tfrac{3}{2}+12\alpha)(1+\alpha)]. \tag{70.22}$$

A table of Γ_1, Γ_2, and Γ_3 for values of $\beta \equiv p_g/p$ ranging from 0 to 1 is given in (**C5**, 59).

The speed of sound in the composite gas of material and thermal radiation can be computed from (48.32), with Γ_1 given by (70.22), p given by (70.2), and the total energy density by

$$\hat{e} = \rho_0 c^2 + (\mathscr{E}/V) \tag{70.23}$$

where $\rho_0 = Nm$ and $\mathscr{E}$ is given by (70.4). It is easy to show that as $\alpha \to 0$ the speed of sound reduces to the adiabatic sound speed of the material, and as $\alpha \to \infty$ it approaches $c/\sqrt{3}$.

Finally, as in (14.31), we can define a variable mean molecular weight μ such that the total pressure (including radiation pressure) is given by $p=\rho kT/\mu m_H$; we can then calculate Q as defined in (14.33). For constant p,

$$(\partial \ln \mu/\partial \ln T)_p = 1+(\partial \ln \rho/\partial \ln T)_p, \tag{70.24}$$

hence

$$Q = -(\partial \ln \rho/\partial \ln T)_p. \tag{70.25}$$

From

$$p = (\rho kT/\mu_0 m_H) + \tfrac{1}{3}a_R T^4, \tag{70.26}$$

one easily finds

$$Q = 1+4\alpha. \tag{70.27}$$

Clearly $Q \to 1$ for a perfect gas ($\alpha \to 0$), and diverges for pure radiation ($\alpha \to \infty$).

71. *Thermodynamics of Equilibrium Radiation Plus an Ionizing Gas*

Let us now consider an equilibrium gas composed of thermal radiation and ionizing hydrogen. Writing x for the ionization fraction, the total pressure is

$$p = p_g + P^* = (1+x)(\mathcal{N}kT/V) + \tfrac{1}{3}a_R T^4, \tag{71.1}$$

and the total internal energy is

$$\mathscr{E} = \tfrac{3}{2}(1+x)\mathcal{N}kT + \mathcal{N}x\varepsilon_H + a_R T^4 V. \tag{71.2}$$

As in §14, x is determined from Saha's equation

$$x^2/(1-x) = \text{Const. } VT^{3/2} \exp(-\varepsilon_H/kT). \tag{71.3}$$

From (71.3) one finds

$$T\left(\frac{\partial x}{\partial T}\right)_v = \frac{x(1-x)}{(2-x)}\left(\frac{3}{2}+\frac{\varepsilon_H}{kT}\right), \tag{71.4}$$

and from (71.1) and (71.3) one can show that

$$T(\partial x/\partial T)_p = \tfrac{1}{2}x(1-x^2)[\tfrac{5}{2}+(\varepsilon_H/kT)+4\alpha] \tag{71.5}$$

and

$$(\partial \ln V/\partial \ln T)_p = 1+4\alpha+\tfrac{1}{2}x(1-x)[\tfrac{5}{2}+(\varepsilon_H/kT)+4\alpha]. \tag{71.6}$$

The heat capacity at constant volume is

$$C_v = \tfrac{3}{2}\mathcal{N}k(1+x)+\mathcal{N}k[\tfrac{3}{2}+(\varepsilon_H/kT)]T(\partial x/\partial T)_v+4a_RT^3V. \tag{71.7}$$

Using (71.4) in (71.7), we find the specific heat per unit mass, $c_v \equiv C_v/(\mathcal{N}m_H)$, is

$$c_v = \left(\frac{k}{m_H}\right)\left[(\tfrac{3}{2}+12\alpha)(1+x)+\frac{x(1-x)}{(2-x)}\left(\frac{3}{2}+\frac{\varepsilon_H}{kT}\right)^2\right], \tag{71.8}$$

which reduces to (14.15) when $\alpha=0$, and to (70.10) when $x=0$. It is evident that both ionization effects and radiation pressure can make a large contribution to c_v.

The heat capacity at constant pressure is obtained from (70.11), with

$$\mathcal{H} = \tfrac{5}{2}(1+x)\mathcal{N}kT+\mathcal{N}x\varepsilon_H+\tfrac{4}{3}a_RT^4V. \tag{71.9}$$

Then

$$\begin{aligned} C_p = {} & \tfrac{5}{2}\mathcal{N}k(1+x)+\mathcal{N}k[\tfrac{5}{2}+(\varepsilon_H/kT)]T(\partial x/\partial T)_p \\ & +\tfrac{4}{3}a_RT^3[4V+T(\partial V/\partial T)_p]. \end{aligned} \tag{71.10}$$

Using (71.5) and (71.6) in $c_p \equiv C_p/(\mathcal{N}m_H)$ we find, after some algebra,

$$c_p = (k/m_H)\{(\tfrac{5}{2}+20\alpha+16\alpha^2)(1+x)+\tfrac{1}{2}x(1-x^2)[\tfrac{5}{2}+(\varepsilon_H/kT)+4\alpha]^2\}, \tag{71.11}$$

which reduces to (14.18) when $\alpha=0$ and to (70.16) when $x=0$.

To compute adiabatic exponents we again require that

$$T\,dS = d\mathcal{E}+p\,dV \equiv 0. \tag{71.12}$$

Then, calculating $d\mathcal{E}$ from (71.2), $p\,dV$ from (71.1), and eliminating dx via the logarithmic derivative of (71.3),

$$\frac{(2-x)}{(1-x)}\frac{dx}{x} = \left(\frac{3}{2}+\frac{\varepsilon_H}{kT}\right)\frac{dT}{T}+\frac{dV}{V}, \tag{71.13}$$

one readily finds

$$\Gamma_3-1=-\left(\frac{\partial \ln T}{\partial \ln V}\right)_s=\frac{1+4\alpha+\frac{1}{2}x(1-x)[\frac{5}{2}+(\varepsilon_H/kT)+4\alpha]}{\frac{3}{2}+12\alpha+\frac{1}{2}x(1-x)\{[\frac{3}{2}+(\varepsilon_H/kT)]^2+\frac{3}{2}+12\alpha\}} \tag{71.14}$$

which reduces to (70.18) when $x=0$ and to (14.30) when $\alpha=0$.

Using the equation of state along with (71.12) and (71.13) one can eliminate both dx and dV in favor of dp and dT. After considerable algebra one obtains

$$\frac{\Gamma_2-1}{\Gamma_2}=\frac{1+4\alpha+\frac{1}{2}x(1-x)[\frac{5}{2}+(\varepsilon_H/kT)+4\alpha]}{\beta\{\frac{5}{2}+20\alpha+16\alpha^2+\frac{1}{2}x(1-x)[\frac{5}{2}+(\varepsilon_H/kT)+4\alpha]^2\}}, \tag{71.15}$$

which reduces to (70.21) when $x=0$ and to (14.28) when $\alpha=0$. Next, from (14.22), (71.14), and (71.15) we find

$$\Gamma_1=\frac{\beta\{\frac{5}{2}+20\alpha+16\alpha^2+\frac{1}{2}x(1-x)[\frac{5}{2}+(\varepsilon_H/kT)+4\alpha]^2\}}{\frac{3}{2}+12\alpha+\frac{1}{2}x(1-x)\{[\frac{3}{2}+(\varepsilon_H/kT)]^2+\frac{3}{2}+12\alpha\}}, \tag{71.16}$$

which reduces to (14.29) when $\alpha=0$ and to (70.22) when $x=0$.

Again, the speed of sound in the composite material-radiation gas can be computed from (48.32) with Γ_1 given by (71.16), p given by (71.1), and the total energy density $\hat{e}$ by (70.27), with $\mathscr{E}$ obtained from (71.2).

Finally, rewriting (70.24) as $Q=(\partial \ln V/\partial \ln T)_p$, we see from (71.6) that

$$Q=1+4\alpha+\tfrac{1}{2}x(1-x)[\tfrac{5}{2}+(\varepsilon_H/kT)+4\alpha], \tag{71.17}$$

which reduces to (14.34) and (70.26) in the appropriate limits.

More general formulae for the thermodynamic properties of a gas composed of thermal radiation and several ionizing species can be found in (**C7**, §9.18), (**K2**), and (**M1**).

The formulae derived in this section and in §70 give an accurate description of the thermodynamic properties of a radiating fluid when the radiation field is thermalized to its equilibrium distribution function and the material is in equilibrium at the same absolute temperature as the radiation. These formulae apply, for example, from the deeper layers of a stellar atmosphere down into the stellar interior. They sometimes can give useful first estimates even for a nonequilibrium radiation field, but in such cases they should be used with caution because not only may they be inaccurate numerically, but the whole conceptual framework of equilibrium thermodynamics on which they are based becomes problematical, or even invalid (§86).

6.3 The Interaction of Radiation and Matter

We now consider how radiation interacts with material. We first set forth formulae for computing rates of absorption, emission, and scattering of radiation in terms of atomic cross sections and level populations. As we will see in §6.4 and §6.5, these quantities, if given, suffice to determine the radiation field via the equation of transfer.

72. *Absorption, Emission, and Scattering*

THE EXTINCTION COEFFICIENT

When radiation passes through material, energy is generally removed from the beam. We describe this loss in terms of an *opacity* or *extinction coefficient* (sometimes loosely called the *total absorption coefficient*) $\chi(\mathbf{x}, t; \mathbf{n}, \nu)$, defined such that an element of material of length dl and cross section dS, oriented normal to a beam of radiation having specific intensity $I(\mathbf{x}, t; \mathbf{n}, \nu)$ propagating along $\mathbf{n}$ into solid angle $d\omega$ in frequency band $d\nu$, removes an amount of energy

$$\delta\mathscr{E} = \chi(\mathbf{x}, t; \mathbf{n}, \nu) I(\mathbf{x}, t; \mathbf{n}, \nu)\, dl\, dS\, d\omega\, d\nu\, dt \tag{72.1}$$

from the beam in a time interval dt. Opacity is the sum, over all states that can absorb at frequency ν, of the product of the occupation numbers of those states (cm^{-3}) times their atomic cross sections (cm^2) at that frequency. The dimensions of χ_ν are cm^{-1}; the quantity $\lambda_\nu \equiv (1/\chi_\nu)$ cm is the *mean free path* of photons of frequency ν in the material.

In the fluid rest frame, the opacity is isotropic, but its frequency spectrum can be complicated, consisting of many overlapping continuum absorption edges, overlaid by thousands to millions of lines, each with a characteristic profile. In the laboratory frame, where the fluid is generally moving, the situation is much more complex. As a result of Doppler shift, a photon moving in direction $\mathbf{n}$ with frequency ν in the lab frame has a frequency

$$\nu_0 = \nu(1 - \mathbf{n} \cdot \mathbf{v}/c) \tag{72.2}$$

in the fluid frame of material moving with velocity $\mathbf{v}$. Hence radiation moving in, say, the direction of the fluid flow interacts with the material at a different fluid-frame frequency than does radiation of the same lab-frame frequency moving in, say, the opposite direction. It is thus absorbed at a different rate because atomic cross sections vary with frequency; the lab-frame opacity therefore becomes anisotropic. (Strictly speaking we should also allow for the effects of *aberration* between the two frames; these can be ignored for our present purposes, but will be accounted for in Chapter 7.)

THE EMISSION COEFFICIENT

The *emission coefficient* (or *emissivity*) $\eta(\mathbf{x}, t; \mathbf{n}, \nu)$ of the material is defined such that the amount of radiant energy released by a material element of length dl and cross section dS, into a solid angle $d\omega$ around a direction $\mathbf{n}$, in frequency interval $d\nu$ in a time dt is

$$\delta\mathscr{E} = \eta(\mathbf{x}, t; \mathbf{n}, \nu)\, dl\, dS\, d\omega\, d\nu\, dt. \tag{72.3}$$

The dimensions of η are ergs cm^{-3} s^{-1} Hz^{-1} sr^{-1}. The emissivity may be isotropic in the rest frame of the material, but is anisotropic in the lab frame when the material moves, for the same reasons χ is.

We will sometimes add a subscript c for "continuum" and l for "line" to both χ and η.

SCATTERING

It is important to distinguish between "true" or "thermal" absorption-emission processes, and the process of *scattering*. In the former case, energy removed from the beam is converted into material thermal energy, and energy is emitted into the beam at the expense of material energy. Examples of "true absorption" processes are these: (1) A photon ionizes an atom; its energy goes into the ionization energy of the atom plus the kinetic energy of the free electron. (2) A photon excites an atom, which is subsequently de-excited by a collision with another particle; the photon's energy goes into the kinetic energy of the collision partners. The inverses of these processes produce "thermal emission" in which energy is extracted from the thermal energy of hot material and converted into radiation. Other examples are given in (**M2**, §2.1).

In contrast, in a scattering process a photon interacts with a scattering center and emerges from the event moving in a different direction, generally with a slightly different frequency. Little or none of the photon's energy goes into (or comes from) the thermal energy of the gas. Examples are as follows: (1) A photon excites an atom from state a to state b; the atom decays radiatively back to state a. (2) A photon collides with a free electron (Thomson or Compton scattering) or with an atom or molecule in which it excites a resonance (Rayleigh or Raman scattering).

It is thus convenient to define a *true absorption coefficient* $\kappa(\mathbf{x}, t; \mathbf{n}, \nu)$ and a *scattering coefficient* $\sigma(\mathbf{x}, t; \mathbf{n}, \nu)$. The extinction coefficient is then

$$\chi(\mathbf{x}, t; \mathbf{n}, \nu) = \kappa(\mathbf{x}, t; \mathbf{n}, \nu) + \sigma(\mathbf{x}, t; \mathbf{n}, \nu). \tag{72.4}$$

Similarly we break the total emissivity into a thermal part η^t and a scattering part η^s:

$$\eta(\mathbf{x}, t; \mathbf{n}, \nu) = \eta^t(\mathbf{x}, t; \mathbf{n}, \nu) + \eta^s(\mathbf{x}, t; \mathbf{n}, \nu). \tag{72.5}$$

In certain simple situations we can write explicit expressions for η^s, which provide useful archetypes for later discussion. We will assume that the scattering is *conservative* so that all of the energy removed from the beam by the process is immediately re-emitted. For example, consider a spectrum line with total scattering cross section σ_l, and profile $\phi(\nu)$ normalized such that in the fluid frame

$$\int_0^\infty \phi(\nu_0)\, d\nu_0 = 1. \tag{72.6}$$

The suffix "0" on any quantity implies that it is measured in the comoving frame. If σ_l is isotropic, the total energy removed from the beam is

$$\sigma_l(\mathbf{x}, t) \int_0^\infty d\nu_0' \phi(\mathbf{x}, t; \nu_0') \oint d\omega_0' I_0(\mathbf{x}, t; \mathbf{n}_0', \nu_0')$$
$$= 4\pi\sigma_l(\mathbf{x}, t) \int_0^\infty \phi(\mathbf{x}, t; \nu_0') J_0(\mathbf{x}, t; \nu_0')\, d\nu_0'. \tag{72.7}$$

In general the re-emission of this energy is described by a *redistribution function* $R(\mathbf{n}', \nu'; \mathbf{n}, \nu)$ giving the joint probability that a photon $(\mathbf{n}', \nu')$ is absorbed and a photon $(\mathbf{n}, \nu)$ is emitted. We will not discuss the complication of *partial redistribution* [cf. (**M2**, Chaps. 2 and 13)], but will assume for simplicity that the photons are emitted *isotropically* in angle and are *randomly redistributed* (also called *complete redistribution*) over the line profile, in which case the fluid-frame emission by scattering is

$$\eta_0^s(\mathbf{x}, t; \nu_0) = \sigma_l(\mathbf{x}, t)\phi(\mathbf{x}, t; \nu_0) \int_0^\infty \phi(\mathbf{x}, t; \nu_0') J_0(\mathbf{x}, t; \nu_0')\, d\nu_0'. \tag{72.8}$$

In view of (72.2), the lab-frame emissivity is then

$$\eta^s(\mathbf{x}, t; \mathbf{n}, \nu) = \sigma_l(\mathbf{x}, t)\phi(\mathbf{x}, t; \mathbf{n}, \nu_0) \int_0^\infty d\nu' \oint d\omega' \phi(\mathbf{x}, t; \mathbf{n}', \nu_0') I(\mathbf{x}, t; \mathbf{n}', \nu'), \tag{72.9}$$

where (again ignoring aberration)

$$\phi(\mathbf{x}, t; \mathbf{n}_0, \nu_0) \equiv \phi[\mathbf{x}, t; \mathbf{n}, \nu(1 - \mathbf{n} \cdot \mathbf{v}/c)]. \tag{72.10}$$

The assumption of complete redistribution is a good approximation in many cases of interest, for example, within the Doppler core of a line (where Doppler shifts efficiently scramble the frequencies of absorbed and emitted photons), or when excited atoms suffer many elastic collisions before a photon is re-emitted (the excited electrons are randomly redistributed over the substates of the upper level, destroying any correlation between absorption and emission frequencies in the line profile). The extreme opposite case occurs when the scattering is isotropic and *coherent*; then the emissivity is

$$\eta_0^s(\mathbf{x}, t; \nu_0) = \sigma_0(\mathbf{x}, t) J_0(\mathbf{x}, t; \nu_0). \tag{72.11}$$

This expression is often used to describe Thomson scattering of continuum photons by free electrons. One can assume coherence because the Thomson cross section σ_e is frequency independent, and the frequency variation of continuum radiation is slow enough that Doppler shifts produced by typical fluid velocities can be ignored. Similarly, isotropy is a good approximation because the angular variation of the (dipole) phase function is weak. With these approximations (72.11) can be used in either the fluid or the lab frame.

We emphasize that the essential characteristic of scattering is that the rate of emission depends mainly on the radiation intensity at $(\mathbf{x}, t)$, and but little (if at all) on the amount of thermal energy there. Because the radiation field may originate mainly from *other* points in the medium, scattering processes are fundamentally *nonlocal*, and decouple the local emission rate from the local thermal pool. We also emphasize that (72.8), (72.9), and (72.11) are meant only to provide archetypes. In general, it is difficult to decide to what extent any particular process (e.g., absorption and emission of photons in a line) is a "true absorption-thermal emission" process or a "scattering" process. Most are a mixture because radiative and collisional processes operate simultaneously; this is true in both lines and continua. The true physics of the situation emerges only when the transfer equation is coupled directly to the equations of statistical equilibrium, which describe explicitly how atomic levels are populated and depopulated (see §6.6).

THE KIRCHHOFF–PLANCK RELATION

An important relation between thermal emission and absorption coefficients exists in strict thermodynamic equilibrium (TE). In an adiabatic enclosure, material (at rest) and radiation equilibrate to a uniform temperature and an isotropic radiation field [cf. (**C5,** 199–206) and (**M5,** 93–96)]. Moreover, in order to achieve a steady state, the amount of energy absorbed by the material in each range $(d\nu, d\omega)$ must exactly equal the amount it emits in that range. Therefore in TE

$$(\eta_\nu^t)^* \equiv (\kappa_\nu I_\nu)^*, \qquad (72.12)$$

where asterisks denote equilibrium values. But $I_\nu^* \equiv B_\nu(T)$, hence we obtain the *Kirchhoff-Planck relation*

$$(\eta_\nu^t)^* \equiv \kappa_\nu^* B_\nu(T). \qquad (72.13)$$

Strictly speaking, (72.13) applies only in TE. But when gradients of physical properties over a photon destruction length are very small, (72.13) is valid to a high degree of approximation at *local* values of the thermodynamic state variables. Hence we often invoke the hypothesis of local thermodynamic equilibrium (LTE) to write (in the comoving frame)

$$\eta_0^t(\mathbf{x}, t; \nu_0) = \kappa_0^*(\mathbf{x}, t; \nu_0) B[\nu_0, T(\mathbf{x}, t)]. \qquad (72.14)$$

Although (72.14) is certainly satisfactory in the diffusion limit (see §80) where the assumptions stated above hold, it cannot be guaranteed true, and may lead to significant errors, when free transport of radiation occurs, because the radiation field then acquires a nonlocal and/or nonequilibrium character that tends to drive the state of the material away from LTE. We shall analyze the meaning of LTE further in §84; in the meantime we regard it as a computational expedient that sometimes must be used, even when of doubtful validity, to render a problem tractable (e.g., in most radiation hydrodynamics applications).

73. *The Einstein Relations*

Consider now radiative transitions between two bound atomic states: a lower level i with statistical weight g_i, and an upper level j with statistical weight g_j, which are separated by an energy $h\nu_{ij} = \varepsilon_j - \varepsilon_i$, where ε_i and ε_j are measured relative to the atom's ground state. Throughout this section all quantities are evaluated in the fluid frame.

The radiative processes that connect i and j are described by three probability coefficients B_{ij}, B_{ji}, and A_{ji} introduced by Einstein. The *absorption probability* B_{ij} is defined such that the number of photons absorbed in the line per unit volume per unit time is

$$r_{ij} = n_i \phi_\nu B_{ij} I_\nu (d\omega/4\pi)\, d\nu, \tag{73.1}$$

where ϕ_ν is the line profile. The rate of energy absorption per unit volume is then

$$a_\nu I_\nu = (B_{ij} h\nu_{ij}/4\pi) n_i \phi_\nu I_\nu. \tag{73.2}$$

Here a_ν is the macroscopic *absorption coefficient*, uncorrected for stimulated emission (see below).

An atom in the upper state can either decay *spontaneously* to the lower state, or be *stimulated* to decay by radiation in the line. The *spontaneous emission probability* A_{ji} is defined such that the rate of energy emission per unit volume is

$$\dot{\mathscr{E}}_\nu \text{ (spontaneous)} = (A_{ji} h\nu_{ij}/4\pi) n_j \phi_\nu. \tag{73.3}$$

Here we have tacitly assumed that the line emission profile is identical to the absorption profile (complete redistribution). The *stimulated* (or *induced*) *emission probability* B_{ji} is defined such that the rate of stimulated energy emission per unit volume is

$$\dot{\mathscr{E}}_\nu \text{ (stimulated)} = (B_{ji} h\nu_{ij}/4\pi) n_j \phi_\nu I_\nu. \tag{73.4}$$

Notice that spontaneous emission is isotropic, whereas stimulated emission has the same angular distribution as I_ν. In an induced emission, the incident photon leads to the emission of an *identical* photon (i.e., two photons emerge from the event). In this sense, induced emission can be viewed as *negative absorption*, and we can subtract (73.4) from (73.2) to obtain a *net absorption coefficient*, corrected for stimulated emission. This procedure is not quite correct because in general the absorption and emission profiles differ; however for complete redistribution they are identical.

The coefficients B_{ij}, B_{ji}, and A_{ji} are related, as can be seen by demanding detailed balancing in thermodynamic equilibrium, which, from (73.2) to (73.4) implies

$$(n_i/n_j)^* B_{ij} I_\nu^* = A_{ji} + B_{ji} I_\nu^*, \tag{73.5}$$

where asterisks denote TE values. But $I_\nu^* \equiv B_\nu$, and by Boltzmann's formula (12.38)

$$(n_j/n_i)^* = (g_j/g_i)\exp(-h\nu_{ij}/kT), \tag{73.6}$$

hence

$$B_\nu = A_{ji}/[(n_i/n_j)^* B_{ij} - B_{ji}] = (A_{ji}/B_{ji})/[(g_i B_{ij}/g_j B_{ji})e^{h\nu_{ij}/kT} - 1]. \tag{73.7}$$

Comparing (73.7) with (67.11) we see that

$$g_i B_{ij} = g_j B_{ji} \tag{73.8}$$

and

$$A_{ji} = (2h\nu_{ij}^3/c^2)B_{ji}. \tag{73.9}$$

Although our argument, for simplicity, invokes thermodynamic equilibrium, both (73.8) and (73.9) hold in general because the Einstein coefficients depend on atomic properties only.

From (73.2), (73.4), and (73.8) we can write the *line absorption coefficient* (in the comoving frame), corrected for stimulated emission as

$$\chi_l(\nu) = n_i(B_{ij}h\nu_{ij}/4\pi)[1-(g_i n_j/g_j n_i)]\phi_\nu; \tag{73.10}$$

in the lab frame we must account for Doppler shifts in ϕ, as in (72.10). In LTE we can use (73.6) in (73.10) to obtain

$$\chi_l^*(\nu) = n_i^*(B_{ij}h\nu_{ij}/4\pi)[1-\exp(h\nu_{ij}/kT)]\phi_\nu, \tag{73.11}$$

where n_i^* is computed from (13.6) using actual values of n_e and n_{ion}. The factor in square brackets in (73.11) is often called "the" correction for stimulated emission; however, this identification is correct only in LTE.

The *line emission coefficient* in the comoving frame is

$$\eta_l(\nu) = n_j(A_{ji}h\nu_{ij}/4\pi)\phi_\nu, \tag{73.12}$$

and the LTE emissivity is obtained by replacing n_j with n_j^*. In writing transfer equations (cf. §77) it is often convenient to use the ratio of emissivity to opacity, which is called the *source function* S_ν. For a line, the source function is

$$S_l = n_j A_{ji}/(n_i B_{ij} - n_j B_{ji}) = (2h\nu_{ij}^3/c^2)/[(g_j n_i/g_i n_j) - 1]. \tag{73.13}$$

Because the frequency variation of the factor ν^3 is weak compared to the variation of ϕ_ν, (73.13) is often called the *frequency-independent* line source function; in contrast, the line source function can have a very strong frequency dependence if we account for the difference between the emission and absorption profiles (partial redistribution). In LTE, S_l reduces to B_ν, as expected from the Kirchhoff–Planck relation.

74. *The Einstein–Milne Relations*

The Einstein relations were generalized to continua by Milne (**M4**), whose treatment we sketch; as in §73 we work in the fluid frame. Suppose an atom is photoionized to produce an ion plus a free electron moving with speed v. Let n_0 be the number density of atoms, n_1 the density of ions, and $n_e(v)\,dv$ the density of electrons with speeds on the range $(v, v+dv)$, assumed Maxwellian. If p_ν is the *photoionization probability* of the atom by radiation in the frequency range $(\nu, \nu+d\nu)$, the *photoionization rate* is $n_0 p_\nu I_\nu\,d\nu$; the energy absorption coefficient is $\alpha_\nu = h\nu p_\nu$. Let $F(v)$ be the *spontaneous recombination probability* and $G(v)$ the *induced recombination probability* for electrons with speeds $(v, v+dv)$ to recombine with the ions. Then the *recombination rate* for electrons with speed v is $n_1 n_e(v)[F(v)+G(v)I_\nu]v\,dv$. The photon energy required to ionize the atom and produce an electron with speed v is

$$h\nu = \varepsilon_{\text{ion}} + \tfrac{1}{2}mv^2, \tag{74.1}$$

whence we have $h\,d\nu = mv\,dv$.

In thermodynamic equilibrium, the number of photoionizations equals the number of recombinations. Therefore

$$n_0^* p_\nu B_\nu = n_1^* n_e(v)[F(v)+G(v)B_\nu](h/m), \tag{74.2}$$

which implies that

$$B_\nu = [F(v)/G(v)]/\{[n_0^* p_\nu m/n_1^* n_e(v)hG(v)]-1\}. \tag{74.3}$$

Comparing (74.3) with (67.11) we see that

$$F(v) = (2h\nu^3/c^2)G(v) \tag{74.4}$$

and

$$p_\nu/G(v) = (h/m)[n_e(v)(n_1/n_0)^*]e^{h\nu/kT}. \tag{74.5}$$

But in TE, $n_e(v)\,dv$ is the Maxwellian distribution

$$n_e(v)\,dv = n_e(m/2\pi kT)^{3/2}\exp(-\tfrac{1}{2}mv^2/kT)4\pi v^2\,dv, \tag{74.6}$$

and the ratio $(n_1 n_e/n_0)^*$ is given by Saha's equation

$$(n_1/n_0)^* = n_e(g_0/2g_1)(h^2/2\pi mkT)^{3/2}\exp(\varepsilon_{\text{ion}}/kT) \equiv n_e\Phi_0(T). \tag{74.7}$$

Using (74.1), (74.6), and (74.7) in (74.5) we obtain

$$p_\nu = (8\pi m^2 v^2 g_1/h^2 g_0)G(v) = (4\pi c^2 m^2 v^2 g_1/h^3 g_0 \nu^3)F(v), \tag{74.8}$$

where the second equality follows from (74.4). Equations (74.4) and (74.8) are the continuum analogues of (73.8) and (73.9); they apply in general, not just in TE.

Using the above results we can write the continuum absorption coefficient, corrected for stimulated emission, as

$$\kappa_\nu = h\nu[n_0 p_\nu - (h/m)n_1 n_e(v)G(v)]. \tag{74.9}$$

Recalling that $\alpha_\nu = h\nu p_\nu$, and using (74.1) and (74.6) to (74.8) we find

$$\kappa_\nu = (n_0 - n_0^* e^{-h\nu/kT})\alpha_\nu, \tag{74.10}$$

where n_0^* is the LTE value of n_0 computed from Saha's equation using actual values of n_e and n_1, that is, $n_0^* = n_1 n_e \Phi_0(T)$. In LTE,

$$\kappa_\nu^* = n_0^*(1 - e^{-h\nu/kT})\alpha_\nu. \tag{74.11}$$

As before the factor $(1 - e^{-h\nu/kT})$ is often called the correction factor for stimulated emission, but this is correct only in LTE. Notice that in the continuum the induced emission rate always has its LTE value, whereas for a spectral line this rate depends on the actual upper level population, and hence may depart from its LTE value. This is not surprising, because recombination, whether spontaneous or induced, results from collisions between ions and electrons; if these particles have an equilibrium (Maxwellian) velocity distribution, recombination must occur at the LTE rate.

The spontaneous continuum emission coefficient is

$$\eta_\nu^t = h\nu n_1 n_e(v) F(v)(h/m) = [h n_1 n_e(v) F(v)/m p_\nu]\alpha_\nu. \tag{74.12}$$

Using (74.6) to (74.8) we find that (74.12) reduces to

$$\eta_\nu^t = (2h\nu^3/c^2) n_0^* \alpha_\nu e^{-h\nu/kT} = n_0^*(1 - e^{-h\nu/kT})\alpha_\nu B_\nu(T) = \kappa_\nu^* B_\nu(T). \tag{74.13}$$

Thus, provided that we define n_0^* in terms of the actual density of electrons and ions, continuum emission occurs at its LTE rate (as predicted by the Kirchhoff-Planck relation) because it is a collisional process. Hence for continua, the general formula for the opacity differs from its LTE form, but that for the emissivity does not; for lines, the general opacity and emissivity both differ from their LTE forms.

75. *Opacity and Emission Coefficients*

In addition to the *bound-bound* (line) and *bound-free* (photoionization) processes described in §§73 and 74, radiation can be absorbed and emitted during collisions between two free particles in *free-free* transitions (e.g., bremsstrahlung). Because this process is collisional, it always occurs at the LTE rate (using actual electron and ion densities). The total opacity (emissivity) at any frequency ν is the sum of the opacities (emissivities) of all processes that occur at the frequency. If we write $\alpha_{ij}(\nu)$, $\alpha_{i\kappa}(\nu)$, and $\alpha_{\kappa\kappa}(\nu)$ for bound-bound, bound-free, and free-free cross sections respectively, then from (73.10) and (74.10) the *total opacity* is

$$\begin{aligned}\chi_\nu &= \sum_i \sum_{j>i} [n_i - (g_i/g_j)n_j]\alpha_{ij}(\nu) + \sum_i (n_i - n_i^* e^{-h\nu/kT})\alpha_{i\kappa}(\nu) \\ &\quad + \sum_\kappa n_e n_\kappa \alpha_{\kappa\kappa}(\nu, T)(1 - e^{-h\nu/kT}) + n_e\sigma_e \qquad (75.1) \\ &\equiv \kappa_\nu + \sigma_\nu,\end{aligned}$$

where the last term represents Thomson scattering by free electrons.

Similarly, from (73.12) and (74.13) the *total thermal emissivity* is

$$\eta_\nu^t = (2h\nu^3/c^2)\Big[\sum_i \sum_{j>i} n_j(g_i/g_j)\alpha_{ij}(\nu) + \sum_i n_i^*\alpha_{i\kappa}(\nu)e^{-h\nu/kT} + \sum_\kappa n_e n_\kappa \alpha_{\kappa\kappa}(\nu, T)e^{-h\nu/kT}\Big] \qquad (75.2)$$

Both (75.1) and (75.2) apply in the fluid frame. If the fluid moves, we must account for Doppler shifts as in (72.2) when calculating lab-frame opacities and emissivities, and in general we have $\chi = \chi(\mathbf{x}, t; \mathbf{n}, \nu)$ and $\eta = \eta(\mathbf{x}, t; \mathbf{n}, \nu)$ in the lab frame.

In the limit of LTE, (75.1) and (75.2) simplify to

$$\chi_\nu^* = \Big\{\sum_i n_i^*\Big[\alpha_{i\kappa}(\nu) + \sum_{j>i}\alpha_{ij}(\nu)\Big] + \sum_\kappa n_e n_\kappa \alpha_{\kappa\kappa}(\nu, T)\Big\}(1 - e^{-h\nu/kT}) + n_e\sigma_e \equiv \kappa_\nu^* + n_e\sigma_e, \qquad (75.3)$$

and

$$(\eta_\nu^t)^* = (2h\nu^3/c^2)e^{-h\nu/kT}\Big\{\sum_i n_i^*\Big[\alpha_{i\kappa}(\nu) + \sum_{j>i}\alpha_{ij}(\nu)\Big] + \sum_\kappa n_e n_\kappa \alpha_{\kappa\kappa}(\nu, T)\Big\}. \qquad (75.4)$$

Clearly $(\eta_\nu^t)^* = \kappa_\nu^* B_\nu$, as expected from the Kirchhoff–Planck relation (72.11). Again (75.3) and (75.4) apply in the comoving frame; in the lab frame both χ^* and η^* depend on $(\mathbf{x}, t; \mathbf{n}, \nu)$ when the fluid moves.

6.4 The Equation of Transfer

76. Derivation of the Transfer Equation

Consider an element of material of length ds and cross section dS, fixed in the laboratory frame. We calculate the change, in a time dt, in the energy of the radiation field contained in a frequency interval $d\nu$, traveling into solid angle $d\omega$ along a direction $\mathbf{n}$ normal to dS, as it passes through the material (see Figure 76.1). The difference between the amount of energy that emerges at position $\mathbf{x}+\Delta\mathbf{x}$ at time $t+\Delta t$ and the amount incident at position $\mathbf{x}$ at time t must equal the difference between the amount of energy created by emission from the material and the amount absorbed. Thus in a Cartesian coordinate system

$$[I(\mathbf{x}+\Delta\mathbf{x}, t+\Delta t; \mathbf{n}, \nu) - I(\mathbf{x}, t; \mathbf{n}, \nu)]\, dS\, d\omega\, d\nu\, dt = [\eta(\mathbf{x}, t; \mathbf{n}, \nu) - \chi(\mathbf{x}, t; \mathbf{n}, \nu)I(\mathbf{x}, t; \mathbf{n}, \nu)]\, ds\, dS\, d\omega\, d\nu\, dt. \qquad (76.1)$$

If we let s be the path length along the ray, $\Delta t = ds/c$, and

$$I(\mathbf{x}+\Delta\mathbf{x}, t+\Delta t; \mathbf{n}, \nu) = I(\mathbf{x}; t; \mathbf{n}, \nu) + [(1/c)(\partial I/\partial t) + (\partial I/\partial s)]\, ds. \qquad (76.2)$$

Substituting (76.2) into (76.1) we obtain the *transfer equation*

$$[(1/c)(\partial/\partial t) + (\partial/\partial s)]I(\mathbf{x}, t; \mathbf{n}, \nu) = \eta(\mathbf{x}, t; \mathbf{n}, \nu) - \chi(\mathbf{x}, t; \mathbf{n}, \nu)I(\mathbf{x}, t; \mathbf{n}, \nu). \qquad (76.3)$$

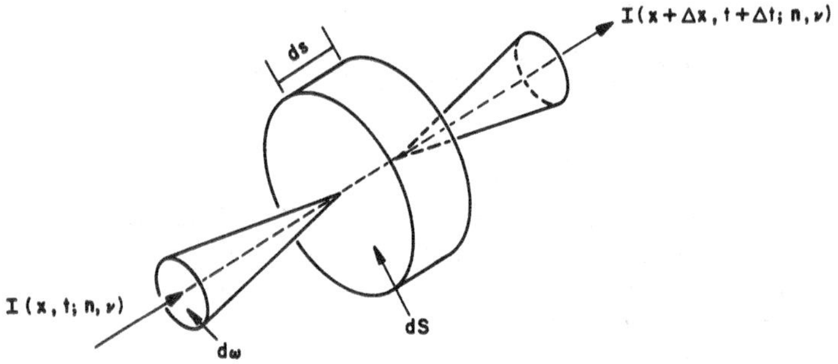

Fig. 76.1 Pencil of radiation passing through a material element.

Because s is a coordinate-independent pathlength, (76.3) applies in arbitrary coordinate systems, provided we use an appropriate expression to evaluate $(\partial/\partial s)$.

The derivation just given of the transfer equation is classical, macroscopic, and phenomenological in character. It omits reference to such important phenomena as polarization, dispersion, coherence, interference, and quantum effects, none of which are correctly described by (76.3). An excellent discussion of the approximations inherent in, and the validity of, the classical radiative transfer equation is given in (**P3,** 47–49). Good discussions of the transfer equation from the point of view of quantum field theory are given in (**H1**), (**L1**), (**L2**), (**L3**), (**O1**).

The mathematical expression for $(\partial/\partial s)$ depends on geometry. In Cartesian coordinates

$$\frac{\partial I}{\partial s}=\left(\frac{\partial x}{\partial s}\right)\frac{\partial I}{\partial x}+\left(\frac{\partial y}{\partial s}\right)\frac{\partial I}{\partial y}+\left(\frac{\partial z}{\partial s}\right)\frac{\partial I}{\partial z}=n_x\frac{\partial I}{\partial x}+n_y\frac{\partial I}{\partial y}+n_z\frac{\partial I}{\partial z}, \tag{76.4}$$

where (n_x, n_y, n_z) are components of the unit vector $\mathbf{n}$ along the direction of propagation. The transfer equation is then

$$[(1/c)(\partial/\partial t)+(\mathbf{n}\cdot\boldsymbol{\nabla})]I(\mathbf{x}, t; \mathbf{n}, \nu)=\eta(\mathbf{x}, t; \mathbf{n}, \nu)-\chi(\mathbf{x}, t; \mathbf{n}, \nu)I(\mathbf{x}, t; \mathbf{n}, \nu). \tag{76.5}$$

For a one-dimensional planar atmosphere, (76.5) reduces to

$$[(1/c)(\partial/\partial t)+\mu(\partial/\partial z)]I(z, t; \mu, \nu) = \eta(z, t; \mu, \nu)-\chi(z, t; \mu, \nu)I(z, t; \mu, \nu), \tag{76.6}$$

and for static media or steady flows the time derivative can be dropped, yielding

$$\mu[\partial I(z; \mu, \nu)/\partial z]=\eta(z; \mu, \nu)-\chi(z; \mu, \nu)I(z; \mu, \nu). \tag{76.7}$$

If the opacity and emissivity are given, (76.7) is an ordinary differential equation, while (76.6) is a partial differential equation. If scattering terms

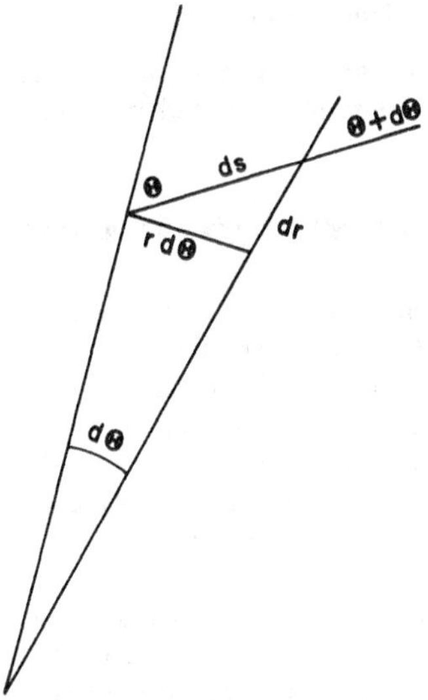

Fig. 76.2 Photon propagation angle in spherical symmetry.

are present, the mean intensity (an integral over angle) appears on the right-hand side, producing *integrodifferential equations.*

In curvilinear coordinates the coordinate basis vectors rotate with respect to the straight-line path determined by a fixed propagation vector $\mathbf{n}$. Therefore to allow for the changes in the components of $\mathbf{n}$ measured along these basis vectors we evaluate $(\partial/\partial s)$ as $(\partial/\partial s)=\mathbf{n}\cdot\nabla+(d\mathbf{n}/ds)\cdot\nabla_{\mathbf{n}}$ where $\nabla_{\mathbf{n}}$ denotes differentiation with respect to the direction cosines of $\mathbf{n}$. For example, in general spherical geometry, the derivative $(\partial/\partial s)$ introduces terms in $\partial/\partial r$, $\partial/\partial\theta$, $\partial/\partial\phi$, $\partial/\partial\Theta$, and $\partial/\partial\Phi$. But for spherical symmetry the terms $\partial/\partial\theta$, $\partial/\partial\phi$, and $\partial/\partial\Phi$ all vanish identically. From Figure 76.2 we see that $dr=\cos\Theta\, ds=\mu\, ds$ and $r\,d\Theta=-\sin\Theta\, ds=-(1-\mu^2)^{1/2}\, ds$, hence

$$\frac{\partial}{\partial s}=\left(\frac{\partial r}{\partial s}\right)\frac{\partial}{\partial r}+\left(\frac{\partial\Theta}{\partial s}\right)\frac{\partial}{\partial\Theta}=\cos\Theta\frac{\partial}{\partial r}-\frac{\sin\Theta}{r}\frac{\partial}{\partial\Theta}=\mu\frac{\partial}{\partial r}+\frac{(1-\mu^2)}{r}\frac{\partial}{\partial\mu}. \tag{76.8}$$

Therefore the transfer equation for a spherically symmetric medium is

$$\left[\frac{1}{c}\frac{\partial}{\partial t}+\mu\frac{\partial}{\partial r}+\frac{(1-\mu^2)}{r}\frac{\partial}{\partial\mu}\right]I(r,t;\mu,\nu)$$
$$=\eta(r,t;\mu,\nu)-\chi(r,t;\mu,\nu)I(r,t;\mu,\nu). \tag{76.9}$$

Notice that even when the $\partial/\partial t$ term is dropped, (76.9) is a partial differential equtaion, or a partial integrodifferential equation.

77. *Optical Depth and Source Function*

The concept of *optical depth* is central to discussions of transfer. If $\mathbf{x}$ and $\mathbf{x}'$ are two points in the medium separated by $l=|\mathbf{x}'-\mathbf{x}|$, the optical depth between them is

$$\tau_\nu(\mathbf{x},\mathbf{x}')=\int_0^l \chi(\mathbf{x}+\mathbf{n}s;\mathbf{n},\nu)\,ds \tag{77.1}$$

where ds is a path-length increment, and $\mathbf{n}$ is a unit vector along the straight line $(\mathbf{x},\mathbf{x}')$. Here we have allowed for the possibility that the material may be moving, in which case τ can depend both on the separation between the two points, and on the direction in which the integration is performed. Recalling that χ_ν^{-1} is the mean free path of a photon of frequency ν, we see that $\tau_\nu(\mathbf{x},\mathbf{x}')$ is equal to the number of photon mean free paths between $\mathbf{x}$ and $\mathbf{x}'$.

For a static planar medium, optical depth is customarily measured vertically downward from the upper boundary at $z=z_{\text{max}}$, and usually provides a more convenient depth variable for transfer calculations than does the geometrical depth z. Remembering that z increases upward we have

$$d\tau_\nu=-\chi_\nu\,dz \tag{77.2}$$

and

$$\tau_\nu(z)=\int_z^{z_{\text{max}}}\chi_\nu(z')\,dz'. \tag{77.3}$$

For a slant ray emerging from the medium with angle-cosine μ relative to the vertical, $ds=dz/\mu$, hence the slant optical depth along the ray down to geometric depth z is $\tau_\nu(z)/\mu$. For static spherical media a similar definition can be written for the *radial optical depth,* that is, the optical depth measured inward along a radius vector.

The transfer equation is often written in terms of the source function

$$S(\mathbf{x},t;\mathbf{n},\nu)\equiv\eta(\mathbf{x},t;\mathbf{n},\nu)/\chi(\mathbf{x},t;\mathbf{n},\nu). \tag{77.4}$$

For example, in a static planar medium the transfer equation assumes its "standard form"

$$\mu(\partial I_\nu/\partial\tau_\nu)=I_\nu-S_\nu; \tag{77.5}$$

we study this equation extensively in §6.5.

The discussion in §6.3 suggests several archetype expressions for S_ν; unless specified otherwise these apply only in the comoving frame of the fluid, or in static media. In LTE, (72.13) implies

$$S_\nu=B_\nu(T). \tag{77.6}$$

If we have a contribution from coherent isotropic scattering then

$$\chi_\nu=\kappa_\nu+\sigma_\nu \tag{77.7a}$$

and

$$\eta_\nu = \kappa_\nu B_\nu + \sigma_\nu J_\nu, \tag{77.7b}$$

hence

$$S_\nu = (\kappa_\nu B_\nu + \sigma_\nu J_\nu)/(\kappa_\nu + \sigma_\nu). \tag{77.8}$$

For a spectrum line with an overlapping LTE continuum

$$\chi_\nu = \kappa_c + \chi_l \phi_\nu. \tag{77.9}$$

If a fraction ε of the line emission is thermal and the remainder is isotropic scattering with complete redistribution, then

$$\eta_\nu = \kappa_c B_\nu + \chi_l \phi_\nu \left[(1-\varepsilon) \int \phi_\nu J_\nu \, d\nu + \varepsilon B_\nu\right], \tag{77.10}$$

hence

$$S_\nu = \left(\frac{r+\varepsilon\phi_\nu}{r+\phi_\nu}\right) B_\nu + \left[\frac{(1-\varepsilon)\phi_\nu}{r+\phi_\nu}\right] \int \phi_\nu J_\nu \, d\nu \equiv \xi_\nu B_\nu + (1-\xi_\nu)\bar{J}, \tag{77.11}$$

where $r \equiv \kappa_c/\chi_l$. We give a physical justification for (77.10) and (77.11) in §87. In the laboratory frame, ϕ_ν becomes $\phi(\mathbf{x}, t; \mathbf{n}, \nu)$ as in (72.10), and $\bar{J}$ becomes a double integral of the specific intensity over both angle and frequency, as in (72.9).

We emphasize that the source functions (77.6), (77.8), and (77.11) are meant only to be illustrative; a more complete discussion is given in §6.7.

78. Moments of the Transfer Equation

Angular moments of the transfer equation are both physically important and mathematically useful. To obtain the *zero-order moment equation* we multiply the time-dependent transfer equation in Cartesian coordinates,

$$[(1/c)(\partial/\partial t) + n^i(\partial/\partial x^i)]I(\mathbf{x}, t; \mathbf{n}, \nu) = \eta(\mathbf{x}, t; \mathbf{n}, \nu) - \chi(\mathbf{x}, t; \mathbf{n}, \nu)I(\mathbf{x}, t; \mathbf{n}, \nu), \tag{78.1}$$

by $(d\omega/4\pi)$ and integrate over all solid angles. Using (64.1), (65.2), and (65.6) we find

$$(1/c)(\partial J_\nu/\partial t) + \nabla \cdot \mathbf{H}_\nu = (1/4\pi) \oint [\eta(\mathbf{x}, t; \mathbf{n}, \nu) - \chi(\mathbf{x}, t; \mathbf{n}, \nu)I(\mathbf{x}, t; \mathbf{n}, \nu)] \, d\omega; \tag{78.2}$$

or, in view of (64.4),

$$(\partial E_\nu/\partial t) + \nabla \cdot \mathbf{F}_\nu = \oint [\eta(\mathbf{x}, t; \mathbf{n}, \nu) - \chi(\mathbf{x}, t; \mathbf{n}, \nu)I(\mathbf{x}, t; \mathbf{n}, \nu)] \, d\omega \tag{78.3}$$

Integrating over all frequencies we have

$$(\partial E/\partial t) + \nabla \cdot \mathbf{F} = \int_0^\infty d\nu \oint d\omega [\eta(\mathbf{x}, t; \mathbf{n}, \nu) - \chi(\mathbf{x}, t; \mathbf{n}, \nu)I(\mathbf{x}, t; \mathbf{n}, \nu)]. \tag{78.4}$$

The reduction of these equations to one-dimensional planar geometry is trivial.

Equations (78.3) and (78.4) are *energy equations* for the radiation field. Integrating them over a fixed volume element and applying the divergence theorem we see that the rate of change of the radiant energy in the volume equals (1) the total rate of energy emission from the material, minus (2) the total rate of energy absorption by the material, minus (3) the net flow of radiant energy through the volume element's boundary surface.

For a spherically symmetric medium, (78.3) and (78.4) become

$$(\partial E_\nu/\partial t)+r^{-2}[\partial(r^2F_\nu)/\partial r]$$
$$=2\pi\int_{-1}^{1}[\eta(r,t;\mu,\nu)-\chi(r,t;\mu,\nu)I(r,t;\mu,\nu)]\,d\mu, \tag{78.5}$$

and

$$(\partial E/\partial t)+r^{-2}[\partial(r^2F)/\partial r]$$
$$=2\pi\int_0^\infty d\nu\int_{-1}^{1}d\mu[\eta(r,t;\mu,\nu)-\chi(r,t;\mu,\nu)I(r,t;\mu,\nu)], \tag{78.6}$$

which can also be obtained by direct integration of (76.9) over $d\omega$ and $d\nu$. The total *luminosity* passing through a spherical shell of radius r is

$$L(r,t)=4\pi r^2F(r,t), \tag{78.7}$$

so (78.6) can be rewritten as

$$(\partial E/\partial t)+(4\pi r^2)^{-1}(\partial L/\partial r)$$
$$=2\pi\int_0^\infty d\nu\int_{-1}^{1}d\mu[\eta(r,t;\mu,\nu)-\chi(r,t;\mu,\nu)I(r,t;\mu,\nu)]. \tag{78.8}$$

In a *static* medium (i.e., no time-dependence or hydrodynamic motions) we must have $(\partial E/\partial t)\equiv 0$. Furthermore, for the material to be in a steady state it must be in *radiative equilibrium* (i.e., it must emit exactly as much energy as it absorbs). Under these conditions the right-hand sides of (78.4) and (78.8) vanish identically, hence $\nabla\cdot\mathbf{F}\equiv 0$. That is, in radiative equilibrium the flux is constant with depth in planar geometry, and the luminosity is constant with radius in spherical geometry. We emphasize that radiative equilibrium occurs only in an absolutely static medium, and represents a limiting form of the radiation energy equation. We will discuss the general radiation energy equation and its coupling to energy equations for radiating fluids in motion in greater detail in Chapter 7.

The *first-order moment equation* for the radiation field is obtained by multiplying (78.1) by $\mathbf{n}$, and integrating against $(d\omega/4\pi)$, which yields

$$c^{-1}(\partial H_\nu^i/\partial t)+(\partial K_\nu^{ij}/\partial x^j)=(1/4\pi)\oint[\eta(\mathbf{x},t;\mathbf{n},\nu)$$
$$-\chi(\mathbf{x},t;\mathbf{n},\nu)I(\mathbf{x},t;\mathbf{n},\nu)]n^i\,d\omega. \tag{78.9}$$

Here $K_\nu^{ij}\equiv(c/4\pi)P_\nu^{ij}$, as defined by (66.2). Multiplying (78.9) by $(4\pi/c)$ we obtain, in tensor notation

$$c^{-2}(\partial\mathbf{F}_\nu/\partial t)+\nabla\cdot\mathsf{P}_\nu=c^{-1}\oint[\eta(\mathbf{x},t;\mathbf{n},\nu)-\chi(\mathbf{x},t;\mathbf{n},\nu)I(\mathbf{x},t;\mathbf{n},\nu)]\mathbf{n}\,d\omega, \tag{78.10}$$

which, integrated over all frequencies, yields

$$c^{-2}(\partial \mathbf{F}/\partial t)+\nabla \cdot \mathsf{P}=c^{-1}\int_0^\infty d\nu \oint d\omega[\eta(\mathbf{x}, t; \mathbf{n}, \nu)-\chi(\mathbf{x}, t; \mathbf{n}, \nu)I(\mathbf{x}, t; \mathbf{n}, \nu)]\mathbf{n}. \tag{78.11}$$

Equations (78.10) and (78.11) are *momentum equations* for the radiation field. To verify this interpretation, recall from (65.7) and (65.8) that the radiative momentum density is c^{-2} times the flux, and from §66 that P is the radiation-momentum flux-density tensor. Furthermore, the momentum of a photon with energy $h\nu$ moving in direction $\mathbf{n}$ is $(h\nu/c)\mathbf{n}$. Thus integrating (78.11) over a fixed volume, and applying the divergence theorem, we find that the rate of change of the radiant momentum in the volume equals (1) the net rate of momentum input into the radiation field by emission from the material, minus (2) the net rate of absorption of radiation momentum by the material, minus (3) the rate of transport of radiative momentum across the boundary surface of the volume. As a by-product we see that the integral

$$\mathbf{f}_R=c^{-1}\int_0^\infty d\nu \oint d\omega \chi(\mathbf{x}, t; \mathbf{n}, \nu)I(\mathbf{x}, t; \mathbf{n}, \nu)\mathbf{n} \tag{78.12}$$

is the *radiation force*, per unit volume, on the material.

In a spherically symmetric medium, (78.10) and (78.11) become

$$c^{-2}(\partial F_\nu/\partial t)+(\partial P_\nu/\partial r)+(3P_\nu-E_\nu)/r$$
$$=(2\pi/c)\int_{-1}^{1} [\eta(r, t; \mu, \nu)-\chi(r, t; \mu, \nu)I(r, t; \mu, \nu)]\mu\, d\mu, \tag{78.13}$$

and

$$c^{-2}(\partial F/\partial t)+(\partial P/\partial r)+(3P-E)/r$$
$$=(2\pi/c)\int_0^\infty d\nu \int_{-1}^{1} d\mu[\eta(r, t; \mu, \nu)-\chi(r, t; \mu, \nu)I(r, t; \mu, \nu)]\mu. \tag{78.14}$$

These results can also be obtained by direct integration (76.9) over $\mu\, d\omega$ and $d\nu$.

Thus far we have allowed for material motions, assuming that χ and η depend on angle and frequency. Considerable simplification is obtained for a static medium ($\mathbf{v}\equiv 0$) when, in addition, the radiation field is time independent. These assumptions provide a good framework for the development of basic methods for solving transfer equations (cf. §6.5). Because both χ and η are isotropic in a static medium, (78.2) in planar geometry reduces to

$$(\partial H_\nu/\partial z)=\eta_\nu-\chi_\nu J_\nu, \tag{78.15a}$$

or

$$(\partial F_\nu/\partial z)=4\pi\eta_\nu-c\chi_\nu E_\nu. \tag{78.15b}$$

Similarly (78.10) reduces to

$$(\partial K_\nu/\partial z) = [\partial(f_\nu J_\nu)/\partial z] = -\chi_\nu H_\nu \tag{78.16a}$$

or

$$(\partial P_\nu/\partial z) = [\partial(f_\nu E_\nu)/\partial z] = -(\chi_\nu/c)F_\nu. \tag{78.16b}$$

The radiation momentum equation now reads

$$(\partial P/\partial z) = -c^{-1}\int_0^\infty \chi_\nu F_\nu\,d\nu; \tag{78.17}$$

the integral over η_ν vanishes because the net momentum loss by the material through isotropic emission is identically zero.

Similarly, in a static spherical medium, (78.5) reduces to

$$r^{-2}[\partial(r^2 F_\nu)/\partial r] = 4\pi\eta_\nu - c\chi_\nu E_\nu, \tag{78.18}$$

and (78.13) becomes

$$(\partial P_\nu/\partial r) + (3P_\nu - E_\nu)/r = -(\chi_\nu/c)F_\nu \tag{78.19a}$$

or

$$\partial(f_\nu E_\nu)/\partial r + (3f_\nu - 1)E_\nu/r = -(\chi_\nu/c)F_\nu. \tag{78.19b}$$

In addition to their physical significance, the moment equations provide powerful tools for solving transfer problems because they eliminate angle variables from the problem and thereby reduce its dimensionality. On the other hand, from (78.15) and (78.16), or (78.18) and (78.19), we see an essential difficulty: the first n moment equations always contain all moments through order $n+1$; thus we have one more unknown to determine than there are equations. This difficulty is known as the *closure problem.* It is very instructive to compare the closure problem for the radiation equations with the corresponding problem for the equations of gas dynamics. Inasmuch as the specific intensity characterizes fully both the angular and energy distribution of the radiation field, our derivation of the radiation energy and momentum equations in terms of moments of the intensity is conceptually identical to the derivation of the fluid equations from kinetic theory as discussed in §30 (and in §43 for relativistic fluids). We saw there that we can write the energy density, heat flux, and stress (including viscous effects) in the fluid in terms of suitable averages over the distribution function. If, as in §31, we assume that the distribution function is isotropic, the system of fluid equations closes exactly, and both the heat flux and the viscous stresses vanish identically. The same is true for radiation; if we assume I_ν is perfectly isotropic, we know that the radiation stress tensor becomes diagonal and isotropic with $P_\nu = \frac{1}{3}E_\nu$, and that $F_\nu \equiv 0$, so no further closure is necessary. On the other hand, if we assume that the distribution function is not isotropic, but that $\lambda/l \ll 1$, where λ is a particle mean free path, we are again able to achieve closure by deriving explicit expressions for the fluid heat flux $\mathbf{q}$ and the viscous stress tensor $\boldsymbol{\sigma}$. As is shown in §§80 and 97, entirely analogous results are possible for radiation in the limit that $\lambda_p/l \ll 1$, where λ_p is the photon mean free path.

The real problem arises near boundary surfaces where a mean free path (photon or particle) may exceed any characteristic structural length in the flow. We must then find other methods for evaluating the averages that appear as the energy flux or as nonisotropic (perhaps even off-diagonal) contributions to the stress tensor, in the fluid and/or radiation energy and momentum equations. We have ignored this problem for ordinary fluids because it becomes important only in extremely rarefied flows [e.g., the interplanetary medium (**H3**)]. But it cannot be ignored for radiation because we always must deal with regions in which $\lambda/l \ll 1$ while $\lambda_p/l \gg 1$; indeed these are the very layers of a radiating flow that we can observe. Here we must face the closure problem squarely.

In one-dimensional problems we have two equations containing the three scalars E_ν, F_ν, and P_ν, and one approach is to close the system with variable Eddington factors f_ν, as in (78.16b) and (78.19b). When solving the moment equations we assume that f_ν is known. We subsequently determine f_ν from a separate angle-by-angle formal solution of the full transfer equation assuming that the radiation energy density (which appears in the source function) is known; we then iterate the two steps to convergence. As the value of f_ν converges, the closure becomes essentially exact. In radiation-hydrodynamics calculations where computational speed is paramount, a yet-simpler procedure is sometimes adopted: one uses approximate analytical formulae to determine f from the geometry of the problem and from that ratio (F/E) [see, e.g., (**F2**), (**S4**)]. G. Minerbo (**M6**) developed an elegant formulation of this kind; Minerbo's formulation is useful also in multidimensional problems where the full Eddington tensor $\mathsf{f} \equiv \mathsf{P}/E$ must be specified.

Alternatively, we can rewrite the transfer equation in terms of angle-dependent mean-intensity-like and flux-like variables (see §83), and obtain exact closure of two coupled angle-dependent equations that strongly resemble the moment equations, and have many of their desirable properties. These equations can be discretized and solved directly by efficient numerical methods.

6.5 Solution of the Transfer Equation

We now address the problem of *solving* the transfer equation. To develop insight we first discuss the formal solution and special solutions for important special cases; we then discuss general numerical techniques. Inasmuch as we now focus mainly on mathematical rather than physical content of the equations, we will usually use the Eddington variables J_ν, H_ν, and K_ν in preference to the dynamical variables E_ν, F_ν, and P_ν.

We concentrate almost entirely on the solution of the *time-independent* transfer equation (the exception is an analytical expression for the time-dependent formal solution). The techniques developed here provide a foundation for later work. Equally important, it turns out that ignoring the time dependence of the transfer eqution itself is actually not a bad approximation for many astrophysical flow problems. To see why this is so, let us examine some characteristic time scales.

To deal with *radiation flow* (e.g., the propagation of a radiation front into material) we must consider very short time scales, $t_R \sim l/c$ or $t_\lambda \sim \lambda_p/c$, corresponding to a photon flight time over a characteristic structural length l, or over a photon mean free path λ_p. For such problems it is obviously necessary to solve the full time-dependent transfer equation in order to describe properly the dynamics of the radiating flow. In Chapter 7 we analyze the relative importance of terms in the transfer equation for moving media in the radiation flow limit, and some of the numerical methods discussed in §7.3 can be used in this regime. Nevertheless, we normally consider such problems to lie outside the scope of this book.

The primary concerns of this book is with problems of *fluid flow* on typical dynamical time scales $t_f \sim l/v$, where l is a characteristic length in the flow, and v is a typical flow velocity; for example, l might be of the order of a scale height H, and v of the order of the material sound speed a. In an optically thin region, the ratio t_R/t_f is $O(v/c)$, which is a small number in the astrophysical flows we will consider; in a stellar pulsation $(v/c) \sim 10^{-4}$, in a stellar wind $(v/c) \sim 10^{-2}$, and even in a supernova explosion $(v/c) \lesssim 3\times 10^{-2}$. (In some flows, e.g., relativistic collapse or laser-fusion experiments, t_f can be very much shorter, approaching t_R.) When $t_R \ll t_f$, the radiation field at any position adjusts essentially instantaneously to changes in physical conditions throughout the flow. This means that on time scales appropriate to a calculation of flow dynamics we can *ignore* the explicit time variation of the radiation field, and can consider it to be in a sequence of quasi-steady states, each of which is consistent with the instantaneous physical structure of the flow. We describe the radiation field in this regime as *quasi static* or *quasi stationary*. Of course we must still account explicitly for time derivatives of the radiation energy and momentum densities in the energy and momentum conservation equations for the radiating fluid, where these quantities appear on an equal footing with the corresponding material terms (see Chapters 7 and 8). The same remark also applies in the cases now to be discussed.

In optically thick regions (e.g., a stellar interior) the situation is different. Here photons *diffuse* by a random-walk process with a mean free path λ_p, and we can essentially discard the transfer equation, replacing it with a simple asymptotic solution (cf. §80). In the diffusion process a photon suffers of the order of $(l/\lambda_p)^2$ interactions with the material as it travels a distance l; hence the characteristic radiation *diffusion time* is $t_d \sim (l^2/c\lambda_p)$. The ratio (t_d/t_f) is thus $O(lv/\lambda_p c)$. In the *true* or (*static*) *diffusion limit*, the dynamical time scale of interest (e.g., a nuclear-evolution time) is so long that $t_f \gg t_d$ [which implies that $(v/c) \ll (\lambda_p/l)$], and random-walk diffusion sets the response time of the radiation field to changes in physical conditions;

because $(l/\lambda_p) \gg 1$, this response can be quite slow. At first sight it might seem that we should adopt timesteps of the order of t_d in our calculation, even though we are not interested in radiation diffusion phenomena per se, but wish only to follow events on the time scale t_f. However, in this regime the radiation field saturates to its thermal equilibrium value (see §80), with high collision rates assuring rapid photon creation and destruction in the dense material. Hence the radiation field is closely *frozen* to instantaneous local conditions as they vary. Therefore, as we computationally follow the time evolution of the fluid properties, we automatically track the time variation of the radiation field, which simply passes through a sequence of states in instantaneous equilibrium with local physical conditions as determined by the flow. In effect the radiation field is again quasi stationary.

If, on the other hand, the material is optically thick enough to trap photons, but now $(v/c) \gtrsim (\lambda_p/l)$ so that $t_f \lesssim t_d$, flow-induced changes in the physical properties of the medium can *drive* changes in the radiation field faster than they could have occurred if only diffusion were operative. We refer to this regime as the *dynamic diffusion* limit. Here we must account for the coupled time variation of the material properties and radiation field by performing a time-dependent solution of both the momentum and energy equations for the radiating fluid, and the radiation energy and momentum equations, simultaneously, on a fluid-flow time scale.

In general we must use radiation moment equations that bridge the optically thick and thin limits correctly, and we will see in Chapter 7 that it is then essential to include terms that are formally of $O(v/c)$. But, in any event, the only reason we need to solve the transfer equation itself is to obtain Eddington factors, required to close the moment equations. In most problems these ratios, which reflect almost purely *geometric* information about source distributions and boundaries, are given with good accuracy by a static *snapshot*, that is, a solution of the time-independent transfer equation using current values of the physical properties in the flow.

In sum, we need to solve the full time-dependent *transfer* equation (in distinction to *moment* equations) only when we wish to treat radiation flow; but generally not if we wish merely to treat fluid flow. We thus see that there is ample motivation for studying the time-independent transfer problem.

79. Formal Solution

BOUNDARY CONDITIONS

The radiation field in any volume consists both of photons emitted by the material within the volume, and of radiation that penetrates the bounding surface of the volume from imposed external sources. The latter are fixed by appropriate *boundary conditions*, which are needed to specify a unique solution. For an arbitrary convex volume V bounded by surface S we must therefore be given

$$I(\mathbf{x}_S, t; \mathbf{n}, \nu) = f(\mathbf{x}_S, t; \mathbf{n}, \nu) \tag{79.1}$$

for all $\mathbf{x}_S$ on S, along all rays $\mathbf{n}$ that penetrate into V (i.e., for $\mathbf{n} \cdot \mathbf{N} \leq 0$, where $\mathbf{N}$ is the outward normal of S), at all frequencies.

In astrophysical problems the generality of (79.1) is usually unnecessary because we consider planar or spherically symmetric geometries. In planar geometry we encounter two classes of problems: (1) a *finite slab* of thickness Z, and (2) a *semi-infinite atmosphere*, which is a medium (such as a stellar atmosphere) that has an open boundary surface separating vacuum from material that is so optically thick that it can be imagined to extend to infinity.

For the finite slab we must specify functions f^+ and f^- describing the incoming radiation on both faces, that is,

$$I(Z, t; \mu, \nu) = f^-(t; \mu, \nu), \qquad \mu \leq 0, \tag{79.2}$$

at the boundary nearest the observer (the *top* of the slab), and

$$I(0, t; \mu, \nu) = f^+(t; \mu, \nu), \qquad \mu \geq 0, \tag{79.3}$$

at the farther boundary (the *bottom* of the slab).

For a semi-infinite atmosphere we have an upper boundary condition of the form (79.2). In posing a lower boundary condition, we note that at very great depth the radiation field must satisfy a *boundedness* condition

$$\lim_{\tau_\nu \to \infty} [e^{-\tau_\nu/\mu} I(\tau_\nu, t; \mu, \nu)] = 0 \tag{79.4}$$

for the solution to be well behaved mathematically. In practice the solution must always be truncated at some large depth, where we impose a boundary condition that expresses f^+ in terms of the local source function and its gradient, or fixes the flux transported across the boundary. These conditions follow naturally from physical considerations in the diffusion limit (cf. §80).

In spherical geometry we deal with *spherical shells* or semi-infinite *spherical envelopes*. For a spherical shell we have equations analogous to (79.2) and (79.3) at $r = R$ (outer radius) and at $r = r_c$ (inner or *core* radius) respectively. In the semi-infinite case, we use (79.2) at $r = R$, and apply the diffusion approximation at $r = r_c$.

For time-dependent problems, the spatial boundary conditions must be augmented by an *initial condition* that gives the radiation field within V at $t = 0$, that is,

$$I(\mathbf{x}, 0; \mathbf{n}, \nu) = g(\mathbf{x}; \mathbf{n}, \nu) \tag{79.5}$$

for all $\mathbf{x}$ within V.

THE TIME-INDEPENDENT FORMAL SOLUTION

Consider now the transfer equation for a static medium (or steady flow), so that $(\partial/\partial t)\equiv 0$. Then in Cartesian coordinates

$$\mathbf{n}\cdot\nabla I=\eta-\chi I \tag{79.6}$$

where all variables are functions of $(\mathbf{x};\mathbf{n},\nu)$. The operator $\mathbf{n}\cdot\nabla=n^i(\partial/\partial x^i)$ is the derivative with respect to path length along the ray whose direction is $\mathbf{n}$. Radiation at $\mathbf{x}$ moving in direction $\mathbf{n}$ consists of photons that were emitted along $\mathbf{n}$ from points along the line $\mathbf{x}-|\mathbf{x}'-\mathbf{x}|\,\mathbf{n}$. It is therefore convenient to let s be the path length *backward* along $\mathbf{n}$, so that (79.6) becomes

$$-\frac{d}{ds}[I(\mathbf{x}-\mathbf{n}s;\mathbf{n},\nu)]+\chi(\mathbf{x}-\mathbf{n}s;\mathbf{n},\nu)I(\mathbf{x}-\mathbf{n}s;\mathbf{n},\nu)=\eta(\mathbf{x}-\mathbf{n}s;\mathbf{n},\nu). \tag{79.7}$$

In what follows we suppress mention of $(\mathbf{n},\nu)$ for brevity.

Equation (79.7) is a linear, first-order differential equation, which has an integrating factor $\exp[-\tau(s_0,s)]$, where

$$\tau(s_0,s)\equiv\int_{s_0}^{s}\chi(\mathbf{x}-\mathbf{n}s'')\,ds'', \tag{79.8}$$

and s_0 is an arbitrary point along $\mathbf{n}$. Notice that τ depends on both $\mathbf{n}$ and ν. Using (79.8) to rewrite (79.7) we have

$$-\frac{d}{ds}[I(\mathbf{x}-\mathbf{n}s)e^{-\tau(s_0,s)}]=e^{-\tau(s_0,s)}\eta(\mathbf{x}-\mathbf{n}s), \tag{79.9}$$

whence

$$I(\mathbf{x}-\mathbf{n}s)e^{-\tau(s_0,s)}-I(\mathbf{x}-\mathbf{n}s_0)=\int_{s}^{s_0}\eta(\mathbf{x}-\mathbf{n}s')e^{-\tau(s_0,s')}\,ds', \tag{79.10}$$

or, finally,

$$I(\mathbf{x}-\mathbf{n}s)=e^{-\tau(s,s_0)}I(\mathbf{x}-\mathbf{n}s_0)+\int_{s}^{s_0}\eta(\mathbf{x}-\mathbf{n}s')e^{-\tau(s,s')}\,ds'. \tag{79.11}$$

To obtain I at a specific $\mathbf{x}$ we set $s=0$, and choose s_0 such that $\mathbf{x}_S=\mathbf{x}-s_0\mathbf{n}$ lies on the boundary surface S. Then

$$\begin{aligned}I(\mathbf{x};\mathbf{n},\nu)=&\int_0^{s_0}\eta(\mathbf{x}-\mathbf{n}s';\mathbf{n},\nu)\exp\left[-\int_0^{s'}\chi(\mathbf{x}-\mathbf{n}s'';\mathbf{n},\nu)\,ds''\right]ds'\\&+f(\mathbf{x}_S;\mathbf{n},\nu)\exp\left[-\int_0^{s_0}\chi(\mathbf{x}-\mathbf{n}s'';\mathbf{n},\nu)\,ds''\right].\end{aligned} \tag{79.12}$$

Equation (79.12) is the general formal solution of the time-independent transfer problem. It states that the intensity of radiation traveling along $\mathbf{n}$ at point $\mathbf{x}$ is the sum of photons emitted from all points along the line segment $\mathbf{x}-\mathbf{n}s$, attenuated by the integrated absorptivity of the intervening material, plus an attenuated contribution from photons entering the boundary surface where it is pierced by that line segment. Note, however, that the apparent simplicity of (79.12) is illusory, for if η contains scattering terms (which depend on I), then (79.12) is an *integral equation* that must be *solved* for I, and therefore brings us no closer to the answer than the original differential equation. But if η is purely thermal, or is given, I can be computed from (79.12) by quadrature; this computation can be done either by direct evaluation of (79.12), or by equivalent, efficient, differential-equation techniques described in §83.

PLANAR MEDIA

Consider (79.12) for a static, planar, semi-infinite medium with no radiation incident at the upper boundary (e.g., a stellar atmosphere). For outgoing radiation at height z in the medium

$$I(z;\mu,\nu)=\int_{-\infty}^{z}\eta(z';\nu)e^{-(\tau_\nu'-\tau_\nu)/\mu}\,dz'/\mu,\qquad(0\le\mu\le 1),\qquad(79.13)$$

or

$$I(\tau_\nu;\mu,\nu)=\int_{\tau_\nu}^{\infty}S_\nu(\tau_\nu')e^{-(\tau_\nu'-\tau_\nu)/\mu}\,d\tau_\nu'/\mu,\qquad(0\le\mu\le 1),\qquad(79.14)$$

where $S_\nu=\eta_\nu/\chi_\nu$ is the source function, and we have invoked (79.4). Similarly, for incoming radiation

$$I(\tau_\nu;\mu,\nu)=\int_{0}^{\tau_\nu}S_\nu(\tau_\nu')e^{(\tau_\nu-\tau_\nu')/\mu}\,d\tau_\nu'/(-\mu),\qquad(-1\le\mu\le 0).\qquad(79.15)$$

Equation (79.14) yields directly the *emergent intensity* seen by an observer outside the atmosphere ($\tau_\nu=0$):

$$I(0;\mu,\nu)=\int_{0}^{\infty}S_\nu(\tau_\nu)e^{-\tau_\nu/\mu}\,d\tau_\nu/\mu.\qquad(79.16)$$

If we assume that near the surface S_ν is given by the linear expansion $S_\nu=a_\nu+b_\nu\tau_\nu$ we find

$$I(0;\mu,\nu)=a_\nu+b_\nu\mu=S(\tau_\nu/\mu=1),\qquad(79.17)$$

which is known as the *Eddington-Barbier relation*. This important result shows that the emergent intensity along a ray is approximately equal to the source function at slant optical depth unity along that ray, that is, at about one photon mean free path from the surface.

THE SCHWARZSCHILD-MILNE RELATIONS

We can also use (79.14) and (79.15) to obtain concise expressions for the mean intensity and flux within the medium. Thus the mean intensity is

$$\begin{aligned}J_\nu(\tau_\nu)&=\tfrac{1}{2}\int_{-1}^{1}I_\nu(\tau_\nu,\mu)\,d\mu\\&=\tfrac{1}{2}\int_{0}^{1}d\mu/\mu\int_{\tau_\nu}^{\infty}d\tau_\nu'S_\nu(\tau_\nu')e^{-(\tau_\nu'-\tau_\nu)/\mu}-\tfrac{1}{2}\int_{-1}^{0}d\mu/\mu\int_{0}^{\tau_\nu}d\tau_\nu'S_\nu(\tau_\nu')e^{(\tau_\nu-\tau_\nu')/\mu}.\end{aligned}\qquad(79.18)$$

Substituting $w \equiv \pm\mu^{-1}$ in the first and second integrals respectively, and interchanging the order of integration, we find

$$J_\nu(\tau_\nu) = \tfrac{1}{2}\left[\int_{\tau_\nu}^{\infty} S_\nu(\tau'_\nu)E_1(\tau'_\nu - \tau_\nu)\, d\tau'_\nu + \int_0^{\tau_\nu} S_\nu(\tau'_\nu)E_1(\tau_\nu - \tau'_\nu)\, d\tau'_\nu\right]$$
$$= \tfrac{1}{2}\int_0^{\infty} S(\tau'_\nu)E_1\,|\tau'_\nu - \tau_\nu|\, d\tau'_\nu, \tag{79.19}$$

which was first derived by K. Schwarzschild, and bears his name. Here $E_1(x)$ is the *first exponential integral* from the family

$$E_n(x) \equiv \int_1^{\infty} y^{-n} e^{-xy}\, dy = x^{n-1}\int_x^{\infty} y^{-n} e^{-y}\, dy, \tag{79.20}$$

whose mathematical properties are discussed in (**A1**, Chap. 5).

By a similar analysis we derive

$$H_\nu(\tau_\nu) = \tfrac{1}{2}\left[\int_{\tau_\nu}^{\infty} S_\nu(\tau'_\nu)E_2(\tau'_\nu - \tau_\nu)\, d\tau'_\nu - \int_0^{\tau_\nu} S_\nu(\tau'_\nu)E_2(\tau_\nu - \tau'_\nu)\, d\tau'_\nu\right] \tag{79.21}$$

and

$$K_\nu(\tau_\nu) = \tfrac{1}{2}\int_0^{\infty} S_\nu(\tau'_\nu)E_3\,|\tau'_\nu - \tau_\nu|\, d\tau'_\nu; \tag{79.22}$$

these expressions were first obtained by E. A. Milne (**M5**).

Equations (79.19), (79.21), and (79.22) are used frequently in radiative transfer theory, and are often abbreviated to an operator notation:

$$\Lambda_\tau[f(x)] \equiv \tfrac{1}{2}\int_0^{\infty} f(x)E_1\,|\tau - x|\, dx, \tag{79.23}$$

$$\Phi_\tau[f(x)] \equiv 2\left[\int_\tau^{\infty} f(x)E_2(x-\tau)\, dx - \int_0^{\tau} f(x)E_2(\tau - x)\, dx\right], \tag{79.24}$$

and

$$X_\tau[f(x)] \equiv \tfrac{1}{2}\int_0^{\infty} f(x)E_3\,|\tau - x|\, dx. \tag{79.25}$$

The mathematical properties of these operators are discussed in detail in (**K1**, Chap. 2).

The exponential integrals all have the asymptotic behavior $E_n(x) \sim e^{-x}/x$ for $x \gg 1$. Thus from (79.19) and (79.22) one sees that the values of $J(\tau)$ and $K(\tau)$ for $\tau \gg 1$ are effectively fixed by the value of S over a range $\Delta\tau \approx \pm 1$ from the point in question. In contrast, the flux is given by a differencing operator, its value being determined by the difference between the amount of emission from deeper layers and that from shallower layers.

To illustrate, consider a linear source function $S(\tau) = a + b\tau$. From

(79.14), (79.15), and (79.23) to (79.25) we find

$$I(\tau, \mu) = (a + b\tau) + b\mu, \qquad (0 \le \mu \le 1), \tag{79.26}$$

$$I(\tau, \mu) = (a + b\tau) + b\mu - (a + b\mu)e^{\tau/\mu}, \qquad (-1 \le \mu \le 0), \tag{79.27}$$

$$J(\tau) = \Lambda_\tau[S(t)] = a + b\tau + \tfrac{1}{2}[bE_3(\tau) - aE_2(\tau)], \tag{79.28}$$

$$H(\tau) = \tfrac{1}{4}\Phi_\tau[S(t)] = \tfrac{1}{3}b + \tfrac{1}{2}[aE_3(\tau) - bE_4(\tau)], \tag{79.29}$$

and

$$K(\tau) = \tfrac{1}{4}X_\tau[S(t)] = \tfrac{1}{3}(a + b\tau) + \tfrac{1}{2}[bE_5(\tau) - aE_4(\tau)]. \tag{79.30}$$

From (79.26) to (79.30) we see that at great depth $I(\tau, \mu)$ contains an isotropic component equal to $S(\tau)$ and an anisotropic component proportional to the gradient of S. Similarly, for $\tau \gg 1$, $J(\tau) \to S(\tau)$ and $K(\tau) \to \frac{1}{3}S(\tau)$ hence $f = K/J \to \frac{1}{3}$, as one expects because for $\tau \gg 1$, $I(\tau, \mu)$ becomes essentially isotropic. Furthermore, $H(\tau) \to \frac{1}{3}b$, which shows explicitly that the flux depends only on the local gradient.

At the surface, boundary effects become important. Noting that $E_n(0) = (n-1)^{-1}$, one finds $J(0) = \frac{1}{2}a + \frac{1}{4}b$, showing that $J(0) < S(0)$ if b is small. In particular for $b = 0$ (isothermal medium), $J(0) = \frac{1}{2}S(0)$ as expected physically because $J(0)$ is then the average of a hemisphere having no radiation ($\mu \le 0$) with one in which $I \equiv S = \text{constant}$. If $b \ge 2a$, then $J(0) \ge S(0)$, and the contribution of photons from deeper, brighter layers outweighs the dilution effects of the hemisphere with no radiation. Similarly $H(0) = \frac{1}{4}a + \frac{1}{2}b$, showing that the surface flux is larger, the faster S increases inward.

Finally, note again that when S contains a scattering term [e.g., equation (77.7)], (79.19) is an integral equation for J.

TIME-DEPENDENT RADIATION FIELD

Suppose now that the material properties and the radiation field are explicitly time dependent. The transfer equation in Cartesian coordinates is then

$$c^{-1}(\partial I/\partial t) + \mathbf{n} \cdot \nabla I = \eta - \chi I. \tag{79.31}$$

Again, the radiation moving in direction $\mathbf{n}$ at $(\mathbf{x}, t)$ consists of photons emitted in direction $\mathbf{n}$ from points along the line $\mathbf{x}' = \mathbf{x} - \mathbf{x}s$ at *retarded times* $t' = t - cs$, where s measures path length backward along the ray. In terms of s, (79.31) can be written

$$-\frac{d}{ds}\left[I\left(\mathbf{x} - \mathbf{n}s, t - \frac{s}{c}; \mathbf{n}, \nu\right)\right] + \chi\left(\mathbf{x} - \mathbf{n}s, t - \frac{s}{c}; \mathbf{n}, \nu\right) I\left(\mathbf{x} - \mathbf{n}s, t - \frac{s}{c}; \mathbf{n}, \nu\right)$$
$$= \eta\left(\mathbf{x} - \mathbf{n}s, t - \frac{s}{c}; \mathbf{n}, \nu\right). \tag{79.32}$$

Equation (79.32) admits an integrating factor, yielding the indefinite integral

$$I(\mathbf{x}, t; \mathbf{n}, \nu) = \int \eta(\mathbf{x}-\mathbf{n}s', t-s'/c; \mathbf{n}', \nu) \times \exp\left[-\int_0^{s'} \chi(\mathbf{x}-\mathbf{n}s'', t-s''/c; \mathbf{n}, \nu)\, ds''\right] ds'. \tag{79.33}$$

The physical interpretation of (79.33) is straightforward: the intensity at $(\mathbf{x}, t)$ is the integrated contribution of photons emitted in direction $\mathbf{n}$ from all $\mathbf{x}'=\mathbf{x}-\mathbf{n}s$, at retarded times $t'=t-s/c$, attenuated by the time-dependent opacity the photons encounter as they travel from $\mathbf{x}'$ to $\mathbf{x}$.

The solution (79.33) is incomplete because it does not match initial conditions imposed on I at $t=0$, or account for photons entering the boundary surface S. To account for boundary conditions we restrict the range of integration along $\mathbf{n}$ from $s=0$ to $s=s_0=|\mathbf{x}_S-\mathbf{x}|$, and add a source term localized to s_0 for all $t>0$. To match the initial condition we add a source term at $t=0$ whose value equals $I(t=0)$ for all $s \leq s_0$, and restrict the range of integration over the emissivity to retarded times $t' \geq 0$ along the ray. Thus in (79.33) we formally replace η by

$$\tilde{\eta}(s) = \eta(\mathbf{x}-\mathbf{n}s, t-s/c; \mathbf{n}, \nu)H(t-s/c) + (1/c)g(\mathbf{x}; \mathbf{n}, \nu)H(s_0-s)\,\delta(t) + f(\mathbf{x}_S, t; \mathbf{n}, \nu)\,\delta(s-s_0)H(t), \tag{79.34}$$

where H is the Heaviside function $[H(x)=0,\ x\leq 0$, and $H(x)=1,\ x>0]$, δ is the Dirac function, and f and g are given boundary conditions as in (79.1) and (79.5).

Subsituting (79.34) into (79.33) we obtain

$$\begin{aligned} I(\mathbf{x}, t; \mathbf{n}, \nu) = &\int_0^{s_0} \eta(\mathbf{x}-\mathbf{n}s', t-s'/c; \mathbf{n}, \nu)H(t-s'/c) \\ &\times \exp\left[-\int_0^{s'} \chi(\mathbf{x}-\mathbf{n}s'', t-s''/c; \mathbf{n}, \nu)\, ds''\right] ds' \\ &+ g(\mathbf{x}-\mathbf{n}ct; \mathbf{n}, \nu) \exp\left[-\int_0^{ct} \chi(\mathbf{x}-\mathbf{n}s'', t-s''/c; \mathbf{n}, \nu)\, ds''\right] H(s_0-ct) \\ &+ f(\mathbf{x}-\mathbf{n}s_0, t-s_0/c; \mathbf{n}, \nu) \exp\left[-\int_0^{s_0} \chi(\mathbf{x}-\mathbf{n}s'', t-s''/c; \mathbf{n}, \nu)\, ds''\right] H(ct-s_0). \end{aligned} \tag{79.35}$$

The first term in (79.35) accounts for all photons emitted by the material, suitably attenuated, from retarded times $t'>0$. The second term matches the solution onto the initial conditions at $t'=0$ unless t is so large that the ray already penetrated the boundary surface at a retarded time $t_0'= t-s_0/c>0$; in this event the third term comes into play and imposes the known boundary condition at $\mathbf{x}=\mathbf{x}_S$ for the correct retarded time t_0'.

80. The Diffusion Limit

In the interior of a star we reach the *diffusion limit* where the optical depth is large, photon mean free paths are small, and photons diffuse through the material in a random walk. Because the radiation is efficiently trapped, and the average temperature gradient in a star is so small, the radiation field is thermal to a high degree of approximation.

Thus for $\tau_\nu \gg 1$, $S_\nu \to B_\nu$, and in the neighborhood of any chosen τ_ν, S_ν can be represented by

$$S_\nu(t_\nu) = \sum_{n=0}^{\infty} B_\nu^{(n)}(\tau_\nu)(t_\nu - \tau_\nu)^n/n! \tag{80.1}$$

where $B_\nu^{(n)} \equiv (\partial^n B_\nu/\partial \tau_\nu^n)$. Here we tacitly assume the medium is planar because photon mean free paths are so small that curvature effects are negligible. If we substitute (80.1) into (79.14), for $0 \le \mu \le 1$ we find

$$I_\nu(\tau_\mu, \mu) = B_\nu(\tau_\nu) + \mu(\partial B_\nu/\partial \tau_\nu) + \mu^2(\partial^2 B_\nu/\partial \tau_\nu^2) + \dots. \tag{80.2}$$

The result for $-1 \le \mu \le 0$ differs from (80.2) only by terms of $O(e^{-\tau_\nu/|\mu|})$, hence for $\tau_\nu \gg 1$ we can use (80.2) for the full range $-1 \le \mu \le 1$. From (64.2), (65.5), and (66.8) we then find

$$J_\nu(\tau_\nu) = B_\nu(\tau_\nu) + \tfrac{1}{3}(\partial^2 B_\nu/\partial \tau_\nu^2) + \dots \tag{80.3}$$

$$H_\nu(\tau_\nu) = \tfrac{1}{3}(\partial B_\nu/\partial \tau_\nu) + \tfrac{1}{5}(\partial^3 B_\nu/\partial \tau_\nu^3) + \dots \tag{80.4}$$

and

$$K_\nu(\tau_\nu) = \tfrac{1}{3}B_\nu(\tau_\nu) + \tfrac{1}{5}(\partial^2 B_\nu/\partial \tau_\nu^2) + \dots. \tag{80.5}$$

We now ask how quickly these series converge. To obtain order-of-magnitude estimates of the derivatives we approximate them by difference quotients: $(\partial^n B_\nu/\partial \tau_\nu^n) \sim B_\nu/\tau_\nu^n$. Then the ratio of successive terms in (80.3) to (80.5) is $O(1/\tau_\nu^2) = O(\lambda_\nu^2/l^2)$ where l is a characteristic structural length. In a stellar envelope a conservative estimate of l is H, the pressure scale height, which in the Sun ranges from 10^2 km near the surface to $\gtrsim 10^3$ km in the interior. Thus representative values for (λ_ν/l) lie on the range 10^{-7} to 10^{-10}, implying a convergence factor of order 10^{-14} to 10^{-20} for the series. Hence we need only the first terms of (80.3) to (80.5), but we must retain two terms in (80.2) because it is the small asymmetry produced by the gradient that yields a nonzero flux.

Thus, well inside an opaque, static medium an asymptotic solution of the transfer equation is

$$J_\nu(\tau_\nu) = 3K_\nu(\tau_\nu) = B_\nu(\tau_\nu) \tag{80.6}$$

and

$$H_\nu = \tfrac{1}{3}(\partial B_\nu/\partial \tau_\nu) = -\tfrac{1}{3}(\partial B_\nu/\partial T)(dT/dr)/\chi_\nu. \tag{80.7}$$

These results are consistent with (79.28) to (79.30) for $\tau \gg 1$; (80.7) also follows immediately from (78.19) for $f_\nu \equiv \tfrac{1}{3}$. Thus far we have shown only that (80.6) and (80.7) apply in a static medium; we will see in §97 that they also apply in the comoving frame of a moving medium when velocity-gradient terms are neglected.

Equations (80.6) show that in the diffusion limit the radiation energy density and radiation pressure have their equilibrium values despite the anisotropy in the radiation field; only if we carry terms of $O(\lambda_p^2/l^2)$ in (80.3) and (80.5) is there a departure from equilibrium, but, as we have seen, such terms are truly negligible. Furthermore, (80.7) shows that the radiation flux is also a local quantity that depends on the local temperature gradient. Indeed, integrating (80.7) over frequency we find

$$F = (L/4\pi r^2) = -(4\pi/3)\left[\int_0^\infty \chi_\nu^{-1}(\partial B_\nu/\partial T)\,d\nu\right](dT/dr), \tag{80.8}$$

which shows that for $\tau_\nu \gg 1$ the radiant heat flux has precisely the same mathematical form, $\mathbf{F} = -K_R \nabla T$, as molecular heat conduction in a gas. In terms of the *Rosseland mean opacity* χ_R (cf. §82) defined by

$$\chi_R^{-1} \equiv \left[\int_0^\infty \chi_\nu^{-1}(\partial B_\nu/\partial T)\,d\nu\right]\Big/\int_0^\infty (\partial B_\nu/\partial T)\,d\nu, \tag{80.9}$$

the effective *radiative conductivity* is

$$K_R = (4\pi/3\chi_R)(dB/dT) = \tfrac{4}{3}c\lambda_R a_R T^3, \tag{80.10}$$

where λ_R is the mean free path corresponding to χ_R. This expression for K_R is exactly what one expects from the mean-free-path arguments of §29: the conductivity is proportional to the product of (1) the energy density associated with the transporting particles (photons) divided by the temperature, (2) the particle speed, and (3) the particle mean free path [cf. (29.23)].

Insofar as it predicts that the energy density and hydrostatic pressure of the radiation field are fixed by the local temperature, while the radiative flux is proportional to the temperature gradient, the diffusion-limit solution of the transfer equation is closely analogous to the first-order Chapman-Enskog solution of the Boltzmann equation for material gases. For moving media the analogy can be pushed even further (cf. §97), and one finds that P contains viscous terms proportional to the rate of strain tensor.

It is important to note that the diffusion-limit flux is a very small leak compared to the radiation energy density. Thus define the *effective temperature* T_{eff} of a star such that

$$L = 4\pi R^2 \sigma_R T_{\text{eff}}^4 \tag{80.11}$$

where L and R are the stellar luminosity and radius; from (68.9), T_{eff} is the temperature of an equivalent black body of radius R that radiates a total luminosity L. The emergent flux is

$$F = \sigma_R T_{\text{eff}}^4; \tag{80.12}$$

this net energy flux (erg cm^{-2} s^{-1}) is transported by particles (photons) that have velocity c, hence the effective energy density associated with the flux is F/c. The total radiation energy density in the field is $E=(4\pi/c)B=4\sigma_R T^4/c$, hence

$$\frac{\text{(Effective energy density in radiation flow)}}{\text{(Total energy density of radiation)}} \sim \frac{(\sigma_R T_{\text{eff}}^4/c)}{(\sigma_R T^4/c)} = \left(\frac{T_{\text{eff}}}{T}\right)^4. \tag{80.13}$$

For a star like the Sun, $T_{\text{eff}} \approx 6\times 10^3$ K, whereas interior temperatures are $\sim 10^7$ K. Thus in the solar interior the leak is about one part in 10^{12}; very small indeed! In contrast, near the surface $T\approx T_{\text{eff}}$, and the energy density associated with freely escaping photons is the energy density of the field itself.

From (80.8) and (80.9) we obtain one of the standard equations of stellar structure, that is,

$$\frac{dT}{dM_r} = \frac{-3\chi_R}{16\sigma_R \rho T^3}\frac{L_r}{(4\pi r^2)^2}, \tag{80.14}$$

where $dM_r = 4\pi r^2 \rho\, dr$ is the mass in a shell of radius r and thickness dr, and L_r is the luminosity passing through a sphere of radius r. In stellar interiors work (80.14) is viewed as an equation that determines the temperature gradient; but it has deeper significance as an asymptotic solution of the radiative transfer equation. Notice that dM_r is a material element, hence M_r can be used as a Lagrangean variable if the material moves (e.g., stellar pulsation). We show in §97 that (80.14) remains valid in moving material provided that we neglect terms of $O(\lambda_p v/lc)$, and we measure *all* quantities, including L_r (or F_r), in the comoving fluid frame.

Similarly the time-independent energy equation (78.8) for a static medium can be rewritten as

$$(dL_r/dM_r) = (4\pi/\rho)\int_0^\infty (\eta_\nu - \chi_\nu J_\nu)\, d\nu = (4\pi/\rho)\int_0^\infty \chi_\nu (S_\nu - J_\nu)\, d\nu. \tag{80.15}$$

If we adopt a source function of the form (77.8), scattering terms cancel identically and we obtain

$$(dL_r/dM_r) = (4\pi/\rho)\int_0^\infty \kappa_\nu (B_\nu - J_\nu)\, d\nu, \tag{80.16}$$

where κ_ν is the thermal absorption coefficient, and in writing η_ν we assumed LTE. One might expect from (80.6) that in the diffusion limit the right-hand side of (80.16) will vanish identically. Physically this would correspond to a state of radiative equilibrium (cf. §78), in which the material emits exactly the amount of energy it absorbs, and the luminosity is independent of radius. However, in the deep interior there can also be an irreversible release of thermonuclear energy at a rate of ε ergs gm^{-1} s^{-1}, which drives a *luminosity gradient.* The thermonuclear energy released in a mass shell is $\varepsilon\, dM_r$, which for a steady state must equal the change in luminosity across that shell; accounting for this additional term (80.16) reduces to its standard stellar interiors form

$$(dL_r/dM_r) = \varepsilon. \tag{80.17}$$

A more revealing derivation of (80.17) in §96 shows that it is only the static limit of a more general energy equation representing the first law of thermodynamics for the composite matter-radiation gas [see (96.10) and (96.11)].

Finally, we show that in the limit of large optical depth, the transfer equation can be manipulated into a time-dependent diffusion equation for the energy density. First, in the diffusion regime we can neglect the term $c^{-2}(\partial \mathbf{F}_\nu/\partial t)$ in (78.10) compared to the right-hand side which, in the fluid frame where χ and η are isotropic, reduces to $(\chi_\nu \mathbf{F}_\nu/c)$, because the time required for photons to random walk a distance l is $\Delta t \sim (l/\lambda_p)^2(\lambda_p/c) = l^2/\lambda_p c$, hence the ratio of the terms in question is

$$c^{-2}(\partial F_\nu/\partial t)/(\chi_\nu F_\nu/c) \sim (1/\chi c\, \Delta t) = (\lambda_p/l)^2 \sim 10^{-18} \tag{80.18}$$

for typical values $l \sim H = 10^3$ km and $\lambda_p \sim 10^{-1}$ cm. Thus we can use the static form of (78.11) restated as

$$\mathbf{F}_\nu = -c\chi_\nu^{-1}\nabla P_\nu = -\tfrac{1}{3}c\chi_\nu^{-1}\nabla E_\nu, \tag{80.19}$$

where we made use of the isotropy of P_ν. Substituting (80.19) into (78.3) we have (for a static medium)

$$\frac{1}{c}\frac{\partial E_\nu}{\partial t} = \frac{1}{3}\nabla\cdot\left(\frac{1}{\chi_\nu}\nabla E_\nu\right) + \left(\frac{4\pi}{c}\right)\eta_\nu - \chi_\nu E_\nu, \tag{80.20}$$

which, as asserted, has the same form as the time-dependent diffusion equation $(\partial f/\partial t) = \nabla^2 f + \mathcal{S}$ where $\mathcal{S}$ is a source-sink term.

81. The Wave Limit

Having seen that, at large optical depth, the transfer equation behaves like a diffusion equation, we now show that in a vacuum ($\chi_\nu = \eta_\nu = 0$) it reduces to the wave equation, which is to be expected from classical electrodynamics. Although one obtains a perfectly unattenuated wave only in true vacuum, the radiation field also approaches this free streaming limit in optically thin media.

In the absence of material the transfer equation (76.3) becomes

$$(\partial I/\partial t) + c(\partial I/\partial s) = 0 \tag{81.1}$$

where s measures the pathlength along the ray $\mathbf{n}$. Defining $I^+ \equiv I(\mathbf{x}, t; \mathbf{n}, \nu)$

and $I^- \equiv I(\mathbf{x}, t; -\mathbf{n}, \nu)$, (81.1) yields

$$(\partial I^+/\partial t) + c(\partial I^+/\partial s) = 0 \tag{81.2}$$

and

$$(\partial I^-/\partial t) - c(\partial I^-/\partial s) = 0. \tag{81.3}$$

Now define the mean-intensity-like quantity

$$j \equiv \tfrac{1}{2}(I^+ + I^-) \tag{81.4}$$

and the fluxlike quantity

$$h \equiv \tfrac{1}{2}(I^+ - I^-); \tag{81.5}$$

then (81.2) and (81.3) can be added and subtracted to produce

$$(\partial j/\partial t) + c(\partial h/\partial s) = 0 \tag{81.6}$$

and

$$(\partial h/\partial t) + c(\partial j/\partial s) = 0. \tag{81.7}$$

Equations (81.6) and (81.7) combine into the wave equations

$$(\partial^2 j/\partial t^2) = c^2(\partial^2 j/\partial s^2) \tag{81.8}$$

and

$$(\partial^2 h/\partial t^2) = c^2(\partial^2 h/\partial s^2), \tag{81.9}$$

which have the solutions

$$j(s, t) = A_1 f_1(s - ct) + A_2 f_2(s + ct) \tag{81.10}$$

and

$$h(s, t) = B_1 f_1(s - ct) + B_2 f_2(s + ct). \tag{81.11}$$

We recognize these as traveling waves moving along $\pm\mathbf{n}$. As usual, the constants $A_1, \ldots, B_2$ are determined by initial and boundary conditions.

Equations (81.10) and (81.11) imply that one can construct a particular solution of the form

$$I(\mathbf{x}, t; \mathbf{n}', \nu') = I_0\, \delta(s - ct)\, \delta(\mathbf{n}' - \mathbf{n})\, \delta(\nu' - \nu), \tag{81.12}$$

that is, a monochromatic plane wave traveling along $\mathbf{n}$ with velocity c. As noted in §66, for such a wave $J_\nu = H_\nu = K_\nu$.

Wave equations also follow from the moment equations once we know that solutions of the form (81.12) exist. Thus if we choose $\mathbf{n} = \mathbf{k}$ in planar geometry, (78.3) and (78.10) become

$$(\partial J_\nu/\partial t) + c(\partial H_\nu/\partial z) = 0 \tag{81.13}$$

and

$$(\partial H_\nu/\partial t) + c(\partial K_\nu/\partial z) = 0, \tag{81.14}$$

and because $J_\nu = K_\nu$ for a plane wave, (81.13) and (81.14) combine to give

$$(\partial^2 J_\nu/\partial t^2) - c^2(\partial^2 J_\nu/\partial z^2) = 0 \tag{81.15}$$

and

$$(\partial^2 H_\nu/\partial t^2) - c^2(\partial^2 H_\nu/\partial z^2) = 0, \qquad (81.16)$$

which are standard wave equations in planar geometry.

In spherical geometry (78.5) becomes

$$(\partial J_\nu/\partial t) + (c/r^2)[\partial(r^2 H_\nu)/\partial r] = 0, \qquad (81.17)$$

while (78.11) reduces to

$$(\partial H_\nu/\partial t) + (c/r^2)[\partial(r^2 J_\nu)/\partial r] = 0 \qquad (81.18)$$

when we recall (66.10) and demand $K_\nu = J_\nu$. Equations (81.17) and (81.18) combine to yield

$$(\partial^2 J_\nu/\partial t^2) = (c^2/r^2)[\partial^2(r^2 J_\nu)/\partial r^2] = c^2\nabla^2 J_\nu \qquad (81.19)$$

and

$$(\partial^2 H_\nu/\partial t^2) = c^2\nabla^2 H_\nu, \qquad (81.20)$$

which are standard wave equations in spherical geometry.

82. The Grey Atmosphere, Mean Opacities, and Multigroup Methods

MOTIVATION AND ASSUMPTIONS

We now consider a highly simplified problem, which provides valuable experience in solving the transfer equation: radiative transfer and energy balance in a static LTE medium composed of *grey material* (one whose absorption coefficient is independent of frequency). This problem can be solved relatively easily and completely, and yields reasonable estimates of the run of the physical properties in the outer layers of stars, thus giving (1) moderately accurate boundary conditions for stellar envelope and interior calculations, and (2) starting solutions for iterative methods for handling more accurate treatments of the physics.

For grey material $\chi_\nu \equiv \chi$, and the transfer equation (77.5) becomes

$$\mu(\partial I_\nu/\partial\tau) = I_\nu - S_\nu. \qquad (82.1)$$

Integrating over frequency we have

$$\mu(\partial I/\partial\tau) = I - S, \qquad (82.2)$$

where quantities such as I, J, H, K, B, and S without subscripts denote frequency-integrated variables, for example,

$$I \equiv \int_0^\infty I_\nu\, d\nu. \qquad (82.3)$$

Because the medium is static, it must be in radiative equilibrium; hence

$$4\pi\int_0^\infty \chi_\nu J_\nu\, d\nu = 4\pi\int_0^\infty \chi_\nu S_\nu\, d\nu, \qquad (82.4)$$

which, for grey material, reduces to $J \equiv S$. Thus the transfer equation to be solved is

$$\mu(\partial I/\partial \tau) = I - J, \tag{82.5}$$

which is a *homogeneous* integrodifferential equation for I, posing what is called *Milne's problem*. The Milne problem is to be solved for $J(\tau)$ [hence $S(\tau)$] from which we compute $I(\tau, \mu)$ and $K(\tau)$. We already know that $H(\tau)$ is constant in radiative equilibrium [cf. (82.7)].

With the additional assumption of LTE, $S_\nu \equiv B_\nu$, hence

$$B[T(\tau)] = \sigma_R T^4/\pi = S(\tau) = J(\tau). \tag{82.6}$$

Therefore if we can determine $J(\tau)$ from (82.5), then (82.6) allows us to associate a temperature with the radiation field at each depth.

GREY MOMENT EQUATIONS

Calculating the zeroth moment of (82.5) we obtain

$$(dH/d\tau) = J - S = J - J \equiv 0, \tag{82.7}$$

which shows that the flux is indeed constant. The first moment yields

$$(dK/d\tau) = H, \tag{82.8}$$

which has the exact integral

$$K(\tau) = H(\tau + C), \tag{82.9}$$

where C is a constant. For $\tau \gg 1$, $J(\tau) \to 3K(\tau)$; hence (82.9) implies that when $\tau \gg 1$, $J(\tau) \approx 3H\tau$. This result suggests a general expression for $J(\tau)$ of the form

$$J(\tau) = 3H[\tau + q(\tau)], \tag{82.10}$$

where $q(\tau)$, the *Hopf function*, is a bounded function, to be determined. In terms of $q(\tau)$, (82.9) becomes

$$K(\tau) = H[\tau + q(\infty)]. \tag{82.11}$$

The solution of the grey problem consists of the calculation of $q(\tau)$. Given this function we can determine the run of temperature with depth from

$$T^4 = \tfrac{3}{4} T_{\text{eff}}^4 [\tau + q(\tau)], \tag{82.12}$$

which follows from (82.6), (82.10), and (80.12).

A wide variety of methods have been developed for determining $q(\tau)$; these are discussed at length in (**C6**) and (**K1**); in fact, it is possible to obtain an exact solution in closed form. We shall not review this large literature here, but will discuss only two methods: the simplest approximation, which yields roughly the right answer, and a second, which yields accurate answers for the grey problem and also provides the basic approach used in solving more complex transfer problems.

THE EDDINGTON APPROXIMATION

To obtain an approximate solution of the grey problem, Eddington made the simplifying assumption that the relation $J(\tau)=3K(\tau)$, valid at great depth, holds throughout the entire medium. Although Eddington's approximation is not exact, it is nevertheless reasonable. For example, the relation $J=3K$ holds (1) in the diffusion regime where $I(\tau,\mu)=I_0(\tau)+I_1(\tau)\mu$; and (2) in the *two-stream approximation* where $I(\tau,\mu)\equiv I^+(\tau)$ for $0\leq\mu\leq 1$ and $I(\tau,\mu)\equiv I^-(\tau)$ for $-1\leq\mu\leq 0$ for arbitrary (but μ-independent) values of I^+ and I^-. The latter provides a rough representation of the radiation field near the boundary of a semi-infinite medium, for we may let $I^-/I^+\rightarrow 0$ as $\tau\rightarrow 0$, and $I^-/I^+\rightarrow 1$ for $\tau\gtrsim 1$.

From (82.10) and (82.11) we see that Eddington's approximation is equivalent to writing

$$J_E(\tau)=3H(\tau+C'),\tag{82.13}$$

where the constant C' is to be determined. To fix C' we use $S=J_E$ in (79.21) to calculate the emergent flux, obtaining

$$H(0)=\tfrac{3}{2}H\int_0^\infty(\tau+C')E_2(\tau)\,d\tau=\tfrac{3}{2}H[E_4(0)+C'E_3(0)]=\tfrac{3}{2}H(\tfrac{1}{3}+\tfrac{1}{2}C').\tag{82.14}$$

Demanding $H(0)\equiv H$ we find $C'=\frac{2}{3}$. Thus in the Eddington approximation $q_E(\tau)\equiv\frac{2}{3}$,

$$J_E(\tau)=3H(\tau+\tfrac{2}{3})\tag{82.15}$$

and

$$T^4=\tfrac{3}{4}T_{\text{eff}}^4(\tau+\tfrac{2}{3}).\tag{82.16}$$

The exact solution of the grey problem gives $q(0)=1/\sqrt{3}=0.577\ldots$, and $q(\infty)=0.710\ldots$, compared with $q_E\equiv 0.666\ldots$. Not surprisingly, the error in q_E and J_E is largest at the surface; one finds $\Delta J(0)/J_{\text{exact}}(0)\approx 0.15$. On the other hand, (82.16) predicts $(T_0/T_{\text{eff}})_E=(\frac{1}{2})^{1/4}\approx 0.841$, which agrees well with $(T_0/T_{\text{eff}})_{\text{exact}}=(\frac{1}{4}\sqrt{3})^{1/4}\approx 0.811$. Thus Eddington's approximation does provide a good first estimate for the temperature structure of a grey atmosphere. Note also that (82.16) predicts $T=T_{\text{eff}}$ at $\tau=\frac{2}{3}$; for this reason $\tau=\frac{2}{3}$ is often considered to be the effective depth of continuum formation in a semi-infinite medium.

Furthermore, using (82.15) in (79.16) we find

$$I_E(\tau=0,\mu)=H(2+3\mu),\tag{82.17}$$

which shows that the radiation field is peaked in the direction of outward flow; indeed $I_E(\mu=1)/I_E(\mu=0)=2.5$. Using (82.17) to calculate J and K at $\tau=0$ we find $J_E(0)=\frac{7}{4}H$ and $K_E(0)=\frac{17}{24}H$, whence the variable Eddington factor at the surface is $f_E(0)=\frac{17}{42}\approx 0.405$; the exact solution yields $f(0)=q(\infty)/3q(0)\approx 0.410$. These results are important because they show that using even a rough estimate ($f\equiv\frac{1}{3}$) of the Eddington factor to solve the transfer equation we obtain a source function that yields a reasonably accurate angular distribution of the intensity, from which we can compute a much better estimate of f throughout the entire atmosphere. We return to this point in §83.

THE METHOD OF DISCRETE ORDINATES

While Eddington's approach gives useful results, it lacks accuracy and generality. We can also solve the grey problem by rewriting (82.6) explicitly as an integrodifferential equation

$$\mu[\partial I(\tau, \mu)/\partial\tau] = I(\tau, \mu) - \tfrac{1}{2}\int_{-1}^{1} I(\tau, \mu)\, d\mu, \tag{82.18}$$

and approximating the integral as a *quadrature sum*. In this procedure a function $f(\mu)$ defined on $-1 \le \mu \le 1$ is sampled at a set of *quadrature points* $\{\mu_m\}$, $(m = \pm 1, \ldots, \pm M)$, where $0 \le \mu_m \le 1$ and $\mu_{-m} = -\mu_m$. Applying standard techniques of numerical analysis, one can generate a set of *quadrature weights* $\{b_m\}$ defined such that the definite integral in (82.18) is represented as a weighted sum over the *discrete ordinates* $\{f(\mu_m)\}$, that is,

$$\int_{-1}^{1} f(\mu)\, d\mu \approx \sum_{m=-M}^{M} b_m f(\mu_m). \tag{82.19}$$

This procedure assumes, in effect, that $f(\mu)$ is the unique interpolating polynomial of order $2M-1$ that passes through the $2M$ ordinates $\{f(\mu_m)\}$.

For the transfer problem we thus represent the angular variation of $I(\tau, \mu)$ by a set of *pencils* of radiation $I_i \equiv I(\tau, \mu_i)$, and replace the integrodifferential equation (82.18) by a coupled set of ordinary differential equations:

$$\mu_i(dI_i/d\tau) = I_i - \tfrac{1}{2}\sum_{m=-M}^{M} b_m I_m, \qquad (m = \pm 1, \ldots, \pm M). \tag{82.20}$$

One thinks of each pencil I_m as giving an average of $I(\mu)$ over a definite range of μ around μ_m. On physical grounds it is reasonable to expect the solution to become increasingly accurate as M increases, and to limit the exact solution as $M \to \infty$.

The *method of discrete ordinates* described above provides an extremely powerful tool for solving transfer problems, and it will be exploited heavily in §§83 and 88, and in Chapter 7. For the grey problem, Chandrasekhar (**C6**) obtained a complete analytical solution of (82.20), which matches the boundary conditions and gives constant flux; furthermore, by studying the limit $M \to \infty$ he deduced many properties of the exact $q(\tau)$. The full exact solution was first obtained by completely different (Laplace transform) methods, discussed in (**K1**) and (**M2,** Chap. 3).

The accuracy of a quadrature formula depends both on the number of quadrature points used, and on their distribution within the interval. A good discussion of methods for constructing quadrature formulae is given in (**C6,** Chap. 2). In *Newton-Cotes formulae* the $\{\mu_m\}$ are equally spaced; for $2M$ points the quadrature is exact if $I(\mu)$ is a polynomial of order $\leq 2M-1$. A more favorable choice is the *Gauss formula,* in which the $\{\mu_m\}$ are the roots of the Legendre polynomial P_{2M}, and which is exact if $I(\mu)$ is a polynomial of order $\leq 4M-1$. An even better choice for transfer problems is the *double-Gauss formula* suggested by Sykes (**S5**), which is now universally used. Here one uses a separate M-point Gauss formula on each of the subintervals $[-1, 0]$ and $[0, 1]$, the points being the roots of P_M suitably shifted and scaled from $[-1, 1]$ to each subinterval. (An important exception is that for $M=1$ one must choose $\mu_{\pm 1} = \pm 1/\sqrt{3}$.) The quadrature is exact if $I(\mu)$ is a polynomial of order $\leq 2M-1$ on each subinterval. Although for a given number of points the formal accuracy of the double-Gauss formula is lower on each subinterval than the ordinary Gauss formula, for transfer problems it is vastly superior because $I(+\mu)$ and $I(-\mu)$ are approximated independently, hence it can account for the fact that as $\tau \rightarrow 0$, $I(-\mu) \rightarrow 0$ while $I(+\mu)$ remains finite. The ordinary Gauss formula spanning $[-1, 1]$ tries, in effect, to integrate through the discontinuity at $\mu = 0$, and naturally loses accuracy in doing so. Quadrature points and weights for double-Gauss formulae are given in (**A1,** 921).

THE NONGREY PROBLEM

The opacity frequency-spectrum of real material is complicated (cf. §72), and in solving realistic transfer problems we confront the difficult question of how best to model this spectrum. Several approaches have been developed; each has strengths and weaknesses, which, for any particular problem, must be weighed carefully. Typically a compromise must be made between accuracy and economy of computation.

One obvious approach is the *direct method* in which a large number of frequency points are chosen so as to represent all the major features of the opacity (e.g., continuum edges and strong lines); one then solves the transfer equation or the moment equations at each of these frequencies. This approach is satisfactory for material with a relatively simple spectrum (e.g., in hot stars where the dominant processes are bound-free and free-free absorption by H, H^+, He, He^+, and He^{++}; Thomson scattering by free electrons; and absorption in a small number of strong lines) and in such cases it can yield accurate results. But for complex spectra (e.g., in cool stars with millions of atomic and molecular lines) the direct method is prohibitively costly, especially for dynamical problems. Let us therefore consider alternatives.

MEAN OPACITIES

The simplest possible solution of the nongrey problem would be obtained if a single *mean opacity* could represent correctly the total transport of radiation through the material; the nongrey problem would then reduce to an equivalent grey problem, whose solution is known. Not surprisingly, such a reduction is not possible in general, basically because opacities enter *nonlinearly* in transport processes. Nevertheless, certain mean opacities have important physical significance and permit considerable simplification of the nongrey problem in some regimes, while providing useful approximations in others where they are not rigorously correct. Here we discuss only planar geometry, but mean opacities can be used in any geometry.

(*a*) *The Rosseland Mean* The key role usually played by radiation in a radiating fluid is the transport of energy. Can we define a mean opacity that guarantees the correct total transport of radiant energy? We will obtain the correct total flux from the first moment equation (78.16) if $\bar{\chi}$ is chosen such that

$$-\int_0^\infty \frac{1}{\chi_\nu}\frac{\partial K_\nu}{\partial z}\,d\nu = \int_0^\infty H_\nu\,d\nu = H \equiv -\frac{1}{\bar{\chi}}\frac{dK}{dz} = -\frac{1}{\bar{\chi}}\int_0^\infty \frac{\partial K_\nu}{\partial z}\,d\nu. \quad (82.21)$$

or

$$\bar{\chi} \equiv \int_0^\infty (\partial K_\nu/\partial z)\,d\nu \bigg/ \int_0^\infty \chi_\nu^{-1}(\partial K_\nu/\partial z)\,d\nu. \quad (82.22)$$

The difficulty in obtaining the correct $\bar{\chi}$ in general is clear from (82.22): we must know K_ν to compute $\bar{\chi}$, but to determine K_ν we must solve the full nongrey problem.

However, in the diffusion regime $K_\nu \to \frac{1}{3}B_\nu$, and $\bar{\chi}$ then reduces to the *Rosseland mean* defined by (80.9). Thus in the diffusion regime we *can* replace an arbitrarily complex opacity spectrum by a single average that guarantees the correct radiative energy transport; this is why the Rosseland mean is universally used in stellar interiors work. Furthermore, in LTE, χ_ν is a function of T and ρ, while B_ν is a function of T only; hence χ_R can be computed once and for all as a function of local state variables.

Note that χ_R is a *harmonic mean*, giving greatest weight to the most transparent regions of the spectrum. Thus opaque features (e.g., strong lines) affect χ_R mainly by reducing the *bandwidth* through which efficient energy transport can occur. A change in the absolute strength of opaque features has little or no effect on χ_R; but the addition of a continuum source and/or many faint lines to an otherwise weakly absorbing spectral region raises the minimum value of χ_ν there, and can increase χ_R significantly.

From (82.21) it follows that, even when the material is not grey, in the diffusion limit $(dK/d\tau_R) = \frac{1}{3}(dB/d\tau_R) = H$, where $H = (\sigma_R T_{\text{eff}}^4/4\pi)$, and $\tau_R = -\int \chi_R\,dz$. Hence the temperature distribution for $\tau_R \gg 1$ in a static nongrey medium in radiative equilibrium is accurately given by the modified grey relation

$$T^4 = \tfrac{3}{4}T_{\text{eff}}^4[\tau_R + q(\tau_R)]. \quad (82.23)$$

Detailed calculations for nongrey radiative-equilibrium atmospheres show that (82.23) is in fact an excellent approximation at depth, and provides a good starting estimate for iterative methods (cf. §88) at the surface. Of course, (82.23) cannot guarantee flux conservation when the diffusion approximation breaks down; thus near a boundary surface the Rosseland mean may seriously underestimate the effective opacity and may yield poor results for radiative energy balance (**C3**).

(*b*) *The Flux Mean* Instead of energy transport we might focus on momentum balance and choose a mean that gives the correct radiation force on the material. From (78.12) and (78.17) the radiation force on a static medium is

$$f_R = (4\pi/c) \int_0^\infty \chi_\nu H_\nu \, d\nu; \tag{82.24}$$

the same expression holds for moving material if all quantities are measured in the comoving frame (cf. §96). Thus the correct radiation force results if $\bar{\chi}$ is defined by

$$\int_0^\infty \chi_\nu H_\nu \, d\nu = \bar{\chi} \int_0^\infty H_\nu \, d\nu = \bar{\chi} H, \tag{82.25}$$

whence

$$\chi_H \equiv \int_0^\infty \chi_\nu H_\nu \, d\nu / H, \tag{82.26}$$

which is called the *flux mean.*

Like the Rosseland mean, χ_H does not allow a complete reduction of the nongrey problem to an equivalent grey problem. Moreover we cannot compute χ_H until we know H_ν, which is obtained only by solving the full nongrey problem. But in the diffusion regime (80.7) implies that χ_H is identical to the Rosseland mean:

$$(\chi_H)_{\text{diffusion}} = \int_0^\infty (\partial B_\nu/\partial T) \, d\nu \Big/ \int_0^\infty \chi_\nu^{-1} (\partial B_\nu/\partial T) \, d\nu \equiv \chi_R. \tag{82.27}$$

Thus in the diffusion limit, the Rosseland mean yields not only the correct energy transport, but the correct momentum balance as well. For this reason the Rosseland mean is often chosen as a representative opacity for use in the momentum equations in problems of radiation hydrodynamics.

(*c*) *The Planck Mean and The Absorption Mean* Other definitions of mean opacities result from requiring correct values for the total energy emitted or absorbed by the material. For a static LTE medium the right-hand side of the radiation energy equation (78.4) reduces to $4\pi \int_0^\infty \kappa_\nu (B_\nu - J_\nu) \, d\nu$ even when scattering terms are present. The same expressions holds for moving media provided that all quantities are measured in the comoving frame (cf. §96). To obtain the correct total emission we thus define a mean opacity $\bar{\kappa}$ such that

$$\int_0^\infty \kappa_\nu B_\nu\,d\nu = \bar{\kappa}\int_0^\infty B_\nu\,d\nu \equiv \kappa_P B(T) \tag{82.28}$$

or

$$\kappa_P \equiv \int_0^\infty \kappa_\nu B_\nu\,d\nu/(\sigma_R T^4/\pi), \tag{82.29}$$

which is called the *Planck mean*. Notice that, like the Rosseland mean, κ_P can be computed once and for all as a function of ρ and T.

To obtain the correct total absorption we must use the *absorption mean* κ_J defined by

$$\kappa_J \equiv \int_0^\infty \kappa_\nu J_\nu\,d\nu \Big/ \int_0^\infty J_\nu\,d\nu. \tag{82.30}$$

But, like χ_H, κ_J cannot be evaluated unless we have solved the full nongrey transfer problem. It is therefore important that we can show that, in the optically thin regime, κ_P provides a reasonable estimate of the total absorption and thus serves as a useful substitute for κ_J, just as χ_R does for χ_H in the diffusion regime.

In particular, to achieve radiative equilibrium we should choose $\bar{\kappa}$ such that

$$\int_0^\infty \kappa_\nu(B_\nu - J_\nu)\,d\nu = 0 \equiv \bar{\kappa}\int_0^\infty (B_\nu - J_\nu)\,d\nu. \tag{82.31}$$

When the material is transparent ($\tau_\nu \ll 1$ at all frequencies), J_ν is essentially fixed, and the integrals in (82.31) are dominated by the frequencies at which $\kappa_\nu \gg \bar{\kappa}$. For $\bar{\tau} \lesssim 1$ we can represent B_ν by a linear expansion

$$B_\nu(t) = B_\nu(\bar{\tau}) + (\partial B_\nu/\partial\bar{\tau})(t - \bar{\tau}) \approx B_\nu(\bar{\tau}) + (\bar{\kappa}/\kappa_\nu)(\partial B_\nu/\partial\bar{\tau})(t_\nu - \tau_\nu), \tag{82.32}$$

whence, by application of the Λ operator, we obtain

$$J_\nu(\bar{\tau}) \approx B_\nu(\bar{\tau})[1 - \tfrac{1}{2}E_2(\bar{\tau})] + \tfrac{1}{2}(\bar{\kappa}/\kappa_\nu)(\partial B_\nu/\partial\bar{\tau})[E_3(\tau_\nu) + \tau_\nu E_2(\tau_\nu)]. \tag{82.33}$$

In the limit $\bar{\tau} \to 0$, $E_2 \to 1$ and $E_3 \to \frac{1}{2}$, and $(B_\nu - J_\nu) \approx \frac{1}{2}B_\nu - \frac{1}{4}(\bar{\kappa}/\kappa_\nu)(\partial B_\nu/\partial\bar{\tau})$. The second term is least important when $\kappa_\nu \gg \bar{\kappa}$, that is, precisely when the first term makes the largest contribution in (82.31). Hence we most nearly achieve energy balance in optically thin material if $\bar{\kappa}$ satisfies

$$\tfrac{1}{2}\bar{\kappa}\int_0^\infty B_\nu\,d\nu = \tfrac{1}{2}\int_0^\infty \kappa_\nu B_\nu\,d\nu, \tag{82.34}$$

that is, if $\bar{\kappa} = \kappa_P$.

Thus κ_P is a good representative opacity for use in the radiation energy equation. On the other hand, use of κ_P in the first moment equation does not yield the correct flux in the diffusion limit, and we again conclude that no one mean opacity completely reduces the nongrey problem to a grey problem.

MEAN-OPACITY REPRESENTATION OF THE MOMENT EQUATIONS

In LTE,

$$(dH/dz) = \kappa_P B - \kappa_J J \tag{82.35}$$

and

$$(dK/dz) = -\chi_H H \tag{82.36}$$

are exact frequency-integrated moment equations. But to solve these equations one must know κ_J and χ_H, which implies solving the nongrey equations. An effective method sometimes used to handle transfer and energy balance in nongrey media is to rewrite (82.35) and (82.36) as

$$(dH/dz) = \kappa_P (B - k_J J) \tag{82.37}$$

and

$$(dK/dz) = -\chi_R k_H H \tag{82.38}$$

where the ratios $k_J \equiv (\kappa_J/\kappa_P)$ and $k_H \equiv (\chi_H/\chi_R)$ are to be determined iteratively, starting from an initial estimate of unity (or values from the previous time-step in a dynamical calculation—see §7.3).

The idea is to use (82.37) and (82.38) along with a constraint of energy balance to determine the temperature distribution, and then perform a frequency-by-frequency formal solution of the transfer equation, using the new temperature distribution, to update (J_ν/J) and (H_ν/H), and then k_J and k_H. With these improved estimates of k_J and k_H we can repeat the first step to determine an improved temperature distribution. Each step of this iteration procedure is relatively cheap. Nevertheless it presupposes that frequency-dependent opacities $\kappa_\nu(\rho, T)$ are available (which may not be true), and that we are willing to calculate the full frequency spectrum of the radiation field; in practice it may prove too costly in dynamical calculations for hundreds or thousands of timesteps. We then have little choice but to adopt $k_J \equiv 1$ and $k_H \equiv 1$; although the results so obtained are not exact, because in general κ_P is not the correct absorption average nor does χ_R equal χ_H except in the diffusion regime, they provide nonetheless a reasonable first approximation.

MULTIGROUP METHODS

The preceding discussion points out that detailed simulation of the opacity spectrum by the direct approach is generally too costly, while replacing it by one or two representative means may be too crude; we therefore seek a middle ground. The weakness of the mean opacity approach is that it averages over the entire spectrum. Given large fluctuations in κ_ν and the possibility of large differences between B_ν and J_ν, it is obvious that κ_P, say, is unlikely to equal κ_J exactly. However it is much easier to define a meaningful opacity in a narrow spectral range, and one asks whether it is possible to represent the physically important features of the opacity spectrum with a few (but more than one or two) astutely chosen parameters. Two such methods have been suggested and widely applied.

In the first, the *multigroup method*, the spectrum is divided into a number of *frequency groups*, each of which spans a definite range (ν_g, ν_{g+1}). Within each group, source terms and radiation quantities are replaced by values that can be viewed either as integrals over the group, for example,

$$B_g \equiv \int_{\nu_g}^{\nu_{g+1}} B_\nu \, d\nu, \tag{82.39}$$

$$J_g \equiv \int_{\nu_g}^{\nu_{g+1}} J_\nu \, d\nu, \tag{82.40}$$

(and similarly for H_g and K_g), or as representative constant values within the group. The zeroth moment equation for group g in a static LTE medium is then

$$(dH_g/dz) = \kappa_{P,g} B_g - \bar{\kappa}_g J_g, \tag{82.41}$$

where the *group Planck mean* is

$$\kappa_{P,g} \equiv \int_{\nu_g}^{\nu_{g+1}} \kappa_\nu B_\nu \, d\nu / B_g. \tag{82.42}$$

We must now decide what value is to be assigned to $\bar{\kappa}_g$ to obtain the correct total absorption in the group. If we were to take literally the picture that J_g is constant within (ν_g, ν_{g+1}), then we should use the straight average opacity

$$\bar{\kappa}_g \equiv \int_{\nu_g}^{\nu_{g+1}} \kappa_\nu \, d\nu / (\nu_{g+1} - \nu_g). \tag{82.43}$$

However, if we use (82.43), then at great depth where $J_\nu \rightarrow B_\nu$, we do not necessarily recover equality between the energy absorbed and emitted within the group (i.e., between $\bar{\kappa}_g J_g$ and $\kappa_{P,g} B_g$) even though $J_g \equiv B_g$, because we have used different weighting schemes in (82.42) and (82.43). Furthermore, in the limit of using only one group, (82.43) is not at all reasonable physically. For these reasons it is usually argued that the multigroup zeroth moment equations should be written as

$$(dH_g/dz) = \kappa_{P,g}(B_g - J_g), \tag{82.44}$$

which reduces to the mean-opacity method for a single group. On the other hand, in the optically thin limit there is no rationale for weighting the absorption term by the Planck function, that is, for replacing $\bar{\kappa}_g$ with $\kappa_{P,g}$, and it may actually be preferable to use $\bar{\kappa}_g$ as defined in (82.43) (e.g., in an optically thin layer illuminated by a perfectly smooth continuum having a radiation temperature markedly different from the local material temperature).

By similar reasoning, the first moment equation for group g can be written

$$(dK_g/dz) = -\bar{\chi}_g H_g. \tag{82.45}$$

Here one argues that to obtain the correct total flux transport we should use a harmonic mean for $\bar{\chi}_g$. Again if we literally take K_g to be constant in a group we would use the simple harmonic mean

$$(\bar{\chi}_g)^{-1} \equiv \int_{\nu_g}^{\nu_{g+1}} \chi_\nu^{-1}\, d\nu/(\nu_{g+1} - \nu_g). \tag{82.46}$$

But this choice of $\bar{\chi}_g$ does not necessarily yield the correct flux in the diffusion limit; to guarantee that we should write

$$(dK_g/dz) = -\chi_{R,g} H_g \tag{82.47}$$

where $\chi_{R,g}$ is the *group Rosseland mean*

$$(\chi_{R,g})^{-1} \equiv \int_{\nu_g}^{\nu_{g+1}} \chi_\nu^{-1}(\partial B_\nu/\partial T)\, d\nu \bigg/ \int_{\nu_g}^{\nu_{g+1}} (\partial B_\nu/\partial T)\, d\nu. \tag{82.48}$$

In most multigroup formulations, (82.47) is used in preference to (82.45). Like (82.44) for the zeroth moment, (82.47) provides a reasonable representation in the limit of one group; (82.45) does not, but may be more realistic for optically thin material.

In summary, the formulation of the multigroup method is not unique, and the results are unavoidably somewhat ambiguous. The method is least accurate for coarse frequency groups, for which there can be significant differences among the various group means, leading to the same problems as in the mean-opacity method. It gives increasingly better results as more, and finer, groups are used, because then the details of the weighting procedure are less important and the various group averages become more nearly equal; of course it also becomes more costly.

OPACITY DISTRIBUTION FUNCTIONS

The second important technique for handling radiative transfer in complex spectra uses *opacity distribution functions* (ODF). The ODF method has been applied extensively in astrophysics and yields excellent results in a wide variety of situations (**C3**), (**D1**). Here the whole spectrum is divided into a number of frequency *intervals*. For a calculation of the total radiative energy and momentum transport in each interval, the exact position of a particular feature (e.g., a spectral line) within the interval is not important; instead, we need to know the fraction of the interval that is relatively transparent (continuum), moderately opaque (weak lines), or very opaque (strong lines). Therefore in each interval we compute the opacity at a large number of uniformly spaced points and bin similar opacity values; we thus construct a distribution function giving the value of the opacity versus the cumulative fraction of the interval covered by opacity less than or equal to this value—see (**K3,** 3–7) or (**M2,** 167–169). This function is smooth and is well represented by a relatively small number of *pickets,* each of which covers a prechosen fraction of the interval with a constant opacity equal to the average of the distribution function in that band; an example is shown in (**K3,** 7).

For each picket we write a transfer equation (or moment equations); in these equations the opacity now has a unique value, which eliminates the ambiguity of the multigroup method. In this respect the ODF is similar to the direct method, and can be viewed as a way of degrading unwanted details of the opacity spectrum (which act as high-frequency "noise") to the minimum level required to give correct energy transport. (The multigroup method did just the opposite: it reduced the overly severe filtering inherent in whole-spectrum means by applying the averaging scheme over smaller intervals.)

THE OPACITY SAMPLING TECHNIQUE

We just mention one other approach: the *opacity sampling technique* (**S3**). Here one solves the transfer equation using actual opacities at a large number of frequencies chosen at random throughout the spectrum. This method has been used successfully to construct static model atmospheres, but it does not seem easily adaptable to dynamical problems so we will not discuss it further.

83. *Numerical Methods*

Analytical solutions of transfer problems are rare, and in most cases of interest we must use numerical methods. We describe here an approach that has proven to be general, flexible, and powerful in treating both radiative transfer and its coupling to the constraints of energy and momentum balance and to the equations of statistical equilibrium. For the present we consider only time-independent transfer in static media; we extend the method to other cases in later chapters.

THE PROBLEM OF SCATTERING

An important obstacle encountered in solving transfer problems is the scattering term, which decouples the radiation field from local sources and sinks, and introduces global transport of photons over large distances. This term permits an open boundary to make itself felt at great depth ($\tau_\nu \gg 1$) in the medium, and allows J_ν to depart significantly from B_ν even at depths where one would have expected them to be identical.

Scattering terms may appear explicitly in the source function. For example, S_ν may have the form [cf. (77.8)]

$$S_\nu = \xi_\nu B_\nu + (1-\xi_\nu) J_\nu \tag{83.1}$$

where ξ_ν is the *thermal coupling parameter*

$$\xi_\nu = \kappa_\nu/(\kappa_\nu + \sigma_\nu). \tag{83.2}$$

In astrophysical media, ξ_ν can sometimes be quite small. For instance, in hot stellar atmospheres Thomson scattering can be the dominant opacity source, and ξ_ν may be of order 10^{-4} into very deep layers of the atmosphere. In spectral lines, the thermalization parameter ε in (77.11) can be very small, and because $r \ll 1$, ξ_ν will also be small, say 10^{-8} or even less.

To see the implications of dominant scattering terms, consider the following simplified problem. Suppose the depth variation of the Planck function is

$$B_\nu = a_\nu + b_\nu \tau_\nu, \tag{83.3}$$

and that ξ_ν is depth independent. Using (83.1) in the zeroth moment equation (78.15) we have

$$(\partial H_\nu/\partial \tau_\nu) = \xi_\nu (J_\nu - B_\nu), \tag{83.4}$$

and the first-order moment equation is

$$(\partial K_\nu/\partial \tau_\nu) = H_\nu. \tag{83.5}$$

Making the Eddington approximation $K_\nu = \frac{1}{3}J_\nu$ we can combine (83.4) and (83.5) into

$$\tfrac{1}{3}(\partial^2 J_\nu/\partial \tau_\nu^2) = \xi_\nu (J_\nu - B_\nu). \tag{83.6}$$

Because $(\partial^2 B_\nu/\partial \tau_\nu^2)$ from (83.3) is zero, we can replace J_ν in the second derivative by $(J_\nu - B_\nu)$. Solving, we obtain

$$J_\nu = B_\nu + \alpha_\nu \exp[-(3\xi_\nu)^{1/2}\tau_\nu] + \beta_\nu \exp[(3\xi_\nu)^{1/2}\tau_\nu]. \tag{83.7}$$

The unknown constants α_ν and β_ν are determined by boundary conditions. At great depth we demand $J_\nu \to B_\nu$, hence $\beta_\nu \equiv 0$. At the surface we use the Eddington–Krook boundary condition $J_\nu(0) = \sqrt{3}\, H_\nu(0)$ (which is consistent with the exact solution of the grey problem). But from (83.5) we can also write $H_\nu(0) = \frac{1}{3}(\partial J_\nu/\partial \tau_\nu)_0$; evaluating the derivative from (83.7) and applying the boundary condition we find α_ν, and thus obtain finally

$$J_\nu(\tau_\nu) = a_\nu + b_\nu \tau_\nu + (b_\nu - \sqrt{3}\, a_\nu) \exp[-(3\xi_\nu)^{1/2}\tau_\nu]/[\sqrt{3}\,(1+\xi_\nu^{1/2})]. \tag{83.8}$$

Equation (83.8) reveals the essential physics of the problem. For simplicity consider an isothermal medium ($b_\nu \equiv 0$). First, we see that J_ν can depart markedly from B_ν at the surface; $J_\nu(0) = \xi_\nu^{1/2} B_\nu/(1+\xi_\nu^{1/2}) \approx \xi_\nu^{1/2} B_\nu$ for $\xi_\nu \ll 1$. Thus for $\xi_\nu \ll 1$, $J_\nu(0) \ll B_\nu(0)$. Second, this departure extends to great depths in the medium. The slow decay of the exponential in (83.8) implies that $J_\nu \to B_\nu$ only when $\tau_\nu \gtrsim (\xi_\nu)^{-1/2}$; the values ξ_ν quoted above show that this *thermalization depth* can be orders of magnitude larger than $\tau_\nu \sim 1$.

We can easily understand why the thermalization depth is so large from physical considerations. The parameter ξ_ν is essentially the probability per scattering event that a photon is thermally destroyed. To assure thermalization, the photon must scatter on the order of $n \sim 1/\xi_\nu$ times. Because the photon executes a random walk, in n scatterings it will travel $n^{1/2} \sim \xi_\nu^{-1/2}$ mean free paths in any direction, in particular toward the boundary surface. Thus when the point of photon emission is at a depth $\tau_\nu \gtrsim \xi_\nu^{-1/2}$ the probability that the photon will be thermalized before it escapes approaches unity, hence $J_\nu \rightarrow B_\nu$.

These results have important implications for developing numerical methods that can handle scattering terms successfully. For example, suppose that instead of solving the problem as we did above, we decided to compute $J_\nu = \Lambda(S_\nu)$; because S_ν contains J_ν, this is an integral equation. To avoid solving the integral equation directly we might try to proceed by *iteration.* Suppose we take $J_\nu = B_\nu$ as a first guess. Using $S_\nu = B_\nu$ we compute a new estimate of J_ν; using this J_ν we then calculate a new value of S_ν, re-evaluate the lambda operator for J_ν, and iterate. We saw in §79 that the kernel of Λ has an exponential falloff, which implies an effective information-propagation range of only $\Delta\tau \sim 1$; in the present problem one would therefore need to perform at least $\xi_\nu^{-1/2}$ iterations (a large number!) to propagate information about the existence of the surface [and the large departure of $J_\nu(0)$ from $B_\nu(0)$] over the entire thermalization depth. This failure also applies to all other equivalent methods that attempt to find S_ν by an iteration like the one just described (which we refer to generically as "Λ iteration").

We were able to obtain the correct solution (83.8) because we solved the problem by an analytical method that dealt *explicitly* with the scattering term. We conclude that any successful numerical method must do likewise, that is, scattering terms must appear explicitly in the source function, and the method must solve the resulting equations directly. In fact, the situation can be even more complex than we have indicated because scattering terms can be "hidden" in other aspects of the problem. For example, if we demand radiative equilibrium, the requirement that $\int \chi_\nu S_\nu \, d\nu = \int \chi_\nu J_\nu \, d\nu$ imposes a coupling of S_ν to J_ν that implies that S_ν will behave like a scattering term; this is seen clearly in the grey problem where $S \equiv J$, and the transfer equation (82.5) is the equivalent of a *pure scattering problem* (i.e., $\xi_\nu \equiv 0$). Furthermore, we will see in §87 that when we drop the assumption of LTE, the equations of statistical equilibrium also imply the presence of scattering terms in the source function.

The discussion leading to (83.7) and (83.8) also shows that in solving transfer problems we must deal with the *two-point boundary conditions* posed by (79.2) to (79.4); the same will be true for any numerical method. One might try to evade this problem by integrating the transfer equation from, say, the deepest point in the medium to the surface, using as initial conditions a guess for $I^-(\tau_{\max}, \mu)$ at the lower boundary, along with $I^+(\tau_{\max}, \mu)$ given by the lower boundary condition. However, unless the guess for I^- is perfect, the values computed for $I^-(0, \mu)$ will not agree with those specified by the upper boundary condition. One must then try to make a new guess for $I^-(\tau_{max})$ that would yield improved agreement between the computed and imposed surface values of I^-.

Not only is this *eigenvalue method* $[I^-(\tau_{max}, \mu_i)$ is to be determined for all $\mu_i < 0]$ inefficient, it is also strongly unstable. Note that the solution (83.7) contains both ascending and descending exponentials; the latter is the true solution, the former is a parasite. Unless we suppress the parasite [as we did analytically in deriving (83.8)], it tends to run away and swamp the true solution. To avoid such problems, our adopted numerical method must explicitly treat the two-point boundary-value nature of the transfer problem from the outset [see (**M2,** 150–151) for further discussion].

SECOND-ORDER FORM OF THE TRANSFER EQUATION

A powerful computational method for solving transfer problems is based on a second-order form of the transfer equation. Consider a static, planar medium. Choose the column mass $dm = -\rho\, dz$ or

$$m(z) = \int_z^{z_{max}} \rho(z')\, dz' \tag{83.9}$$

as independent variable; this is a Lagrangean variable suitable for dynamical calculations. Note that m increases downward into the medium; the opposite convention is used in stellar structure calculations where M_r increases outward with radius, cf. §80. Let

$$\omega(\nu) \equiv \chi(\nu)/\rho \tag{83.10}$$

be the opacity per gram, in terms of which we write optical depth increments as $d\tau_\nu = \omega_\nu\, dm$.

For radiation moving in two antiparallel pencils $\pm\mu$, we have two transfer equations

$$\pm\mu[\partial I(\pm\mu, \nu)/\partial m] = \omega(\nu)[I(\pm\mu, \nu) - S(\nu)], \tag{83.11}$$

where for brevity we suppress explicit mention of m dependence. As in §81 define a mean-intensity-like variable

$$j(\mu, \nu) \equiv \tfrac{1}{2}[I(+\mu, \nu) + I(-\mu, \nu)], \qquad (0 \le \mu \le 1), \tag{83.12}$$

and a fluxlike variable

$$h(\mu, \nu) \equiv \tfrac{1}{2}[I(+\mu, \nu) - I(-\mu, \nu)], \qquad (0 \le \mu \le 1). \tag{83.13}$$

Then adding the two equations (83.11) for $\pm\mu$, we obtain

$$\mu[\partial h(\mu, \nu)/\partial m] = \omega(\nu)[j(\mu, \nu) - S(\nu)], \tag{83.14}$$

and subtracting them we have

$$\mu[\partial j(\mu, \nu)/\partial m] = \omega(\nu) h(\mu, \nu). \tag{83.15}$$

These equations, first derived by P. Feautrier (**F1**), strongly resemble the zeroth and first moments of the transfer equation, but with two important differences: (1) they contain only j and h (i.e., the system closes), and (2) they are angle dependent.

Using (83.15) we can eliminate h from (83.14), obtaining the second-order equation

$$\frac{\mu^2}{\omega_\nu}\frac{\partial}{\partial m}\left[\frac{1}{\omega_\nu}\frac{\partial j(\mu, \nu)}{\partial m}\right] = j(\mu, \nu) - S(\nu). \tag{83.16}$$

or, in abbreviated notation,

$$\mu^2(\partial^2 j_{\mu\nu}/\partial \tau_\nu^2) = j_{\mu\nu} - S_\nu, \qquad (0 \le \mu \le 1). \tag{83.17}$$

These equations, supplemented by boundary conditions, can be solved by efficient numerical algorithms to be discussed shortly. We stress that in solving (83.16) and (83.17), any scattering terms in S_ν arc to be written out *explicitly*; this introduces integrals of $j_{\mu\nu}$ over angle (and sometimes over frequency) on the right-hand side. Having solved (83.17) for $j(\mu, \nu)$ we can find $h(\mu, \nu)$ from (83.15).

An important property of (83.14) to (83.17) is that they are accurate in the diffusion regime. Thus if we integrate (83.17) over μ we find

$$J_\nu = S_\nu + (\partial^2 K_\nu/\partial \tau_\nu^2), \tag{83.18}$$

which in the diffusion regime yields

$$J_\nu = B_\nu + \tfrac{1}{3}(\partial^2 B_\nu/\partial \tau_\nu^2), \tag{83.19}$$

agreeing with (80.3) to second order. Similarly the integral over angle of μ times (83.15) yields the first term of (80.4) in the diffusion limit; as the next term is of third order, this implies the flux is accurate to second order. Experience shows that (83.14) to (83.17) behave correctly at both small and large optical depths.

BOUNDARY CONDITIONS

To obtain a unique solution of (83.17) we must impose boundary conditions at $\tau = 0$ and $\tau = \tau_{\max}$. At $\tau_\nu = 0$, we usually set $I(-\mu, \nu) \equiv 0$; hence from (83.12) and (83.13) $h_{\mu\nu}(0) \equiv j_{\mu\nu}(0)$, and (83.15) then yields

$$\mu(\partial j_{\mu\nu}/\partial \tau_\nu)_0 = j_{\mu\nu}(0). \tag{83.20}$$

In a finite slab, at $\tau = \tau_{\max}$ we specify $I(\tau_{\max}, +\mu, \nu) = I^+_{\mu\nu}$; from the identity $h_{\mu\nu}(\tau_{\max}) \equiv I^+_{\mu\nu} - j_{\mu\nu}(\tau_{\max})$, (83.15) then becomes

$$\mu(\partial j_{\mu\nu}/\partial \tau_\nu)_{\tau_{\max}} = I^+_{\mu\nu} - j_{\mu\nu}(\tau_{\max}). \tag{83.21}$$

Equations (83.17), (83.20), and (83.21) are sufficient to specify the run of $j_{\mu\nu}$ with depth in the slab.

In a semi-infinite medium we can impose the diffusion approximation at $\tau_{\max}$ and take $I(\tau_{\max}, \mu, \nu) = B_\nu(\tau_{\max}) + \mu(\partial B_\nu/\partial \tau_\nu)_{\tau_{\max}}$, whence $j_{\mu\nu} = B_\nu(\tau_{\max})$

and $h_{\mu\nu}(\tau_{\max}) = \mu(\partial B_\nu/\partial \tau_\nu)_{\tau_{\max}}$; (83.21) then reduces to

$$(\partial j_{\mu\nu}/\partial \tau_\nu)_{\tau_{\max}} = (\partial B_\nu/\partial \tau_\nu)_{\tau_{\max}} = [(\partial B_\nu/\partial T)(dT/dm)/\omega_\nu]_{\tau_{\max}}. \tag{83.22}$$

Alternatively we may wish to specify the total flux H transported across the lower boundary. In the diffusion limit we have

$$H = \int_0^\infty d\nu \int_0^1 d\mu\, \mu h_{\mu\nu} = (dT/dm) \int_0^\infty d\nu \omega_\nu^{-1}(\partial B_\nu/\partial T) \int_0^1 d\mu\, \mu^2, \tag{83.23a}$$

or

$$(dT/dm) = 3H \Big/ \int_0^\infty \omega_\nu^{-1}(\partial B_\nu/\partial T)\, d\nu; \tag{83.23b}$$

hence from (83.22)

$$\left.\frac{\partial j_{\mu\nu}}{\partial \tau_\nu}\right|_{\tau_{\max}} = \left[\frac{3H}{\omega_\nu}\left(\frac{\partial B_\nu}{\partial T}\right) \Big/ \int_0^\infty \frac{1}{\omega_\nu}\frac{\partial B_\nu}{\partial T}\, d\nu\right]_{\tau_{\max}} \tag{83.24}$$

Equations (83.17), (83.20), and either (83.22) or (83.24) are sufficient to specify the run of $j_{\mu\nu}$ with depth in a semi-infinite medium.

DISCRETIZATION

We now convert the differential equation (83.17) and its accompanying boundary conditions into *difference equations* by a discretization of all variables. We choose a discrete set of angle points $\{\mu_m\}$, $(m = 1, \ldots, M)$, and frequency points $\{\nu_n\}$, $(n = 1, \ldots, N)$, spanning the ranges $0 \leq \mu \leq 1$ and $0 \leq \nu \leq \infty$. We divide the medium into a set of D mass shells whose boundaries are specified by the mesh $\{m_d\}$, $(d = 1, \ldots, D+1)$; each cell has a mass $m_{d+(1/2)} = m_{d+1} - m_d$. In general, the mass cells will be of unequal size. Variables whose values are specified at cell centers are given half-integral indices [e.g., $j_{d+(1/2),mn} \equiv j(m_{d+(1/2)}, \mu_m, \nu_n)$]. Variables specified on cell surfaces are given integer indices [e.g., $h_{dmn} \equiv h(m_d, \mu_m, \nu_n)$]; see Figure 83.1.

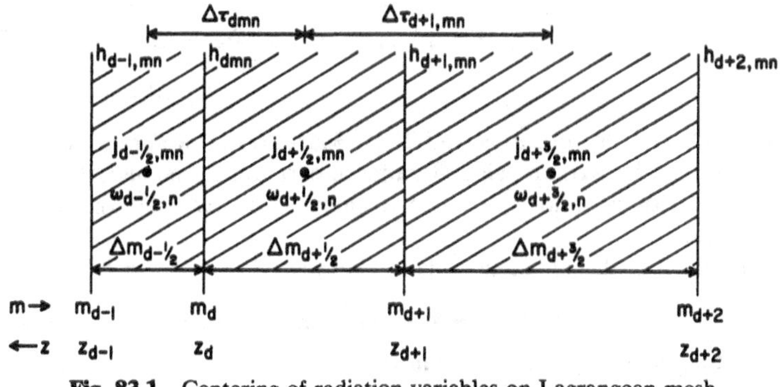

Fig. 83.1 Centering of radiation variables on Lagrangean mesh.

Integrals are replaced by quadrature sums, for example,

$$\int_0^\infty y(\nu)\,d\nu \to \sum_{n=1}^{N} a_n y(\nu_n) \tag{83.25a}$$

and

$$\int_0^1 y(\mu)\,d\mu \to \sum_{m=1}^{M} b_m y(\mu_m). \tag{83.25b}$$

For example, source functions such as (77.8) and (77.11) are represented as

$$S_{d+(1/2),n} = \alpha_{d+(1/2),n} \sum_{k=1}^{K} w_k j_{d+(1/2),k} + \beta_{d+(1/2),n}. \tag{83.26}$$

The first term in (83.26) is the scattering integral, and the second is the thermal term. Here we have grouped all combinations of angles and frequencies into a single serial set with index $k = 1, \ldots, K = MN$ [e.g., $(\mu_k, \nu_k) \equiv (\mu_m, \nu_n)$ where $k = m + (n-1)M$], and w_k is a combined weight for both angle and frequency quadratures, which may include a profile function.

Similarly, derivatives are replaced by difference formulae, for example,

$$(d^2x/d\tau^2)_{d+(1/2)} = [(dx/d\tau)_{d+1} - (dx/d\tau)_d]/\Delta\tau_{d+(1/2)}, \tag{83.27}$$

where

$$(dx/d\tau)_d = [x_{d+(1/2)} - x_{d-(1/2)}]/\Delta\tau_d. \tag{83.28}$$

Thus defining

$$\Delta\tau_{dk} \equiv \tfrac{1}{2}[\omega_{d-(1/2),k}\,\Delta m_{d-(1/2)} + \omega_{d+(1/2),k}\,\Delta m_{d+(1/2)}], \tag{83.29}$$

and

$$\Delta\tau_{d+(1/2),k} \equiv \tfrac{1}{2}(\Delta\tau_{dk} + \Delta\tau_{d+1,k}), \tag{83.30}$$

a second-order accurate difference representation of (83.17) is

$$\frac{1}{\Delta\tau_{d+(1/2),k}}\left[\left(\frac{\mu_k^2}{\Delta\tau_{d+1,k}}\right)j_{d+(3/2),k} - \mu_k^2\left(\frac{1}{\Delta\tau_{dk}} + \frac{1}{\Delta\tau_{d+1,k}}\right)j_{d+(1/2),k} + \left(\frac{\mu_k^2}{\Delta\tau_{dk}}\right)j_{d-(1/2),k}\right] = j_{d+(1/2),k} - S_{d+(1/2),k}, \tag{83.31}$$

$$(k = 1, \ldots, K),\ (d = 2, \ldots, D-1)$$

where $S_{d+(1/2),k}$ is given by (83.26). [Fourth-order accurate difference equations can be written using Hermite integration formulae (**A6**).] These $(D-2)$ sets of equations must be augmented by two sets of boundary conditions.

Consider first the upper boundary. From (83.14) we have

$$h_{2k} = h_{1k} + [\Delta\tau_{(3/2)k}/\mu_k][j_{(3/2)k} - S_{(3/2)k}], \tag{83.32}$$

where $\Delta\tau_{(3/2)k} \equiv \omega_{(3/2)k}\,\Delta m_{3/2}$. From (83.15),

$$h_{2k} = \mu_k[j_{(5/2)k} - j_{(3/2)k}]/\Delta\tau_{2k}. \tag{83.33}$$

Applying (83.20) from the surface of the uppermost cell to its center we have

$$j_{1k} \equiv h_{1k} = \mu_k[j_{(3/2)k} - j_{1k}]/\tfrac{1}{2}\,\Delta\tau_{(3/2)k}, \tag{83.34}$$

whence

$$j_{1k} = h_{1k} = j_{(3/2)k}/(1 + \tfrac{1}{2}\,\Delta\tau_{(3/2)k}/\mu_k). \tag{83.35}$$

Using (83.33) and (83.35) in (83.32) we obtain the desired boundary condition

$$\mu_k[j_{(5/2)k} - j_{(3/2)k}]/\Delta\tau_{2k} = j_{(3/2)k}/[1 + \tfrac{1}{2}\Delta\tau_{(3/2)k}/\mu_k] + [\Delta\tau_{(3/2)k}/\mu_k][j_{(3/2)k} - S_{(3/2)k}]. \tag{83.36}$$

To obtain a lower boundary condition, we can use (83.21), (83.22), or (83.24). In particular, (83.24) implies

$$\frac{[j_{D+(1/2),k} - j_{D-(1/2),k}]}{\Delta\tau_{Dk}} = \frac{3H}{\omega_{D+(1/2),k}}\left(\frac{\partial B_k}{\partial T}\right)_{D+(1/2)} \Bigg/ \left[\sum_{n=1}^{N} \frac{a_n}{\omega_{D+(1/2),n}}\left(\frac{\partial B_n}{\partial T}\right)_{D+(1/2)}\right]. \tag{83.37}$$

Equation (83.37) is only of first-order accuracy, but this is usually sufficient in the diffusion regime. Using Hermite formulae it is possible to write third-order accurate boundary conditions (**A6**).

Equations (83.31), (83.36), and (83.37) comprise DK equations in the same number of unknowns $\{j_{d+(1/2),k}\}$; let us now consider how they are to be solved.

THE FORMAL SOLUTION

The simplest transfer problem, called the *formal solution*, is to calculate the radiation field at all depths, angles, and frequencies when the source function is known [e.g., in an LTE medium ($S_\nu \equiv B_\nu$)] whose temperature structure is given. The formal solution is also used to evaluate variable Eddington factors in certain iterative procedures discussed below.

Represent the depth variation of j for a particular angle and frequency (μ_k, ν_k) by the column vector

$$\mathbf{j}_k \equiv [j_{(3/2)k}, j_{(5/2)k}, \ldots, j_{D+(1/2),k}], \tag{83.38}$$

and the (known) depth variation of $S(\nu_k)$ by

$$\mathbf{S}_k \equiv [S_{(3/2)k}, S_{(5/2)k}, \ldots, S_{D+(1/2),k}]. \tag{83.39}$$

Then the transfer equation (83.31) and boundary conditions (83.36) and (83.37) for $j_{d+(1/2),k}$ $(d = 1, \ldots, D)$ are of the form

$$\mathbf{T}_k\mathbf{j}_k = \mathbf{S}_k, \tag{83.40}$$

where $\mathbf{T}_k$ is a $(D \times D)$ *tridiagonal* matrix.

The solution of (83.40) is effected by a standard Gaussian elimination scheme [cf. (83.52) to (83.54)], which requires of the order of cD operations, where c is a numerical constant of order unity. If we have a total of K angle-frequency choices, the computational effort to obtain the full radiation field scales as $cDK = cDMN$, which is irreducible because we wish to determine DK values of j. On vector computers, several systems of the form (83.40) can be solved in parallel, resulting in an enormous increase in efficiency.

Having determined $j_{d+(1/2),k}$ for all d and k we can calculate the moments

$$J_{d+(1/2),n} \equiv \sum_m b_m j_{d+(1/2),mn} \tag{83.41}$$

and

$$K_{d+(1/2),n} = \sum_m b_m \mu_m^2 j_{d+(1/2),mn}, \tag{83.42}$$

and hence the Eddington factor $f_{d+(1/2),n} \equiv [K_{d+(1/2),n}/J_{d+(1/2),n}]$ for $d = 1, \ldots, D$. Using the difference representation of (83.15) we can calculate h_{dk} and hence fluxcs on the cell boundaries

$$H_{dn} = \sum_m b_m \mu_m h_{dmn}, \tag{83.43}$$

for $d = 2, \ldots, D$. The flux at the lower boundary is fixed by the lower boundary condition. At the upper boundary we calculate j_{1k} using (83.35), and can then evaluate

$$J_{1n} = \sum_m b_m j_{1mn} \equiv \mathscr{j}_{1n} J_{(3/2)n}, \tag{83.44}$$

$$H_{1n} = \sum_m b_m \mu_m j_{1mn} \equiv \mathscr{h}_{1n} J_{1n}, \tag{83.45}$$

and

$$K_{1n} = \sum_m b_m \mu_m^2 j_{1mn} = f_{1n} J_{1n}. \tag{83.46}$$

The geometrical factors $\mathscr{j}_{1n}$ and $\mathscr{h}_{1n}$ will be used to pose boundary conditions for the radiation moment equations [see (83.61)].

THE FEAUTRIER METHOD

A more realistic transfer problem is to calculate the radiation field when the source function explicitly contains scattering terms as in (83.26). We solve such problems by two different methods. In the *Feautrier method* (**F1**), one defines vectors

$$\mathbf{j}_{d+(1/2)} \equiv [j_{d+(1/2),1}, j_{d+(1/2),2}, \ldots, j_{d+(1/2),K}], \qquad (d = 1, \ldots, D), \tag{83.47}$$

containing all angle-frequency components of the radiation field at a single depth, and vectors

$$\mathbf{L}_{d+(1/2)} \equiv [\beta_{d+(1/2),1}, \beta_{d+(1/2),2}, \ldots, \beta_{d+(1/2),K}], \qquad (d = 1, \ldots, D), \tag{83.48}$$

containing the angle-frequency components of the thermal source term.

The transfer equation (83.31) plus boundary conditions (83.36) and (83.37) can then be written as a *block tridiagonal* system of the form

$$-\mathbf{A}_{d+(1/2)}\mathbf{j}_{d-(1/2)}+\mathbf{B}_{d+(1/2)}\mathbf{j}_{d+(1/2)}-\mathbf{C}_{d+(1/2)}\mathbf{j}_{d+(3/2)} = \mathbf{L}_{d+(1/2)}, \qquad (d=1,\ldots,D), \tag{83.49}$$

The matrices **A**, **B**, and **C** are all of dimension $(K\times K)$; **A** and **C** are diagonal, containing portions of the finite-difference operator for all angle-frequency points down the diagonal. **B** is a full matrix containing terms from the difference operator down the diagonal, plus diagonal and off-diagonal terms coupling radiation at each angle-frequency point to all others through the quadrature sum for the scattering integral in (83.26). [If Hermite formulae (**A6**) are used, **A** and **C** are also full.] The upper boundary condition implies that $\mathbf{A}_{3/2}\equiv 0$, and the lower boundary condition implies $\mathbf{C}_{D+(1/2)}\equiv 0$.

Equations (83.49) are solved by Gaussian elimination. At level $d+\frac{1}{2}$ we express $\mathbf{j}_{d+(1/2)}$ in terms of $\mathbf{j}_{d+(3/2)}$ and use this expression to eliminate $\mathbf{j}_{d+(1/2)}$ from the next equation. Starting at the upper boundary we have

$$\mathbf{j}_{3/2}=(\mathbf{B}_{3/2}^{-1}\mathbf{C}_{3/2})\mathbf{j}_{5/2}+(\mathbf{B}_{3/2}^{-1}\mathbf{L}_{3/2})\equiv\mathbf{D}_{3/2}\mathbf{j}_{5/2}+\boldsymbol{\nu}_{5/2}. \tag{83.50}$$

Substituting (83.50) into (83.49) for $d=2$ we find

$$\mathbf{j}_{5/2}=(\mathbf{B}_{5/2}-\mathbf{A}_{5/2}\mathbf{D}_{3/2})^{-1}\mathbf{C}_{5/2}\mathbf{j}_{7/2}+(\mathbf{B}_{5/2}-\mathbf{A}_{5/2}\mathbf{D}_{3/2})^{-1}(\mathbf{L}_{5/2}+\mathbf{A}_{5/2}\boldsymbol{\nu}_{3/2}) \equiv\mathbf{D}_{5/2}\mathbf{j}_{7/2}+\boldsymbol{\nu}_{5/2}, \tag{83.51}$$

and in general we have

$$\mathbf{j}_{d+(1/2)}=\mathbf{D}_{d+(1/2)}\mathbf{j}_{d+(3/2)}+\boldsymbol{\nu}_{d+(1/2)}, \tag{83.52}$$

where

$$\mathbf{D}_{d+(1/2)}\equiv[\mathbf{B}_{d+(1/2)}-\mathbf{A}_{d+(1/2)}\mathbf{D}_{d-(1/2)}]^{-1}\mathbf{C}_{d+(1/2)}, \tag{83.53}$$

and

$$\boldsymbol{\nu}_{d+(1/2)}\equiv[\mathbf{B}_{d+(1/2)}-\mathbf{A}_{d+(1/2)}\mathbf{D}_{d-(1/2)}]^{-1}[\mathbf{L}_{d+(1/2)}+\mathbf{A}_{d+(1/2)}\boldsymbol{\nu}_{d-(1/2)}]. \tag{83.54}$$

We compute $\mathbf{D}_{d+(1/2)}$ and $\boldsymbol{\nu}_{d+(1/2)}$ for $d=1$ through $d=D-1$. At $d=D$, $\mathbf{C}_{D+(1/2)}=0$, and $\mathbf{j}_{D+(1/2)}\equiv\boldsymbol{\nu}_{D+(1/2)}$. Having found $\mathbf{j}_{D+(1/2)}$ we obtain all other $\mathbf{j}_{d+(1/2)}$ for $d=D-1, D-2,\ldots,1$ from (83.52) by successive back substitution. From $\mathbf{j}_{d+(1/2)}$, one can evaluate $J_{d+(1/2),n}$ and $S_{d+(1/2),n}$ using appropriate angle (or angle-frequency) quadratures. The forward-backward sweep enforces the two-point boundary conditions, and the explicit appearance of scattering terms guarantees correct thermalization. The method is computationally robust. To estimate the computational effort, we note that a solution of a full linear system of order n requires $O(n^3)$ operations; there are D such systems, hence the total effort scales as $cDK^3=cDM^3N^3$.

VARIABLE EDDINGTON FACTORS

In the Feautrier method, the scaling of the computational effort as the cube of the product of the number of angles and frequencies is very unfavorable.

Clearly it is essential to eliminate any angle-frequency information about the radiation field that is not absolutely necessary. In this vein, we note that typical scattering kernels in static media (or in the fluid frame of moving media) are isotropic, hence only J_ν, not $j_{\mu\nu}$, enters. This simplification holds even when the transfer equation is coupled to constraints of energy and momentum balance, and to the statistical equilibrium equations (cf. §88). We therefore eliminate the superfluous angular information by using moments of the transfer equation, closing the system by use of variable Eddington factors.

Thus, integrating (83.17), (83.20), and (83.24) over angle we obtain

$$\partial^2(f_\nu J_\nu)/\partial\tau_\nu^2 = J_\nu - S_\nu, \tag{83.55}$$

$$[\partial(f_\nu J_\nu)/\partial\tau_\nu]_0 = H_\nu(0), \tag{83.56}$$

and

$$[\partial(f_\nu J_\nu)/\partial\tau_\nu]_{\tau_{max}} = \left[(H/\omega_\nu)(\partial B_\nu/\partial T)\Big/\int_0^\infty \omega_\nu^{-1}(\partial B_\nu/\partial T)\,d\nu\right]_{\tau_{max}}. \tag{83.57}$$

The finite difference form of (83.55) is

$$\frac{1}{\Delta\tau_{d+(1/2),n}}\left[\frac{f_{d-(1/2),n}J_{d-(1/2),n}}{\Delta\tau_{dn}} - \left(\frac{1}{\Delta\tau_{dn}}+\frac{1}{\Delta\tau_{d+1,n}}\right) f_{d+(1/2),n}J_{d+(1/2),n} + \frac{f_{d+(3/2),n}J_{d+(3/2),n}}{\Delta\tau_{d+1,n}}\right] = J_{d+(1/2),n} - S_{d+(1/2),n}, \qquad (d=2,\ldots,D-1). \tag{83.58}$$

The lower boundary condition can be represented as

$$\frac{f_{D+(1/2),n}J_{D+(1/2),n} - f_{D-(1/2),n}J_{D-(1/2),n}}{\Delta\tau_{Dn}} = \frac{3H}{\omega_{D+(1/2),n}}\left(\frac{\partial B_n}{\partial T}\right)_{D+(1/2)}\Big/\sum_n \frac{a_n}{\omega_{D+(1/2),n}}\left(\frac{\partial B_n}{\partial T}\right)_{D+(1/2)}. \tag{83.59}$$

By an analysis similar to that leading from (83.32) to (83.36) one finds that

$$H_{1n} = h_{1n}J_{1n} = h_{1n}f_{(3/2)n}J_{(3/2)n}/(f_{1n} + \tfrac{1}{2}\Delta\tau_{(3/2)n}h_{1n}); \tag{83.60}$$

hence the upper boundary condition can be written as

$$[f_{(5/2)n}J_{(5/2)n} - f_{(3/2)n}J_{(3/2)n}]/\Delta\tau_{2n} = h_{1n}f_{(3/2)n}J_{(3/2)n}/[f_{1n} + \tfrac{1}{2}\Delta\tau_{(3/2)n}h_{1n}] + \Delta\tau_{(3/2)n}[J_{(3/2)n} - S_{(3/2)n}]. \tag{83.61}$$

Equations (83.58) to (83.61) are of the same form as (83.49) and are solved by the same Gaussian elimination scheme. But now the computational effort scales as cDN^3, lower by a factor of M^3; in typical calculations $M=3$ or 4.

To solve (83.58) to (83.61) we must know f_ν at all depths and $h_{1\nu}$ at the surface. We proceed iteratively: (1) From a first guess for S_ν (e.g., $S_\nu = B_\nu$)

we carry out a formal solution, which yields $j_{\mu\nu}$ at all angles, frequencies, and depths. (2) We then calculate f_ν at all depths from (83.41) and (83.42), and $\mathscr{j}_{1\nu}$ and $\mathscr{k}_{1\nu}$ from (83.44) and (83.45). The essential point is that the Eddington factors are determined with substantially better accuracy than the radiation field itself because they are *shape factors* that depend only on the *ratio* of radiation moments. For instance, local scale-factor errors in J_ν and K_ν simply drop out. (3) Given f_ν and $\mathscr{k}_{1\nu}$, we solve (83.58) to (83.61) using expressions for S_ν in which scattering terms appear explicitly. In this step, one obtains the correct thermalization properties of J_ν. (4) We then re-evaluate S_ν using the new value of J_ν; this updated S_ν will, in general, differ from the original S_ν. We therefore recalculate new f_ν's via step (1), and iterate to convergence.

If I iterations are required to achieve convergence, the total computing effort scales as $I(cDMN+c'DN^3) \ll c''DM^3N^3$ for moderate values of M and I. Experience shows that the Eddington-factor iteration generally converges rapidly ($I \sim 3$), so that substantial savings are realized.

THE RYBICKI METHOD

An alternative to the Feautrier method was devised by Rybicki (**R1**). In Feautrier's method, all frequency-dependent information is grouped together at each depth, and the solution proceeds depth by depth; this method can handle an explicit frequency dependence of the scattering terms, such as those that arise in partial redistribution problems. But in many problems the scattering term is independent of frequency; for example, in line formation with complete redistribution only the quantity $\bar{J} \equiv \int \phi_\nu J_\nu \, d\nu$ appears. In such cases the frequency information retained by the Feautrier method is redundant, and Rybicki showed that the system can be reorganized in a way that has more favorable computing-time requirements.

Assume that the source function has the form $S_\nu = \alpha_\nu \bar{J} + \beta_\nu$, and let

$$\bar{\mathbf{J}} \equiv [\bar{J}_{3/2}, \bar{J}_{5/2}, \ldots, \bar{J}_{D+(1/2)}] \tag{83.62}$$

represent the run of $\bar{J}$ with depth. Then at each angle-frequency point k the transfer equation has the form [cf. (83.40)]

$$\mathbf{T}_k \mathbf{j}_k + \mathbf{U}_k \bar{\mathbf{J}} = \mathbf{K}_k, \qquad (k = 1, \ldots, K), \tag{83.63}$$

where $\mathbf{T}_k$ is a $(D \times D)$ tridiagonal matrix representing the differential operator, $\mathbf{U}_k$ is a $(D \times D)$ diagonal matrix containing the depth variation of the coefficient $\alpha_{d+(1/2),k}$ of the scattering term, and $\mathbf{K}_k$ is a vector of length D containing the depth variation of the thermal source term $\beta_{d+(1/2),k}$. In addition we have D equations that define $\bar{J}_{d+(1/2)}$, namely

$$\bar{J}_{d+(1/2)} = \sum_{k=1}^{K} w_{d+(1/2),k} j_{d+(1/2),k}, \qquad (d = 1, \ldots, D). \tag{83.64}$$

which are equivalent to the matrix equation

$$\bar{\mathbf{J}} = \sum_{k=1}^{K} \mathbf{V}_k \mathbf{j}_k \tag{83.65}$$

where each $\mathbf{V}_k$ is a $(D \times D)$ diagonal matrix containing the depth variation of the quadrature weight for angle-frequency point k in (83.64).

The system comprising (83.63) and (83.64) can be solved efficiently. For each k we solve the tridiagonal system (83.63) to find the vector $\mathbf{A}_k$ and the matrix $\mathbf{B}_k$ in

$$\mathbf{j}_k = \mathbf{T}_k^{-1}\mathbf{K}_k - (\mathbf{T}_k^{-1}\mathbf{U}_k)\bar{\mathbf{J}} \equiv \mathbf{A}_k - \mathbf{B}_k\bar{\mathbf{J}}. \tag{83.66}$$

Substituting (83.66) into (83.65) for all k we develop the final system $\mathbf{C}\bar{\mathbf{J}} = \mathbf{D}$, where $\mathbf{C}$ is the full $(D \times D)$ matrix

$$\mathbf{C} = \mathbf{I} + \sum_{k=1}^{K} \mathbf{V}_k \mathbf{B}_k, \tag{83.67}$$

and $\mathbf{D}$ is a vector of length D,

$$\mathbf{D} = \sum_{k=1}^{K} \mathbf{V}_k \mathbf{A}_k. \tag{83.68}$$

We solve the final system for $\bar{\mathbf{J}}$, from which we calculate $\mathbf{S}_k$, the run of the source function at angle-frequency point k over depth. One can then find any desired $\mathbf{j}_k$ from (83.66).

The calculation of each $\mathbf{B}_k$ requires $O(D^2)$ operations, as does the multiplication by $\mathbf{V}_k$ and summation into $\mathbf{C}$. Solution of the final full system requires $O(D^3)$ operations. Hence the total computing effort scales as $cD^2K + c'D^3 = cD^2MN + c'D^3$; in practice the first term usually dominates. We now see the advantage of the Rybicki scheme over the Feautrier scheme: the computing effort varies *linearly* with the number of angle-frequency points, rather than as the cube; therefore Rybicki's method is preferable for problems with large numbers of angles and frequencies. On the other hand, Rybicki's method works only if the scattering term is frequency independent, whereas Feautrier's method can handle partial redistribution. Furthermore, with Feautrier's scheme it is relatively easy to impose constraints of energy and momentum balance, and to couple the transfer equation to the equations of statistical equilibrium (cf. §88), whereas imposition of these additional constraints makes Rybicki's scheme prohibitively costly. One should analyze each problem to determine the relative cost of the two methods, and choose the one that is optimum [see (**M2**, 161) and (**M2**, §12–3)].

SPHERICAL GEOMETRY

While planar geometry is often adequate in astrophysical transfer problems, to study extended envelopes or the structure of a star as a whole we must work in spherical geometry. Using variable Eddington factors, one can write the transfer equation for spherical media in a form that closely resembles the planar equation.

On a radial optical depth scale $d\tau_\nu = -\chi_\nu \, dr$, the moment equations (78.18) and (78.19) are

$$\partial(r^2 H_\nu)/\partial\tau_\nu = r^2(J_\nu - S_\nu) \tag{83.69}$$

and

$$[\partial(f_\nu J_\nu)/\partial\tau_\nu] - (3f_\nu - 1)J_\nu/\chi_\nu r = H_\nu. \tag{83.70}$$

One cannot derive a simple second-order equation by substituting (83.70) for H_ν directly into (83.69). But we can cast the left-hand side of (83.70) into a more convenient form by introducing a *sphericality factor* q_ν defined (**A5**) such that

$$\frac{1}{q_\nu}\frac{\partial(f_\nu q_\nu J_\nu)}{\partial\tau_\nu} \equiv \frac{\partial(f_\nu J_\nu)}{\partial\tau_\nu} - \frac{(3f_\nu - 1)}{\chi_\nu r} J_\nu, \tag{83.71}$$

whence it follows that

$$\ln q_\nu = \int_{r_c}^{r} [(3f_\nu - 1)/r' f_\nu] \, dr', \tag{83.72}$$

where r_c is the *core radius*, that is, the inner boundary of the medium. Note that q_ν is a geometrical factor on the same footing as f_ν, and is determined if f_ν is given.

Using (83.71) we can rewrite (83.70) as

$$\partial(f_\nu q_\nu J_\nu)/\partial\tau_\nu = q_\nu H_\nu, \tag{83.73}$$

which, when substituted into (83.69) yields the *combined moment equation*

$$\frac{1}{q_\nu}\frac{\partial}{\partial\tau_\nu}\left[\frac{r^2}{q_\nu}\frac{\partial(f_\nu q_\nu J_\nu)}{\partial\tau_\nu}\right] = \frac{r^2}{q_\nu}(J_\nu - S_\nu). \tag{83.74}$$

Defining the new variable $dX_\nu \equiv (q_\nu/r^2)\, d\tau_\nu$, we rewrite (83.74) in the second-order form

$$\partial^2(f_\nu q_\nu J_\nu)/\partial X_\nu^2 = (r^4/q_\nu)(J_\nu - S_\nu). \tag{83.75}$$

To obtain an outer boundary condition at $r = R$, define the geometrical factor

$$\hbar_\nu \equiv H_\nu(R)/J_\nu(R) = \int_0^1 I(R, \mu, \nu)\mu \, d\mu \Big/ \int_0^1 I(R, \mu, \nu)\, d\mu. \tag{83.76}$$

Then from (83.73) we have

$$\partial(f_\nu q_\nu J_\nu)/\partial X_\nu \,|_{r=R} = \hbar_\nu R^2 J_\nu(R). \tag{83.77}$$

At the inner boundary we apply the diffusion approximation, fixing the temperature gradient by demanding the correct total flux transport as in

(83.23) and (83.24). We then have

$$\left.\frac{\partial(f_\nu q_\nu J_\nu)}{\partial X_\nu}\right|_{r=r_c} = r_c^2 H_c\left[\frac{1}{\chi_\nu}\left(\frac{\partial B_\nu}{\partial T}\right)\bigg/\int_0^\infty \frac{1}{\chi_\nu}\left(\frac{\partial B_\nu}{\partial T}\right) d\nu\right]_{r=r_c}, \tag{83.78}$$

where $H_c = L/(16\pi^2 r_c^2)$.

Equations (83.75), (83.77), and (83.78) can be discretized on a spherical mesh and solved by the Feautrier scheme. The mesh may represent either a set of surfaces of constant radii, $\{r_i\}$, $(i=1,\ldots,I+1)$, with $r_1 = r_c$ and $r_{I+1} = R$, or a set of mass shells $\{M_i\}$, where M_i is the total mass contained inside r_i. These choices are convenient for Eulerian and Lagrangean calculations, respectively. In the latter, the radiation flux is placed on cell surfaces, and the mean intensity, radiation pressure, and material properties are located at cell centers. The computing effort to solve the system for N frequencies scales as cIN if there is no frequency coupling in the source function, and as cIN^3 if there is.

To carry out the computations just described we must know the Eddington factors. These are determined from a frequency-by-frequency formal solution for given values of S_ν using either of two methods. One method is to solve the transfer equation (76.9) along *rays* tangent to a set of spherical shells (see Figure 83.2). If s measures the path length from the symmetry axis along such a ray, it is easy to show that the differential operator $\mu(\partial/\partial r) + r^{-1}(1-\mu^2)(\partial/\partial\mu)$ is identically $(\partial/\partial s)$. Hence if $I^\pm$ denotes the intensity traveling along $\pm s$, (76.9) becomes

$$\pm[\partial I^\pm(s, p, \nu)/\partial s] = \chi(r, \nu)[S(r, \nu) - I^\pm(s, p, \nu)], \tag{83.79}$$

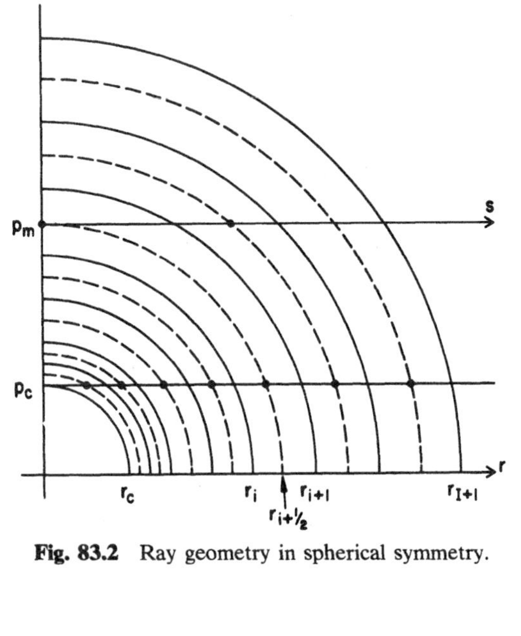

Fig. 83.2 Ray geometry in spherical symmetry.

where p is the impact parameter of the ray and $r=r(s,p)=(s^2+p^2)^{1/2}$. If we define

$$d\tau(s,p,\nu)\equiv-\chi[r(s,p),\nu]\,ds, \tag{83.80}$$

$$j(s,p,\nu)\equiv\tfrac{1}{2}[I^+(s,p,\nu)+I^-(s,p,\nu)], \tag{83.81}$$

and

$$h(s,p,\nu)\equiv\tfrac{1}{2}[I^+(s,p,\nu)-I^-(s,p,\nu)]; \tag{83.82}$$

then the sum and difference of equations (83.79) yield

$$[\partial h(s,p,\nu)/\partial\tau(s,p,\nu)]=j(s,p,\nu)-S(s,p,\nu) \tag{83.83}$$

and

$$[\partial j(s,p,\nu)/\partial\tau(s,p,\nu)]=h(s,p,\nu), \tag{83.84}$$

hence

$$[\partial^2 j(s,p,\nu)/\partial\tau^2(s,p,\nu)]=j(s,p,\nu)-S(s,p,\nu). \tag{83.85}$$

To discretize we choose a set of rays defined by impact parameters $\{p_m\}$, $(m=1,\ldots,M=I+C)$, where the first C rays intersect the core, with $p_C=r_1$, and the remainder are tangent to the spherical surfaces through cell centers [i.e., $p_m=r_{(m-C)+(1/2)}$]. The cell centers are defined such that $r_{i+(1/2)}$ contains half the volume (or mass) between r_i and r_{i+1}, that is,

$$r^3_{i+(1/2)}\equiv\tfrac{1}{2}(r_i^3+r^3_{i+1}). \tag{83.86}$$

The mth ray intersects both cell surfaces, inducing a mesh $\{s_{lm}\}$, and the spherical surfaces through cell centers, inducing a second mesh $s_{l+(1/2),m}$, where

$$s_{lm}=(r_l^2-p_m^2)^{1/2}, \tag{83.87a}$$

and

$$s_{l+(1/2),m}=[r^2_{l+(1/2)}-p_m^2]^{1/2}. \tag{83.87b}$$

For rays inside the core $(p_m\le p_C)$, $\{s_{lm}\}$ is defined for $l=1,\ldots,I+1$, and $\{s_{l+(1/2),m}\}$ for $l=1,\ldots,I$. For rays outside the core, $\{s_{lm}\}$ is defined for $l=m-C+1,\ldots,I+1$, and $\{s_{l+(1/2),m}\}$ for $l=m-C,\ldots,I$. By symmetry, we need to calculate the solution only to the right of the vertical axis in Figure 83.2, and to the right of the core.

We write a difference-equation representation of (83.85) in terms of $j_{i+(1/2),mn}\equiv j[s_{i+(1/2),m},p_m,\nu_n]$, similar to (83.31) with $\mu\equiv1$. We define

$$\Delta\tau_{imn}\equiv\chi_{i-(1/2),n}[s_{im}-s_{i-(1/2),m}]+\chi_{i+(1/2),n}[s_{i+(1/2),m}-s_{im}] \tag{83.88a}$$

and

$$\Delta\tau_{i+(1/2),mn}\equiv\tfrac{1}{2}(\Delta\tau_{imn}+\Delta\tau_{i+1,mn}). \tag{83.88b}$$

For rays outside the core $(m>C)$, we can apply (83.85) at $l=m-C,\ldots,I$. At the axis of symmetry we require that the fictitious quantity $j_{[m-C-(1/2)],mn}\equiv j_{[m-C+(1/2)],mn}$ to obtain an equation containing only $j_{[m-C+(1/2)],mn}$ and $j_{[m-C+(3/2)],mn}$. At the outer boundary we manipulate difference representations of (83.83) and (83.84), using $j(R)\equiv h(R)$ (no incident radiation), as in (83.32) to (83.36) to obtain an equation containing only $j_{I+(1/2),mn}$ and $j_{I-(1/2),mn}$. For rays that intersect the core, we obtain an inner boundary condition by imposing a known flux from the core [see (**M2**, 252–254) for further details].

Along each ray we have a tridiagonal system of order L, where L ranges from 1 to I. Summing over all rays, for all frequencies, we find the total computing effort scales as $c'I^2N$ (assuming $I \gg C$). On vector computers, the equations for all frequencies on a given ray can be solved in parallel.

An alternative way of doing the formal solution in spherical geometry is to develop a finite difference representation of (76.9) with r and μ as independent variables. Suppose we start with the transfer equation in conservation form:

$$3\mu \frac{\partial(r^2 I_\nu)}{\partial(r^3)} + \frac{1}{r}\frac{\partial}{\partial\mu}[(1-\mu^2)I_\nu] = \eta_\nu - \chi_\nu I_\nu. \tag{83.89}$$

Writing (83.89) for $\pm\mu$, and taking the sum and difference of these equations, we have

$$3\mu \frac{\partial(r^2 h_\nu)}{\partial(r^3)} + \frac{1}{r}\frac{\partial}{\partial\mu}[(1-\mu^2)h_\nu] = \eta_\nu - \chi_\nu j_\nu \tag{83.90}$$

and

$$3\mu \frac{\partial(r^2 j_\nu)}{\partial(r^3)} + \frac{1}{r}\frac{\partial}{\partial\mu}[(1-\mu^2)j_\nu] = -\chi_\nu h_\nu, \tag{83.91}$$

where j_ν and h_ν are defined as in (83.12) and (83.13), for $0 \leq \mu \leq 1$.

We discretize (83.90) and (83.91) on a radial mesh with h_ν located at cell surfaces $\{r_i\}$, $(i = 1, \ldots, I+1)$, and j_ν at cell centers $\{r_{i+(1/2)}\}$, $(i = 1, \ldots, I)$, and introduce both a frequency mesh $\{\nu_n\}$ and angular mesh $\{\mu_m\}$; a typical choice for $\{\mu_m\}$ is a double-Gauss quadrature of order M. Equations (83.90) and (83.91) then become

$$3\mu_m(r_{i+1}^2 h_{i+1,mn} - r_i^2 h_{imn})/(r_{i+1}^3 - r_i^3) + [2r_{i+(1/2)}]^{-1} \sum_{m'} D_{mm'}(h_{im'n} + h_{i+1,m'n})$$
$$= \eta_{i+(1/2),n} - \chi_{i+(1/2),n} j_{i+(1/2),mn}, \qquad (m = 1, \ldots, M), \tag{83.92}$$

and

$$3\mu_m[r_{i+(1/2)}^2 j_{i+(1/2),mn} - r_{i-(1/2)}^2 j_{i-(1/2),mn}]/[r_{i+(1/2)}^3 - r_{i-(1/2)}^3]$$
$$+ (2r_i)^{-1} \sum_{m'} D_{mm'}[j_{i-(1/2),m'n} + j_{i+(1/2),m'n}]$$
$$= -\chi_{in} h_{imn}, \qquad (m = 1, \ldots, M), \tag{83.93}$$

where

$$\chi_{in} \equiv \{\chi_{i-(1/2)}[r_i - r_{i-(1/2)}] + \chi_{i+(1/2)}[r_{i+(1/2)} - r_i]\}/[r_{i+(1/2)} - r_{i-(1/2)}]. \tag{83.94}$$

The matrix $D_{mm'}$ is a discrete representation of the angle-derivative terms in (83.90) and (83.91), evaluated at μ_m; it may be constructed in several different ways. One approach (**C1**), (**C2**), (**C6**, §91), (**N1**) is to assume that $j(\mu)$ and $h(\mu)$ are represented by (unique) interpolating polynomials of order $(M-1)$ determined by the angle mesh $\{\mu_m\}$. These polynomials can be written conveniently as

$$j(\mu) \approx \sum_{i=1}^{M} L_i(\mu) j(\mu_i), \tag{83.95}$$

and similarly for $h(\mu)$. Here $L_i(\mu)$ is the Lagrange polynomial

$$L_i(\mu) \equiv \frac{(\mu-\mu_1)\ldots(\mu-\mu_{i-1})(\mu-\mu_{i+1})\ldots(\mu-\mu_M)}{(\mu_i-\mu_1)\ldots(\mu_i-\mu_{i-1})(\mu_i-\mu_{i+1})\ldots(\mu_i-\mu_M)} \tag{83.96}$$

which is defined such that $L_i(\mu_j) = \delta_{ij}$. Then

$$(\partial j/\partial\mu)_m = \sum_{i=1}^{M} (dL_i/d\mu)_{\mu_m} j(\mu_i) \tag{83.97}$$

whence

$$D_{mm'} = (1-\mu_m^2)(dL_{m'}/d\mu)_{\mu_m} - 2\mu_m \,\delta_{mm'}. \tag{83.98}$$

The derivatives $(dL_i/d\mu)$ are easily calculated analytically from (83.96). A related approach is to represent $(1-\mu^2)j(\mu)$ and $(1-\mu^2)h(\mu)$ instead of $j(\mu)$ and $h(\mu)$ by the interpolating polynomials.

The **D** matrix given by (83.98) is full, and for large M (needed for accuracy) has large elements of alternating sign. This is a characteristic of all high-order differentiation formulae, reflecting the fact that high-order polynomials can oscillate wildly over their interval of definition; numerical noise or other errors in the solution then tend to be amplified, and the computation may become unstable.

The difficulty just described can be circumvented by using a *discrete-space method* similar to Carlson's S_N *method* (**C4**), (**L4**), (**L5**). Here, one assumes that the discrete ordinates $j(\mu_m)$ and $h(\mu_m)$ represent j and h within an angular cell $[\mu_{m-(1/2)}, \mu_{m+(1/2)}]$. The cell boundaries are taken to be $\mu_{1/2} \equiv 0$, and

$$\mu_{m+(1/2)} \equiv \mu_{m-(1/2)} + b_m, \qquad (m = 1, \ldots, M), \tag{83.99}$$

where b_m is the quadrature weight associated with μ_m. The properties of quadrature formulae assure that $\mu_{m-(1/2)} < \mu_m < \mu_{m+(1/2)}$, and that $\mu_{M+(1/2)} \equiv 1$.

Thus, integrating (83.90) over $[\mu_{m-(1/2)}, \mu_{m+(1/2)}]$, and suppressing depth and frequency subscripts, we obtain

$$3b_m\mu_m \frac{\partial(r^2 h_m)}{\partial(r^3)} + \frac{1}{r}\{[1-\mu_{m+(1/2)}^2]h_{m+(1/2)} - [1-\mu_{m-(1/2)}^2]h_{m-(1/2)}\} = b_m(\eta - \chi j_m), \qquad (m = 1, \ldots, M). \tag{83.100}$$

Noticing that by symmetry $h_{1/2} = h(\mu = 0) \equiv 0$, we see that (83.100) has the desirable property of being strictly conservative: when summed over all m, it yields the zeroth moment equation (83.69) exactly.

We must now represent $h_{n\pm(1/2)}$ in terms of the discrete ordinates $\{h_m\}$; we then discretize the spatial operator as before, and obtain an equation of the form (83.92). To maximize stability of the solution, one may represent $h(\mu)$ by a linear spline (**P2**), so that

$$h_{m+(1/2)} = \{[\mu_{m+1} - \mu_{m+(1/2)}]h_m + [\mu_{m+(1/2)} - \mu_m]h_{m+1}\}/(\mu_{m+1} - \mu_m),$$
$$(m = 1, \ldots, M-1). \quad (83.101)$$

We need not specify $h_{M+(1/2)}$ as its coefficient in (83.100) vanishes because $\mu_{M+(1/2)} \equiv 1$. Substituting (83.101) into (83.100), and recalling $h_{1/2} \equiv 0$, we find that $\mathbf{D}$ is a tridiagonal matrix with elements

$$b_m D_{m,m-1} = -[1 - \mu^2_{m-(1/2)}][\mu_m - \mu_{m-(1/2)}]/(\mu_m - \mu_{m-1}), \qquad (m = 2, \ldots, M), \quad (83.102a)$$

$$b_m D_{mm} = \frac{[1 - \mu^2_{m+(1/2)}][\mu_{m+1} - \mu_{m+(1/2)}]}{(\mu_{m+1} - \mu_m)} - \frac{[1 - \mu^2_{m-(1/2)}][\mu_{m-(1/2)} - \mu_{m-1}]}{(\mu_m - \mu_{m-1})}, \qquad (m = 2, \ldots, M), \quad (83.102b)$$

$$b_m D_{m,m+1} = [1 - \mu^2_{m+(1/2)}][\mu_{m+(1/2)} - \mu_m]/(\mu_{m+1} - \mu_m),$$
$$(m = 1, \ldots, M-1), \quad (83.102c)$$

and

$$b_1 D_{11} = (1 - \mu^2_{3/2})(\mu_2 - \mu_{3/2})/(\mu_2 - \mu_1). \quad (83.102d)$$

In developing a discrete-space representation of (83.91) we first rewrite it as

$$3\mu^2 \frac{\partial(r^2 j_\nu)}{\partial(r^3)} + \frac{(\mu^2 - 1)}{r} j_\nu + \frac{1}{r}\frac{\partial}{\partial\mu}[\mu(1-\mu^2)j_\nu] = -\chi_\nu \mu h_\nu, \quad (83.103)$$

which when integrated over $[\mu_{m-(1/2)}, \mu_{m+(1/2)}]$ becomes

$$3b_m\mu_m^2 \frac{\partial(r^2 j_m)}{\partial r^3} + b_m \frac{(\mu_m^2 - 1)}{r} j_m + \frac{1}{r}\{\mu_{m+(1/2)}[1 - \mu^2_{m+(1/2)}]j_{m+(1/2)} - \mu_{m-(1/2)}[1 - \mu^2_{m-(1/2)}]j_{m-(1/2)}\} = -\chi b_m \mu_m h_m, \qquad (m = 1, \ldots, M). \quad (83.104)$$

Recalling that $\mu_{1/2} \equiv 0$ and $\mu_{M+(1/2)} \equiv 1$, we see that (83.104) has the desirable property that when summed over m it yields the first-moment equation (83.70) exactly. Discretizing (83.104) we obtain

$$3\mu_m^2[r^2_{i+(1/2)} j_{i+(1/2),mn} - r^2_{i-(1/2)} j_{i-(1/2),mn}]/[r^3_{i+(1/2)} - r^3_{i-(1/2)}] + (2r_i)^{-1}\sum_{m'} D'_{mm'}[j_{i-(1/2),mn} + j_{i+(1/2),mn}] = -\chi_{in}\mu_m h_{imn}, \qquad (m = 1, \ldots, M). \quad (83.105)$$

Representing $j(\mu)$ by a linear spline as in (83.101), one finds that $\mathbf{D}'$ is tridiagonal with elements

$$b_m D'_{m,m-1} = -\mu_{m-(1/2)}[1-\mu^2_{m-(1/2)}][\mu_m - \mu_{m-(1/2)}]/(\mu_m - \mu_{m-1}), \qquad (m=2,\ldots,M), \quad (83.106a)$$

$$b_m D'_{mm} = (\mu^2_m - 1) + \frac{\mu_{m+(1/2)}[1-\mu^2_{m+(1/2)}][\mu_{m+1} - \mu_{m+(1/2)}]}{(\mu_{m+1} - \mu_m)} - \frac{\mu_{m-(1/2)}[1-\mu^2_{m-(1/2)}][\mu_{m-(1/2)} - \mu_{m-1}]}{(\mu_m - \mu_{m-1})} \qquad (m=1,\ldots,M), \quad (83.106b)$$

and

$$b_m D'_{m,m+1} = \mu_{m+(1/2)}[1-\mu^2_{m+(1/2)}][\mu_{m+(1/2)} - \mu_m]/(\mu_{m+1} - \mu_m), \qquad (m=1,\ldots,M-1).$$

If the discrete-space equations are formulated in terms of the intensity $I^{\pm}$, instead of j and h, the fact that a photon trajectory always has a larger value of μ at the point of absorption than at the point of creation argues (**L6**) that the angle derivative should be represented by an upstream difference, that is, $I_{m+(1/2)} \equiv I_m$. This argument does not apply, however, to j and h, which mix information at $\pm\mu$.

Equations (83.92) and (83.93), with suitable boundary conditions, pose a coupled system of the general form

$$-\mathbf{A}_i \mathbf{j}_{i+(1/2)} + \mathbf{B}_i \mathbf{h}_i - \mathbf{C}_i \mathbf{j}_{i-(1/2)} = \mathbf{D}_i, \qquad (i=I+1,\ldots,1), \quad (83.107)$$

and

$$-\mathbf{E}_{i+(1/2)} \mathbf{h}_{i+1} + \mathbf{F}_{i+(1/2)} \mathbf{j}_{i+(1/2)} - \mathbf{G}_{i+(1/2)} \mathbf{h}_i = \mathbf{H}_{i+(1/2)}, \qquad (i=I,\ldots,1). \quad (83.108)$$

Here $\mathbf{j}_{i+(1/2)}$ and $\mathbf{h}_i$ are vectors of length M containing the angle components $\mathbf{j}_{i+(1/2),mn}$ and $\mathbf{h}_{imn}$; the matrices **A**, **B**, **C**, and **E**, **F**, **G** are all of dimension $M \times M$. The boundary conditions imply that $\mathbf{A}_{I+1} \equiv 0$ and $\mathbf{C}_1 \equiv 0$. Because the equations are already angle coupled, scattering terms in the source function can be written out explicitly as in (83.26) at no extra computational cost. Equations (83.107) and (83.108) are solved by Gaussian elimination, starting with the forward elimination

$$\mathbf{h}_i = \mathbf{K}_i \mathbf{j}_{i-(1/2)} + \mathbf{L}_i \qquad (83.109)$$

and

$$\mathbf{j}_{i-(1/2)} = \mathbf{M}_{i-(1/2)} \mathbf{h}_{i-1} + \mathbf{N}_{i-(1/2)}, \qquad (83.110)$$

for $i = I+1, \ldots, 2$, where

$$\mathbf{K}_i \equiv [\mathbf{B}_i - \mathbf{A}_i \mathbf{M}_{i+(1/2)}]^{-1} \mathbf{C}_i, \qquad (83.111)$$

$$\mathbf{L}_i \equiv [\mathbf{B}_i - \mathbf{A}_i \mathbf{M}_{i+(1/2)}]^{-1} [\mathbf{D}_i + \mathbf{A}_i \mathbf{N}_{i+(1/2)}], \qquad (83.112)$$

$$\mathbf{M}_{i-(1/2)} \equiv [\mathbf{F}_{i-(1/2)} - \mathbf{E}_{i-(1/2)} \mathbf{K}_i]^{-1} \mathbf{G}_{i-(1/2)}, \qquad (83.113)$$

and

$$\mathbf{N}_{i-(1/2)} \equiv [\mathbf{F}_{i-(1/2)} - \mathbf{E}_{i-(1/2)} \mathbf{K}_i]^{-1} [\mathbf{H}_{i-(1/2)} + \mathbf{E}_{i-(1/2)} \mathbf{L}_i]. \qquad (83.114)$$

The forward sweep is completed by applying (83.109) at $i=1$, where $\mathbf{C}_1 \equiv 0$ implies $\mathbf{h}_1 \equiv \mathbf{L}_1$. We then back-substitute into (83.110) and (83.109) to determine $\mathbf{j}_{i+(1/2)}$ and $\mathbf{h}_{i+1}$ for $i=1, \ldots, I$. The computational effort for I radial shells, M angles, and N frequencies scales as cIM^3N.

Both the tangent-ray and discrete-space schemes have advantages and disadvantages. The tangent-ray method is expensive for large I because of the large number of tangent rays and mesh points per ray. However the solution along each ray is a true formal solution and is relatively cheap. Furthermore, this method provides good angular resolution of the radiation field (particularly when $R/r_c \gg 1$ and the radiation field becomes strongly peaked towards $\mu = 1$) because it samples all source shells. The discrete-space scheme uses M fixed angles and is economical when M is small. But because it is no longer a true formal solution (the equations are angle coupled) the computational expense rises rapidly with M. For small M the angular resolution may be inadequate when $R/r_c \gg 1$ and/or the radiation field is strongly forward peaked. But the method may be completely adequate, with small M, for problems in which a geometrically thin transport layer (e.g., a stellar atmosphere) surrounds a large diffusion region (e.g., a stellar interior) within which sphericity effects are important [because (r/r_c) varies over a large range] but the radiation field is nearly isotropic.

Having found j_ν at all depths and frequencies by either method, we can evaluate the f_ν's and q_ν's needed in the moment equations. We iterate between the moment equations and the formal solution as in the planar case. If J iterations are required for convergence, the total computing effort scales as $J(cIN^3 + c'I^2N)$ or $J(cIN^3 + c'IM^3N)$ for the tangent-ray and discrete-space methods respectively.

6.6 Statistical Equilibrium in the Presence of a Radiation Field

The occupation numbers of atomic levels in a radiating fluid are not, in general, those predicted by equilibrium statistical mechanics for local values of the temperature and density (LTE). In some cases the departures from LTE are severe; generally they are driven by radiative processes that deviate from equilibrium values in regions where photons escape efficiently through a boundary. Using a stellar atmosphere as an example, we first examine the microscopic requirements of LTE, and show that they are unfulfilled. We then formulate rate equations that determine the state of the material from local values of the temperature and density and the radiation field.

84. The Microscopic Implications of LTE

The hypothesis of LTE makes several tacit assumptions; let us examine some of these critically.

DETAILED BALANCE

In thermodynamic equilibrium, *all* processes are in detailed balance, that is, every transition, of any kind, is exactly canceled by its inverse. Detailed balance holds for collisional processes so long as the particle distribution function is Maxwellian; collisions then occur at their equilibrium rates (per particle) at the local kinetic temperature. If these were the only processes operating, we would have LTE.

However, radiative processes compete with collisions. When they occur at their equilibrium rates (e.g., the radiation field is Planckian), they also are in detailed balance, and help drive the material toward LTE. But if the radiation field has a nonequilibrium distribution the radiative rates can be out of balance and will tend to drive the material away from LTE. In the interior (or deep in the atmosphere) of a star the radiation field does, in fact, thermalize to the Planck function, and LTE obtains. However, at the surface of a star the radiation field is out of equilibrium in two important respects. First, there are no incoming photons, hence the field is anisotropic, and is *dilute* because the mean intensity (which sets absorption rates) averages over a hemisphere containing no radiation. More important, the outward-moving photons at the surface originate mainly from layers at unit optical depth in the atmosphere; these layers in general have physical properties (e.g., temperature) substantially different from the surface layers. Hence even for Planckian emission from the deeper layers, at the surface the characteristic radiation temperature T_R can be quite different from the local kinetic temperature T_k. The effect is largest at frequencies where the material is most transparent and the unit optical-depth surface lies deepest. Moreover, when $h\nu/kT \gg 1$, $B_\nu(T)$ varies as $e^{-h\nu/kT}$, hence $B_\nu(T_R)$ may differ from $B_\nu(T_k)$ by orders of magnitude even for a modest difference between T_R and T_k. Thus the frequency spectrum of the intensity near the surface will generally be strongly non-Planckian, and radiative rates will be far from detailed balance.

We note in passing that even when the material is not in LTE we can use detailed-balancing arguments to determine *rate coefficients*; for example, we can express collisional de-excitation rates in terms of excitation rates. We shall use this device repeatedly.

THE PARTICLE VELOCITY-DISTRIBUTION FUNCTION

In equilibrium, all particles have Maxwellian velocity distributions at a single kinetic temperature; this distribution is the unique result of elastic collisions among particles (cf. §§8 and 9). However, in general both radiative and inelastic collisional processes can perturb the equilibrium; for example, a collisional excitation of an atom by an electron lowers the electron's energy by a discrete amount, while a recombination prevents further elastic collisions. Thus establishment of a Maxwellian distribution hinges on whether the elastic collisions occur much more frequently than inelastic or radiative processes.

Characteristic times for these competing processes are the self-collision time t_c for electrons [cf. (10.26)], the radiative recombination time t_{rr}, and the inelastic collision time t_{ic}. For conditions in stellar atmospheres, conservative estimates show that $t_c/t_{rr} \sim 10^{-7}$ and $t_c/t_{ic} \lesssim 10^{-3}$, so the elastic collisions always dominate, and the velocity distribution should be Maxwellian (**B2**), (**M2,** §5–3). Moreover, demanding steady state in an atmosphere of hydrogen atoms, protons, electrons, and radiation while allowing energy exchange among all components, one finds (**B1**) that $|T_{\text{ion}} - T_{\text{elec}}| \lesssim 10^{-3} T_{\text{elec}}$ provided that the electron density $n_e \gtrsim 10^{10}$. Thus in what follows we can safely assume that all particles are, in fact, Maxwellian at a single temperature.

EXCITATION AND IONIZATION EQUILIBRIUM

When collisional excitation and ionization rates (and their inverses) dominate the corresponding radiative rates, LTE should prevail; this is the case in dense laboratory plasmas and stellar interiors, but not in a stellar atmosphere. One can show (**B2**) that in the Sun ($T \sim 6000$ K) the ratio of radiative to collisional ionization rates ranges from about 2 to about 10^3 for atoms with ionization potentials $\varepsilon_{\text{ion}} \sim 1$ eV and ~ 8 eV respectively. For an O-star ($T \sim 3 \times 10^4$ K) these ratios are 0.2 and 20; but atoms with $\varepsilon_{\text{ion}} \sim 1$ eV are no longer important because they are completely ionized. Hence photoionization rates generally vastly exceed collisional ionization rates; similarly, radiative recombinations outweigh collisional recombinations.

Similarly, radiative excitation rates in the Sun are from 6 to 300 times larger than collision rates for transitions in the spectral range 3000 Å $\leq \Lambda \leq$ 9000 Å. For an O-star the radiative rates exceed collisional only for $\Lambda \lesssim 5000$ Å; but the bulk of the radiation emerges in the far ultraviolet for such stars, thus for all practical purposes radiative processes far outweigh collisions.

We thus see that in stellar atmospheres both excitation and ionization equilibria can be driven away from LTE by radiative processes. We next must ask "over what depth-range can radiative rates depart markedly from their equilibrium values?". Early discussions [e.g., (**U1**), (**B2**)] concluded that equilibrium radiative rates in any transition are attained at optical depths greater than unity in that transition. However, this conclusion is false [see (**T2,** 141–147) or (**M2,** §5–3)] because of the effects of scattering, which allow J_ν to depart from B_ν (hence departures from LTE to occur) to a depth equal to a photon destruction length (cf. §83). Furthermore, because level populations are determined by *all* rates within the transition array, departure of even one transition from equilibrium tends to drive the population of *all* levels away from LTE. Only at depths where the radiation field is thermalized at *all* frequencies (i.e., $\tau_\nu \gtrsim \xi_\nu^{-1/2}$ for all ν) is the recovery of LTE guaranteed.

In summary, in a stellar atmosphere (whence the photons we observe originate!) one must, in general, allow for departures from LTE. The machinery to do so is provided by non-LTE rate equations, which we now consider.

85. Non-LTE Rate Equations

GENERAL FORM

We derive the rate equations for the general case of a moving medium. Let n_{ik} be the number density of particles in level i of chemical species k in a fixed volume. The rate of change of n_{ik} is determined by the net rate of flow of particles into the volume and the net rate of transitions into level i from all other levels j by atomic processes. Thus

$$(\partial n_{ik}/\partial t) = -\nabla \cdot (n_{ik}\mathbf{v}) + \sum_{j\neq i} n_{jk}P_{ji}^k - n_{ik}\sum_{j\neq i} P_{ij}^k; \tag{85.1}$$

here P_{ij}^k is the total (radiative plus collisional) transition rate from level i to level j. If we sum (85.1) over all states of species k, all atomic rates cancel term by term, and, writing $N_k \equiv \sum_i n_{ik}$, we obtain a continuity equation for species k:

$$(\partial N_k/\partial t) = -\nabla \cdot (N_k\mathbf{v}). \tag{85.2}$$

Multiplying (85.2) by m_k, the mass of species k, summing over species, and noting that $\rho = \sum_k m_k N_k$, we recover the continuity equation

$$(\partial \rho/\partial t) = -\nabla \cdot (\rho\mathbf{v}). \tag{85.3}$$

Using (85.3) to eliminate $\nabla \cdot \mathbf{v}$ we can rewrite (85.1) in the Lagrangean form

$$(Dn_{ik}/Dt) - (n_{ik}/\rho)(D\rho/Dt) = \sum_{j\neq i} n_{jk}P_{ji}^k - n_{ik}\sum_{j\neq i} P_{ij}^k, \tag{85.4}$$

or, more instructively, as

$$\rho[D(n_{ik}/\rho)Dt] = \sum_{j\neq i} n_{jk}P_{ji}^k - n_{ik}\sum_{j\neq i} P_{ij}^k. \tag{85.5}$$

Equation (85.5) states that in a material element, the rate of change of the number of particles, per unit mass, in a particular level equals the net number, per unit mass, entering that level via atomic transitions. The Lagrangean form of (85.2) is

$$D(N_k/\rho)/Dt = 0, \tag{85.6}$$

which states that the total number of particles of a chemical species, per unit mass, in a material element is constant.

For steady flow (85.1) simplifies to

$$\sum_{j\neq i} n_{jk}P_{ji}^k - n_{ik}\sum_{j\neq i} P_{ij}^k - \nabla \cdot (n_{ik}\mathbf{v}) = 0, \tag{85.7}$$

and for a static medium it becomes

$$\sum_{j\neq i} n_{jk}P_{ji}^k - n_{ik}\sum_{j\neq i} P_{ij}^k = 0. \tag{85.8}$$

In what follows we concentrate mainly on formulating and solving (85.8), which illustrates all the essential physics; we generalize to time-dependent flows at the end of the section. In all cases, all rates are to be evaluated in the comoving fluid frame.

COLLISION RATES

In an ionized plasma, atomic collisions with charged particles dominate all others because of the long range of Coulomb interactions. Furthermore, because the collision rate is proportional to the flux (hence velocity) of the incident particles, we usually need consider only electrons, whose velocities are a factor of $(Am_{\rm H}/m_e)^{1/2} \approx 43A^{1/2}$ greater than those of ions of atomic weight A.

Let $\sigma_{ij}(v)$ be the cross section for transitions $(i \rightarrow j)$ produced by collisions with electrons moving at speed v. The number of collisional excitations $(i \rightarrow j)$ is

$$n_iC_{ij} = n_in_e\int_{v_0}^{\infty} \sigma_{ij}(v)f(v)v\,dv, \tag{85.9}$$

where $f(v)$ is the Maxwellian velocity distribution and v_0 is the speed corresponding to the threshold energy E_{ij} of the transition (i.e., $\frac{1}{2}m_ev_0^2 = E_{ij}$). In equilibrium we have $n_i^*C_{ij} = n_j^*C_{ji}$, which allows us to compute the collisional deexcitation rate $(j \rightarrow i)$ as

$$n_jC_{ji} = n_j(n_i/n_j)^*C_{ij} = n_j(g_ie^{h\nu_{ij}/kT}/g_j)C_{ij}. \tag{85.10}$$

Similarly, the collisional ionization rate from level i is

$$n_iC_{i\kappa} = n_in_e\int_{v_0}^{\infty} \sigma_{i\kappa}(v)f(v)v\,dv, \tag{85.11}$$

and the collisional recombination rate is

$$n_\kappa C_{\kappa i} = n_\kappa(n_i/n_\kappa)^*C_{i\kappa} = n_i^*C_{i\kappa} \tag{85.12}$$

where n_i^* is computed using actual electron and ion densities, that is, $(n_i/n_\kappa)^* \equiv n_e\Phi_{i\kappa}(T)$.

RADIATIVE RATES

The number of upward radiative bound-bound transitions $(i \rightarrow j)$ is

$$\begin{aligned} n_iR_{ij} &= n_iB_{ij}\int \phi_\nu J_\nu\,d\nu \equiv n_iB_{ij}\bar{J}_{ij} \\ &= n_i(4\pi\alpha_{ij}/h\nu_{ij})\bar{J}_{ij} = n_i4\pi\int (h\nu)^{-1}\alpha_{ij}(\nu)J_\nu\,d\nu. \end{aligned} \tag{85.13}$$

The last form will prove useful later. We emphasize that (85.13) applies when all quantities are measured in the comoving frame. In the laboratory frame $\bar{J}$ is the much more complicated expression

$$\bar{J}_{ij} = (4\pi)^{-1} \int d\nu \oint d\omega \phi[\nu(1-\mathbf{n}\cdot\mathbf{v}/c)]I(\mathbf{n}, \nu). \tag{85.14}$$

The total number (spontaneous plus induced) of downward radiative transitions $(j \rightarrow i)$ is

$$n_j R'_{ji} = n_j(A_{ji} + B_{ji}\bar{J}_{ij}) = n_j(4\pi\alpha_{ij}/h\nu_{ij})(g_i/g_j)[(2h\nu_{ij}^3/c^2) + \bar{J}_{ij}]. \tag{85.15}$$

A prime has been added to R'_{ji} to reserve the unadorned symbol for a different use below. We cast (85.15) into a more useful form by factoring out the quantity $(n_i/n_j)^* = (g_i/g_j)e^{h\nu_{ij}/kT}$ from the right-hand side to obtain

$$n_j R'_{ji} \equiv n_j\left(\frac{n_i}{n_j}\right)^* R_{ji} = n_j\left(\frac{n_i}{n_j}\right)^*\left[4\pi \int \frac{\alpha_{ij}(\nu)}{h\nu}\left(\frac{2h\nu^3}{c^2} + J_\nu\right)e^{-h\nu/kT}\, d\nu\right]. \tag{85.16}$$

In (85.16) we can take ν-dependent factors inside the integral because the line profile ϕ_ν varies swiftly. Despite its apparently cumbersome form, this expression will allow us to systematize notation effectively.

Now consider bound-free transitions. The photoionization rate is

$$n_i R_{i\kappa} = n_i 4\pi \int_{\nu_0}^{\infty} [\alpha_{i\kappa}(\nu)J_\nu/h\nu]\, d\nu. \tag{85.17}$$

To calculate the spontaneous recombination rate we invoke a detailed balancing argument. In TE the number of spontaneous recombinations equals the number of photoionizations corrected for stimulated emissions, that is,

$$(n_\kappa R'_{\kappa i})^*_{\text{spon}} = n_i^* 4\pi \int_{\nu_0}^{\infty} [\alpha_{i\kappa}(\nu)/h\nu]B_\nu(1-e^{-h\nu/kT})\, d\nu. \tag{85.18}$$

The corresponding nonequilibrium rate is obtained by calculating n_i^* using the actual ion density n_κ; thus

$$\begin{aligned}(n_\kappa R'_{\kappa i})_{\text{spon}} &= n_\kappa(n_i/n_\kappa)^* 4\pi \int_{\nu_0}^{\infty} [\alpha_{i\kappa}(\nu)/h\nu]B_\nu(1-e^{-h\nu/kT})\, d\nu \\ &= n_\kappa(n_i/n_\kappa)^* 4\pi \int_{\nu_0}^{\infty} [\alpha_{i\kappa}(\nu)/h\nu](2h\nu^3/c^2)e^{-h\nu/kT}\, d\nu.\end{aligned} \tag{85.19}$$

The equilibrium number of stimulated emissions is the term containing $e^{-h\nu/kT}$ in (85.18); to obtain the nonequilibrium number we replace B_ν by J_ν, and calculate n_i^* using the actual ion density. Thus

$$(n_\kappa R'_{\kappa i})_{\text{stim}} = n_\kappa(n_i/n_\kappa)^* 4\pi \int_{\nu_0}^{\infty} [\alpha_{i\kappa}(\nu)/h\nu]J_\nu e^{-h\nu/kT}\, d\nu. \tag{85.20}$$

Combining (85.19) and (85.20) we get the total recombination rate

$$n_\kappa R'_{\kappa i} \equiv n_\kappa (n_i/n_\kappa)^* R_{\kappa i} = n_\kappa \left(\frac{n_i}{n_\kappa}\right)^* 4\pi \int_{\nu_0}^{\infty} \frac{\alpha_{i\kappa}(\nu)}{h\nu} \left(\frac{2h\nu^3}{c^2} + J_\nu\right) e^{-h\nu/kT}\, d\nu. \tag{85.21}$$

Comparing (85.13) with (85.17), and (85.16) with (85.21), shows that the notation is systematized by writing all upward radiative rates ($i \to j$), whether j is bound or free, as $n_i R_{ij}$ where

$$R_{ij} \equiv 4\pi \int_{\nu_0}^{\infty} [\alpha_{ij}(\nu) J_\nu / h\nu]\, d\nu, \tag{85.22}$$

and all downward radiative rates ($j \to i$) as $n_j (n_i/n_j)^* R_{ji}$, where

$$R_{ji} \equiv 4\pi \int_{\nu_0}^{\infty} [\alpha_{ij}(\nu)/h\nu][(2h\nu^3/c^2) + J_\nu] e^{-h\nu/kT}\, d\nu. \tag{85.23}$$

Notice that in equilibrium $R_{ij}^* \equiv R_{ji}^*$.

SOLUTION OF THE RATE EQUATIONS

We now assemble individual rates into equations from which occupation numbers can be computed. For the present we take as given the temperature T, the total particle density N, and the mean intensity J_ν at all frequencies. One can regard the temperature and gas pressure ($p = NkT$) as coming from a solution of the momentum and energy equations, and J_ν from a solution of the transfer equation. For simplicity we consider a gas of pure hydrogen.

For each bound level we have a *rate equation* of the form

$$\sum_{j<i} n_j (R_{ji} + C_{ji}) - n_i \left[\sum_{j<i} (n_j/n_i)^* (R_{ij} + C_{ji}) + \sum_{j>i}^{\kappa} (R_{ij} + C_{ij}) \right]$$
$$+ \sum_{j>i}^{\kappa} n_j (n_i/n_j)^* (R_{ji} + C_{ij}) = 0, \qquad (i = 1, \ldots, L), \tag{85.24}$$

where L is the number of bound levels. We can also write a total ionization equation

$$\sum_{i=1}^{L} n_i (R_{i\kappa} + C_{i\kappa}) - n_p \sum_{i=1}^{L} (n_i/n_p)^* (R_{\kappa i} + C_{i\kappa}) = 0, \tag{85.25}$$

where n_p is the proton density. However (85.25) contains no new information because it is merely the sum of (85.24) over all bound levels.

We thus have L equations in $L+2$ variables: $n_1, \ldots, n_L, n_p, n_e$. To close the system we invoke *charge conservation*,

$$n_p = n_e, \tag{85.26}$$

and *number conservation*,

$$\sum_{i=1}^{L} n_i + n_p + n_e = N. \tag{85.27}$$

Clearly (85.26) is trivial and could be used to eliminate n_e or n_p; we retain it because in a gas mixture several species may contribute electrons.

The system composed of (85.24) and (85.26) can be written

$$\mathbf{A}\mathbf{n}=\mathbf{B}, \tag{85.28}$$

where $\mathbf{n}$ is the column vector

$$\mathbf{n}=(n_1, \ldots, n_L, n_p), \tag{85.29}$$

$\mathbf{A}$ is a matrix of dimension $(L+1)\times(L+1)$ containing the transition rates, and $\mathbf{B}$ is the column vector

$$\mathbf{B}=(0, 0, \ldots, 0, n_e). \tag{85.30}$$

If T, n_e, and J_ν are given, all elements of $\mathbf{A}$ and $\mathbf{B}$ are known, and we can solve the linear system (85.28) for $\mathbf{n}$ by standard numerical methods. But if we are given N, not n_e, the unknown n_e must also be determined from the complete system comprising (85.28) and (85.27); this system is *nonlinear* because n_e appears in collision and recombination rates in elements of $\mathbf{A}$, which in turn multiply the level populations $n_1, \ldots, n_p$. The nonlinear system is solved by an iterative linearization scheme.

Suppose we have an estimate n_e^0 of the electron density (say from assuming LTE), which we use to calculate estimates $\mathbf{A}^0$ and $\mathbf{B}^0$ of $\mathbf{A}$ and $\mathbf{B}$, and we solve $\mathbf{A}^0\mathbf{n}^0=\mathbf{B}^0$ for an approximate set of occupation numbers $\mathbf{n}^0$. In general, n_e^0 and $\mathbf{n}^0$ will fail to satisfy (85.27), and we must improve these initial estimates. To do so, we set $\mathbf{n}=\mathbf{n}^0+\delta\mathbf{n}$ and $n_e=n_e^0+\delta n_e$, and expand (85.28) to first order, obtaining

$$\mathbf{A}^0(\mathbf{n}^0+\delta\mathbf{n})+\mathbf{n}^0\cdot(\partial\mathbf{A}/\partial n_e)^0\,\delta n_e=\mathbf{B}^0+(\partial\mathbf{B}/\partial n_e)^0\,\delta n_e. \tag{85.31}$$

Here

$$[\mathbf{n}^0\cdot(\partial\mathbf{A}/\partial n_e)^0]_i=\sum_j(\partial A_{ij}/\partial n_e)^0 n_j^0, \tag{85.32}$$

and

$$(\partial\mathbf{B}/\partial n_e)=(0, \ldots, 0, 1). \tag{85.33}$$

The derivatives $(\partial A_{ij}/\partial n_e)$ can be computed analytically. Using the fact that $\mathbf{A}^0\mathbf{n}^0=\mathbf{B}^0$, we rewrite (85.31) as

$$\mathbf{A}^0\,\delta\mathbf{n}+[\mathbf{n}^0\cdot(\partial\mathbf{A}/\partial n_e)^0-(\partial\mathbf{B}/\partial n_e)^0]\,\delta n_e=0. \tag{85.34}$$

Further, from (85.27) we have

$$\sum_{i=1}^{L}\delta n_i+\delta n_p+\delta n_e=N-\left(\sum_{i=1}^{L}n_i^0+n_p^0+n_e^0\right). \tag{85.35}$$

Equations (85.34) and (85.35) provide a linear system yielding $\delta\mathbf{n}$ and δn_e. Solving them, we revise all populations and recompute $\mathbf{A}$ and $\mathbf{B}$ using the new value of n_e, and iterate until $|\delta n_i/n_i|$ is less than some small number for all n_i and n_e. Convergence is quadratic, giving precise results in a few iterations if the original estimate of n_e^0 is reasonable. In some formulations one uses *departure coefficients*

$$b_i \equiv n_i/n_i^* \tag{85.36}$$

instead of occupation numbers per se.

As we will see in §6.7 and Chapters 7 and 8, the linearization procedure just described is a restrictive example of the general procedure used to solve the coupled equations of hydrodynamics, radiative transfer, and statistical equilibrium. In general we do not know T, N, and J_ν, but have only preliminary estimates; we then linearize (85.28) as

$$\delta\mathbf{n} = (\partial\mathbf{n}/\partial n_e)\,\delta n_e + (\partial\mathbf{n}/\partial T)\,\delta T + \sum_k (\partial\mathbf{n}/\partial J_k)\,\delta J_k \tag{85.37}$$

where the derivatives all have the form

$$(\partial\mathbf{n}/\partial x) = \mathbf{A}^{-1}[(\partial\mathbf{B}/\partial x) - \mathbf{n}\cdot(\partial\mathbf{A}/\partial x)]. \tag{85.38}$$

Analytical expressions can be written for $(\partial\mathbf{A}/\partial n_e)$, $(\partial\mathbf{A}/\partial T)$, and $(\partial\mathbf{A}/\partial J_k)$; see (**A7**), (**M2**), and (**M3**). In (85.38) all terms on the right-hand side are evaluated using current estimates of $\mathbf{A}$, $\mathbf{B}$, and $\mathbf{n}$, the latter obtained from $\mathbf{An} = \mathbf{B}$. Equations (85.37) and (85.35) are then solved simultaneously with the linearized transfer and dynamical equations (see §88).

Finally, we outline the method for solving rate equations in a dynamical atmosphere. Assuming that all quantities are known at t^n, we wish to solve for new level populations at time $t^{n+1} = t^n + \Delta t^{n+(1/2)}$. In (85.24) we replace the zero right-hand side with the time derivative on the left-hand side of (85.5), represented by a finite difference, obtaining

$$[(\mathbf{n}^{n+1}/\rho^{n+1}) - (\mathbf{n}^n/\rho^n)]/\Delta t^{n+(1/2)} = \theta\mathbf{a}^{n+1}(\mathbf{n}^{n+1}/\rho^{n+1}) + (1-\theta)\mathbf{a}^n(\mathbf{n}^n/\rho^n). \tag{85.39}$$

where the superscripts indicate the time level at which quantities are evaluated. The parameter θ determines the centering of the time derivative. Intuitively one expects $\theta = \frac{1}{2}$ to be optimal because it weighs information at t^n and t^{n+1} equally in the integration over $\Delta t^{n+(1/2)}$. However, for most astrophysical problems the microscopic-process equilibrium times are orders of magnitude shorter than a dynamical timestep, hence level populations evolve through a sequence of quasistatic equilibria. Equations (85.39) are therefore very *stiff*, and for numerical stability one must choose $\theta = 1$, that is, a backward Euler or fully implicit scheme. Equations (85.39) can then be rewritten as

$$\tilde{\mathbf{a}}^{n+1}\mathbf{n}^{n+1} = \mathbf{b}^{n+1}, \tag{85.40}$$

where $b_i^{n+1} = (-\rho^{n+1}/\rho^n)n_i^n/\Delta t^{n+(1/2)}$, $(i = 1, \ldots, L)$, and $\tilde{\mathbf{a}}^{n+1}$ is the rate matrix with the term $-1/\Delta t^{n+(1/2)}$ added to the diagonal elements. Equations (85.26) and (85.27) remain unaltered, and are also evaluated at t^{n+1}. The system (85.40) plus (85.26) is again of the general form $\mathbf{An} = \mathbf{B}$, and the linearization scheme proceeds exactly as before, yielding equations of the general form (85.37). These are solved simultaneously with (85.35) and the linearized transfer and hydrodynamic equations (see §§88 and 104).

86. *Thermal Properties of a Nonequilibrium Gas*

In LTE, all occupation numbers and thermodynamic properties of the material are determined by two local state variables, say N and T (cf. §14). But for non-LTE, the rate equations imply that $n_i = n_i(N, T, J_\nu)$, where J_ν denotes the complete frequency spectrum of the mean intensity. Thus we cannot determine the particle distribution function until we know the photon distribution function; that is, we have as many new "state variables" as are required to specify the frequency distribution of the radiation field. Moreover, the radiation field is fundamentally nonlocal.

Thus when there are departures from LTE we cannot hope to express thermodynamic properties of the material as a function of local state variables. At best we can write some properties in terms of particle densities. For example, we can express the internal energy of the material (assumed to be pure hydrogen) as

$$\rho e = \sum_{i=1}^{L} n_i \varepsilon_i + n_p \varepsilon_I + \tfrac{3}{2} NkT \qquad (86.1)$$

where ε_I is the ionization energy of hydrogen, and $\varepsilon_i = (1 - i^{-2})\varepsilon_I$. Similarly the specific enthalpy is

$$\rho h = \sum_{i=1}^{L} n_i \varepsilon_i + n_p \varepsilon_I + \tfrac{5}{2} NkT. \qquad (86.2)$$

We used this approach in §75 to write formulae for the opacity and emissivity.

For other quantities, such as entropy, we can go back to basic definitions such as (11.1) and (11.2), but their usefulness is problematical. In yet other cases it is not clear how a quantity can be defined; for example, we cannot write formulae for c_v and c_p because the gas responds differently when the energy input is radiative than when it is thermal energy of microscopic motions. Some quantities are simply meaningless; for example, the adiabatic exponents no longer make sense because, in the presence of a non-equilibrium radiation field, the material is inherently nonadiabatic owing to radiative energy exchange from one point in the medium to another.

It is sometimes useful to have rough models for the nonequilibrium thermodynamic properties of material, for example, in analyses of linear wave propagation or local stability. For example, one might assume an *imposed* radiation field and fixed radiative rates, and then perturb the rate equations with respect to, say, δT and δp, and use the δn's to evaluate such quantities as $\delta\chi_\nu$ and $\delta\eta_\nu$ (useful in the linearized transfer equation), or δe (for studies of energy balance). A cruder model would be to assume all the b factors for bound levels to be fixed, and then calculate δn_i as $b_i\, \delta n_i^*$ for perturbations δT and δp. But we stress that these approaches are not rigorous, and the only completely satisfactory treatment is to enforce a self-consistent nonlocal coupling among the fluid equations, rate equations, and transfer equations from the outset.

6.7 Solution of the Coupled Transfer and Statistical Equilibrium Equations in Static Media

We are now in a position to combine the transfer and statistical equilibrium equations with the goal of obtaining consistency between the radiation field and material properties, subject to constraints of momentum and energy balance, which is central to several classes of astrophysical problems: (1) In the static non-LTE *model atmospheres problem* we attempt to determine simultaneously the radiation field (from the transfer equation), the state of the material (from the statistical equilibrium equations), and the structure of the medium (from constraints of hydrostatic and radiative equilibrium). We focus closely on this problem in this chapter because it provides a good framework for the development of the basic formalism. A simpler problem is to assume LTE, replacing the statistical equilibrium equations with the Saha-Boltzmann relation. (2) In the *statistical equilibrium problem* we solve the combined transfer and statistical equilibrium equations assuming that the atmospheric structure is known. Here one includes many levels and transitions so as to simulate an observed spectrum. We will not discuss this class of problem; see (**M2,** Chaps. 10–12). (3) In the *dynamical model atmospheres problem* we solve the transfer equation, time-dependent statistical equilibrium equations, and the hydrodynamical equations. Again the problem can be simplified by assuming LTE. Such problems are considered in Chapters 7 and 8. (4) Finally, one may do a statistical equilibrium problem in a time-varying atmosphere, for example, to calculate the spectrum of a variable star. Such problems are outside the scope of this book.

87. The Two-Level Atom

The two characteristic features of non-LTE transfer problems are that (1) the source function (for both lines and continua) contains a dominant scattering term and (2) the radiation field in one transition affects the radiation field in other transitions. To gain physical insight we consider a highly simplified atomic model, the *two-level atom*, consisting of two bound levels and a continuum, for which we can obtain an analytical expression for the source function that displays the scattering term explicitly, and shows the nature of the other source-sink terms. We can then see how atomic processes affect the thermalization of the radiation field in the line joining the two levels, and armed with this insight we can appreciate the implications of the computational strategy developed in §88. To supplement our brief discussion see (**M2,** Chap. 11).

THE LINE SOURCE FUNCTION

Consider the spectrum line formed between the lower and upper levels, l and u, of an atom. Assume that the line is so strong (e.g., a resonance line) that we need consider only the opacity and emissivity of the line itself. From (73.10), the average line opacity is

$$\chi_{lu} = (B_{lu}h\nu_{lu}/4\pi)[n_l - (g_l/g_u)n_u]. \tag{87.1}$$

Let $d\tau \equiv -\chi_{lu}\,dz$; the transfer equation is then

$$\mu(dI_\nu/d\tau) = \phi_\nu(I_\nu - S_l), \tag{87.2}$$

where the source function is [cf. (73.13)]

$$S_l = (2h\nu_{lu}^3/c^2)[(g_u n_l/g_l n_u) - 1]^{-1}. \tag{87.3}$$

The occupation numbers of levels l and u are determined by the two rate equations

$$n_l(B_{lu}\bar{J}_{lu} + C_{lu} + R_{l\kappa} + C_{l\kappa}) = n_u(A_{ul} + B_{ul}\bar{J}_{lu} + C_{ul}) + n_l^*(R_{\kappa l} + C_{l\kappa}) \tag{87.4}$$

and

$$n_u(A_{ul} + B_{ul}\bar{J}_{lu} + C_{ul} + R_{u\kappa} + C_{u\kappa}) = n_l(B_{lu}\bar{J}_{lu} + C_{lu}) + n_u^*(R_{\kappa u} + C_{u\kappa}), \tag{87.5}$$

where $\bar{J}_{lu} \equiv \int \phi_\nu J_\nu\,d\nu$. From (87.4) and (87.5) we derive an explicit analytical expression for S_l by solving for the ratio (n_l/n_u), substituting into (87.3), and using the Einstein relations (73.8) and (73.9); after some algebra one obtains

$$S_l = [\bar{J}_{lu} + \varepsilon B_\nu(T_e) + \theta]/(1 + \varepsilon + \eta), \tag{87.6}$$

where

$$\varepsilon \equiv C_{ul}(1 - e^{-h\nu_{lu}/kT})/A_{ul}, \tag{87.7}$$

$$\eta \equiv \frac{(R_{u\kappa} + C_{u\kappa})n_l^*(R_{\kappa l} + C_{l\kappa}) - g_l(R_{l\kappa} + C_{l\kappa})n_u^*(R_{\kappa u} + C_{u\kappa})/g_u}{A_{ul}[n_l^*(R_{\kappa l} + C_{l\kappa}) + n_u^*(R_{\kappa u} + C_{u\kappa})]}, \tag{87.8}$$

and

$$\theta \equiv \left(\frac{2h\nu_{lu}^3}{c^2}\right)\left(\frac{g_l}{g_u A_{ul}}\right)\frac{(R_{l\kappa} + C_{l\kappa})n_u^*(R_{\kappa u} + C_{u\kappa})}{[n_l^*(R_{\kappa l} + C_{l\kappa}) + n_u^*(R_{\kappa u} + C_{u\kappa})]}. \tag{87.9}$$

Equation (87.6), first derived by R. N. Thomas (**T1**), has played an important role in line-formation theory.

Each term in (87.6) has a simple physical interpretation. Consider first the denominator, which contains *sink terms*, all normalized to the spontaneous emission rate. The first term corresponds to scattered photons. The second is the rate of collisional de-excitation of the upper level, hence accounts for collisional destruction and thermalization of line photons. The third is proportional to the total ionization rate from the upper level times the fraction of recombinations into the lower level, hence accounts for destruction of line photons by processes that transfer electrons from the upper to the lower state via the continuum.

The numerator contains *source terms*. The first is a noncoherent scattering reservoir of line photons resulting from the cumulative contributions of sources and sinks over an interaction volume. The second is the thermal source, representing photons created by collisional excitation of an electron to the upper level, followed by radiative decay; this term depends explicitly on the local electron kinetic temperature T_e. The third term is proportional to the total ionization rate from the ground level times the fraction of recombinations into the upper level, hence accounts for line photons created by continuum processes that transfer electrons from the lower to upper level, whence they radiatively decay. The quantity θ can be rewritten as $\theta = \eta B^*$, where

$$B^* = \left(\frac{2h\nu_{lu}^3}{c^2}\right)\left\{\left(\frac{n_l^* g_u}{n_u^* g_l}\right)\left[\frac{(R_{u\kappa}+C_{u\kappa})(R_{\kappa l}+C_{l\kappa})}{(R_{l\kappa}+C_{l\kappa})(R_{\kappa u}+C_{u\kappa})}\right]-1\right\}^{-1}. \qquad (87.10)$$

We can then express B^* as $B_\nu(T_R)$ where T_R is a characteristic radiation temperature that can exceed, or be less than, T_e depending on the relative sizes of photoionization and recombination rates.

At great depth in an atmosphere, large densities imply high collision rates, ε becomes large, and S_l thermalizes. If the continua are also opaque so that $J_\nu \to B_\nu$, then $R_{l\kappa} \to R_{l\kappa}^* = R_{\kappa l}^*$, while $R_{u\kappa} \to R_{u\kappa}^* = R_{\kappa u}^*$, and $B^* \to B_\nu(T_e)$. Hence at depth $S_l \to B_\nu(T_e)$. The behavior of S_l near the boundary depends on the relative sizes of the source-sink terms; in general both ε and η will be much smaller than unity. If $\varepsilon > \eta$ and $\varepsilon B > \eta B^*$, the line is *collision dominated*; S_l then tends to couple to the local electron temperature T_e. If $\eta > \varepsilon$ and $\eta B^* > \varepsilon B$, the line is *photoionization dominated*; here S_l tends to couple to the radiation temperature T_R. *Mixed domination* lines have $\varepsilon > \eta$ but $\eta B^* > \varepsilon B$ or vice versa. These three cases can behave quite differently in a stellar chromosphere, see (**M2**, Chap. 11).

THE LINE TRANSFER EQUATION

Given a model atmosphere, one can compute the depth variation of ε, η, $B_\nu(T_e)$, and B^*. Substituting (87.6) into (87.2) we have a noncoherent scattering problem with known coefficients, which can readily be solved by the numerical techniques described in §83. While such a solution certainly yields the answer, it reveals little about the underlying physics. To gain further insight let us therefore consider a simpler problem, and suppose that $\eta = \theta \equiv 0$, and that both ε and B are constant. The transfer equation is then

$$\mu(\partial I_\nu/\partial\tau) = \phi_\nu[I_\nu - (1-\tilde{\varepsilon})\bar{J} - \tilde{\varepsilon}B], \qquad (87.11)$$

where $\tilde{\varepsilon} \equiv \varepsilon/(1+\varepsilon)$. Equation (87.11) bears a strong resemblance to the archetype scattering problem (83.1) to (83.6) except that now the scattering is noncoherent. We thus expect that the radiation field in a line will behave as described in §83, that is, that $\bar{J}$ will depart markedly from B at the surface, and that this departure will extend over a thermalization depth $\mathcal{L} \gg 1$ in line optical-depth units.

Detailed analysis [see e.g., (**M2**, §11–2)] shows that these expectations are met. In particular one finds that

$$S_l(0) = \varepsilon^{1/2} B, \tag{87.12}$$

in close analogy with (83.8). The thermalization depth differs from our earlier result because the noncoherence of the scattering process implies that photons that would have been trapped in the line core at depths where the core is opaque can instead get redistributed into the transparent line wings, whence they escape. We can estimate the effects of this escape mechanism on S_l by the following argument.

THE THERMALIZATION DEPTH

Suppose the line has a Doppler profile. Measure frequencies within the line in Doppler units, that is, $x \equiv (\nu - \nu_0)/\Delta\nu_D$ where ν_0 is the line-center frequency, and the *Doppler width* is $\Delta\nu_D = (\nu_0/c)(2kT/m)^{1/2}$. Then the profile function is

$$\phi(x) = \pi^{-1/2} e^{-x^2}. \tag{87.13}$$

The escape probability for a photon of frequency x traveling along a ray with angle-cosine μ from a point at line optical depth τ is $\exp(-\tau\phi_x/\mu)$. Weighting escapes by ϕ_x, the probability of emission at x, we find the total *photon escape probability*

$$\begin{aligned} P_e(\tau) &= \tfrac{1}{2}\int_{-\infty}^{\infty} dx\,\phi_x \int_0^1 d\mu\, e^{-\tau\phi_x/\mu} = \tfrac{1}{2}\int_{-\infty}^{\infty} dx\,\phi_x \int_1^{\infty} dy\, y^{-2} e^{-\tau\phi_x y} \\ &= \tfrac{1}{2}\int_{-\infty}^{\infty} E_2(\tau\phi_x)\phi_x\, dx. \end{aligned} \tag{87.14}$$

Define x_1 such that $\tau\phi(x_1) = 1$. Then for $\tau \gg 1$, $E_2(\tau\phi_x)$ is approximately zero when $|x| \lesssim x_1$ because then $\tau\phi_x \gg 1$, and is approximately unity when $|x| \gtrsim x_1$ because $\tau\phi_x \ll 1$. We can thus approximate P_e as

$$P_e(\tau) \approx \int_{x_1}^{\infty} \phi_x\, dx, \tag{87.15}$$

which shows that P_e is determined by photon escapes in the line wing.

When $\tau \gg 1$, $x_1 \gg 1$. In this limit (87.15) yields

$$P_e(\tau) = \tfrac{1}{2}\,\mathrm{erfc}\,(x_1) \approx e^{-x_1^2}/(2\pi^{1/2} x_1) \tag{87.16}$$

for a Doppler profile, for which, from (87.13)

$$x_1 = [\ln(\tau/\pi^{1/2})]^{1/2}, \tag{87.17}$$

so that

$$P_e(\tau) \approx k/[\tau(\ln\tau)^{1/2}], \tag{87.18}$$

where k is a constant of order unity.

The escape probability is to be compared with P_d, the *photon destruction probability* per scattering. Ignoring continuum terms, $P_d = (C_{ul}/A_{ul}) = \varepsilon$ for a two-level atom. If $P_e \ll P_d$, the photon will surely be thermalized before it escapes and therefore $S_l \to B$; if $P_e \gg P_d$ photons escape freely, hence S_l can depart from B, perhaps drastically. It is thus reasonable to estimate the *thermalization length* $\mathscr{L}$ as the depth at which $P_e(\mathscr{L}) \approx P_d$. For a Doppler profile this gives

$$\mathscr{L} \sim k/\varepsilon \tag{87.19}$$

where k is a factor of order unity. We again find that the range over which S_l can differ markedly from B can be enormous when ε is small, as it usually is near a boundary surface. For coherent scattering we found earlier that $\mathscr{L} \sim \varepsilon^{-1/2}$; the larger value predicted by (87.19) results from the effects of noncoherence. For other line profiles $\mathscr{L}$ can be even larger; for example, for a Lorentz profile $\mathscr{L} \sim \varepsilon^{-2}$.

OVERLAPPING CONTINUUM

We assumed above that all opacity and emissivity comes from the line itself. When there is an overlapping continuum, which we assume is in LTE, the source function is given by (77.11), and the transfer equation becomes

$$\mu(\partial I_\nu/\partial\tau) = \phi_\nu[I_\nu - (1-\xi_\nu)\bar{J} - \xi_\nu B_\nu]. \tag{87.20}$$

Detailed analysis shows [see e.g., (**M2**, 350–354)] that the behavior of S_l is then determined by the average thermal coupling parameter

$$\bar{\xi} \equiv \int_{-\infty}^{\infty} \phi_x \xi_x \, dx = \delta + \varepsilon(1-\delta), \tag{87.21}$$

where

$$\delta \equiv r \int_{-\infty}^{\infty} \phi_x(\phi_x + r)^{-1} \, dx. \tag{87.22}$$

Physically, δ is the continuum destruction probability: it is the profile-weighted average of the probability $r/(\phi + r)$ that a line photon will be absorbed by the continuum, times the (unit) probability that the absorbed photon is thermalized. The continuum processes set a floor on $\bar{\xi}$, often much larger than ε itself, and guarantee that S_l will be thermalized when the continuum optical depth is unity or larger, even if the line thermalization processes represented by ε would not by themselves force $S_l \to B$.

TRANSFER IN A RESONANCE CONTINUUM

The preceding analysis can be adapted to continua, say the Lyman continuum of hydrogen. We adopt the simplest possible atomic model: one bound state plus a continuum. We assume that the continuum is so opaque that we can ignore all other sources of opacity and emissivity. Then from

(75.1) and (75.2)

$$\chi_\nu = (n_1 - n_1^* e^{-h\nu/kT})\alpha_\nu = (b_1 - e^{-h\nu/kT}) n_1^* \alpha_\nu, \tag{87.23}$$

while

$$\eta_\nu = n_1^* \alpha_\nu B_\nu (1 - e^{-h\nu/kT}). \tag{87.24}$$

For the Lyman continuum $h\nu/kT \gg 1$ (otherwise the hydrogen is mostly ionized and the assumptions just stated fail), hence we can neglect $e^{-h\nu/kT}$. We then have

$$S_\nu = \eta_\nu/\chi_\nu = B_\nu/b_1, \tag{87.25}$$

and the transfer equation to be solved is

$$\mu(\partial I_\nu/\partial\tau) = \phi_\nu[I_\nu - (B_\nu/b_1)]; \tag{87.26}$$

here we have written $\alpha_\nu \equiv \alpha_0\phi_\nu$ and $d\tau \equiv -\chi_0\, dz = -n_1\alpha_0\, dz$ where α_0 is the cross section at the continuum head.

From the rate equation for a one-level atom, we find that b_1 is given by

$$b_1 = \left[4\pi \int_{\nu_0}^{\infty} (\alpha_\nu/h\nu) B_\nu\, d\nu + C_{1\kappa}\right] \Big/ \left[4\pi \int_{\nu_0}^{\infty} (\alpha_\nu/h\nu) J_\nu\, d\nu + C_{1\kappa}\right], \tag{87.27}$$

where we ignored stimulated emission. Because $(h\nu/kT) \gg 1$, the factor $\exp(-h\nu/kT)$ in B_ν decays away rapidly from its value at $\nu = \nu_0$. Thus the two integrands in (87.27) peak sharply at $\nu \approx \nu_0$, and to a good approximation we can replace the integrals by $4\pi w_0(\alpha_0/h\nu_0)J_0$ and $4\pi w_0(\alpha_0/h\nu_0)B_0$, where w_0 is an appropriate quadrature weight. We then consider the transfer problem only at the continuum head. If we write $\varepsilon = C_{1\kappa}/[(4\pi w_0\alpha_0 B_0/h\nu_0) + C_{1\kappa}]$, the source function at ν_0 is

$$S_0 = (1-\varepsilon)J_0 + \varepsilon B_0. \tag{87.28}$$

and the transfer equation at ν_0, in the Eddington approximation, becomes

$$\tfrac{1}{3}(d^2J_0/d\tau_0^2) = \varepsilon(J_0 - B_0). \tag{87.29}$$

Equation (87.29) is identical to the archetype problem (83.6), hence we know that the thermalization depth will be $\sim\varepsilon^{-1/2}$, while at the boundary $S_0 \sim \varepsilon^{1/2}B_0$, which implies $b_1 \sim \varepsilon^{-1/2}$. In a stellar atmosphere $\varepsilon \ll 1$, hence in the Lyman continuum there are large departures from LTE, which persist to great optical depth. Thus both lines and continua are subject to major non-LTE effects.

THE MULTILEVEL PROBLEM

The atomic models discussed above are drastically oversimplified; to achieve realism we must consider multilevel atoms. In the multilevel case we might try to follow the approach used above, manipulating the rate equation to obtain an analytical expression of the form

$$S_{ij} = (\bar{J}_{ij} + \alpha_{ij})/(1 + \beta_{ij}) \tag{87.30}$$

for the source function in each transition $i \rightarrow j$. This approach is called the *equivalent two-level atom* (ETLA) *method* because (87.30) is of the same general form as (87.6). There are many different formulations of this method, which use different groupings of terms within α_{ij} and β_{ij}, and different methods of solving the equations [see, e.g., (**A3**, Chap. 3), (**J1**, Chaps. 6 and 8), (**M2**, §12–1), and (**T2**, Chap. 3)]. Given source functions of the form (87.30), it is straightforward to solve the transfer equation in each transition.

Superficially this procedure looks much simpler than it is. The fundamental problem is that the coefficients α_{ij} and β_{ij} contain terms that depend on the radiation fields in all transitions other than $(i \rightarrow j)$ [e.g., for a two-level atom (87.6) explicitly displays the dependence of S_l on the radiative rates in the continua $(l \rightarrow \kappa)$ and $(u \rightarrow \kappa)$]. Thus we cannot in fact specify the source-sink terms for any S_{ij} until we know the radiation field in every other transition. But we can't compute these radiation fields until we know the relevant source functions. We must therefore use some kind of iteration scheme [see e.g., (**A3**, Chaps. 4 and 7), (**T2**, Chaps. 5 and 6)]. In some cases the iteration will converge; in others it fails, forcing one to seek clever schemes (often based on trial and error) to accelerate convergence.

In the face of the difficulties just described, Jefferies (**J1**, Chap. 8) stressed the conceptual importance of viewing radiation in the entire transition array of an atom as belonging to a *collective photon pool*, thus recognizing that photons do not belong uniquely to any one transition, but interlock all transitions in a manner prescribed by the equations of statistical equilibrium. From this standpoint one sees that it is essential to treat all transitions *simultaneously*; let us now address this problem in the context of constructing a static non-LTE model atmosphere.

88. *The Complete Linearization Method*

In the non-LTE model atmospheres problem we wish to determine, at each point in the medium, the frequency distribution of the radiation field, the distribution of atoms and ions over their bound states, and the temperature and density. The imposed constraints are momentum balance, energy balance, steady-state statistical equilibrium, and charge and number conservation. Our discussion will be brief; further details are given in (**M2**, Chaps. 7 and 12).

BASIC EQUATIONS

The pressure distribution is determined by the equation of hydrostatic equilibrium $\boldsymbol{\nabla} p = \mathbf{f}$, where $\mathbf{f}$ is the total force per unit volume acting on the material; cf. (23.25). For a one-dimensional planar medium

$$(dp/dm) = g - (f_R/\rho), \tag{88.1}$$

where $dm \equiv -\rho\, dz$ and, from (78.12),

$$(f_R/\rho) = (4\pi/c) \int_0^\infty \omega_\nu H_\nu \, d\nu, \tag{88.2}$$

where $\omega_\nu \equiv \chi_\nu/\rho$. Using (78.17) we can rewrite (88.1) and (88.2) as

$$(dp/dm) + (dP/dm) = g, \tag{88.3}$$

where

$$P = \int_0^\infty P_\nu \, d\nu = (4\pi/c) \int_0^\infty f_\nu J_\nu \, d\nu. \tag{88.4}$$

The temperature distribution is determined by the requirement of radiative equilibrium:

$$4\pi \int_0^\infty (\eta_\nu - \chi_\nu J_\nu) \, d\nu = 0. \tag{88.5}$$

Assuming thermal emission and coherent electron scattering, $\chi_\nu = \kappa_\nu + n_e\sigma_e$ and $\eta_\nu = \eta_\nu^t + n_e\sigma_e J_\nu$, hence (88.5) becomes

$$4\pi \int_0^\infty (\eta_\nu^t - \kappa_\nu J_\nu) \, d\nu = 0, \tag{88.6}$$

where η_ν^t and κ_ν are given by (75.1) and (75.2).

The state of the material is determined by the equations of statistical equilibrium, charge conservation, and number conservation, as in (85.25) to (85.27).

The only radiation quantity appearing in these equations is J_ν, hence we use the transfer equation in angle-integrated form:

$$\frac{\partial^2(f_\nu J_\nu)}{\partial \tau_\nu^2} = J_\nu - S_\nu = \left(1 - \frac{n_e\sigma_e}{\chi_\nu}\right) J_\nu - \frac{\eta_\nu^t}{\chi_\nu}. \tag{88.7}$$

DISCRETIZATION

Equations (88.3), (88.6), (88.7), and (85.25) to (85.27) completely specify the problem. We discretize them for a set of frequencies $\{\nu_k\}$, and mass shells with boundaries at $\{m_d\}$ as in §83. The discretized form of (88.3) is

$$\begin{aligned} k[N_{d+(1/2)}T_{d+(1/2)} - N_{d-(1/2)}T_{d-(1/2)}] \\ + (4\pi/c) \sum_{k=1}^{K} w_k [f_{d+(1/2),k} J_{d+(1/2),k} - f_{d-(1/2),k} J_{d-(1/2),k}] \\ = g\, \Delta m_d, \qquad (d = 2, \ldots, D), \end{aligned} \tag{88.8}$$

where $\Delta m_d \equiv \frac{1}{2}[\Delta m_{d-(1/2)} + \Delta m_{d+(1/2)}]$, and w_k is a frequency-quadrature weight. To obtain a starting value at $d = 1$ we use (88.1), assuming that the material is transparent (so that H_ν, hence f_R, is fixed from the boundary outward). We have

$$N_{3/2} k T_{3/2} = m_1 \left[g - (4\pi/c) \sum_{k=1}^{K} w_k \omega_{3/2k} h_k j_k J_{3/2k} \right], \tag{88.9}$$

where the geometric factors h_k and j_k are defined by (83.44) and (83.45).

The discrete representation of (88.6) is

$$\sum_{k=1}^{K} w_k[\eta^t_{d+(1/2),k} - \kappa_{d+(1/2),k} J_{d+(1/2),k}] = 0, \qquad (d = 1, \ldots, D), \tag{88.10}$$

and the transfer equation and its boundary conditions are given by (83.58) to (83.61). The material equations can be used exactly as written in §85.

METHOD OF SOLUTION

In solving the system posed above we must bear in mind several important points. First, no variable is more fundamental than any other because all interact; thus within the slab $\Delta m_{d+(1/2)}$ the solution vector to be found is

$$\boldsymbol{\psi}_{d+(1/2)} = [J_{d+(1/2),1}, \ldots, J_{d+(1/2),K}, N, T, n_e, n_1, \ldots, n_L], \tag{88.11}$$

where L is the total number of levels including ions. Some of the information in $\boldsymbol{\psi}_{d+(1/2)}$ is, strictly speaking, redundant; in particular the J's determine the n's via the rate equations. But because we wish to know all these quantities we retain them in the formulation.

Second, all variables are strongly coupled globally via radiative transfer. A change in *any* variable at *any* point in the medium implies changes in *all* variables at *all* other points. The method of solution must account for this global coupling.

Third, the system is nonlinear, and must be solved by iteration.

Assume that the desired solution $\boldsymbol{\psi}_{d+(1/2)}$ can be written $\boldsymbol{\psi}_{d+(1/2)} = \boldsymbol{\psi}^0_{d+(1/2)} + \delta\boldsymbol{\psi}_{d+(1/2)}$ where $\boldsymbol{\psi}^0_{d+(1/2)}$ is the current (approximate) solution. The system to be solved at each depth point is of the form $\mathbf{f}_{d+(1/2)}(\boldsymbol{\psi}) = 0$, so we demand that $\delta\boldsymbol{\psi}$ be such that $\mathbf{f}_{d+(1/2)}(\boldsymbol{\psi}^0 + \delta\boldsymbol{\psi}) = 0$. We linearize this system and demand that

$$\mathbf{f}_{d+(1/2)}(\boldsymbol{\psi}^0) + \sum_j [\partial \mathbf{f}_{d+(1/2)}/\partial \psi_j]\, \delta\psi_j = 0, \qquad (d = 1, \ldots, D), \tag{88.12}$$

where j runs over all variables. These equations determine the $\delta\boldsymbol{\psi}$'s; let us examine them in more detail.

The linearized equation of hydrostatic equilibrium is

$$\begin{aligned} k[T_{d+(1/2)}\, \delta N_{d+(1/2)} + N_{d+(1/2)}\, \delta T_{d+(1/2)} \\ - T_{d-(1/2)}\, \delta N_{d-(1/2)} - N_{d-(1/2)}\, \delta T_{d-(1/2)}] \\ + (4\pi/c) \sum_{k=1}^{K} w_k[f_{d+(1/2),k}\, \delta J_{d+(1/2),k} - f_{d-(1/2),k}\, \delta J_{d-(1/2),k}] \\ = g\, \Delta m_d - N_{d+(1/2)} k T_{d+(1/2)} + N_{d-(1/2)} k T_{d-(1/2)} \\ - (4\pi/c) \sum_{k=1}^{K} w_k[f_{d+(1/2),k} J_{d+(1/2),k} - f_{d-(1/2),k} J_{d-(1/2),k}], \end{aligned} \tag{88.13}$$

with a similar equation for the linearized boundary condition (88.9). The linearized radiative equilibrium equation is

$$\sum_{k=1}^{K} w_k[\kappa_{d+(1/2),k}\,\delta J_{d+(1/2),k}+J_{d+(1/2),k}\,\delta\kappa_{d+(1/2),k}-\delta\eta^t_{d+(1/2),k}]$$
$$=\sum_{k=1}^{K} w_k[\eta^t_{d+(1/2),k}-\kappa_{d+(1/2),k}J_{d+(1/2),k}] \qquad (88.14)$$

where from (75.1) we express $\delta\kappa$ as

$$\delta\kappa=(\partial\kappa/\partial T)\,\delta T+(\partial\kappa/\partial n_e)\,\delta n_e+\sum_{l=1}^{L}(\partial\kappa/\partial n_l)\,\delta n_l, \qquad (88.15)$$

and similarly for $\delta\eta^t$.

The linearized transfer equations are

$$\begin{aligned}
&\frac{f_{d-(1/2),k}\,\delta J_{d-(1/2),k}}{\Delta\tau_{dk}\,\Delta\tau_{d+(1/2),k}}+\frac{f_{d+(3/2),k}\,\delta J_{d+(3/2),k}}{\Delta\tau_{d+(1/2),k}\,\Delta\tau_{d+1,k}}\\
&\quad-\left\{\frac{f_{d+(1/2),k}}{\Delta\tau_{d+(1/2),k}}\left(\frac{1}{\Delta\tau_{dk}}+\frac{1}{\Delta\tau_{d+1,k}}\right)+\left[1-\frac{n_{e,d+(1/2)}\sigma_e}{\chi_{d+(1/2),k}}\right]\right\}\delta J_{d+(1/2),k}\\
&\quad+a_{d+(1/2),k}\,\delta\omega_{d-(1/2),k}+b_{d+(1/2),k}\,\delta\omega_{d+(1/2),k}+c_{d+(1/2),k}\,\delta\omega_{d+(3/2),k}\\
&\quad-[\eta^t_{d+(1/2),k}+n_{e,d+(1/2)}\sigma_e J_{d+(1/2),k}]\frac{\delta\chi_{d+(1/2),k}}{\chi^2_{d+(1/2),k}}\\
&\quad+\frac{\delta\eta^t_{d+(1/2),k}}{\chi_{d+(1/2),k}}+\frac{\sigma_e J_{d+(1/2),k}}{\chi_{d+(1/2),k}}\,\delta n_{e,d+(1/2)}\\
&=\beta_{d+(1/2),k}+J_{d+(1/2),k}-\frac{[n_{e,d+(1/2)}\sigma_e J_{d+(1/2),k}+\eta^t_{d+(1/2),k}]}{\chi_{d+(1/2),k}}
\end{aligned} \qquad (88.16)$$

where

$$\alpha_{d+(1/2),k}\equiv[f_{d+(1/2),k}J_{d+(1/2),k}-f_{d-(1/2),k}J_{d-(1/2),k}]/[\Delta\tau_{d+(1/2),k}\,\Delta\tau_{dk}], \qquad (88.17)$$

$$\gamma_{d+(1/2),k}\equiv[f_{d+(1/2),k}J_{d+(1/2),k}-f_{d+(3/2),k}J_{d+(3/2),k}]/[\Delta\tau_{d+(1/2),k}\,\Delta\tau_{d+1,k}], \qquad (88.18)$$

$$\beta_{d+(1/2),k}\equiv\alpha_{d+(1/2),k}+\gamma_{d+(1/2),k}, \qquad (88.19)$$

$$a_{d+(1/2),k}\equiv\{\alpha_{d+(1/2),k}+\tfrac{1}{2}\beta_{d+(1/2),k}[\Delta\tau_{dk}/\Delta\tau_{d+(1/2),k}]\}/[\omega_{d-(1/2),k}+\omega_{d+(1/2),k}], \qquad (88.20)$$

$$c_{d+(1/2),k}\equiv\{\gamma_{d+(1/2),k}+\tfrac{1}{2}\beta_{d+(1/2),k}[\Delta\tau_{d+1,k}/\Delta\tau_{d+(1/2),k}]\}/[\omega_{d+(1/2),k}+\omega_{d+(3/2),k}], \qquad (88.21)$$

and

$$b_{d+(1/2),k}\equiv a_{d+(1/2),k}+c_{d+(1/2),k}. \qquad (88.22)$$

The density is

$$\rho = (N - n_e)m_{\rm H} \sum_i \alpha_i A_i \equiv (N - n_e)\bar{m} \tag{88.23}$$

where α_i is the fractional abundance, by number, of element i, having atomic weight A_i; hence $\delta\rho$, which appears in $\delta\omega$, can be written $\delta\rho = \bar{m}(\delta N - \delta n_e)$.

Equations (88.13) to (88.23) plus the linearized statistical equilibrium, number conservation, and charge conservation equations (85.37) and (85.35) can be assembled into a system of the form

$$-\mathbf{A}_{d+(1/2)}\,\delta\boldsymbol{\psi}_{d-(1/2)} + \mathbf{B}_{d+(1/2)}\,\delta\boldsymbol{\psi}_{d+(1/2)} - \mathbf{C}_{d+(1/2)}\,\delta\boldsymbol{\psi}_{d+(3/2)} = \mathbf{L}_{d+(1/2)}, \tag{88.24}$$

which is solved by the Feautrier method (cf. §83). Here $\mathbf{L}_{d+(1/2)}$ is the residual error in the transfer and constraint equations for the current solution $\boldsymbol{\psi}^0_{d+(1/2)}$; as $\mathbf{L} \to 0$, the corrections $\delta\boldsymbol{\psi} \to 0$.

Mathematically the complete linearization method is merely a multidimensional Newton-Raphson iteration; at each stage the equations are internally consistent to first order. But it is extremely important to appreciate the system's physical content. Thus the linearized rate equations provide an algorithm that describes how photons are shuffled about within the collective photon pool in response to a change in material properties, or, reciprocally, how the material responds to a change in the radiation field. The linearized hydrostatic and radiative equilibrium equations describe how the pressure and temperature respond to local changes in the radiation field or material properties. Local changes are coupled to all other points in the medium via the tridiagonal linearized hydrostatic and transfer equations. This linearization correctly handles both local and global coupling, and experience shows it to be an effective tool for solving the problem.

The iteration scheme starts from an initial model constructed assuming LTE and adopting the grey temperature distribution on the Rosseland mean scale [cf. (82.23)]. The hydrostatic equation is integrated step by step for this $T(\tau_R)$, and from N and T one determines n_i^* for all atomic levels at all depths. Using the linearization procedure, the model is adjusted to give strict radiative equilibrium for the actual nongrey opacity, still assuming LTE. To obtain a non-LTE model, the Saha-Boltzmann relations are replaced by the statistical equilibrium equations.

After each linearization step, all occupation numbers are updated using the new temperature, density, and radiation field; new Eddington factors are then determined from a formal solution, and the procedure is iterated to convergence. If the Eddington factors were known and fixed, convergence would be quadratic; in practice the need to update Eddington factors slows the convergence rate, but errors usually diminish substantially at each iteration. Further details are given in (**M2,** Chap. 7), (**M3**), and the references cited therein. Note that for LTE models the Rybicki scheme is more efficient than the Feautrier scheme [see (**G1**) or (**M2,** 180–185)].

We now have all of the ingredients needed to describe radiative transfer and the interaction between radiation and material. In the next chapter we apply this knowledge to the dynamics of radiating fluids.

References

(A1) Abramowitz, M. and Stegun, I. (1964) *Handbook of Mathematical Functions.* Washington, D.C.: U.S. Department of Commerce.
(A2) Aller, L. H. (1963) *Astrophysics: The Atmospheres of the Sun and Stars.* (2nd ed.) New York: Ronald.
(A3) Athay, R. G. (1972) *Radiation Transport in Spectral Lines.* Dordrecht: Reidel.
(A4) Auer, L. H. (1967) *Astrophys. J. Letters,* **150,** L53.
(A5) Auer, L. H. (1971) *J. Quant. Spectrosc. Rad. Transf.,* **11,** 573.
(A6) Auer, L. H. (1976) *J. Quant. Spectrosc. Rad. Transf.,* **16,** 931.
(A7) Auer, L. H. and Mihalas, D. (1969) *Astrophys. J.,* **158,** 641.
(B1) Bhatnagar, P., Krook, M., Menzel, D. H., and Thomas, R. N. (1955) *Vistas in Astron.,* **1,** 296.
(B2) Böhm, K.-H. (1960) in *Stellar Atmospheres,* ed. J. L. Greenstein, Chapter 3. Chicago: University of Chicago.
(C1) Campbell, P. M. (1965) *Lawrence Livermore Laboratory Report No. UCRL-12411.* Livermore: University of California.
(C2) Campbell, P. M. (1969) *Int. J. Heat Mass Transf.,* **12,** 497.
(C3) Carbon, D. (1974) *Astrophys. J.,* **187,** 135.
(C4) Carlson, B. G. (1963) in *Methods in Computational Physics,* ed. B. Alder, S. Fernbach, and M. Rotenberg, Vol. **1,** Chapter 1. New York: Academic.
(C5) Chandrasekhar, S. (1939) *An Introduction to the Study of Stellar Structure.* Chicago: University of Chicago.
(C6) Chandrasekhar, S. (1950) *Radiative Transfer.* Oxford: Oxford University Press.
(C7) Cox, J. P. and Giuli, R. T. (1968) *Principles of Stellar Structure.* New York: Gordon and Breach.
(D1) Dragon, J. N. and Mutschlecner, J. P. (1980) *Astrophys. J.,* **239,** 1045.
(F1) Feautrier, P. (1964) *C. R. Acad. Sci. Paris,* **258,** 3189.
(F2) Freeman, B. E., Hauser, L. E., Palmer, J. T., Pickard, S. O., Simmons, G. M., Williston, D. G., and Zerkle, J. E. (1968). *DASA Report No. 2135,* Volume I. La Jolla: Systems, Science, and Software, Inc.
(G1) Gustafsson, B. and Nissen, P. (1972) *Astron. and Astrophys.,* **19,** 261.
(H1) House, L. L. and Steinitz, R. (1975) *Astrophys. J.,* **195,** 235.
(H2) Huang, K. (1963) *Statistical Mechanics.* New York: Wiley.
(H3) Hundhausen, A. J. (1972) *Coronal Expansion and Solar Wind.* New York: Springer.
(J1) Jefferies, J. T. (1968) *Spectral Line Formation.* Waltham: Blaisdell.
(K1) Kourganoff, V. (1952) *Basic Methods in Transfer Problems.* Oxford: Oxford University Press.
(K2) Krishna-Swamy, K. S. (1961) *Astrophys. J.,* **134,** 1017.

(K3) Kurucz, R. L., Peytremann, E., and Avrett, E. H. (1974) *Blanketed Models for Early-Type Stars.* Washington, D.C.: Smithsonian Institution.
(L1) Landi Degl'innocenti, E. (1983) *Solar Physics,* **85,** 3.
(L2) Landi Degl'innocenti, E. and Landi Degl'innocenti, M. (1972) *Solar Phys.,* **27,** 319.
(L3) Landi Degl'innocenti, E. and Landi Degl'innocenti, M. (1975) *Nuovo Cimento,* **27B,** 134.
(L4) Lathrop, K. D. and Carlson, B. G. (1967) *J. Comp. Phys.,* **2,** 173.
(L5) Lathrop, K. D. and Carlson, B. G. (1971) *J. Quant. Spectrosc. Rad. Transf.,* **11,** 921.
(L6) Lund, C. M. and Wilson, J. R. (1980) *Lawrence Livermore Laboratory Report No. UCRL-84678.* Livermore: University of California.
(M1) Mihalas, D. (1965) *Astrophys. J.,* **141,** 564.
(M2) Mihalas, D. (1978) *Stellar Atmospheres.* (2nd ed.) San Francisco: Freeman.
(M3) Mihalas, D., Auer, L. H., and Heasley, J. N. (1975) *NCAR Technical Note No. TN/STR-104.* Boulder: National Center for Atmospheric Research.
(M4) Milne, E. A. (1924) *Phil. Mag.,* **47,** 209.
(M5) Milne, E. A. (1966) in *Selected Papers on Physical Processes in Ionized Plasmas,* ed. D. H. Menzel, p. 77. New York: Dover.
(M6) Minerbo, G. (1978) *J. Quant. Spectrosc. Rad. Transf.,* **20,** 541.
(N1) Nelson, R. (1967) *Lawrence Livermore Laboratory Report No. TPNU 66–8.* Livermore: University of California.
(O1) Osborne, R. K. and Klevans, E. H. (1961) *Ann. Phys.,* **15,** 105.
(P1) Pai, S.-I. (1966) *Radiation Gas Dynamics.* New York: Springer.
(P2) Peraiah, A. and Grant, I. P. (1973) *J. Inst. Maths. Applics.,* **12,** 75.
(P3) Pomraning, G. C. (1973) *The Equations of Radiation Hydrodynamics.* Oxford: Pergamon.
(R1) Rybicki, G. B. (1971) *J. Quant. Spectrosc. Rad. Transf.,* **11,** 589.
(S1) Sampson, D. H. (1965) *Radiative Contributions to Energy and Momentum Transport in a Gas.* New York: Interscience.
(S2) Slater, J. C. (1960) *Quantum Theory of Atomic Structure.* New York: McGraw-Hill.
(S3) Sneden, C., Johnson, H. R., and Krupp, B. M. (1976) *Astrophys. J.,* **204,** 281.
(S4) Spillman, G. R. (1968) *Formulation of the Eddington Factor for Use in One-Dimensional Nonequilibrium Diffusion Calculations.* Los Alamos Scientific Laboratory Office Memorandum, January 25, 1968.
(S5) Sykes, J. B. (1951) *Mon. Notices Roy. Astron. Soc.,* **111,** 377.
(T1) Thomas, R. N. (1957) *Astrophys. J.,* **125,** 260.
(T2) Thomas, R. N. (1965) *Some Aspects of Non-Equilibrium Thermodynamics in the Presence of a Radiation Field.* Boulder: University of Colorado.
(U1) Unsöld, A. (1955) *Physik der Sternatmosphären.* (2nd ed.) Berlin: Springer.
(W1) Woolley, R. v.d. R. and Stibbs, D. W. N. (1953) *The Outer Layers of a Star.* Oxford: Oxford University Press.

7

The Equations of Radiation Hydrodynamics

In astrophysical flows, radiation often contains a large fraction of the energy density, momentum density, and stress (i.e., pressure) in the radiating fluid. Furthermore, radiative transfer is usually the most effective energy-exchange mechanism within the fluid. To describe the behavior of such flows we need conservation laws that account accurately for both the material and the radiative contributions to the flow dynamics.

To estimate the importance of radiation in fixing the *local* properties of a radiating fluid, consider the ratio R of the material internal energy density $\hat{e}$ to the radiation energy density E; for a perfect gas and equilibrium radiation

$$R \equiv \hat{e}/E = (3k/2a_R)(N/T^3) = 2.8 \times 10^{-2} N/T^3.$$

R also gives a measure of the relative importance of gas and radiation pressure because $p = \frac{2}{3}\hat{e}$ for a perfect gas, and $P = \frac{1}{3}E$ for radiation. Clearly radiation is most important at high temperatures and/or low densities. The two energy densities are about equal when $T_{\text{keV}} \approx 2\rho^{1/3}$, where T_{keV} is the temperature in kilovolts (1.2×10^7 K) and ρ is the material density in g cm^{-3}. Therefore, when temperatures reach a few keV (e.g., in X-ray sources or stellar interiors), radiation dominates the energy and pressure in the radiating fluid even at high densities.

In astrophysical systems R has a large range. For example $R \sim 10^4$ in the solar atmosphere, so radiative contributions to the energy density can be ignored; in an O-star atmosphere $R \sim 0.1$, and radiation is overwhelmingly important. This striking difference reflects both factor-of-ten larger temperatures and much lower densities in the O-star atmosphere compared to the Sun. Similarly, at the Sun's center $R \sim 500$, but at the center of an O-star $R \sim 1$. The large value of R in the solar interior reflects high densities (~ 100 g cm^{-3}) and temperatures of only about a kilovolt; in contrast the central temperature of an O-star is a few kilovolts and densities are a few g cm^{-3}.

The situation for energy *transport* in radiating flows is quite different. Radiative energy transfer usually dominates all other mechanisms even when temperatures are only about 1 eV and $E \ll \hat{e}$. In particular, radiative transport usually greatly exceeds thermal conduction because in equilibrium the photons and material particles have the same average energy, but photons travel at the speed of light, whereas material particles move only at about the sound speed; moreover, photons usually have much longer mean free paths than particles.

A semiquantitative measure of the relative importance of radiative and material energy transport in a radiating flow is given by the dimensionless *Boltzmann number*

$$\text{Bo} \equiv (\rho c_p T v)/(\sigma_R T^4),$$

which is the ratio of the material enthalpy flux to the radiative flux from a free surface at temperature T. The Boltzmann number plays the same role for radiating fluids as the Peclet number does for nonradiating fluids [cf. (28.4)]. Recalling that $\sigma_R = \frac{1}{4} a_R c$, one sees that near a radiating surface Bo is of the order of (v/c) times the ratio defined above. In the solar atmosphere $(v/c) \sim 2 \times 10^{-5}$, and in an O-star atmosphere $(v/c) \sim 10^{-4}$, whence we conclude that radiative transport is dominant in the outer layers of most stars. In the interior of a star we must replace $\sigma_R T^4$ with $\sigma_R T_{\text{eff}}^4$, the *net* radiative flux; here energy transport by convection can dominate if the fluid moves at even a small fraction of its sound speed. If, on the other hand, the material is stable against convection (**C5**, Chap. 13), then radiative transport dominates in the interior as well.

Thus far we have discussed radiation as if it plays only an incidental role in a flow. But, in some cases, radiation can *drive* flows. For example, in the outer layers of a star radiative energy and momentum transport can drive or damp waves, drive stellar winds, and inhibit gravitational accretion. Furthermore, the temperature and density response of the opacity in the envelope of some stars allows radiation to drive stellar pulsations.

The equations of radiation hydrodynamics can be formulated in a variety of ways; each has advantages and disadvantages. One fundamental issue is whether to write the equations in an inertial frame fixed relative to an external observer (or the center of the star), or in the comoving fluid frame. Another concerns how best to describe the dynamical behavior of the radiation field. Thus in a stellar interior the radiation and material are in equilibrium, and we can treat the radiating fluid as a composite gas whose total energy, pressure, etc. are simple sums of the radiative and material contributions. But such an approach is virtually useless in the outer layers of a star where the radiation field has a strongly nonlocal character; here we must couple the dynamical equations to a full radiation transport equation.

In a moving fluid, the equation of transfer contains $O(v/c)$ frame-dependent terms that lead to similar terms in the dynamical equations for the radiating fluid. In contrast, the frame-dependent terms for a nonradiating fluid are only $O(v^2/c^2)$ (cf. §42). One can understand how $O(v/c)$ effects arise in a radiating fluid from simple classical considerations. First, there is an *advection* effect: a fluid element tends to "sweep up" ("leave behind") photons traveling against (along) its velocity vector, thus increasing (decreasing) the radiation energy density with which it can interact. Second, *Doppler shifts* affect the spectral distribution of the radiation field incident on the material. Consider a reference state with two fluid elements at rest, between which a certain energy and momentum exchange occurs. Now move one element relative to the other. Then, in addition to the change in the photon number density produced by advection, each photon will be blue (red) shifted, hence will have higher (lower) energy, when the two elements approach (recede from) one another. Both of these $O(v/c)$ effects can significantly affect the energy and momentum balance in a radiating fluid when the radiation field is intense.

The arguments advanced above are qualitative, and only serve to motivate a thorough mathematical analysis. In this work we will be guided by two precepts. First, we will pay close attention to the frame in which the equations are being written. In the past, failure to discriminate carefully between frames has led to confusion in the formulation of the dynamical equations, to misapplication of results valid in one frame to others in which they are not, and to serious conceptual errors. Second, we will retain mathematical consistency among various sets of equations to $O(v/c)$. The analysis is sometimes tedious, and may test our readers' patience. We assure them that this effort is not merely a quixotic obsession, but is essential to achieve equivalence among different forms of the radiating-fluid dynamical equations, both in a given frame, and between frames. The effort is vindicated by the surprising result that in certain regimes of interest, terms that are formally only $O(v/c)$ actually dominate over all others in the equations.

For didactic simplicity we ignore scattering and assume LTE. Though these restrictions afford considerable simplification, the resulting equations are complicated, and methods for solving them are not yet fully developed. Nevertheless it is essential to derive physically accurate equations, for it is clearly more useful to solve the correct equations, however approximately, than to solve incorrect equations, even exactly.

We first discuss (§7.1) the Lorentz transformation properties of quantities appearing in the transfer equation. In §7.2 we first write the transfer equation for moving media, then derive the energy and momentum equations for the radiating fluid (i.e., material plus radiation). We treat inertial-frame equations first because the derivation of the comoving-frame transfer equation is more complicated. We next discuss (§7.3) methods for solving these equations in one-dimensional flows. Here we consider first the important limiting case of diffusion, which offers penetrating insight into the dynamical behavior of the radiation field. We then discuss the comoving-frame equations, which are ideal for one-dimensional Lagrangean hydrodynamics calculations. Finally we consider two important versions of the inertial-frame equations.

7.1 Lorentz Transformation of the Transfer Equation

In order to write the transfer equation in different frames, we must determine the Lorentz transformation properties of its constituents: the specific intensity, opacity, emissivity, and photon directions and energies. In the formulae below the affix "0" denotes the comoving frame, in which material properties are isotropic.

89. The Photon Four-Momentum

In §37 we showed that the *photon four-momentum* is

$$M^\alpha = (h\nu/c)(1, \mathbf{n}) \tag{89.1}$$

where ν, $h\nu$, and $\mathbf{n}$ are the frequency, energy, and direction of propagation of the photon. The photon propagation four vector is

$$K^\alpha = (2\pi\nu/c)(1, \mathbf{n}). \tag{89.2}$$

Both K^α and M^α are null vectors.

The components of M^α in (89.1) are in Cartesian coordinates, hence are physical components. Later we will also need the contravariant components of M^α in spherical coordinates having a line element

$$ds^2 = -c^2\,dt^2 + dr^2 + r^2(d\theta^2 + \sin^2\theta\,d\phi^2). \tag{89.3}$$

Using equation (A3.41) we find

$$M^0 = h\nu/c, \tag{89.4a}$$

$$M^1 = (h\nu/c)\mu, \tag{89.4b}$$

$$M^2 = (h\nu/c)[(1-\mu^2)^{1/2}\cos\Phi]/r, \tag{89.4c}$$

and

$$M^3 = (h\nu/c)[(1-\mu^2)^{1/2}\sin\Phi]/(r\sin\theta), \tag{89.4d}$$

where $\Theta \equiv \cos^{-1}\mu$ and Φ are the polar and azimuthal angles of $\mathbf{n}$ relative to $\hat{\mathbf{r}}$.

If a photon has frequency ν and travels in direction $\mathbf{n}$ as measured in the lab frame, it will have some other frequency ν_0 and direction $\mathbf{n}_0$ as measured by an observer attached to a fluid element moving with velocity $\mathbf{v}$ relative to lab frame. Because M^α is a four-vector, its components in the two frames are related by the general Lorentz transformations (35.33) and (35.34), whence we obtain

$$\nu_0 = \gamma\nu(1 - \mathbf{n}\cdot\mathbf{v}/c) \tag{89.5}$$

and

$$\mathbf{n}_0 = (\nu/\nu_0)\{\mathbf{n} - \gamma(\mathbf{v}/c)[1 - (\gamma\mathbf{n}\cdot\mathbf{v}/c)/(\gamma+1)]\}, \tag{89.6}$$

or, equivalently,

$$\nu = \gamma\nu_0(1 + \mathbf{n}_0\cdot\mathbf{v}/c) \tag{89.7}$$

and

$$\mathbf{n} = (\nu_0/\nu)\{\mathbf{n}_0 + \gamma(\mathbf{v}/c)[1 + (\gamma\mathbf{n}_0\cdot\mathbf{v}/c)/(\gamma+1)]\}. \tag{89.8}$$

For the special case of motion along the z axis, (89.5) and (89.6) simplify to

$$(\nu_0, \nu_0 n_{x0}, \nu_0 n_{y0}, \nu_0 n_{z0}) = [\gamma\nu(1-n_z\beta), \nu n_x, \nu n_y, \gamma\nu(n_z-\beta)], \quad (89.9)$$

which implies

$$[\nu_0; \mu_0; (1-\mu_0^2)^{1/2}; \Phi_0] = [\gamma\nu(1-\beta\mu); (\mu-\beta)/(1-\beta\mu); (1-\mu^2)^{1/2}/\gamma(1-\beta\mu); \Phi]. \quad (89.10)$$

Similarly the inverse transformation gives

$$[\nu; \mu; (1-\mu^2)^{1/2}] = [\gamma\nu_0(1+\beta\mu_0); (\mu_0+\beta)/(1+\beta\mu_0); (1-\mu_0^2)^{1/2}/\gamma(1+\beta\mu_0)]. \quad (89.11)$$

Equations (89.10) and (89.11) describe the Doppler shift and aberration of light between frames in relative motion; the classical formulae are obtained by retaining terms only to $O(v/c)$, that is, by setting $\gamma \equiv 1$. These equations also apply to radial flow in spherical geometry.

From (89.10) one finds $d\nu_0 = (\nu_0/\nu)\, d\nu$, $d\mu_0 = (\nu/\nu_0)^2\, d\mu$, and $d\Phi = d\Phi_0$. Then recalling that $d\omega = d\mu\, d\Phi$ we see that $\nu\, d\nu\, d\omega$ is a Lorentz invariant:

$$\nu\, d\nu\, d\omega = \nu_0\, d\nu_0\, d\omega_0, \quad (89.12)$$

a result we will use repeatedly. Equation (89.12) has a deeper physical significance. In §43 we showed that for particles of any kind

$$d^3p/\tilde{e} = p^2\, dp\, d\omega/\tilde{e} \quad (89.13)$$

is an invariant. In particular, for photons $p = h\nu/c$ and $\tilde{e} = h\nu = cp$, hence the invariance of (89.13) implies (89.12).

90. Transformation Laws for the Specific Intensity, Opacity, and Emissivity

To determine the transformation properties of the specific intensity, we follow L. H. Thomas (**T1**) and calculate the number of photons N in a frequency interval $d\nu$, passing through an element of area dS oriented perpendicular to the z axis, into a solid angle $d\omega$ along an angle $\Theta = \cos^{-1}\mu$ to the z axis in a time interval dt. Let dS be stationary in the lab frame. Then

$$N = [I(\mu, \nu)/h\nu](d\omega\, d\nu)(dS \cos\Theta\, dt). \quad (90.1)$$

To an observer in a frame moving with velocity v along the z axis, dS appears to be moving with a velocity v in the negative z direction. This observer would therefore count

$$N_0 = [I_0(\mu_0, \nu_0)/h\nu_0](d\omega_0\, d\nu_0)[dS \cos\Theta_0\, dt_0 + (v/c)\, dS\, dt_0] \quad (90.2)$$

photons passing through dS; the first term gives the number of photons that would have been counted if dS had been stationary, while the second is the photon number density $\psi_0 = (I_0/ch\nu_0)$ times the volume $(dS\, v\, dt_0)$

swept out by dS in a time $dt_0 = \gamma\, dt$. But both observers must count the same *number* of photons passing through dS, hence $N = N_0$. Equating (90.1) and (90.2), and using (89.11) and (89.12) we find

$$I(\mu, \nu) = (\nu/\nu_0)^3 I_0(\mu_0, \nu_0). \tag{90.3}$$

That is, the quantity

$$\mathcal{I}(\mu, \nu) \equiv I(\mu, \nu)/\nu^3 \tag{90.4}$$

is a Lorentz invariant, called the *invariant intensity*.

We can obtain the same result by applying to the photon distribution function f_R the general arguments of §43, which led to the conclusion that the particle distribution function $f(\mathbf{x}, \mathbf{p}, t)$ is Lorentz invariant. From (63.4) we then immediately see that $\mathcal{I} = I/\nu^3 = \text{const.} \times f_R$ is an invariant.

Now consider the emissivity. Observers in all frames will count the same *number* of photons emitted from a definite volume element into a particular solid angle and frequency interval in a specified time interval. Hence

$$\eta(\mu, \nu)\, d\omega\, d\nu\, dV\, dt/h\nu = \eta_0(\nu_0)\, d\omega_0\, d\nu_0\, dV_0\, dt_0/h\nu_0. \tag{90.5}$$

Then using (89.12) and recalling that $dV\, dt$ is an invariant we find

$$\eta(\mu, \nu) = (\nu/\nu_0)^2 \eta_0(\nu_0), \tag{90.6}$$

where we noted that η is isotropic in the comoving frame.

Similarly, observers in all frames will count the same number of photons absorbed by a definite material element from a particular frequency interval and solid angle in a specified time interval. Hence

$$\chi(\mu, \nu) I(\mu, \nu)\, d\nu\, d\omega\, dV\, dt/h\nu = \chi_0(\nu_0) I_0(\mu_0, \nu_0)\, d\nu_0\, d\omega_0\, dV_0\, dt_0/h\nu_0, \tag{90.7}$$

whence

$$\chi(\mu, \nu) = (\nu_0/\nu)\chi_0(\nu_0). \tag{90.8}$$

We can also derive (90.8) from (90.3) and (90.6) by arguing that to achieve energy balance in equilibrium we must be able to equate the number of emissions and absorptions by a material element in all frames.

In deriving (90.3), (90.4), (90.6), and (90.8) we made use of the special Lorentz transformation for simplicity. The same results apply for arbitrary relative motion of the two frames provided that μ is replaced by $\mathbf{n}$, and we use (89.5) to (89.8) to relate $(\nu, \mathbf{n})$ to $(\nu_0, \mathbf{n}_0)$.

91. *The Radiation Stress-Energy Tensor and Four-Force Vector*

THE STRESS-ENERGY TENSOR

We now seek an expression for the *radiation stress-energy tensor* R, the spacetime generalization of the radiation stress tensor P defined in §65. We can infer the form of R by requiring that the space components R^{ij} be the rate of transport of the ith component of the radiative momentum per unit volume through a unit area oriented perpendicular to the jth coordinate axis. Thus we write

$$R^{ij} = \int f_R M^i c n^j \, d^3M \tag{91.1}$$

which is the integral of (number of particles per cm^3 per unit phase volume) × (momentum in i direction per particle) × (velocity component in j direction) over all phase space. But for a photon $cn^j = c^2 M^j/\tilde{e}$, so we tentatively generalize (91.1) to

$$R^{\alpha\beta} = c^2 \int f_R M^\alpha M^\beta \frac{d^3M}{\tilde{e}}. \tag{91.2}$$

$R^{\alpha\beta}$ is obviously a four-tensor because it is the integral of the outer product of the four-vector M^α with itself, times the invariants f_R and $d^3M/\tilde{e}$.

We have already seen that the space components of (91.2) are the radiative stress. The component

$$R^{00} = \int f_R h\nu \, d^3M \tag{91.3}$$

is the integral of (number of particles per cm^3 per unit phase volume) × (energy per particle) over all phase space, and hence equals the radiation energy density. Likewise

$$R^{0i} = (1/c) \int f_R h\nu c n^i \, d^3M \tag{91.4}$$

equals $(1/c)$ times the energy flux density in the ith direction, while

$$R^{i0} = c \int f_R n^i (h\nu/c) \, d^3M \tag{91.5}$$

equals c times the momentum density in the ith direction. Thus R as given by (91.2) is a one-to-one analogue, for radiation, of the material stress-energy tensor defined in §40.

Note that (91.2) can also be applied to material particles, for which $p^i = mv^i$ and $\tilde{e} = mc^2$, where m is the relative mass of the particle. Thus (91.2) is the covariant generalization of the particle momentum flux density tensor (43.45), and provides a general expression for the stress-energy tensor in kinetic theory. The discussion above is purposely heuristic; a much deeper analysis that emphasizes the geometric aspects of the problem can be found in (**S6,** Chaps. 1–3).

Using (63.4) to replace f_R with the specific intensity, and noting that $p^2 \, dp \, d\omega = h^3\nu^2 \, d\nu \, d\omega/c^3$, we can write a continuum version of (91.2) as

$$R^{\alpha\beta} = c^{-1} \int_0^\infty d\nu \oint d\omega I(\mathbf{n}, \nu) n^\alpha n^\beta, \tag{91.6}$$

where we define $n^0 \equiv 1$ as in (89.2). $R^{\alpha\beta}$ as given by (91.6) is manifestly covariant because it is the outer product of the photon four-momentum with itself, times the invariants $I\nu^{-3}$ and $\nu\, d\nu\, d\omega$, integrated over all angles and frequencies. An equivalent form of (91.6) is

$$\mathsf{R} = \begin{pmatrix} E & c^{-1}\mathbf{F} \\ c^{-1}\mathbf{F} & \mathsf{P} \end{pmatrix}, \tag{91.7}$$

where E, $\mathbf{F}$, and P are the radiation energy density, flux, and stress tensor as defined in §§64 to 66. The elements of (91.7) can obviously be interpreted in exactly the same way as (91.3) to (91.5).

Using (66.6), one finds that in planar geometry (91.7) reduces to

$$\mathsf{R} = \begin{pmatrix} E & 0 & 0 & c^{-1}F \\ 0 & \frac{1}{2}(E-P) & 0 & 0 \\ 0 & 0 & \frac{1}{2}(E-P) & 0 \\ c^{-1}F & 0 & 0 & P \end{pmatrix} \tag{91.8}$$

where $i = 1, 2, 3$ denote (x, y, z) respectively. The components in (91.8) are physical components, and are identical to the components measured with respect to an orthonormal tetrad in a curvilinear (e.g., spherical) coordinate system. Using the transformation rules (A3.47) we can write the contravariant components of R in spherical symmetry as

$$R^{\alpha\beta} = \begin{pmatrix} E & c^{-1}F & 0 & 0 \\ c^{-1}F & P & 0 & 0 \\ 0 & 0 & \frac{1}{2}(E-P)/r^2 & 0 \\ 0 & 0 & 0 & \frac{1}{2}(E-P)/r^2\sin^2\theta \end{pmatrix}, \tag{91.9}$$

where now $i = 1, 2, 3$ denote (r, θ, ϕ). Equations (91.7) to (91.9) also give the comoving-frame radiation stress-energy tensor R_0 if all quantities are measured in that frame.

The connection between R and R_0 is obtained from the Lorentz transformations (35.41) and (35.42). One finds

$$E = \gamma^2(E_0 + 2c^{-2}v_iF_0^i + c^{-2}v_iv_jP_0^{ij}), \tag{91.10}$$

$$F^i = \gamma\{F_0^i + \gamma E_0 v^i + v_jP_0^{ij} + [\gamma/c^2(\gamma+1)][(2\gamma+1)v_jF_0^j + \gamma v_jv_kP_0^{jk}]v^i\}, \tag{91.11}$$

and

$$\begin{aligned} P^{ij} = P_0^{ij} &+ \gamma c^{-2}(v^iF_0^j + v^jF_0^i) + \gamma^2c^{-2}E_0v^iv^j \\ &+ [\gamma^2/c^2(\gamma+1)](v^jv_kP_0^{ik} + v^iv_kP_0^{kj} + 2\gamma c^{-2}v_kF_0^kv^iv^j) \\ &+ [\gamma^2/c^2(\gamma+1)]^2(v_kv_lP_0^{kl})v^iv^j. \end{aligned} \tag{91.12}$$

For one-dimensional flow in planar geometry, (91.10) to (91.12) reduce to

$$E = \gamma^2(E_0 + 2\beta c^{-1}F_0 + \beta^2 P_0), \tag{91.13}$$

$$F = \gamma^2[(1+\beta^2)F_0 + vE_0 + vP_0], \tag{91.14}$$

and

$$P = \gamma^2(P_0 + 2\beta c^{-1}F_0 + \beta^2 E_0). \tag{91.15}$$

These equations also apply in spherical symmetry for radial flow. We can further reduce (91.13) to (91.15) to $O(v/c)$, obtaining

$$E = E_0 + 2\beta c^{-1}F_0, \tag{91.16}$$

$$F = F_0 + vE_0 + vP_0, \tag{91.17}$$

and

$$P = P_0 + 2\beta c^{-1}F_0. \tag{91.18}$$

The corresponding inverse transformations are

$$(E_0, F_0, P_0) = [E - 2\beta c^{-1}F, F - v(E+P), P - 2\beta c^{-1}F]. \tag{91.19}$$

Equations (91.16) to (91.19) can also be derived by using (89.11), (89.12), and (90.3) expanded to first order in v/c. Thus $I_\nu\, d\nu\, d\omega = (\nu/\nu_0)^2 I^0_{\nu_0}\, d\nu_0\, d\omega_0 \approx (1+2\beta\mu_0)I^0_{\nu_0}\, d\nu_0\, d\omega_0$, from which (91.16) follows by integrating over solid angle and frequency. Similarly, $\mu I_\nu\, d\nu\, d\omega = (\mu_0+\beta)(1+\beta\mu_0)I^0_{\nu_0}\, d\nu_0\, d\omega_0 \approx [\mu_0 + \beta(1+\mu_0^2)]I^0_{\nu_0}\, d\nu_0\, d\omega_0$ leads to (91.17), while $\mu^2 I_\nu\, d\nu\, d\omega = (\mu_0+\beta)^2 I^0_{\nu_0}\, d\nu_0\, d\omega_0 \approx (\mu_0^2 + 2\beta\mu_0)I^0_{\nu_0}\, d\nu_0\, d\omega_0$ leads to (91.18). Note that (91.10) to (91.19) apply only to frequency-integrated moments.

THE FOUR-FORCE DENSITY VECTOR

By analogy with (42.1) we expect the dynamical equations for the radiation field to have the general form

$$R^{\alpha\beta}_{;\beta} = -G^\alpha, \tag{91.20}$$

where G^α is the *radiation four-force density* acting on the material. Thus the time component G^0 equals c^{-1} times the net rate of radiative energy input, per unit volume, into the matter, while the space components G^i equal the net rate of radiative momentum input. From these physical interpretations it is easy to write G^α in terms of macroscopic absorption and emission coefficients as

$$G^0 = c^{-1}\int_0^\infty d\nu \oint d\omega[\chi(\mathbf{n}, \nu)I(\mathbf{n}, \nu) - \eta(\mathbf{n}, \nu)] \tag{91.21a}$$

and

$$G^i = c^{-1}\int_0^\infty d\nu \oint d\omega[\chi(\mathbf{n}, \nu)I(\mathbf{n}, \nu) - \eta(\mathbf{n}, \nu)]n^i. \tag{91.21b}$$

G^α is manifestly a four-vector, being the integral of the four-vector

$\nu(1, \mathbf{n})$, times the invariants $(\chi I/\nu^2)$ or (η/ν^2) and $\nu\, d\nu\, d\omega$, over all angles and frequencies. Thus (91.20), with R given by (91.7) and G^α by (91.21), is indeed a covariant conservation relation for the radiation field. For example, (91.20) in Cartesian coordinates yields the moment equations (78.4) and (78.11) derived in Chapter 6, consistent with the physical interpretation of those equations.

The relationship between G^α and G_0^α is obtained by Lorentz transformation. For one-dimensional flow in planar or spherical geometry,

$$G^0 = \gamma(G_0^0 + \beta G_0^1) \tag{91.22a}$$

and

$$G^1 = \gamma(G_0^1 + \beta G_0^0), \tag{91.22b}$$

or equivalently,

$$G_0^0 = \gamma(G^0 - \beta G^1) \tag{91.23a}$$

and

$$G_0^1 = \gamma(G^1 - \beta G^0). \tag{91.23b}$$

Here

$$cG^0 = 2\pi \int_0^\infty d\nu \int_{-1}^1 d\mu [\chi(\mu, \nu) I(\mu, \nu) - \eta(\mu, \nu)], \tag{91.24a}$$

$$cG^1 = 2\pi \int_0^\infty d\nu \int_{-1}^1 d\mu [\chi(\mu, \nu) I(\mu, \nu) - \eta(\mu, \nu)]\mu, \tag{91.24b}$$

$$cG_0^0 = \int_0^\infty [c\chi_0(\nu_0) E_0(\nu_0) - 4\pi\eta_0(\nu_0)]\, d\nu_0, \tag{91.25a}$$

and

$$cG_0^1 = \int_0^\infty \chi_0(\nu_0) F_0(\nu_0)\, d\nu_0. \tag{91.25b}$$

92. *Covariant Form of the Transfer Equation*

THE PHOTON BOLTZMANN EQUATION

For convenience, in this section we use units in which $h = c = 1$ and work in Cartesian coordinates. The standard Boltzmann equation for particles is

$$(\partial f/\partial t) + v^i(\partial f/\partial x^i) + p^i(\partial f/\partial p^i) = (Df/Dt)_{\text{coll}}. \tag{92.1}$$

An obvious covariant generalization of (92.1) is

$$\left(\frac{dx^\alpha}{d\tau}\right)\frac{\partial f}{\partial x^\alpha} + \left(\frac{dp^\alpha}{d\tau}\right)\frac{\partial f}{\partial p^\alpha} = \left(\frac{\delta f}{\delta \tau}\right)_{\text{coll}}, \tag{92.2}$$

where $(\delta/\delta\tau)$ is the intrinsic derivative with respect to proper time. Because photon world lines lie on the null cone, proper time is not a useful variable for the photon Boltzmann equation, so we replace τ by a new affine path-length variable ℓ defined such that

$$p^\alpha \equiv (dx^\alpha/d\ell). \tag{92.3}$$

[See (**S7,** §2.4) for a similar approach for geodesics, to which we return in §95.] We can then rewrite (92.2) as

$$p^\alpha(\partial f/\partial x^\alpha)+\dot{p}^\alpha(\partial f/\partial p^\alpha)=(\delta f/\delta \ell)_{\text{coll}}, \tag{92.4}$$

where

$$\dot{p}^\alpha \equiv (dp^\alpha/d\ell). \tag{92.5}$$

For photons we identify p^α with M^α, and write the right-hand side in terms of a source ε and a sink $-\alpha f_R$, representing photon emission and absorption by the material. Thus the *photon Boltzmann equation* is

$$M^\alpha(\partial f_R/\partial x^\alpha)+\dot{M}^\alpha(\partial f_R/\partial M^\alpha)=\varepsilon-\alpha f_R, \tag{92.6}$$

or, in terms of the invariant intensity,

$$M^\alpha(\partial \mathscr{I}/\partial x^\alpha)+\dot{M}^\alpha(\partial \mathscr{I}/\partial M^\alpha)=\mathscr{e}-\mathscr{a}\mathscr{I}. \tag{92.7}$$

Equation (92.7) applies in all frames, in particular in inertial frames. In the absence of general relativistic effects, photon trajectories in inertial frames are straight lines, hence $\dot{M}^\alpha \equiv 0$ (i.e., the photon four-momentum is conserved). Thus in an inertial frame (92.7) reduces to

$$M^\alpha \mathscr{I}_{,\alpha}=\mathscr{e}-\mathscr{a}\mathscr{I}. \tag{92.8}$$

Substituting $\mathscr{I}=I/\nu^3$ and noting that ν is now a constant, we find that the left-hand side of (92.8) is ν^{-2} times the left-hand side of the time-dependent transfer equation (76.5). Therefore on the right-hand side we can identify

$$\mathscr{e}=\eta_\nu/\nu^2 \tag{92.9a}$$

and

$$\mathscr{a}=\nu\chi_\nu. \tag{92.9b}$$

That is, $\mathscr{e}$ and $\mathscr{a}$ are just the invariant emissivity and invariant opacity discussed in §90.

LORENTZ INVARIANCE OF THE TRANSFER EQUATION

Let us now show that the transfer equation is covariant under Lorentz transformation between inertial frames. We stress that this statement holds only between frames moving *uniformly* relative to one another (see below and §95).

One approach is to argue that because $\mathscr{I}$ is a Lorentz invariant, $\mathscr{I}_{,\alpha}$ must be a covariant four-vector, hence $M^\alpha\mathscr{I}_{,\alpha}$ is an invariant. Thus between two inertial frames we can write

$$\begin{aligned}\frac{1}{\nu^2}[\eta(\mathbf{n},\nu)-\chi(\mathbf{n},\nu)I(\mathbf{n},\nu)]&=\frac{1}{\nu^2}\left[\frac{1}{c}\frac{\partial I(\mathbf{n},\nu)}{\partial t}+\mathbf{n}\cdot\nabla I(\mathbf{n},\nu)\right]\\ &=M^\alpha\mathscr{I}_{,\alpha}\equiv M'^\alpha\mathscr{I}'_{,\alpha}=\frac{1}{\nu'^2}\left[\frac{1}{c}\frac{\partial I'(\mathbf{n}',\nu')}{\partial t'}+\mathbf{n}'\cdot\nabla' I'(\mathrm{n}',\nu')\right].\end{aligned} \tag{92.10}$$

Equating the left- and right-most expressions in (92.10), and applying

(90.3), (90.6), and (90.8) we have

$$\frac{1}{c}\frac{\partial I'(\mathbf{n}', \nu')}{\partial t'} + \mathbf{n}' \cdot \nabla' I'(\mathbf{n}', \nu') = \eta'(\mathbf{n}', \nu') - \chi'(\mathbf{n}', \nu') I'(\mathbf{n}', \nu'), \quad (92.11)$$

which is identical in form to the transfer equation in the unprimed frame, as asserted.

Alternatively, we can use equations (35.39) and (35.12) to infer the transformation properties of the four-gradient (a covariant vector); for the special Lorentz transformation

$$\left(\frac{1}{c}\frac{\partial}{\partial t}, \frac{\partial}{\partial x}, \frac{\partial}{\partial y}, \frac{\partial}{\partial z}\right) = \left[\gamma\left(\frac{1}{c}\frac{\partial}{\partial t'} - \beta\frac{\partial}{\partial z'}\right), \frac{\partial}{\partial x'}, \frac{\partial}{\partial y'}, \gamma\left(\frac{\partial}{\partial z'} - \frac{\beta}{c}\frac{\partial}{\partial t'}\right)\right]. \quad (92.12)$$

Combining (92.12) and (89.9) we then have

$$c^{-1}(\partial/\partial t) + (\mathbf{n} \cdot \nabla) \equiv (\nu'/\nu)[c^{-1}(\partial/\partial t') + (\mathbf{n}' \cdot \nabla')]. \quad (92.13)$$

Therefore

$$c^{-1}(\partial I_\nu/\partial t) + (\mathbf{n} \cdot \nabla) I_\nu = \eta_\nu - \chi_\nu I_\nu \quad (92.14)$$

transforms to

$$(\nu'/\nu)[c^{-1}(\partial/\partial t') + (\mathbf{n}' \cdot \nabla')][(\nu/\nu')^3 I'(\mathbf{n}', \nu')] = (\nu/\nu')^2[\eta'(\mathbf{n}', \nu') - \chi'(\mathbf{n}', \nu') I'(\mathbf{n}', \nu')], \quad (92.15)$$

and because ν/ν' is constant for *uniformly* moving frames, we recover (92.11).

NONINERTIAL FRAMES

When we transform from the lab frame to a *noninertial* frame such as the comoving frame of a fluid whose velocity varies in position and time, we can no longer take the ratio $(\nu/\nu')^3$ to be constant and remove it from the differential operator as we did in (92.15). Instead, new terms appear that account for changes in the Lorentz transformation from one point in the flow to another.

Put another way, a photon moving on a straight line with constant frequency in the lab frame suffers differing amounts of aberration and Doppler shift as measured in different fluid elements. Thus, in the ensemble of frames composing the comoving frame, we do *not* have $\dot{M}^\alpha \equiv 0$, and (92.8) ceases to be valid. Instead, we must start from (92.7) and generalize the transfer equation to an equation of the form

$$M^\alpha \mathscr{I}_{|\alpha} = e - a\mathscr{I} \quad (92.16)$$

where the operation $_{|\alpha}$ denotes a derivative taken subject to the constraint that photon paths remain on the null cone in the fluid frame. Equation (92.16) is the *Lagrangean transfer equation*, which we discuss in detail in §95.

7.2 The Dynamical Equations for a Radiating Fluid

We are now in a position to derive the dynamical equations for a radiating fluid. As our interest centers primarily on radiative effects, we will assume, for simplicity, that the material component is an ideal fluid; the effects of viscosity and heat conduction in the material can be included by using the results of Chapters 3 and 4.

We first develop an Eulerian formulation, in which all radiation quantities are measured in the laboratory frame, and both radiation and material properties are considered to be functions of $(\mathbf{x}, t)$. The Eulerian equations are conservation relations for the total (material plus radiation) energy and momentum in a *fixed* volume element. We can cast these equations into "quasi-Lagrangean" or "modified Eulerian" form by grouping time and space derivatives into the Lagrangean time derivative (D/Dt). However, the resulting equations are not truly Lagrangean because radiation quantities are measured in the lab, rather than comoving, frame; we develop the fully Lagrangean view in §§95 and 96.

The Eulerian equations are easier to apply in multidimensional flows; indeed, except in the diffusion approximation the Lagrangean equations have been used only for one-dimensional flows. On the other hand, complexities in the physics of the material properties and/or the radiation-material interaction are most easily handled in the Lagrangean frame; moveover the Lagrangean formulation often affords deeper physical insight.

93. *The Inertial-Frame Transfer Equation for a Moving Fluid*

Consider now the inertial-frame transfer equation for a moving medium, from which we will derive inertial-frame radiation energy and momentum equations. The main question that arises is how best to account for the Doppler shift and aberration of photons from the lab frame into the moving fluid frame, where they interact with the material.

In most astrophysical flows, v/c is so small that it is tempting to ask whether we could simply *ignore* velocity-dependent effects in calculating the radiation-material interaction (at least in the continuum where cross sections change slowly). This procedure has often been used; nevertheless we will shortly see that the answer is actually "no", and that we must retain the distinction between χ and χ_0, and η and η_0 to $O(v/c)$, and solve the transfer equation to this level of accuracy.

In principle we could solve the lab-frame transfer equation by brute force, using a large number of angles and frequencies and transforming these into the comoving frame via (89.5) to (89.11) when computing material absorption and emission coefficients. But this approach is unsatisfactory for two reasons. (1) The interaction terms are cumbersome double integrals over both angle and frequency [cf. (91.21)] that are costly to evaluate. (2) It obscures important underlying physics. We therefore seek other methods of treating the matter-radiation interaction.

FORMULATION

The simplest way to handle the lab-frame angle-frequency dependence of the absorption and emission terms is to use first-order expansions to evaluate the material coefficients at the appropriate fluid-frame frequency. That is, writing

$$\nu/\nu_0 = 1 + (\mathbf{n}\cdot\mathbf{v}/c), \tag{93.1}$$

equation (90.8) expanded to $O(v/c)$ yields

$$\chi(\mathbf{n}, \nu) = \chi_0(\nu) - (\mathbf{n}\cdot\mathbf{v}/c)[\chi_0(\nu) + \nu(\partial\chi_0/\partial\nu)], \tag{93.2}$$

and (90.6) yields

$$\eta(\mathbf{n}, \nu) = \eta_0(\nu) + (\mathbf{n}\cdot\mathbf{v}/c)[2\eta_0(\nu) - \nu(\partial\eta_0/\partial\nu)]. \tag{93.3}$$

Notice that in (93.2) and (93.3), ν is the *lab*-frame frequency of the radiation.

The *transfer equation* in Cartesian coordinates can then be written

$$\begin{aligned}\frac{1}{c}\frac{\partial I(\mathbf{n}, \nu)}{\partial t} + \mathbf{n}\cdot\boldsymbol{\nabla} I(\mathbf{n}, \nu) &= \eta_0(\nu) - \chi_0(\nu) I(\mathbf{n}, \nu) \\ &+ \left(\frac{\mathbf{n}\cdot\mathbf{v}}{c}\right)\left\{2\eta_0(\nu) - \nu\frac{\partial\eta_0}{\partial\nu} + \left[\chi_0(\nu) + \nu\frac{\partial\chi_0}{\partial\nu}\right] I(\mathbf{n}, \nu)\right\}.\end{aligned} \tag{93.4}$$

The advantage gained in this approach is that both χ_0 and η_0 are isotropic, which simplifies the calculation of angular moments of (93.4). While it is reasonable to expect (93.4) to be satisfactory for smooth continua, it will not be adequate for spectral lines because a first-order expansion in $\Delta\nu$ cannot accurately track the rapid variation of χ and η over a line profile, unless the velocity-induced frequency shifts are smaller than a line width (which is not the case for most problems of interest).

Integrating (93.4) over $d\omega$ we obtain the *monochromatic radiation energy equation*

$$(\partial E_\nu/\partial t) + (\partial F_\nu^i/\partial x^i) = 4\pi\eta_0(\nu) - c\chi_0(\nu)E_\nu + (v_i F_\nu^i/c)[\chi_0(\nu) + \nu(\partial\chi_0/\partial\nu)]. \tag{93.5}$$

Integrating (93.4) against $\mathbf{n}\, d\omega$ we obtain the *monochromatic radiation momentum equation*

$$\begin{aligned}c^{-2}(\partial F_\nu^i/\partial t) + (\partial P_\nu^{ij}/\partial x^j) &= -c^{-1}\chi_0(\nu)F_\nu^i + \tfrac{4}{3}\pi c^{-2} v^i[2\eta_0(\nu) - \nu(\partial\eta_0/\partial\nu)] \\ &+ c^{-1} v_j[\chi_0(\nu) + \nu(\partial\chi_0/\partial\nu)]P_\nu^{ij}.\end{aligned} \tag{93.6}$$

Here we noted that

$$\oint n^i n^j\, d\omega = \tfrac{4}{3}\pi\, \delta^{ij}. \tag{93.7}$$

Finally, integrating (93.5) and (93.6) over frequency we obtain the *radiation energy equation*

$$E_{,t} + F^i_{,i} = \int_0^\infty [4\pi\eta_0(\nu) - c\chi_0(\nu)E_\nu]\, d\nu \\ + c^{-1}v_i \int_0^\infty [\chi_0(\nu) + \nu(\partial\chi_0/\partial\nu)]F^i_\nu \, d\nu = -cG^0 \tag{93.8}$$

and the *radiation momentum equation*

$$c^{-2}F^i_{,t} + P^{ij}_{,j} = -c^{-1}\int_0^\infty \chi_0(\nu)F^i_\nu \, d\nu + 4\pi c^{-2}v^i \int_0^\infty \eta_0(\nu)\, d\nu \\ + c^{-1}v_j \int_0^\infty [\chi_0(\nu) + \nu(\partial\chi_0/\partial\nu)]P^{ij}_\nu \, d\nu = -G^i. \tag{93.9}$$

It is important to notice that the first terms on the right-hand sides of (93.8) and (93.9) are *not* cG^0_0 and G^i_0 as defined in (91.25), despite their superficial resemblance. In G^0_0 and G^i_0 *all* quantities are evaluated in the comoving frame; in contrast, in (93.8) and (93.9) the material coefficients are in the comoving frame while radiation quantities and frequencies are in the inertial frame. To call attention to this combination of frames we refer to (93.4) to (93.9) as *mixed-frame equations.*

To obtain the corresponding equations in spherical symmetry we merely replace the left-hand sides of (93.4) to (93.6), (93.8), and (93.9) with the left-hand sides of (76.9), (78.5), (78.6), (78.13), and (78.14) because only the interaction terms are affected by the expansion procedure. Scattering terms are complicated in the mixed-frame equations; we therefore ignore them and set $\chi \equiv \kappa$ for the remainder of §93. A detailed discussion of scattering is given in (**F2**) [see also (**M8**)].

ON THE IMPORTANCE OF $O(v/c)$ TERMS

Let us now examine the physical importance of the v/c terms in (93.8) and (93.9). To simplify the discussion we specialize to grey material:

$$E_{,t} + F^i_{,i} = \kappa(4\pi B - cE) + (\kappa/c)v_iF^i = -cG^0 \tag{93.10}$$

and

$$c^{-2}F_{,t} + P^{ij}_{,j} = (\kappa/c)[-F^i + v^i(4\pi B/c) + v_jP^{ij}] = -G^i. \tag{93.11}$$

Consider first the energy equation. We instantly see that if we omit the $O(v/c)$ terms, we lose a term equal to the rate of work done by the radiation force on the material, a serious error when the radiation field is intense. Furthermore, in the diffusion regime $E_0 \rightarrow (4\pi B/c)$, hence from (91.16) and (91.17) $4\pi B - cE = -2\mathbf{v} \cdot \mathbf{F}/c + O(v^2/c^2)$. Equation (93.10) then becomes

$$E_{,t} + F^i_{,i} = -(\kappa/c)v_iF^i, \tag{93.12}$$

which is essentially the first law of thermodynamics for the radiation field. It states that the rate of change of the radiation energy density in a fixed volume plus the rate of work done by the radiation force on the material equals the net rate of (radiant) heat influx through the boundary surface of the volume. Thus in the diffusion regime we reach three important conclusions. (1) Omission of the $O(v/c)$ terms from the radiation energy equation produces an error equal in size to ignoring the net absorption-emission term, which is unacceptable. (2) We arrive at the physically correct statement (93.12) only by retaining $O(v/c)$ terms. (3) Dimensional analysis suggests that $\kappa\mathbf{v}\cdot\mathbf{F}/c$ is $O(lv/\lambda_p c)$ relative to $\nabla\cdot\mathbf{F}$; hence the velocity-dependent term may actually dominate the energy balance in the dynamic diffusion regime where $v/c \gtrsim \lambda_p/l$.

Next consider the momentum equation in the diffusion regime. On a fluid-flow time scale the time derivative is only $O(\lambda_p v/lc)$ relative to $\kappa F/c$, hence is negligible. We can therefore write

$$F^i = -(c/\kappa)P^{ij}_{,j} + v^i(4\pi B/c) + v_j P^{ij}, \tag{93.13}$$

In §97 we will show that in the diffusion limit $E_0 \to (4\pi B/c)$, $P^{ij} \to P_0^{ij} + O(\lambda_p v/lc) = \frac{1}{3}E_0\,\delta^{ij} + O(\lambda_p v/lc)$, and $\mathbf{F}_0 \to -(c/\kappa)\nabla\cdot \mathsf{P}_0$. Thus to $O(v/c)$ equation (93.13) reduces to

$$F^i = F_0^i + v^i E_0 + v_j P_0^{ij} = F_0^i + \tfrac{4}{3}v^i E_0, \tag{93.14}$$

which is just the Lorentz transformation from $\mathbf{F}_0$ to $\mathbf{F}$, cf. (91.17). Hence if we were to omit $O(v/c)$ terms in (93.11) we would fail to discriminate between the inertial-frame (Eulerian) and the comoving-frame (Lagrangean) radiation flux. To appreciate the importance of this point, recall from §80 that in a stellar interior $(vE_0/F_0) \sim (v/c)(T/T_{\text{eff}})^4 \sim 10^{12}(v/c)$, which implies that even a minuscule velocity produces a huge difference between $\mathbf{F}$ and $\mathbf{F}_0$. In short, the $O(v/c)$ terms in (93.11) are *essential* if we are to obtain the correct lab-frame flux in a moving fluid.

RELATIVE SIZES OF TERMS

The thrust of the discussion above is that terms that are formally $O(v/c)$, and which therefore appear, at first sight, to be negligible can sometimes dominate all others in the equation. Hence we must undertake a detailed analysis of the relative sizes of terms in (93.4) to (93.9) in all regimes of interest. In the streaming limit we consider both radiation-flow and fluid-flow time scales; in the diffusion limit we consider both static and dynamic diffusion.

We assume that $(v/c) \ll 1$, and agree that terms that are *always* of $O(v/c)$ or smaller relative to the dominant terms can be dropped. The key word here is "always" because terms that are negligible in one regime may dominate in another, and because any real flow spans both the optically thin and thick limits. As we desire our calculations to be accurate in both limits and successfully bridge the gap between, *any term found to be essential in one regime must be retained in all regimes.*

In the streaming limit $\lambda_p/l \gtrsim 1$, $E \approx P$, and $F \approx cE$. In the diffusion limit $E = 3P$. For static diffusion $t_f \gg t_d$ and $(v/c) \ll (\lambda_p/l)$; in this case the first term on the right-hand side of (91.17) dominates and $F \to F_0$, hence F/cE is $O(\lambda_p/l)$. For dynamic diffusion $t_f \lesssim t_d$ and $(v/c) \gtrsim (\lambda_p/l)$; in this case the last two terms in (91.17) dominate, and F/cE is $O(v/c)$. Similarly the net absorption-emission term [i.e., $\kappa(cE - 4\pi B)$] is $O(c\lambda_p/l^2)E$ for static diffusion (cf. §80), and $O(v/l)E$ for dynamic diffusion (cf. §97).

Consider first the transfer equation (93.4). In the streaming regime, dimensional analysis suggests that on a fluid-flow time scale the five terms in the equation scale as $(v/c):1:(l/\lambda_p):(l/\lambda_p):(v/c)(l/\lambda_p)$. Here we can drop both the time derivative (the radiation field is quasi static) and the velocity-dependent term on the right-hand side, retaining only the spatial operator and the absorption-emission terms. For radiation flow on a time scale t_R, the $(\partial/\partial t)$ term becomes $O(1)$ and must be retained. Now consider the diffusion regime, grouping the net emission $\eta - \kappa I$ into a single term. For static diffusion the terms scale as $(v/c):1:(\lambda_p/l):(v/c)(l/\lambda_p)$; for dynamic diffusion they scale as $(v/c):1:(v/c):(v/c)(l/\lambda_p)$. In both cases the time derivative can be dropped. For dynamic diffusion the velocity-dependent term may actually dominate all others in the equation. Even for static diffusion it will dominate the net absorption-emission term if $(v/c) \gtrsim (\lambda_p/l)^2$. Inasmuch as we always retain the absorption-emission terms, we must retain the velocity-dependent term as well. In short, *to obtain a correct solution of the inertial-frame transfer equation on a fluid-flow time scale we must retain the spatial operator on the left-hand side of* (93.4), *and all terms on the right-hand side.* To follow radiation flow on a time scale t_R, we must also retain the time derivative.

Next consider the radiation energy equation (93.10), starting with the streaming limit. Dimensional analysis suggests that on a fluid-flow time scale the five terms in the equation scale as $(v/c):1:(l/\lambda_p):(l/\lambda_p):(v/c)(l/\lambda_p)$; thus we need retain only $\nabla \cdot \mathbf{F}$ and the absorption-emission terms. To follow radiation flow, we also need to retain $(\partial/\partial t)$, which becomes $O(1)$ on a time scale t_R. An exceptional case arises if the medium is nearly in radiative equilibrium; here the absorption-emission terms may cancel almost exactly, and $(\partial/\partial t)$ and the velocity-dependent terms can then fix the energy blance. In this event we must retain all terms in the equation. Now consider the static diffusion limit. Here the terms scale as $(v/c)(l/\lambda_p):1:1:(v/c)(l/\lambda_p)$, where the net absorption-emission terms are grouped together. In this regime we can drop both the $(\partial/\partial t)$ and velocity-dependent terms because $(v/c) \ll (\lambda_p/l)$. But when $(v/c) \to (\lambda_p/l)$, all terms in the equation become of the same order and must be retained. In the dynamic diffusion limit the terms scale as $1:1:1:(v/c)(l/\lambda_p)$; here the velocity-dependent term may dominate all others.

Finally, consider the radiation momentum equation (93.9), starting with the streaming limit. On a fluid-flow time scale the terms scale as $(v/c):1:(l/\lambda_p):(v/c)(l/\lambda_p):(v/c)(l/\lambda_p)$. We need retain only $\nabla \cdot \mathsf{P}$ and the integral over $\mathbf{F}$, all other terms being at most $O(v/c)$. On a radiation-flow time scale we must also retain $(\partial/\partial t)$. In the static diffusion limit, the terms scale as $(v/c)(\lambda_p/l):1:1:(v/c)(l/\lambda_p):(v/c)(l/\lambda_p)$. In this regime we can drop both the time-derivative and velocity-dependent terms. But as $(v/c) \rightarrow (\lambda_p/l)$, the velocity-dependent terms become of the same order as $\nabla \cdot \mathsf{P}$ and must be retained, while $(\partial/\partial t)$ is only $O(v^2/c^2)$. Finally, in the dynamic diffusion limit, the terms scale as $(v^2/c^2):1:(v/c)(l/\lambda_p):(v/c)(l/\lambda_p):(v/c)(l/\lambda_p)$. Here we can drop $(\partial/\partial t)$, but must retain $\nabla \cdot \mathsf{P}$, and all three terms on the right-hand side, which are of the same size and may actually dominate the solution [cf. discussion of (93.14)].

In summary, *to solve the inertial-frame radiation energy and momentum equations correctly on a fluid-flow time scale we must retain all terms in both equations except* $(\partial/\partial t)$ *in the momentum equation, which can be dropped.* To follow radiation flow we must retain $(\partial/\partial t)$ in the momentum equation as well. Unfortunately, these requirements make the equations cumbersome to solve.

94. *Inertial-Frame Equations of Radiation Hydrodynamics*

The radiation energy and momentum equations discussed in §93 are to be solved simultaneously with conservation equations for the material, which we now derive.

GENERAL FORM

The dynamical equations for the radiation field can be written (cf. §91)

$$R^{\alpha\beta}_{;\beta} = -G^{\alpha}. \tag{94.1}$$

This expression is manifestly covariant and applies in all frames. In an *inertial* frame the covariant derivative can be evaluated immediately in any coordinate system, using the formulae in §A3. In a *noninertial* frame, we must first construct the spacetime metric before we can compute the Christoffel symbols needed to evaluate the covariant derivative of the stress-energy tensor (see §95).

In Cartesian coordinates, substitution of (91.7) into (94.1) immediately yields the radiation energy equation

$$E_{,t} + F^i_{,i} = -cG^0 \tag{94.2}$$

and the radiation momentum equation

$$c^{-2}F^i_{,t} + P^{ij}_{,j} = -G^i, \tag{94.3}$$

where G^0 and G^i are given in general by (91.21), or to $O(v/c)$ by (93.8) and (93.9). In spherical symmetry we can apply equation (A3.89) to (91.9) or (A3.91) to (91.8), noting that only $(\partial/\partial t)$ and $(\partial/\partial r)$ are nonvanishing, to obtain

$$(\partial E/\partial t)+r^{-2}[\partial(r^2F)/\partial r]=-cG^0 \tag{94.4}$$

and

$$c^{-2}(\partial F/\partial t)+(\partial P/\partial r)+(3P-E)/r=-G^1. \tag{94.5}$$

To obtain dynamical equations for a radiating fluid we use a similar approach, adopting either of two equivalent physical pictures. On one hand, we can consider the radiation field as providing an additional four-force acting on the material, and modify the dynamical equations for the material to read

$$M^{\alpha\beta}_{;\beta}=F^\alpha+G^\alpha. \tag{94.6}$$

Alternatively we can consider the externally imposed four-force F^α to act on a radiating fluid, comprising matter plus radiation, which has a total stress-energy tensor

$$S^{\alpha\beta}=M^{\alpha\beta}+R^{\alpha\beta}; \tag{94.7}$$

we then obtain the dynamical equations

$$(M^{\alpha\beta}+R^{\alpha\beta})_{;\beta}=F^\alpha. \tag{94.8}$$

In view of (94.1), equations (94.6) and (94.8) are mathematically equivalent. As we will see, (94.8) provides a conceptually more satisfying formulation in the diffusion regime, whereas (94.6) is more natural in the streaming limit.

Writing (94.6) and (94.8) in Cartesian coordinates for an ideal material fluid plus radiation, we obtain the relativistically correct equations

$$(\rho_1c^2-p)_{,t}+(\rho_1c^2v^j)_{,j}=v_jf^j+cG^0 \tag{94.9a}$$

and

$$(\rho_1v_i)_{,t}+(\rho_1v_iv^j)_{,j}=f_i-p_{,i}+G_i, \tag{94.10a}$$

or

$$(\rho_1c^2-p+E)_{,t}+(\rho_1c^2v^j+F^j)_{,j}=v_jf^j \tag{94.9b}$$

and

$$(\rho_1v^i+c^{-2}F^i)_{,t}+(\rho_1v^iv^j+P^{ij})_{,j}=f^i-\delta^{ij}p_{,j}. \tag{94.10b}$$

Here $\rho_1\equiv\gamma^2\rho_{000}$, ρ_{000} is defined by (40.9), and f^i is the Newtonian force density. Comparable expressions for general three-dimensional flows in spherical coordinates are given in (**P3**, 230–231).

If we subtract c^2 times the continuity equation (39.8) from (94.9) we obtain

$$[(\gamma-1)\rho c^2+\gamma\rho e+(\gamma^2-1)p]_{,t}+\{[(\gamma-1)\rho c^2+\gamma\rho e+\gamma^2p]v^i\}_{,i}=v_if^i+cG^0 \tag{94.11a}$$

or

$$[(\gamma-1)\rho c^2+\gamma\rho e+(\gamma^2-1)p+E]_{,t}+\{[(\gamma-1)\rho c^2+\gamma\rho e+\gamma^2 p]v^i+F^i\}_{,i}=v_i f^i, \tag{94.11b}$$

which will prove useful below; here, as in (39.5), $\rho\equiv\gamma\rho_0$.

The flows with which we deal are nonrelativistic; let us therefore reduce (94.9) to (94.11) to expressions correct to $O(v/c)$.

THE MOMENTUM EQUATION

We can cast the momentum equations (94.10) into a simpler form (as we did for a nonradiating fluid in §42) by multiplying (94.9) by v^i/c^2 and subtracting from (94.10) to obtain the relativistically correct equations (**W2**):

$$\rho_*(D\mathbf{v}/D\tau)=\mathbf{f}-\nabla p-c^2\mathbf{v}(p_{,t}+\mathbf{v}\cdot\mathbf{f})+\mathbf{G}-c^{-1}\mathbf{v}G^0, \tag{94.12a}$$

or

$$\rho_*\frac{D\mathbf{v}}{D\tau}=\mathbf{f}-\nabla p-\frac{\mathbf{v}}{c^2}\left(\frac{\partial p}{\partial t}+\mathbf{v}\cdot\mathbf{f}\right)-\left(\nabla\cdot\mathsf{P}+\frac{1}{c^2}\frac{\partial\mathbf{F}}{\partial t}\right)+\frac{\mathbf{v}}{c^2}\left(\frac{\partial E}{\partial t}+\nabla\cdot\mathbf{F}\right). \tag{94.12b}$$

Here $\rho_*\equiv\gamma\rho_{000}$, and $\mathbf{G}$ denotes the space components of G^α.

In §42 we saw that the distinctions between t and τ, and ρ and ρ_*, are $O(v^2/c^2)$, as is $\mathbf{v}(p_{,t}+\mathbf{v}\cdot\mathbf{f})/c^2$ relative to other terms. Hence for a non-radiating fluid, the Newtonian momentum equation is correct to $O(v/c)$. In contrast, for a radiating fluid the frame-dependent term $\mathbf{v}G^0/c$ can be $O(v/c)$ relative to the radiation force $\mathbf{G}$ in the streaming limit, hence the radiating-fluid momentum equation correct to $O(v/c)$ is

$$\rho(D\mathbf{v}/Dt)=\mathbf{f}-\nabla p+\mathbf{G}-(\mathbf{v}/c)G^0 \tag{94.13a}$$

or

$$\rho(D\mathbf{v}/Dt)=\mathbf{f}-\nabla p-[c^{-2}(\partial\mathbf{F}/\partial t)+\nabla\cdot\mathsf{P}]+c^{-2}\mathbf{v}[(\partial E/\partial t)+\nabla\cdot\mathbf{F}]. \tag{94.13b}$$

On a fluid-flow time scale the term containing $(\partial E/\partial t)$ is $O(v^2/c^2)$ relative to $\nabla\cdot\mathsf{P}$ and can be dropped. These equations are quasi-Lagrangean in the sense defined earlier.

The first two terms on the right-hand side of (94.13) account for externally imposed and pressure gradient forces. The third term accounts for the radiation force, expressed either as the momentum absorbed by the material from the radiative flux, or as the divergence of the radiation pressure tensor. The last term accounts for changes in the equivalent mass density of the material, as measured in the lab frame, resulting from any net gain or loss of energy by the material through its interaction with the radiation field. This term has often been omitted in discussions of radiation hydrodynamics [see, e.g., equation (9.83) in (**P3**)], but at a sacrifice in logical consistency. In particular, we will see in §96 that it is essential to retain this term in order to make an exact correspondence between the inertial-frame and comoving-frame momentum equations for a radiating fluid.

While granting the *logical* importance of the $O(v/c)$ terms in (94.13), we have agreed that because $(v/c) \ll 1$ we can drop terms that are always of this order, or smaller, for practical computations. In (94.13a), $\mathbf{v}G^0/c$ is at most $O(v/c)$ relative to $\mathbf{G}$ in the streaming limit, and even smaller if the material is in radiative equilibrium. In the diffusion limit $\mathbf{v}G^0/c$ is $O(\lambda_p v/lc)$ or $O(v^2/c^2)$ relative to $\mathbf{G}$ in the static and dynamic diffusion regimes respectively. Thus in all cases this term may be dropped. Similarly, in (94.13b), F/c is $O(1)$ relative to P in the streaming limit, and is $O(\lambda_p/l)$ or $O(v/c)$ relative to P in the static or dynamic diffusion limits; hence on a fluid-flow time scale both terms containing $\mathbf{F}$ are at most $O(v/c)$ relative to $\nabla \cdot \mathsf{P}$ and can be dropped. Thus the inertial-frame momentum equations suited to practical computation are

$$\rho(D\mathbf{v}/Dt) = \mathbf{f} - \nabla p + \mathbf{G} \tag{94.14a}$$

or

$$\rho(D\mathbf{v}/Dt) = \mathbf{f} - \nabla p - \nabla \cdot \mathsf{P}, \tag{94.14b}$$

that is, the standard Newtonian equations of motion including a radiative force.

Expressions for (94.13b) in spherical geometry, with the v/c term omitted, are given in (**P3**, 231).

THE TOTAL ENERGY EQUATION

To obtain the total energy equation for a nonrelativistic radiating fluid we simply let $\gamma \rightarrow 1$ and $(\gamma - 1) \rightarrow \frac{1}{2}v^2/c^2$ in (94.11). We then have

$$(\rho e + \tfrac{1}{2}\rho v^2)_{,t} + \{[\rho(e + \tfrac{1}{2}v^2) + p]v^i\}_{,i} = v_i f^i + cG^0, \tag{94.15a}$$

or

$$(\rho e + \tfrac{1}{2}\rho v^2 + E)_{,t} + \{[\rho(e + \tfrac{1}{2}v^2) + p]v^i + F^i\}_{,i} = v_i f^i. \tag{94.15b}$$

These Eulerian equations are correct to $O(v/c)$. Equation (94.15a) states that the rate of change of the material energy (internal plus kinetic) in a fixed volume equals the rate of work done by external forces and fluid stresses, plus the net rate of energy input to the material by absorption and emission of radiation, minus the net flux of material energy through the surface bounding the volume. Similarly, integrating (94.15b) over a fixed volume element and applying the divergence theorem, we obtain the statement that the rate of change of the total energy (internal, kinetic, and radiative) in the volume equals the rate of work done on the element by external forces and fluid stresses, minus the flux of total energy (material plus radiative) out of the volume. Detailed expressions for (94.15b) in spherical coordinates are given in (**P3**, 232).

Using (19.13) we can rewrite (94.15) in the quasi-Lagrangean form

$$\rho D(e+\tfrac{1}{2}v^2)/Dt+\nabla\cdot(p\mathbf{v})=\mathbf{v}\cdot\mathbf{f}+cG^0 \qquad (94.16a)$$

or

$$\rho D(e+\tfrac{1}{2}v^2)/Dt+(\partial E/\partial t)+\nabla\cdot(p\mathbf{v}+\mathbf{F})=\mathbf{v}\cdot\mathbf{f}. \qquad (94.16b)$$

These equations will prove useful later.

THE MECHANICAL ENERGY EQUATION

To obtain a mechanical energy equation for a radiating fluid, we form the dot product of (94.13) with $\mathbf{v}$ and drop terms of $O(v^2/c^2)$, which yields

$$\rho D(\tfrac{1}{2}v^2)/Dt=-\mathbf{v}\cdot(\nabla p)+\mathbf{v}\cdot(\mathbf{f}+\mathbf{G}) \qquad (94.17a)$$

or

$$\rho D(\tfrac{1}{2}v^2)/Dt=-\mathbf{v}\cdot(\nabla p)+\mathbf{v}\cdot\mathbf{f}-\mathbf{v}\cdot[c^{-2}(\partial\mathbf{F}/\partial t)+\nabla\cdot\mathsf{P}]. \qquad (94.17b)$$

These (quasi-Lagrangean) equations state that the rate of change of the kinetic energy per unit mass in a material element equals the rate of work, per unit mass, done by applied external and radiative forces, minus the work done against fluid stresses.

On a fluid-flow time scale the $(\partial\mathbf{F}/\partial t)$ term in (94.17b) is at most $O(v/c)$ relative to $\nabla\cdot\mathsf{P}$; hence this term is $O(v^2/c^2)$ overall and can be dropped. On a radiation-flow time scale this term is of the same order as $\nabla\cdot\mathsf{P}$ in the streaming limit.

THE GAS-ENERGY EQUATION

In §42 we derived the relativistically correct gas-energy equation for a nonradiating fluid. By exactly the same analysis, using (94.6), (94.9a), and (94.10a) we find that the relativistic gas-energy equation for a radiating fluid is

$$\rho_0\left[\frac{De}{D\tau}+p\frac{D}{D\tau}\left(\frac{1}{\rho_0}\right)\right]=-V_\alpha F^\alpha-V_\alpha G^\alpha. \qquad (94.18)$$

As before, $V_\alpha F^\alpha\equiv 0$, while $-V_\alpha G^\alpha=\gamma(cG^0-\mathbf{v}\cdot\mathbf{G})$. The inner product $V_\alpha G^\alpha$ is not zero for radiation as it is for ordinary body forces because the radiant energy absorbed by the material produces a change in its total proper energy (cf. §37). Recalling that $(dt/d\tau)=\gamma$, we see that the lab-frame gas-energy equation for a radiating fluid is

$$\rho_0\left[\frac{De}{Dt}+p\frac{D}{Dt}\left(\frac{1}{\rho_0}\right)\right]=cG^0-\mathbf{v}\cdot\mathbf{G} \qquad (94.19a)$$

or

$$\rho_0\left[\frac{De}{Dt}+p\frac{D}{Dt}\left(\frac{1}{\rho_0}\right)\right]=-\left(\frac{\partial E}{\partial t}+\nabla\cdot\mathbf{F}\right)+\mathbf{v}\cdot\left(\frac{1}{c^2}\frac{\partial\mathbf{F}}{\partial t}+\nabla\cdot\mathsf{P}\right). \qquad (94.19b)$$

Recalling (91.23a), we see that (94.19a) is the first law of thermodynamics for matter in the presence of radiation. It states that the rate of change of the internal energy per unit mass in a material element plus the rate of mechanical work done by the material in expansion, equals the net rate, per unit mass, of "heat" input from the radiation field, evaluated in the comoving fluid frame (cf. §96); compare with (93.12).

We emphasize that in (94.19) all radiation quantities are measured in the lab frame, while the material properties e, p, and ρ_0 are all measured in the comoving frame. But for the latter the distinction between frames is $O(v^2/c^2)$ and hence can be ignored to $O(v/c)$. Thus (94.19) could also be derived simply by taking the difference between the $O(v/c)$ equations (94.16) and (94.17).

Dimensional analysis suggests that in (94.19a) $\mathbf{v}\cdot\mathbf{G}$ is of the same order as cG^0 in the dynamic diffusion regime, and may exceed cG^0 in the streaming limit if the material is approximately in radiative equilibrium. Hence both terms on the right-hand side must be retained. In (94.19b), $c^{-2}\mathbf{v}\cdot(\partial\mathbf{F}/\partial t)$ is $O(v^2/c^2)$ relative to $\nabla\cdot\mathbf{F}$ on a fluid-flow time scale, and hence can be dropped. The remaining three terms are all of the same order in the dynamic diffusion regime, hence all must be retained. Thus for practical calculations the inertial-frame gas-energy equation is

$$\rho\left[\frac{De}{Dt}+p\frac{D}{Dt}\left(\frac{1}{\rho}\right)\right]=cG^0-\mathbf{v}\cdot\mathbf{G} \tag{94.20a}$$

or

$$\rho\left[\frac{De}{Dt}+p\frac{D}{Dt}\left(\frac{1}{\rho}\right)\right]=-\nabla\cdot\mathbf{F}-\frac{\partial E}{\partial t}+\mathbf{v}\cdot\nabla\cdot\mathsf{P}. \tag{94.20b}$$

Equation (94.20b) can be rewritten in either the Eulerian form

$$(\rho e+E)_{,t}+\nabla\cdot[(\rho e+p)\mathbf{v}+\mathbf{F}]=\mathbf{v}\cdot(\nabla p+\nabla\cdot\mathsf{P}) \tag{94.21}$$

or, using (19.13), in the quasi-Lagrangean form

$$\rho\left[\frac{D}{Dt}\left(e+\frac{E}{\rho}\right)+p\frac{D}{Dt}\left(\frac{1}{\rho}\right)\right]+\nabla\cdot(\mathbf{F}-\mathbf{v}E)=\mathbf{v}\cdot\nabla\cdot\mathsf{P}. \tag{94.22}$$

By straightforward manipulation (94.22) can be recast as

$$\rho\left[\frac{D}{Dt}\left(e+\frac{E}{\rho}\right)+p\frac{D}{Dt}\left(\frac{1}{\rho}\right)+\frac{1}{\rho}\,\mathsf{P}:\nabla\mathbf{v}\right]+\nabla\cdot(\mathbf{F}-\mathbf{v}E-\mathbf{v}\cdot\mathsf{P})=0; \tag{94.23}$$

compare with (96.9). Here $\mathsf{P}:\nabla\mathbf{v}$ denotes the contraction $P^{ij}v_{i,j}$.

COUPLING TO THE RADIATION EQUATIONS

The radiating-fluid momentum and energy equations written above are to be solved simultaneously with the radiation energy and momentum equations of §93 [i.e. (93.8) and (93.9) or perhaps (93.10) and (93.11)]. When simplified for practical computations on fluid-flow time scales, the fluid momentum, total energy, and mechanical energy equations are all standard Newtonian equations which include radiative terms in exactly the way one would expect from heuristic arguments. Only the gas energy equation contains a velocity-dependent radiation term that would be unanticipated from simple Newtonian arguments; this term, often ignored in inertial-frame formulations of the equations of radiation hydrodynamics, is required to convert the net rate of radiant energy input into the material to its value in the comoving fluid frame (cf. §96). Thus the *fluid* equations contain few surprises (the exception being the gas energy equation) and can be handled in the usual way. In contrast, it is in the *radiation* energy and momentum equations that special care is required, for, as we have seen, it is essential that *all* velocity-dependent terms be retained if we are to obtain the correct radiation energy and momentum balance. It is at this juncture that most Eulerian-frame treatments of radiation hydrodynamics are flawed, for the velocity-dependent terms are usually dropped, and the radiation equations are treated as if the material is at rest, which is simply incorrect.

95. *The Comoving-Frame Equation of Transfer*

RATIONALE FOR THE COMOVING FRAME

In radiation hydrodynamics the *comoving frame* of a fluid parcel comprises a *set* of inertial frames, each of which has a velocity that instantaneously coincides with that of the parcel. Clearly this frame is identical to the *Lagrangean frame* of fluid dynamics, and further is the *proper frame* in the relativistic sense, and is therefore the frame in which microscopic descriptions of material properties by thermodynamics and statistical mechanics apply. It is also the frame in which details of the interaction between radiation and matter (e.g., partial redistribution by scattering) are most easily handled (**M13**). Moreover, it offers computational advantages because it is the frame in which material properties are isotropic, and in which the frequency mesh can be tailored to describe accurately the absorption spectrum of the material; the latter point is especially important in line-formation problems (**M11**). Thus the comoving frame is the natural frame for one-dimensional flow problems such as stellar pulsations, and is the frame always used (whether explicitly or implicitly) in stellar evolution calculations that invoke the diffusion-limit solution of the transfer equation.

Because the velocity field in a flow is, in general, a function of both position and time, the comoving frame associated with any particular fluid element is a noninertial frame. Photon trajectories in the comoving frame are therefore not Euclidian straight lines, but are *geodesics* whose shapes are determined by the metric of the curved (i.e., non-Minkowskian) spacetime through which the photons move. In addition, photon frequencies are not constant in this spacetime. As a result, the comoving-frame transfer equation is more complicated than the lab-frame equation, and contains derivatives with respect to angle and frequency in addition to space coordinates and time.

There are two routes by which the comoving-frame equation of transfer can be derived, each having certain advantages. In the first we use special relativity in an inertial spacetime to derive an equation correct to all orders in (v/c); the results can then be reduced to $O(v/c)$. At this point one can safely invoke Galilean relativity because all $O(v/c)$ terms have been accounted for, and all remaining special relativistic terms are $O(v^2/c^2)$ or higher; hence a fullly Lagrangean formulation can be constructed simply by grouping terms to form the Lagrangean time derivative (D/Dt). Alternatively we can derive the equation in a noninertial Lagrangean frame from the outset, using the techniques of general relativity; here we obtain results accurate only to $O(v/c)$, but enjoy a more direct hold on the physics and deeper insight into the geometrical aspects of the problem. We will develop both approaches, limiting the discussion to one-dimensional spherically symmetric flows.

The main goal of §§95 and 96 is to obtain equations in which all *physical* variables, for both radiation and matter, are expressed in the Lagrangean frame. But we emphasize that this choice of frame is critical only for the *dependent* variables, and that the choice of *grid* (i.e., *independent* variables) on which the equations are to be solved is a matter of complete indifference. Indeed we may choose Eulerian coordinates fixed in space, Lagrangean coordinates fixed in the fluid (§98), or a *freely moving* coordinate system that is neither [e.g., an *adaptive mesh* that moves both in inertial space and with respect to fluid elements (**T3**), (**W3**)]. In practice the adaptive-mesh schemes have proven to be extraordinarily powerful tools in solving astrophysical radiation-hydrodynamics problems.

SPECIAL RELATIVISTIC FORMULATION

In deriving relativistic equations of hydrodynamics, we expressed the material stress-energy tensor in terms of proper quantities and calculated derivatives in an inertial spacetime. We can do the same for radiation, obtaining a transfer equation containing intensities, material properties, angles, and frequencies in the comoving frame only.

The inertial-frame transfer equation for spherically symmetric flow is

$$\frac{1}{c}\frac{\partial I(\mu,\nu)}{\partial t}+\mu\frac{\partial I(\mu,\nu)}{\partial r}+\frac{(1-\mu^2)}{r}\frac{\partial I(\mu,\nu)}{\partial \mu}=\eta(\mu,\nu)-\chi(\mu,\nu)I(\mu,\nu). \tag{95.1}$$

Using (90.3), (90.6), and (90.8) we can rewrite (95.1) as

$$\begin{aligned}\left(\frac{\nu}{\nu_0}\right)&\left[\frac{1}{c}\frac{\partial I_0(\mu_0,\nu_0)}{\partial t}+\mu\frac{\partial I_0(\mu_0,\nu_0)}{\partial r}+\frac{(1-\mu^2)}{r}\frac{\partial I_0(\mu_0,\nu_0)}{\partial \mu}\right]\\ &-3\left(\frac{\nu}{\nu_0^2}\right)\left[\frac{1}{c}\frac{\partial \nu_0}{\partial t}+\mu\frac{\partial \nu_0}{\partial r}+\frac{(1-\mu^2)}{r}\frac{\partial \nu_0}{\partial \mu}\right]I_0(\mu_0,\nu_0)\\ &=\eta_0(\nu_0)-\chi_0(\nu_0)I_0(\mu_0,\nu_0).\end{aligned} \tag{95.2}$$

When the derivatives in (95.2) are calculated, it is assumed that both μ and ν are held constant (with the exception of $\partial/\partial\mu$, of course). Because the fluid velocity varies in space and time, the comoving-frame quantities μ_0 and ν_0 are not constant, and we must account for their variations.

To calculate the derivatives of $I_0(\mu_0, \nu_0)$ we apply the chain rules

$$\left.\frac{\partial}{\partial t}\right|_{r\mu\nu} = \left.\frac{\partial}{\partial t}\right|_{r\mu_0\nu_0} + \left.\frac{\partial \mu_0}{\partial t}\right|_{r\mu\nu}\frac{\partial}{\partial \mu_0} + \left.\frac{\partial \nu_0}{\partial t}\right|_{r\mu\nu}\frac{\partial}{\partial \nu_0}, \tag{95.3}$$

$$\left.\frac{\partial}{\partial r}\right|_{t\mu\nu} = \left.\frac{\partial}{\partial r}\right|_{t\mu_0\nu_0} + \left.\frac{\partial \mu_0}{\partial r}\right|_{t\mu\nu}\frac{\partial}{\partial \mu_0} + \left.\frac{\partial \nu_0}{\partial r}\right|_{t\mu\nu}\frac{\partial}{\partial \nu_0}, \tag{95.4}$$

and

$$\left.\frac{\partial}{\partial \mu}\right|_{rt\nu} = \left.\frac{\partial \mu_0}{\partial \mu}\right|_{rt\nu}\frac{\partial}{\partial \mu_0} + \left.\frac{\partial \nu_0}{\partial \mu}\right|_{rt\nu}\frac{\partial}{\partial \nu_0}. \tag{95.5}$$

By repeated use of equations (89.10) and (89.11) one can evaluate all the derivatives written above in terms of comoving-frame quantities only; one finds

$$(\partial\mu_0/\partial t) = -\gamma^2(1-\mu_0^2)(\partial\beta/\partial t), \tag{95.6a}$$

$$(\partial\nu_0/\partial t) = -\gamma^2\mu_0\nu_0(\partial\beta/\partial t), \tag{95.6b}$$

$$(\partial\mu_0/\partial r) = -\gamma^2(1-\mu_0^2)(\partial\beta/\partial r), \tag{95.7a}$$

$$(\partial\nu_0/\partial r) = -\gamma^2\mu_0\nu_0(\partial\beta/\partial r), \tag{95.7b}$$

$$(\partial\mu_0/\partial\mu) = \gamma^2(1+\beta\mu_0)^2, \tag{95.8a}$$

and

$$(\partial\nu_0/\partial\mu) = -\beta\gamma^2(1+\beta\mu_0)\nu_0. \tag{95.8b}$$

Substituting (95.3) to (95.8) into (95.2) we find, after some reduction, the *comoving-frame transfer equation*

$$\begin{aligned}
&\frac{\gamma}{c}(1+\beta\mu_0)\frac{\partial I_0(\mu_0,\nu_0)}{\partial t} + \gamma(\mu_0+\beta)\frac{\partial I_0(\mu_0,\nu_0)}{\partial r}\\
&\quad+\frac{\partial}{\partial\mu_0}\left\{\gamma(1-\mu_0^2)\left[\frac{(1+\beta\mu_0)}{r} - \gamma^2(\mu_0+\beta)\frac{\partial\beta}{\partial r} - \frac{\gamma^2}{c}(1+\beta\mu_0)\frac{\partial\beta}{\partial t}\right]I_0(\mu_0,\nu_0)\right\}\\
&\quad-\frac{\partial}{\partial\nu_0}\left\{\gamma\nu_0\left[\frac{\beta(1-\mu_0^2)}{r} + \gamma^2\mu_0(\mu_0+\beta)\frac{\partial\beta}{\partial r} + \frac{\gamma^2}{c}\mu_0(1+\beta\mu_0)\frac{\partial\beta}{\partial t}\right]I_0(\mu_0,\nu_0)\right\}\\
&\quad+\gamma\left\{\frac{2\mu_0+\beta(3-\mu_0^2)}{r} + \gamma^2(1+\mu_0^2+2\beta\mu_0)\frac{\partial\beta}{\partial r}\right.\\
&\quad\left.+\frac{\gamma^2}{c}[2\mu_0+\beta(1+\mu_0^2)]\frac{\partial\beta}{\partial t}\right\}I_0(\mu_0,\nu_0) = \eta_0(\nu_0) - \chi_0(\nu_0)I_0(\mu_0,\nu_0).
\end{aligned} \tag{95.9}$$

We have grouped terms so that the angle and frequency derivatives are in *conservative form* (i.e., such that they vanish when integrated over their full ranges). Equation (95.9) is valid for $0 \leq |\beta| < 1$, and hence can be used in relativistic flows.

Integrating (95.9) over comoving-frame angles we obtain frequency-dependent moment equations. Define

$$Q_0(\nu_0) \equiv 2\pi \int_{-1}^{1} I_0(\mu_0, \nu_0)\mu_0^3 \, d\mu_0. \tag{95.10}$$

Then integrating (95.9) against $d\omega_0/4\pi$ we obtain the *monochromatic radiation energy equation*

$$\begin{aligned}
&\gamma\left[\frac{\partial E_0(\nu_0)}{\partial t} + \frac{v}{c^2}\frac{\partial F_0(\nu_0)}{\partial t}\right] + \gamma\left[\frac{\partial F_0(\nu_0)}{\partial r} + v\frac{\partial E_0(\nu_0)}{\partial r}\right] \\
&\quad + \gamma\left\{\frac{1}{r}[2F_0(\nu_0) + 3vE_0(\nu_0) - vP_0(\nu_0)] + \gamma^2\frac{\partial v}{\partial r}\left[E_0(\nu_0) + P_0(\nu_0) + \frac{2v}{c^2}F_0(\nu_0)\right]\right. \\
&\quad \left. + \frac{\gamma^2}{c^2}\frac{\partial v}{\partial t}[2F_0(\nu_0) + vE_0(\nu_0) + vP_0(\nu_0)]\right\} \\
&\quad - \frac{\partial}{\partial \nu_0}\left[\!\!\left[\gamma\nu_0\left\{\frac{v}{r}[E_0(\nu_0) - P_0(\nu_0)] + \gamma^2\frac{\partial v}{\partial r}\left[P_0(\nu_0) + \frac{v}{c^2}F_0(\nu_0)\right]\right.\right.\right. \\
&\quad \left.\left.\left. + \frac{\gamma^2}{c^2}\frac{\partial v}{\partial t}[F_0(\nu_0) + vP_0(\nu_0)]\right\}\right]\!\!\right] = 4\pi\eta_0(\nu_0) - c\chi_0(\nu_0)E_0(\nu_0).
\end{aligned} \tag{95.11}$$

Integrating (95.9) against $\mu_0 \, d\omega_0/4\pi$, we obtain the *monochromatic radiation momentum equation*

$$\begin{aligned}
&\frac{\gamma}{c^2}\left[\frac{\partial F_0(\nu_0)}{\partial t} + v\frac{\partial P_0(\nu_0)}{\partial t}\right] + \gamma\left[\frac{\partial P_0(\nu_0)}{\partial r} + \frac{v}{c^2}\frac{\partial F_0(\nu_0)}{\partial r}\right] \\
&\quad + \gamma\left\{\frac{1}{r}\left[3P_0(\nu_0) - E_0(\nu_0) + \frac{2v}{c^2}F_0(\nu_0)\right] + \frac{\gamma^2}{c^2}\frac{\partial v}{\partial r}[2F_0(\nu_0) + vE_0(\nu_0) + vP_0(\nu_0)]\right. \\
&\quad \left. + \frac{\gamma^2}{c^2}\frac{\partial v}{\partial t}\left[E_0(\nu_0) + P_0(\nu_0) + \frac{2v}{c^2}F_0(\nu_0)\right]\right\} \\
&\quad - \frac{\partial}{\partial \nu_0}\left[\!\!\left[\gamma\nu_0\left\{\frac{v}{c^2 r}[F_0(\nu_0) - Q_0(\nu_0)] + \frac{\gamma^2}{c^2}\frac{\partial v}{\partial r}[Q_0(\nu_0) + vP_0(\nu_0)]\right.\right.\right. \\
&\quad \left.\left.\left. + \frac{\gamma^2}{c^2}\frac{\partial v}{\partial t}\left[P_0(\nu_0) + \frac{v}{c^2}Q_0(\nu_0)\right]\right\}\right]\!\!\right] = -\frac{\chi_0(\nu_0)}{c}F_0(\nu_0).
\end{aligned} \tag{95.12}$$

Note that these equations contain *four* moments, unlike the inertial-frame equations in which Q does not appear.

Integrating (95.11) and (95.12) over comoving-frame frequency we obtain the *radiation energy equation*

$$\gamma\left(\frac{\partial E_0}{\partial t}+\frac{v}{c^2}\frac{\partial F_0}{\partial t}\right)+\gamma\left(\frac{\partial F_0}{\partial r}+v\frac{\partial E_0}{\partial r}\right)+\gamma\left[\frac{1}{r}(2F_0+3vE_0-vP_0)\right.$$
$$\left.+\gamma^2\frac{\partial v}{\partial r}\left(E_0+P_0+\frac{2v}{c^2}F_0\right)+\frac{\gamma^2}{c^2}\frac{\partial v}{\partial t}(2F_0+vE_0+vP_0)\right] \quad (95.13)$$
$$=\int_0^\infty [4\pi\eta_0(\nu_0)-c\chi_0(\nu_0)E_0(\nu_0)]\,d\nu_0,$$

and the *radiation momentum equation*

$$\frac{\gamma}{c^2}\left(\frac{\partial F_0}{\partial t}+v\frac{\partial P_0}{\partial t}\right)+\gamma\left(\frac{\partial P_0}{\partial r}+\frac{v}{c^2}\frac{\partial F_0}{\partial r}\right)+\gamma\left[\frac{1}{r}\left(3P_0-E_0+\frac{2v}{c^2}F_0\right)\right.$$
$$\left.+\frac{\gamma^2}{c^2}\frac{\partial v}{\partial r}(2F_0+vE_0+vP_0)+\frac{\gamma^2}{c^2}\frac{\partial v}{\partial t}\left(E_0+P_0+\frac{2v}{c^2}F_0\right)\right] \quad (95.14)$$
$$=-\frac{1}{c}\int_0^\infty \chi_0(\nu_0)F_0(\nu_0)\,d\nu_0.$$

Notice that the third moment Q_0 has vanished from these equations.

To check (95.13) and (95.14), start with the inertial-frame radiation energy and momentum equations

$$(\partial E/\partial t)+(\partial F/\partial r)+2F/r=-cG^0 \quad (95.15)$$

and

$$c^{-2}(\partial F/\partial t)+(\partial P/\partial r)+(3P-E)/r=-G^1, \quad (95.16)$$

and use (91.13) to (91.15) to eliminate (E, F, P) in favor of (E_0, F_0, P_0), and (91.22) to express G^0 and G^1 in terms of G_0^0 and G_0^1. One then finds that (95.15) equals (95.13) plus β times (95.14), and that (95.16) equals (95.14) plus β times (95.13) [cf. (**M4**)]. We are thus assured of exact consistency between the inertial- and comoving-frame equations.

Equations (95.9) to (95.14) apply in the high-velocity limit and hence can be used to describe radiative transfer in, say, the cosmic expansion, supernova blast waves, and other high-velocity flows. But for most flows $(v/c)\ll 1$ and it suffices to work only to $O(v/c)$. To first order in β, the transfer equation reduces to

$$\frac{1}{c}\frac{DI_0(\mu_0,\nu_0)}{Dt}+\frac{\mu_0}{r^2}\frac{\partial}{\partial r}[r^2 I_0(\mu_0,\nu_0)]$$
$$+\frac{\partial}{\partial\mu_0}\left\{(1-\mu_0^2)\left[\frac{1}{r}+\frac{\mu_0}{c}\left(\frac{v}{r}-\frac{\partial v}{\partial r}\right)-\frac{a}{c^2}\right]I_0(\mu_0,\nu_0)\right\}$$
$$-\frac{\partial}{\partial\nu_0}\left\{\nu_0\left[(1-\mu_0^2)\frac{v}{cr}+\frac{\mu_0^2}{c}\frac{\partial v}{\partial r}+\frac{\mu_0 a}{c^2}\right]I_0(\mu_0,\nu_0)\right\} \quad (95.17)$$
$$+\left[(3-\mu_0^2)\frac{v}{cr}+\frac{(1+\mu_0^2)}{c}\frac{\partial v}{\partial r}+\frac{2\mu_0 a}{c^2}\right]I_0(\mu_0,\nu_0)$$
$$=\eta_0(\nu_0)-\chi_0(\nu_0)I_0(\mu_0,\nu_0).$$

Here $a=(\partial v/\partial t)$, the fluid acceleration. In (95.17) we have grouped terms to form the Lagrangean time derivative (D/Dt) and have written the spatial derivative in conservative form.

Similarly, the monochromatic radiation energy equation to $O(v/c)$ is

$$\begin{aligned}\frac{DE_0(\nu_0)}{Dt}&+\frac{1}{r^2}\frac{\partial}{\partial r}[r^2F_0(\nu_0)]+\frac{v}{r}[3E_0(\nu_0)-P_0(\nu_0)]\\&+\frac{\partial v}{\partial r}[E_0(\nu_0)+P_0(\nu_0)]+\frac{2a}{c^2}F_0(\nu_0)\\&-\frac{\partial}{\partial \nu_0}\left[\!\left[\nu_0\left\{\frac{v}{r}[E_0(\nu_0)-P_0(\nu_0)]+\frac{\partial v}{\partial r}P_0(\nu_0)+\frac{a}{c^2}F_0(\nu_0)\right\}\right]\!\right]\\&=4\pi\eta_0(\nu_0)-c\chi_0(\nu_0)E_0(\nu_0),\end{aligned}\tag{95.18}$$

and the monochromatic radiation momentum equation is

$$\begin{aligned}\frac{1}{c^2}\frac{DF_0(\nu_0)}{Dt}&+\frac{\partial P_0(\nu_0)}{\partial r}+\frac{3P_0(\nu_0)-E_0(\nu_0)}{r}\\&+\frac{2}{c^2}\left(\frac{\partial v}{\partial r}+\frac{v}{r}\right)F_0(\nu_0)+\frac{a}{c^2}[E_0(\nu_0)+P_0(\nu_0)]\\&-\frac{\partial}{\partial \nu_0}\left[\!\left[\nu_0\left\{\frac{v}{c^2r}[F_0(\nu_0)-Q_0(\nu_0)]+\frac{1}{c^2}\frac{\partial v}{\partial r}Q_0(\nu_0)+\frac{a}{c^2}P_0(\nu_0)\right\}\right]\!\right]\\&=-\frac{1}{c}\chi_0(\nu_0)F_0(\nu_0).\end{aligned}\tag{95.19}$$

Finally, the radiation energy equation to $O(v/c)$ is

$$\begin{aligned}\frac{DE_0}{Dt}+\frac{1}{r^2}\frac{\partial}{\partial r}(r^2F_0)+\frac{v}{r}(3E_0-P_0)+\frac{\partial v}{\partial r}(E_0+P_0)+\frac{2a}{c^2}F_0\\=\int_0^\infty[4\pi\eta_0(\nu_0)-c\chi_0(\nu_0)E_0(\nu_0)]\,d\nu_0,\end{aligned}\tag{95.20}$$

and the radiation momentum equation is

$$\begin{aligned}\frac{1}{c^2}\frac{DF_0}{Dt}+\frac{\partial P_0}{\partial r}+\frac{3P_0-E_0}{r}+\frac{2}{c^2}\left(\frac{\partial v}{\partial r}+\frac{v}{r}\right)F_0+\frac{a}{c^2}(E_0+P_0)\\=-\frac{1}{c}\int_0^\infty\chi_0(\nu_0)F_0(\nu_0)\,d\nu_0.\end{aligned}\tag{95.21}$$

Equations (95.17) to (95.21) are equivalent to those derived by Castor (**C3**) and Buchler (**B2**), except that Castor omits the acceleration terms. On a fluid-flow time scale these terms are $O(v/c)$ compared to those in (v/r) or $(\partial v/\partial r)$, hence $O(v^2/c^2)$ overall and can be dropped; however if the velocity evolves on a radiation-flow time scale they should be retained. Moreover, as we will see in §97, these terms have an interesting physical significance.

The planar limits of (95.17) to (95.21) and (95.9) to (95.14) are obtained by letting $(1/r)\to 0$. Buchler (**B2**), (**B3**) also gives results for cylindrical geometry.

Finally, to demonstrate explicitly the consistency between the inertial- and comoving-frame dynamical equations for radiation, consider a grey, planar, pure-absorbing medium in LTE. Equations (95.20) and (95.21) become (omitting acceleration terms)

$$(\partial E_0/\partial t)+(\partial F_0/\partial z)+v(\partial E_0/\partial z)+(\partial v/\partial z)(E_0+P_0)=\kappa_0(4\pi B_0-cE_0), \tag{95.22}$$

and

$$c^{-2}(\partial F_0/\partial t)+(\partial P_0/\partial z)+(v/c^2)(\partial F_0/\partial z)+(2/c^2)(\partial v/\partial z)F_0=-c^{-1}\kappa_0 F_0. \tag{95.23}$$

On the other hand, using (91.16) to (91.18), we can rewrite the inertial-frame equation (93.10)

$$(\partial E/\partial t)+(\partial F/\partial z)=\kappa_0(4\pi B_0-cE)+(v/c)\kappa_0 F \tag{95.24}$$

as

$$\frac{\partial E_0}{\partial t}+\frac{\partial F_0}{\partial z}+v\left(\frac{\partial E_0}{\partial z}+\frac{\partial P_0}{\partial z}\right)+\frac{\partial v}{\partial z}(E_0+P_0)=\kappa_0(4\pi B_0-cE_0)-\frac{v}{c}\kappa_0 F_0+O\left(\frac{v^2}{c^2}\right). \tag{95.25}$$

Regrouping terms and using (95.23) we find

$$\begin{aligned}\frac{DE_0}{Dt}+\frac{\partial F_0}{\partial z}+\frac{\partial v}{\partial z}(E_0+P_0)&=\kappa_0(4\pi B_0-cE_0)-v\left(\frac{\kappa_0}{c}F_0+\frac{\partial P_0}{\partial z}\right)\\&=\kappa_0(4\pi B_0-cE_0)+O\left(\frac{v^2}{c^2}\right),\end{aligned} \tag{95.26}$$

which is identical to (95.22). Similarly the inertial-frame equation (93.11)

$$c^{-2}(\partial F/\partial t)+(\partial P/\partial z)=(\kappa_0/c)[-F+(4\pi v/c)B_0+vP] \tag{95.27}$$

becomes

$$\frac{1}{c^2}\frac{\partial F_0}{\partial t}+\frac{\partial P_0}{\partial z}+\frac{2v}{c^2}\frac{\partial F_0}{\partial z}+\frac{2}{c^2}\frac{\partial v}{\partial z}F_0=-\frac{\kappa_0 F_0}{c}+\frac{v}{c^2}\kappa_0(4\pi B_0-cE_0)+O\left(\frac{v^2}{c^2}\right). \tag{95.28}$$

Regrouping terms and using (95.22) we find

$$\begin{aligned}\frac{1}{c^2}\frac{DF_0}{Dt}+\frac{\partial P_0}{\partial z}+\frac{2}{c^2}\frac{\partial v}{\partial z}F_0&=-\frac{\kappa_0 F_0}{c}+\frac{v}{c^2}\left[\kappa_0(4\pi B_0-cE_0)-\frac{\partial F_0}{\partial z}\right]\\&=-\frac{\kappa_0 F_0}{c}+O\left(\frac{v^2}{c^2}\right),\end{aligned} \tag{95.29}$$

which is identical to (95.23).

NONINERTIAL FRAME FORMULATION

Following Lindquist (**L5**) and Castor (**C3**), we now derive the comoving-frame transfer equation directly in a noninertial Lagrangean frame. Again for convenience we use units in which $h=c=1$, converting to physical units at a later stage. The photon Boltzmann equation in a noninertial frame is (cf. §92)

$$M^\alpha(\partial\mathcal{I}/\partial x^\alpha)+\dot{M}^\alpha(\partial\mathcal{I}/\partial M^\alpha)=e-a\mathcal{I}=(\delta\mathcal{I}/\delta\ell)_{\text{coll}} \tag{95.30}$$

where $\mathcal{I}\equiv I/\nu^3$, $e\equiv\eta_\nu/\nu^2$, and $a\equiv\nu\chi_\nu$. Furthermore, $\dot{M}^\alpha\equiv(dM^\alpha/d\ell)$ where ℓ is an affine path-length parameter chosen to satisfy (92.3).

Photon trajectories are *geodesics* in the curved spacetime of the comoving frame. Therefore, the intrinsic derivative $(\delta M^\alpha/\delta\ell)$ is identically zero along a photon trajectory, and from equation (A3.100) we have

$$(\delta M^\alpha/\delta\ell)=(dM^\alpha/d\ell)+\begin{Bmatrix}\alpha\\\beta\gamma\end{Bmatrix}M^\beta(dx^\gamma/d\ell)\equiv 0, \tag{95.31}$$

or, in light of (92.3),

$$\dot{M}^\alpha=(dM^\alpha/d\ell)=-\begin{Bmatrix}\alpha\\\beta\gamma\end{Bmatrix}M^\beta M^\gamma. \tag{95.32}$$

Hence we can rewrite (95.30) as

$$M^\alpha(D\mathcal{I}/Dx^\alpha)=e-a\mathcal{I}=(\delta\mathcal{I}/\delta\ell)_{\text{coll}} \tag{95.33}$$

where the operator

$$(D/Dx^\alpha)\equiv(\partial/\partial x^\alpha)-\begin{Bmatrix}\gamma\\\alpha\beta\end{Bmatrix}M^\beta(\partial/\partial M^\gamma). \tag{95.34}$$

The Christoffel symbols in (95.34) are to be derived from the spacetime metric, and in general will not vanish in the (noninertial) comoving frame even in Cartesian coordinates. We must now recast (95.33) and (95.34) into a more useful form.

In writing (95.33) we have tacitly assumed that the invariant intensity is defined for all possible four-momenta. But M^α is a null vector, hence $\mathcal{I}(x^\alpha, M^\alpha)$ is actually defined only for those arguments M^α that lie on the null cone. We must therefore calculate (D/Dx^α) in such a way as to assure that M^α remains on the null cone as a photon propagates. One way of proceeding is to treat the contravariant space components M^i as independent coordinates, and to calculate (D/Dx^α) as an operator for the subset of vectors $\mathbf{M}$ of constant (null) length. But this approach is cumbersome, especially for systems having special symmetries (e.g., spherical symmetry) where simplifications are often possible. For such systems it is much more convenient to work in an *orthonormal* coordinate frame, using variables adapted to the symmetries in the problem.

Thus let M^α denote the contravariant components of $\mathbf{M}$ with respect to some coordinate system x^α that has a general metric $g_{\alpha\beta}$. Then

$$\mathbf{M}=M^\alpha\boldsymbol{\varepsilon}_\alpha \tag{95.35}$$

where the $\boldsymbol{\varepsilon}_\alpha$ are basis vectors of the coordinate system. In the neighborhood of any point $\mathbf{x}$, introduce an orthonormal *tetrad frame* $\boldsymbol{\varepsilon}_a(\mathbf{x})$, $(a = 0, 1, 2, 3)$, such that

$$\boldsymbol{\varepsilon}_a(\mathbf{x}) \cdot \boldsymbol{\varepsilon}_b(\mathbf{x}) = \eta_{ab} \tag{95.36}$$

where η_{ab} is the Lorentz metric. Relative to this frame, we can express $\mathbf{M}$ in terms of its *tetrad components* M^a as

$$\mathbf{M} = M^a \boldsymbol{\varepsilon}_a. \tag{95.37}$$

Write the transformation between the two coordinate systems as

$$\boldsymbol{\varepsilon}_a = \varepsilon_a^{\alpha} \boldsymbol{\varepsilon}_\alpha \tag{95.38a}$$

and

$$\boldsymbol{\varepsilon}_\alpha = \varepsilon_\alpha^{a} \boldsymbol{\varepsilon}_a. \tag{95.38b}$$

Then clearly

$$M^a = \varepsilon_\alpha^a M^\alpha \tag{95.39a}$$

and

$$M^\alpha = \varepsilon_a^\alpha M^a. \tag{95.39b}$$

Now suppose we choose the particular coordinate transformation $x^\alpha \to x'^\alpha \equiv x^\alpha$ and $M^\alpha \to M^a = \varepsilon_\alpha^a(\mathbf{x}) M^\alpha$ that leaves the coordinate system unchanged, but expresses the photon momentum in terms of tetrad components. Then if we regard $\mathscr{I}$ as a function of (x^α, M^a), the transfer equation can be written

$$\begin{aligned} M^a(D\mathscr{I}/Dx^a) &= (\delta\mathscr{I}/\delta\ell)_{\text{coll}} = (\partial\mathscr{I}/\partial x^\alpha)(dx^\alpha/d\ell) + (\partial\mathscr{I}/\partial M^b)(dM_b/d\ell) \\ &= M^a \varepsilon_a^\alpha(\partial\mathscr{I}/\partial x^\alpha) + (\partial\mathscr{I}/\partial M)(dM^b/d\ell). \end{aligned} \tag{95.40}$$

The operator $\partial_a \equiv \varepsilon_a^\alpha(\partial/\partial x^\alpha)$ is known as the *Pfaffian derivative.*

To calculate $(dM^b/d\ell)$, we recall that photon trajectories are geodesics in the original coordinate system, hence

$$\begin{aligned} \frac{\delta M^\beta}{\delta\ell} &= M^\beta_{,\alpha}\frac{dx^\alpha}{d\ell} + \left\{ {\beta \atop \alpha\gamma} \right\} M^\gamma \frac{dx^\alpha}{d\ell} = (\varepsilon_c^\beta M^c)_{,\alpha}\frac{dx^\alpha}{d\ell} + \left\{ {\beta \atop \alpha\gamma} \right\} M^\alpha M^\gamma \\ &= \varepsilon_c^\beta(dM^c/d\ell) + M^c \varepsilon^\beta_{c,\alpha} M^\alpha + \left\{ {\beta \atop \alpha\gamma} \right\} M^\alpha M^\gamma \equiv 0. \end{aligned} \tag{95.41}$$

Therefore

$$\begin{aligned} \varepsilon_\beta^b \varepsilon_c^\beta \frac{dM^c}{d\ell} = \delta_c^b \frac{dM^c}{d\ell} = \frac{dM^b}{d\ell} &= -\varepsilon_a^\alpha \varepsilon_\beta^b \left(\varepsilon^\beta_{c,\alpha} + \left\{ {\beta \atop \alpha\gamma} \right\} \varepsilon_c^\gamma \right) M^a M^c \\ &= -\varepsilon_a^\alpha \varepsilon_\beta^b \varepsilon^\beta_{c;\alpha} M^a M^c. \end{aligned} \tag{95.42}$$

Then defining the *Ricci rotation coefficient* to be

$$\Gamma^b_{ac} \equiv \varepsilon_a^\alpha \varepsilon_\beta^b \varepsilon^\beta_{c;\alpha}, \tag{95.43}$$

the transfer equation becomes

$$M^a(D\mathscr{I}/Dx^a) = M^a[\partial_a - \Gamma^b_{ac} M^c(\partial/\partial M^b)]\mathscr{I} = e - a\mathscr{I}. \tag{95.44}$$

Notice that, unlike Christoffel symbols, the rotation coefficients are not symmetric in the two lower indices.

Because **M** is a null vector, only three of its tetrad components can be independent, as any three suffice to determine the fourth. Therefore in the evaluation of (95.44) we need differentiate only with three components of **M**, which we take to be the three space components M^a, $(a=1,2,3)$.

We now specialize (95.44) to spherical symmetry. Choose a comoving-frame metric of the general form

$$ds^2 = -e^{2\Psi}\,d\tau^2 + e^{2\Lambda}\,d\imath^2 + R^2(d\theta^2 + \sin^2\theta\,d\phi^2), \tag{95.45}$$

where $\imath$ is a generalized Lagrangean radial coordinate, and Ψ, Λ, and R are functions of $\imath$ and τ only. In spherical symmetry the derivatives $(\partial/\partial\theta)$ and $(\partial/\partial\phi)$ are identically zero, so we need calculate only terms containing $(\partial/\partial\tau)$ and $(\partial/\partial\imath)$. From straightforward calculation one finds that the nonzero Christoffel symbols for (95.45) are:

$$\left.\begin{aligned}
&\begin{Bmatrix}0\\00\end{Bmatrix} = (\partial\Psi/\partial\tau), \qquad \begin{Bmatrix}0\\11\end{Bmatrix} = \exp[2(\Lambda-\Psi)](\partial\Lambda/\partial\tau),\\
&\begin{Bmatrix}0\\22\end{Bmatrix} = \exp(-2\Psi)R(\partial R/\partial\tau),\\
&\begin{Bmatrix}0\\33\end{Bmatrix} = \exp(-2\Psi)R(\partial R/\partial\tau)\sin^2\theta, \qquad \begin{Bmatrix}0\\10\end{Bmatrix} = (\partial\Psi/\partial\imath),\\
&\begin{Bmatrix}1\\00\end{Bmatrix} = \exp[2(\Psi-\Lambda)](\partial\Psi/\partial\imath), \qquad \begin{Bmatrix}1\\11\end{Bmatrix} = (\partial\Lambda/\partial\imath),\\
&\begin{Bmatrix}1\\22\end{Bmatrix} = -\exp(-2\Lambda)R(\partial R/\partial\imath),\\
&\begin{Bmatrix}1\\33\end{Bmatrix} = -\exp(-2\Lambda)R(\partial R/\partial\imath)\sin^2\theta,\\
&\begin{Bmatrix}1\\01\end{Bmatrix} = (\partial\Lambda/\partial\tau), \qquad \begin{Bmatrix}2\\33\end{Bmatrix} = -\sin\theta\cos\theta, \qquad \begin{Bmatrix}2\\02\end{Bmatrix} = R^{-1}(\partial R/\partial\tau),\\
&\begin{Bmatrix}2\\12\end{Bmatrix} = R^{-1}(\partial R/\partial\imath), \qquad \begin{Bmatrix}3\\03\end{Bmatrix} = R^{-1}(\partial R/\partial\tau), \qquad \begin{Bmatrix}3\\13\end{Bmatrix} = R^{-1}(\partial R/\partial\imath),\\
&\text{and}\\
&\begin{Bmatrix}3\\23\end{Bmatrix} = \cot\theta
\end{aligned}\right\} \tag{95.46}$$

At the event $(\tau, \imath, \theta, \phi)$ introduce the orthonormal basis

$$\boldsymbol{\varepsilon}_0 = e^{-\Psi}\boldsymbol{\varepsilon}_\tau, \qquad \boldsymbol{\varepsilon}_1 = e^{-\Lambda}\boldsymbol{\varepsilon}_\imath, \qquad \boldsymbol{\varepsilon}_2 = R^{-1}\boldsymbol{\varepsilon}_\theta, \qquad \text{and} \qquad \boldsymbol{\varepsilon}_3 = (R\sin\theta)^{-1}\boldsymbol{\varepsilon}_\phi. \tag{95.47}$$

One then sees that the transformation matrix ε_a^α is diagonal:

$$\varepsilon_a^\alpha = \begin{pmatrix} e^{-\Psi} & 0 & 0 & 0 \\ 0 & e^{-\Lambda} & 0 & 0 \\ 0 & 0 & R^{-1} & 0 \\ 0 & 0 & 0 & (R\sin\theta)^{-1} \end{pmatrix}, \tag{95.48}$$

whence we have

$$\varepsilon_\alpha^a \equiv (\varepsilon_a^\alpha)^{-1} = \begin{pmatrix} e^{\Psi} & 0 & 0 & 0 \\ 0 & e^{\Lambda} & 0 & 0 \\ 0 & 0 & R & 0 \\ 0 & 0 & 0 & R\sin\theta \end{pmatrix}, \tag{95.49}$$

Furthermore, write $\mathbf{M}$ in terms of spherical coordinates with $\boldsymbol{\varepsilon}_1$ taken to be the polar axis:

$$M^0 = \nu, \qquad M^1 = \nu\cos\Theta, \qquad M^2 = \nu\sin\Theta\cos\Phi, \qquad M^3 = \nu\sin\Theta\sin\Phi, \tag{95.50}$$

where ν is the photon's energy. We can then compute the Jacobian $J(M^1, M^2, M^3/\nu, \Theta, \Phi)$ and its inverse

$$J^{-1} = \frac{\partial(\nu, \Theta, \Phi)}{\partial(M^1, M^2, M^3)}$$

$$= \begin{pmatrix} \cos\Theta & -(\sin\Theta)/\nu & 0 \\ \sin\Theta\cos\Phi & (\cos\Theta\cos\Phi)/\nu & -(\sin\Phi)/\nu\sin\Theta \\ \sin\Theta\sin\Phi & (\cos\Theta\sin\Phi)/\nu & (\cos\Phi)/\nu\sin\Theta \end{pmatrix}, \tag{95.51}$$

whence we have

$$(\partial/\partial M^1) = \mu(\partial/\partial\nu) + \nu^{-1}(1-\mu^2)(\partial/\partial\mu), \tag{95.52}$$

$$(\partial/\partial M^2) = (1-\mu^2)^{1/2}\cos\Phi[(\partial/\partial\nu) - \nu^{-1}\mu(\partial/\partial\mu)], \tag{95.53}$$

and

$$(\partial/\partial M^3) = (1-\mu^2)^{1/2}\sin\Phi[(\partial/\partial\nu) - \nu^{-1}\mu(\partial/\partial\mu)], \tag{95.54}$$

where $\mu \equiv \cos\Theta$. Here we have dropped $(\partial/\partial\Phi)$ because of azimuthal symmetry.

We now must compute the Ricci rotation coefficients. Because ε_a^α and ε_α^a are diagonal, (95.43) reduces to

$$\Gamma^b_{ac} = \varepsilon_a^a(\varepsilon^{-1})_b^b\left(\varepsilon_c^c\begin{Bmatrix} b \\ ac \end{Bmatrix} + \varepsilon^b_{c,a}\,\delta_c^b\right), \tag{95.55}$$

where there is no sum on repeated indices. We can ignore terms with $b=0$ because in (95.44) we differentiate only with respect to space components of M^b. Following (**L5**), define the operators

$$D_\tau \equiv e^{-\Psi}(\partial/\partial\tau) \tag{95.56a}$$

and

$$D_{\imath} \equiv e^{-\Lambda}(\partial/\partial \imath), \tag{95.56b}$$

and the auxiliary variables

$$U \equiv D_\tau R \tag{95.57a}$$

and

$$\Gamma \equiv D_{\imath} R. \tag{95.57b}$$

Using (95.46), (95.48), and (95.49) in (95.55) we find that the nonzero Ricci coefficients are

$$\left.\begin{aligned}&\Gamma^1_{00} = D_{\imath}\Psi, \qquad \Gamma^1_{22} = \Gamma^1_{33} = -\Gamma/R, \qquad \Gamma^1_{10} = D_\tau\Lambda,\\ &\Gamma^2_{33} = -R^{-1}\cot\theta, \qquad \Gamma^2_{20} = \Gamma^3_{30} = U/R,\\ &\Gamma^2_{21} = \Gamma^3_{31} = \Gamma/R, \qquad \text{and} \qquad \Gamma^3_{32} = R^{-1}\cot\theta.\end{aligned}\right\} \tag{95.58}$$

In the transfer equation (95.44) we then have

$$M^a\partial_a = M^a\varepsilon^\alpha_a(\partial/\partial x^\alpha) = \nu D_\tau + \mu\nu D_{\imath}, \tag{95.59}$$

while

$$\begin{aligned}M^aM^c\Gamma^b_{ac}(\partial/\partial M^b) = {}&(M^0M^0\Gamma^1_{00} + M^2M^2\Gamma^1_{22} + M^3M^3\Gamma^1_{33} + M^0M^1\Gamma^1_{10})(\partial/\partial M^1)\\ &+ (M^3M^3\Gamma^2_{33} + M^1M^2\Gamma^2_{21} + M^0M^2\Gamma^2_{20})(\partial/\partial M^2)\\ &+ (M^1M^3\Gamma^3_{31} + M^0M^3\Gamma^3_{30} + M^2M^3\Gamma^3_{32})(\partial/\partial M^3).\end{aligned} \tag{95.60}$$

Substituting (95.50) and (95.52) to (95.54) into (95.60), collecting terms, and using the results along with (95.59) in (95.44) we obtain finally the comoving-frame transfer equation

$$\begin{aligned}&D_\tau\mathcal{I} + \mu D_{\imath}\mathcal{I} - \nu[\mu D_{\imath}\Psi + \mu^2 D_\tau\Lambda + (1-\mu^2)(U/R)](\partial\mathcal{I}/\partial\nu)\\ &+(1-\mu^2)\{(\Gamma/R) - D_{\imath}\Psi + \mu[(U/R) - D_\tau\Lambda]\}(\partial\mathcal{I}/\partial\mu) = \nu^{-1}(e - a\mathcal{I}).\end{aligned} \tag{95.61}$$

Equation (95.61) is exact for the general metric (95.45). To apply it to a particular flow we must obtain explicit expressions for the coefficients in the metric; it is at this point that we must forsake exactness if we wish to obtain analytical results. One sees that some kind of approximation must be made by realizing that in general the acceleration field $\mathbf{a}(\mathbf{r}, t)$ can be arbitrarily complicated, and by recalling that the principle of equivalence implies that this field can be viewed as resulting from the gravitational field of an arbitrarily complex distribution of masses. Thus an attempt to obtain an exact analytical metric for an arbitrary flow field is as difficult as solving exactly the field equations of general relativity for an arbitrary mass distribution, which is not possible by known methods. In practice, it is feasible to work analytically only to $O(v/c)$. An alternative is to construct the metric numerically; but by doing so we forsake having explicit analytical expressions for the metric and the transfer equation. See (**G1**) for a discussion of the numerical approach in the context of radiative transfer.

For one-dimensional spherically symmetric flows, Castor (**C3**) adopted inertial-frame coordinates (t', r, θ, ϕ) and Lagrangean coordinates (t, M_r, θ, ϕ), and related them by the coordinate transformation

$$M_r(r, t') = \int_0^r 4\pi(r')^2\rho(r', t')\, dr' \tag{95.62}$$

and

$$t(r, t') = t' - c^{-2}\int_0^r v(r', t')\, dr', \tag{95.63}$$

where $v = (\partial r/\partial t') = -(4\pi r^2\rho)^{-1}(\partial M_r/\partial t')_r$ is the fluid velocity, Equations (95.62) and (95.63) provide an $O(v/c)$ approximation to a local Lorentz transformation between the inertial and comoving frames in the neighborhood of the event (r, t'). From these equations one readily finds

$$dx \equiv (dM_r/4\pi r^2\rho) = dr - v\, dt' \tag{95.64}$$

and

$$dt = (1 - I/c^2)\, dt' - (v/c^2)\, dr, \tag{95.65}$$

where

$$I \equiv \int_0^r [\partial v(r', t')/\partial t']\, dr'. \tag{95.66}$$

Solving for dr and dt' we have

$$dr = [(1 - I/c^2)/D]\, dx + (v/D)\, dt \tag{95.67}$$

and

$$dt' = (v/c^2 D)\, dx + D^{-1}\, dt \tag{95.68}$$

where

$$D \equiv 1 - (I + v^2)/c^2. \tag{95.69}$$

Substituting (95.67) and (95.68) into the inertial-frame metric

$$ds^2 = dr^2 + r^2(d\theta^2 + \sin^2\theta\, d\phi^2) - c^2(dt')^2 \tag{95.70}$$

we obtain the comoving-frame metric

$$ds^2 = F(dM_r/4\pi r^2\rho)^2 + r^2(d\theta^2 + \sin^2\theta\, d\phi^2) - G\, dt^2 - 2H\, dM_r\, dt \tag{95.71}$$

where [see (**M6**)]

$$F = [(1 - I/c^2)^2 - (v^2/c^2)]/D^2, \tag{95.72}$$

$$G = (c^2 - v^2)/D^2, \tag{95.73}$$

and

$$H = vI/(4\pi r^2\rho c^2 D^2). \tag{95.74}$$

Inasmuch as we are interested in final results correct to $O(v/c)$, we may now discard terms of $O(v^2/c^2)$. We see by inspection that H is $O(v^2/c^2)$, and hence can be dropped, while $F = 1 + O(v^2/c^2)$. For G we have

$$G = c^2/(1 - 2I/c^2) + O(v^2/c^2) = c^2 + 2I + O(v^2/c^2); \tag{95.75}$$

I/c^2 can be $O(v/c)$ for radiation-flow time scales $t_R \sim \Delta r/c$ or when the fluid acceleration is comparable to (cv/r) or $c(dv/dr)$, and hence should be retained.

Comparing (95.71) with (95.45) in which $d\tau \equiv dt$ and $d\imath \equiv dM_r$ we can make the identifications

$$R \equiv r, \qquad \Lambda \equiv -\ln(4\pi r^2\rho), \qquad \text{and} \qquad \Psi \equiv \tfrac{1}{2}\ln(c^2+2I), \quad (95.76)$$

whence we find, to $O(v/c)$,

$$D_\tau \equiv c^{-1}(\partial/\partial t) = c^{-1}(D/Dt) \qquad \text{and} \qquad D_\imath \equiv (4\pi r^2\rho)(\partial/\partial M_r) \equiv (\partial/\partial r). \quad (95.77)$$

Here we noted that the time derivative calculated in the comoving frame is identical to the customary Lagrangean (D/Dt). From (95.57) and (95.77) we find $U=(v/c)$, $\Gamma \equiv 1$,

$$D_\imath\Psi = \frac{1}{c^2+2I}\frac{\partial}{\partial r}\left(\int_0^r \frac{\partial v}{\partial t'}\,dr'\right) = \frac{1}{c^2}\frac{\partial v}{\partial t'} + O\left(\frac{v^2}{c^2}\right) = \frac{a}{c^2}, \quad (95.78)$$

and

$$D_\tau\Lambda = -c^{-1}[D(\ln\rho)/Dt + (2v/r)]. \quad (95.79)$$

Using (95.76) to (95.79) in (95.61) and expressing $\mathcal{I}$, α, and e in terms of $I_0(\mu_0, \nu_0)$, $\chi_0(\nu_0)$, and $\eta_0(\nu_0)$, we find, after some elementary reductions, the comoving-frame transfer equation

$$\begin{aligned}
&\frac{1}{c}\frac{DI_0(\mu_0,\nu_0)}{Dt} + 4\pi\rho\mu_0\frac{\partial}{\partial M_r}[r^2 I_0(\mu_0,\nu_0)] \\
&+\frac{\partial}{\partial\mu_0}\left\{(1-\mu_0^2)\left[\frac{1}{r}+\frac{\mu_0}{c}\left(\frac{3v}{r}+\frac{D\ln\rho}{Dt}\right)-\frac{a}{c^2}\right]I_0(\mu_0,\nu_0)\right\} \\
&-\frac{\partial}{\partial\nu_0}\left\{\nu_0\left[(1-3\mu_0^2)\frac{v}{cr}-\frac{\mu_0^2}{c}\frac{D\ln\rho}{Dt}+\frac{\mu_0 a}{c^2}\right]I_0(\mu_0,\nu_0)\right\} \qquad (95.80)\\
&+\left[(1-3\mu_0^2)\frac{v}{cr}-\frac{(1+\mu_0^2)}{c}\frac{D\ln\rho}{Dt}+\frac{2\mu_0 a}{c^2}\right]I_0(\mu_0,\nu_0) \\
&=\eta_0(\nu_0)-\chi(\nu_0)I_0(\mu_0,\nu_0).
\end{aligned}$$

This equation is fully Lagrangean in the sense that *all* radiation and material properties are in the comoving frame, the independent variable M_r is Lagrangean, and the time derivatives (D/Dt) are evaluated in a moving fluid element. Recalling the equation of continuity

$$(D\ln\rho/Dt) = -r^{-2}[\partial(r^2v)/\partial r] = -(\partial v/\partial r)-(2v/r), \quad (95.81)$$

one easily sees that (95.80) is identical to (95.17). We thus have two logically independent derivations of the result.

Taking angular moments of (95.80) we obtain the monochromatic radiation energy equation

$$\begin{aligned}\frac{DE_0(\nu_0)}{Dt}&+4\pi\rho_0\frac{\partial}{\partial M_r}[r^2F_0(\nu_0)]-\frac{v}{r}[3P_0(\nu_0)-E_0(\nu_0)]\\&-\frac{D\ln\rho}{Dt}[E_0(\nu_0)+P_0(\nu_0)]+\frac{2a}{c^2}F_0(\nu_0)\\&+\frac{\partial}{\partial\nu_0}\left[\!\!\left[\nu_0\left\{\frac{v}{r}[3P_0(\nu_0)-E_0(\nu_0)]+\frac{D\ln\rho}{Dt}P_0(\nu_0)-\frac{a}{c^2}F_0(\nu_0)\right\}\right]\!\!\right]\\&=4\pi\eta_0(\nu_0)-c\chi_0(\nu_0)E_0(\nu_0),\end{aligned}\tag{95.82}$$

and the monochromatic radiation momentum equation

$$\begin{aligned}\frac{1}{c^2}\frac{DF_0(\nu_0)}{Dt}&+4\pi r^2\rho\frac{\partial P_0(\nu_0)}{\partial M_r}+\frac{3P_0(\nu_0)-E_0(\nu_0)}{r}\\&-\frac{2}{c^2}\left(\frac{v}{r}+\frac{D\ln\rho}{Dt}\right)F_0(\nu_0)+\frac{a}{c^2}[E_0(\nu_0)+P_0(\nu_0)]\\&+\frac{\partial}{\partial\nu_0}\left[\!\!\left[\nu_0\left\{\frac{v}{c^2r}[3Q_0(\nu_0)-F_0(\nu_0)]+\frac{1}{c^2}\frac{D\ln\rho}{Dt}Q_0(\nu_0)-\frac{a}{c^2}P_0(\nu_0)\right\}\right]\!\!\right]\\&=-\frac{\chi_0(\nu_0)}{c}F_0(\nu_0),\end{aligned}\tag{95.83}$$

which are equivalent to (95.18) and (95.19).

Integrating over frequency we obtain the radiation energy equation

$$\begin{aligned}\frac{DE_0}{Dt}+4\pi\rho\frac{\partial(r^2F_0)}{\partial M_r}-\frac{v}{r}(3P_0-E_0)-\frac{D\ln\rho}{Dt}(E_0+P_0)+\frac{2aF_0}{c^2}\\=\int_0^\infty[4\pi\eta_0(\nu_0)-c\chi_0(\nu_0)E_0(\nu_0)]\,d\nu_0\end{aligned}\tag{95.84}$$

and the radiation momentum equation

$$\begin{aligned}\frac{1}{c^2}\frac{DF_0}{Dt}+4\pi r^2\rho\frac{\partial P_0}{\partial M_r}+\frac{3P_0-E_0}{r}-\frac{2}{c^2}\left(\frac{v}{r}+\frac{D\ln\rho}{Dt}\right)F_0+\frac{a}{c^2}(E_0+P_0)\\=-\frac{1}{c}\int_0^\infty\chi_0(\nu_0)F_0(\nu_0)\,d\nu_0,\end{aligned}\tag{95.85}$$

which are equivalent to (95.20) and (95.21). These equations also follow directly from

$$R_{0;\beta}^{\alpha\beta}=-G_0^\alpha\tag{95.86}$$

where $R_0^{\alpha\beta}$ is given by (91.9) with all radiation quantities evaluated in the comoving frame, G_0^α is given by (91.25), and the covariant derivatives are evaluated in the curved spacetime of the fluid frame. Using (A3.89) with

Christoffel symbols calculated in the metric (95.71), one can show that (95.86) does, in fact, yield (95.84) and (95.85).

Equations (95.84) and (95.85) apply in spherical symmetry. Buchler has shown (**B2**) that tensorial forms of these equations, applicable in any geometry, are

$$\rho \frac{D}{Dt}\left(\frac{E_0}{\rho}\right) + \nabla \cdot \mathbf{F}_0 + \mathsf{P}_0 : \nabla \mathbf{v} + \frac{2}{c^2} \mathbf{a} \cdot \mathbf{F}_0 + cG_0^0 = 0, \qquad (95.87)$$

and

$$\frac{\rho}{c^2} \frac{D}{Dt}\left(\frac{\mathbf{F}_0}{\rho}\right) + \nabla \cdot \mathsf{P}_0 + \frac{1}{c^2} \mathbf{F}_0 \cdot \nabla \mathbf{v} + \frac{1}{c^2}(E_0 \mathbf{a} + \mathbf{a} \cdot \mathsf{P}_0) + \mathbf{G}_0 = 0. \quad (95.88)$$

The term $\mathsf{P}_0 : \nabla \mathbf{v}$ in (95.87) is dyadic notation for the contraction of P_0 with $\nabla \mathbf{v}$. Buchler also gives tensorial forms for the monochromatic moment equations [see his equations (9) and (10)].

IMPORTANCE OF $O(v/c)$ TERMS

In §93 we showed that in order to solve correctly the inertial-frame transfer equation and its moments one must retain terms that are formally $O(v/c)$ (cf. §93). Building on the discussion by Castor (**C3**), we now show that the same conclusion applies to the comoving-frame radiation and momentum equations. In making estimates of the relative sizes of terms we shall ignore the acceleration terms [which are never larger than $O(v/c)$], and consider $(\partial v/\partial r)$, (v/r), and $(D \ln \rho/Dt)$ to be $O(v/l)$. In the diffusion regime, we shall use results to be derived in §97 for estimating the sizes of the net absorption-emission terms, $\mathbf{F}_0$, and $(3P_0 - E_0)$.

Consider first the radiation energy equation (95.84); group the net absorption-emission into a single term. In the streaming limit, dimensional analysis suggests that on a fluid-flow time-scale the five terms in (95.84) scale as $(v/c):1:(v/c):(v/c):(l/\lambda_p)$, hence we need retain only the flux divergence and the absorption-emission terms; the radiation field is quasi-static. On a radiation-flow time scale we must also retain the (D/Dt) term. If the material is essentially in radiative equilibrium, the absorption-emission terms cancel almost exactly, and the (D/Dt) and velocity-dependent terms, although small, may significantly affect the energy balance; we should then retain all terms. In the static diffusion limit, the terms scale as $(v/c)(l/\lambda_p):1:(v/c)^2:(v/c)(l/\lambda_p):1$, hence only the flux-divergence and absorption-emission terms need be retained. As $(v/c) \to (\lambda_p/l)$, all terms except the one containing $(3P_0 - E_0)$ are of the same order, and all must be kept. In the dynamic diffusion regime the scaling is $1:(c/v)(\lambda_p/l):(v/c)(\lambda_p/l):1:1$. The dominant terms are the rate of change of the energy density, the rate of work done by radiation pressure, and the net absorption-emission terms; the flux divergence is of less importance than in other regimes, and again we can drop $(3P_0 - E_0)$. In summary, *to guarantee the correct radiation energy balance in all regimes, we must retain all terms in* (95.84) *except the acceleration term.*

Now consider the radiation momentum equation (95.85). In the streaming limit, dimensional analysis suggests that on a fluid-flow time scale the terms scale as $(v/c):1:1:(v/c):(l/\lambda_p)$. Hence we need retain only $\nabla \cdot \mathsf{P}_0$ and the integral of $\chi_0 F_0/c$. If we follow radiation flow, the (D/Dt) term must also be kept. In the diffusion regime the terms scale as $(v/c)(\lambda_p/l):1:(v/c)(\lambda_p/l):(v/c)(\lambda_p/l):1$, hence we can drop (D/Dt), $(3P_0 - E_0)$, and the velocity-dependent terms. This result contrasts strongly with that for the inertial-frame radiation momentum equation (where it is essential to retain all the velocity-dependent terms to obtain the correct inertial-frame flux), and reveals an important advantage of the Lagrangean formulation. *In summary, in solving the comoving-frame radiation momentum equation* (95.85) *on a fluid-flow time scale we can drop the time derivative and all velocity-dependent terms.*

Castor (**C3**) arrives at the same conclusions for a pulsating star where (D/Dt) is of the order of ω, the pulsation frequency.

96. Comoving-Frame Equations of Radiation Hydrodynamics

We are now in a position to write the Lagrangean equations of radiation hydrodynamics. We consider one-dimensional spherically symmetric flows; the corresponding planar equations are obtained by taking the limit $(1/r) \rightarrow 0$. We ignore the acceleration terms in the radiation energy and momentum equations, which are $O(v^2/c^2)$ on fluid-flow time scales (but see §97).

THE MOMENTUM EQUATION

The simplest way to obtain the comoving-frame momentum equation is to reduce the relativistically correct equation (94.12a) to the proper frame, in which $\mathbf{v} = 0$ instantaneously. We then have, to $O(v/c)$,

$$\rho_{000}(D\mathbf{v}/Dt) = \mathbf{f} - \nabla p + \mathbf{G}_0. \tag{96.1}$$

For nonrelativistic fluids $(p + \rho_0 e) \ll \rho_0 c^2$, and we can ignore the difference between ρ_{000} and ρ. Specializing (96.1) to one-dimensional spherically symmetric flow we find

$$\rho(Dv/Dt) = -(GM_r\rho/r^2) - (\partial p/\partial r) + (1/c)\int_0^\infty \chi_0(\nu_0)F_0(\nu_0)\, d\nu_0, \tag{96.2}$$

which states that a fluid element accelerates in response to applied external forces (e.g., gravity), the pressure gradient, and the force exerted by the radiation on the material as measured in its rest frame. The velocity-dependent terms in the inertial-frame momentum equation vanish in the Lagrangean frame.

To obtain the comoving-frame analogue of (94.12b), we use (95.85) to eliminate the integral in (96.2), which yields

$$\rho\frac{Dv}{Dt}+\frac{1}{c^2}\frac{DF_0}{Dt}=-\frac{GM_r\rho}{r^2}-\left(\frac{\partial p}{\partial r}+\frac{\partial P_0}{\partial r}+\frac{3P_0-E_0}{r}\right)+\frac{2}{c^2}\left(\frac{v}{r}+\frac{D\ln\rho}{Dt}\right)F_0. \tag{96.3}$$

We can also derive (96.3) by evaluating (94.12b) directly in the comoving frame provided that we replace $[c^{-2}(\partial\mathbf{F}/\partial t)+\nabla\cdot\mathsf{P}]$ with $(R_0^{1\beta})_{;\beta}$ and calculate the covariant derivative using the (nonzero) Christoffel symbols obtained from the metric (95.71). Regrouping terms in (96.3), we can write it in the more instructive form

$$\rho\frac{D}{Dt}\left[v+\left(\frac{F_0}{c^2\rho}\right)\right]=-\frac{GM_r\rho}{r^2}-\frac{\partial(p+P_0)}{\partial r}-\frac{3P_0-E_0}{r}-\frac{1}{c^2}\left(\frac{\partial v}{\partial r}\right)F_0, \tag{96.4}$$

which states that the rate of change of the total (material plus radiative) momentum density in a radiating fluid equals the applied force minus the divergence of the total stress, minus an additional (relativistic) term that arises because the radiant energy flux has inertia (cf. §97).

On a fluid-flow time scale both terms containing F_0 in (96.4) are $O(v/c)$ in the streaming limit, and $O(\lambda_p v/lc)$ in the diffusion limit, relative to $(\partial P_0/\partial r)$, and can be dropped in practical calculations. Hence another useful form of the Lagrangean momentum equation is

$$\rho(D\mathbf{v}/Dt)=\mathbf{f}-\nabla p-\nabla\cdot\mathsf{P}_0. \tag{96.5}$$

Equation (96.5) is slightly more approximate than (96.2), but assumes a particularly simple form in the diffusion limit, where $\nabla\cdot\mathsf{P}_0$ reduces to ∇P_0, so that the fluid acceleration depends on the total (gas plus radiation) pressure gradient.

THE GAS-ENERGY EQUATION

The comoving-frame gas-energy equation follows directly from the relativistic equation (94.18) by evaluating $V_\alpha F^\alpha$ and $V_\alpha G^\alpha$ in the proper frame. We obtain

$$\rho_0\{(De/D\tau)+p[D(1/\rho_0)/D\tau]\}=c(F_0^0+G_0^0), \tag{96.6}$$

where G_0^0 is given by (91.25a). For ordinary body forces $cF_0^0=(\mathbf{v}\cdot\mathbf{f})_0=0$. But in the presence of nonmechanical energy sources cF_0^0 equals the rate, per unit volume, of energy input to the material, as measured in the fluid frame (cf. §37). For example, in stellar interiors thermonuclear reactions irreversibly release ε ergs $\mathrm{g}^{-1}\,\mathrm{s}^{-1}$ into the material. In this case

$$\rho_0\left[\frac{De}{D\tau}+p\frac{D}{D\tau}\left(\frac{1}{\rho_0}\right)\right]=\int_0^\infty[c\chi_0(\nu_0)E_0(\nu_0)-4\pi\eta_0(\nu_0)]\,d\nu_0+\rho_0\varepsilon. \tag{96.7}$$

Equation (96.7) is the first law of thermodynamics for the material; it states that the rate of change of the material energy density plus the rate of work done by the material pressure equals the net rate of energy input from the radiation field and thermonuclear sources, all per unit mass. In what follows we work to $O(v/c)$, hence in (96.7) we replace ρ_0 by ρ and $(D/D\tau)$ by (D/Dt).

THE RADIATION-ENERGY EQUATION

By rearranging terms we can write (95.84) in a form that makes its physical content more apparent:

$$\rho\left[\frac{D}{Dt}\left(\frac{E_0}{\rho}\right)+P_0\frac{D}{Dt}\left(\frac{1}{\rho}\right)-(3P_0-E_0)\frac{v}{\rho r}\right] = \int_0^\infty [4\pi\eta_0(\nu_0)-c\chi_0(\nu_0)E_0(\nu_0)]\,d\nu_0-\frac{1}{r^2}\frac{\partial}{\partial r}(r^2F_0). \tag{96.8}$$

The second and third terms on the left-hand side of (96.8) reduce to $P_0^{ij}v_{i;j}$, the contraction of the radiation-pressure and fluid-velocity tensors, hence equal the rate of work done by the radiation stress [cf. (27.7)]. Thus (96.8) is the first law of thermodynamics for the radiation field; it states that the rate of change of the radiation energy density, plus the rate of work done by radiation pressure, equals the net rate of energy input into the radiation field from the material, minus the net rate of radiant energy flow out of a fluid element by transport [again cf. (27.7)], all per unit mass.

THE FIRST LAW OF THERMODYNAMICS FOR THE RADIATING FLUID

Taking the sum of (96.7) and (96.8) we obtain the first law of thermodynamics for the radiating fluid:

$$\frac{D}{Dt}\left(e+\frac{E_0}{\rho}\right)+p\frac{D}{Dt}\left(\frac{1}{\rho}\right)+\left[P_0\frac{D}{Dt}\left(\frac{1}{\rho}\right)-(3P_0-E_0)\frac{v}{\rho r}\right]=\varepsilon-\frac{\partial}{\partial M_r}(4\pi r^2F_0), \tag{96.9}$$

which states that the rate of change of the total (material plus radiation) energy density in a fluid element plus the rate of work done by the total pressure in the element equals the rate of thermonuclear energy input into the element minus the rate of radiant energy loss by transport to adjacent fluid elements.

When the radiation field is isotropic (e.g., in the diffusion regime), (96.9) simplifies to

$$\frac{D}{Dt}\left(e+\frac{E_0}{\rho}\right)+(p+P_0)\frac{D}{Dt}\left(\frac{1}{\rho}\right)=\varepsilon-\frac{\partial L_r^0}{\partial M_r}, \tag{96.10}$$

where L_r^0 is the luminosity at radius r, measured in the comoving frame. In this limit, the radiating fluid behaves like a gas whose total energy density and pressure are the simple sums of the contributions from the radiation and material components.

In the equilibrium diffusion limit, (96.10) is the standard energy equation used in dynamical stellar evolution calculations [cf. (97.7)]. For a *static* medium, it reduces to one of the standard equations of stellar structure

$$(\partial L_r^0/\partial M_r) = \varepsilon, \tag{96.11}$$

which apply to stable stars evolving on a *nuclear time scale* t_N, which is so long compared to dynamical times of interest (e.g., the free-fall time or a pulsation period) that the evolution is quasi-stationary and fluid motions can be neglected.

THE MECHANICAL ENERGY EQUATION

To obtain the fluid-frame mechanical energy equation we multiply the momentum equation (96.2) by v, which yields

$$\rho D(\tfrac{1}{2}v^2)/Dt = -(GM_r v\rho/r^2) - v(\partial p/\partial r) + (v/c)\int_0^\infty \chi_0(\nu_0)F_0(\nu_0)\,d\nu_0, \tag{96.12}$$

which is identical to (24.8) if we lump the radiative force into f, and to (94.17a) except that here the radiation force is evaluated in the comoving frame.

THE TOTAL ENERGY EQUATION

To obtain a total energy equation we first rewrite (96.12) as

$$\frac{D}{Dt}\left(\tfrac{1}{2}v^2 - \frac{GM_r}{r}\right) + \frac{\partial}{\partial M_r}(4\pi r^2 vp) = p\frac{D}{Dt}\left(\frac{1}{\rho}\right) + \frac{v}{c\rho}\int_0^\infty \chi_0(\nu_0)F_0(\nu_0)\,d\nu_0. \tag{96.13}$$

Next, substituting from (95.85) for the radiation force, and ignoring terms of $O(v^2/c^2)$ we obtain

$$\frac{D}{Dt}\left(\tfrac{1}{2}v^2 - \frac{GM_r}{r}\right) + \frac{\partial}{\partial M_r}[4\pi r^2 v(p+P_0)] = (p+P_0)\frac{D}{Dt}\left(\frac{1}{\rho}\right) - \frac{v}{\rho r}(3P_0 - E_0). \tag{96.14}$$

Finally, adding (96.14) to (96.9) we have

$$\frac{D}{Dt}\left(e + \frac{E_0}{\rho} + \tfrac{1}{2}v^2 - \frac{GM_r}{r}\right) + \frac{\partial}{\partial M_r}\{4\pi r^2[v(p+P_0) + F_0]\} = \varepsilon, \tag{96.15}$$

which is clearly a statement of overall energy conservation for the radiating fluid. All radiation quantities in (96.15) are to be evaluated in the comoving frame.

Equation (96.15) is essentially identical to equation (27.4), written in spherical coordinates, for an inviscid but conducting (via radiation) fluid whose internal energy density is the sum of the gas and radiation energy densities, and whose pressure equals the sum of the gas and radiation pressures, with the term on the right-hand side accounting for "external" energy input from thermonuclear reactions. This equation, with $4\pi r^2 F_0$ replaced by L_r, and E_0 and P_0 given their thermal equilibrium values, is the total energy equation used in dynamical stellar structure calculations [see, for example, (**C4**, eq. 6); (**F1**, eq. 3); (**K7**, eq. 15); or (**L2**, eq. 51.3)].

We can rewrite (96.15) in Eulerian coordinates as

$$\frac{\partial}{\partial t}\left(\rho e + E_0 + \tfrac{1}{2}\rho v^2 - \frac{GM_r\rho}{r}\right) + \frac{1}{r^2}\frac{\partial}{\partial r}\left[\!\left[r^2\left\{\left[\rho\left(e+\tfrac{1}{2}v^2-\frac{GM_r}{r}\right)+p+P_0+E_0\right]v+F_0\right\}\right]\!\right]=\rho\varepsilon. \tag{96.16}$$

Then using (91.17a) and ignoring $O(v^2/c^2)$ terms in converting E_0 to E in the time derivative, we obtain

$$\frac{\partial}{\partial t}\left(\rho e + E + \tfrac{1}{2}\rho v^2 - \frac{GM_r\rho}{r}\right) + \frac{1}{r^2}\frac{\partial}{\partial r}\left[\!\left[r^2\left\{\left[\rho\left(e+\tfrac{1}{2}v^2-\frac{GM_r}{r}\right)+p\right]v+F\right\}\right]\!\right]=\rho\varepsilon, \tag{96.17}$$

which is identical to the Eulerian result (94.15b) when thermonuclear energy release is allowed. In (96.17), radiation quantities are now measured in the laboratory frame.

Assuming that $\dot{\mathcal{M}}$ is so small that we can neglect the time variation of M_r, we can write an explicit integral of (96.17) for the case of steady flow [cf. (24.22) for a nonradiating fluid]. We find

$$\dot{\mathcal{M}}[h+\tfrac{1}{2}v^2-(GM_r/r)]+L_r = 4\pi\int_0^r \rho\varepsilon x^2\,dx. \tag{96.18}$$

That is, the total energy flux passing through a surface of radius r, consisting of the material energy flux (i.e., the mass flux times the enthalpy plus kinetic plus potential energy per unit mass) plus the luminosity radiated by the surface (measured in the lab frame) equals the total thermonuclear energy release in the volume bounded by the surface. In physical terms, (96.18) states that all the energy contained in radiation and in fluid motions in a star originates ultimately from thermonuclear energy release in the star's interior.

CONSISTENCY OF VARIOUS FORMS OF THE COMOVING-FRAME ENERGY AND MOMENTUM EQUATIONS

We now show that $O(v/c)$ terms must also be retained in order to obtain consistency among various forms of the comoving-frame energy equation, and between the comoving-frame and inertial-frame energy and momentum equations. Our discussion summarizes and extends a penetrating analysis of these issues by Castor (**C3**). An earlier, but incomplete, treatment was given by Wendroff (**W2**).

In an optically thin medium, or near a radiating surface of an opaque medium, the radiation field departs strongly from thermal equilibrium, hence J can differ markedly from B, the flux is large, and the radiation pressure tensor is anisotropic. In this regime, it is natural to describe the energy exchange between the material and radiation in terms of direct gains and losses, as in (96.7), and momentum exchange in terms of radiation forces acting on the material, as in (96.2).

In contrast, in the diffusion regime $J \rightarrow B$, so that the net absorption-emission term in (96.7) vanishes to high order, and the flux becomes a very small leak from the large reservoir of radiant energy. The radiation energy density and pressure both approach their equilibrium values, and the radiation pressure becomes isotropic. It is then natural to calculate the total energy content and pressure of the radiating fluid by adding the material and radiative contributions, and to use (96.9) as the energy equation and (96.5) as the momentum equation.

In any practical computation we must choose *one* form of the fluid energy equation even when the flow spans both the optically thin and thick limits. If the $O(v/c)$ terms are retained in the radiation energy equation (95.84), and this equation is solved simultaneously with either fluid energy equation, the choice is immaterial because exact consistency between the two is guaranteed. But suppose we *drop* the $O(v/c)$ terms from (95.84). Then if we use (96.7), we will obtain satisfactory results in the optically thin regime, but will make serious errors in the optically thick regime, where $J \rightarrow B$ and the right-hand side vanishes almost identically, because we have not accounted explicitly for either the rate of change of the internal energy in the radiation or the rate of work done by radiation pressure. Castor concludes (**C3**) that in the diffusion regime the temperature determined from (96.7) with the $O(v/c)$ terms omitted from (95.84) can be in error by an amount of $O(P/p)$. If, instead, we use (96.9) the difficulty is reversed. We then obtain an accurate solution at great depth, but will make serious errors in the optically thin regime where the gas decouples from the radiation; Castor finds that the error in the temperature is again $O(P/p)$. In short, it is *essential* to retain $O(v/c)$ terms in (95.84) in order to bridge the transition between the optically thick and thin limits.

The situation for the momentum equation is different. Here (DF_0/Dt) and the velocity-dependent terms multiplying F_0 in (96.3) are never larger than $O(v/c)$, and are much smaller in the diffusion limit. We can therefore drop these terms, which means that we will obtain consistency with (96.2) even if we drop the time-derivative and velocity-dependent terms from the radiation momentum equation (95.85). Moreover, in the derivation of the mechanical energy equation (96.12), which when combined with (96.9), leads to the total energy equation (96.15), all $O(v/c)$ terms in (95.85) become $O(v^2/c^2)$, and hence can be dropped from the outset. In short, we do not adversely affect consistency among various forms of the energy or momentum equations by dropping all $O(v/c)$ terms from (95.85).

CONSISTENCY OF THE INERTIAL-FRAME AND COMOVING-FRAME ENERGY AND MOMENTUM EQUATIONS FOR A RADIATING FLUID

Let us now examine the mutual consistency of the inertial-frame and comoving-frame energy and momentum equations. Consider first the inertial-frame gas-energy equation (94.19b). On a fluid-flow time scale the $(\partial\mathbf{F}/\partial t)$ term is $O(v^2/c^2)$ relative to $\nabla\cdot\mathbf{F}$ and hence can be dropped. Similarly the (v/c) terms in the transformations of (E, P) into (E_0, P_0) will produce terms of $O(v^2/c^2)$; we thus need to retain $O(v/c)$ terms only to transform $\mathbf{F}$ to $\mathbf{F}_0$. In particular, for one-dimensional spherically symmetric flow we have

$$\begin{aligned}\nabla\cdot\mathbf{F} &= \frac{1}{r^2}\frac{\partial}{\partial r}[r^2(F_0+vE_0+vP_0)]\\ &= \frac{\partial F_0}{\partial r}+\frac{2F_0}{r}+v\left(\frac{\partial E_0}{\partial r}+\frac{\partial P_0}{\partial r}\right)+\left(\frac{\partial v}{\partial r}+\frac{2v}{r}\right)(E_0+P_0).\end{aligned} \tag{96.19}$$

Furthermore, from (66.10)

$$\nabla\cdot\mathsf{P}_0=(\partial P_0/\partial r)+(3P_0-E_0)/r. \tag{96.20}$$

Using these results in (94.19b) we find

$$\rho\left[\frac{De}{Dt}+p\frac{D}{Dt}\left(\frac{1}{\rho}\right)\right]=-\left[\frac{DE_0}{Dt}+\frac{1}{r^2}\frac{\partial}{\partial r}(r^2F_0)+\frac{v}{r}(3E_0-P_0)+(E_o+P_0)\frac{\partial v}{\partial r}\right], \tag{96.21}$$

which, by virtue of (95.84) is identical to the comoving-frame gas-energy equation (96.7). If the velocity-dependent term on the right-hand side of (94.19b) had been omitted, we would be left with an extra term in (96.21) of the form $v(\partial P/\partial r)$, that is, the rate of work done by the fluid against the radiation pressure gradient. For fluids with intense radiation fields, this term is large and would lead to serious errors. By a similar analysis, one readily shows that (94.22) is consistent with (96.9).

Alternatively, consider the inertial-frame equation (94.19a), which for grey material reduces to

$$\rho\left[\frac{De}{Dt}+p\frac{D}{Dt}\left(\frac{1}{\rho}\right)\right]=\kappa_0(cE-4\pi B_0-2\mathbf{v}\cdot\mathbf{F}/c)+O(v^2/c^2). \tag{96.22}$$

Then using (91.16) we have

$$\rho\left[\frac{De}{Dt}+p\frac{D}{Dt}\left(\frac{1}{\rho}\right)\right]=\kappa_0(cE_0-4\pi B_0), \tag{96.23}$$

which is identical to the comoving-frame equation (96.7) for grey material. Had the $O(v/c)$ terms been omitted from (94.19a), from (93.10) and (93.11), or from (91.16), this exact reduction would not be achieved; the error would equal $\kappa_0\mathbf{v}\cdot\mathrm{F}/c$, the rate of work done by radiation forces on the material.

In summary, consistency between the inertial-frame and comoving-frame equations requires that all $O(v/c)$ terms be retained in both gas-energy equations, in the radiation energy equation, and in the transformation laws between frames [see also (**P4**)]. In contrast, all $O(v/c)$ terms can be omitted from the radiation momentum equation without loss of consistency.

Finally, consider the inertial-frame momentum equation (94.13b), which for spherically symmetric flow reduces to

$$\rho\frac{Dv}{Dt}=\frac{-GM_r\rho}{r^2}-\frac{\partial p}{\partial r}-\left[\frac{1}{c^2}\frac{\partial F}{\partial t}+\frac{\partial P}{\partial r}+\frac{(3P-E)}{r}-\frac{v}{c^2}\left(\frac{\partial E}{\partial t}+\frac{\partial F}{\partial r}+\frac{2F}{r}\right)\right]. \tag{96.24}$$

On a fluid-flow time scale the term containing $(\partial E/\partial t)$ is $O(v^2/c^2)$ relative to $(\partial P/\partial r)$, and therefore can be dropped. Similarly all terms containing F are at most $O(v/c)$ relative to the terms in E and P. Hence to obtain a final result accurate to $O(v/c)$ it is sufficient to set $F=F_0$, but all terms must be retained in transforming from (E, P) to (E_0, P_0). Making these conversions we find

$$\rho\frac{Dv}{Dt}=\frac{-GM_r\rho}{r^2}-\frac{\partial p}{\partial r}-\left[\frac{1}{c^2}\frac{\partial F_0}{\partial t}+\frac{v}{c^2}\frac{\partial F_0}{\partial r}+\frac{\partial P_0}{\partial r}+\frac{(3P_0-E_0)}{r}+\frac{2}{c^2}\left(\frac{\partial v}{\partial r}+\frac{v}{r}\right)F_0\right], \tag{96.25}$$

which is identical to the comoving-frame equation (96.3). Thus consistency of the momentum equation between frames is assured if, and only if, one accounts for $O(v/c)$ terms in both frames.

Similarly, in light of (93.10) and (93.11) the inertial-frame momentum equation (94.13a) for a spherically symmetric flow of grey material is

$$\rho(Dv/Dt)=-(GM_r\rho/r^2)-(\partial p/\partial r)+(\kappa_0/c)[F-(v/c)(E+P)]+O(v^2/c^2), \tag{96.26}$$

which, from (91.19), is identical to the comoving-frame equation (96.2) for grey material. Again we see that the $O(v/c)$ terms are essential for consistency.

7.3 Solution of the Equations of Radiation Hydrodynamics

MATHEMATICAL STRUCTURE OF THE PROBLEM

In §§93 to 96 we formulated the equations of radiation hydrodynamics in both the Eulerian and Lagrangean frames; we now ask how to solve them. In this connection it is instructive to count the number of variables to be determined and the number of equations available to determine them, as in §24. As before we must find seven fluid variables: ρ, p, T, e, and three components of $\mathbf{v}$; in addition we must now find ten radiation variables: E, the three components of $\mathbf{F}$, and the six nonredundant components of P.

These seventeen variables are related by nine partial differential equations: the equation of continuity, the material energy equation including radiation terms, three components of the material momentum equation, the radiation energy equation, and three components of the radiation momentum equation. In addition we have two material constitutive relations: the pressure and caloric equations of state. (We assume that the material opacity and emissivity are given as functions of, say, ρ and T.) We are thus short by six equations, which in effect are closure relations relating P_{ij} to E. These relations can be specified a priori in the diffusion regime, but in general they must be determined either iteratively, or from some ad hoc prescription (cf. §78). In addition we must specify appropriate boundary and initial conditions.

The equations of radiation hydrodynamics in two- and three-dimensional flows are truly formidable; indeed they have never been solved for nontrivial problems except in the limit of radiation diffusion. We shall therefore confine our attention to one-dimensional flows, in which case we need to determine only eight variables (ρ, p, T, e, v_r, E, F, P) from a total of five differential equations and two constitutive relations; here we need to specify only one Eddington factor $f = P/E$ in order to complete the system.

NUMERICAL APPROACH

To handle the coupled system described above we replace the differential equations by suitable discrete approximations and solve these numerically. At this juncture it suffices to describe the procedure verbally, reserving a presentation of difference equations for a specific example in §98.

Suppose we wish to solve the Lagrangean equations of radiation hydrodynamics in one-dimensional planar or spherical geometry. We divide the medium into discrete cells, locating velocities at cell interfaces and material properties (density, pressure, etc.) at cell centers, as we did in the absence of radiation (cf. Chapter 5). In the momentum equation we need to know the radiation force (hence the flux) at the same locations as fluid velocities and accelerations (i.e., the interfaces). On the other hand, to apply the first law of thermodynamics to the radiating fluid we need to know the radiation energy density and pressure at the same locations as their material counterparts (i.e., cell centers). Hence we solve the time-dependent radiation moment equations on the same mesh as used for the static transfer equation in Chapter 6.

Next consider the time centering of the variables. For simplicity we assume that, as in Chapter 5, we use an explicit form of the momentum equation. As before we time center material properties at t^n and velocities at $t^{n-(1/2)} \equiv \frac{1}{2}(t^{n-1} + t^n)$. To advance the velocity from $t^{n-(1/2)}$ to $t^{n+(1/2)}$ we need to know the external forces, the pressure gradient, and the radiation force, all at t^n; thus radiation forces, whether computed from the flux as in (96.2) or from the gradient of the radiation pressure as in (96.5), should be centered at t^n. Given velocities $v^{n+(1/2)}$ we can advance the interface positions from t^n to t^{n+1}, and by continuity update the material density to t^{n+1}. For a nonradiating fluid, we determine T, hence e and p, at t^{n+1} from an implicit form of the material energy equation. For a radiating fluid, we must solve implicit forms of the material energy equation, including radiation terms, simultaneously with the radiation energy and momentum equations to determine T, e, p, E, P, and (if needed) F at t^{n+1}. Clearly the radiation quantities must be centered at the same time level as the material properties (i.e., at t^n and t^{n+1}).

Integration of the momentum equation including radiation forces differs only trivially from the cases considered in Chapter 5, hence we focus here primarily on the question of how to formulate and solve the coupled material and radiation energy equations at the advanced time t^{n+1}. We discuss first the diffusion limit, then the Lagrangean equations, and finally two Eulerian or mixed-frame approaches. We apply some of these methods in Chapter 8.

97. Radiation Diffusion Methods

OVERVIEW

We saw in §80 that in static material the radiation field in the *equilibrium diffusion regime* is determined by local fluid properties and their gradients. We will now show that one can also develop an asymptotic solution of the transfer equation and obtain an explicit analytical expression for the radiation stress-energy tensor in an opaque *moving* medium. These results afford deep insight into the dynamical effects of radiation in a radiating flow. One can use the analytical form of the stress-energy tensor directly in the energy and momentum equations for the radiating fluid, and thus, in effect, dispense with the transfer equation altogether. We shall treat only nonrelativistic flows; the diffusion approximation in relativistic flows is discussed in (**G1**). Furthermore we assume pure absorption and ignore scattering; generalizations to include scattering in the Thomson limit can be found in (**H2**) and (**M1**), and in the Compton limit in (**M2**).

The basic assumption made in radiation diffusion theory is that the material is extremely opaque, so that $(\lambda_p/l) \ll 1$. If the material is moving, we must also recognize a second *independent* small parameter, namely $(v/c) \ll 1$. Our goal is to obtain an expression for the radiation stress-energy tensor by solving the transfer equation through an expansion in terms of these small parameters. As we will see, in the comoving frame (but *not* the inertial frame) the lowest-order terms that appear in the solution are $O(\lambda_p/l)$, terms of $O(v/c)$ being entirely absent. Thus we can develop a first-order diffusion theory for the comoving-frame radiation field by ignoring fluid motions even in moving material.

The next-higher-order terms will be $O(\lambda_p v/lc)$, $O(\lambda_p^2/l^2)$, and $O(v^2/c^2)$. Of these, the $O(\lambda_p v/lc)$ terms are the most important, being of first order in each of the physically interesting small parameters; we call the expression containing these terms the "second-order" solution. Terms of $O(\lambda_p^2/l^2)$ have never been treated; those of $O(v^2/c^2)$ are contained implicitly in the covariant expression for the stress-energy tensor $R^{\alpha\beta}$ given later. While one expects intuitively the $O(\lambda_p v/lc)$ terms to be more important than the other second-order terms, it must be admitted that no rigorous analysis has ever been made to justify this assumption.

One should note that the *relative* size of the two independent expansion parameters is important because one has *static diffusion* when $(v/c) \ll (\lambda_p/l)$, so that $t_d \ll t_f$ and photon diffusion limits the rate of energy flow, and *dynamic diffusion* when $(v/c) \gtrsim (\lambda_p/l)$ and advection of energy by the moving fluid sets the effective rate of energy transport (cf. §§93 and 96).

The assumption of equilibrium is restrictive, and certainly will fail near a boundary surface from which radiation escapes freely. To extend the range of applicability of the theory, one can drop the strong assumption of thermal equilibrium and construct a nonequilibrium diffusion theory that invokes simple relations among the radiation moments, but does not presume that the intensity equals the Planck function at the local material temperature.

In practice it is found that diffusion theory predicts too rapid an energy transport where photon mean free paths become comparable to characteristic structural lengths in the flow. One way to overcome this problem is to introduce *flux limiters* that restrict the energy transport to physically allowable values. This approach is convenient and has been applied extensively, but lacks the accuracy of a consistent solution of the full transport equation (cf. §98).

THE ZERO-ORDER AND FIRST-ORDER EQUILIBRIUM DIFFUSION APPROXIMATIONS

In a static medium, the radiation field thermalizes to the Planck function as soon as the material is effectively thick. In moving material, the radiation field can achieve local thermal equilibrium only if a photon is destroyed in essentially the same physical environment as it was created, before local conditions are modified significantly by fluid flow. That is, the mean time between absorptions, $t_\lambda \approx \lambda_p/c$, must be much smaller than a fluid-flow time $t_f \approx l/v$, which implies that we must have $(\lambda_p v/lc) \ll 1$, a condition that obviously will be met at great depth in, say, a stellar envelope where both $(\lambda_p/l) \ll 1$ and $(v/c) \ll 1$.

The simplest physical situation is when the material is so homogeneous within an interaction volume that we can neglect all gradients; formally this regime corresponds to the limit $(\lambda_p/l) \to 0$. Thus, dividing both sides of (95.84) and (95.85) (for grey material) by χ_0 and letting $\chi_0 \to \infty$ we find that $E_0 = (4\pi/c)(\eta_0/\chi_0) = (4\pi/c)B(T) = a_R T^4$, where T is the *material temperature*, and that $F_0 \to 0$. Moreover, in the absence of gradients the radiation field is isotropic so $P_0 = \frac{1}{3}E_0 = \frac{1}{3}a_R T^4$. Similar arguments applied to (95.82) and (95.83) show that $E_0(\nu_0) = 3P_0(\nu_0) = (4\pi/c)B(\nu_0, T)$, and $F_0(\nu_0) \to 0$. Hence a consistent *zero-order*, comoving-frame radiation stress-energy tensor in the equilibrium diffusion limit is

$$\mathsf{R}_0 = \begin{pmatrix} a_R T^4 & \mathbf{0} \\ \mathbf{0} & \frac{1}{3} a_R T^4 \mathsf{I} \end{pmatrix}. \tag{97.1}$$

Here I denotes the 3×3 unit matrix. From (97.1) we see that for radiation, just as for an ordinary material gas, transport effects vanish when $(\lambda_p/l) \to 0$.

To obtain a first-order expression for R_0 we now assume that terms of $O(\lambda_p/l)$ are nonvanishing, but drop all terms of higher order. Consider first the monochromatic radiation momentum equation (95.83) in the isotropic limit, and ignore acceleration terms. Dimensional analysis suggests that on a fluid-flow time scale, all terms containing F_0 and Q_0 on the left-hand side are $O(\lambda_p v/lc)$ relative to the term on the right-hand side, and thus can be dropped in a first-order solution. We then obtain

$$\mathbf{F}_0(\nu_0) = -[c/\chi_0(\nu_0)]\,\boldsymbol{\nabla}\cdot\mathsf{P}_0(\nu_0) = -[4\pi/3\chi_0(\nu_0)]\boldsymbol{\nabla}B(\nu_0, T), \tag{97.2}$$

hence

$$\mathbf{F}_0 = -(c/3\chi_R^0)\boldsymbol{\nabla}(a_R T^4) = -\tfrac{4}{3} a_R c \lambda_R T^3 \boldsymbol{\nabla} T \equiv -K_R \boldsymbol{\nabla} T, \tag{97.3}$$

where χ_R^0 is the Rosseland mean opacity evaluated in the comoving frame, and $\lambda_R = 1/\chi_R^0$ is the Rosseland mean free path for photons. From (97.3) it is clear that $c^{-1}F_0$ is $O(\lambda_p/l)$ relative to E_0 or P_0.

It is noteworthy that (97.3), which applies in a *moving* fluid, is identical in mathematical form to (80.8) for a static medium. Physically this result states that because $(\lambda_p/l) \ll 1$, each fluid element is essentially "unaware" that it is moving, because the radius of the horizon from which it "sees" photons is minuscule compared to the scale of the flow; hence the flux measured in the fluid frame saturates to its static value as determined by the local temperature gradient within the fluid element.

Next consider the radiation energy equation (95.84). Given (97.3), dimensional analysis suggests that all terms on the left-hand side are either $O(\lambda_p^2/l^2)$ or $O(\lambda_p v/lc)$ relative to E_0; hence the departure of E_0 from $a_R T^4$ is at most second-order. Therefore a consistent *first-order* expression for the comoving-frame radiation stress-energy tensor in the equilibrium diffusion regime is

$$\mathsf{R}_0 = \begin{pmatrix} a_R T^4 & -\dfrac{1}{3\chi_R}\boldsymbol{\nabla}(a_R T^4) \\ -\dfrac{1}{3\chi_R}\boldsymbol{\nabla}(a_R T^4) & \frac{1}{3} a_R T^4 \mathsf{I} \end{pmatrix} = \begin{pmatrix} a_R T^4 & -(K_R/c)\boldsymbol{\nabla} T \\ -(K_R/c)\boldsymbol{\nabla} T & \frac{1}{3} a_R T^4 \mathsf{I} \end{pmatrix}. \tag{97.4}$$

We emphasize that (97.4), which contains the standard expressions used in stellar interiors work, applies in the *comoving frame only*. In particular, (97.3) should *not* be used as an expression for the inertial-frame flux F, which differs from F_0 by terms that are often much larger than F_0 itself [cf. (91.17) and (93.13)].

Equation (97.4) reveals the interesting fact that to first order the radiation stress-energy tensor contains a "conductive" energy flux, but no viscous stress, in contrast to ordinary gas dynamics where these terms are of the same order. We will see later that radiative viscosity is $O(\lambda_p v/lc)$ relative to P_0.

By using (97.2) in (96.2), or by writing $P_0 = \frac{1}{3}E_0 = \frac{1}{3}a_R T^4$ and discarding $O(\lambda_p v/lc)$ terms in (96.3), we see that the comoving-frame momentum equation in the first-order equilibrium diffusion approximation is

$$\rho(D\mathbf{v}/Dt) = \mathbf{f} - \nabla(p + \tfrac{1}{3}a_R T^4); \tag{97.5}$$

the radiating fluid thus behaves dynamically like an ideal gas whose total pressure is the sum of the gas and radiation pressure (both isotropic). Equation (97.5) is the standard momentum equation used in dynamical stellar structure calculations [see, e.g., (**C4**, eq. 2), (**C5**, eq. 27.15), (**F1**, eq. 2), (**K7**, eq. 2), (**L1**, eq. 2.55), or (**S3**, eq. 3.24)].

The most useful form of the energy equation in the equilibrium diffusion regime is (96.9), the first law of thermodynamics for the radiating fluid. Using the first-order solution (97.4) in (96.9) we find

$$\frac{D}{Dt}\left(e + \frac{a_R T^4}{\rho}\right) + (p + \tfrac{1}{3}a_R T^4)\frac{D}{Dt}\left(\frac{1}{\rho}\right) = \frac{1}{\rho}\nabla\cdot(K_R \nabla T) + \varepsilon, \tag{97.6}$$

which for one-dimensional spherically symmetric flow reduces to

$$\frac{D}{Dt}\left(e + \frac{a_R T^4}{\rho}\right) + (p + \tfrac{1}{3}a_R T^4)\frac{D}{Dt}\left(\frac{1}{\rho}\right) = \frac{\partial}{\partial M_r}\left[\frac{(4\pi r^2)^2 c\rho}{3\chi_R}\frac{\partial(a_R T^4)}{\partial M_r}\right] + \varepsilon. \tag{97.7}$$

At the risk of tedious repetition we again emphasize that all quantities in these equations are measured in the comoving frame. Equation (97.7) is the standard energy equation used in dynamical stellar-structure calculations [see, e.g., (**C4**, eq. 4), (**C5**, eq. 27.16), (**F1**, eq. 4), (**K7**, eq. 3), (**L2**, eq. 52.1), or (**S3**, eq. 3.19)].

Within the framework of one-dimensional Lagrangean hydrodynamics we may regard ρ and $(D\rho/Dt)$ as known in the discrete representation of (97.7) spanning the interval (t^n, t^{n+1}). Therefore, (97.7) basically has the mathematical form of a heat conduction equation with nonlinear coefficients. Accordingly, to avoid stringent timestep limitations, we use an implicit (e.g., backward Euler) time-differencing scheme. Then, (97.7) (including artifical viscosity terms—cf. §59— and with suitable boundary conditions) provides a tridiagonal system of nonlinear equations for $T^{n+1}_{i+(1/2)}$ at cell centers. As in §88 we linearize this system around some current estimate $T^*_{i+(1/2)}$, obtaining a system of the form

$$-A_{i+(1/2)}\,\delta T_{i-(1/2)} + B_{i+(1/2)}\,\delta T_{i+(1/2)} - C_{i+(1/2)}\,\delta T_{i+(3/2)} = R_{i+(1/2)}, \qquad (i = 1, \ldots, I), \tag{97.8}$$

which is solved by Gaussian elimination. This process is iterated to convergence. Having obtained T^{n+1}, we know e, p, E_0, and P_0 at t^{n+1}, and can then integrate the momentum equation (97.5) from $t^{n+(1/2)}$ to $t^{n+(3/2)}$. In short, in the equilibrium diffusion regime the dynamical equations for a radiating fluid are relatively easy to solve.

Finally, we can obtain the inertial-frame radiation stress-energy tensor in the first-order equilibrium diffusion limit by applying the transformations (91.10)–(91.12) to (97.4). To $O(v/c)$ the main effect is to replace the Lagrangean flux by the Eulerian flux, cf. (93.14). Thus using (97.4) and (91.17) in (94.21) one finds that the Eulerian energy equation, correct to $O(v/c)$, for a radiating fluid in the equilibrium diffusion regime is

$$(\rho e + a_R T^4)_{,t} + \nabla \cdot [(\rho e + a_R T^4)\mathbf{v}] + (p + \tfrac{1}{3}a_R T^4)\nabla \cdot \mathbf{v} = \nabla \cdot (K_R \nabla T) + \rho\varepsilon. \tag{97.9}$$

Because the material properties are functions of (ρ, T), (97.9) suffices to determine T at t^{n+1}. A simple rearrangement reduces (97.9) to (97.6); the same result is obtained by starting from (94.22) and making $O(v/c)$ transformations of inertial-frame radiation quantities into the fluid frame [see also (**P4**)].

THE "SECOND-ORDER" EQUILIBRIUM DIFFUSION APPROXIMATION

In first-order diffusion theory we dropped time derivatives and velocity-dependent terms and obtained a radiation stress-energy tensor comprising an energy density, an isotropic pressure, and an energy flux proportional to the local temperature gradient. We now develop the next level of approximation, retaining terms of $O(\lambda_p v/lc)$; the radiation stress-energy tensor then contains dissipative terms corresponding to *radiative viscosity*. The main purpose of the discussion is to exhibit the complete one-to-one correspondence that exists between the dynamical behavior of radiation in the diffusion regime and that of a viscous, heat-conducting, relativistic material fluid as described in §4.3.

Radiative viscosity was first discussed by Jeans (**J1**), (**J2**) and Milne (**M14**), (**M15**), who concluded that it provides an efficient mechanism for angular momentum transport in stellar interiors, and promotes solid-body rotation of stars. Their analyses are not relativistically correct, and some of their formulae are flawed. The correct formulae were first derived in a penetrating paper by L. H. Thomas (**T1**), and later, in manifestly covariant analyses, by Hazelhurst and Sargent (**H1**) and by Simon (**S4**).

The approach followed by Thomas is to use (89.5), (90.6), and (90.8) to write the mixed-frame transfer equation in Cartesian coordinates as

$$\begin{aligned}[c^{-1}(\partial/\partial t) + \mathbf{n} \cdot \nabla]I(\mathbf{n}, \nu) &= (\nu/\nu_0)^2 \eta_0(\nu_0) - (\nu_0/\nu)\kappa_0(\nu_0) I(\mathbf{n}, \nu) \\ &= [(1 - v^2/c^2)/(1 - \mathbf{n} \cdot \mathbf{v}/c)^2]\eta_0(\nu_0) \\ &\quad - [(1 - \mathbf{n} \cdot \mathbf{v}/c)/(1 - v^2/c^2)^{1/2}]\kappa_0(\nu_0) I(\mathbf{n}, \nu),\end{aligned} \tag{97.10}$$

and then, starting from the LTE solution $I = \eta_0/\kappa_0$, to solve (97.10) by iteration, which yields

$$I(\mathbf{n}, \nu) = \frac{(1-v^2/c^2)^{3/2}}{(1-\mathbf{n}\cdot\mathbf{v}/c)^3}\frac{\eta_0}{\kappa_0} - \frac{(1-v^2/c^2)^{1/2}}{(1-\mathbf{n}\cdot\mathbf{v}/c)}\frac{1}{\kappa_0}\left(\frac{1}{c}\frac{\partial}{\partial t}+\mathbf{n}\cdot\nabla\right)\frac{(1-v^2/c^2)^{3/2}}{(1-\mathbf{n}\cdot\mathbf{v}/c)^3}\frac{\eta_0}{\kappa_0}+\cdots. \tag{97.11}$$

One can then evaluate R by direct integration of $I(\mathbf{n}, \nu)$ over ω and ν. The calculation is straightforward, but lengthy and cumbersome, and the rearrangement of terms in the final expression for R (**T1**, eq. 7 and the unnumbered equation on p. 248) into a covariant form (**T1**, eqs. 7.1 and 8) is rather tricky.

The most physically appealing approach is Simon's, who used the Eckart decomposition theorem to express R in terms of quantities that are easily evaluated in the comoving frame; we follow Simon's analysis here. Applying (44.14) to the tensor $W^{\alpha\beta} \equiv c^2 R^{\alpha\beta}$ we can write

$$R^{\alpha\beta} = \mathcal{P}^{\alpha\beta} + c^{-2}(\mathcal{F}^{\alpha}V^{\beta} + V^{\alpha}\mathcal{F}^{\beta} + \mathcal{E}V^{\alpha}V^{\beta}) \tag{97.12}$$

where

$$\mathcal{E} = c^{-2}V_{\alpha}V_{\beta}R^{\alpha\beta}, \tag{97.13a}$$

$$\mathcal{F}^{\alpha} = -S^{\alpha}_{\beta}R^{\beta\gamma}V_{\gamma}, \tag{97.13b}$$

and

$$\mathcal{P}^{\alpha\beta} = S^{\alpha}_{\gamma}S^{\beta}_{\delta}R^{\gamma\delta}. \tag{97.13c}$$

Here S^{α}_{β} is the projection tensor defined by (44.2); we use a different letter for it here to avoid confusion with the radiation pressure tensor P.

Because V_{α} is orthogonal to S^{α}_{β}, it is orthogonal to both $\mathcal{F}^{\alpha}$ and $\mathcal{P}^{\alpha\beta}$, which implies that

$$\mathcal{F}^{0} = v_i\mathcal{F}^{i}/c, \tag{97.14}$$

and that

$$\mathcal{P}^{i0} = \mathcal{P}^{ij}v_j/c \tag{97.15a}$$

and

$$\mathcal{P}^{00} = \mathcal{P}^{ij}v_iv_j/c^2. \tag{97.15b}$$

In the comoving frame where $V^{\alpha}_0 = (c, 0, 0, 0)$, we thus have

$$\mathcal{P}^{\alpha 0}_0 = 0, \tag{97.16a}$$

and

$$\mathcal{F}^{0}_0 = 0 \tag{97.16b}$$

and hence from (97.12)

$$\mathcal{P}^{ij}_0 = P^{ij}_0 = R^{ij}_0, \tag{97.17a}$$

$$\mathcal{F}^{i}_0 = F^{i}_0 = R^{0i}_0, \tag{97.17b}$$

and

$$\mathcal{E}_0 = E_0 = R^{00}_0. \tag{97.17c}$$

These results are completely general.

Suppose now that in the comoving frame LTE obtains, so that $\eta_0(\nu_0) = \kappa_0(\nu_0)B(\nu_0, T)$ where T is the *material temperature* measured in the comoving frame. Then the comoving-frame emissivity can be written in terms of invariants as

$$e_0 = a_0 g_R^0, \tag{97.18}$$

where g_R^0 is the invariant photon distribution function for blackbody radiation at rest relative to the observer, that is,

$$g_R^0 = (c^2/h^4\nu_0^3)B(\nu_0, T) = (2/h^3)[\exp(h\nu_0/kT) - 1]^{-1}, \tag{97.19}$$

and a_0 is the invariant opacity

$$a_0(\nu_0) = h\nu_0\kappa_0(\nu_0) = h\nu_0 K_0, \tag{97.20}$$

where K_0 is a world scalar.

In order to generalize (97.18) to an arbitrary frame, we notice that

$$M_\alpha V^\alpha \equiv -h\nu_0; \tag{97.21}$$

hence the appropriate covariant generalization of g_R^0 is

$$g_R = (2/h^3)[\exp(-M_\alpha V^\alpha/kT) - 1]^{-1} \tag{97.22}$$

and the covariant generalization of a_0 is

$$a = (-M_\alpha V^\alpha)K_0. \tag{97.23}$$

Thus for LTE in the fluid frame the photon Boltzmann equation (92.6) in an arbitrary inertial frame is

$$M^\alpha f_{R,\alpha} = (-M_\alpha V^\alpha/c)K_0(g_R - f_R), \tag{97.24}$$

where for convenience we have temporarily adopted Cartesian coordinates.

To obtain an approximation for f_R, we assume that in the diffusion regime f_R does not differ much from g_R, and use (97.24) to develop the expansion

$$f_R = g_R + (c/K_0 M_\alpha V^\alpha)M^\alpha g_{R,\alpha} + \dots. \tag{97.25}$$

In (97.25), g_R depends on x^α because both T and V^α are functions of x^α. We could now follow Thomas and use (97.25) in (91.2) to calculate $R^{\alpha\beta}$ directly. It is much simpler, however, to carry out the calculation in the comoving frame, and then reconstruct R in the inertial frame via (97.12) to (97.17).

Because $V_\alpha V^\alpha_{,\beta} \equiv 0$, $(V^{(0)}_{,\alpha})_0 \equiv 0$ in the comoving frame. Therefore in this frame

$$\begin{aligned}(M^\alpha g_{R,\alpha})_0 &= M_0^\alpha[(\partial g_R/\partial T)T_{,\alpha} + (\partial g_R/\partial V^\beta)V^\beta_{,\alpha}]_0 \\ &= M_0^\alpha[(\partial g_R^0/\partial T)(\partial T/\partial x_0^\alpha) + (\partial g_R/\partial V^i)_0(\partial V^i/\partial x^\alpha)_0].\end{aligned} \tag{97.26}$$

From (97.20)

$$(\partial g_R^0/\partial T) = (c^2/h^4\nu_0^3)[\partial B(\nu_0, T)/\partial T], \tag{97.27}$$

and from (97.22)

$$\frac{\partial g_R}{\partial V^\alpha}=\left(\frac{2}{h^3}\right)\frac{(M_\alpha/kT)\exp(-M_\alpha V^\alpha/kT)}{[\exp(-M_\alpha V^\alpha/kT)-1]^2}, \tag{97.28}$$

whence

$$(\partial g_R/\partial V^i)_0=(Tn_i/c)(\partial g_R^0/\partial T). \tag{97.29}$$

Substituting (97.26) to (97.29) into (97.25), evaluated in the comoving frame, we obtain

$$f_R^0=g_R^0-\left[\frac{c^2}{h^4\nu_0^3\kappa_0(\nu_0)}\frac{\partial B(\nu_0,T)}{\partial T}\right]\frac{M_0^\alpha}{|M_0^\alpha|}\left(\frac{\partial T}{\partial x_0^\alpha}+\frac{Tn_i^0}{c}\frac{\partial V^i}{\partial x_0^\alpha}\right). \tag{97.30}$$

We can now evaluate R_0 by substituting (97.30) into (91.2). Recalling that $d^3M=M^2\,dM\,d\omega=h^3c^{-3}\nu^2\,d\nu\,d\omega$, we see from (97.17c) and (91.3) that

$$\begin{aligned}\mathscr{E}_0&=\int_0^\infty d\nu_0\oint d\omega_0(h^4\nu_0^3/c^3)f_R^0\\&=\frac{4\pi}{3}\left[B(T)-\frac{1}{\kappa_R^0}\frac{\partial B(T)}{\partial T}\left(\frac{\partial T}{\partial x_0^0}+\frac{T}{c}\langle n^\alpha n_i\rangle_0\frac{\partial V_0^i}{\partial x_0^\alpha}\right)\right].\end{aligned} \tag{97.31}$$

Here

$$B(T)=(c/4\pi)a_RT^4 \tag{97.32}$$

and κ_R^0 is the comoving-frame Rosseland mean

$$\frac{1}{\kappa_R^0}\equiv\int_0^\infty\frac{1}{\kappa_0(\nu_0)}\frac{\partial B(\nu_0,T)}{\partial T}\,d\nu_0\Bigg/\int_0^\infty\frac{\partial B(\nu_0,T)}{\partial T}\,d\nu_0, \tag{97.33}$$

a world scalar. Noting that

$$\langle n^\alpha n_i\rangle_0\equiv(4\pi)^{-1}\oint n_0^\alpha n_i^0\,d\omega_0=\tfrac{1}{3}\delta_i^\alpha, \tag{97.34}$$

we can rewrite (97.31) as

$$\mathscr{E}_0=a_RT^4-\frac{4}{3}\left(\frac{a_RT^4}{c\kappa_R^0}\right)\left(\frac{\partial V_0^i}{\partial x_0^i}+\frac{3c}{T}\frac{\partial T}{\partial x_0^0}\right). \tag{97.35}$$

Similarly, using (97.30) in (97.17b) and (91.4) we find

$$\mathscr{F}_0^i=-\frac{4}{3}\left(\frac{a_RT^4}{c\kappa_R^0}\right)\left(\frac{c^2}{T}\frac{\partial T}{\partial x_0^i}+c\frac{\partial V_0^i}{\partial x_0^0}\right). \tag{97.36}$$

Finally, from (97.17a) and (91.1) we obtain

$$\mathscr{P}_0^{ij}=\tfrac{1}{3}a_RT^4\,\delta^{ij}-\frac{4\pi}{c\kappa_R^0}\frac{\partial B(T)}{\partial T}\left(\langle n^in^j\rangle\frac{\partial T}{\partial x_0^0}+\langle n^in^jn_kn^l\rangle\frac{T}{c}\frac{\partial V_0^k}{\partial x_0^l}\right). \tag{97.37}$$

But

$$\langle n^in^jn_kn^l\rangle=\tfrac{1}{15}(\delta^{ij}\,\delta_k^l+\delta_k^i\,\delta^{jl}+\delta^{il}\,\delta_k^j), \tag{97.38}$$

hence

$$\mathcal{P}_0^{ij} = \tfrac{1}{3} a_R T^4 \,\delta^{ij} - \frac{4}{15}\left(\frac{a_R T^4}{c\kappa_R^0}\right)\left(\frac{\partial V_0^i}{\partial x_0^k}\,\delta^{kj} + \frac{\partial V_0^j}{\partial x_0^k}\,\delta^{ki} + \frac{\partial V_0^k}{\partial x_0^k}\,\delta^{ij} + \frac{5c}{T}\frac{\partial T}{\partial x_0^0}\,\delta^{ij}\right). \tag{97.39}$$

Equations (97.35), (97.36), and (97.39) apply in the comoving frame. To generalize these expressions to an arbitrary frame, in particular the lab frame, we merely cast them into covariant forms that reduce to the correct results in the comoving frame. Let

$$\mu_R \equiv \tfrac{4}{15}(a_R T^4/c\kappa_R^0) = (T/5c^2)K_R \tag{97.40}$$

be the *coefficient of radiative viscosity*, a world scalar; K_R is the radiative conductivity as defined in (97.3). From (97.35) we see by inspection that a covariant expression for $\mathscr{E}$ is

$$\mathscr{E} = a_R T^4 - 5\mu_R[V^\alpha_{;\alpha} + 3V^\alpha(\ln T)_{,\alpha}]. \tag{97.41}$$

To write a covariant expression for $\mathscr{F}^\alpha$ we note from (97.16b) and (97.36) that the comoving-frame flux has nonvanishing space components but a vanishing time component. We can assure this behavior by writing $\mathscr{F}^\alpha$ in terms of the projection tensor as

$$\mathscr{F}^\alpha = -K_R(T_{,\beta} + c^{-2}TA_\beta)(g^{\alpha\beta} + c^{-2}V^\alpha V^\beta). \tag{97.42}$$

This expression is exactly analogous to Eckart's covariant material heat-conduction vector (46.22); it differs from Simon's (**S4**) equation (82), which is not. By similar reasoning, we replace the Kronecker deltas in (97.39) with projection tensors, and noting that

$$(V^\alpha V^\beta V^\gamma)_{;\gamma} = V^\alpha V^\beta V^\gamma_{;\gamma} + V^\alpha V^\gamma V^\beta_{;\gamma} + V^\beta V^\gamma V^\alpha_{;\gamma}, \tag{97.43}$$

we find that (97.39) can be written covariantly as

$$\begin{aligned}\mathcal{P}^{\alpha\beta} = \tfrac{1}{3} a_R T^4(g^{\alpha\beta} + c^{-2}V^\alpha V^\beta) - \mu_R[&g^{\alpha\beta}V^\gamma_{;\gamma} + g^{\beta\gamma}V^\alpha_{;\gamma} + g^{\alpha\gamma}V^\beta_{;\gamma}\\ &+ c^{-2}(V^\alpha V^\beta V^\gamma)_{;\gamma} + 5V^\gamma(\ln T)_{,\gamma}(g^{\alpha\beta} + c^{-2}V^\alpha V^\beta)].\end{aligned} \tag{97.44}$$

Finally, substituting (97.41), (97.42), and (97.44) into (97.12) we find that the covariant expression for the diffusion-limit radiation stress-energy tensor in an arbitrary frame is

$$\begin{aligned}R^{\alpha\beta} = \tfrac{1}{3} a_R T^4(g^{\alpha\beta} + 4c^{-2}V^\alpha V^\beta) - \mu_R[&g^{\alpha\beta}V^\gamma_{;\gamma} + g^{\beta\gamma}V^\alpha_{;\gamma} + g^{\alpha\gamma}V^\beta_{;\gamma}\\ &+ 6c^{-2}(V^\alpha V^\beta V^\gamma)_{;\gamma}\\ &+ 5(\ln T)_{;\gamma}(g^{\alpha\beta}V^\gamma + g^{\beta\gamma}V^\alpha + g^{\alpha\gamma}V^\beta + 6c^{-2}V^\alpha V^\beta V^\gamma)].\end{aligned} \tag{97.45}$$

Aside from minor differences in notation, this result is identical to Thomas's (**T1**). Equation (97.45) is the direct analogue, for radiation, of (45.3) for matter.

Equations (97.41), (97.42), (97.44), and (97.45) are relativistically correct, and apply in all frames. To gain physical insight it is most instructive to work in the comoving frame, developing a solution that is internally consistent to $O(\lambda_p v/lc)$. Thus, reducing (97.41) to (97.44) to the comoving frame, we find [cf. (97.34) to (97.39)]

$$E_0 = a_R T^4 - 5\mu_R[3(D \ln T/Dt) + \nabla \cdot \mathbf{v}], \tag{97.46}$$

$$(F_0)_i = -K_R(T_{,i} + c^{-2}Ta_i), \tag{97.47}$$

and

$$P_0^{ij} = [\tfrac{1}{3}a_R T^4 - 5\mu_R(D \ln T/Dt)]\,\delta^{ij} - \mu_R(v^i_{,k}\,\delta^{kj} + v^j_{,k}\,\delta^{ik} + v^k_{,k}\,\delta^{ij}). \tag{97.48}$$

We can cast (97.48) into the same form as the stress tensor for a Newtonian fluid [cf. (25.3)], that is

$$P_0^{ij} = \mathscr{P}\,\delta^{ij} - 2\mu_R D^{ij} - \zeta_R v^k_{,k}\,\delta^{ij} \tag{97.49}$$

where

$$\mathscr{P} \equiv \tfrac{1}{3}a_R T^4 - 5\mu_R(D \ln T/Dt) \tag{97.50}$$

is the isotropic component of the radiation pressure, D^{ij} is the traceless rate-of-strain tensor (32.34), and

$$\zeta_R \equiv \tfrac{5}{3}\mu_R. \tag{97.51}$$

It is obvious from (97.49) why μ_R is called the viscosity coefficient for radiation.

The corresponding results in curvilinear coordinates are obtained by replacing δ_{ij} with g_{ij} and ordinary derivatives with covariant derivatives.

A dimensional analysis of (97.46) and (97.50) suggests that the departures of E_0 from $a_R T^4$ and of $\mathscr{P}$ from $\frac{1}{3}a_R T^4$ are both $O(\lambda_p v/lc)$ relative to the dominant terms; hence when $v/c \gg \lambda_p/l$ these departures can be larger than in a static medium [where they are $O(\lambda_p^2/l^2)$, cf. §80]. The radiative viscous terms in (97.49) are also $O(\lambda_p v/lc)$ relative to the leading term. Furthermore, from (97.46) we can now verify the result cited in §§93 and 95 that the net absorption-emission term $\kappa(4\pi B - cE)$ in the radiation energy equation is $O(v/l)E$ for dynamic diffusion [compared to $O(c\lambda_p/l^2)E$ for static diffusion]. Analysis of (97.47) suggests that the acceleration term is only $O(v^2/c^2)$ relative to the leading term of the flux. Normally we would drop a term of this order, but we retain it here for reasons that will emerge below.

Radiative viscosity arises because photons deposit their momentum in the fluid element in which they are absorbed; it is thus a direct analogue of molecular viscosity. In both cases, momentum exchange via particle motions within the fluid produces a frictional force, with faster-moving elements tending to drag along slower elements, and vice versa. Indeed, it is easy to derive the coefficient of radiative viscosity from simple mean-free-path arguments. Consider photons emitted from the origin O, traveling along the positive x axis until they are absorbed at P, one mean free path λ_p from O. Let the material have a velocity shear (dv/dx) parallel to the y axis. Then, owing to aberration, photons of frequency ν emitted from O will have a y component of momentum $p_y = -(h\nu/c)\sin\theta$, where $\sin\theta = v_p/c = \lambda_p(dv/dx)/c$, when they are absorbed at P. Hence the radiative viscous force on a fluid element at P with cross section δA perpendicular to the x axis is $f_y(\nu) = -(cn_\nu\,\delta A)\times(h\nu/c)\times(\lambda_p/c)(dv/dx)$, where n_ν is the photon number density. Summing over all energies we find $f_y = -(a_R T^4 \lambda_p/c)(dv/dx)$, whence we identify $\mu_R \sim (a_R T^4 \lambda_p/c)$, which, aside from a numerical factor, is (97.40).

One can define a *radiative Reynolds number* Re_R by substituting μ_R into (28.1). Dimensional analysis suggests that $\text{Re}_R \sim (t_f/t_\lambda)\times$(material kinetic energy density/radiation energy density). Thus the radiative Reynolds number can be small, hence radiative viscous effects important, when the radiant energy density is large compared to the material energy density (i.e., at high temperatures) and/or when the photon mean free path is long.

From (97.49) and (97.51) it also appears that radiation has a substantial bulk viscosity ζ_R. As emphasized by Weinberg (**W1**) it is necessary to make a deeper inquiry about this quantity. In particular he notes the proof by Tisza (**T2**) that bulk viscosity is absent in any gas for which the trace of the total stress-energy tensor is expressible as a function of only $\rho_{00}c^2$, the total energy density, and/or n, the particle number density. This is the case for a gas of structureless point particles in both the nonrelativistic and extreme relativistic limits [cf. (43.50) and (43.53)], hence for a gas of photons. Therefore one would expect photons to have zero bulk viscosity when interacting with relativistic material particles, in apparent contradiction to (97.51). Moreover, according to (97.46), the comoving-frame energy density contains dissipative terms, in contradiction to the general result from Eckart's theory (cf. §§44 and 45) that

$$(M^{\alpha\beta}_{\text{viscous}} + M^{\alpha\beta}_{\text{conduction}})V_\alpha V_\beta \equiv 0. \tag{97.52}$$

Weinberg demonstrates that these apparent contradictions can be resolved by noticing that while Thomas and Eckart define particle number density in the same way, they use different definitions for the temperature. Thomas uses the *material* temperature T measured by a comoving observer, whereas Eckart defines a temperature T_* by the requirement that the comoving-frame energy density $c^{-2}V_\alpha V_\beta M^{\alpha\beta}$ equal the total energy density $\hat{e} = c^2\rho_{00}(n, T_*)$ in *thermal equilibrium* at temperature T_*. Weinberg shows that

$$T - T^* = 15\mu_R\left(\frac{\partial\hat{e}}{\partial T}\right)_n^{-1}\left[\frac{1}{3} - \frac{(\partial p/\partial T)_n}{(\partial\hat{e}/\partial T)_n}\right]\nabla\cdot\mathbf{v}, \tag{97.53}$$

that when this difference is taken into account, Thomas's results are consistent with Eckart's general theory, and that in (97.49) one now has

$$\mu_R = \tfrac{4}{15}(a_R T_*^4/c\kappa_R^0) \tag{97.54}$$

and

$$\zeta_R = 15\mu_R[(\partial p/\partial\hat{e})_n - \tfrac{1}{3}]^2. \tag{97.55}$$

From (43.50) and (43.53) we see that (97.55) yields zero radiative bulk viscosity when the radiation interacts with extremely relativistic material, and reduces to Thomas's expression when the material is nonrelativistic.

The physical essence of the preceding analysis is that in the diffusion regime a radiating fluid behaves, in the comoving frame, like a viscous, heat-conducting gas that has a total "internal" energy density (per gram)

$$e_{\text{tot}} = e + (E_0/\rho_0) \tag{97.56}$$

a total hydrostatic pressure

$$p_{\text{tot}} = p + \mathsf{P}, \tag{97.57}$$

a total energy flux

$$\mathbf{q}_{\text{tot}} = \mathbf{q} + \mathbf{F}_0 = -(K + K_R)(\nabla T + c^{-2} T\mathbf{a}), \tag{97.58}$$

and a total viscous stress tensor

$$\sigma^{ij}_{\text{tot}} = (\mu + \mu_R)(v^i_{,k}\,\delta^{jk} + v^j_{,k}\,\delta^{ik} - \tfrac{2}{3} v^k_{,k}\,\delta^{ij}) + \zeta_R v^k_{,k}\,\delta^{ij}. \tag{97.59}$$

Adopting this view, we should be able to obtain valid momentum and gas-energy equations for the radiating fluid using the analysis of §§46 and 47 for a nonideal relativistic fluid. For example, reduction of the momentum equation (47.3) to $O(v/c)$ in the comoving frame in a spherically symmetric flow gives

$$\begin{aligned}\left[\rho + \frac{1}{c^2}(\rho e + p + E_0 + \mathscr{P})\right] a_r &= f_r - \frac{\partial}{\partial r}(p + \mathscr{P}) \\ &+ \frac{\partial}{\partial r}\left[\tfrac{4}{3}(\mu + \mu_R) r \frac{\partial}{\partial r}\left(\frac{v}{r}\right) + \zeta_R\left(\frac{\partial v}{\partial r} + \frac{2v}{r}\right)\right] + 4(\mu + \mu_R)\frac{\partial}{\partial r}\left(\frac{v}{r}\right) \\ &- \frac{1}{c^2}\frac{D}{Dt}(F_0 + q) - \frac{2}{c^2}\left(\frac{\partial v}{\partial r} + \frac{v}{r}\right)(F_0 + q).\end{aligned} \tag{97.60}$$

Here we ignored material bulk viscosity, and dropped one term, $-(c^{-2}\zeta\nabla\cdot\mathbf{v})a_r$, which is formally $O(v^2/c^2)$.

We can check (97.60) by comparing it to (96.1) (to which material viscous and conduction terms are added) in which the comoving-frame radiative force term $\mathbf{G}_0$ is evaluated using (95.88) or (95.85), with E_0, F_0, and P_0 obtained from (97.46) to (97.50). Thus computing the rr component of P_0 from (97.49) we find

$$P_0 = \mathscr{P} - \tfrac{4}{3}\mu_R[(\partial v/\partial r) - (v/r)] - \zeta_R \nabla\cdot\mathbf{v}, \tag{97.61}$$

whence, from (97.46) and (97.50), we have

$$(3P_0 - E_0)/r = -4\mu_R[\partial(v/r)/\partial r]. \tag{97.62}$$

Using (97.61) and (97.62) in (95.85) we reproduce the radiative terms in (97.60) exactly, which is very satisfying.

Furthermore, we can now give a precise physical interpretation of the acceleration terms in (95.88): they account for the equivalent inertia of the radiation enthalpy density, playing the same role as the material enthalpy terms in ρ_{000} in (96.1). Similarly the terms $c^{-2}[\mathbf{F}_0 \cdot (\nabla \mathbf{v}) + \mathbf{F}_0(\nabla \cdot \mathbf{v})]$ in (95.88) account for the interaction of the radiation momentum density with the shear-flow field, playing the same role as the last three terms in (47.3), which account for the inertia of the material heat flux. Finally, $c^{-2}(DF_0/Dt)$ accounts for the rate of change of the momentum density of the radiation field [cf. (96.4)], and plays essentially the same role as ρa_r for the matter.

Dimensional analysis of (97.60) suggests that $c^{-2}(E_0 + \mathcal{P})a_r$ is $O(v^2/c^2)$ relative to $\partial \mathcal{P}/\partial r$, while all of the radiation viscosity and flux terms on the right-hand side of (97.60) are $O(\lambda_p v/lc)$. The former can thus be dropped (along with the corresponding material terms) because we have already omitted all other terms that are formally $O(v^2/c^2)$, while *all* of the latter must be retained if the treatment of radiation viscosity effects in the momentum equation is to be consistent.

Similarly, reducing the gas energy equation (46.15) to $O(v/c)$ in the comoving frame with the help of (46.7) and adding a thermonuclear energy-release term we find that

$$\rho\left[\frac{D}{Dt}\left(e + \frac{E_0}{\rho}\right) + (p + \mathcal{P})\frac{D}{Dt}\left(\frac{1}{\rho}\right)\right]$$
$$= \rho\varepsilon_0 + \Phi - \nabla \cdot (\mathbf{q} + \mathbf{F}_0) - \frac{2}{c^2}\mathbf{a} \cdot (\mathbf{q} + \mathbf{F}_0) + \frac{(K + K_R)}{c^2}\nabla \cdot \mathbf{v}\frac{DT}{Dt} \tag{97.63}$$

where Φ is the total dissipation function

$$\Phi \equiv 2(\mu + \mu_R)D_{ij}D^{ij} + \zeta_R(\nabla \cdot \mathbf{v})^2. \tag{97.64}$$

Again, we have ignored the bulk viscosity of the material.

We can check (97.63) for spherically symmetric flow by comparing it to (96.6), to which we add material viscous and conduction terms, and calculate the radiation energy-deposition rate cG_0^0 from (95.87) or (95.84). Thus using (97.61) and (97.62) we recover all the radiative terms in (97.63) except the last term on the right-hand side. The origin of the discrepancy is obscure, but dimensional analysis suggests that the term in question is formally only $O(v^2/c^2)$ relative to $\nabla \cdot \mathbf{F}_0$ and could therefore be neglected (but see below). The term may correspond to a high-order term omitted from (95.87), which is formally accurate only to $O(v/c)$.

From (97.63) and (97.64) we see that radiation generates entropy in a radiating fluid both by energy transport down a temperature gradient and by viscous dissipation, in complete parallelism with the corresponding material processes. Furthermore, we can now interpret the $\mathbf{a} \cdot \mathbf{F}_0$ term in the comoving-frame radiation energy equation physically as accounting for the equivalent inertia of the radiant energy flow, playing the same role as the inertia of thermal heat conduction in material (cf. §46).

Dimensional analysis of (97.63) suggests that the terms containing μ_R in (97.46) and (97.48), and the acceleration term in (97.47), all give rise to terms in (97.63) that are of the same order as Φ and the term containing $\mathbf{a} \cdot \mathbf{F}_0$. Thus if we wish to study the effects of radiative viscous dissipation we should retain *all* terms in (97.46) to (97.48) and in (97.63) for consistency. However, at this point we must note the disturbing fact that the dimensional analysis suggest that all of the radiative dissipative terms in (97.63) are formally only $O(v^2/c^2)$ relative to the leading term $\nabla \cdot \mathbf{F}_0$!

This result may indicate that (97.63) is inadequate to describe radiative viscous effects because neither the comoving-frame radiation energy equation, nor (97.46) to (97.48) for R_0, are formally accurate to $O(v^2/c^2)$ and, furthermore, $O(v^2/c^2)$ effects have also been omitted from the material terms. Actually the situation is completely analogous to that discussed in §51 for the damping of acoustic waves by material viscosity and thermal conduction: viscous terms affect the momentum balance (whereas conduction terms do not) while conduction terms dominate the energy dissipation (whereas viscous terms are negligible because they are of second order in the velocity perturbation).

The dimensional analysis used above may not be accurate in every case, and for some flows the radiative dissipative terms may be larger, see (**M7**). But in any event, we conclude that radiative viscous dissipation is generally small for nonrelativistic fluids, in harmony with the conclusions of Cox (**C5**, §27.6d) and Kopal (**K5**), (**K6**) that the effects of both radiative and material viscosity on stellar pulsation are negligible compared to "turbulent viscosity" produced by convective motions in stellar envelopes.

Equations (97.46) to (97.51) provide an accurate radiation stress-energy tensor in the diffusion limit, as needed for, say, dynamical stellar evolution calculations. But it would be laborious to implement the full equations in a computation, and it is natural to ask whether any terms can be dropped in practical applications. A variety of simplified equations for treating radiative viscous effects have been proposed. For example, Simon (**S4**) suggested that the (D/Dt) terms can be dropped in both (97.46) and (97.48). This omission is unsatisfactory in general because dimensional analysis suggests that all the terms containing μ_R in these equations are of the same order. Therefore, if we choose to retain the viscous stress in (97.49), we must retain (DT/Dt) in (97.50); we then must retain all terms in (97.46) to guarantee that trace $\mathsf{P}_0 = E_0$. More revealing, using (97.46) to (97.48) in the comoving-frame radiation energy equation (95.84) for a planar, grey medium we find that we must retain all terms in (97.46) in order to obtain internal consistency to $O(\lambda_p v/lc)$; in particular, omission of the (D/Dt) term in (97.46) is tantamount to dropping (DE_0/Dt) in (95.84), which is obviously unacceptable.

Similar criticisms may be leveled at simplified forms of the momentum and energy equations that have been derived by Newtonian reasoning. For example, in the two standard works on stellar pulsation and stability by

Ledoux (**L1**) and Ledoux and Walraven (**L2**), the momentum equation for the radiating fluid is taken to be

$$\rho(Dv_i/Dt) = f_i + (p + p_R)_{,i} + g_{ij}\sigma^{jk}_{;k} \tag{97.65}$$

where σ^{ij} is the total viscous stress tensor (97.59) [cf. eqs. (2.14) and (2.55) of (**L1**) and eqs. (48.4) and (50.1) of (**L2**)]. The first law of thermodynamics for the radiating fluid is written

$$\rho\left[\frac{D}{Dt}\left(e+\frac{E_R}{\rho}\right)+(p+p_R)\frac{D}{Dt}\left(\frac{1}{\rho}\right)\right] = \rho\varepsilon - \boldsymbol{\nabla}\cdot(\mathbf{q}+\mathbf{F}_R)+\Phi \tag{97.66}$$

where Φ is given by (97.64) [see the equation following (2.62) in (**L1**) and eq. (52.1) in (**L2**)]. In (**L2**), $E_R \equiv a_R T^4$ [cf. their eqs. (49.37) and (51.1)], while p_R includes dissipative terms [cf. their eq. (49.43)]. In (**L1**), $E_R \equiv a_R T^4$ and $p_R \equiv \frac{1}{3} a_R T^4$ [cf. eqs. (2.4) and (2.14) of that reference]. In both (**L1**) and (**L2**) the acceleration term in (97.47) is dropped from $\mathbf{F}_R$.

Unfortunately, all of the simplifications just described neglect terms that are of the same order as the viscous terms that are retained. For example, (97.65) omits the rate of change of the radiation momentum density and the dynamical interaction of the radiation momentum with the velocity gradient, even though these terms are formally of the same order as the viscous stress; it is not, therefore, a consistent equation of motion, and (97.60) should be used instead. Similarly, (97.66) omits terms from E_R, F_R, and p_R, as well as $\mathbf{a}\cdot\mathbf{F}_R$, which are all formally of the same order as Φ, and this equation is not consistent at the level at which radiative dissipative effects enter.

The important conclusion that can be drawn from the discussion above is that all dynamical effects of radiation, including radiative viscosity, inertia of the radiant heat flux and enthalpy density, and radiation momentum density are correctly described by the comoving-frame radiation momentum and energy equations (95.87) and (95.88), provided that all terms in those equations are retained. Thus one can account for these dynamical effects by solving the (full) comoving-frame equations directly, without resort to the second-order diffusion approximation, which supplies analytical results valid only in the limit of small photon mean free paths. In practice, it may be difficult to retain sufficient numerical accuracy to calculate the dissipation terms with any significance at great depth, but it is just then that the departures from adiabaticity they produce are small, perhaps $O(v^2/c^2)$. On the other hand, in the transport regime the comoving-frame moment equations provide a direct means of handling radiative momentum input into, and dissipation in, the material in radiation-dominated flows, just when these effects are large. Analytical expressions for viscous terms cannot be written in the transport regime because they depend on the *global* structure of the velocity field over a photon mean free path (now comparable to a characteristic structural length) instead of on *local* velocity gradients. The "viscous" effects are now just part of the nonlocal momentum and energy transport calculated from the numerical solution of the radiation moment equations.

THE NONEQUILIBRIUM DIFFUSION APPROXIMATION

One of the assumptions made by equilibrium diffusion theory is that the radiation field and the material are in thermal equilibrium, which implies that the radiation has a Planckian distribution at the material temperature. This assumption is unnecessarily restrictive because problems arise in which the material is opaque, but the matter and radiation are not in equilibrium, for example when the material energy balance is driven by hydrodynamical processes faster than it can relax radiatively, or when the radiation field varies too rapidly for the material to follow instantaneously. It is therefore productive to construct a *nonequilibrium diffusion theory* in which the radiation field can have an arbitrary spectral distribution and energy density. Interesting early discussions of this approach appear in (**C1**), (**C2**), and (**F3**).*

It is most natural to formulate the theory in the Lagrangean frame, for that is the frame in which the radiation field isotropizes and $P_0(\nu_0) \rightarrow \frac{1}{3}E_0(\nu_0)$ when $\lambda_p \rightarrow 0$. Thus, assuming isotropy in (95.83) and dropping all terms of $O(\lambda_p v/lc)$ and higher, we have the nonequilibrium diffusion flux

$$\mathbf{F}_0(\nu_0) = -[c/3\chi_0(\nu_0)]\boldsymbol{\nabla} E_0(\nu_0). \tag{97.67}$$

Using this expression in (95.82) we obtain the nonequilibrium-diffusion-limit monochromatic radiation-energy equation

$$\begin{aligned}\rho\left[\frac{D}{Dt}\left[\frac{E_0(\nu_0)}{\rho}\right]+\frac{1}{3}\left\{E_0(\nu_0)-\frac{\partial}{\partial\nu_0}[\nu_0 E_0(\nu_0)]\right\}\frac{D}{Dt}\left(\frac{1}{\rho}\right)\right]\\ =\boldsymbol{\nabla}\cdot\left[\frac{c}{3\chi_0(\nu_0)}\boldsymbol{\nabla} E_0(\nu_0)\right]+\kappa_0(\nu_0)[4\pi B(\nu_0, T)-cE_0(\nu_0)].\end{aligned} \tag{97.68}$$

Here we have assumed that χ_0 and η_0 are given by (77.7).

Integrating over all frequencies we obtain the total flux

$$\mathbf{F}_0 = -(c/3\bar{\chi})\boldsymbol{\nabla} E_0, \tag{97.69}$$

and the nonequilibrium radiation diffusion equation

$$\rho\left[\frac{D}{Dt}\left(\frac{E_0}{\rho}\right)+\tfrac{1}{3}E_0\frac{D}{Dt}\left(\frac{1}{\rho}\right)\right]=\boldsymbol{\nabla}\cdot\left(\frac{c}{3\bar{\chi}}\boldsymbol{\nabla} E_0\right)+c(\kappa_P a_R T^4-\kappa_E E_0), \tag{97.70}$$

* What we call "equilibrium diffusion" is called *radiation heat conduction* by some authors (**Z1**, 151–153), and "radiation diffusion" by others. What we call "nonequilibrium diffusion" is also sometimes called "radiation diffusion" (**Z1**, 154–156) (which can lead to confusion) or the "Eddington approximation" (which is imprecise because one can invoke isotropy, $P = \frac{1}{3}E$, *without* dropping the $\partial F/\partial t$ term, which destroys the wavelike character of the radiation moment equations in transparent material and forces them to yield a diffusion equation). We recommend the terminology used here because it is descriptive and specific.

an expression first derived by Castor (**C3**). Here $\bar{\chi}$ is defined by

$$\nabla E_0/\bar{\chi} \equiv \int_0^\infty [\nabla E_0(\nu_0)/\chi_0(\nu_0)]\, d\nu_0, \tag{97.71}$$

[cf. (82.22)], taking different values along different coordinate axes if necessary; κ_P is the Planck mean; and κ_E is the absorption mean [cf. (82.30)]

$$\kappa_E \equiv \int_0^\infty e(\nu_0)\kappa_0(\nu_0)\, d\nu_0, \tag{97.72}$$

where $e(\nu_0) \equiv E_0(\nu_0)/E_0$ is the radiation energy spectral profile in the comoving frame.

Equations (97.69) and (97.70) are extensions of first-order equilibrium diffusion theory. One can make similar extensions of the second-order theory (**H2**); in essence one replaces $a_R T^4$ with E_0 in the expressions for μ_R and P_0.

To emphasize the similarity of (97.70) to the equilibrium diffusion equation (97.7) we parameterize the total radiation energy density in terms of a *radiation temperature* T_R, which in general is distinct from the material temperature T. Thus, if we define

$$a_R T_R^4 \equiv E_0, \tag{97.73}$$

equation (97.70) becomes

$$\rho\left[\frac{D}{Dt}\left(\frac{a_R T_R^4}{\rho}\right) + \tfrac{1}{3} a_R T_R^4 \frac{D}{Dt}\left(\frac{1}{\rho}\right)\right] = \nabla \cdot \left(\frac{4}{3}\frac{a_R c T_R^3}{\bar{\chi}} \nabla T_R\right) + a_R c(\kappa_P T^4 - \kappa_E T_R^4). \tag{97.74}$$

We stress that T_R is only a parameter describing the total radiation energy density, and that $e(\nu_0)$ need not be Planckian at T_R (although it may be useful to *assume* that it is—see below). In this *two-temperature description* we must determine two variables, T and T_R, hence must solve (97.74) along with the material energy equation

$$\rho\left[\frac{De}{Dt} + p\frac{D}{Dt}\left(\frac{1}{\rho}\right)\right] = a_R c(\kappa_E T_R^4 - \kappa_P T^4) + \rho\varepsilon. \tag{97.75}$$

Discrete versions of (97.74) and (97.75) are to be solved simultaneously for T^{n+1} and T_R^{n+1} at t^{n+1}. Both equations are linearized around current estimates T^* and T_R^*; for one-dimensional flows one obtains a block tridiagonal system with 2×2 blocks. This system is solved by Gaussian elimination, and the linearization procedure is iterated to convergence.

Thus far we have assumed that the coefficients $\bar{\chi}$ and κ_E are given. In grey material, $\kappa_P \equiv \kappa_E \equiv \kappa$ and $\bar{\chi} \equiv \kappa + \sigma$. For nongrey material the simplest approximation is to set, by analogy with equilibrium diffusion, $\bar{\chi} = \chi_R$ and $\kappa_E = \kappa_P$, with all opacities evaluated at the material temperature. This approach assumes that the radiation field is not strongly out of equilibrium with the material. A simple alternative is to take the two-temperature description literally, and assume that the spectral distribution of $E_0(\nu_0)$ is actually $B(\nu_0, T_R)$. Then, following (**F3**), we can define the *two-temperature mean opacities*

$$\kappa_E(T, T_R) \equiv \int_0^\infty \kappa_0(\nu_0, T) B(\nu_0, T_R)\, d\nu_0/(a_R c T_R^4/4\pi), \tag{97.76}$$

and

$$\bar{\chi}(T, T_R) \equiv \left[\int_0^\infty \frac{1}{\chi_0(\nu_0, T)} \left(\frac{\partial B}{\partial T}\right)_{T_R} d\nu_0 \bigg/ (a_R c T_R^3/\pi)\right]^{-1}. \tag{97.77}$$

This approach accounts approximately for the nonequilibrium character of the radiation field, and reduces to standard equilibrium diffusion when $T_R = T$.

To improve on the schemes described above we must determine the actual spectrum $e(\nu_0)$ by solving the monochromatic diffusion equation (97.68). This equation is complicated mathematically by the frequency derivative, which accounts for the varying Doppler shifts experienced by photons as they travel through moving material. Nonetheless, this term has been written in conservative form, and vanishes when (97.68) is integrated over frequency. One might therefore argue that we could simply drop the frequency derivative, knowing that the correct total energy density would still be obtained. The fallacy of this argument is revealed by a thought experiment devised by Buchler (**B3**).

Consider an adiabatic enclosure containing an extremely opaque, homogeneous medium (which implies $\mathbf{F} \equiv 0$), whose sole opacity is Thomson scattering ($\kappa \equiv 0$). Then, without the frequency derivative, (97.68) reduces to

$$\frac{D \ln E_0(\nu_0)}{Dt} - \frac{4}{3}\frac{D \ln \rho}{Dt} = 0, \tag{97.78}$$

which predicts that

$$E_0(\nu_0) \propto V^{-4/3}. \tag{97.79}$$

This result is correct for the *total* energy density in an adiabatic enclosure [cf. (69.71)], but not for the *monochromatic* energy density. Indeed, the fact that $E \propto V^{-4/3}$ in equilibrium implies that $T \propto V^{-1/3}$, so that the spectral distribution $B(\nu_0, T)$ *must* change when the enclosure contracts or expands. We will obtain the correct radiation energy spectral profile only if we retain the frequency derivative in (97.68). This term accounts for the progressive redshift (blueshift) of radiation undergoing adiabatic expansion (compression). Moreover, in the nonequilibrium case it allows redistribution of radiation over frequency to produce a non-Planckian spectrum, which may be of critical importance, for example when energy cascades from the spectral peak into a high-energy tail in material undergoing compression. In short, the $(\partial/\partial\nu)$ term must be retained in order to obtain the correct spectral distribution $e(\nu_0)$ (which is the whole point of doing a frequency-dependent calculation!).

Introducing frequency groups (or ODF pickets) as in §82, in one-dimensional flows we must solve G diffusion equations

$$\frac{D}{Dt}\left(\frac{E_g}{\rho}\right)+\tfrac{1}{3}[E_g-\Delta_g(\nu E_\nu)]\frac{D}{Dt}\left(\frac{1}{\rho}\right)$$
$$=\frac{\partial}{\partial M_r}\left[\frac{c(4\pi r^2)^2}{3(\bar{\chi}_g/\rho)}\frac{\partial E_g}{\partial M_r}\right]+4\pi\left(\frac{\kappa_{P,g}}{\rho}\right)B_g-c\left(\frac{\bar{\kappa}_g}{\rho}\right)E_g, \tag{97.80}$$

simultaneously with the material energy equation

$$\frac{De}{Dt}+p\frac{D}{Dt}\left(\frac{1}{\rho}\right)=\sum_g\left[c\left(\frac{\bar{\kappa}_g}{\rho}\right)E_g-4\pi\left(\frac{\kappa_{P,g}}{\rho}\right)B_g\right], \tag{97.81}$$

where

$$E_g\equiv\int_{\nu_g}^{\nu_{g+1}}E_0(\nu_0)\,d\nu_0. \tag{97.82}$$

Here $\kappa_{P,g}$ is the group Planck mean defined by (82.42), and $\bar{\kappa}_g$ and $\bar{\chi}_g$ are, respectively, appropriate direct and harmonic averages of the opacity within the group.

There are several possible choices for $\bar{\kappa}_g$ and $\bar{\chi}_g$ (cf. §82). For example, one might choose $\bar{\kappa}_g=\kappa_{P,g}(T)$ and $\bar{\chi}_g=\chi_{R,g}(T)$ computed at the material temperature T. Or, generalizing (97.76) and (97.77), one can introduce *two-temperature group means* (**F3**) and set $\bar{\kappa}_g=\kappa_{P,g}(T, T_R)$ and $\bar{\chi}_g=\chi_R(T, T_R)$ where

$$\kappa_{P,g}(T, T_R)\equiv\int_{\nu_g}^{\nu_{g+1}}\kappa_0(\nu_0, T)B(\nu_0, T_R)\,d\nu_0/B_g \tag{97.83}$$

and

$$\chi_{R,g}(T, T_R)\equiv\left[\int_{\nu_g}^{\nu_{g+1}}\frac{1}{\chi_0(\nu_0, T)}\left(\frac{\partial B_\nu}{\partial T}\right)_{T_R}d\nu_0\Big/\int_{\nu_g}^{\nu_{g+1}}\left(\frac{\partial B_\nu}{\partial T}\right)_{T_R}d\nu_0\right]^{-1}. \tag{97.84}$$

In either case $\kappa_{P,g}$ in the emission term remains a function of T only. Another possibility is to use an ODF, in which case $\bar{\kappa}_g=\kappa_{\text{picket}}(T)$ and $\bar{\chi}_g=\chi_{\text{picket}}(T)$, which are constants within each picket; we then also use κ_{picket} for $\kappa_{P,g}$ in the emission term. For expository ease we will use the generic notation $\bar{\kappa}_g$ and $\bar{\chi}_g$ to represent any of the choices just described.

The term $\Delta_g(\nu E_\nu)$ in (97.80) denotes the integral of $\partial[\nu_0E_0(\nu_0)]/\partial\nu_0$ over group g. We use upstream differencing, remembering that photons are redshifted (blueshifted) if the material expands (is compressed). hence for expansion we couple group g to g+1 and write

$$\Delta_g(\nu E_\nu)=(\nu E_\nu)_{g+1}-(\nu E_\nu)_g=\nu_{g+(3/2)}(E_{g+1}/\Delta\nu_{g+1})-\nu_{g+(1/2)}(E_g/\Delta\nu_g), \tag{97.85a}$$

and for compression we couple group g to group g − 1:

$$\Delta_g(\nu E_\nu) = \nu_{g+(1/2)}(E_g/\Delta\nu_g) - \nu_{g-(1/2)}(E_{g-1}/\Delta\nu_{g-1}). \qquad (97.85b)$$

Here $\nu_{g+(1/2)} \equiv \frac{1}{2}(\nu_g + \nu_{g+1})$, and the divisor $\Delta\nu_g \equiv (\nu_{g+1} - \nu_g)$ enters to convert E_g, an integral over the frequency band, back to a spectral density E_ν. When the discrete form of (97.80) is summed over all groups, the $\Delta(\nu E_\nu)$ terms telescope, and their sum vanishes identically.

The multigroup diffusion equations and the material energy equation are discretized in space and time in the usual way. If one uses two-temperature opacities, it is convenient to adjoin the definition

$$\sum_g E_g = a_R T_R^4 \qquad (97.86)$$

at each depth point. Assuming that all material properties are $f(\rho, T)$ or $f(\rho, T, T_R)$, and that ρ^{n+1} is known from the explicit hydrodynamics, the goal is to determine the solution vectors

$$\boldsymbol{\psi}_d \equiv (E_1, \ldots, E_G, T, T_R)_d^{n+1}, \qquad (d = 1, \ldots, D), \qquad (97.87)$$

of the nonlinear system.

Consider first a direct solution of the problem. We linearize all equations around current values $\boldsymbol{\psi}_d^*$; material properties are linearized in terms of δT (and, if appropriate, δT_R). The resulting linear system in $(\delta E_1, \ldots, \delta E_G, \delta T, \delta T_R)$ is block tridiagonal, and can be solved using the Feautrier technique described in §88 and iterated to convergence. The total computational effort scales as $cD(G+2)^3$; the solution is thus costly for a large number of frequency groups, and an alternative procedure may be preferable.

An efficient iteration scheme is provided by the *multifrequency/grey method*, which was developed in the *VERA code* [see (**F4**) and §§98 and 99]. Here we break the solution into two parts, each of which can be done relatively cheaply. In the first step, we assume that we know the radiation energy spectral profile $e_g \equiv E_g/E_0$. We then compute

$$\kappa_E \equiv \sum_g e_g \bar{\kappa}_g \equiv k_E \kappa_P \qquad (97.88)$$

and

$$(1/\bar{\chi}) \equiv \sum_g (1/\bar{\chi}_g)(\partial E_g/\partial M_r)/(\partial E_0/\partial M_r) \equiv k_F \chi_R, \qquad (97.89)$$

and use these means in the integrated diffusion equation (97.74) and the material energy equation (97.75). We linearize these equations (holding the ratios k_E and k_F fixed), solve for δT and δT_R, and update T and T_R at all depths; this step requires only $cD(2)^3$ operations. In the second step, we assume that we know the run of T and T_R. We can then solve the multigroup equations (97.80) for all E_g's using an "inner iteration" procedure. To avoid coupling among the equations, we put the $\Delta_g(\nu E_\nu)$ terms on the right-hand side along with the source term B_g. To start, we use the old

e_g scaled to the current value of E_0 to estimate the Δ_g's. We then have G independent tridiagonal systems of order D, which can be solved in parallel with a computational effort that scales as cDG. We then update the Δ_g's on the right-hand side, and iterate to convergence, thus determining e_g. We then return to the first step to compute more accurate values of T and T_R, and carry this "outer iteration" to convergence.

The multifrequency/grey method is efficient because in the first step even a rough estimate of the spectral distribution yields reasonable values for κ_E and $\bar{\chi}$, hence for T and T_R. Similarly, in the second step the computed spectral *distribution* will be reasonably accurate even if T and T_R (hence e and E_0) contain local errors.

Before leaving nonequilibrium diffusion theory, it is worthwhile to critique earlier formulations [e.g., (**C2**) and (**F3**)] which, unfortunately, are not physically consistent. In essence, these analyses start from the lab-frame equations (93.10) and (93.11), assume that $P^{ij} = \frac{1}{3}E\delta^{ij}$, and drop all $O(v/c)$ terms on the right-hand side, obtaining an energy equation of the form

$$(\partial E/\partial t) = \kappa(4\pi B - cE) + \nabla \cdot [(c/3\chi)\nabla E]. \tag{97.90}$$

The problem is that (97.90) is not correct in either the Lagrangean or Eulerian frame. Although the right-hand side of (97.90) looks like the right-hand side of the Lagrangean equation (97.70) (if we ignore the distinction between E and E_0), (97.90) is nevertheless not Lagrangean because on the left-hand side the distinction between $(\partial E/\partial t)$ and $\rho[D(E/\rho)/Dt]$ is not made, and the rate of work done by radiation pressure, $\frac{1}{3}\rho E[D(1/\rho)/Dt]$, is missing. On the other hand (97.90) is not a correct Eulerian equation because on comparing with (93.10) we see that the right-hand side lacks a term $c^{-1}\chi\mathbf{v} \cdot \mathbf{F}$ equal to the rate of work done by radiation forces on the material, and furthermore that in the flux divergence the comoving-frame flux has been used instead of the lab-frame flux, thus omitting the dominant term $\frac{4}{3}E_0\mathbf{v}$ that discriminates $\mathbf{F}$ (Eulerian) from $\mathbf{F}_0$ (Lagrangean).

Indeed, the derivation of the correct Eulerian nonequilibrium diffusion equation is a bit tricky. Not only must we discriminate between $\mathbf{F}$ and $\mathbf{F}_0$, but we also must be careful how we relate the inertial-frame energy density E to the parameter T_R. If we naively write $E = \frac{1}{3}a_R T_R^4$, the radiation-energy and gas-energy equations contain extra terms and do not reduce correctly to their Lagrangean counterparts (although the result for the radiating fluid—radiation plus material—is correct). Instead, we must define T_R by (97.73); then (91.16) implies that

$$E = a_R T_R^4 - (8a_R T_R^3/3c\chi_R)\mathbf{v} \cdot \nabla T_R. \tag{97.91}$$

The second term in (97.91) is important only when we compute the net absorption-emission. Using (97.91) and (93.14) in (93.10), we find that the Eulerian nonequilibrium-diffusion radiation energy equation correct to

$O(v/c)$ is

$$(a_R T_R^4)_{,t} = a_R c(\kappa_P T^4 - \kappa_E T_R^4) + \nabla \cdot [(4a_R c T_R^3 / 3\chi_R)\nabla T_R - \tfrac{4}{3} a_R T_R^4 \mathbf{v}] + \tfrac{4}{3} a_R T_R^3 \mathbf{v} \cdot \nabla T_R, \qquad (97.92)$$

which reduces to the more revealing form

$$(a_R T_R^4)_{,t} + \nabla \cdot (a_R T_R^4 \mathbf{v}) + \tfrac{1}{3} a_R T_R^4 \nabla \cdot \mathbf{v} = a_R c(\kappa_P T^4 - \kappa_E T_R^4) + \nabla \cdot [(4a_R c T_R^3 / 3\chi_R)\nabla T_R] \qquad (97.93)$$

which is identical to (97.74). Similarly, using (97.92) in (94.20) we find that the Eulerian gas-energy equation, correct to $O(v/c)$, in the nonequilibrium diffusion limit is

$$(\rho e)_{,t} + \nabla \cdot (\rho e \mathbf{v}) + p\nabla \cdot \mathbf{v} = a_R c(\kappa_E T_R^4 - \kappa_P T^4) + \rho\varepsilon, \qquad (97.94)$$

which is identical to (97.75).

In summary, (97.90) is unsatisfactory because either a term equal to four times the radiation work term is omitted if it is used as a Lagrangean equation, or the wrong flux is computed and a term equal to the rate of work done by radiation is omitted if it is used as an Eulerian equation.

THE PROBLEM OF FLUX LIMITING IN DIFFUSION THEORY

Inasmuch as the basic assumption of diffusion theory is that $\lambda_p \ll l$, there is no reason to expect the time-dependent radiation diffusion equation to yield accurate results in transparent media. Nevertheless, because of its simplicity, diffusion theory is often used throughout the flow, in both opaque and transparent regions; typically one then finds (**B1**), (**C2**) too large an energy transport in the optically thin material. Moreover, diffusion theory usually gives a serious overestimate of the energy deposited by a radiation front penetrating into cold material, particularly at early times. Nonequilibrium diffusion generally gives better results than equilibrium diffusion, but both are significantly in error.

In the most extreme cases, the flux out of an optically thin zone predicted by diffusion theory may exceed the energy density times the velocity of light,

$$|\mathbf{F}_{\text{diffusion}}| > cE, \qquad (97.95)$$

implying that the effective speed of energy propagation,

$$V_E \equiv |\mathbf{F}|/E = |\tfrac{1}{3} c\lambda_p \nabla E|/E \sim c(\lambda_p/3l), \qquad (97.96)$$

exceeds the velocity of light.

Both of these results are physically absurd, and clearly reflect a breakdown of the theory. The root of the problem is that the diffusion equation tacitly assumes that photons *always* travel a distance of the order of λ_p, even if λ_p exceeds the free-flight distance $c\,\Delta t$ corresponding to the timestep Δt. To follow a radiation front we choose $\Delta t \sim \Delta x/c$, hence we can expect diffusion theory to break down in regions where $\lambda_p > \Delta x$. The problem should come as no surprise, for it is well known that the linear diffusion equation has a formally *infinite* signal speed (**M16,** 862–865), whereas the signal speed for transport of radiation in transparent material is limited to the speed of light (§81).

One way to overcome the problem is to use a *flux limiter,* first suggested by J. R. Wilson (unpublished); see also (**W4**). The idea is to alter the diffusion-theory formula for the flux in such a way as to yield the standard result in the high-opacity limit, while simulating free streaming in transparent regions. For example, we might use an expression of the form

$$\mathbf{F} = -c\,\nabla E/[(3/\lambda_p) + |\nabla E|/E], \tag{97.97}$$

which yields $\mathbf{F}_{\text{diffusion}}$ when $\lambda_p/l \ll 1$, and limits to

$$\mathbf{F} = cE\mathbf{n} \tag{97.98}$$

in transparent regions; here $\mathbf{n}$ is a unit vector opposite to ∇E (i.e., down the gradient). Equation (97.97) is only illustrative; numerous other expressions have been proposed (**K2**), (**L3**), (**L4**), the one most often used in astrophysics appearing in (**A1**).

While flux-limited diffusion has been widely used, this approach provides only an ad hoc "fixup" of generally unknown accuracy; it gives the correct limits, and has qualitatively correct behavior in between, but it could be quantitatively wrong (perhaps seriously) in the intermediate regime. Fundamentally the flux-limiting problem results from dropping the time derivative from the radiation momentum equation, which precludes recovery of the wave-equation character of the coupled radiation energy and momentum equations in the optically thin limit. Indeed, dimensional analysis suggests that on a radiation-flow time scale $c^{-2}(\partial F/\partial t)$ is $O(\lambda_p/l)$ relative to $\chi F/c$, and will dominate the solution for a radiation front in transparent material.

If we retain the time derivative in the radiation momentum equation, we can, in fact, recover the wave equation in transparent material, and the solution is automatically flux limited (**M10**). We examine this assertion briefly here to motivate the developments of §§98 and 99. To make the point while avoiding unnecessary complications, we consider time-dependent transport in planar, static material. The radiation energy and momentum equations are

$$(\partial E/\partial t) + (\partial F/\partial z) = \kappa(4\pi B - cE) \tag{97.99}$$

and

$$c^{-2}(\partial F/\partial t) + [\partial(fE)/\partial z] = -(\chi/c)F, \tag{97.100}$$

where f is the variable Eddington factor. We first replace the time derivative in (97.100) by a finite difference, while leaving the spatial derivative in continuous form; for stability we use a fully implicit scheme, which gives

$$F^{n+1} = \frac{-c}{\gamma + \chi^{n+1}} \frac{\partial(f^{n+1}E^{n+1})}{\partial z} + \frac{\gamma}{\gamma + \chi^{n+1}} F^n, \tag{97.101}$$

where

$$\gamma \equiv 1/c\,\Delta t. \tag{97.102}$$

Equation (97.101) provides an analytical expression for the flux, which can be used in (97.99); but let us first examine its physical content.

Define λ such that

$$\lambda^{-1} \equiv \gamma + \chi = (c\,\Delta t)^{-1} + \lambda_p^{-1} = \lambda_t^{-1} + \lambda_p^{-1}. \tag{97.103}$$

Clearly the effective mean free path λ is the harmonic mean of the optical mean free path λ_p and the free-flight distance λ_t through which a photon can travel in a timestep Δt. We then rewrite (97.101) as

$$F^{n+1} = -c\lambda^{n+1}[\partial(f^{n+1}E^{n+1})/\partial z] + (\lambda/\lambda_t)F^n. \tag{97.104}$$

In opaque material $\chi \gg 1$, $\lambda_p \ll \lambda_t$, and $\lambda \to \lambda_p$, while $f \to \frac{1}{3}$. Equation (97.104) then reduces to the standard diffusion result

$$F^{n+1} = -\tfrac{1}{3}c\lambda_p^{n+1}(\partial E^{n+1}/\partial z). \tag{97.105}$$

In transparent material $\chi \to 0$, $\lambda_p \to \infty$, hence $\lambda \to \lambda_t$, which shows that (97.103) correctly limits the effective mean free path to the photon flight distance instead of allowing it to become arbitrarily large. In this regime $f \to 1$, $F \to cE$, and (97.104) reduces to

$$F^{n+1} = F^n - \lambda_t(\partial F^{n+1}/\partial z), \tag{97.106}$$

which makes the physically correct statement that in the optically thin limit a change in the local value of the flux results from information communicated about the flux gradient within a photon free-flight distance λ_t.

If we now replace the time derivative in the energy equation with a backwards time difference and eliminate the flux via (97.104), we obtain the *combined moment equation*

$$(\gamma + \kappa^{n+1})E^{n+1} - \frac{\partial}{\partial z}\left[\lambda^{n+1}\frac{\partial(f^{n+1}E^{n+1})}{\partial z}\right] = \left(\frac{4\pi}{c}\right)\kappa^{n+1}B^{n+1} + \gamma E^n - \left(\frac{\gamma}{c}\right)\frac{\partial}{\partial z}(\lambda^{n+1}F^n). \tag{97.107}$$

The mathematical structure of (97.107) is similar to the nonequilibrium diffusion equation (97.70); its physical content, however, is much larger. In the limit of high opacity and/or long timesteps, $(\kappa/\gamma) \gg 1$, and (97.107) reduces to

$$\frac{1}{3}\frac{\partial}{\partial z}\left(\lambda_p^{n+1}\frac{\partial E^{n+1}}{\partial z}\right) = \kappa^{n+1}\left[E^{n+1} - \left(\frac{4\pi}{c}\right)B^{n+1}\right], \tag{97.108}$$

which is a quasi-static nonequilibrium diffusion equation at the advanced time level. In the limit of low opacity and/or short timesteps, $(\kappa/\gamma) \ll 1$, and, setting $f=1$, we find

$$E^{n+1} - (c\,\Delta t)^2(\partial^2 E^{n+1}/\partial z^2) = E^n - (\gamma c)^{-1}(\partial F^n/\partial z) = 2E^n - E^{n-1}, \tag{97.109}$$

where the second equality follows from a backward Euler representation of (97.99) at t^n for $\kappa=0$. Regrouping terms we have

$$c^2(\partial^2 E^{n+1}/\partial z^2) = (E^{n+1} - 2E^n + E^{n-1})/\Delta t^2 \approx (\partial^2 E/\partial t^2)^n, \tag{97.110}$$

which is an approximation to the wave equation; the miscentering of the time derivative results from use of fully implicit difference formulae in both moment equations. If we were to use $f=\frac{1}{3}$ in the transparent limit we would still obtain the wave equation, but with a propagation speed of only $c/\sqrt{3}$ instead of c, a result that also follows rigorously from (97.99) and (97.100) when $\kappa = \chi \equiv 0$ and $f=\frac{1}{3}$. Thus it is essential to use accurate Eddington factors in order to obtain the correct propagation speed.

In summary, if we solve the full time-dependent radiation moment equations we recover both the diffusion and wave limits, and avoid the problem of flux limiting, which is a mere artifact of the approximations inherent in the diffusion equation. Furthermore, we must use accurate Eddington factors if we are to recover the correct streaming limit. Let us therefore now consider how to solve the full radiation transport problem in moving media.

98. Transport Solution in the Comoving Frame

In this section we discuss methods for solving the Lagrangean equations of radiation hydrodynamics in the transport regime. Because the complete set of equations is complicated, we tailor the discussion to situations of astrophysical interest. As before we consider only one-dimensional spherically symmetric flows of a single material. We will emphasize fluid-flow time scales, but write equations that generally behave correctly on radiation-flow time scales as well. To close the system of moment equations we use variable Eddington factors; these are to be evaluated in a subsidiary angle-frequency-dependent formal solution. Similarly, the spectral distributions required to form mean absorption coefficients for the radiation energy and momentum equations are to be obtained from a subsidiary solution of multigroup equations.

RADIATION ENERGY AND MOMENTUM EQUATIONS

For economy of notation we omit the affix "0" with the understanding that *in this section all physical variables are measured in the comoving frame.* As discussed in §95, we retain all $O(v/c)$ terms in the radiation energy equation; in the radiation momentum equation we drop them all except the

(D/Dt) term, which is kept to assure flux limiting, as discussed in §97. Thus the radiation equations to be solved are the monochromatic radiation energy equation

$$\frac{D}{Dt}\left(\frac{E_\nu}{\rho}\right)+\left[f_\nu \frac{D}{Dt}\left(\frac{1}{\rho}\right)-(3f_\nu-1)\frac{v}{\rho r}\right]E_\nu-\frac{\partial}{\partial \nu}\left\{\left[f_\nu \frac{D}{Dt}\left(\frac{1}{\rho}\right)-(3f_\nu-1)\frac{v}{\rho r}\right]\nu E_\nu\right\}$$
$$=\frac{\kappa_\nu}{\rho}(4\pi B_\nu-cE_\nu)-\frac{\partial(4\pi r^2 F_\nu)}{\partial M_r}, \quad (98.1)$$

and the monochromatic radiation momentum equation

$$\frac{1}{c^2}\frac{DF_\nu}{Dt}+4\pi r^2\rho\frac{\partial(f_\nu E_\nu)}{\partial M_r}+\frac{(3f_\nu-1)}{r}E_\nu=-\frac{\chi_\nu F_\nu}{c}. \quad (98.2)$$

As in §83 we define the sphericity factor

$$\ln q_\nu \equiv \int_{r_c}^{r}[(3f_\nu-1)/f_\nu r']\,dr', \quad (98.3)$$

which allows us to rewrite (98.2) as

$$\frac{1}{c^2}\frac{DF_\nu}{Dt}+\frac{4\pi r^2\rho}{q_\nu}\frac{\partial(f_\nu q_\nu E_\nu)}{\partial M_r}=-\frac{\chi_\nu F_\nu}{c}. \quad (98.4)$$

Integrating (98.1) over frequency we obtain the radiation energy equation

$$\frac{D}{Dt}\left(\frac{E}{\rho}\right)+\left[f\frac{D}{Dt}\left(\frac{1}{\rho}\right)-(3f-1)\frac{v}{\rho r}\right]E=\frac{1}{\rho}(4\pi\kappa_P B-c\kappa_E E)-\frac{\partial(4\pi r^2 F)}{\partial M_r}. \quad (98.5)$$

Here

$$f\equiv\int_0^\infty f_\nu e_\nu\,d\nu \quad (98.6)$$

and

$$\kappa_E\equiv\int_0^\infty \kappa_\nu e_\nu\,d\nu, \quad (98.7)$$

where the radiation energy spectral profile $e_\nu \equiv E_\nu/E$ is presumed known. Similarly, integrating (98.2) over frequency we obtain the radiation momentum equation

$$\frac{1}{c^2}\frac{DF}{Dt}+\frac{4\pi r^2\rho}{q}\frac{\partial(fqE)}{\partial M_r}=-\frac{\chi_F F}{c}, \quad (98.8)$$

where now

$$\ln q\equiv\int_{r_c}^{r}[(3f-1)/fr']\,dr', \quad (98.9)$$

and

$$\chi_F\equiv\int_0^\infty \chi_\nu \mathscr{f}_\nu\,d\nu. \quad (98.10)$$

Again the radiation flux spectral profile $\mathscr{f}_\nu \equiv F_\nu/F$ is assumed known.

We can use (98.5) and (98.8) as a coupled system on an interleaved grid with E defined at cell centers and F defined on cell boundaries. Alternatively we can use (98.8) to eliminate F from the radiation energy equation, thereby producing a combined moment equation for E. To obtain such an equation we replace the time derivative in (98.8) by a finite difference while leaving the spatial operator in continuous form. For stability we use a backwards difference, obtaining

$$F = -\frac{4\pi c r^2 \rho \lambda}{q} \frac{\partial (fqE)}{\partial M_r} + \left(\frac{\lambda}{\lambda_t}\right) F^n, \tag{98.11}$$

where λ is the effective mean free path defined in (97.103), and $\lambda_t \equiv c\,\Delta t$. In (98.11), all variables are evaluated at a time Δt beyond some arbitrary reference time t^n; notice that as $\Delta t \to 0$, $\lambda \to 0$, $(\lambda/\lambda_t) \to 1$, and $F \to F^n$. Substituting (98.11) into (98.5) we have

$$\begin{aligned} \frac{D}{Dt}\left(\frac{E}{\rho}\right) + \left[f \frac{D}{Dt}\left(\frac{1}{\rho}\right) - (3f-1)\frac{v}{\rho r}\right] E &= \frac{1}{\rho}(4\pi\kappa_P B - c\kappa_E E) \\ + \frac{\partial}{\partial M_r}\left[\frac{(4\pi r^2)^2 c\rho\lambda}{q} \frac{\partial (fqE)}{\partial M_r}\right] &- \frac{\partial}{\partial M_r}\left[4\pi r^2 \left(\frac{\lambda}{\lambda_t}\right) F^n\right]. \end{aligned} \tag{98.12}$$

If we ignore the time dependence of the radiation momentum equation (equivalent to letting $\lambda_t \to \infty$, $\lambda \to \lambda_p$) we obtain the combined moment equation

$$\begin{aligned} \frac{D}{Dt}\left(\frac{E}{\rho}\right) + \left[f \frac{D}{Dt}\left(\frac{1}{\rho}\right) - (3f-1)\frac{v}{\rho r}\right] E & \\ = \frac{1}{\rho}(4\pi\kappa_P B - c\kappa_E E) &+ \frac{\partial}{\partial M_r}\left[\frac{(4\pi r^2)^2 c\rho}{q\chi_F} \frac{\partial (fqE)}{\partial M_r}\right], \end{aligned} \tag{98.13}$$

which was first derived by Castor (**C3**) (who made the additional approximations that $\kappa_E = \kappa_P$ and $\chi_F = \chi_R$). This equation can be used in a variety of astrophysical problems [see e.g., (**K3**)] but (98.12) is preferable if we wish to model phenomena on a radiation-flow time scale. If we set $f \equiv \frac{1}{3}$ and $q \equiv 1$ in (98.13) we recover the nonequilibrium diffusion equation (97.70) or (97.74); if, instead, we set $(D/Dt) \equiv 0$ and $v \equiv 0$, we recover (83.65), the combined moment equation for a static medium.

FLUID EQUATIONS

The radiation equations are to be solved simultaneously with the fluid momentum equation

$$(Dv/Dt) = -(GM_r/r^2) - 4\pi r^2[\partial(p+Q)/\partial M_r] + (\chi_F/\rho c)F, \tag{98.14a}$$

or

$$(Dv/Dt) = -(GM_r/r^2) - 4\pi r^2\{[\partial(p+Q)/\partial M_r] + (1/q)[\partial(fqE)/\partial M_r]\}, \tag{98.14b}$$

and the material energy equation

$$(De/Dt)+(p+Q)[D(1/\rho)/Dt]=\varepsilon+(c\kappa_E E-4\pi\kappa_P B)/\rho\equiv\dot{e}. \tag{98.15}$$

Here Q is the pseudoviscous pressure, and ε is the rate of thermonuclear energy generation; all material properties are assumed to be known functions of ρ and T.

COMPUTATIONAL STRATEGY

The problem is to solve (98.14) and (98.15) coupled to either (98.5) and (98.8), or to (98.12), as a function of time. As an *example* we outline the computational procedure in a Lagrangean coordinate system; but we remind the reader that any coordinate grid may be used, with adaptive-mesh schemes (**T3**), (**W3**) being the preferred choice for many problems.

In an explicit Lagrangean scheme (cf. §59) we may use either (98.14a) or (98.14b) to advance v from $t^{n-(1/2)}$ to $t^{n+(1/2)}$ because p, Q, E, F, f, and q are all known at t^n. Equation (98.14a) is essentially exact, whereas (98.14b) omits both time- and velocity-dependent terms in accordance with the analysis of §96. The omitted terms are at most $O(v/c)$, hence their effect on energy balance is at most $O(v^2/c^2)$.

Having obtained $v^{n+(1/2)}$ we update radii and densities to t^{n+1}. We must then solve (98.14) simultaneously with either (98.12) for T and E, or with (98.5) and (98.8) for T, E, and F at t^{n+1}. Here we assume that f, q, κ_E, and χ_F are known at t^{n+1}, which implies that we are given f_ν, q_ν, e_ν, and $\not{f}_\nu$ at t^{n+1}. But of course these quantities depend on T^{n+1} and E^{n+1}, and the whole set must be determined self-consistently.

The computationally most economical method proceeds in three steps, which are iterated to consistency. (1) Given estimates of the geometric factors and spectral profiles we solve the coupled material energy and radiation moment equations for T^{n+1}, E^{n+1}, and F^{n+1}. (2) Given these values of T and E we evaluate the source-sink terms in the angle-frequency-dependent transfer equation and perform a formal solution for f_ν and q_ν. Because the Eddington factor is only a *ratio* of radiation moments and is primarily *geometry dependent*, reasonable distributions of T and E yield relatively accurate values of f_ν and q_ν. (3) Using the current estimates of T^{n+1}, f_ν, and q_ν, we solve the monochromatic moment equations to obtain new estimates of the spectral profiles e_ν and $\not{f}_\nu$. We then update the frequency-integrated quantities f and q, re-evaluate χ_E and χ_F, and return to step (1).

Aside from the additional step of updating Eddington factors, the above procedure is identical to the multifrequency/grey method of solving the multigroup diffusion equation. Indeed, if the Eddington factors can be determined reasonably cheaply (perhaps being updated only every few timesteps), little more effort is needed to solve the full transport equations than the corresponding diffusion equations. Inasmuch as a transport solution is inherently much more accurate in transparent regions near boundaries, it may be false economy to use a diffusion calculation at all.

DIFFERENCE EQUATIONS

An explicit difference equation for the fluid momentum equation is

$$[v_i^{n+(1/2)} - v_i^{n-(1/2)}]/\Delta t^n = -GM_i/(r_i^n)^2 \\ -4\pi (r_i^n)^2 [p_{i+(1/2)}^n - p_{i-(1/2)}^n + Q_{i+(1/2)}^{n-(1/2)} - Q_{i-(1/2)}^{n-(1/2)}]/\Delta M_i + (\chi_i^n/\rho_i^n c) F_i^n, \tag{98.16a}$$

or

$$[v_i^{n+(1/2)} - v_i^{n-(1/2)}]/\Delta t^n = -GM_i/(r_i^n)^2 \\ -4\pi (r_i^n)^2 \{p_{i+(1/2)}^n - p_{i-(1/2)}^n + Q_{i+(1/2)}^{n-(1/2)} - Q_{i-(1/2)}^{n-(1/2)} \\ + [f_{i+(1/2)}^n q_{i+(1/2)}^n E_{i+(1/2)}^n - f_{i-(1/2)}^n q_{i-(1/2)}^n E_{i-(1/2)}^n]/q_i^n\}/\Delta M_i. \tag{98.16b}$$

Here $i = 2, \ldots, I$, and we have dropped the subscript "F" on χ. In the case of unequal timesteps, one may wish to center both $p_{i+(1/2)}$ and r_i at $t^{n+\lambda}$, as in (59.64) and (59.87). The pseudoviscous pressure Q is computed according to the prescriptions in §59.

At the lower boundary we assume that $v_1 = f(t)$, a known function of time; for example $v_1 \equiv 0$ at the center of a star or at the fixed inner boundary of a pulsating envelope. At the upper boundary we assume that the material and pseudoviscous pressures are zero; we further assume that any material outside of r_{I+1} is optically thin so that radiation quantities are invariant beyond that point. We then have

$$[v_{I+1}^{n+(1/2)} - v_{I+1}^{n-(1/2)}]/\Delta t^n = -GM_{I+1}/(r_{I+1}^n)^2 \\ +4\pi (r_{I+1}^n)^2 [p_{I+(1/2)}^n + Q_{I+(1/2)}^{n-(1/2)}]/\Delta M_{I+1} + (\chi_{I+1}^n/\rho_{I+1}^n c) F_{I+1}^n, \tag{98.17a}$$

or

$$[v_{I+1}^{n+(1/2)} - v_{I+1}^{n-(1/2)}]/\Delta t^n \\ = -GM_{I+1}/(r_{I+1}^n)^2 + 4\pi (r_{I+1}^n)^2 \{[p_{I+(1/2)}^n + Q_{I+(1/2)}^{n-(1/2)}]\,\Delta M_{I+1} \\ + 2[f_{I+(1/2)}^n q_{I+(1/2)}^n - f_{I+1}^n q_{I+1}^n \jmath_{I+1}^n] E_{I+(1/2)}^n / q_I^n \,\Delta M_{I+(1/2)}\}. \tag{98.17b}$$

Here $\Delta M_{I+1} \equiv \frac{1}{2}\Delta M_{I+(1/2)} + \Delta M_{I+(3/2)}$, where $\Delta M_{I+(3/2)}$ represents any mass assumed to lie outside of r_{I+1}, and $\jmath_{I+1} \equiv E_{I+1}/E_{I+(1/2)}$ is a geometrical factor determined from the formal solution [cf. (83.44)].

Having advanced $v_i^{n-(1/2)}$ to $v_i^{n+(1/2)}$ we update shell positions

$$r_i^{n+1} = r_i^n + v_i^{n+(1/2)}\,\Delta t^{n+(1/2)}, \tag{98.18}$$

and densities

$$\rho_{i+(1/2)}^{n+1} = (3/4\pi)\,\Delta M_{i+(1/2)}/[(r_{i+1}^{n+1})^3 - (r_i^{n+1})^3]. \tag{98.19}$$

The material energy equation is represented as

$$e^{n+1}_{i+(1/2)} - e^{n}_{i+(1/2)} + \{\tfrac{1}{2}[p^{n}_{i+(1/2)} + p^{n+1}_{i+(1/2)}] + Q^{n+(1/2)}_{i+(1/2)}\}\{[1/\rho^{n+1}_{i+(1/2)}] - [1/\rho^{n}_{i+(1/2)}]\} = \Delta t^{n+(1/2)}[(1-\theta)\dot{e}^{n}_{i+(1/2)} + \theta\dot{e}^{n+1}_{i+(1/2)}], \qquad (i=1,\ldots,I), \tag{98.20}$$

where $\frac{1}{2} \leq \theta \leq 1$, and

$$\dot{e}^{m}_{i+(1/2)} \equiv \varepsilon^{m}_{i+(1/2)} + [c\kappa^{m}_{E,i+(1/2)}E^{m}_{i+(1/2)} - 4\pi\kappa^{m}_{P,i+(1/2)}B^{m}_{i+(1/2)}]/\rho^{m}_{i+(1/2)}. \tag{98.21}$$

To (98.20) we must adjoin difference-equation representations of the radiation energy and momentum equations. We will use a fully implicit (backward Euler) scheme for both equations. This choice has several advantages. (1) It is physically sensible on fluid-flow time scales, because when $c\,\Delta t \gg \Delta r$, the radiation field is essentially quasistatic at the advanced time level. Indeed, experience shows (**M8**) that the fully implicit scheme should be used even on radiation-flow time scales because time-centered differencing tends to produce large, unphysical oscillations of the solution. (2) It maximizes stability; a von Neumann local stability analysis shows that the fully implicit equations are unconditionally stable. (3) It is algebraically simple; other choices lead to more complex equations (**M8**), (**M10**).

The fully implicit representation of the radiation momentum equation (98.8) is

$$F^{n+1}_{i} = -\beta^{n+1}_{i,i+(1/2)}E^{n+1}_{i+(1/2)} + \beta^{n+1}_{i,i-(1/2)}E^{n+1}_{i-(1/2)} + \alpha^{n+1}_{i}F^{n}_{i}, \qquad (i=2,\ldots,I),$$

$$(i=2,\ldots,I), \tag{98.22}$$

where, writing $\gamma \equiv 1/c\,\Delta t$,

$$\alpha^{n+1}_{i} \equiv \gamma/(\gamma + \chi^{n+1}_{i}) \tag{98.23}$$

and

$$\beta^{n+1}_{i,i\pm(1/2)} \equiv 4\pi c(r^{n+1}_{i})^2\rho^{n+1}_{i}f^{n+1}_{i\pm(1/2)}q^{n+1}_{i\pm(1/2)}/q^{n+1}_{i}(\gamma + \chi^{n+1}_{i})\,\Delta M_{i}. \tag{98.24}$$

Here ρ^{n+1}_{i} and χ^{n+1}_{i} are suitable averages across the interface r_i. For ρ_i it is reasonable to adopt the mass-weighted average

$$\rho_i \equiv [\rho_{i-(1/2)}\,\Delta M_{i-(1/2)} + \rho_{i+(1/2)}\,\Delta M_{i+(1/2)}]/[\Delta M_{i-(1/2)} + \Delta M_{i+(1/2)}]. \tag{98.25}$$

Similarly, to obtain the correct optical-depth increment between cell centers one might adopt

$$(\chi/\rho)_i \equiv [(\chi/\rho)_{i-(1/2)}\,\Delta M_{i-(1/2)} + (\chi/\rho)_{i+(1/2)}\,\Delta M_{i+(1/2)}]/[\Delta M_{i-(1/2)} + \Delta M_{i+(1/2)}]. \tag{98.26}$$

For certain classes of problems, however, (98.26) is unsatisfactory. For example, in stellar pulsation calculations the temperature sensitivity of the opacity is so steep ($\chi \propto T^{12}$) that there can be enormous (one or two orders of magnitude) jumps in the opacity between successive zones. In this case,

Christy has shown **(C4)** [see also **(S5)**] that reasonably accurate flux transport is obtained in the diffusion limit by using the energy-weighted harmonic average

$$(\rho/\chi)_i \equiv \frac{\{[T_{i-(1/2)}]^4(\rho/\chi)_{i-(1/2)}+[T_{i+(1/2)}]^4(\rho/\chi)_{i+(1/2)}\}}{\{[T_{i-(1/2)}]^4+[T_{i+(1/2)}]^4\}}. \tag{98.27}$$

For an inner boundary condition we fix the incident flux. If r_1 is the center of a star, $F_1 \equiv 0$. If r_1 is the radius of a static core inside a dynamical envelope, we set $F_1 = L_1/4\pi r_1^2$ where L_1 is assumed given. To obtain an outer boundary condition we apply (98.22) on the half-shell from $r_{I+(1/2)}$ to R_{I+1}; then

$$F_{I+1}^{n+1} = -\tilde{\beta}_{I+1,I+1}^{n+1}E_{I+1}^{n+1} + \tilde{\beta}_{I+1,I+(1/2)}^{n+1}E_{I+(1/2)}^{n+1} + \tilde{\alpha}_{I+1}^{n+1}F_{I+1}^{n}, \tag{98.28}$$

where $\tilde{\alpha}$ and $\tilde{\beta}$ are defined by (98.23) and (98.24) with ΔM_i replaced by $\frac{1}{2}\Delta M_{I+(1/2)}$. Invoking a geometrical closure of the form

$$F_{I+1}^{n+1} = \hbar_{I+1}^{n+1}cE_{I+1}^{n+1} \tag{98.29}$$

where $\hbar$ is given by the formal solution, as in (83.45), we find

$$F_{I+1}^{n+1} = \beta_{I+1,I+(1/2)}^{n+1}E_{I+(1/2)}^{n+1} + \alpha_{I+1}^{n+1}F_{I+1}^{n}, \tag{98.30}$$

where

$$\alpha_{I+1}^{n+1} \equiv c\hbar_{I+1}^{n+1}\tilde{\alpha}_{I+1}^{n+1}/(ch_{I+1}^{n+1}+\tilde{\beta}_{I+1,I+1}^{n+1}) \tag{98.31}$$

and

$$\beta_{I+1,I\pm(1/2)}^{n+1} \equiv c\hbar_{I+1}^{n+1}\tilde{\beta}_{I+1,I\pm(1/2)}^{n+1}/(ch_{I+1}^{n+1}+\tilde{\beta}_{I+1,I+1}^{n+1}). \tag{98.32}$$

The fully implicit representation of the radiation energy equation (98.8) is

$$\begin{aligned}
&[E_{i+(1/2)}^{n+1}/\rho_{i+(1/2)}^{n+1}] + [\![f_{i+(1/2)}^{n+1}\{[1/\rho_{i+(1/2)}^{n+1}]-[1/\rho_{i+(1/2)}^{n}]\}\\
&\quad -[3f_{i+(1/2)}^{n+1}-1](v/r)_{i+(1/2)}^{n+(1/2)}\,\Delta t^{n+(1/2)}/\rho_{i+(1/2)}^{n+1}]\!]E_{i+(1/2)}^{n+1}\\
&= [E_{i+(1/2)}^{n}/\rho_{i+(1/2)}^{n}] + \Delta t^{n+(1/2)}\{[4\pi\kappa_{P,i+(1/2)}^{n+1}B_{i+(1/2)}^{n+1} - c\kappa_{E,i+(1/2)}^{n+1}E_{i+(1/2)}^{n+1}]/\rho_{i+(1/2)}^{n+1}\\
&\quad -4\pi[(r_{i+1}^{n+1})^2F_{i+1}^{n+1}-(r_i^{n+1})^2F_i^{n+1}]/\Delta M_{i+(1/2)}\}, \qquad (i=1,\ldots,I).
\end{aligned} \tag{98.33}$$

Equations (98.22), (98.30), and (98.33) are a coupled system for F_i^{n+1}, $(i=1,\ldots,I+1)$, and $E_{i+(1/2)}^{n+1}$, $(i=1,\ldots,I)$. Alternatively we can use (98.22) and (98.30) in (98.33) to eliminate the F_i's analytically, obtaining a finite difference representation of the combined moment equation (98.12). Thus for interior shells $(i=2,\ldots,I-1)$ we have

$$\begin{aligned}
&[E_{i+(1/2)}^{n+1}/\rho_{i+(1/2)}^{n+1}] + [\![f_{i+(1/2)}^{n+1}\{[1/\rho_{i+(1/2)}^{n+1}]-[1/\rho_{i+(1/2)}^{n}]\}\\
&\quad -[3f_{i+(1/2)}^{n+1}-1](v/r)_{i+(1/2)}^{n+(1/2)}\,\Delta t^{n+(1/2)}/\rho_{i+(1/2)}^{n+1}]\!]E_{i+(1/2)}^{n+1}\\
&= [E_{i+(1/2)}^{n}/\rho_{i+(1/2)}^{n}] + \Delta t^{n+(1/2)}[\![[4\pi\kappa_{P,i+(1/2)}^{n+1}B_{i+(1/2)}^{n+1} - c\kappa_{E,i+(1/2)}^{n+1}E_{i+(1/2)}^{n+1}]/\rho_{i+(1/2)}^{n+1}\\
&\quad +4\pi\{(r_{i+1}^{n+1})^2[\beta_{i+1,i+(3/2)}^{n+1}E_{i+(3/2)}^{n+1} - \beta_{i+1,i+(1/2)}^{n+1}E_{i+(1/2)}^{n+1} - \alpha_{i+1}^{n+1}F_{i+1}^{n}]\\
&\quad -(r_i^{n+1})^2[\beta_{i,i+(1/2)}^{n+1}E_{i+(1/2)}^{n+1} - \beta_{i,i-(1/2)}^{n+1}E_{i-(1/2)}^{n+1} - \alpha_i^{n+1}F_i^{n}]\}/\Delta M_{i+(1/2)}]\!].
\end{aligned} \tag{98.34}$$

For the innermost shell ($i=1$) we have

$$\begin{aligned}(E^{n+1}_{3/2}/\rho^{n+1}_{3/2}) + \{f^{n+1}_{3/2}[(1/\rho^{n+1}_{3/2}) - (1/\rho^{n}_{3/2})] \\ - (3f^{n+1}_{3/2} - 1)(v/r)^{n+(1/2)}_{3/2}\,\Delta t^{n+(1/2)}/\rho^{n+1}_{3/2}\}E^{n+1}_{3/2} \\ = (E^{n}_{3/2}/\rho^{n}_{3/2}) + \Delta t^{n+(1/2)}\{(4\pi\kappa^{n+1}_{P,3/2}B^{n+1}_{3/2} - c\kappa^{n+1}_{E,3/2}E^{n+1}_{3/2})/\rho^{n+1}_{3/2} \\ + [4\pi(r^{n+1}_{2})^2(\beta^{n+1}_{2,5/2}E^{n+1}_{5/2} - \beta^{n+1}_{2,3/2}E^{n+1}_{3/2} - \alpha^{n+1}_{2}F^{n}_{2}) + L_1]/\Delta M_{3/2}\},\end{aligned} \tag{98.35}$$

and for the outermost shell ($i=I$),

$$\begin{aligned}[E^{n+1}_{I+(1/2)}/\rho^{n+1}_{I+(1/2)}] + [\![f^{n+1}_{I+(1/2)}\{[1/\rho^{n+1}_{I+(1/2)}] - [1/\rho^{n}_{I+(1/2)}]\} \\ - [3f^{n+1}_{I+(1/2)} - 1](v/r)^{n+(1/2)}_{I+(1/2)}\,\Delta t^{n+(1/2)}/\rho^{n+1}_{I+(1/2)}]\!]E^{n+1}_{I+(1/2)} \\ = [E^{n}_{I+(1/2)}/\rho^{n}_{I+(1/2)}] \\ + \Delta t^{n+(1/2)}[\![[4\pi\kappa^{n+1}_{P,i+(1/2)}B^{n+1}_{i+(1/2)} - c\kappa^{n+1}_{E,I+(1/2)}E^{n+1}_{i+(1/2)}]/\rho^{n+1}_{I+(1/2)} \\ - 4\pi\{(r^{n+1}_{I+1})^2[\beta^{n+1}_{I+1,I+(1/2)}E^{n+1}_{I+(1/2)} + \alpha^{n+1}_{I+1}F^{n}_{I+1}] \\ + (r^{n}_{I})^2[\beta^{n+1}_{I,I+(1/2)}E^{n+1}_{I+(1/2)} - \beta^{n+1}_{I,I-(1/2)}E^{n+1}_{I-(1/2)} - \alpha^{n+1}_{I}F^{n}_{I}]\}/\Delta M_{I+(1/2)}]\!].\end{aligned} \tag{98.36}$$

To evaluate the cell-centered quantity $(v/r)_{i+(1/2)}$ in (98.33) to (98.36) we define $r_{i\pm(1/2)}$ as in (83.86), whence by differentiation with respect to time we have

$$(v/r)_{i+(1/2)} = (r_i^2 v_i + r_{i+1}^2 v_{i+1})/(r_i^3 + r_{i+1}^3). \tag{98.37}$$

For calculations on a nuclear-evolutionary or an acoustic time scale, $\gamma\,\Delta r \ll 1$ and $\gamma\lambda_p \ll 1$, hence we can usually set $\gamma = 0$ in the radiation momentum equation. For phenomena on much shorter time scales (e.g., accretion onto compact objects), $\gamma\,\Delta r$ may approach unity, and $\gamma\lambda_p$ may greatly exceed unity in transparent regions of the flow; the time derivative in the radiation momentum equation must then be kept.

Equations (98.22), (98.30), and (98.33), plus the material energy equation (98.20) can be assembled into a system of nonlinear equations for the unknowns $T^{n+1}_{i+(1/2)}$, $E^{n+1}_{i+(1/2)}$, and F^{n+1}_{i+1}, $(i=1,\ldots,I)$. We linearize the system around trial estimates $[T^*_{i+(1/2)}, E^*_{i+(1/2)}, F^*_{i+1}]$ to obtain a block tridiagonal system of (3×3) matrices coupling a set of solution vectors $\boldsymbol{\delta\psi}_i \equiv [\delta T_{i+(1/2)}, \delta E_{i+(1/2)}, \delta F_{i+1}]$, $(i=1,\ldots,I)$; the linearized system is solved by Gaussian elimination and iterated to consistency. In this procedure we might write $\kappa_E \equiv k_E\kappa_P$ and $\chi_F \equiv k_F\chi_R$, and linearize κ_P and χ_R as functions of T, assuming that the ratios k_E and k_F determined in the multigroup step remain fixed. Alternatively we might calculate $(\partial\kappa_E/\partial T)$ from $(\partial\kappa_g/\partial T)$ in each group weighted by the spectral profile e_g, and similarly for $(\partial\chi_F/\partial T)$.

If the combined moment equation is used instead of the individual radiation energy and momentum equations, the unknowns are reduced to $(T, E)^{n+1}_{i+(1/2)}$, $(i=1,\ldots,I)$. The F_i^{n+1}'s can be calculated from the $E^{n+1}_{i+(1/2)}$'s using (98.22) and (98.30). The method of solution is the same, but the matrices are now only (2×2). The advantage of the smaller matrix size may, however, be offset by the more complicated structure of the linearized equations.

If the hydrodynamic equations are coupled to the radiation equations in implicit, rather than explicit, form, we must determine simultaneously r_i^{n+1}, v_i^{n+1}, $\rho_{i+(1/2)}^{n+1}$, $T_{i+(1/2)}^{n+1}$, $E_{i+(1/2)}^{n+1}$, and $F_{i+(1/2)}^{n+1}$. When linearized, this set of equations produces a block tridiagonal system of (6×6) matrices coupling solution vectors $\delta\boldsymbol{\psi}_i \equiv [\delta r_{i+1}, \delta v_{i+1}, \delta\rho_{i+(1/2)}, \delta T_{i+(1/2)}, \delta E_{i+(1/2)}, \delta F_{i+1}]$, $(i = 1, \ldots, I)$. The higher cost of solving this larger system at each timestep may be offset by having to take many fewer timesteps because we no longer need to observe the Courant condition imposed on the explicit scheme.

ENERGY CONSERVATION

Equation (96.15) is an expression of total energy conservation for a radiating fluid. Integrating over mass we have

$$\frac{D}{Dt}\int\left(e+\frac{E}{\rho}+\tfrac{1}{2}v^2-\frac{GM_r}{r}\right)dM_r+4\pi r_{I+1}^2 v_{I+1}P_{I+1}+L_{I+1}-L_1=\int \varepsilon\, dM_r. \tag{98.38}$$

Here, for simplicity, we have assumed that the inner boundary is fixed $(v_1 \equiv 0)$, and that the gas and pseudoviscous pressure vanish at the outer boundary. Introducing discretized variables, and replacing integrals over mass and time by sums, we obtain the total energy conservation law

$$\mathscr{E}^{n+1}=\sum_i \left[\!\left[\{e_{i+(1/2)}^{n+1}+[E_{i+(1/2)}^{n+1}/\rho_{i+(1/2)}^{n+1}]\}\,\Delta M_{i+(1/2)}+[\tfrac{1}{2}(v_i^{n+1})^2-GM_i/r_i^n]\,\Delta M_i\right]\!\right] + \mathscr{W}_{I+1}^{n+1}+\mathscr{L}^{n+1}-q^{n+1}=\text{constant}, \tag{98.39}$$

where

$$\mathscr{W}_{I+1}^{n+1}\equiv\sum_{k=0}^{n}\Delta t^{k+(1/2)}(\tfrac{4}{3}\pi)[(r_{I+1}^{k+1})^3-(r_{I+1}^{k})^3]\tfrac{1}{2}(P_{I+1}^{k}+P_{I+1}^{k+1}), \tag{98.40}$$

$$\mathscr{L}^{n+1}\equiv\sum_{k=0}^{n}\Delta t^{k+(1/2)}[(1-\theta')^{k+(1/2)}(L_{I+1}^{k}-L_1^{k})+\theta'^{k+(1/2)}(L_{I+1}^{k+1}-L_1^{k+1})], \tag{98.41}$$

and

$$q^{n+1}\equiv\sum_{k=0}^{n}\Delta t^{k+(1/2)}\sum_i[(1-\theta)^{k+(1/2)}\varepsilon_{i+(1/2)}^{k}+\theta^{k+(1/2)}\varepsilon_{i+(1/2)}^{k+1}]\,\Delta M_{i+(1/2)}. \tag{98.42}$$

The difference equations do not explicitly guarantee (98.39). Instead, we evaluate $\mathscr{E}^{n+1}$ after each integration step, and use its constancy as a check on the quality of the solution.

FORMAL SOLUTION

The Eddington factors needed in the radiation energy and momentum equations are obtained from a formal solution of the angle-frequency-dependent transfer equation using current estimates of source-sink terms.

To obtain a solution accurate to $O(v/c)$, we would need to solve (95.17) or (95.80), which would be difficult because these are partial differential equations in four independent variables, hence costly to solve [cf. (**M4**), (**M12**)]. We thus seek a simpler procedure.

The simplest approach is to drop both (D/Dt) and all velocity-dependent terms from the transfer equation. We then solve the static transfer problem along straight rays, as in §83, using the instantaneous positions of mass shells and current values of all material properties. We thus obtain a *snapshot* of the radiation field, from which we can calculate E_ν and P_ν, hence f_ν and q_ν, at all depths and frequencies. For some problems it may even be possible to ignore curvature effects. For example, in a pulsating star the radiation field is anisotropic only in the atmosphere, which is usually so thin compared to a stellar radius that it can be assumed to be plane parallel; curvature effects become important deeper in the envelope (a significant fraction of the radius in), but here we can simply set $f_\nu \equiv \frac{1}{3}$.

Intuitively one expects static snapshots to be adequate for most astrophysical applications because the ratio f_ν should vary less rapidly than the individual values of E_ν and P_ν. On the other hand, this approach is likely to be inadequate for flows that contain radiation fronts, that evolve on time scales of the same order as t_R, or that are transparent over such enormous distances that retardation effects are important (e.g., the cosmic expansion). Here we must at least retain information about the time variation of the radiation field.

For the purpose of computing Eddington factors we therefore propose a *model Lagrangean transfer equation* that (1) accounts for time dependence, and (2) is consistent with the radiation energy and momentum equations, but (3) omits inessential complications arising from ray curvature and Doppler shifts. We obtain such an equation by rewriting (95.80) as

$$\frac{\rho}{c}\frac{D}{Dt}\left(\frac{I_\nu}{\rho}\right)+4\pi r^2\rho\mu\frac{\partial I_\nu}{\partial M_r}+\frac{(1-\mu^2)}{r}\frac{\partial I_\nu}{\partial \mu}$$
$$+\frac{1}{c}\left\{\left(\frac{3v}{r}+\frac{D\ln\rho}{Dt}\right)\frac{\partial}{\partial\mu}[\mu(1-\mu^2)I_\nu]-\left[(1-3\mu^2)\frac{v}{r}-\mu^2\frac{D\ln\rho}{Dt}\right]\frac{\partial(\nu I_\nu)}{\partial\nu}\right\} \tag{98.43}$$
$$=\eta_\nu-\left[\chi_\nu+(1-3\mu^2)\frac{v}{cr}-\frac{\mu^2}{c}\frac{D\ln\rho}{Dt}\right]I_\nu,$$

and then *dropping* all of the terms in the braces which (1) are only $O(v/c)$, and (2) vanish identically when integrated over angle and frequency. The remaining terms describe time-dependent transfer along straight rays, and include two velocity-dependent terms that account for the rate of work done by radiation pressure.

Forming symmetric and antisymmetric averages of (98.43) for $\pm\mu$ we have

$$\frac{\rho}{c}\frac{D}{Dt}\left(\frac{j_\nu}{\rho}\right)+4\pi r^2\rho\mu\frac{\partial h_\nu}{\partial M_r}+\frac{(1-\mu^2)}{r}\frac{\partial h_\nu}{\partial \mu}=\eta_\nu-\left[\chi_\nu+(1-3\mu^2)\frac{v}{cr}-\frac{\mu^2}{c}\frac{D\ln\rho}{Dt}\right]j_\nu, \tag{98.44}$$

and

$$\frac{\rho}{c}\frac{D}{Dt}\left(\frac{h_\nu}{\rho}\right)+4\pi r^2\rho\mu\frac{\partial j_\nu}{\partial M_r}+\frac{(1-\mu^2)}{r}\frac{\partial j_\nu}{\partial \mu}=-\left[\chi_\nu+(1-3\mu^2)\frac{v}{cr}-\frac{\mu^2}{c}\frac{D\ln\rho}{Dt}\right]h_\nu. \tag{98.45}$$

We drop the velocity-dependent terms in (98.45), which are never larger than $O(v/c)$ (cf. §95), and use

$$\frac{1}{c}\frac{Dh_\nu}{Dt}+4\pi r^2\rho\mu\frac{\partial j_\nu}{\partial M_r}+\frac{(1-\mu^2)}{r}\frac{\partial j_\nu}{\partial \mu}=-\chi_\nu h_\nu. \tag{98.46}$$

In (98.44) to (98.46) we regard j and h as functions of $(M_r, t; \mu, \nu)$.

When (98.44) is integrated over angle and frequency we recover the Lagrangean radiation energy equation (98.5); similarly (98.46) yields the Lagrangean radiation momentum equation (98.8). These equations thus meet all the desiderata stated above.

One approach to solving these equations is to rewrite them along tangent rays through cell centers, as in §83. Then

$$\frac{\rho}{c}\frac{D}{Dt}\left(\frac{j_\nu}{\rho}\right)+\frac{\partial h_\nu}{\partial s}=\eta_\nu-\left[\chi_\nu+(1-3\mu^2)\frac{v}{cr}-\frac{\mu^2}{c}\frac{D\ln\rho}{Dt}\right]j_\nu, \tag{98.47}$$

and

$$\frac{1}{c}\frac{Dh_\nu}{Dt}+\frac{\partial j_\nu}{\partial s}=-\chi_\nu h_\nu. \tag{98.48}$$

The computational effort to solve (98.47) and (98.48) for I shells and G groups scales as cI^2G. A complication with this method is that the angles $\{\mu_m\}$ at which the tangent rays intersect mass shells change over $\Delta t^{n+(1/2)}$ because the shells move. To form time differences we need $j^n[M_{i+(1/2)}, \mu^{n+1}_{i+(1/2),m}, \nu_g]$, not $j^n[M_{i+(1/2)}, \mu^{n}_{i+(1/2),m}, \nu_g]$. In principle one should interpolate j and h as functions of angle at t^n. However if the motion of the shells is small over $\Delta t^{n+(1/2)}$, $(v/c \ll 1)$, the interpolation can be ignored, the resulting error being of the same order as made by neglecting the $O(v/c)$ terms dropped in (98.44) and (98.46). The problem does not arise in planar geometry, where motions of the mass zones do not affect the direction of rays when aberration is neglected.

The alternative is to rewrite (98.44) and (98.46) in conservative form,

$$\frac{\rho}{c}\frac{D}{Dt}\left(\frac{j_\nu}{\rho}\right)+4\pi\rho\mu\frac{\partial}{\partial M_r}(r^2h_\nu)+\frac{1}{r}\frac{\partial}{\partial\mu}[(1-\mu^2)h_\nu]$$
$$=\eta_\nu-\left[\chi_\nu+(1-3\mu^2)\frac{v}{cr}-\frac{\mu^2}{c}\frac{D\ln\rho}{Dt}\right]j_\nu \tag{98.49}$$

and

$$\frac{\mu}{c}\frac{Dh_\nu}{Dt}+4\pi\rho\mu^2\frac{\partial}{\partial M_r}(r^2 j_\nu)+\frac{(\mu^2-1)}{r}j_\nu+\frac{1}{r}\frac{\partial}{\partial\mu}[\mu(\mu^2-1)j_\nu]=-\chi_\nu\mu h_\nu, \tag{98.50}$$

and then use a discrete representation on a radial mesh $\{r_i\}$, angle mesh $\{\mu_m\}$, and frequency mesh $\{\nu_g\}$, as in §83. The computational effort required to solve these equations for I shells, M angles, and G groups scales as cIM^3G.

Before considering the multigroup step it is worth noting that the comoving-frame transfer problem for spectral lines is rather different. Here Doppler shifts dominate all other $O(v/c)$ terms because the radiation field in a line changes markedly over a Doppler width $\Delta\nu_D$. Dimensional analysis shows that the $(\partial/\partial\nu)$ term is effectively amplified to $O(v/v_{th})$ where v_{th} is of the order of the sound speed in the material; hence aberration and advection terms can be ignored. The comoving-frame line-transfer equation is then a partial differential equation in (s, t, ν) along straight rays, and is relatively easily solved for steady expansion (**M9**), (**M11**), (**M13**); however, for nonmonotonic flows the problem is much more complex (**N1**), and it may be preferable to work in the inertial frame (**M5**). Unfortunately, present prospects for a consistent treatment of the effects of line blanketing on energy and momentum balance in dynamical media are dim, despite their probable importance.

THE MULTIGROUP EQUATIONS

The formal solution yields estimates of E_g and F_g for all groups as a by-product. However these results may not yield satisfactory spectral profiles e_g and f_g because frequency derivatives were ignored in the formal solution. Instead, we should calculate E_g and F_g from the monochromatic moment equations including the frequency derivatives.

Because we do not require energy densities and fluxes simultaneously in this step it is computationally more efficient to work with the combined moment equation. By the same steps leading to (98.22) and (98.30) we can write (98.4) as

$$F_{ig}^{n+1}=-\beta_{i,i+(1/2),g}^{n+1}E_{i+(1/2),g}^{n+1}+\beta_{i,i-(1/2),g}^{n+1}E_{i-(1/2),g}^{n+1}+\alpha_{ig}^{n+1}F_{ig}^{n}, \quad (i=2,\ldots,I), \tag{98.51}$$

and

$$F_{I+1,g}^{n+1}=\beta_{I+1,I+(1/2),g}^{n+1}E_{I+(1/2),g}^{n+1}+\alpha_{I+1,g}^{n+1}F_{I+1,g}^{n}, \tag{98.52}$$

where the α's and β's are defined in (98.23), (98.24), (98.31), and (98.32), except that frequency (group) subscripts are appended to f, q, and χ. At the inner boundary we can apply the planar diffusion because $\lambda_p \ll r_1$.

$$F_{1g}^{n+1}=\left(\frac{\chi_R^{n+1}}{\bar{\chi}_g^{n+1}}\right)_1\frac{(\partial B_\nu/\partial T)_{1g}^{n+1}}{(dB/dT)_1^{n+1}}\left(\frac{L_1}{4\pi r_1^2}\right). \tag{98.53}$$

Using (98.51) to (98.53) to eliminate fluxes from the discrete representation of (98.1) we obtain the multigroup combined moment equation

$$
\begin{aligned}
&[E^{n+1}_{i+(1/2),g}/\rho^{n+1}_{i+(1/2)}] + [\![f^{n+1}_{i+(1/2),g}\{[1/\rho^{n+1}_{i+(1/2)}] - [1/\rho^{n}_{i+(1/2)}]\}\\
&\quad - [3f^{n+1}_{i+(1/2),g} - 1](v/r)^{n+(1/2)}_{i+(1/2)}\,\Delta t^{n+(1/2)}/\rho^{n+1}_{i+(1/2)}]\!]E^{n+1}_{i+(1/2),g} = [E^{n}_{i+(1/2),g}/\rho^{n}_{i+(1/2)}]\\
&\quad + \Delta t^{n+(1/2)}[\![[4\pi\kappa^{n+1}_{P,i+(1/2),g}B^{n+1}_{i+(1/2),g} - c\bar{\kappa}^{n+1}_{i+(1/2),g}E^{n+1}_{i+(1/2),g}]/\rho^{n+1}_{i+(1/2)}\\
&\quad + 4\pi\{(r^{n+1}_{i+1})^2[\beta^{n+1}_{i+1,i+(3/2),g}E^{n+1}_{i+(3/2),g} - \beta^{n+1}_{i+1,i+(1/2),g}E^{n+1}_{i+(1/2),g} - \alpha^{n+1}_{i+1,g}F^{n}_{i+1,g}]\\
&\quad - (r^{n+1}_{i})^2[\beta^{n+1}_{i,i+(1/2),g}E^{n+1}_{i+(1/2),g} - \beta^{n+1}_{i,i-(1/2),g}E^{n+1}_{i-(1/2),g} - \alpha^{n+1}_{ig}F^{n}_{ig}]\}/\Delta M_{i+(1/2)}]\!]\\
&\quad + \{[1/\rho^{n+1}_{i+(1/2)}] - [1/\rho^{n}_{i+(1/2)}]\}\,\Delta_g(\nu f_\nu E_\nu)^{n+1}_{i+(1/2)}\\
&\quad - (v/r)^{n+(1/2)}_{i+(1/2)}\,\Delta t^{n+(1/2)}\,\Delta_g[\nu(3f_\nu - 1)E_\nu]^{n+1}_{i+(1/2)}/\rho^{n+1}_{i+(1/2)},\\
&\quad (i = 2, \ldots, I-1;\ g = 1, \ldots, G).
\end{aligned}
\tag{98.54}
$$

The Δ_g's in (98.54) are to be written using upstream differencing, as in (97.86). Analysis of (98.1) shows that the effective "advection velocity" of photons in frequency space is

$$
(D\nu/Dt) = [f_\nu(D \ln \rho/Dt) + (3f_\nu - 1)(v/r)]\nu. \tag{98.55}
$$

If $(D\nu/Dt) > 0$, photons are shifted to higher frequencies during the timestep, and writing x_ν for either f_ν or $(3f_\nu - 1)$ we choose

$$
\Delta_g(\nu x_\nu E_\nu) = [\nu_{g+(1/2)}/\Delta\nu_g]x_g E_g - [\nu_{g-(1/2)}/\Delta\nu_g]x_{g-1}E_{g-1}; \tag{98.56a}
$$

if $(D\nu/Dt) < 0$ we choose

$$
\Delta_g(\nu x_\nu E_\nu) = [\nu_{g+(3/2)}/\Delta\nu_{g+1}]x_{g+1}E_{g+1} - [\nu_{g+(1/2)}/\Delta\nu_g]x_g E_g. \tag{98.56b}
$$

Equation (98.55) shows that the sign of the frequency shift is not necessarily the same as $(D\rho/Dt)$, as it is in the diffusion limit, and may even be different for different frequencies at a given position in the flow. This property implies that the difference equations (98.54) may not be exactly conservative when summed over frequency.

Boundary conditions are obtained by applying the discrete representation of (98.1) at $i = 1$ and $i = I$, using (98.52) and (98.53) to eliminate F^{n+1}_{1g} and $F^{n+1}_{I+1,g}$.

To handle the frequency coupling economically, we put the Δ_g terms on the right-hand side along with known source terms, and solve the equations iteratively. As a first estimate we set E_g in Δ_g (only) to $E^{n+1}_g \approx e^n_g E^{n+1}$, and solve the resulting G systems in parallel. We use the new E_g's to reevaluate the Δ_g's, and iterate the solution to convergence. For the continuum this procedure converges quickly because the frequency-derivative terms result only in a minor redistribution of energy among bins. For spectral lines (e.g., included in an ODF) this is not the case, but we cannot pursue this point further here.

If J iterations are required to achieve convergence, the computational effort in this step scales as $cJIG$. If one performs a direct solution to handle the coupling between groups instead of the iterative procedure just described, the computational effort would scale as cIG^3, which is probably prohibitively expensive for dynamical calculations (as was also true for multigroup diffusion).

Having found $E^{n+1}_{i+(1/2),g}$, we calculate F^{n+1}_{ig} for all i and g from (98.51) and (98.52). We can then update the spectral profiles e_g and f_g, and use these to re-evaluate χ_F and κ_E. Finally, given the radiation field and source-sink terms in the comoving fluid frame, one must remember to transform back to the inertial frame when calculating the specific intensity or flux seen by an external observer; this transformation is particularly important in spectral lines.

99. *Transport Solution by Mixed-Frame and VERA-Code Methods*

In the mixed-frame method (cf. §93) the specific intensity, angles, and frequencies are measured in the lab frame, while material properties are evaluated in the fluid frame by an expansion procedure. All velocity-dependent terms then appear only on the right-hand side of the equations. The method is fundamentally Eulerian, though it is possible to cast it into a quasi-Lagrangean form.

In the VERA (Variable Eddington Radiation Approximation) method, which was developed before the Lagrangean equations of §95 had been derived, the approach was to use a lab-frame spacetime operator, and fixed *lab-frame* angles and frequencies, while evaluating both the radiation field and material properties in the comoving frame. The method thus entails expansions of both radiation and material quantities, and the resulting equations are complicated, containing velocity-dependent terms both on the right-hand side and inside the differential operator. The physical meaning of the equations is often obscure, and in retrospect a pure Lagrangean formulation is clearly preferable. Nevertheless the method merits discussion because for many years VERA was the only code that handled $O(v/c)$ terms, and was the source of many innovative techniques such as variable Eddington factors and the multifrequency/grey method.

THE MIXED-FRAME METHOD

The mixed-frame transfer equation for one-dimensional spherically symmetric flow is [cf. (93.4)]

$$c^{-1}(\partial I_\nu/\partial t)+\mu(\partial I_\nu/\partial r)+r^{-1}(1-\mu^2)(\partial I_\nu/\partial\mu)=\eta_\nu-\kappa_\nu I_\nu+(\mu v/c)(\tilde{\kappa}_\nu I_\nu+\tilde{\eta}_\nu) \tag{99.1}$$

where

$$\tilde{\kappa}_\nu\equiv\kappa_\nu+\nu(\partial\kappa_\nu/\partial\nu) \tag{99.2}$$

and

$$\tilde{\eta}_\nu\equiv 3\eta_\nu-[\partial(\nu\eta_\nu)/\partial\nu]. \tag{99.3}$$

Here we have omitted the affix "0" on material quantities, with the understanding that all material properties are evaluated in the comoving frame; furthermore we have ignored scattering ($\sigma_\nu \equiv 0$) so that $\chi_\nu \equiv \kappa_\nu$.

For simplicity we assume that the opacity is represented by an opacity distribution function so that $\kappa_\nu \equiv \kappa_g$ on (ν_g, ν_{g+1}). Then the multigroup version of (99.1) is

$$c^{-1}(\partial I_g/\partial t)+\mu(\partial I_g/\partial r)+r^{-1}(1-\mu^2)(\partial I_g/\partial \mu)=\eta_g-\kappa_g I_g+(\mu v/c)(\kappa_g I_g+\tilde{\eta}_g), \tag{99.4}$$

where $\eta_g = \kappa_g B_g$, B_g is given by (82.39), and

$$\tilde{\eta}_g \equiv \int_{\nu_g}^{\nu_{g+1}} \tilde{\eta}_\nu \, d\nu = 3\kappa_g B_g + [\nu_{g+1}\eta(\nu_{g+1}) - \nu_g \eta(\nu_g)] \equiv 3\kappa_g \tilde{B}_g. \tag{99.5}$$

Equation (99.3) is written in conservative form so that the term in square brackets vanishes identically when integrated over $(0, \infty)$. To assure this property in the multigroup equations one can either (1) drop the term in square brackets in (99.5), which will be small if the frequency spectrum of the opacity is smooth and $(\nu_{g+1} - \nu_g)$ is not too large, or (2) choose an appropriate definition of $\eta(\nu_g)$, for example, $\eta(\nu_g) \equiv \frac{1}{2}(\kappa_{g-1}B_{g-1}+\kappa_g B_g)$.

Taking symmetric and antisymmetric averages of (99.4) for $\pm\mu$ we find

$$c^{-1}(\partial j_g/\partial t)+\mu(\partial h_g/\partial r)+r^{-1}(1-\mu^2)(\partial h_g/\partial \mu)=\kappa_g(B_g-j_g)+(\mu v/c)\kappa_g h_g, \tag{99.6}$$

and

$$c^{-1}(\partial h_g/\partial t)+\mu(\partial j_g/\partial r)+r^{-1}(1-\mu^2)(\partial j_g/\partial \mu)=-\kappa_g h_g+(\mu v/c)\kappa_g(j_g+3\tilde{B}_g). \tag{99.7}$$

Integrating these equations over angle we obtain the multigroup energy equation

$$(\partial E_g/\partial t)=-(1/r^2)[\partial(r^2F_g)/\partial r]+\kappa_g(4\pi B_g-cE_g)+(v/c)\kappa_g F_g \equiv R_g, \tag{99.8}$$

and momentum equation

$$c^{-2}(\partial F_g/\partial t)+(1/q_g)[\partial(f_g q_g E_g)/\partial r]=-(\kappa_g/c)F_g+(v/c)\kappa_g[f_g E_g+(4\pi/c)\tilde{B}_g], \tag{99.9}$$

where $f_g \equiv P_g/E_g$, and q_g is defined as in (98.3).

Integrating (99.8) and (99.9) over frequency we obtain the radiation energy equation

$$(\partial E/\partial t)=-(1/r^2)[\partial(r^2F)/\partial r]+4\pi\kappa_P B-c\kappa_E E+(v/c)\kappa_F F\equiv R \tag{99.10}$$

and the radiation momentum equation

$$c^{-2}(\partial F/\partial t)+(1/q)[\partial(fqE)/\partial r]=-(\kappa_F/c)F+(v/c)[\kappa_K fE+(4\pi/c)\kappa_P B], \tag{99.11}$$

where f is defined by (98.6), q by (98.9), κ_E by (98.7), κ_F by (98.10), and

$$\kappa_K \equiv \left(\int \kappa_\nu K_\nu \, d\nu/K\right) = \left(\sum_g e_g f_g \kappa_g\right)\Big/ f. \tag{99.12}$$

All terms in (99.10) and (99.11) must be kept in order to account for the work done by radiation forces, and to get the correct flux.

These radiation equations can be coupled to the Eulerian equation of continuity (19.9), momentum equation [cf. (94.14)]

$$\frac{\partial(\rho v)}{\partial t} + \frac{\partial}{\partial r}(\rho v^2 + p) + \frac{1}{q}\frac{\partial(fqE)}{\partial r} = -\frac{GM_r\rho}{r^2}, \tag{99.13}$$

and total energy equation [cf. (94.16)]

$$\frac{\partial}{\partial t}(\rho e + \tfrac{1}{2}\rho v^2 + E) + \frac{1}{r^2}\frac{\partial}{\partial r}[r^2(\rho e + \tfrac{1}{2}\rho v^2 + p)v + r^2 F] = \frac{-GM_r\rho v}{r^2}, \tag{99.14}$$

The computational strategy is essentially the same as for the Lagrangean scheme (§98) or for multigroup diffusion (§97). We solve the coupled nonlinear hydrodynamic and radiation energy and momentum equations, assuming that Eddington factors and spectral distributions are known. We then update the Eddington factors in a formal solution and obtain new spectral distributions from the multigroup moment equations. The process is iterated to convergence.

A wide variety of methods can be used to solve the Eulerian hydrodynamic equations, including special techniques to guarantee conservation and to handle shocks. These are discussed thoroughly in (**R1**, Chap. 12) and (**R2**, Chap. 5). In the absence of radiation one can use effectively explicit techniques like the Lax–Wendroff method; with radiation it is better to use implicit schemes such as outlined in (**K4**) [see also (**T3**)]. The momentum and total energy equations given in (**K4**) are easily generalized to include all the radiation terms written in (99.13) and (99.14), but these authors assume that the radiation field is quasi-static at the advanced time level, and ignore O(v/c) terms in the radiation equations. It is therefore worthwhile to write difference representations of the radiation equations here.

As before, we center fluxes on spherical shells $\{r_i\}$, ($i = 1, \ldots, I+1$) and energy densities at cell centers $\{r_{i+(1/2)}\}$, ($i = 1, \ldots, I$), defined by (83.86). For the radiation energy equation we have

$$\begin{aligned}
[E^{n+1}_{i+(1/2)} - E^n_{i+(1/2)}]/\Delta t^{n+(1/2)} &= 3[(r^{n+1}_{i+1})^2 F^{n+1}_{i+1} - (r^{n+1}_i)^2 F^{n+1}_i]/[(r^{n+1}_{i+1})^3 - (r^{n+1}_i)^3] \\
&\quad + 4\pi\kappa^{n+1}_{P,i+(1/2)} B^{n+1}_{i+(1/2)} - c\kappa^{n+1}_{E,i+(1/2)} E^{n+1}_{i+(1/2)} \qquad (99.15) \\
&\quad + \tfrac{1}{2}(v^{n+1}_i \kappa^{n+1}_{F,i} F^{n+1}_i + v^{n+1}_{i+1}\kappa^{n+1}_{F,i+1} F^{n+1}_{i+1})/c, \qquad (i = 1, \ldots, I),
\end{aligned}$$

and for the radiation momentum equation we write

$$(\gamma+\kappa_{F,i}^{n+1})F_i^{n+1}$$
$$=-c[f_{i+(1/2)}^{n+1}q_{i+(1/2)}^{n+1}E_{i+(1/2)}^{n+1}-f_{i-(1/2)}^{n+1}q_{i-(1/2)}^{n+1}E_{i-(1/2)}^{n+1}]/\{q_i[r_{i+(1/2)}^{n+1}-r_{i-(1/2)}^{n+1}]\}$$
$$+v_i^{n+1}[\langle\kappa_K fE\rangle_i^{n+1}+(4\pi/c)\langle\kappa_P B\rangle_i^{n+1}]+\gamma F_i^n, \quad (i=2,\ldots,I). \quad (99.16)$$

In (99.16), $x\rangle_i\equiv\frac{1}{2}[x_{i-(1/2)}+x_{i+(1/2)}]$, and $\gamma\equiv 1/[c\,\Delta t^{n+(1/2)}]$.

Assembling the difference representations of the hydrodynamic and radiation equations we get a tridiagonal nonlinear system, which is linearized and solved for corrections $[\delta\rho_{i+(1/2)}, \delta v_{i+1}, \delta T_{i+(1/2)}, \delta E_{i+(1/2)}, \delta F_{i+1}]$, $(i=1,\ldots,I)$ to current estimates of these quantities. Using the current estimate of the temperature distribution we calculate source-sink terms and carry out a formal solution at each frequency. As in the Lagrangean case we can either do a solution along tangent rays with a system of the form

$$c^{-1}(\partial j_g/\partial t)+(\partial h_g/\partial s)=\kappa_g[(B_g-j_g)+(\mu v/c)h_g] \quad (99.17)$$

and

$$c^{-1}(\partial h_g/\partial t)+(\partial j_g/\partial s)=\kappa_g[-h_g+(\mu v/c)(j_g+3\tilde{B}_g)], \quad (99.18)$$

or for a fixed set of angles $\{\mu_m\}$, with a system of the form

$$\frac{1}{c}\left(\frac{\partial j_{mg}}{\partial t}\right)+3\mu_m\frac{\partial(r^2h_{mg})}{\partial(r^3)}+\frac{1}{r}\sum_{m'}D_{mm'}h_{m'g}=\kappa_g\left[(B_g-j_{mg})+\left(\frac{\mu_m v}{c}\right)h_{mg}\right], \quad (99.19)$$

and

$$\frac{\mu_m}{c}\left(\frac{\partial h_{mg}}{\partial t}\right)+3\mu_m^2\frac{\partial(r^2j_{mg})}{\partial(r^3)}+\frac{1}{r}\sum_{m'}D'_{mm'}j_{m'g}=\kappa_g\left[-\mu_m h_{mg}+\left(\frac{\mu_m^2 v}{c}\right)(j_{mg}+3\tilde{B}_g)\right]. \quad (99.20)$$

Either set may be differenced as described in §83. For I radial shells, G groups, and M angles the computational effort for the tangent-ray method scales as cI^2G, and for the angle-differenced method as $cIGM^3$.

An advantage of the mixed-frame formulation is that in the formal solution one can solve the full transfer equation, which is *exactly* consistent with both the multigroup and integrated moment equations. Furthermore, because the spatial mesh is fixed, the radiation quantities are computed at the same set of positions and angles at each time level; no interpolations are required.

Given updated Eddington factors we can calculate revised spectral distributions from the multigroup moment equations. These equations have no frequency coupling (in contrast to their Lagrangean counterparts) hence can be solved directly with a computational effort that scales as cIG.

The advantages of the mixed-frame scheme are that (1) the equations have a simple structure, (2) a solution of the full transfer equation is possible, and (3) the multigroup equations are uncoupled. These advantages must be weighed against two serious disadvantages.

(1) Frequency derivatives of the material opacity and emissivity are required. We have avoided the issue here by assuming a constant opacity within groups. For plasmas where the opacity is dominated by light ions and is relatively smooth, reasonable estimates of the necessary derivatives may be obtainable; but for complex, jagged spectra it is extremely difficult to estimate meaningful opacity derivatives. In particular the expansion procedure used in this approach is unsatisfactory for spectral lines, which can be of great importance in media with velocity gradients. A spectral line is smeared over frequency by a velocity gradient, hence the effective widths of lines increase and continuum windows between lines are filled in, producing a substantial increase in the Rosseland mean. An interesting discussion of the problem using Sobolev theory [see (**M3,** Chap. 14)] is given in (**K1**), where velocity effects are found to be major. The problems just described are less serious in the Lagrangean frame, where the opacity is always that measured by an observer at rest relative to the material. Here the complication to be faced is the proper treatment of the $\partial(\nu E_\nu)/\partial\nu$ term, which is fundamentally easier because the frequency variation of E_ν is usually smooth (perhaps even Planckian) even when the opacity spectrum is jagged.

(2) We obtain the correct lab-frame flux only by including the two v/c terms on the right-hand side of (99.11), which dominate $\nabla\cdot\mathsf{P}$ in the dynamic diffusion limit. Thus the accuracy of the computed flux hangs on obtaining an accurate representation of these terms in the difference equations, which is nontrivial because of the interleaving of the radiation variables on the grid. In contrast, in the Lagrangean formulation we obtain the correct comoving-frame flux in both the streaming and diffusion limits without any essential difficulty.

The mixed-frame equations can also be written in quasi-Lagrangean form. To illustrate the approach with a minimum of complication, we consider planar geometry, using the column mass $dm=-\rho\,dz$ as independent variable. Using the definition $(D/Dt)=(\partial/\partial t)+\mathbf{v}\cdot\nabla$ and the equation of continuity, one finds

$$(\partial I_\nu/\partial t)\equiv\rho[D(I_\nu/\rho)/Dt]-\nabla\cdot(I_\nu\mathbf{v}). \tag{99.21}$$

Thus the mixed-frame transfer equation in Cartesian coordinates can be written in quasi-Lagrangean form

$$\frac{1}{c}\frac{D}{Dt}\left(\frac{I_\nu}{\rho}\right)-\frac{\partial}{\partial m}\left[\left(\mu-\frac{v}{c}\right)I_\nu\right]=\left(\frac{\kappa_\nu}{\rho}\right)(B_\nu-I_\nu)+\left(\frac{\mu v}{c\rho}\right)(\tilde{\kappa}_\nu I_\nu+\eta_\nu), \tag{99.22}$$

whence we obtain the multigroup equations

$$\frac{1}{c}\frac{D}{Dt}\left(\frac{j_g}{\rho}\right)-\frac{\partial}{\partial m}\left(\mu h_g-\frac{v}{c}j_g\right)=\left(\frac{\kappa_g}{\rho}\right)(B_g-j_g)+\left(\frac{\mu v}{c}\right)\left(\frac{\kappa_g}{\rho}\right)h_g \tag{99.23}$$

and

$$\frac{1}{c}\frac{D}{Dt}\left(\frac{h_g}{\rho}\right)-\frac{\partial}{\partial m}\left(\mu j_g-\frac{v}{c}h_g\right)=-\left(\frac{\kappa_g}{\rho}\right)h_g+\left(\frac{\mu v}{c}\right)\left(\frac{\kappa_g}{\rho}\right)(j_g+3\tilde{B}_g), \tag{99.24}$$

the multigroup moment equations

$$\frac{D}{Dt}\left(\frac{E_g}{\rho}\right)-\frac{\partial}{\partial m}(F_g-vE_g)=\left(\frac{\kappa_g}{\rho}\right)\left(4\pi B_g-cE_g+\frac{v}{c}F_g\right) \tag{99.25}$$

and

$$\frac{1}{c^2}\frac{D}{Dt}\left(\frac{F_g}{\rho}\right)-\frac{\partial}{\partial m}\left(f_gE_g-\frac{v}{c^2}F_g\right)=\frac{1}{c}\left(\frac{\kappa_g}{\rho}\right)[-F_g+v(f_gE_g+\tilde{B}_g)], \tag{99.26}$$

and the radiation energy and momentum equations

$$\frac{D}{Dt}\left(\frac{E}{\rho}\right)-\frac{\partial}{\partial m}(F-vE)=\frac{1}{\rho}\left(4\pi\kappa_P B-c\kappa_E E+\frac{v}{c}\kappa_F F\right) \tag{99.27}$$

and

$$\frac{1}{c^2}\frac{D}{Dt}\left(\frac{F}{\rho}\right)-\frac{\partial}{\partial m}\left(fE-\frac{v}{c^2}F\right)=\frac{1}{c\rho}[-\kappa_F F+v(\kappa_K fE+\kappa_P B)]. \tag{99.28}$$

According to the dimensional arguments of §93, (99.28) can be simplified to

$$c^{-2}(DF/Dt)-[\partial(fE)/\partial m]=(1/c\rho)[-\kappa_F F+v(\kappa_K fE+\kappa_P B)]; \tag{99.29}$$

equation (99.26) can be similarly simplified.

Equations (99.28) and (99.29) are coupled to the quasi-Lagrangean momentum equation (94.14)

$$\frac{Dv}{Dt}=-g+\frac{\partial}{\partial m}(p+P)=-g+\frac{\partial p}{\partial m}+\left(\frac{1}{c\rho}\right)[\kappa_F F-v(\kappa_K fE+\kappa_P B)] \tag{99.30}$$

and gas energy equation (94.19)

$$(De/Dt)+p[D(1/\rho)/Dt]=\varepsilon+[c\kappa_E F-4\pi\kappa_P B-2(v/c)\kappa_F F]/\rho. \tag{99.31}$$

The solution of (99.27) and (99.29) to (99.31) proceeds essentially as outlined in §98. The material equations are now complicated by the presence of velocity-dependent terms. The momentum equation presents no difficulty if the first form in (99.30) is used, but in the energy equation cancellation among terms on the right-hand side may be troublesome in the diffusion regime. The quasi-Lagrangean equations (99.27) and (99.29) suffer the same disadvantages as their Eulerian counterparts (99.10) and (99.11). Moreover, in differencing $\partial(vE)/\partial m$ in (99.27) the centering is bad, and one may lose accuracy. The equations in spherical geometry are even messier and harder to handle well. The quasi-Lagrangean transfer equations (99.23) and (99.24) contain similar awkward terms, and the formal solution becomes complicated because now the mesh moves. One either can ignore the motions, or else must interpolate quantities at the old time level; both options introduce inaccuracies.

In summary, the mixed-frame method suffers a number of disadvantages, and while neither formulation is completely free of problems, we judge the Lagrangean approach of §98 to be physically more appealing, and both simpler and more accurate in application.

THE VERA-CODE METHOD

In the VERA-code method (**F4**), (**P2**), (**S2**), spacetime, angles, and frequencies are measured in the lab frame while the specific intensity and material properties are computed in the comoving frame. We will derive the equations in spherical symmetry, and then restate important results in tensor notation so that they apply in general geometries.

We start from the mixed-frame equation

$$\begin{aligned}\frac{1}{c}\frac{\partial I_{\mu\nu}}{\partial t}+\frac{\mu}{r^2}\frac{\partial}{\partial r}(r^2 I_{\mu\nu})+\frac{1}{r}\frac{\partial}{\partial\mu}[(1-\mu^2)I_{\mu\nu}]\\ =\eta_\nu^0-\kappa_\nu^0 I_{\mu\nu}+\frac{\mu v}{c}\left[3\eta_\nu^0-\frac{\partial}{\partial\nu}(\nu\eta_\nu^0)+I_{\mu\nu}\frac{\partial}{\partial\nu}(\nu\kappa_\nu^0)\right],\end{aligned}\tag{99.32}$$

and transform $I_{\mu\nu}\equiv I(r,t;\mu,\nu)$ to the comoving frame by a first-order expansion. Thus from (89.10), (89.11), and (90.3) we have

$$\begin{aligned}I_{\mu\nu}&=(\nu/\nu_0)^3 I^0(\mu_0,\nu_0)=(1+\beta\mu_0)^3[I^0(\mu,\nu)+(\nu_0-\nu)(\partial I^0/\partial\nu)\\&\quad+(\mu_0-\mu)(\partial I^0/\partial\mu)]+O(v^2/c^2)\\&=(1+3\beta\mu)I_{\mu\nu}^0-\beta\mu\nu(\partial I^0/\partial\nu)-\beta(1-\mu^2)(\partial I^0/\partial\mu)+O(v^2/c^2).\end{aligned}\tag{99.33}$$

More generally,

$$I(\mathbf{n},\nu)=[1+(3\mathbf{v}\cdot\mathbf{n}/c)]I^0(\mathbf{n},\nu)-(\mathbf{v}\cdot\mathbf{n}/c)\nu(\partial I^0/\partial\nu)-(\mathbf{v}/c)\cdot\nabla_{\mathbf{n}}I^0,\tag{99.34}$$

where $\nabla_{\mathbf{n}}$ denotes the gradient with respect to the direction cosines of the propagation vector.

Integrating (99.33) over solid angle we obtain transformation laws for the monochromatic radiation moments:

$$E_\nu=E_\nu^0+(2v/c^2)F_\nu^0-(v/c^2)[\partial(\nu F_\nu^0)/\partial\nu],\tag{99.35}$$

$$F_\nu=F_\nu^0+v(E_\nu^0+P_\nu^0)-v[\partial(\nu P_\nu^0)/\partial\nu],\tag{99.36}$$

and

$$P_\nu=P_\nu^0+(2v/c^2)F_\nu^0-(v/c^2)[\partial(\nu Q_\nu^0)/\partial\nu].\tag{99.37}$$

Or, in tensor form

$$E_\nu = E_\nu^0 + (2/c^2)\mathbf{v}\cdot\mathbf{F}_\nu^0 - c^{-2}[\partial(\nu\mathbf{v}\cdot\mathbf{F}_\nu^0)/\partial\nu], \tag{99.38}$$

$$\mathbf{F}_\nu = \mathbf{F}_\nu^0 + \mathbf{v}E_\nu^0 + \mathbf{v}\cdot\mathsf{P}_\nu^0 - [\partial(\nu\mathbf{v}\cdot\mathsf{P}_\nu^0)/\partial\nu], \tag{99.39}$$

and

$$\mathsf{P}_\nu = \mathsf{P}_\nu^0 + c^{-2}(\mathbf{v}\mathbf{F}_\nu^0 + \mathbf{F}_\nu^0\mathbf{v}) - c^{-2}[\partial(\nu\mathbf{v}\cdot\mathsf{Q}_\nu^0)/\partial\nu], \tag{99.40}$$

where

$$\mathsf{Q}_\nu^0 \equiv \oint I^0(\mathbf{n},\nu)\mathbf{nnn}\,d\omega. \tag{99.41}$$

One sees by inspection that (99.35) to (99.40), when integrated over frequency yield the standard $O(v/c)$ transformations of the radiation stress-energy tensor [cf. (91.10) to (91.12) and (91.16) to (91.18)].

Substituting (99.33) into (99.32), and using (99.21), we obtain the transfer equation

$$\begin{aligned}\frac{\rho}{c}\frac{D}{Dt}\left(\frac{I_{\mu\nu}^0}{\rho}\right)+\frac{\mu}{r^2}\frac{\partial}{\partial r}(r^2I_{\mu\nu}^0)+\frac{1}{r}\frac{\partial}{\partial\mu}[(1-\mu^2)I_{\mu\nu}^0]=\eta_\nu^0-\kappa_\nu^0I_{\mu\nu}^0+\frac{1}{cr^2}\frac{\partial}{\partial r}(r^2vI_{\mu\nu}^0)\\ -\frac{1}{c}\left[\!\!\left[\left[\kappa_\nu^0+\frac{1}{c}\frac{\partial}{\partial t}+\frac{\mu}{r^2}\frac{\partial}{\partial r}r^2+\frac{1}{r}\frac{\partial}{\partial\mu}(1-\mu^2)\right]\left\{\mu v\left[2I_{\mu\nu}^0-\frac{\partial}{\partial\nu}(\nu I_{\mu\nu}^0)\right]\right.\right.\right.\\ \left.\left.\left.-v\frac{\partial}{\partial\mu}[(1-\mu^2)I_{\mu\nu}^0]\right\}-\mu v\left[3\eta_\nu^0-\frac{\partial}{\partial\nu}(\nu\eta_\nu^0)+I_{\mu\nu}^0\frac{\partial}{\partial\nu}(\nu\kappa_\nu^0)\right]\right]\!\!\right]+O(v^2/c^2).\end{aligned} \tag{99.42}$$

Taking the zeroth angular moment of (99.42), we obtain, after some reduction,

$$\begin{aligned}\rho\frac{D}{Dt}\left(\frac{E_\nu^0}{\rho}\right)+\frac{1}{r^2}\frac{\partial}{\partial r}[r^2(F_\nu^0+vP_\nu^0)]-4\pi\eta_\nu^0+c\kappa_\nu^0E_\nu^0+\frac{v}{c}\kappa_\nu^0F_\nu^0\\ =\frac{\partial}{\partial\nu}\left\{\nu\left[\frac{v}{c}\kappa_\nu^0F_\nu^0+\frac{1}{c^2}\frac{\partial}{\partial t}(vF_\nu^0)+\frac{1}{r^2}\frac{\partial}{\partial r}(r^2vP_\nu^0)\right]\right\}-\frac{2}{c^2}\frac{\partial}{\partial t}(vF_\nu^0),\end{aligned} \tag{99.43}$$

or, more generally,

$$\begin{aligned}\rho[D(E_\nu^0/\rho)/Dt]+\nabla\cdot(\mathbf{F}_\nu^0+\mathbf{v}\cdot\mathsf{P}_\nu^0)-4\pi\eta_\nu^0+c\kappa_\nu^0E_\nu^0+(\kappa_\nu^0/c)\mathbf{v}\cdot\mathbf{F}_\nu^0\\ =\frac{\partial}{\partial\nu}\left\{\nu\left[\frac{\kappa_\nu^0}{c}\mathbf{v}\cdot\mathbf{F}_\nu^0+\frac{1}{c^2}\frac{\partial}{\partial t}(\mathbf{v}\cdot\mathbf{F}_\nu^0)+\nabla\cdot(\mathbf{v}\cdot\mathsf{P}_\nu^0)\right]\right\}-\frac{2}{c^2}\frac{\partial}{\partial t}(\mathbf{v}\cdot\mathbf{F}_\nu^0).\end{aligned} \tag{99.44}$$

Similarly, after rearrangement the first moment of (99.42) can be written

$$\begin{aligned}\frac{\rho}{c^2}\frac{D}{Dt}\left(\frac{F_\nu^0}{\rho}\right)+\frac{\partial P_\nu^0}{\partial r}+\frac{3P_\nu^0-E_\nu^0}{r}+\frac{\kappa_\nu^0F_\nu^0}{c}=-\frac{1}{c^2}\left[E_\nu^0\frac{\partial v}{\partial t}+\frac{\partial}{\partial t}(vP_\nu^0)+F_\nu^0\frac{\partial v}{\partial r}\right]\\ +\frac{1}{c^2}\frac{\partial}{\partial\nu}\left\{\nu\left[\frac{\partial}{\partial t}(vP_\nu^0)+\frac{\partial}{\partial r}(vQ_\nu^0)+\frac{v(3Q_\nu^0-F_\nu^0)}{r}+v(c\kappa_\nu^0P_\nu^0-\tfrac{4}{3}\pi\eta_\nu^0)\right]\right\},\end{aligned} \tag{99.45}$$

or, in tensor form,

$$\frac{\rho}{c^2}\frac{D}{Dt}\left(\frac{\mathbf{F}_\nu^0}{\rho}\right)+\nabla\cdot\mathsf{P}_\nu^0+\frac{\kappa_\nu^0}{c}\mathbf{F}_\nu^0=-\frac{1}{c^2}\left[E_\nu^0\frac{\partial\mathbf{v}}{\partial t}+\frac{\partial(\mathbf{v}\cdot\mathsf{P}_\nu^0)}{\partial t}+\mathbf{F}_\nu^0\cdot\nabla\mathbf{v}\right]$$
$$+\frac{1}{c^2}\frac{\partial}{\partial\nu}\left\{\nu\left[\frac{\partial}{\partial t}(\mathbf{v}\cdot\mathsf{P}_\nu^0)+\nabla\cdot(\mathbf{v}\cdot\mathsf{Q}_\nu^0)+c\kappa_\nu^0\mathbf{v}\cdot\mathsf{P}_\nu^0-\frac{4\pi}{3}\eta_\nu^0\mathbf{v}\right]\right\}. \quad (99.46)$$

In deriving (99.45) and (99.46) the first term on the right was reduced by use of (99.43) and (99.44).

Integrating (99.44) and (99.46) over frequency we obtain the radiation energy equation

$$\rho[D(E^0/\rho)/Dt]+\nabla\cdot(\mathbf{F}^0+\mathbf{v}\cdot\mathsf{P}^0)-4\pi\kappa_P^0B^0+c\kappa_E^0E^0+(\kappa_F^0/c)\mathbf{v}\cdot\mathbf{F}^0$$
$$=-(2/c^2)\,\partial(\mathbf{v}\cdot\mathbf{F}^0)/\partial t \quad (99.47)$$

and the radiation momentum equation

$$(\rho/c^2)[D(\mathbf{F}^0/\rho)/Dt]+\nabla\cdot\mathsf{P}^0+(\kappa_F^0/c)\mathbf{F}^0$$
$$=-c^{-2}[E^0(\partial\mathbf{v}/\partial t)+\partial(\mathbf{v}\cdot\mathsf{P}^0)/\partial t+\mathbf{F}^0\cdot\nabla\mathbf{v}]. \quad (99.48)$$

Despite its rather different appearance, (99.47) can be reduced to the Lagrangean radiation energy equation (95.87). To see this we expand

$$\partial(\mathbf{v}\cdot\mathbf{F}^0)/\partial t=\mathbf{a}\cdot\mathbf{F}+\mathbf{v}\cdot(\partial\mathbf{F}^0/\partial t) \quad (99.49)$$

and

$$\nabla\cdot(\mathbf{v}\cdot\mathsf{P}^0)=\mathsf{P}^0:\nabla\mathbf{v}+\mathbf{v}\cdot(\nabla\cdot\mathsf{P}^0), \quad (99.50)$$

whence we find, after rearrangement,

$$\rho\frac{D}{Dt}\left(\frac{E_0}{\rho}\right)+\nabla\cdot\mathbf{F}^0+\frac{\mathbf{v}}{c^2}\cdot\frac{\partial\mathbf{F}^0}{\partial t}+\mathsf{P}^0:\nabla\mathbf{v}+\frac{2}{c^2}\mathbf{a}\cdot\mathbf{F}^0$$
$$=4\pi\kappa_P^0B^0-c\kappa_E^0E^0-\mathbf{v}\cdot\left(\frac{1}{c^2}\frac{\partial\mathbf{F}^0}{\partial t}-\nabla\cdot\mathsf{P}^0+\frac{\kappa_F^0}{c}\mathbf{F}^0\right). \quad (99.51)$$

The last term on the right-hand side of (99.51) vanishes to $O(v^2/c^2)$ by virtue of (99.48). Thus (99.51) differs from (95.87) only by the term $c^{-2}\mathbf{v}\cdot(\partial\mathbf{F}^0/\partial t)$ on the left-hand side. This term is present because the operator ∇ in (99.51) is in the Eulerian frame, whereas in (95.87) it is in the Lagrangean frame. From (92.12) [see also (**B2,** equation 2)] one has, to $O(v/c)$,

$$\nabla_E=\nabla_L-c^{-2}\mathbf{v}(\partial/\partial t), \quad (99.52)$$

whence we see that (99.51) and (95.87) are physically equivalent. Nevertheless, one should note that although (99.47) reduces algebraically to (95.87), from a computational point of view it is quite different, being more difficult to handle well, and probably yielding less accurate results.

By a similar rearrangement of (99.48) we find

$$\frac{\rho}{c^2}\frac{D}{Dt}\left(\frac{\mathbf{F}^0}{\rho}\right)+\nabla\cdot\mathsf{P}^0+\frac{\mathbf{v}}{c^2}\cdot\frac{\partial\mathsf{P}^0}{\partial t}+\mathbf{F}^0\cdot\nabla\mathbf{v}+\frac{1}{c^2}\left(E^0\frac{\partial\mathbf{v}}{\partial t}+\mathsf{P}^0\cdot\frac{\partial\mathbf{v}}{\partial t}\right)+\frac{\kappa_F\mathbf{F}^0}{c}=0, \tag{99.53}$$

which, by virtue of (99.52), reduces to (95.88).

Equations (99.47) and (99.48) are closed with variable Eddington factors and then coupled to Lagrangean material momentum and energy equations such as (98.14) and (98.15); the combined system is solved by the multifrequency/grey technique as described in §98. (Indeed, the VERA code is the original source of this method.) Difference representations of the VERA equations are thoroughly documented in (**F4**), (**P2**), and (**S2**); here we merely critique the computational procedure.

The VERA equations are more complicated than their Lagrangean counterparts, and many approximations were made in implementing them. Thus in solving (99.47) and (99.48) the VERA code drops all the terms on the right-hand sides of these equations. Analysis shows that these terms are at most $O(v/c)$ relative to the dominant terms, hence their omission is justified. Equation (99.48) is then essentially identical to our Lagrangean equation (98.8). In (99.47) the term $\mathbf{v}\cdot\mathsf{P}^0$ has an awkward centering relative to $\mathbf{F}^0$, as does $\mathbf{v}\cdot\mathbf{F}^0$ relative to E^0. In retrospect one realizes that these terms should have been expanded and canceled as in (99.51), which would have yielded a simpler, fully Lagrangean equation.

The situation for the multigroup equations is worse; all terms on the right-hand sides of (99.44) and (99.46) were dropped. These omissions are harmless in (99.46). Most of the terms omitted from (99.44) can be dropped without ill effect, but it is important to retain $\partial(\nu\nabla\cdot\mathbf{v}\cdot\mathsf{P}^0_\nu)/\partial\nu$ in order to obtain the correct spectral distribution (cf. §97). Again the Lagrangean equation (98.1) is both simpler and more accurate.

Finally, in the formal solution, it is clear that a rigorous treatment of (99.42) is hopeless. Early versions of the VERA code (**F4**) used formulae of the kind mentioned in §78 to evaluate the Eddington factor from E^0 and F^0. Later versions [(**F5**), (**P2**), (**S2**)] evaluated f from either time-retarded solutions or static snapshots along tangent rays, omitting all velocity-dependent terms. As was argued in §98, this approach should be adequate in many applications, but it is less reliable and satisfying than the Lagrangean methods sketched in (98.50) to (98.58), or the mixed-frame approach outlined in (99.17) to (99.20).

In summary, the VERA equations resemble the Lagrangean equations, but are more complicated and more difficult to solve to a high level of internal consistency; the Lagrangean equations are to be preferred [see also (**B2**, pp. 298–299)].

References

(A1) Alme, M. L. and Wilson, J. R. (1974) *Astrophys. J.*, **194,** 147.

(B1) Barfield, W. D., von Holdt, R., and Zachariasen, F. (1954) *Los Alamos Scientific Laboratory Report No. LA-1709.* Los Alamos: University of California.

(B2) Buchler, J. R. (1979) *J. Quant. Spectrosc. Rad. Transf.*, **22,** 293.

(B3) Buchler, J. R. (1983) *J. Quant. Spectrosc. Rad. Transf.*, **30,** 395.

(C1) Campbell, P. M. (1965) *Lawrence Radiation Laboratory Report No. UCRL-12411.* Livermore: University of California.

(C2) Campbell, P. M. and Nelson, R. G. (1964) *Lawrence Radiation Laboratory Report No. UCRL-7838.* Livermore: University of California.

(C3) Castor, J. I. (1972) *Astrophys. J.*, **178,** 779.

(C4) Christy, R. F. (1964) *Rev. Mod. Phys.*, **36,** 555.

(C5) Cox, J. P. and Giuli, R. T. (1968) *Principles of Stellar Structure.* New York: Gordon and Breach.

(E1) Epstein, R. I. (1981) *Astrophys. J. Letters*, **244,** L89.

(F1) Falk, S. W. and Arnett, W. D. (1977) *Astrophys. J. Supp.*, **33,** 515.

(F2) Fraser, A. R. (1966) *Atomic Weapons Research Establishment Report No. O-82/65.* Aldermaston: U.K. Atomic Energy Authority.

(F3) Freeman, B. E. (1965) *Los Alamos Scientific Laboratory Report No. LA-3377,* Los Alamos: University of California.

(F4) Freeman, B. E., Hauser, L. E., Palmer, J. T., Pickard, S. O., Simmons, G. M., Williston, D. G., and Zerkle, J. E. (1968) *Defense Atomic Support Agency Report No. DASA 2135,* Vol. 1. La Jolla: Systems, Science, and Software, Inc.

(F5) Freeman, B. E., Palmer, J. T., Simmons, G. M., Schaibly, J. H., and Zerkle, J. E. (1969) *Defense Atomic Support Agency Report No. DASA 2258-III.* La Jolla: Systems, Science, and Software, Inc.

(G1) Glaviano, M. C. and Raymond, D. J. (1981) *Astrophys. J.*, **243,** 271.

(H1) Hazelhurst, J. and Sargent, W. L. W. (1959) *Astrophys. J.*, **130,** 276.

(H2) Hsieh, S.-H. and Spiegel, E. A. (1976) *Astrophys. J.*, **207,** 244.

(J1) Jeans, J. H. (1926) *Mon. Not. Roy. Astron. Soc.*, **86,** 328.

(J2) Jeans, J. H. (1926) *Mon. Not. Roy. Astron. Soc.*, **86,** 444.

(K1) Karp, A. H., Lasher, G., Chan, K. L., and Salpeter, E. E. (1977) *Astrophys. J.*, **214,** 161.

(K2) Kershaw, D. S. (1976) *Lawrence Livermore Laboratory Report No. UCRL-78378.* Livermore: University of California.

(K3) Klein, R. I., Stockman, H. S., and Chevalier, R. A. (1980) *Astrophys. J.*, **237,** 921.

(K4) Kneer, F. and Nakagawa, Y. (1976) *Astron. and Astrophys.*, **47,** 65.

(K5) Kopal, Z. (1964) *Astrophysica Norvegica,* **9,** 239.

(K6) Kopal, Z. (1965) *Z. für Astrophys.*, **61,** 156.

(K7) Kutter, G. S. and Sparks, W. M. (1972) *Astrophys. J.*, **175,** 407.

(L1) Ledoux, P. (1965) in *Stellar Structure,* ed. L. H. Aller and D. B. McLaughlin, Chapter 10. Chicago: University of Chicago.

(L2) Ledoux, P. and Walraven, Th. (1958) in *Handbuch der Physik,* Vol. 51, *Astrophysics II: Stellar Structure,* ed. S. Flügge, p. 353. Berlin: Springer.

(L3) Levermore, C. D. (1979) *Lawrence Livermore Laboratory Report No. UCID-18229.* Livermore: University of California.

(L4) Levermore, C. D. and Pomraning, G. C. (1981) *Astrophys. J.* **248,** 321.

(L5) Lindquist, R. W. (1966) *Ann. Phys.*, **37,** 487.

(M1) Masaki, I. (1971) *Pub. Astron. Soc. Japan*, **23,** 425.

(M2) Masaki, I. (1981) *Pub. Astron. Soc. Japan*, **33,** 77.

(M3) Mihalas, D. (1978) *Stellar Atmospheres.* (2nd ed.) San Francisco: Freeman.

(M4) Mihalas, D. (1980) *Astrophys. J.*, **237,** 574.

(M5) Mihalas, D. (1980) *Astrophys. J.*, **238,** 1042.

(M6) Mihalas, D. (1981) *Astrophys. J.*, **250,** 373.

(M7) Mihalas, D. (1983) *Astrophys. J.*, **266,** 242.

(M8) Mihalas, D. and Klein, R. I. (1982) *J. Comp. Phys.*, **46,** 97.

(M9) Mihalas, D. and Kunasz, P. B. (1978) *Astrophys. J.*, **219,** 635.

(M10) Mihalas, D. and Weaver, R. P. (1982) *J. Quant. Spectrosc. Rad. Transf.*, **28,** 213.

(M11) Mihalas, D., Kunasz, P. B., and Hummer, D. G. (1975) *Astrophys. J.*, **202,** 465.

(M12) Mihalas, D., Kunasz, P. B., and Hummer, D. G. (1976) *Astrophys. J.*, **206,** 515.

(M13) Mihalas, D., Kunasz, P. B., and Hummer, D. G. (1976) *Astrophys. J.*, **210,** 419.

(M14) Milne, E. A. (1929) *Mon. Not. Roy. Astron. Soc.*, **89,** 519.

(M15) Milne, E. A. (1930) *Quart. J. of Math. (Oxford)*, **1,** 1.

(M16) Morse, P. M. and Feshbach, H. (1953) *Methods of Theoretical Physics.* New York: McGraw-Hill.

(N1) Noerdlinger, P. D. (1981) *Astrophys. J.*, **245,** 682.

(P1) Pai, S. I. (1966) *Radiation Gas Dynamics.* New York: Springer.

(P2) Palmer, J. T., Freeman, B. E., and Schaibly, J. H. (1969) *Defense Atomic Support Agency Report No. DASA 2420-1.* La Jolla: Systems, Science, and Software, Inc.

(P3) Pomraning, G. C. (1973) *The Equations of Radiation Hydrodynamics.* Oxford: Pergamon.

(P4) Pomraning, G. C. (1974) *J. Quant. Spectrosc. Rad. Transf.*, **14,** 657.

(R1) Richtmyer, R. D. and Morton, K. W. (1967) *Difference Methods for Initial-Value Problems.* (2nd ed.) New York: Interscience.

(R2) Roache, P. J. (1976) *Computational Fluid Dynamics.* Albuquerque: Hermosa.

(S1) Sampson, D. H. (1965) *Radiative Contributions to Energy and Momentum Transport in a Gas.* New York: Interscience.

(S2) Schaibly, J. and Wilson, A. R. (1972) *Systems, Science, and Software Report No. 3SCR-1449.* La Jolla: Systems, Science, and Software, Inc.

(S3) Sears, R. L. and Brownlee, R. R. (1965) in *Stellar Structure*, ed. L. H. Aller and D. B. McLaughlin, Chapter 11. Chicago: University of Chicago.

(S4) Simon, R. (1963) *J. Quant. Spectrosc. Rad. Transf.*, **3,** 1.

(S5) Stobie, R. S. (1969) *Mon. Not. Roy. Astron. Soc.*, **144,** 461.

(S6) Synge, J. L. (1957) *The Relativistic Gas.* Amsterdam: North Holland.

(S7) Synge, J. L. and Schild, A. R. (1949) *Tensor Calculus.* Toronto: University of Toronto.

(T1) Thomas, L. H. (1930) *Quart. J. of Math. (Oxford)*, **1,** 239.

(T2) Tisza, L. (1942) *Phys. Rev.*, **61,** 531.

(T3) Tscharnuter, W. and Winkler, K.-H. (1969) *Computer Phys. Comm.*, **18,** 171.

(W1) Weinberg, S. (1971) *Astrophys. J.*, **168,** 175.

(W2) Wendroff, B. (1963) *Los Alamos Scientific Laboratory Report No. LAMS-2795.* Los Alamos: University of California.

(W3) Winkler, K.-H. and Norman, M. (1984) in *Astrophysical Radiation Hydrodynamics,* ed. K.-H. Winkler and M. Norman. Dordrecht: Reidel.

(W4) Winslow, A. M. (1968) *Nucl. Sci. and Eng.*, **32,** 101.

(Z1) Zel'dovich, Ya. B. and Raizer, Yu. P. (1966) *Physics of Shock Waves and High-Temperature Hydrodynamic Phenomena.* New York: Academic.

8

Radiating Flows

In a radiating fluid, radiation affects the energy and momentum balance in the flow, and can drive the thermodynamic state of the material out of equilibrium. In this chapter we consider a few interesting examples of radiating-flow problems in both the linear and nonlinear regimes.

For small-amplitude disturbances we focus on radiative energy and momentum exchange; we first examine the radiative smoothing of temperature fluctuations in a static medium, then the effects of radiation on acoustic waves in a homogeneous medium, and finally the effects of radiation on acoustic-gravity waves in a stratified medium. We shall draw our examples primarily from astrophysics where considerable attention has been given to radiative effects on the propagation and dissipation of waves in the atmospheres of the Sun and stars.

For nonlinear disturbances we meet a much richer variety of phenomena. We consider first the conceptually simple problem of penetration of radiation into a passive static medium as a thermal wave. We then examine the effects of radiative transport across steady shocks and in propagating shocks in both the weak- and strong-shock limits, including the case of a propagating non-LTE shock, where radiation determines the state of the material. We next examine the interplay of radiation and hydrodynamics in propagating ionization fronts. Finally, we consider the dynamics of radiation-driven stellar winds, where the primary effect of radiation is on the momentum balance in the flow.

The reader should note that in most of the applications to be discussed the treatment of radiation falls far below the standards set in Chapter 7, although reasonably complete and consistent solutions are obtained for one or two simple problems. We make this remark not as a criticism of the existing literature, but rather to call attention to the rewarding opportunities that exist for new research exploiting the more complete formulation of the dynamical behavior of radiation that is now available.

8.1 Small-Amplitude Disturbances

100. Radiative Damping of Temperature Fluctuations

Valuable insight into the effects of radiative energy exchange on small-amplitude disturbances can be obtained by examining the smoothing of temperature fluctuations in a radiating fluid.

QUASI-STATIC RADIATION TRANSPORT

Consider a field of temperature perturbations imposed on a static, grey, LTE ambient medium, initially in radiative equilibrium. We assume there are no fluid motions ($\mathbf{v} \equiv 0$), and that heat is exchanged only radiatively. Under these assumptions the gas energy equation (96.7) reduces to

$$\rho(\partial e/\partial t) = 4\pi\kappa(J - B), \tag{100.1}$$

where, from (52.21), $(\partial e/\partial t) = c_v(\partial T/\partial t)$ in a static medium. The thermal source is $B = (a_R c/4\pi)T^4$, and the mean intensity is $4\pi J = \oint I\, d\omega$. We assume that the characteristic time scale associated with the disturbances is so long that the radiation field can be taken to be quasi-static. Then I is given by the static transfer equation

$$(\partial I/\partial s) = \kappa(B - I), \tag{100.2}$$

where s is the path length along a ray. In using (100.2) we neglect all dynamical effects of the radiation field.

For small disturbances we linearize, writing

$$T(\mathbf{x}, t) = T_0(\mathbf{x}) + T_1(\mathbf{x}, t), \tag{100.3}$$

$$B(\mathbf{x}, t) = B_0(\mathbf{x}) + (a_R c T_0^3/\pi)T_1 \equiv B_0 + B_1, \tag{100.4}$$

and

$$\kappa(\mathbf{x}, t) = \kappa_0(\mathbf{x}) + (\partial\kappa_0/\partial T)T_1 \equiv \kappa_0 + \kappa_1. \tag{100.5}$$

The linearized energy equation is then

$$\rho c_v(\partial T_1/\partial t) = 4\pi\kappa_0(J_1 - B_1) + 4\pi\kappa_1(J_0 - B_0), \tag{100.6}$$

where J_1 is the local perturbation of the mean intensity induced by perturbations in the source–sink terms throughout the medium. Because we assume that the material is initially in radiative equilibrium, $J_0 \equiv B_0$, hence the term containing κ_1 in (100.6) vanishes identically.

J_1 is the angle average of I_1, the local change in the specific intensity, which can be calculated from the linearized transfer equation

$$(\partial I_1/\partial s) = \kappa_0(B_1 - I_1) + \kappa_1(B_0 - I_0). \tag{100.7}$$

If we now make the simplifying assumptions that the ambient medium is *homogeneous* and of *infinite extent* (appropriate for a study of, say, pure acoustic waves), then the unperturbed radiation field will be *isotropic*, which implies that $I_0 \equiv J_0 \equiv B_0 = \text{constant}$. Hence the term containing κ_1 in (100.7) vanishes identically. Thus in a homogeneous medium, both the energy equation and the transfer equation are unaffected (to first order) by a perturbation in the opacity.

I_1 is found directly from the formal solution of (100.7):

$$I_1(\mathbf{x}_0, \mathbf{n}) = \int_0^\infty B_1(\mathbf{x}_0 - \mathbf{n}s)e^{-\kappa_0 s}\kappa_0\, ds. \tag{100.8}$$

The energy equation for a field of temperature perturbations in an infinite, homogeneous, static medium is thus

$$[\partial T_1(\mathbf{x}_0)/\partial t] = -\nu\left[T_1(\mathbf{x}_0) - (4\pi)^{-1}\oint d\omega \int_0^\infty T_1(\mathbf{x}_0 - \mathbf{n}s)e^{-\kappa_0 s}\kappa_0\, ds\right], \tag{100.9}$$

where

$$\nu \equiv 4a_R c\kappa_0 T_0^3/\rho c_v = 16\sigma_R \kappa_0 T_0^3/\rho c_v \tag{100.10}$$

is an inverse time scale characterizing the rate of energy loss by radiative emission in the absence of reabsorption.

Suppose now we have a field of planar temperature disturbances varying as $e^{i\mathbf{k}\cdot\mathbf{x}}$. Choose $\mathbf{k}$ as a preferred direction defining the x axis (which is otherwise arbitrary in a homogeneous medium), and let $\cos^{-1}\mu$ be the angle between $\mathbf{k}$ and $\mathbf{n}$. We can then rewrite (100.9) as

$$[\partial T_1(x_0)/\partial t] = -\nu\left[T_1(x_0) - \tfrac{1}{2}\int_{-1}^{1} d\mu \int_0^\infty T_1(x_0 \mp \xi)e^{-\kappa_0\xi/|\mu|}\kappa_0\, d\xi/|\mu|\right], \tag{100.11}$$

where the sign in the argument of the integrand is chosen opposite to the sign of μ.

Equation (100.11) admits separable solutions of the form

$$T_1(x, t) = \phi(k, t)e^{ik(x-x_0)}. \tag{100.12}$$

Note that because (100.11) is linear, any linear combination of solutions of the general form (100.12) will satisfy (100.11), hence we can synthesize the behavior of an arbitrary field of fluctuations by a suitable superposition of its Fourier components. Using (100.12) in (100.11) we have

$$(\partial\phi/\partial t) = -n(k)\phi, \tag{100.13}$$

where, by virtue of symmetry considerations that simplify the integral,

$$n(k) = \nu\left[1 - \int_0^1 d\mu \int_0^\infty \cos(\mu k y/\kappa_0)e^{-y}\, dy\right]. \tag{100.14}$$

Thus a spatially harmonic temperature disturbance with wavenumber k decays exponentially from its initial value $\phi(k, 0)$ according to

$$\phi(k, t) = \phi(k, 0)\exp[-t/t_{RR}(k)] \tag{100.15}$$

where the *radiative relaxation time* is $t_{RR}(k) = 1/n(k)$. It is clear on physical grounds that n must always be positive because in the situation we are considering $|J_1|$ is always less than or equal to $|B_1|$ (equality occurring only in the limit of infinite optical thickness), and therefore regions of enhanced temperature always tend to cool while cooler regions tend to heat, thus damping the disturbance.

Table 100.1. Radiative Relaxation Rates and Eddington Factors

κ_0/k	$n(k)/\nu$	$n_E(k)/\nu$	$f(k)$
0.0	1.000	1.000	0.000
0.2	0.725	0.893	0.106
0.4	0.524	0.676	0.176
0.6	0.382	0.481	0.222
0.8	0.283	0.342	0.253
1.0	0.215	0.250	0.273
1.5	0.118	0.129	0.301
2.0	0.073	0.076	0.314
5.0	0.013	0.013	0.330
∞	0.000	0.000	0.333

From standard tables one has

$$\int_0^\infty \cos\,(\mu k y/\kappa_0)e^{-y}\,dy = [1+(k/\kappa_0)^2\mu^2]^{-1}, \tag{100.16}$$

which gives the angular distribution of $I_1(\mu)$ for a given (k/κ_0). Using (100.16) in (100.14) we obtain the dispersion relation for the *thermal relaxation mode* of a radiating fluid:

$$n(k) = \nu[1-(\kappa_0/k)\cot^{-1}(\kappa_0/k)], \tag{100.17}$$

a result first obtained by Spiegel (**S18**).

The ratio $\kappa_0/k = \kappa_0\Lambda/2\pi = \tau_\Lambda/2\pi$, where τ_Λ is the optical thickness of one wavelength of the perturbation. As shown in Table 100.1, $n(k)$ decreases monotonically from ν to zero as τ_Λ varies from zero to infinity; that is, optically thin disturbances damp rapidly whereas optically thick disturbances damp slowly. To understand this result intuitively one notes that in an optically thin perturbation $I_1 \to 0$ because positive and negative fluctuations of T_1 contribute equally along each line of sight, and average to zero. In this case $J_1 \to 0$, hence the damping rate is set entirely by local emission losses, independently of k. In contrast, in an optically thick perturbation $J_1 \approx B_1$, the difference between the two being set by radiation diffusion; therefore thermal emission is closely balanced by reabsorption, and the cooling rate is slow. By expanding (100.17) we find that as $(\kappa_0/k) \to \infty$, $n(k) \to \nu k^2/3\kappa_0^2$; therefore $t_{RR} \to 3\kappa_0^2/k^2\nu \propto (\rho c_v T_0/a_R T_0^4) \times (l^2/c\lambda_p) = (\hat{e}/E)t_d$ where $\hat{e}$ is the material energy density, E is the radiation energy density, and t_d is the radiation diffusion time.

Unno and Spiegel (**U9**) clarified the physical significance of the exact solution obtained above by using moments of the radiation field and invoking the Eddington approximation. In the limit that both hydrodynamic motions and the dynamics of the radiation field (specifically the rate of change of the radiant energy density) can be ignored, the first law of thermodynamics for the radiating fluid (96.9) yields an alternative energy equation:

$$\rho c_v(\partial T/\partial t) = -\nabla \cdot \mathbf{F}. \tag{100.18}$$

It is important to remember the restrictive assumptions on which (100.18) is based. Next, making the Eddington approximation $K_{ij} = \frac{1}{3}J\,\delta_{ij}$ and dropping the time-dependent term in the radiation momentum equation, we obtain an explicit expression for $\mathbf{F}$ [cf. (97.68)], by means of which we can rewrite (100.18) as

$$\rho c_v(\partial T/\partial t) = \nabla \cdot [(4\pi/3\kappa)\,\nabla J]. \tag{100.19}$$

The approximations made here are the same as those used in the nonequilibrium diffusion approximation (cf. §97). Then substituting for J from (100.1) we have

$$\rho c_v(\partial T/\partial t) = \nabla \cdot \{(1/3\kappa)\,\nabla[a_R c T^4 + (\rho c_v/\kappa)(\partial T/\partial t)]\}, \tag{100.20}$$

which describes the thermal behavior of a static radiating medium, in the Eddington approximation, when time evolution of the radiation field is ignored. Finally, linearizing (100.20) and using the fact that $\nabla T_0 \equiv 0$ (homogeneous medium) we find

$$(\nabla^2 - 3\kappa^2)(\partial T_1/\partial t) = -\nu\,\nabla^2 T_1, \tag{100.21}$$

where ν is defined by (100.10).

The essential physics emerges when we examine (100.21) in the opaque and transparent regimes. In opaque material $\kappa l \to \infty$, where l is a characteristic length, and (100.21) limits to the diffusion equation

$$(\partial T_1/\partial t) = (\nu/3\kappa^2)\,\nabla^2 T_1, \tag{100.22}$$

which shows that radiative relaxation occurs on a characteristic time scale $\propto 3\kappa^2 l^2/\nu$, as found above. For transparent material $\kappa l \to 0$ and (100.21) limits to *Newton's law of cooling*

$$(\partial T_1/\partial t) = -\nu T_1, \tag{100.23}$$

according to which the rate of cooling is linearly proportional to the size of the temperature fluctuation, and has a characteristic time scale $t_{RR} = 1/\nu$.

Using a trial solution of the form (100.12) in (100.21) we recover (100.13), but with the exact $n(k)$ replaced by

$$n_E(k) = \nu/[1 + 3(\kappa_0/k)^2]. \tag{100.24}$$

Equation (100.24) has the same limiting behavior as (100.17) when $(\kappa_0/k) \to 0$ and $(\kappa_0/k) \to \infty$, and, as shown in Table 100.1, provides a reasonably good approximation in between.

When the effects of scattering are included, κ_0 is replaced by $(\kappa + \sigma)_0$ in

(100.19), hence (100.24) becomes

$$n_E(k) = \nu/\{1+3[\kappa_0(\kappa+\sigma)_0/k^2]\}. \qquad (100.25)$$

From (100.25) one sees that for a given total opacity $\chi_0 = (\kappa+\sigma)_0$, increasing σ_0 relative to κ_0 always increases the relaxation time. Indeed in the limit of pure scattering ($\kappa_0 = 0$) we have the noteworthy result that $n_E(k) \equiv \nu = 0$, hence the medium behaves adiabatically, as one would expect because photons are conserved by a pure scattering process. When the effects of energy exchange by thermal conduction are included (**A6**), (**D1**), the relaxation rate becomes $n = n_{\text{rad}} + n_{\text{cond}}$, where $n_{\text{cond}} = (K/\rho c_v)k^2$; here K is the material thermal conductivity. Comparison of n_{rad} and n_{cond} shows that radiation dominates only in long-wavelength disturbances, specifically when

$$k^2 < (16\sigma_R\kappa_0 T_0^3/K) - 3\kappa_0(\kappa+\sigma)_0. \qquad (100.26)$$

From the fairly close agreement of $n_E(k)$ and $n(k)$ many authors have concluded that the Eddington approximation is valid for the perturbed radiation field in both the optically thick and thin limits. We can examine this conclusion critically by calculating the Eddington factor directly from (100.16), obtaining

$$f(k) = \int_0^1 \mu^2[1+(k/\kappa_0)^2\mu^2]^{-1}\,d\mu \bigg/ \int_0^1 [1+(k/\kappa_0)^2\mu^2]^{-1}\,d\mu$$
$$= (\kappa_0/k)[1-(\kappa_0/k)\cot^{-1}(\kappa_0/k)]/\cot^{-1}(\kappa_0/k). \qquad (100.27)$$

Numerical values for $f(k)$ are given in Table 100.1. For optically thick disturbances $f \to \frac{1}{3}$. But for optically thin disturbances f actually *vanishes*, indicating that the perturbed radiation field is far from isotropic. In fact, the perturbed radiation field has a "pancake-shaped" distribution around the normal to the plane of the disturbance, because in the plane ($\mu = 0$), $I_1(\mu) = B_1$, but as $\mu \to 1$ (i.e., along $\mathbf{k}$) $I_1 \to 0$ because contributions from the sinusoidal variation of B sum to zero when there is no attenuation along the ray.

The real reason that $n_E = n$ when $\kappa_0/k = 0$ is *not* that the Eddington approximation is valid in this limit, but rather that $J_1 \propto (\kappa_0/k) \to 0$ in an optically thin disturbance. Hence the absorption term $\kappa_0 J_1$ in the energy equation vanishes, and the relaxation rate is set solely by the emission term $\kappa_0 B_1$, which is independent of both k and f. Indeed, retracing the derivation of (100.24), one finds that $n_E \equiv \nu$ when $\kappa_0/k = 0$ no matter what numerical value is chosen for the closure ratio K/J; that is, the Eddington approximation is *irrelevant* in the optically thin limit.

Spiegel's formula for t_{RR} has been extensively applied in estimating the effects of radiative damping on waves in stellar atmospheres (cf. §§101 and 102). But it is well to emphasize the restrictive assumptions on which it rests: an infinite, grey, homogeneous medium in LTE; initial radiative equilibrium; and no dynamics of either the matter or the radiation field. It entirely neglects *boundary* and *nonlocal transport* effects arising from inhomogeneities in the medium. The conditions under which the formula is known to be valid are therefore very limited, and one must be careful not to misapply it.

As an example of the consequences of dropping some of the assumptions made in deriving (100.17), consider an ambient homogeneous, static, LTE medium not initially in radiative equilibrium, but in a steady state under the action of radiative energy exchange and a constant nonradiative (e.g., magnetic) energy input (or loss) $\dot{q}$. Then

$$(\partial e/\partial t)_0 = 4\pi\kappa_0(J_0 - B_0) + \dot{q} = 0, \tag{100.28}$$

which implies $J_0 \neq B_0$. (Such a state can be realized only for a *finite* homogeneous medium, for otherwise J_0 would inevitably saturate to B_0.) In this case we cannot omit the term $4\pi\kappa_1(J_0 - B_0)$ from the linearized energy equation, nor the term $\kappa_1(I_0 - B_0)$ from the linearized transfer equation. Retracing the analysis we find that the effect is to replace B_1 in both (100.6) and (100.8) by an equivalent source

$$\tilde{B}_1 = B_1 + (\partial \ln \kappa_0/\partial T)(B_0 - J_0)T_1, \tag{100.29}$$

and therefore ν in (100.10) et seq. by

$$\tilde{\nu} = (4\pi\kappa_0/\rho c_v)[(a_R c/\pi)T_0^3 + (\partial \ln \kappa_0/\partial T)(B_0 - J_0)]. \tag{100.30}$$

When $\dot{q} > 0$, then $B_0 > J_0$, and the nonradiative energy input is balanced by excess emission. Then if $(\partial\kappa_0/\partial T) > 0$, we have $\tilde{\nu} > \nu$, as one expects because an increase in opacity produces an increased rate of emission. If, on the other hand, $(\partial\kappa_0/\partial T) < 0$, so that the material radiates less efficiently as it is heated, then $\tilde{\nu} < \nu$, and the relaxation time increases. Indeed if the second term in (100.30) is sufficiently negative, $\tilde{\nu}$ can become negative and an initial fluctuation will grow rather than decay; in this case the material is *thermally unstable.*

TIME-DEPENDENT RADIATION TRANSPORT

To extend the analysis we now allow for the finite propagation speed of light and consider time-dependent radiation transport; we thereby allow the radiation field to have a dynamical character. As before, we assume no material motions, which means that the material will respond only passively to the radiation field. Intuitively we expect to find again a thermal relaxation mode, but modified by the finite photon flight time $t_\lambda \equiv \lambda_p/c = 1/c\kappa$, and in addition, other new modes arising from the dynamical nature of the radiation field, including attenuated propagating radiation waves that correspond to the flow of radiation through an absorbing medium.

We adjoin the radiation energy and momentum equations to the gas energy equation and for steadily driven disturbances in an infinite homogeneous medium derive a dispersion relation (which is independent of global initial-boundary conditions) for the coupled set (**A6**), (**D1**). To close the system of moments we invoke the Eddington approximation, encouraged by the good results it gives in the quasi-static case.

For an infinite, homogeneous, grey, LTE medium initially in radiative equilibrium the perturbation equations to be solved are (100.6) and

$$(1/c\kappa)(\partial J_1/\partial t)+(1/\kappa)(\partial H_1/\partial x)=B_1-J_1 \tag{100.31}$$

and

$$(1/c\kappa)(\partial H_1/\partial t)+(1/3\kappa)(\partial J_1/\partial x)=-H_1, \tag{100.32}$$

where we noted that $J_0 \equiv B_0$ and $H_0 \equiv 0$. Assuming plane-wave perturbations of the form $\phi_1 = \Phi e^{ikx}e^{-nt}$ we obtain the system

$$\begin{pmatrix} nt_\lambda-1 & -ik/\kappa & 1 \\ -ik/3\kappa & nt_\lambda-1 & 0 \\ \nu & 0 & n-\nu \end{pmatrix}\begin{pmatrix} J_1 \\ H_1 \\ B_1 \end{pmatrix}=0 \tag{100.33}$$

which has a nontrivial solution only if the determinant of coefficients is zero. From this requirement we obtain the dispersion relation

$$z^3-(\alpha+2)z^2+(\alpha+\beta+1)z-\alpha\beta=0, \tag{100.34}$$

where $z \equiv nt_\lambda$, $\alpha \equiv \nu t_\lambda$, and $\beta \equiv k^2/3\kappa^2$.

In general, (100.34) has either three real roots or one real root and two conjugate complex roots. Given the roots n_i, $(i=1,2,3)$, one finds that the *eigenvectors* of the system have components

$$\mathbf{V}_i(k)=(J_1, H_1, B_1)_i=[1, (in_it_\lambda\kappa/k)(n_i-\nu-t_\lambda^{-1})/(\nu-n_i), \nu/(\nu-n_i)]\times J_1. \tag{100.35}$$

To gain physical insight we examine (100.34) in various limiting cases. Suppose first that $t_\lambda \to 0$, which implies that $c\to\infty$, hence *quasi-static radiation*. We then recover (100.24) and thus have the same thermal relaxation mode as before. We find

$$J_1 = B_1/[1+(k^2/3\kappa^2)] \tag{100.36a}$$

and

$$H_1 = -(ik/3\kappa)J_1. \tag{100.36b}$$

Note that $J_1 \to B_1$ and $H_1 \to 0$ as $\tau_\Lambda \to \infty$; and $J_1 \to 0$, $H_1 \to 0$ as $\tau_\Lambda \to 0$; H_1 lags J_1 by $\pi/2$ as a function of x.

Next, suppose that $\nu \to 0$, which corresponds to material with *infinite heat capacity*. Here the state of the matter is frozen, and a disturbance in the radiating fluid can propagate only by radiation. The dispersion relation reduces to

$$z[(z-1)^2+\beta]=0, \tag{100.37}$$

which yields roots $z_1=0$ and $z_{2,3}=1\pm ik/\sqrt{3}\,\kappa$. For $n=0$ we can impose an arbitrary B_1, which does not decay in time; J_1 and H_1 are again given by

(100.36), and merely represent the static adjustment of the radiation field to an imposed source perturbation (constant in time). The other two roots are more interesting; we find $B_1 \equiv 0$ and

$$H_1 = \mp J_1/\sqrt{3} \tag{100.38a}$$

where

$$J_1 \propto e^{-t/t_\lambda} e^{ik[x \pm (c/\sqrt{3})t]}. \tag{100.38b}$$

These are *damped radiation waves* propagating along the $\pm x$ axis with a phase and group speed $c/\sqrt{3}$, attenuating in time at a given spatial position on a time scale t_λ, or in space following a particular phase crest with a spatial scale $(\sqrt{3}\,\kappa)^{-1}$. These modes were rejected by previous authors (**A6**), (**D1**), (**F5**), but are, in fact, legitimate modes in a radiating fluid. The propagation speed is $c/\sqrt{3}$ instead of c because we have used the Eddington approximation and therefore obtain the *telegrapher's equation* in the optically thin limit instead of the exact radiation wave equation [cf. the discussion following (97.110)]. To obtain a better solution we would need to calculate an accurate Eddington factor for the time-dependent radiation field.

Next consider a *homogeneous disturbance* ($k \equiv 0$). From (100.33) we find the dispersion relation

$$z(z-1)[z-(\alpha+1)] = 0 \tag{100.39}$$

which has roots $z_1 = 0$, $z_2 = 1$, and $z_3 = \alpha + 1$. For the nondecaying mode $n_1 = 0$ we find that $J_1 \equiv B_1$ and $H_1 \equiv 0$. Here we have merely reached a *new equilibrium* in the radiation field by making identical, constant changes in J and B; H_1 is zero by symmetry. For the root $n_2 t_\lambda = 1$ we find that $J_1 = B_1 = 0$, whereas H_1 is arbitrary. Thus we may apply any nonzero perturbation in the specific intensity of the form $I_1(\mu) = \Sigma a_i P_i(\mu)$ as long as $a_0 = a_2 = 0$, which implies that $J_1 = K_1 = 0$ (Eddington approximation); a_1 and a_i for $i \geq 3$ may be arbitrary. Alternatively we may impose an azimuthal anisotropy such that J_1 and K_1 are zero (**F5**). Here we have an *isotropization mode,* in which an initial angular anisotropy of the radiation field is removed by absorption and isotropic re-emission on a radiation-flow time scale t_λ; the process is analogous to the establishment of an isotropic velocity distribution function for material particles in a deflection time t_D (cf. §10). Finally, for the root $n_3 = \nu + t_\lambda^{-1}$ we find $J_1 = -B_1/\nu t_\lambda$, which yields exact energy conservation in the linearized gas energy equation, and $H_1 = 0$. Here we have an *exchange mode* in which a given amount of energy is removed from the radiation field and temporarily deposited in the material (or vice versa) thus conserving the total fluid energy, but destroying radiative equilibrium. The disturbance decays back to equilibrium at a rate even faster than the relaxation rate of a transparent disturbance because we have simultaneously increased B (hence the emissivity) and decreased J (hence the rate of absorption), or vice versa, which results in a larger temperature imbalance (hence relaxation rate) than in an optically thin disturbance where B is altered but the ambient J_0 is unperturbed.

Consider next the *opaque limit*, $\beta \ll 1$, where (100.34) yields three real roots. We find that to first order in β the smallest root of (100.34) is

$$n_1(k) \approx \nu(k^2/3\kappa^2)/(1+\nu t_\lambda), \tag{100.40}$$

which is just the thermal relaxation mode, but with an effective relaxation rate $\nu/(1+\nu t_\lambda)$. The decreased relaxation rate is what one would expect intuitively because the radiation now takes a finite time to flow. In this mode J_1 and B_1 always have the same sign. For $\beta \ll 1$, J_1 and B_1 are nearly equal and $|H_1| \ll |J_1|$; and $J_1 \to B_1$ while $H_1 \to 0$ when $k \to 0$. The isotropization-mode root in the limit of small β is

$$n_2(k) \approx \nu t_\lambda\{1-[(k^2/3\kappa^2)(\nu t_\lambda - 1)/\nu t_\lambda]\}. \tag{100.41}$$

Again we have both $|J_1| \ll |H_1|$ and $|B_1| \ll |H_1|$. Finally, the exchange-mode root in the limit of small β is

$$n_3(k) \approx \nu + t_\lambda^{-1} - [(k^2/3\kappa^2)/\nu t_\lambda^2(1+\nu t_\lambda)]. \tag{100.42}$$

In this mode J_1 and B_1 always have opposite signs, and H_1 is small and 90° out of phase with J_1.

Finally, consider the *transparent limit* $\beta \to \infty$. We find that (100.34) has one real root

$$n_1(\infty) \approx \nu \tag{100.43}$$

and two complex roots

$$n_{2,3}(\infty) \approx c\kappa \pm ick/\sqrt{3}. \tag{100.44}$$

The real root corresponds to a pure damped disturbance; as we will shortly see, the mode by which it decays depends on the value of α. The complex roots correspond to two damped radiation waves propagating with phase and group speed $c/\sqrt{3}$. In these modes the sign parity of J_1 relative to B_1 is opposite to that in the surviving mode corresponding to n_1. This fact and the conjugate relation of the complex roots guarantees that with the three modes we can always synthesize an imposed perturbation in which J_1 and B_1 have arbitrary relative amplitude and phase.

The analytical results discussed above are represented in Figure 100.1, which shows $n(k)$ obtained from numerical solutions of (100.34). For optically thick disturbances we always find three real roots, corresponding to the exchange, isotropization, and thermal-relaxation modes. For small optical thickness we always find one real root and two complex roots corresponding to damped radiation waves. When $\alpha < 1$ the real root corresponds to the thermal relaxation mode; when $\alpha > 1$ it corresponds to the exchange mode. The connectivity of the various branches changes abruptly at $\alpha = 1$, as illustrated. For the special case $\alpha = 1$ the dispersion

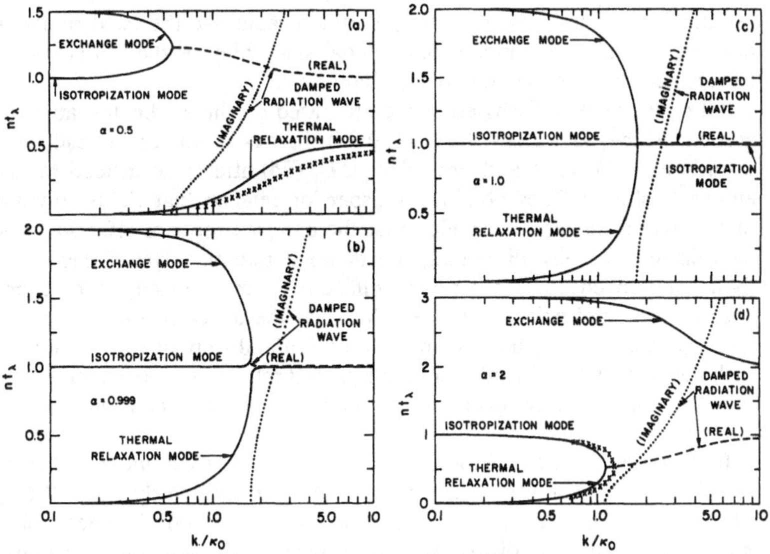

Fig. 100.1 Decay rates, as a function of optical thickness, of temperature fluctuations in a static medium.

relation

$$(z-1)(z^2-2z+\beta)=0 \tag{100.45}$$

can be solved exactly; one finds $z_1=1$ and $z_{2,3}=1\pm(1-\beta)^{1/2}$. The first root corresponds to the isotropization mode, which survives into the transparent limit in this case; the other roots yield damped radiation waves when $\beta>1$, and the thermal-relaxation and exchange modes when $\beta<1$.

For a specific choice of k/κ and νt_λ, an arbitrary disturbance can be projected onto the three eigenvectors of the system, yielding three components of the perturbation. The ith component decays away on a time scale $1/\text{Re}\,(n_i)$, hence the mode with the smallest root dominates the long-term time evolution of the radiating fluid. For example, in the opaque limit, local imbalances in radiative equilibrium will be removed first by direct exchange between radiation and material energy, next the radiation field will isotropize, and finally the residual perturbation will be smoothed by thermal relaxation. For optically thin disturbances energy transport in radiation waves becomes efficient, and two of the modes that exist for optically thick disturbances will be replaced by these modes. For small α, $t_\lambda<t_{RR}$, hence the radiation field will adjust essentially instantaneously to the state of the material via damped radiation waves; the final rate of relaxation (in the thermal relaxation mode) is set by the heat capacity of the material. In contrast, for large α, $t_\lambda>t_{RR}$, hence the radiation is essentially "frozen" and the material adjusts rapidly to the *local* radiation field via the exchange mode; final global smoothing of the initial disturbance then proceeds via damped radiation waves.

All of the results discussed above are based on the Eddington approximation, which, as we saw from (100.27), breaks down for optically thin disturbances. Delache and Froeschle (**D1**), (**F5**) attempted instead to find an exact solution. They obtained a dispersion relation that yields only one or two real roots, and concluded that the complex roots of (100.34) must be rejected, thereby discarding the damped radiation waves. However, their solution encounters a severe difficulty in the optically thin regime because they find only one mode, which has the unacceptable implication that one is not free to impose an arbitrary initial disturbance, but only one with the correct relationship (both in sign and relative size) between J_1 and B_1. This lack of a complete set of modes indicates a deficiency in the analysis.

In fact, the formal solution in (**D1**) is invalid when $nt_\lambda \geq 1$ because initial-boundary conditions are not accounted for correctly [cf. (79.35)]. The mathematical symptom is that a certain integral diverges unless $nt_\lambda < 1$; physically the divergence occurs because the integral is swamped by an exponentially divergent source when the integration is extended to $t \rightarrow -\infty$ instead of being truncated at $t = 0$ (the instant when the initial perturbation was imposed). The solution in (**F5**) suffers from a similar problem. Furthermore, when correct limits are applied in the formal solution it is no longer possible to use a separable solution of the form (100.12). Thus the proposed "exact" solution appears to have only limited applicability.

In our opinion the Eddington approximation should always yield results that are at least qualitatively correct. For example, suppose $nt_\lambda < 1$. Here we expect the Eddington approximation to be valid in the opaque limit, and to become irrelevant in the transparent limit. As a test we replace the factor $\frac{1}{3}$ in (100.32) to (100.34) with the quasi-static $f(k)$ given by (100.27). As shown in Figure 100.1 we find little change in the modes with $nt_\lambda < 1$. The same remark holds for modes with $nt_\lambda \geq 1$, but in this case we cannot guarantee that (100.27) is valid. However, we know that for a mode with $nt_\lambda > 1$ the material equilibrates to the local radiation field via *isotropic* emission and absorption processes on a time scale shorter than that required for radiation to flow to (or from) adjacent regions. Therefore if we assume (legitimately) that the initial radiation perturbation is isotropic, we can argue that it must remain isotropic during the lifetime of the mode, hence that the Eddington approximation will apply. Similarly, for $nt_\lambda = 1$ the effect of the isotropization mode is to isotropize an initially anisotropic distribution. Likewise the damped radiation modes will propagate an initial isotropic disturbance isotropically. In all cases the Eddington approximation appears reasonable.

The radiative relaxation of a medium comprising non-LTE two-level atoms, radiation, and an ambient LTE gas is discussed in (**L7**), (**F5**), and (**F6**). Appropriate rate equations are adjoined to the gas energy and radiation transport equations. The resulting dispersion relation is more complicated and yields a richer spectrum of modes, which now depend on characteristic time scales governing the kinetics of statistical equilibrium (e.g., radiative and collisional rates) in addition to the parameters entering in the cases discussed above.

Virtually all astrophysical discussions of the radiative damping of temperature fluctuations and/or waves are based on (100.17) or (100.24), and thus ignore the time dependence of the radiation field. This simplification is usually justified because the dynamical time scales of many astrophysical phenomena are enormously longer than a photon flight time, hence the radiation field is indeed quasi-static and the fast exchange, isotropization, and damped-radiation-wave modes are of little interest. A similar situation is encountered in fluid dynamics where in order to follow the evolution of flow phenomena having long time scales, one can make the *anelastic approximation* and adopt modified equations of hydrodynamics that suppress sound waves, thereby filtering out variations on short time scales that otherwise are a nuisance computationally.

THERMAL RESPONSE OF THE SOLAR ATMOSPHERE

In attempting to apply (100.17) to estimate the radiative relaxation time of temperature fluctuations (or waves) in a stellar atmosphere, one must account for two important effects: (1) the variation of material properties, hence ν, with height, and (2) the presence of an open boundary.

Estimates of the relaxation time for an optically thin disturbance, $t_{RR}(\infty)$, have been made by several authors [e.g., (**B6**, 326), (**S17**), (**S24**), (**U3**)] using realistic model solar atmospheres. Allowing for continuum opacities only, one finds that $t_{RR}(\infty)$ in the photosphere ($\tau_{\text{cont}} \approx 1$) is about 1 s, and rises rapidly with height, reaching a maximum of about 800 s at about 700 km above the photosphere. At greater heights the relaxation time begins to drop because of rising temperature, then passes through a secondary maximum as hydrogen ionizes, and finally plunges sharply. These results are modified drastically when radiative losses in spectral lines are included (**G5**); one then finds that $t_{RR}(\infty)$ rises to about 500 s just above the temperature minimum, then falls to only 90 s in the mid-chromosphere where line losses are large, before rising again to about 400 s when hydrogen ionizes. Unfortunately it is difficult to allow properly for self-absorption in the spectral lines, hence to estimate accurately their net cooling rate, and the line-loss term is uncertain by at least a factor of 2.

Below $\tau_{\text{cont}} \approx 1$, $t_{RR}(\infty)$ drops rapidly as κ rises sharply. But for a disturbance of finite wavenumber k, the increase of τ_Λ with increasing κ implies that $t_{RR}(k)$ increases rapidly, in accordance with (100.17). Ultimately, $t_{RR}(k)$ becomes so large that a time-periodic disturbance behaves essentially adiabatically ($\omega t_{RR} \gg 1$). Because the atmosphere has an open boundary the effective relaxation rate of a disturbance depends on its optical depth τ in the atmosphere, as well as on τ_Λ. Thus a horizontal perturbation will relax by horizontal radiative exchange between crests and troughs if $\tau_\Lambda \ll \tau$, but if $\tau_\Lambda \gg \tau$ it relaxes more efficiently as a result of vertical radiative losses through the open boundary. Ulrich (**U8**) suggested that in calculating t_{RR} from (100.17) we use an effective optical thickness given by

$$\tau_{\text{eff}}^{-2} \equiv \tau_\Lambda^{-2} + \tau^{-2}. \qquad (100.46)$$

In physical terms (100.46) gives the harmonic mean of the number of absorptions a photon requires to cross one wavelength of the disturbance, and the number to escape from the atmosphere. For a vertical disturbance one can use (100.46) or simply choose $\tau_{\text{eff}} = \min(\tau, \tau_\Lambda)$. In an exponential atmosphere $\tau_\Lambda : \tau = \Lambda : H$, hence τ_Λ sets the relaxation rate only for short-wavelength disturbances (or for long-wavelength disturbances deep in the envelope where H becomes large).

The thermal response of the solar atmosphere to periodic time variations of the radiative flux incident from below is examined numerically in (**W5**). The atmosphere is assumed to be motionless, but allowance is made for an inhomogeneous vertical structure. A sinusoidal variation with a 10 percent amplitude in the radiative flux is imposed at the lower boundary. Initial transients (the subjects of study elsewhere in this section) are allowed to die out, and the final periodic solution driven by the boundary condition is obtained. From the numerical results one finds that (1) the amplitude of the temperature fluctuation decreases with increasing height, (2) there is a phase lag between the imposed flux and the temperature response, (3) the lag increases with height, and (4) the lag is an increasing fraction of a period as the period decreases.

To understand these results qualitatively, consider the optically thin part of the atmosphere, and assume that

$$T(z, t) = T_0[1 + \xi(z)e^{i\omega t}] \qquad (100.47)$$

and

$$J(t) = J_0(1 + \varepsilon e^{i\omega t}), \qquad (100.48)$$

where T_0 is a suitable average and $J_0 = B_0 = \sigma_R T_0^4/\pi$. The perturbation ε is constant because the region considered is optically thin. The linearized energy equation then reduces to

$$i\omega\xi(z) = (4\sigma_R \langle\kappa\rangle T_0^3/c_v)[\varepsilon - 4\xi(z)]. \qquad (100.49)$$

In general both ε and ξ are complex, but we can choose the time coordinate so that ε is real, whence we have

$$\xi_I = -\tfrac{1}{4}\nu\varepsilon\omega/(\omega^2 + \nu^2) \qquad (100.50a)$$

and

$$\xi_R/\xi_I = -\nu/\omega. \qquad (100.50b)$$

Therefore

$$|\xi(z)| = \tfrac{1}{4}\nu(z)\varepsilon/[\omega^2 + \nu^2(z)]^{1/2} \tag{100.51a}$$

and

$$\phi(z) = -\tan^{-1}[\omega/\nu(z)]. \tag{100.51b}$$

Here ϕ is the phase angle between J_1 and T_1; negative ϕ indicates that T lags J. In the high-frequency limit, $\omega/\nu \to \infty$,

$$|\xi| \to \tfrac{1}{4}\nu\varepsilon/\omega \to 0 \tag{100.52a}$$

and

$$\phi \to -\frac{\pi}{2}, \tag{100.52b}$$

so the thermal response lags the input by 90° and its amplitude vanishes. In the low-frequency limit, $\omega/\nu \to 0$,

$$|\xi| \to \tfrac{1}{4}\varepsilon \tag{100.53a}$$

and

$$\phi \to 0, \tag{100.53b}$$

hence the atmosphere passes through a series of quasi-equilibria with vanishing phase lag. Equation (100.51a) shows that the phase lag must increase with height because ν decreases outward through the photosphere; the approximate results given by (100.51) are in good agreement with the numerical results.

The radiative relaxation of a two-dimensional checkerboard distribution of temperature and density fluctuations simulating a hydrodynamic model of convection cells in the solar atmosphere is discussed in (**L8**). The relaxation rate is found to depend on the cell size, the size of the velocity field, and the amplitude of the initial temperature fluctuation.

101. Propagation of Acoustic Waves in a Radiating Fluid

In this section we examine the effects of radiation on acoustic waves propagating in an infinite, homogeneous medium in LTE, initially in radiative equilibrium. We first consider wave damping by radiative energy exchange, which is generally very efficient, especially near boundary surfaces (e.g., the solar photosphere) where radiative relaxation times are very short compared to typical wave periods. By comparison, wave damping by viscosity and thermal conduction is negligible in most situations of astrophysical interest. We treat the spatial damping of driven harmonic disturbances (i.e., real ω and complex k) as opposed to the time decay of a transient initial disturbance (i.e., real k and complex ω). We then consider the more fundamental role played by radiation through its contributions to the total energy density and pressure in the fluid; we find that radiation can radically alter the dynamical properties of wave modes in the fluid. Finally, we consider briefly the effects of radiation forces on acoustic waves.

NEWTONIAN COOLING (OPTICALLY THIN PERTURBATIONS)

Consider first an optically thin disturbance in which radiative energy exchange is adequately described by the Newtonian cooling approximation (**S16**), (**S25**). We assume that the radiation field is quasi-static and ignore the dynamical behavior of the radiation component of the radiating fluid. From (100.23) the net heat input to the gas is then

$$(Dq/Dt) = -\rho c_v \nu T_1 \tag{101.1}$$

where ν is given by (100.10). Using (101.1) in (52.19) we can write the gas energy equation as

$$(Dp/Dt) - a^2(D\rho/Dt) = -16(\Gamma_3 - 1)\sigma_R \kappa_0 T_0^4 (T_1/T_0). \tag{101.2}$$

For the special case of a perfect gas with constant specific heats, we can rewrite the right-hand side of (101.2) in a more convenient form. For such a gas $\Gamma_3 = \gamma$, $(\gamma - 1)c_v = R$, $a^2 = \gamma p/\rho$, and $T_1/T_0 = (p_1/p_0) - (\rho_1/\rho_0)$, and the linearized version of (101.2) reduces to

$$(\partial p_1/\partial t) = a^2(\partial \rho_1/\partial t) + \mathbf{v} \cdot (a^2 \nabla \rho_0 - \nabla p_0) - (a^2 \rho_0/\gamma t_{RR})[(p_1/p_0) - (\rho_1/\rho_0)], \tag{101.3}$$

where for brevity we write $t_{RR} \equiv t_{RR}(\infty) = \nu^{-1}$. Note in passing that for an ionizing gas the linearized energy equation is

$$(\partial p_1/\partial t) = a^2(\partial \rho_1/\partial t) + \mathbf{v} \cdot (a^2 \nabla \rho_0 - \nabla p_0) - a^2 \rho_0[\alpha_2(p_1/p_0) - \alpha_1(\rho_1/\rho_0)], \tag{101.4}$$

where, from (54.84a), α_1 and α_2 for a pure hydrogen gas are

$$\alpha_1 \equiv [16(\Gamma_3 - 1)\sigma_R \kappa_0 T_0^4/\Gamma_1 p_0]/\{1 + \tfrac{1}{2}x(1-x)[\tfrac{5}{2} + (\varepsilon_I/kT)]\} \tag{101.5}$$

and

$$\alpha_2 \equiv [1 + \tfrac{1}{2}x(1-x)]\alpha_1. \tag{101.6}$$

For a homogeneous medium the gradient terms in (101.3) vanish, hence the dynamics of a radiatively damped acoustic wave is determined (in the Newtonian cooling approximation) by

$$(\partial \rho_1/\partial t) = -\rho_0 \nabla \cdot \mathbf{v}_1, \tag{101.7}$$

$$\rho_0(\partial \mathbf{v}_1/\partial t) = -\nabla p_1, \tag{101.8}$$

and

$$(\partial p_1/\partial t) - a^2(\partial \rho_1/\partial t) = -(a^2 \rho_0/\gamma t_{RR})[(p_1/p_0) - (\rho_1/\rho_0)]. \tag{101.9}$$

Taking $(\partial^2/\partial t^2)$ of (101.9) and using (48.6), which follows from (101.7) and

(101.8), we obtain the wave equation

$$\{[(\partial^2/\partial t^2) - a^2\,\nabla^2](\partial/\partial t) + t_{RR}^{-1}[(\partial^2/\partial t^2) - (a^2/\gamma)\,\nabla^2]\}p_1 = 0. \quad (101.10)$$

For plane waves (101.10) yields the dispersion relation

$$k^2 = \left(\frac{\gamma\omega^2}{a^2}\right)\left[\frac{(1+\gamma\omega^2 t_{RR}^2) - i(\gamma-1)\omega t_{RR}}{1+\gamma^2\omega^2 t_{RR}^2}\right]. \quad (101.11)$$

Because k is complex we find damped progressive waves varying as $e^{i(\omega t - k_R x)}e^{-k_I x}$.

In the high-frequency limit, $\omega t_{RR} \gg 1$, and

$$k \approx (\omega/a)\{1 - i[\tfrac{1}{2}(\gamma-1)/\gamma\omega t_{RR}]\}, \quad (101.12)$$

which corresponds to acoustic waves traveling with the adiabatic sound speed, having a characteristic damping length

$$L \approx [2\gamma/(\gamma-1)]a t_{RR}. \quad (101.13)$$

Note that $L/\Lambda \sim \omega t_{RR} \gg 1$, hence in the limit of high frequencies and/or long damping times, acoustic waves behave essentially adiabatically and suffer only a small damping *per cycle*. Note, however, that (101.13) shows that the geometrical distance over which the wave damps can be made arbitrarily small by making the relaxation time sufficiently short; in this sense high-frequency waves can be heavily damped.

In the low-frequency limit, $\omega t_{RR} \ll 1$, and

$$k \approx (\omega/a_T)[1 - i\tfrac{1}{2}(\gamma-1)\omega t_{RR}], \quad (101.14)$$

where a_T is the isothermal sound speed $a/\gamma^{1/2}$. We now have acoustic waves traveling with speed a_T, with a damping length

$$L \approx [2/(\gamma-1)]a_T t_{RR}/(\omega t_{RR})^2, \quad (101.15)$$

which implies that $L/\Lambda \sim (\omega t_{RR})^{-1} \gg 1$. Thus according to Newtonian cooling theory, in the limit of low frequencies and/or short radiative relaxation times radiative exchange obliterates temperature fluctuations in the gas, and acoustic waves propagate isothermally with negligible spatial damping.

For arbitrary ωt_{RR} one solves (101.10) numerically. As shown in Figure 101.1, the phase speed $v_p = \omega/k_R$ rises abruptly from a_T to a near $\omega t_{RR} \approx 1$. At the same time, the damping length $L/\Lambda = |k_R/2\pi k_I|$ passes through a minimum. Indeed, near $\omega t_{RR} \sim 1$, $L/\Lambda \approx 1$, so these waves decay after traveling only a few wavelengths.

It must be emphasized that all of the above results apply only for optically thin disturbances, $\kappa/k \ll 1$, $\Lambda/\lambda_p \ll 1$. The significance of this remark will become clear shortly.

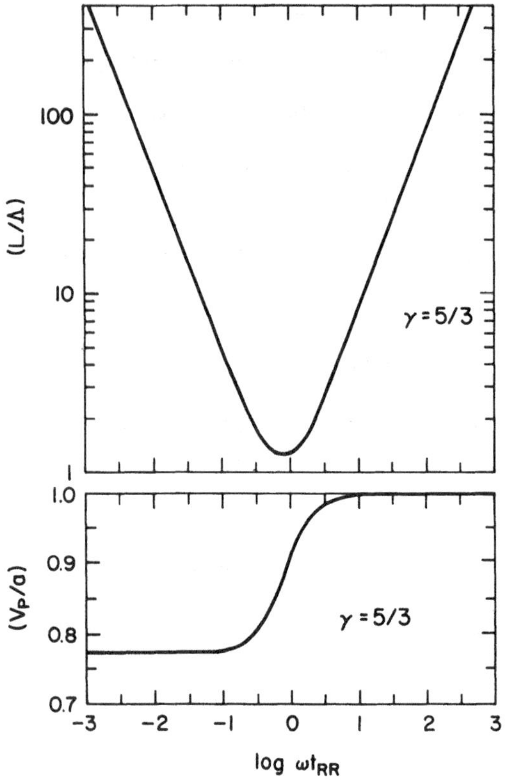

Fig. 101.1 Damping length and phase speed of acoustic mode in Newtonian cooling approximation.

EQUILIBRIUM DIFFUSION (OPTICALLY THICK DISTURBANCES)

In extremely opaque material (e.g., inside a star), radiation comes into thermal equilibrium with the matter, and energy exchange proceeds by diffusion; in this regime radiation can have important dynamical effects on wave propagation. In treating the radiation field we can apply the equilibrium diffusion approximation provided that the disturbance is sufficiently optically thick, that is, that $\kappa/k \gg 1$, and $\Lambda/\lambda_p \gg 1$.

The dynamical equations for a radiating fluid in the equilibrium diffusion regime were derived in §97. From (97.5) and (97.6) the momentum equation is

$$\rho(D\mathbf{v}/Dt) = -\nabla\bar{p} \tag{101.16}$$

and the energy equation is

$$\rho\{(D\bar{e}/Dt) + \bar{p}[D(1/\rho)/Dt]\} = \nabla \cdot (\bar{K}\,\nabla T). \tag{101.17}$$

Here $\bar{p}$, $\bar{e}$, and $\bar{K}$ are the *total* pressure, energy density, and conductivity of the fluid:

$$\bar{p} = p_{gas} + p_{rad} = \rho RT + \tfrac{1}{3}a_R T^4 \equiv (1+\alpha)\rho RT, \tag{101.18}$$

$$\bar{e} = e_{gas} + (a_R T^4/\rho) = \tfrac{3}{2}RT + (a_R T^4/\rho), \tag{101.19}$$

and

$$\bar{K} = K_{thermal} + K_{rad}, \tag{101.20}$$

where K_{rad} is defined by (97.3). Using (52.19) we rewrite (101.17) as

$$(D\bar{p}/Dt) - a^2(D\rho/Dt) = (\Gamma_3 - 1)\,\nabla \cdot (\bar{K}\,\nabla T) \tag{101.21}$$

where

$$a^2 \equiv \Gamma_1 \bar{p}/\rho = (1+\alpha)\Gamma_1(a^2{}_{gas}/\gamma), \tag{101.22}$$

and Γ_3 and Γ_1 are given by (70.18) and (70.22).

For a small disturbance we linearize these equations. The linearized continuity and momentum equations are the same as (101.7) and (101.8) (with p_1 replaced by $\bar{p}_1$), and therefore again yield (48.6). The linearized energy equation is

$$(\partial \bar{p}_1/\partial t) - a^2(\partial \rho_1/\partial t) = (\Gamma_3 - 1)\bar{K}\,\nabla^2 T_1. \tag{101.23}$$

We can eliminate T_1 in favor of $\bar{p}_1$ and ρ_1 by means of the linearized equation of state

$$T_1/T_0 = [(1+\alpha)(\bar{p}_1/\bar{p}_0) - (\rho_1/\rho_0)]/(1+4\alpha). \tag{101.24}$$

Thus using (101.24) in (101.23) and making use of (70.16), (70.18), and (70.22) we find

$$(\partial \bar{p}_1/\partial t) - a^2(\partial \rho_1/\partial t) = \Gamma\chi[\nabla^2 \bar{p}_1 - (a^2/\Gamma)\,\nabla^2 \rho_1] \tag{101.25}$$

where we defined an effective Γ as

$$\Gamma \equiv (1+\alpha)\Gamma_1 \tag{101.26}$$

and the thermal diffusivity is

$$\chi \equiv \bar{K}/\rho_o c_p. \tag{101.27}$$

With these definitions (101.25) is formally identical to (51.5) for a heat-conducting gas. Taking $(\partial^2/\partial t^2)$ of (101.25) and using (48.6) to eliminate ρ_1 we obtain the wave equation

$$\{[(\partial^2/\partial t^2) - a^2\,\nabla^2](\partial/\partial t) - \Gamma\chi[(\partial^2/\partial t^2) - (a^2/\Gamma)\,\nabla^2]\,\nabla^2\}\bar{p}_1 = 0. \tag{101.28}$$

For a plane wave (101.28) yields the dispersion relation

$$(ak/\omega)^4 - [\Gamma - i(a^2/\chi\omega)](ak/\omega)^2 - i(a^2/\chi\omega) = 0, \tag{101.29}$$

which is quadratic in $(ak/\omega)^2$, and contains two dimensionless numbers: Γ and $\chi\omega/a^2$. This dispersion relation is formally identical to (51.16) for a thermally conducting inviscid material gas. Hence we obtain the same modes and physical interpretation as before: an adiabatic (isothermal) radiation-modified acoustic wave, and a slow (fast) radiation-diffusion wave in the low (high) frequency limits, respectively. These waves have the same propagation characteristics as described in §51 (cf. Figures 51.1 and 51.2); only the thermodynamic parameters (χ, Γ, a) differ from their earlier definitions. Thus in the equilibrium diffusion regime radiation is an inseparable part of the radiating fluid, with photons behaving dynamically like "honorary material particles".*

It should be noted that as $\alpha \rightarrow \infty$, Γ, a, and c_p diverge, but this is an artifact of having ignored the rest energy of the fluid in our analysis [cf. (48.32) and (70.27) for a]. The divergence occurs only at extremely high temperatures where relativistic effects are major and a relativistic analysis is required. In contrast, it follows from (101.22) and (101.26) that $a^2/\Gamma \equiv (a_T^2)_{\text{gas}}$, hence the phase speed of the isothermal acoustic wave depends on gas properties only.

The analysis can be extended to include the effects of electromagnetic fields in an ionized plasma; see (**P1,** §8.2).

TIME-INDEPENDENT TRANSPORT (EDDINGTON APPROXIMATION)

The Newtonian cooling and equilibrium diffusion approximations conflict with each other in that they predict opposite variations of the propagation speed (i.e., adiabatic versus isothermal) of the acoustic mode in going from low to high frequency. This contradiction arises because each scheme breaks down in one or the other limit. Thus at a sufficiently low frequency the wavelength of a disturbance is so long that it becomes optically thick (no matter how transparent the material), and the Newtonian cooling approximation no longer applies. Conversely, at very high frequencies the wavelength of a disturbance becomes so small that it is optically thin (no matter how opaque the material) and the diffusion approximation is no longer valid because a photon mean free path exceeds the characteristic spatial scale of gradients in the disturbance (recall the discussion of flux limiting in §97).

These considerations show that it is imperative to account for *transport effects* arising from finite photon mean free paths in the disturbance. Our qualitative expectation based on the diffusion approximation is that waves should be adiabatic at very low frequencies, and become isothermal above some critical frequency; but then at some sufficiently high frequency the waves should again become adiabatic, as predicted by the Newtonian cooling approximation. Precisely this behavior was found by Stein and Spiegel (**S23**) in their analysis of the time decay of an initial disturbance, allowing for transport effects (but ignoring the time dependence and dynamical behavior of the radiation field). In keeping with the rest of the

* We are indebted to Dr. J. I. Castor for this felicitous expression.

discussion of this section, we consider instead the spatial damping of a driven disturbance along the lines explored by Vincenti and Baldwin (**V6**), (**V7**, §§12.5–12.8).

The dynamical behavior of the material is governed by the continuity equation, the material momentum equation

$$\rho(D\mathbf{v}/Dt) = -\nabla p, \tag{101.30}$$

and the gas energy equation

$$(Dp/Dt) - a^2(D\rho/Dt) = 4\pi(\gamma-1)\kappa(J-B). \tag{101.31}$$

Here p refers to the gas pressure only, and $a^2 = \gamma p/\rho$. In (101.30) we have neglected the radiation force on the material.

We assume that the radiation field is quasi-static and make the Eddington approximation. The radiation energy equation is then

$$(\partial H/\partial x) = \kappa(B-J) \tag{101.32}$$

and the radiation momentum equation is

$$\tfrac{1}{3}(\partial J/\partial x) = -\kappa H. \tag{101.33}$$

Equations (101.32) and (101.33) provide a significant improvement over the Newtonian cooling approximation because they apply in both the optically thick and thin limits. They are also an improvement over equilibrium diffusion because they discriminate between J and B, which is crucial when the disturbance becomes optically thin. However they sacrifice some of the logical consistency inherent in the equilibrium diffusion analysis because they ignore the dynamics of the radiation field; we will remedy that flaw later.

In linearizing the radiation equations we note that $J_0 \equiv B_0$ and $H_0 \equiv 0$, and introduce nondimensional radiation variables $j_1 \equiv J_1/B_0$, $h_1 \equiv H_1/B_0$, and $B_1/B_0 = 4T_1/T_0 \equiv 4\theta_1$. Combining the linearized forms of (101.32) and (101.33) we have

$$(3\kappa^2)^{-1}(\partial^2 j_1/\partial x^2) = j_1 - 4\theta_1. \tag{101.34}$$

In linearizing the continuity and material momentum and energy equations we use a velocity potential $u_1 \equiv (\partial\phi_1/\partial x)$ which implies that $(\partial\rho_1/\partial t) = -\rho_0(\partial^2\phi_1/\partial x^2)$ and, from (101.30), $p_1 = -\rho_0(\partial\phi_1/\partial t)$. Using these expressions in the linearized gas energy equation we find

$$(\partial^2\phi_1/\partial t^2) - a^2(\partial^2\phi_1/\partial x^2) = (4a^3\kappa/\text{Bo})(4\theta_1 - j_1) \tag{101.35}$$

where Bo is the Boltzmann number obtained by setting the characteristic flow speed equal to the sound speed:

$$\text{Bo} \equiv \rho_0 c_p a/\sigma_R T_0^3. \tag{101.36}$$

Similarly the linearized equation of state for the material can be written

$$(\partial^2\phi_1/\partial t^2) - (a^2/\gamma)(\partial^2\phi_1/\partial x^2) + (a^2/\gamma)(\partial\theta_1/\partial t) = 0. \tag{101.37}$$

For a plane wave (101.34), (101.35), and (101.37) imply

$$\begin{pmatrix} 0 & -4 & 1+(k^2/3\kappa^2) \\ (a^2k^2/\omega^2)-\gamma & ia^2/\omega & 0 \\ (a^2k^2/\omega^2)-1 & -16a^3\kappa/\omega^2\text{Bo} & 4a^3\kappa/\omega^2\text{Bo} \end{pmatrix} \begin{pmatrix} \phi_1 \\ \theta_1 \\ j_1 \end{pmatrix} = 0. \quad (101.38)$$

Setting the determinant of (101.38) equal to zero we obtain the dispersion relation

$$[1-i(16\tau_a/\text{Bo})](ak/\omega)^4-[1-3\tau_a^2-i(16\gamma\tau_a/\text{Bo})](ak/\omega)^2-3\tau_a^2=0. \quad (101.39)$$

Here

$$\tau_a \equiv a\kappa/\omega = \tau_\Lambda/2\pi, \quad (101.40)$$

where τ_Λ is the optical thickness of one wavelength of a disturbance of frequency ω traveling with the adiabatic sound speed a. Like (101.29), (101.39) is quadratic in (a^2k^2/ω^2), hence we again get two distinct wave modes. One is a radiation-modified acoustic wave; the other is a nonequilibrium radiation diffusion wave analogous to a thermal wave.

The importance of radiation to the behavior of these waves is measured by Bo. In the limit $\text{Bo} \rightarrow \infty$ radiative energy exchange with the material ceases. In this special case the dispersion relation factors into

$$[(ak/\omega)^2-1][(ak/\omega)^2+3\tau_a^2]=0, \quad (101.41)$$

and we obtain (1) an undamped adiabatic acoustic wave in which θ_1 and v_1 are related by (48.24b), and J_1 and B_1 are related by (100.36), and (2) a radiation-field perturbation j_1 decaying as $\exp(-\sqrt{3}\,\kappa x)$ [as one expects for the Eddington approximation, cf. (83.7)], while $\phi_1=\theta_1\equiv 0$.

Solving (101.39) for $\tau_a \ll 1$ (small optical thickness and/or high frequency) we find a weakly damped acoustic wave with

$$k \approx (\omega/a)[1-i8\tau_a(\gamma-1)/\text{Bo}], \quad (101.42)$$

which implies that $v_p = a$ and $L/\Lambda = [\text{Bo}/16\pi\tau_a(\gamma-1)] \gg 1$, and a fast, strongly damped radiation diffusion wave with

$$k \approx (\omega/a)[(8\sqrt{3}\,\gamma\tau_a^2/\text{Bo})-i\sqrt{3}\,\tau_a], \quad (101.43)$$

which implies that $v_p/a = \text{Bo}/8\sqrt{3}\,\gamma\tau_a^2$ and $L = 1/\sqrt{3}\,\kappa$ or $L/\Lambda = 4\gamma\tau_a/\pi\text{Bo}$. Note that as $\tau_a \rightarrow 0$, the damping length for the acoustic mode becomes infinite, whereas the diffusion mode has an infinite phase speed and a fixed geometrical damping length, while $L/\Lambda \rightarrow 0$. The infinite propagation speed of the diffusion mode reflects the failure of the quasi-static radiation equations to provide flux limiting in optically thin material.

For $\tau_a \gg 1$ (i.e., large optical thickness and/or low frequency) we find a damped acoustic wave with

$$k \approx (\omega/a)[1-i8(\gamma-1)/3\tau_a\text{Bo}], \quad (101.44)$$

which implies that $v_p = a$ and $L/\Lambda = [3\tau_a \mathrm{Bo}/16\pi(\gamma-1)] \gg 1$, and a slow, heavily damped diffusion wave with

$$k \approx (\omega/a)(3\tau_a \mathrm{Bo}/32)^{1/2}(1-i), \qquad (101.45)$$

which implies that $v_p/a = (32/3\tau_a \mathrm{Bo})^{1/2} \ll 1$ and $L/\Lambda = 1/2\pi$. The prediction by (101.44) that the acoustic-mode speed always equals the material sound speed independent of the radiation energy density (i.e., Bo) is in contradiction with equilibrium diffusion theory (which is valid when $\tau_a \gg 1$) and reflects the failure of (101.32) and (101.33) to account for the dynamics of the radiation field.

In general, (101.39) must be solved numerically. Results for various values of Bo are shown in Figures 101.2 and 101.3. For context, Bo is of order 10 in the solar photosphere and at the Sun's center, 10^{-2} at the center of an O-star, 10^{-5} in an X-ray source, and 10^{-14} or smaller in a solar flare. Figure 101.2 shows that the acoustic mode is indeed adiabatic at high and low frequencies, and is isothermal over a range approximately inversely proportional to Bo. Furthermore, we see that the damping length is large when the phase speed is constant, but drops sharply where v_p makes a transition between a and a_T. Figure 101.3 shows that v_p in the

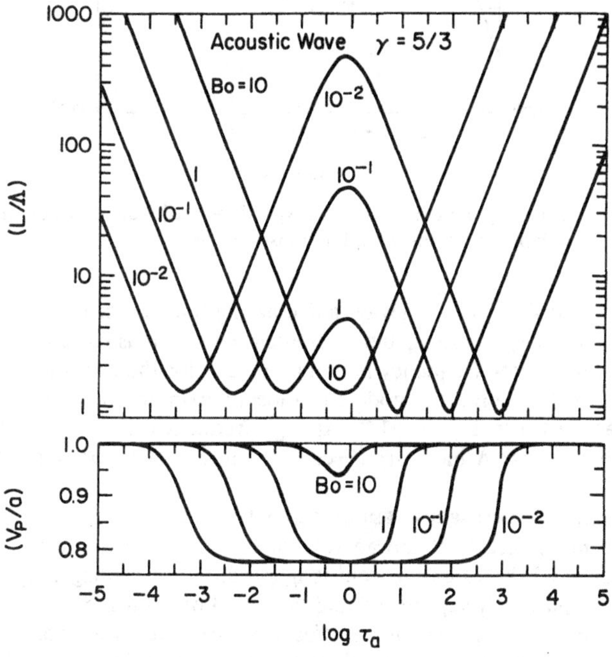

Fig. 101.2 Damping length and phase speed of acoustic mode for quasi-static radiation field, allowing for transport effects.

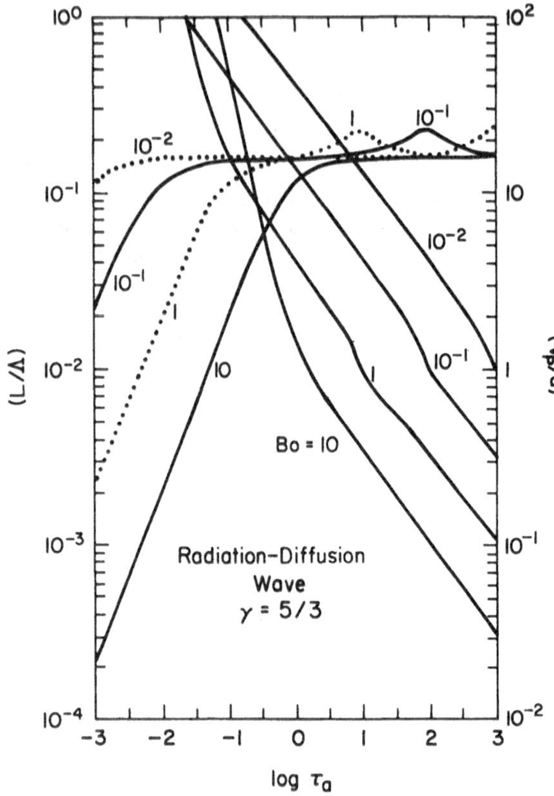

Fig. 101.3 Damping length and phase speed of radiation diffusion mode for quasi-static radiation field, allowing for transport effects.

radiation diffusion wave is always a decreasing function of τ_a, and increases with Bo when τ_a is small, but decreases with increasing Bo when τ_a is large. Near the low-frequency minimum of L/Λ for the acoustic mode, L/Λ for the radiation-diffusion mode has a local maximum like that of thermal waves as shown in Figure 51.2. At high frequencies, L is fixed but L/Λ decreases because Λ increases [because v_p increases as ω^2, see (101.43)].

TIME-DEPENDENT TRANSPORT (EDDINGTON APPROXIMATION)

The two main defects of the analysis just presented are that (1) the time dependence of the radiation field is ignored, hence propagating radiation waves are spuriously suppressed and the radiation diffusion wave is not flux limited, and (2) the dynamical effects of the radiation (work done by radiation pressure and the rate of change of the radiation energy density in the radiating fluid) are neglected. Therefore, to complete the physical picture we account for these phenomena by including the radiation pressure gradient in the material momentum equation

$$\rho(D\mathbf{v}/Dt) = -\nabla p - \nabla P \tag{101.46}$$

and by using the Lagrangean radiation energy and momentum equations

$$c^{-1}(DJ/Dt) - (4J/3c\rho)(D\rho/Dt) + (\partial H/\partial x) = \kappa(B - J) \tag{101.47}$$

and

$$c^{-1}(DH/Dt) + \tfrac{1}{3}(\partial J/\partial x) = -\kappa H. \tag{101.48}$$

The gas-energy equation (101.31) remains the same as before.

In (101.47) and (101.48), we have made the Eddington approximation, so in (101.46) $P = \frac{1}{3}E = (4\pi/3c)J$. In (101.46) we have neglected the time derivative of H, which is permissible because that term is at most $O(a/c)$ relative to ∇P, which in turn produces terms that are only $O(a/c)$ relative to the dominant terms in the dispersion relation (except at very small Boltzmann numbers).

The linearized continuity equation again yields $(\partial\rho_1/\partial t) = -\rho_0(\partial^2\phi_1/\partial x^2)$, while the linearized material momentum equation is

$$\rho_0(\partial u_1/\partial t) = -(\partial p_1/\partial x) - (4\pi B_0/3c)(\partial j_1/\partial x), \tag{101.49}$$

which implies

$$p_1 = -\rho_0(\partial\phi_1/\partial t) - (4\pi B_0/3c)j_1. \tag{101.50}$$

Using these expressions in the linearized gas-energy equation and material equation of state we find

$$(\partial^2\phi_1/\partial t^2) - a^2(\partial^2\phi_1/\partial x^2) + [4a^3/3c(\gamma-1)\text{Bo}](\partial j_1/\partial t) = (4a^3\kappa/\text{Bo})(4\theta_1 - j_1) \tag{101.51}$$

and

$$(\partial^2\phi_1/\partial t^2) - (a^2/\gamma)(\partial^2\phi_1/\partial x^2) + (a^2/\gamma)(\partial\theta_1/\partial t) + [4a^3/3c(\gamma-1)\text{Bo}](\partial j_1/\partial t) = 0. \tag{101.52}$$

For a plane wave, (101.51) and (101.52) become

$$(a^2k^2 - \omega^2)\phi_1 - (16a^3\kappa/\text{Bo})\theta_1 + (4a^3\kappa/\text{Bo})[1 + i\tfrac{1}{3}(\gamma-1)^{-1}\tau_c^{-1}]j_1 = 0, \tag{101.53}$$

and

$$(a^2k^2 - \gamma\omega^2)\phi_1 + ia^2\omega\theta_1 + (4a^3\kappa/\text{Bo})[i\tfrac{1}{3}\gamma(\gamma-1)^{-1}\tau_c^{-1}]j_1 = 0. \tag{101.54}$$

By analogy with (101.40) we have defined

$$\tau_c \equiv c\kappa/\omega, \tag{101.55}$$

the optical thickness associated with a disturbance of frequency ω traveling at the speed of light (not sound). Notice that $\tau_c : \tau_a = c : a$, hence in an acoustic wave of any appreciable optical thickness $\tau_c \gg 1$.

The linearized radiation equations are

$$(c\kappa)^{-1}(\partial j_1/\partial t)+(4/3c\kappa)(\partial^2\phi_1/\partial x^2)+\kappa^{-1}(\partial h_1/\partial x)=4\theta_1-j_1 \quad (101.56)$$

and

$$(c\kappa)^{-1}(\partial h_1/\partial t)+(3\kappa)^{-1}(\partial j_1/\partial x)=-h_1, \quad (101.57)$$

where we noted that $J_0 \equiv B_0$ and $H_0 \equiv 0$. For plane waves, (101.56) and (101.57) become

$$-(4k^2/3c\kappa)\phi_1-4\theta_1+(1+i\tau_c^{-1})j_1-i(k/\kappa)h_1=0 \quad (101.58)$$

and

$$-i(k/3\kappa)j_1+(1+i\tau_c^{-1})h_1=0, \quad (101.59)$$

which, when combined, yield

$$-(4k^2/3c\kappa)(1+i\tau_c^{-1})\phi_1-4(1+i\tau_c^{-1})\theta_1+[(1+i\tau_c^{-1})^2+(k^2/3\kappa^2)]j_1=0. \quad (101.60)$$

Thus we have

$$\begin{pmatrix} -(4k^2/3c\kappa)(1+i\tau_c^{-1}) & -4(1+i\tau_c^{-1}) & (1+i\tau_c^{-1})^2+(k^2/3\kappa^2) \\ (a^2k^2/\omega^2)-\gamma & ia^2/\omega & i(4a^3\kappa/\omega^2\text{Bo})\tfrac{1}{3}\gamma(\gamma-1)^{-1}\tau_c^{-1} \\ (a^2k^2/\omega^2)-1 & -16a^3\kappa/\omega^2\text{Bo} & (4a^3\kappa/\omega^2\text{Bo})[1+i\tfrac{1}{3}(\gamma-1)^{-1}\tau_c^{-1}] \end{pmatrix} \times \begin{pmatrix} \phi_1 \\ \theta_1 \\ j_1 \end{pmatrix}=0. \quad (101.61)$$

From the determinant of (101.61) we obtain, after some reduction, the dispersion relation

$$\begin{aligned} &[1-i(16\tau_a/\text{Bo})]z^4 \\ &\quad+\{3\tau_a^2(1+i\tau_c^{-1})^2-1+i(16\gamma\tau_a/\text{Bo})+(16a/c\text{Bo})\tau_a^2(1+i\tau_c^{-1}) \\ &\quad\times[5+i\tfrac{1}{3}(\gamma-1)^{-1}\tau_c^{-1}+(16a/3c\text{Bo})\gamma(\gamma-1)^{-1}]\}z^2 \\ &\quad-3\tau_a^2[(1+i\tau_c^{-1})^2+(16\gamma a/c\text{Bo})(1+i\tau_c^{-1})]=0, \end{aligned} \quad (101.62)$$

where $z \equiv ak/\omega$. Equation (101.62) is more complicated than (101.39) and admits a richer variety of wave modes. It is easy to study analytically only in limiting cases. Notice that (101.62) contains yet another dimensionless parameter $r \equiv a/c\text{Bo}$; we consider the cases of small and large r separately.

In most laboratory experiments and familiar stellar astrophysical regimes, temperatures are low enough to guarantee that $r \ll 1$ because $a/c \ll 1$, even though Bo may be much smaller than unity and radiation makes a significant contribution to the energy-momentum balance in the fluid. For example, at the center of the Sun Bo ~ 10, $a/c \sim 10^{-3}$, hence $r \sim 10^{-4}$; at the center of an O-star Bo $\sim 10^{-2}$, $a/c \sim 2\times10^{-3}$, hence $r \sim 0.2$.

In the small-r regime we drop terms in r and r^2 from (101.62), and analyze

$$[1-i(16\tau_a/\mathrm{Bo})]z^4+[3\tau_a^2(1+i\tau_c^{-1})^2-1+i(16\gamma\tau_a/\mathrm{Bo})]z^2-3\tau_a^2(1+i\tau_c^{-1})^2=0. \tag{101.63}$$

It is evident that for large τ_a, (101.63) reduces to (101.39) because $\tau_c \gg \tau_a$. Hence for $\tau_a \gg 1$ the behavior of the modes is essentially the same as discussed above for quasi-static radiation. On the other hand, for $\tau_a \ll \tau_c \lesssim 1$, the time dependence of the radiation field becomes important. In this limit (101.63) factors approximately into

$$(z^2-1)[z^2+3\tau_a^2(1+i\tau_c^{-1})^2]\approx 0. \tag{101.64}$$

Equation (101.64) has two roots: $k \approx \omega/a$, corresponding (formally) to an adiabatic acoustic wave, and

$$k \approx \sqrt{3}\,[(\omega/c)-i\kappa], \tag{101.65}$$

corresponding to a damped radiation wave propagating with speed $c/\sqrt{3}$ (Eddington approximation). This (flux-limited) radiation wave displaces the radiation diffusion wave at moderate-to-small values of τ_c. The geometrical damping length of this mode remains fixed at $L=1/\sqrt{3}\,\kappa$, whereas $L/\Lambda=(1/2\pi\tau_c)\rightarrow\infty$ as $\tau_c\rightarrow 0$. The acoustic mode is also damped; analysis of (101.63) shows that to first order in τ_a we recover (101.42). This result is, of course, only formal, as an acoustic wave cannot exist at frequencies characteristic of light waves because internal processes in the gas invalidate the inviscid continuum description of the fluid at much lower frequencies.

As the temperature of the fluid is raised, (a/c) increases and Bo decreases. Thus the ratio r may eventually become of order unity or greater; for example, in an X-ray source $(a/c)\sim 2\times 10^{-3}$ while $\mathrm{Bo}\sim 10^{-5}$, hence $r\sim 200$. In this regime we must therefore analyze the full dispersion relation (101.62). The analysis shows that for $\tau_a \ll 1$ we recover (101.42) and (101.65), so we again have an attenuated radiation wave and (formally) an acoustic wave propagating at the sound speed of the gas component of the fluid.

The limit $\tau_a \gg 1$ is more interesting. Here we find a weakly damped radiation-dominated acoustic wave with

$$k\approx(\omega/a)\tfrac{4}{3}[a/(\gamma-1)c\mathrm{Bo}]^{1/2}[1-i\tfrac{3}{32}(\gamma-1)(c^2\mathrm{Bo}/a^2\tau_a)], \tag{101.66}$$

which implies

$$v_p/a\approx\tfrac{4}{3}[a/(\gamma-1)c\mathrm{Bo}]^{1/2} \tag{101.67}$$

and $L/\Lambda\approx[16/3\pi(\gamma-1)](a^2/c^2\mathrm{Bo})\tau_a\gg 1$; and a strongly damped, slow, radiation diffusion wave with

$$k\approx(\omega/a)(a/c\mathrm{Bo})[8\gamma\tau_a\mathrm{Bo}/3(\gamma-1)]^{1/2}(1-i), \tag{101.68}$$

which implies $v_p/a\approx(c\mathrm{Bo}/a)[3(\gamma-1)/8\gamma\tau_a\mathrm{Bo}]^{1/2}\ll 1$ and $L/\Lambda=1/2\pi$.

To appreciate (101.67) physically, recall from (101.22) that the sound speed in a radiating fluid is $a_{\text{fluid}} = [(1+\alpha)\Gamma_1/\gamma]^{1/2} a_{\text{gas}}$, where $\alpha \equiv p_{\text{rad}}/p_{\text{gas}} = [4\gamma/3(\gamma-1)](a/c\text{Bo})$. For large r, $\alpha \gg 1$ and $\Gamma_1 \to \frac{4}{3}$; hence (101.67) simply states that the radiation-dominated acoustic mode propagates at the sound speed appropriate for a radiating fluid whose pressure and energy density are dominated by radiation. The acoustic-mode phase speed obtained from numerical solutions of (101.62) for large τ_a does, in fact, agree precisely with a_{fluid} as computed from (101.22).

As shown in Figures 101.4 and 101.5, for $r \ll 1$ the material dominates the dynamical behavior of the fluid, with only one difference from the results given by the quasi-static theory: at small τ_a the fast radiation diffusion mode, found before, is transformed into a propagating radiation

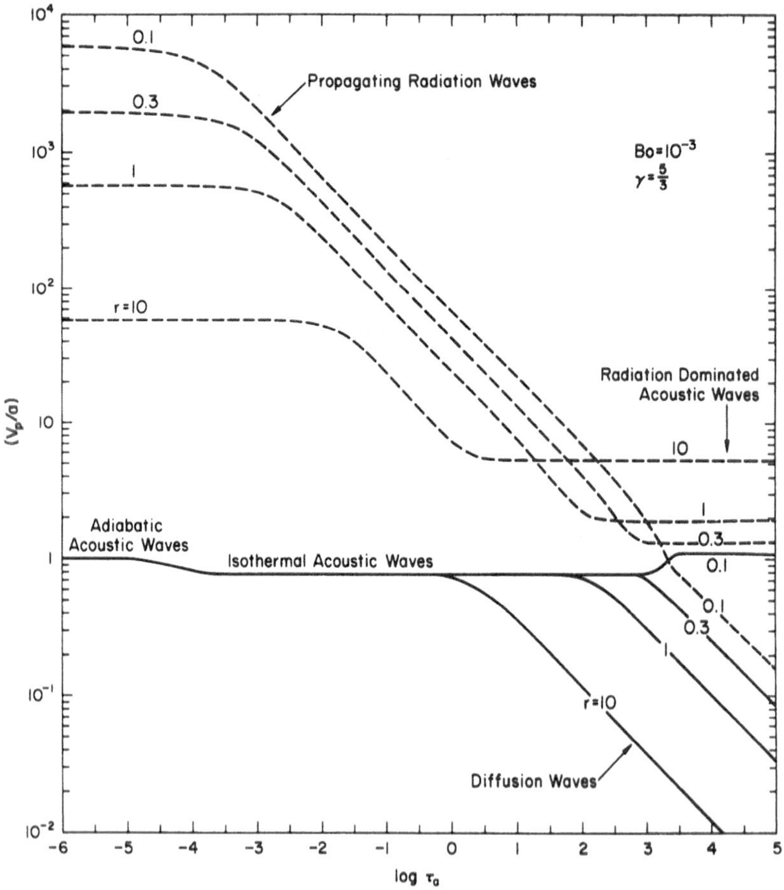

Fig. 101.4 Phase speed for acoustic, diffusion, and propagating radiation modes, allowing for time dependence and dynamical behavior of radiation field.

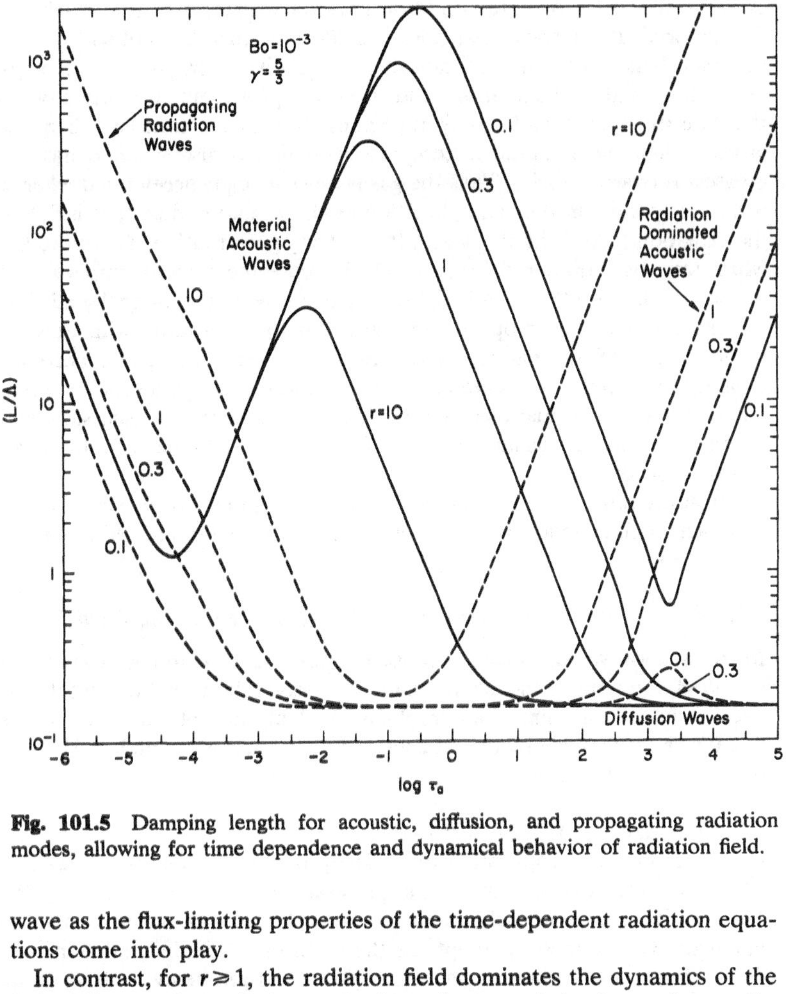

Fig. 101.5 Damping length for acoustic, diffusion, and propagating radiation modes, allowing for time dependence and dynamical behavior of radiation field.

wave as the flux-limiting properties of the time-dependent radiation equations come into play.

In contrast, for $r \gtrsim 1$, the radiation field dominates the dynamics of the fluid. When τ_a is small the modes have the same behavior as for small r because the radiation and material are essentially uncoupled. But as τ_a increases it is the propagating radiation wave that merges continuously into the radiation-dominated acoustic mode, while the material acoustic mode first changes from adiabatic to isothermal and then merges continuously into the slow radiation diffusion mode.

RADIATIVE AMPLIFICATION OF ACOUSTIC WAVES

The propagation of optically thin acoustic waves in a homogeneous medium with a large radiation flux has been analyzed by Hearn (**H1**). He finds that under certain conditions the waves can be amplified by the work done by the radiation force on wave-induced variations of the opacity of the material. If the wave frequency is sufficiently low (but not so low that the disturbances become optically thick), radiative energy exchange obliterates temperature fluctuations and the wave propagates isothermally. In this case the opacity varies only in response to changes in density, hence is largest when the material is compressed, which is also when it has the greatest forward velocity. Thus the gas is most strongly accelerated when it is moving fastest in the same direction as the radiation flux, that is, when the radiation force is in phase with the velocity perturbation. Therefore the work done by radiation forces tends to increase the velocity amplitude of the wave. On the other hand, high-frequency waves are essentially adiabatic, and the decrease in opacity with increasing temperature (which occurs in hot, e.g., stellar, material) more than offsets the density-induced increase; hence these waves are damped by radiative energy losses. Unfortunately many approximations were made in this exploratory discussion, and a complete analysis using consistent Lagrangean radiation equations remains to be done.

An approximate theory describing the development of waves into the nonlinear regime under the action of this mechanism is contained in (**H2**).

102. Propagation of Acoustic-Gravity Waves in a Radiating Fluid

In this section we consider the propagation of acoustic-gravity waves in a stratified radiating atmosphere. Unfortunately relatively little work has been done on this important problem, and at present the state of the analysis is far less complete and consistent than that presented in §101 for pure acoustic waves.

WAVE DAMPING BY NEWTONIAN COOLING

Consider first the propagation of optically thin acoustic-gravity waves in which the radiative energy exchange produces Newtonian cooling. We assume the radiation field is quasi-static, hence ignore its dynamical behavior. Under these assumptions the main effect of the radiation is to damp the waves. An additional effect, as we will see below, is that we no longer obtain either pure progressive or pure standing (evanescent) waves separated crisply into distinct regions in the diagnostic diagram as in the adiabatic case discussed in §53.

(*a*) *Isothermal Atmosphere* Following Souffrin (**S16**), (**S17**) we first assume a planar isothermal atmosphere composed of a perfect gas having constant specific heats. The linearized gas energy equation (101.3) then reduces to

$$(\partial p_1/\partial t)+w_1(dp_0/dz)-a^2[(\partial \rho_1/\partial t)+w_1(d\rho_0/dz)]=-t_{RR}^{-1}[p_1-(a^2/\gamma)\rho_1]. \tag{102.1}$$

For simplicity we assume that the radiative relaxation time t_{RR} is constant with height. Assuming that ρ_1, p_1, and w_1 are of the form (53.30) we can reduce (102.1) to

$$i\omega P[1-(i/\omega t_{RR})]-ia^2\omega R[1-(i/\gamma\omega t_{RR})]+a^2(\omega_g^2/g)W=0, \quad (102.2)$$

which differs from (53.31d) only by the imaginary terms in the coefficients of $i\omega P$ and $ia^2\omega R$. Here we used the fact that $\omega_g^2=(\gamma-1)g/\gamma H$ for a perfect isothermal gas.

Although we can again use (53.30) for P, R, W, U, and Θ, the vertical wavenumber k_z will now be complex. For this reason we replace, for the time being, ik_zW and ik_zP in (53.31a) and (53.31c) by $-(dW/dz)$ and $-(dP/dz)$, and combine those equations with (102.2) to obtain the following differential equation for W:

$$\begin{aligned}\{(\omega^2/a^2)-[1-(\omega_g^2/\omega^2)]k_x^2-(1/4H^2)+(d^2/dz^2)\\ -(i/\gamma\omega t_{RR})[(\gamma\omega^2/a^2)-k_x^2-(1/4H^2)+(d^2/dz^2)]\}W=0\end{aligned} \quad (102.3)$$

or

$$\{h_0+(d^2/dz^2)-(i/\gamma\omega t_{RR})[h_0+(d^2/dz^2)+(\gamma-1)(\omega^2/a^2)-(\omega_g^2/\omega^2)k_x^2]\}W=0. \quad (102.4)$$

Identical equations hold for P or R. In the isothermal, adiabatic limit $h_0=k_z^2$ [see (54.89)].

Again following Souffrin we note that in general we can write $k_z = k_R+ik_I$ and, assuming $W\propto\exp(-ik_zz)$,

$$(d^2W/dz^2)\equiv-(h_R+ih_I)W=-(k_R+ik_I)^2W=-[(k_R^2-k_I^2)+2ik_Ik_R]W. \quad (102.5)$$

Then substituting $-(h_R+ih_I)W$ for (d^2W/dz^2) in (102.4) we find that h_R and h_I are given by

$$h_I=\gamma\omega t_{RR}(1+\gamma^2\omega^2t_{RR}^2)^{-1}[(\omega_g^2/\omega^2)k_x^2-(\gamma-1)(\omega^2/a^2)] \quad (102.6)$$

and

$$\begin{aligned}h_R=[(\omega_g^2/\omega^2)-1]k_x^2+a^{-2}(\omega^2-\omega_a^2)\\ -(1+\gamma^2\omega^2t_{RR}^2)^{-1}[(\omega_g^2/\omega^2)k_x^2-(\gamma-1)(\omega^2/a^2)],\end{aligned} \quad (102.7)$$

where, as in §52, $\omega_a\equiv a/2H$. The real and imaginary parts of k_z are determined from $h_I=2k_Rk_I$ and $h_R=k_R^2-k_I^2$, which yield

$$k_R^2=\tfrac{1}{2}[h_R+(h_R^2+h_I^2)^{1/2}] \quad (102.8)$$

and

$$k_I^2=\tfrac{1}{2}[-h_R+(h_R^2+h_I^2)^{1/2}]. \quad (102.9)$$

The positive sign was chosen for the radical to make both k_R^2 and k_I^2 positive (i.e., k_R and k_I real) whether h_R is positive or negative.

There remains an ambiguity about which sign of $(k_R^2)^{1/2}$ and $(k_I^2)^{1/2}$ to choose. Souffrin imposes the requirement that the energy flux be positive in the positive z direction, that is, he requires that energy be carried upward from a source. The vertical component of the energy flux is given by

$$(\phi_w)_z = \tfrac{1}{4}(P^*W + PW^*) = \frac{WW^*\omega k_R(\omega^4 - g^2k_x^2)}{\omega^4\{k_R^2 + [k_I - (1/2H) + (gk_x^2/\omega^2)]^2\}} \tag{102.10}$$

which is positive if and only if

$$\omega k_R(\omega^4 - g^2k_x^2) > 0. \tag{102.11}$$

If we multiply both sides of $h_I = 2k_Rk_I$ by $\omega k_R(\omega^4 - g^2k_x^2)$ and use (102.6) we find

$$\begin{aligned} k_Rk_I\omega(\omega^4 - g^2k_x^2) &= \tfrac{1}{2}h_I\omega(\omega^4 - g^2k_x^2) \\ &= -[\gamma t_{RR}\omega_g^2/2g^2(1+\gamma^2\omega^2t_{RR}^2)](\omega^4 - g^2k_x^2)^2 < 0, \end{aligned} \tag{102.12}$$

which is negative because both factors in the right-most expression are intrinsically positive. Comparing (102.11) and (102.12) we conclude that $k_I \leq 0$. Moreover we see that gravity waves, for which $g^2k_x^2 > \omega^4$, have $k_R < 0$ for upward propagation of energy, whereas acoustic waves, for which $\omega^4 > g^2k_x^2$, have positive k_R, which was also the case for adiabatic acoustic-gravity waves as discussed in §54.

From (102.8) and (102.9) we see that when $h_R = 0$, $|k_R| = |k_I|$; when $h_R > 0$, $|k_R| > |k_I|$; and when $h_R < 0$, $|k_R| < |k_I|$. Thus when $h_R < 0$, the waves are heavily damped over a single vertical wavelength, and were classified by Souffrin as *mainly damped* or *mainly evanescent*, whereas waves with $h_R > 0$ are classified as *mainly propagating*. The boundaries separating these propagating and evanescent regions in the diagnostic diagram are defined by the curves $h_R = 0$, and are shown in Figure 102.1 for several values of t_{RR}, ranging from 0 (isothermal) to ∞ (adiabatic). Damped, propagating acoustic waves lie above the upper curve, which asymptotes at small values of k_x to

$$\omega_{aN}^2(t_{RR}) = \tfrac{1}{2}[[\omega_a^2 - (1/\gamma t_{RR}^2) + \{[\omega_a^2 - (1/\gamma t_{RR}^2)]^2 + (4\omega_a^2/\gamma^2 t_{RR}^2)\}^{1/2}]], \tag{102.13}$$

which is the effective acoustic cutoff frequency in the Newtonian cooling approximation. Note that ω_{aN} varies with t_{RR}. The curve bounding the region of propagating low-frequency waves asymptotes at large k_x to

$$\omega_{gN}^2(t_{RR}) = \omega_g^2 - (1/\gamma^2 t_{RR}^2), \tag{102.14}$$

the effective gravity-wave cutoff frequency in the Newtonian cooling approximation. Thus if $\omega_g < (1/\gamma t_{RR})$, gravity waves cannot propagate even when the atmosphere is convectively stable. Equation (102.14) is modified

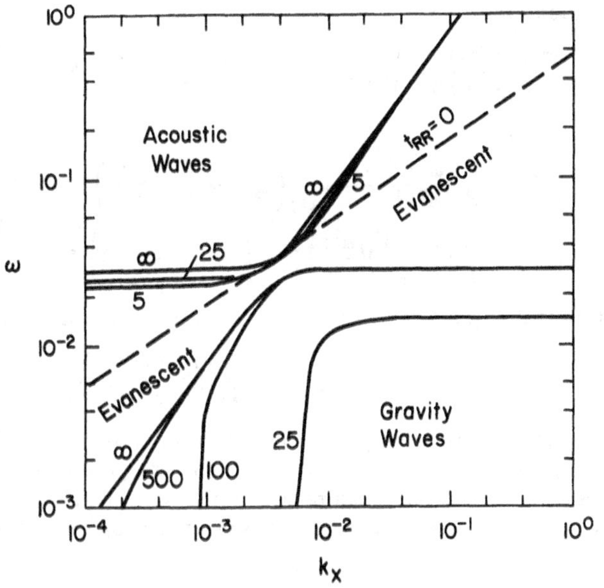

Fig. 102.1 Diagnostic diagram showing boundary curves between mostly propagating acoustic waves (upper left), mainly evanescent waves, and mostly propagating gravity waves (lower right), labeled by the value of t_{RR}.

when ω is allowed to be complex, and Stix finds (**S24**) that an impulsively generated packet of gravity waves can propagate if $\omega_g > (1/2\gamma t_{RR})$, a slightly less stringent condition.

In the expression for the energy flux, equation (102.10), WW^* is not constant with z but contains the factor $\exp(-2|k_I|z)$, hence the wave energy flux diminishes with increasing height. When $|k_I|$ is large, the decrease is very rapid. In the limit of instantaneous temperature smoothing, that is, when $t_{RR} \to 0$, we find $h_I \to 0$ and

$$h_R \to (\gamma-1)(\omega^2/a^2) - (\omega_a^2/a^2) - k_x^2, \tag{102.15}$$

which yields the dispersion relation for propagating isothermal sound waves; in this limit gravity waves are absent altogether.

Physically these results are not surprising, because (in the absence of gradients in the composition of the gas) the buoyancy force that drives gravity waves arises solely from horizontal temperature fluctuations, which vanish when $t_{RR} \to 0$. In contrast, acoustic waves are driven by pressure gradients whether there are associated temperature perturbations or not.

The polarization relations (53.32) are also modified by radiative energy exchange, and can change drastically when t_{RR} is small. Defining $\alpha \equiv (\gamma\omega t_{RR})^{-1}$ and $r \equiv (\gamma\omega^2 - a^2k_x^2)/(\omega^2 - a^2k_x^2)$, we find the polarization relations for waves in an isothermal atmosphere under the Newtonian cooling approximation are

$$P=\frac{\omega a^2(1+ir\alpha)}{(\omega^2-a^2k_x^2)(1+r^2\alpha^2)}\left\{k_R+\alpha\left(k_I-\frac{1}{2H}\right)+i\left[\left(\frac{\gamma-2}{2\gamma H}\right)+k_I-\alpha k_R\right]\right\}W, \tag{102.16}$$

$$R=\frac{\omega(1+ir\alpha)}{(\omega^2-a^2k_x^2)(1+r^2\alpha^2)}\left\{k_R+\gamma\alpha\left(k_I-\frac{1}{2H}\right)\right. \\ \left.+\frac{i}{H}\left[\left(\frac{\gamma-1}{\gamma}\right)\left(\frac{a^2k_x^2}{\omega^2}\right)-\frac{1}{2}\right]+i(k_I-\gamma\alpha k_R)\right\}W, \tag{102.17}$$

and, noting that $(T_1/T_0)=(p_1/p_0)-(\rho_1/\rho_0)$ implies $\Theta=(\gamma P/a^2)-R$,

$$\Theta=\frac{\omega(\gamma-1)(1+ir\alpha)}{(\omega^2-a^2k_x^2)(1+r^2\alpha^2)}\left\{k_R+ik_I+\frac{i}{H}\left[\frac{1}{2}-\frac{1}{\gamma}\left(\frac{a^2k_x^2}{\omega^2}\right)\right]\right\}W. \tag{102.18}$$

The leading real and imaginary terms in (102.16) to (102.18) are the same as in (53.32). The quantity r is of the order unity except when ω^2 or $\gamma\omega^2$ is nearly equal to $a^2k_x^2$, while α can range from very small values (for nearly adiabatic propagation) to very large values when $2\pi\gamma t_{RR}$ is much less than a wave period.

(*b*) *Solar Model Atmosphere* In a nonisothermal atmosphere, use of the Newtonian cooling approximation provides a simple but, unfortunately, inconsistent method for studying the interaction between linear waves and the radiation field. Some of the inconsistency arises from the fact that the model atmospheres [e.g., HSRA (**G3**) or VAL (**V4**), (**V5**)] chosen to represent the ambient medium in which the waves propagate are not in radiative equilibrium. Because the physical mechanisms that determine the temperature structure of the solar atmosphere are not actually known, we have little choice but to include an unspecified nonradiative source–sink term in the gas energy equation and write

$$\frac{Dp}{Dt}-a^2\frac{D\rho}{Dt}=(\Gamma_3-1)\left[4\pi\int_0^\infty \kappa_\nu(J_\nu-S_\nu)\,d\nu-\nabla\cdot\mathbf{F}_{\rm nr}\right] \tag{102.19}$$

Here $\mathbf{F}_{\rm nr}$ represents some sort of nonradiative energy flux chosen such that $\nabla\cdot\mathbf{F}_{\rm nr}$ exactly balances the net radiative gains and losses in the static atmosphere, that is,

$$4\pi\left[\int_0^\infty \kappa_\nu(J_\nu-S_\nu)\,d\nu\right]_0=(\nabla\cdot\mathbf{F}_{\rm nr})_0. \tag{102.20}$$

Because we do not know how to write $\mathbf{F}_{\rm nr}$, we cannot do more than guess at how a wave-induced perturbation $(\mathbf{F}_{\rm nr})_1$ would depend on T_1 and ρ_1. Therefore, in the linearized gas energy equation we have no choice but to ignore this term altogether.

A second problem is that whenever the departure from radiative equilibrium in the ambient atmosphere is large, the term $\int \kappa_{\nu 1}(J_\nu-S_\nu)_0\,d\nu$, which is ignored in the Newtonian cooling approximation, can be large and important (cf. §100). Furthermore, in the Newtonian cooling formulation, the net radiative gain term $4\pi \int \kappa_{\nu 0}(J_\nu - S_\nu)_1 \, d\nu$ is approximated in terms of the *local* cooling time, which is derived as if at each height the atmosphere were optically thin over a wavelength and infinite, isotropic, homogeneous, and isothermal at the local temperature. The cooling time is then calculated from (100.17) and (100.10) using the local values of $T(z)$, $\rho(z)$, $\kappa(z)$, etc. at each height; except in the low photosphere $(\kappa_0/k) \approx 0$. However as can be seen from Figure 54.1, each of the assumptions underlying (100.17) is poor in some region of the atmosphere, with the most serious errors occurring at continuum optical depths $\tau_c \sim 10^{-2}$ to 1, where both $T(z)$ and $\kappa(z)$ vary rapidly with height, and where the gas is neither very optically thick nor optically thin. Unfortunately this is also the region where the radiative damping effects are the most important.

Retracing the arguments of §100 it is clear that (100.17) is a poor approximation to the solution of (100.7) for a highly inhomogeneous and anisotropic medium. In particular, in the solar photosphere and chromosphere the perturbations of J are essentially determined by the perturbations in S at an optical depth of about unity, not locally. Hence the nonlocally driven part of J_1 (which is lost in the local cooling-time formulation) may dominate over the locally driven part.

In treating the propagation of linear acoustic-gravity waves in the solar atmosphere we can take into account the variations of temperature, density, sound speed, buoyancy frequency, and ionization properties of the atmosphere. The resulting variations with height of the real and imaginary parts of the vertical "wavenumber" of a wave of given ω and k_x imply height-dependent variations in all properties of the wave. To describe radiative-exchange effects the best treatments available all use the Newtonian cooling approximation despite the criticisms we have just leveled at it; we merely caution the reader to remember the caveats expressed above when evaluating the results of this work. Whether the results obtained are even qualitatively correct can be determined ultimately only by computations that treat the radiation field self-consistently with the fluid equations.

If we assume the density to be fixed by the requirement of hydrostatic equilibrium, then, as in §54, the density is given by (54.75) with $H(z)$ defined by (54.68). [See (**M8**) for a discussion about $\rho(z)$, H, ω_{BV}^2, and a when a "turbulent pressure" is included in the model.] The amplitude functions are again as in (54.77) with $E(z)$ defined by (54.76), and the Brunt–Väisälä frequency $\omega_{\mathrm{BV}}(z)$ is given by (54.67), $H_p(z)$ by (54.69), and $\Gamma_1(z)$ by (14.19).

The linearized continuity and momentum equations are unchanged from their adiabatic forms (54.78a) and (54.78b). The linearized energy equation, written in terms of amplitude functions, becomes

$$i\omega[1-(i/\omega t_{RR})]P(z) - i\omega a^2[1-(i/\Gamma\omega t_{RR})]R(z) + (a^2\omega_{\mathrm{BV}}^2/g)W(z) = 0, \tag{102.21}$$

or

$$i\omega(1-i\alpha\Gamma)P(z)-i\omega a^2(1-i\alpha)R(z)+(a^2\omega_{BV}^2/g)W(z)=0; \qquad (102.22)$$

here $\alpha\equiv(\Gamma\omega t_{RR})^{-1}$ and $\Gamma\equiv c_p/c_v$ including ionization effects as in (14.15) and (14.18). Combining (102.22) with (54.78a) and (54.78b) yields the following rather unwieldy equation for $P(z)$:

$$\begin{aligned}&\left\{\left(\frac{\omega_{BV}^2}{\omega^2}-1\right)k_x^2+\frac{\omega^2}{a^2}-\frac{1}{4H^2}+\left(\frac{d^2}{dz^2}\right)\right.\\&\left.-i\alpha\left[\frac{\Gamma\omega^2}{a^2}-k_x^2-\frac{1}{4H^2}+\left(\frac{d^2}{dz^2}\right)+\left(\frac{1}{H}-\frac{1}{H_p}\right)\left(\frac{d}{dz}+\frac{1}{2H}\right)\right]\right\}P(z)\\&+\left\{\frac{d}{dz}\left(\frac{1}{2H}\right)-\frac{d\ln a^2}{dz}\left(\frac{d}{dz}-\frac{1}{2H}\right)\right.\\&-i\alpha\left[\frac{d}{dz}\left(\frac{1}{2H}\right)-\left(\frac{d\ln a^2}{dz}+\frac{d\ln\alpha}{dz}\right)\left(\frac{d}{dz}-\frac{1}{2H}\right)\right.\\&\left.\left.-\frac{ig\Gamma}{a^2}\left(\frac{d\ln\alpha}{dz}+\frac{d\ln\Gamma}{dz}\right)\right]\right\}P(z)=0.\end{aligned} \qquad (102.23)$$

The expression in the first set of braces is identical to that in (102.3) except that ω_{BV}^2 and Γ replace ω_g^2 and γ, and a new term, which is identically zero in an isothermal medium where $H\equiv H_p$, has appeared. The second set of braces contains the derivative terms that arise because all the atmospheric properties now vary with height. The terms containing derivatives of a^2, ω_{BV}^2, Γ, and $(1/2H)$ are not exactly the same in the corresponding equations for $W(z)$ and $R(z)$.

To obtain numerical results for a realistic model atmosphere it is easier not to use (102.23) but to either (1) solve (102.22), (54.78a), and (54.78b) simultaneously at all depth points or (2) regard the atmosphere as a set of thin layers, chosen such that no atmospheric property changes much through a layer, and that the layer thickness is small compared to a vertical wavelength of the wave being studied. Within each layer all atmospheric properties are taken to be constant, but the derivatives (dT/dz) and $(d\mu/dz)$ obtained from the model are used to calculate H and ω_{BV}^2 (which in turn are taken to be constant within a layer).

We will describe results obtained by using method (2). All terms in the first set of braces in (102.23) were calculated as they would be for a continuous model; the value for each layer was then chosen to be either the midpoint of the layer or at an interface. All terms in the second set of braces were assumed to be negligible and were dropped. Because the approximations inherent in the original equations imply that at best only qualitative results will be obtained, the small inaccuracies incurred in this implementation of the layer method are not important.

We will discuss only gravity waves, for which the region of propagation lies above about 100 km, where $t_{RR}>1/\Gamma\omega_{BV}$ [cf. (102.14)]. The adopted radiative relaxation times are given by the linear relation shown in Figure 102.2, which closely approximates the relaxation times computed by Stix

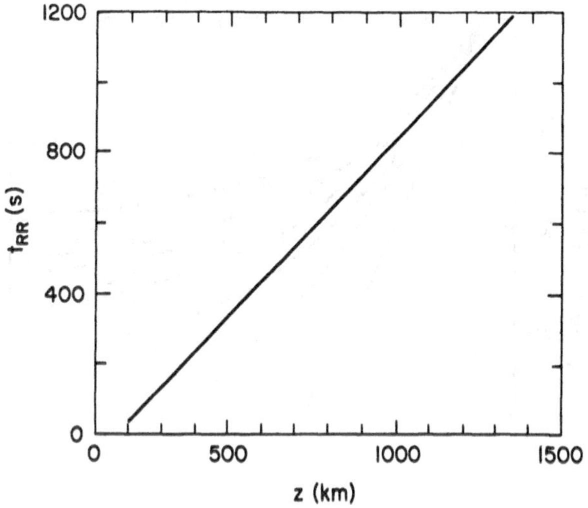

Fig. 102.2 Representative run of radiative relaxation time in a model solar atmosphere.

(**S24**) in the region 100 to 500 km, and provides a continuous extension into the chromosphere where radiative relaxation times are not known very accurately (cf. §100). For the smallest values of t_{RR} that still exceed $(\Gamma\omega_{BV})^{-1}$, the real part of k_z is much smaller than its adiabatic value, hence the vertical wavelength of the wave increases and the group velocity decreases. The change is greatest for waves that already have small values of k_z in the adiabatic limit.

Energy loss occurs most rapidly from gravity waves with large k_z, that is, from waves with large k_x for a given ω, or with small ω for a given k_x. This fact is evident in Figure 102.3, where waves A to D all have $\Lambda_x = 2000$ km and form a sequence of decreasing frequency, whereas waves D to G have a fixed period of 500 s, and form a sequence with decreasing k_x. The least damped waves are A, B, and G, while the energy flux decreases most rapidly with height for D.

Recalling (53.24b) for the vertical energy flux, we see that the energy flux can decrease both as a result of a decrease in the amplitudes of p_1 and w_1, and from an increase in the phase lag $|\delta_{PW}|$ toward $\pi/2$. The phase lags and relative amplitudes of the perturbations can be found from the polarization relations, which are now given by

$$P = \frac{ia^2\omega(\kappa_1 + i\alpha\kappa_2)}{(\kappa_1^2 + \alpha^2\kappa_2^2)}\left[\frac{\omega_{BV}^2}{g} - (1 - i\alpha)\left(\frac{1}{2H} - \frac{d}{dz}\right)\right]W, \qquad (102.24)$$

$$R = \frac{i\omega(\kappa_1 + i\alpha\kappa_2)}{(\kappa_1^2 + \alpha^2\kappa_2^2)}\left[\frac{a^2\omega_{BV}^2 k_x^2}{g\omega^2} - (1 - i\Gamma\alpha)\left(\frac{1}{2H} - \frac{d}{dz}\right)\right]W, \quad (102.25)$$

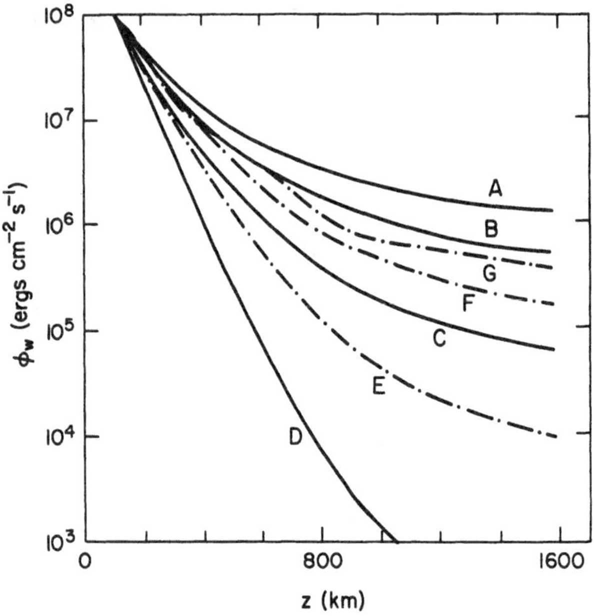

Fig. 102.3 Height dependence of energy flux for radiatively damped gravity waves with horizontal wavelength 2000 km and periods of 700, 800, 1000, and 1500 s (curves A to D), and for gravity waves with period 1500 s and horizontal wavelengths 2000, 3000, 5000, and 7000 km (curves D to G).

and

$$\Theta = \frac{1}{Q}\left(\frac{\Gamma}{a^2}P - R\right) = \frac{i\omega(\kappa_1 + i\alpha\kappa_2)}{(\kappa_1^2 + \alpha^2\kappa_2^2)}\left[\frac{\omega_{BV}^2\kappa_2}{g\omega^2} + (\Gamma-1)\left(\frac{d}{dz} - \frac{1}{2H}\right)\right]W, \quad (102.26)$$

where $\kappa_1 \equiv \omega^2 - a^2k_x^2$, $\kappa_2 \equiv \Gamma\omega^2 - a^2k_x^2$, and Q is given by (14.33).

When $\alpha > 1$, the terms $(\kappa_1 + i\alpha\kappa_2)$ and $(1 - i\alpha)$ are dominated by the imaginary part. In the low photosphere (102.14) implies that t_{RR} must be greater than about 25 s for gravity waves to propagate, and we find that $\alpha > 1$ for $25\text{ s} < t_{RR} < 100\text{ s}$ if the wave period is 600 s, and $\alpha > 1$ for $25\text{ s} < t_{RR} \leq 170\text{ s}$ if the period is 1000 s. Each of the complex terms, and $(d/dz) \rightarrow -ik_{zR} + k_{zI}$, can thus show a phase change from the adiabatic value that approaches $\pi/2$ when $t_{RR} \approx 25$ s, particularly if the wave period is large.

The polarization relation for Θ is the simplest of the three, especially when the vertical wavenumber is small; then the phase lag is determined mainly by $i\omega(\kappa_1 + i\alpha\kappa_2)\kappa_2$. For $\omega^2 \ll a^2k_x^2$, $\kappa_1 \approx \kappa_2 \approx -a^2k_x^2$ and $\Theta \approx |C|\, i(1 + i\alpha)W$, hence $\sin\delta_{TW} \approx (1+\alpha^2)^{-1}$, $\cos\delta_{TW} \approx -\alpha(1+\alpha^2)^{-1}$, and $\tan\delta_{TW} \approx -1/\alpha$. Thus δ_{TW} tends to $\pi/2$ as $\alpha \rightarrow 0$ (adiabatic) and $\delta_{TW} \rightarrow \pi$

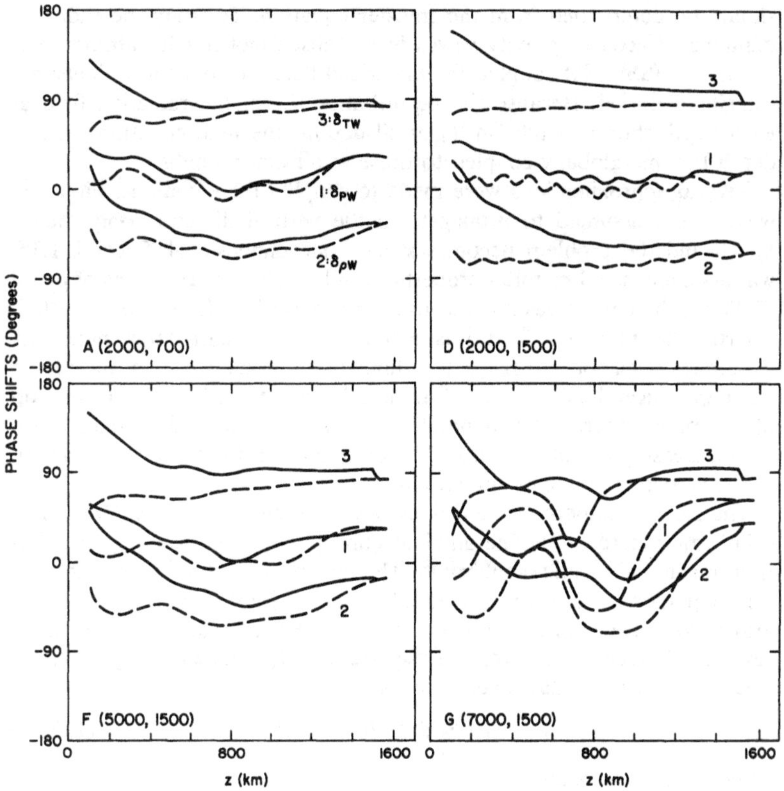

Fig. 102.4 Phase shifts for gravity waves in model solar atmosphere. (1) δ_{PW}. (2) $\delta_{\rho W}$. (3) δ_{TW}. *Solid curves* allow for radiative damping in Newtonian cooling approximation. *Dashed curves* are for adiabatic waves.

as $\alpha \to \infty$ (strong radiative damping). Precisely this behavior is seen in Figure 102.4, where the adiabatic and radiatively damped phase lags are compared for several waves.

The polarization relations for P and R are more complicated, each containing three terms that are real in the adiabatic limit and complex when radiative damping occurs. The values of δ_{RW} and δ_{PW} in Figure 102.4 for radiatively damped waves both reflect the interplay of the three complex terms, as well as some interference (much less than in the adiabatic case) between upward-propagating and reflected waves.

WAVE DAMPING BY RADIATIVE TRANSPORT

Near the open boundary of a radiating medium such as a stellar atmosphere the radiation terms in the energy equation are strongly nonlocal, and should be determined from the transfer equation. To study the radiative damping of acoustic-gravity waves in a realistic model solar atmosphere, Schmieder (**S5**), (**S6**) coupled the linearized fluid equations to a linearized transfer equation. Because the perturbation $J_{\nu 1}$ of the radiation field at each depth then depends on T_1 at all depths, the fluid equations at one depth become globally coupled to those at all other depths.

Several approximations were made to simplify the equations. First, the waves were assumed to propagate in the vertical direction only, hence $k_x \equiv 0$ and the problem becomes strictly one dimensional. Second, LTE was assumed, so all material properties can be written as functions of ρ and T. Third, the perfect gas law was used and ionization effects are neglected. Fourth, the radiation field was assumed to be quasi-static, hence its dynamical behavior was ignored. Schmieder analyzed waves in the region from an optical depth of about 3 up to a height of 500 km, that is, the top of the photosphere. The assumption of linearity gradually breaks down with increasing height, but should be quite good in the lower layers where radiative damping is most important.

The run of temperature and density was taken from the HSRA model solar atmosphere (**G3**). The ambient atmosphere is thus neither in radiation nor in hydrostatic equilibrium. The linearized continuity and momentum equations are given by (52.24) and (52.25) (ignoring the nongravitational force term that is implicit in the HSRA model). These can be rewritten in terms of the vertical displacement ζ_1, defined by $w_1 = (\partial \zeta_1 / \partial t)$. Thus integrating (52.24) over time we have

$$\rho_1 + \zeta_1 (d\rho_0/dz) + \rho_0(\partial \zeta_1/\partial z) = 0, \tag{102.27}$$

while (52.25) becomes

$$\rho_0(\partial^2 \zeta_1/\partial t^2) = -(\partial p_1/\partial z) - \rho_1 g. \tag{102.28}$$

The gas-energy equation (ignoring the nonradiative term that drives the ambient atmosphere out of radiative equilibrium) becomes

$$\begin{aligned} &\frac{p_0}{\rho_0 T_0}\left(\frac{\partial T_1}{\partial t} + \frac{\partial \zeta_1}{\partial t}\frac{dT_0}{dz}\right) - \frac{(\gamma-1)p_0}{\rho_0}\frac{\partial}{\partial z}\left(\frac{\partial \zeta_1}{\partial t}\right) \\ &\quad = 4\pi(\gamma-1)\left[\int_0^\infty \kappa_{\nu 0}(B_{\nu 1} - J_{\nu 1})\, d\nu + \int_0^\infty \kappa_{\nu 1}(B_{\nu 0} - J_{\nu 0})\, d\nu\right]. \end{aligned} \tag{102.29}$$

The first term on the right-hand side results from the wave-induced perturbations of the mean intensity and the local Planck function. The second term arises from two effects: (1) the perturbation of the opacity, and (2) the departure of the ambient atmosphere from radiative equilibrium, which implies that $J_0 \neq B_0$. When $J_0 - B_0$ is large, the second term, which is strongly model dependent, may also be large; thus errors in the assumed $T(z)$ of the unperturbed model may produce important errors in the radiative damping calculations.

The mean intensity is approximated by the average of the incoming and outgoing intensity calculated from the formal solution (79.14) and (79.15) for a representative ray, using the Planck function from the model atmosphere. The radiative energy exchange term in (102.29) is approximated by a quadrature sum over a few representative frequencies. Perturbed intensities are computed from the linearized formal solution, which is constructed recursively between successive depth points in a discrete mesh. To pose boundary conditions for the transfer equation, it is assumed that there is no incoming radiation at the top of the atmosphere, whereas the bottom of the atmosphere is taken deep enough that the diffusion approximation can be applied. The hydrodynamic boundary conditions are imposed at the upper boundary where the value of ζ_1 is fixed; all perturbations are normalized to this value. Initial relations between T_1 and ζ_1 at the uppermost two grid points are obtained by assuming adiabatic motion and a pure outgoing wave at the top of the atmosphere. The resulting matrix equation is solved numerically and iterated to consistency.

In Schmieder's solution the general form of the radiative exchange term at depth point i is

$$\dot{q}_i = \sum_j A_{ij} T_{1,j} + \sum_j B_{ij} \rho_{1,j}. \tag{102.30}$$

The matrix B_{ij} is almost diagonal. The matrix A_{ij}, however, clearly reveals the strong nonlocal effect of the radiation-field perturbations produced by temperature perturbations up to about one photon mean free path away from the chosen depth i. For heights above about 200 km the dominant contributions to $\dot{q}_i$ from A_{ij} come from (1) a region between about -40 km and $+140$ km, and (2) the local temperature perturbation, which produces the diagonal elements. The departure of the ambient atmosphere from radiative equilibrium also makes an important contribution to the diagonal elements. Below 200 km, the opacity increases exponentially, and the largest off-diagonal elements in A_{ij} become confined to a small range of depths immediately above and below the point z_i. At large optical depths, a photon mean free path becomes small compared to the grid spacing and A_{ij} becomes essentially diagonal.

Schmieder finds that the wave amplitudes increase less rapidly with height for radiatively damped waves than for adiabatic waves. Evanescent waves are less affected by radiative damping than propagating waves; therefore although the amplitude of an adiabatic evanescent wave grows more slowly with height than that of a propagating wave, the difference in growth rates between radiatively damped evanescent and propagating waves is not large. For example, for a 140-s propagating nonadiabatic wave, the relative temperature perturbation ($|T_1|/T_0\,|w_1|$) is distinctly suppressed compared to that of an adiabatic wave in the region of strong damping. For a 300 s evanescent wave, however, the relative temperature perturbation is actually larger in the nonadiabatic case than in the adiabatic case.

In the isothermal Newtonian-cooling approximation, one can show that $|T_1|/T_0|w_1|$ is always decreased by radiative damping, whether the wave propagates or is evanescent. Schmieder's result for the 300 s wave thus arises from a nonlocal effect. The wavelength of an evanescent wave is very long, hence the phase changes little from the bottom to the top of the photosphere. Therefore transfer of radiation from around optical depth unity to greater heights tends to enhance $|T_1|/T_0$ relative to $|w_1|$.

In a second paper (**S6**) Schmieder discusses the possibility of determining an "equivalent damping time", which could be used in the Newtonian cooling formalism, at each height in the atmosphere. She first derives at each height the damping time that would cause the displacement amplitude to decrease locally at the same rate as given by the nonlocal computation; these are shown as curves "A" in Figure 102.5 for waves of different periods. However, as we saw above for Newtonian cooling in an isothermal atmosphere, radiative damping also alters the phase relations among the perturbation variables. Schmieder derives another equivalent damping time for each wave from the changes in phase lags; these are shown as curves "B" in Figure 102.5, with error bars resulting from the uncertainties in

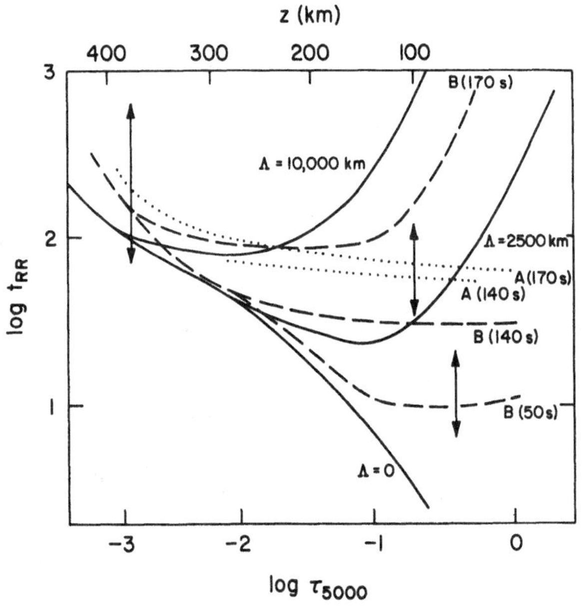

Fig. 102.5 Equivalent damping times derived from amplitude variations (dotted curves) and phase variations (dashed curves) of acoustic waves of indicated periods. Radiative damping is calculated from a solution of the transfer equations; error bars (vertical arrows) indicate uncertainties in phase curves. solid curves show local damping times computed from Spiegel's formula. From (**S6**), by permission.

assigning phases in waves that vary strongly with height. The two sets of curves give markedly different results. Had damping times been estimated from amplitude ratios, yet different results would have been obtained.

For comparison, damping times computed for waves of the same periods using Spiegel's formula (100.17) are shown in Figure 102.5 as solid curves. The important conclusion to be drawn is that nonlocal effects resulting from radiative transfer produce *qualitatively* different propagation characteristics in a wave, and that it simply is not possible to reproduce these effects self-consistently with a single equivalent damping time.

STABILITY OF ACOUSTIC-GRAVITY WAVES IN A RADIATING FLUID

The possibility of radiation-driven instabilities in a stratified radiating atmosphere has been discussed by Berthomieu et al. (**B4**) and by Spiegel (**S19**). In the former paper, it is shown that above a certain critical frequency, isothermal, optically thin perturbations in an isothermal slab of an atmosphere traversed by a radiation field can be amplified by radiation in a drift instability. In the latter paper it is shown that under certain circumstances radiation forces can drive instabilities in a stratified radiating fluid. A detailed analysis is presented for quasi-adiabatic *photoacoustic* and *photogravity* modes. The results are intriguing, but it would take us too far afield to discuss them here; the interested reader should consult the original paper.

8.2 Nonlinear Flows

103. Thermal Waves

Thermal waves result from conductive energy-transport processes within a fluid, which give rise to an energy flux $\mathbf{q} = -K\,\nabla T$. For nonradiating neutral gases it is usually satisfactory to assume *linear conduction* (K independent of T) because the conductivity depends only weakly on temperature (cf. §33). But in ionized plasmas where $K \propto T^{5/2}$, and in opaque radiating fluids where the radiation conduction coefficient depends strongly on T, we must treat *nonlinear conduction.* The distinction is important because thermal waves behave qualitatively differently in the two cases.

A problem of some interest in radiation hydrodynamics is the penetration of radiation from a hot source into cold material, a process that is reasonably well described by treating the radiation field in the diffusion approximation. Practical examples are the penetration of stellar radiation into the interstellar medium at the instant of star formation or of a supernova explosion, or the irradiation of a fusion pellet by intense laser beams. Such propagating radiation fronts are called *Marshak waves* (**M4**) or *radiation diffusion waves.*

Because radiative energy exchange is very efficient, significant radiation penetration and energy deposition can occur in a time much too short for the fluid to be set into motion. In this section we therefore consider the penetration of radiation into a *static medium*. Eventually, of course, the material becomes hot, pressure gradients build, and the fluid flows; we examine the penetration of radiation into a moving medium in §106. In order to gain insight we will emphasize simple problems for which analytical solutions are possible, and then compare with numerical results obtained using techniques discussed in §97.

BEHAVIOR OF LINEAR CONDUCTION WAVES

In a static conducting fluid, the energy equation is

$$\rho c_v(\partial T/\partial t) = \nabla \cdot (K \nabla T). \tag{103.1}$$

Ignoring the spatial and temporal variation of K we can rewrite (103.1) as

$$(\partial T/\partial t) = \chi \nabla^2 T \tag{103.2}$$

where $\chi \equiv K/\rho c_v$ is the thermal diffusivity of the material. Equation (103.2) is to be solved subject to given initial and boundary conditions.

A classic problem is to solve (103.2) in one-dimensional planar geometry for an instantaneous heat pulse from a plane source in an infinite, cold ($T = 0$) medium. Thus if we release $\mathscr{E}$ ergs cm^{-2} in the plane $x = 0$ at $t = 0$, we solve (103.2) with the initial condition

$$T(x, 0) = Q\,\delta(x) \tag{103.3}$$

and boundary conditions

$$T(\pm\infty, t) = 0. \tag{103.4}$$

By energy conservation

$$\int_{-\infty}^{\infty} T(x, t)\, dx = Q \equiv \mathscr{E}/\rho c_v \tag{103.5}$$

at all later times.

The well-known solution (**L2,** §51), (**M13,** 862) of this problem is

$$T(x, t) = [Q/(4\pi\chi t)^{1/2}] \exp(-x^2/4\chi t), \tag{103.6}$$

which is sketched in Figure 103.1a for various values of $\tau \equiv 4\chi t$. Equation (103.6) exhibits two important properties characteristic of linear conduction: (1) the bulk of the energy is contained within the region $|\Delta x| \lesssim (4\chi t)^{1/2}$, which shows that the range of energy penetration grows as $t^{1/2}$ (appropriate for a random-walk process). The peak temperature decreases as $t^{-1/2}$, as one also would expect from (103.5), which implies that $T_{\text{peak}}\,\Delta x \approx$ constant. (2) The process has an infinite signal speed in the sense that even at an infinitesimal time the temperature is nonzero for all x. The long-range tail, which contains only an infinitesimal amount of energy, results because the conductivity is finite everywhere, hence always admits a nonzero energy propagation no matter how low the temperature and how shallow the gradient.

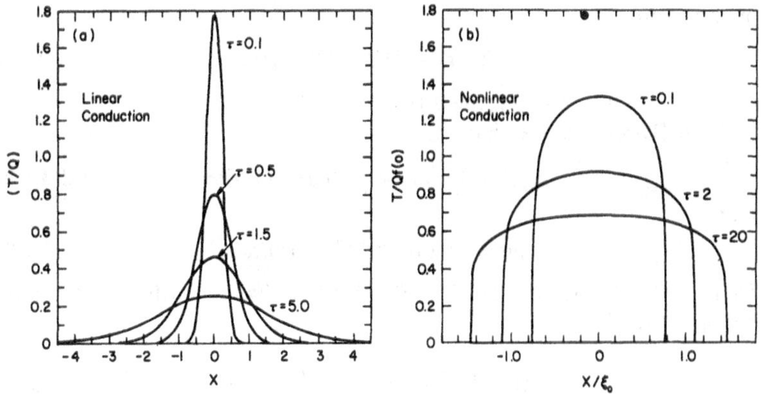

Fig. 103.1 Time development of temperature pulse for (a) linear and (b) nonlinear conduction.

BEHAVIOR OF NONLINEAR CONDUCTION WAVES

Suppose now that the conductivity is a function of temperature; in particular assume a power-law variation $K = K_0 T^n$, which allows K to increase sharply with rising temperature and to vanish at zero temperature. We must now solve

$$(\partial T/\partial t) = (\rho c_v)^{-1} \nabla \cdot (K_0 T^n \nabla T) \equiv \alpha \nabla \cdot (T^n \nabla T). \qquad (103.7)$$

Again suppose we apply an instantaneous heat pulse in the plane $x = 0$ of an infinite, cold medium. In this case heat cannot penetrate instantaneously to infinity because a finite amount of heat must first be deposited in the material to raise its conductivity above zero. The point at which the temperature first rises from zero, defining the position of the thermal wave, propagates into the medium with a finite velocity. Moreover because the conductivity, hence efficiency of heat transport, rises with rising temperature, we have a positive feedback mechanism that allows the temperature to rise very sharply from zero at a well-defined *thermal front*, behind which there is a nearly isothermal region extending back to the origin, as sketched in Figure 103.1b.

We can derive an exact solution for the problem just posed from a similarity analysis. The variables entering the problem are x [cm], t [s], and T [K]; in addition there are two parameters, Q [K cm] and α [$\text{cm}^2\,\text{s}^{-1}\,\text{K}^{-n}$]. The only dimensionless combination that can be formed from x, t, α, and Q is

$$\xi \equiv x/(\alpha Q^n t)^{1/(n+2)} \qquad (103.8)$$

which we adopt as our similarity variable. It follows from (103.8) that the position of the front varies as $x_f \sim t^{1/(n+2)}$. The quantity $Q/(\alpha Q^n t)^{1/(n+2)} = (Q^2/\alpha t)^{1/(n+2)}$ has the dimensions of temperature, hence we seek a solution of the form

$$T(x, t) = (Q^2/\alpha t)^{1/(n+2)} f(\xi) \tag{103.9}$$

where $f(\xi)$ is to be determined from (103.7).

From (103.8) we see that

$$(\partial f/\partial t) = -(n+2)^{-1}(\xi/t)(df/d\xi) \tag{103.10a}$$

and

$$(\partial f/\partial x) = (\alpha Q^n t)^{-1/(n+2)}(df/d\xi). \tag{103.10b}$$

Using (103.9) and (103.10) in (103.7) we derive an ordinary differential equation for f:

$$(n+2)\frac{d}{d\xi}\left(f^n \frac{df}{d\xi}\right) + \xi\frac{df}{d\xi} + f = 0. \tag{103.11}$$

The solution of this equation is (**L2**, 196), (**Z3**, 665)

$$f(\xi) = [n\xi_0^2/2(n+2)]^{1/n}[1-(\xi/\xi_0)^2]^{1/n} \tag{103.12}$$

for $\xi < \xi_0$, and $f(\xi) = 0$ for $\xi \geq \xi_0$. Here ξ_0 is a constant of integration that can be determined from the energy-conservation requirement (103.5), which, in view of (103.8) and (103.9), becomes

$$\int_{-\infty}^{\infty} f(\xi)\, d\xi = \int_{-\xi_0}^{\xi_0} f(\xi)\, d\xi = 1. \tag{103.13}$$

Evaluation of the integral gives

$$\xi_0^{n+2} = [(n+2)^{(n+1)} 2^{(1-n)}/n\pi^{(n/2)}][\Gamma(\tfrac{1}{2}+1/n)/\Gamma(1/n)]. \tag{103.14}$$

From (103.8) the position of the thermal front is given by

$$x_f = \xi_0(\alpha Q^n t)^{1/(n+2)}. \tag{103.15}$$

The nonlinear conduction solution shown in Figure 103.1b is a plot of $T/Qf(0)$ for $n=6$, as a function of x/ξ_0, for various values of $\tau \equiv \alpha Q^n t$.

MARSHAK WAVES

Let us now consider nonlinear radiation diffusion. We choose certain boundary conditions of interest in astrophysical applications and reserve the name "Marshak wave" for this particular class of problems. For static material, the energy equation in the equilibrium diffusion approximation is

$$\rho c_v(\partial T/\partial t) = \nabla \cdot (K_R \nabla T). \tag{103.16}$$

The rate of change of the radiation energy density has been neglected on the left-hand side, so the theory applies only when the fluid is not strongly radiation dominated.

From (97.3) the radiation conduction coefficient is

$$K_R = \tfrac{4}{3}a_R c\lambda_p T^3 = \tfrac{4}{3}a_R c T^3/\chi_R \tag{103.17}$$

where λ_p is the photon mean free path and χ_R is the Rosseland mean extinction coefficient. K_R is a very strong function of T because, for typical astrophysical materials, the Rosseland mean opacity scales as

$$\chi_R = \chi_0 \rho^\alpha T^{-n} \tag{103.18}$$

where n is about 3.5 [cf. (**C23,** 378–380), (**S10,** 68–70)]. Hence K_R typically varies as T^6 or T^7, and even more rapidly under certain circumstances.

A complete discussion of one-dimensional radiation diffusion for the case of a constant driving temperature at the plane $x = 0$ has been given by Petschek, Williamson, and Wooten (**P2**); see also (**K4**). In their notation we wish to solve

$$(\partial T/\partial t) = \mathcal{K}(\partial^2 T^m/\partial x^2) \tag{103.19}$$

where

$$\mathcal{K} \equiv (4a_R c/3m)/(\chi_0 c_v \rho^{\alpha+1}) \tag{103.20}$$

and

$$m \equiv n + 4, \tag{103.21}$$

subject to the condition $T \equiv T_0$ at $x = 0$.

Adopting the similarity variable

$$\xi \equiv x/(2\mathcal{K}T_0^{m-1}t)^{1/2} \equiv Ax/t^{1/2}, \tag{103.22}$$

which implies

$$(\partial f/\partial t) = -\tfrac{1}{2}(\xi/t)(df/d\xi) \tag{103.23a}$$

and

$$(\partial^2 f/\partial x^2) = (A^2/t)(d^2 f/d\xi^2), \tag{103.23b}$$

and adopting the scaled temperature

$$\tau(x, t) \equiv T(x, t)/T_0$$

as dependent variable, we can rewrite (103.19) as

$$(d^2\tau^m/d\xi^2) = -\xi(d\tau/d\xi). \tag{103.24}$$

The boundary conditions are $\tau \equiv 1$ at $\xi = 0$, and $\tau = (d\tau^m/d\xi) = 0$ at some point ξ_0 yet to be determined. The latter implies that the flux vanishes at ξ_0 because

$$F = -K_R(\partial T/\partial x) = -\rho c_v(\mathcal{K}T_0^{m+1}/2t)^{1/2}(d\tau^m/d\xi). \tag{103.25}$$

The total energy in the wave is

$$\mathcal{E} = \rho c_v \int_0^{x_0} T\, dx = \rho c_v(2\mathcal{K}T_0^{m+1}t)^{1/2}\varepsilon \tag{103.26}$$

where

$$\varepsilon \equiv \int_0^{\xi_0} \tau\, d\xi. \tag{103.27}$$

Table 103.1. Properties of Marshak Waves

		ξ_0		ε	
n	m	Exact	Approximate	Exact	Approximate
0.0	4.0	1.231	1.118	0.940	0.894
1.0	5.0	1.177	1.095	0.952	0.913
2.0	6.0	1.143	1.080	0.960	0.926
2.5	6.5	1.130	1.074	0.963	0.931
3.0	7.0	1.120	1.069	0.965	0.935
3.5	7.5	1.111	1.065	0.968	0.939
4.0	8.0	1.103	1.061	0.970	0.943
5.0	9.0	1.091	1.054	0.973	0.949
10.0	14.0	1.057	1.035	0.982	0.966

Energy conservation implies that $(\partial \mathscr{E}/\partial t) = F(x=0, t)$, which is equivalent to

$$\varepsilon = -(d\tau^m/d\xi)_{\xi=0}, \tag{103.28}$$

a result needed below.

After appropriate transformation, (103.24) is readily integrated numerically. For a given n, hence m, the solution yields ξ_0, ε, and $\tau(\xi)$; extensive results are tabulated in (**P2**). A few representative values are summarized in Table 103.1. Petschek et al. also develop a hierarchy of approximate analytical solutions starting from the zeroth approximation $\tau(\xi) \equiv 1$ for $0 \leq \xi \leq \xi_0$ and zero for $\xi > \xi_0$, which implies that $\xi_0 = 1$ and $\varepsilon = 1$. Another approximate solution is obtained by making the *constant flux approximation*, demanding that the net flux be constant behind the wave front so that

$$(d\tau^m/d\xi) = -C \tag{103.29}$$

for $0 \leq \xi \leq \xi_0$. Integrating from ξ to ξ_0 we have

$$\tau^m = C\xi_0[1 - (\xi/\xi_0)]. \tag{103.30}$$

But $\tau(0) \equiv 1$, hence $C = 1/\xi_0$. Furthermore, from (103.27)

$$\varepsilon = \int_0^{\xi_0} [1 - (\xi/\xi_0)]^{1/m} \, d\xi = \xi_0 m/(m+1), \tag{103.31}$$

which, when used in (103.28) with $(d\tau^m/d\xi) = -1/\xi_0$ from (103.29), yields

$$\xi_0 = [(m+1)/m]^{1/2}. \tag{103.32}$$

The values of ξ_0 and ε obtained from (103.32) and (103.31) are compared with the exact values in Table 103.1; the error in ξ_0 is at most 10 percent and in ε at most 5 percent, so the approximation is useful. Petschek et al. call this the "one-halfth" approximation because it is intermediate in accuracy between the zeroth and first approximations of their hierarchy.

COMPARISON WITH NONEQUILIBRIUM DIFFUSION AND TRANSPORT

Comparisons of the Marshak-wave similarity solution with nonequilibrium diffusion and full transport calculations are reported in (**B1**) and (**C1**). In these calculations the temperature dependence of the opacity is ignored but the rate of change of the radiation energy density is included, so the equation solved is

$$\frac{\partial}{\partial t}(\rho c_v T + a_R T^4) = \tfrac{1}{3} a_R c \lambda_p \frac{\partial^2 T^4}{\partial x^2}. \qquad (103.33)$$

If both x and ct are measured in units of λ_p, the problem depends on only one dimensionless parameter, $\eta \equiv a_R T_0^3/\rho c_v$. Equation (103.33) is solved by transforming to the new variable $\xi \equiv x/(ct)^{1/2}$ and by integrating the resulting ordinary differential equation for $\tau \equiv \eta (T/T_0)^4$ numerically. The value of ξ_0 at the front depends only on η. The transport solution is effected by a discrete-ordinate solution of the time-dependent transfer equation in (**B1**) and by a Monte Carlo simulation in (**C1**).

Results obtained from imposing the boundary condition $T(x=0, t) \equiv T_0$ are shown in Figure 103.2 for a case with $\eta = 1.69$ (which implies $\xi_0 = 1.06$). In this plot the Marshak wave is represented by a single curve, whereas the shape of the transport solution evolves in time. The Marshak wave always deposits too much energy into the material and significantly

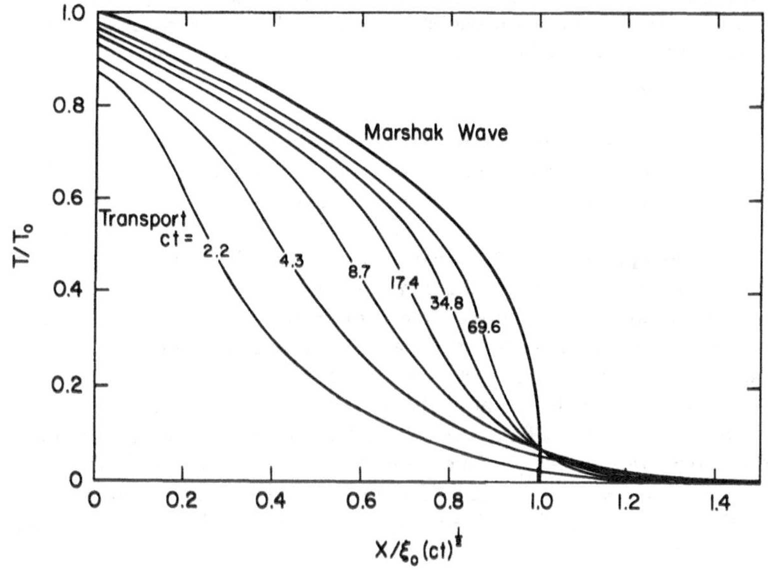

Fig. 103.2 Comparison of Marshak wave with transport solution for radiation penetration into a cold medium. From (**B1**).

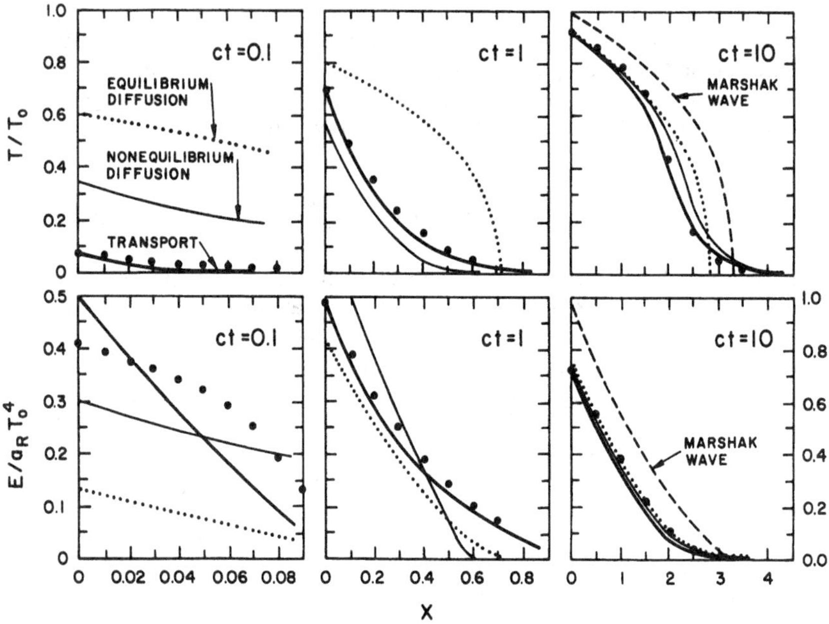

Fig. 103.3 Comparison of equilibrium diffusion, nonequilibrium diffusion, and transport solutions for radiation penetrating into cold material. From (**C1**). Heavy dots show results from an unpublished flux-limited diffusion calculation, kindly communicated by Dr. C. D. Levermore.

outruns the real radiation front (except for the exponentially attenuated tail of photons that have entered the slab at normal incidence and have penetrated to a depth $x = ct$). The too-rapid heating of the material is shown vividly in Figure 103.3 where the Marshak front can be plotted on the graphs only for $ct = 10$; at earlier times it is off the right edge of the graph: at $x = 0.335$ for $ct = 0.1$ and $x = 1.06$ at $ct = 1$.

The solution can be improved considerably by choosing a better boundary condition. In reality the material at $x = 0$ does not equilibrate instantaneously to $T = T_0$, but reaches this temperature only after a finite time. Thus our constant-temperature boundary condition forces the radiative conductivity to be too large from the outset, thereby admitting too large a flux, hence rate of energy deposition. A more accurate description of the physics is to assume a *constant imposed radiation field* $I(x = 0, \mu) \equiv I^- = B(T_0) = (a_R c/4\pi)T_0^4$ for $-1 \leq \mu \leq 0$. Now from (83.12), (83.13), and (83.15) we have

$$\mu^2(\partial j/\partial \tau)|_{x=0^+} = \mu h(x = 0^+) = \mu[j(x = 0^+) - I^-], \qquad (103.34)$$

which, when integrated over angle ($0 \le \mu \le 1$) yields

$$-\tfrac{1}{3}\lambda_p(\partial J/\partial x)|_{x=0^+} = \tfrac{1}{2}[J(x=0^+) - I^-]. \qquad (103.35)$$

But $J(x=0) = (c/4\pi)E(x=0) = (a_R c/4\pi)[T(x=0)]^4$, hence we have the boundary condition

$$[\tfrac{1}{2}T^4 - \tfrac{1}{3}\lambda_p(\partial T^4/\partial x)]_{x=0} = \tfrac{1}{2}T_0^4. \qquad (103.36)$$

With (103.36) as the boundary condition, it is no longer possible to solve the problem with a similarity transformation like that used earlier. Instead, (103.33) and (103.36) are solved numerically using the techniques described in §97; the results are denoted "equilibrium diffusion approximation" in Figure 103.3.

Although more accurate than the Marshak wave, the equilibrium diffusion solution still yields too much material heating and too low a radiation density at early times. The results can be improved further by using the nonequilibrium diffusion approximation of §97, which allows the material and the radiation field to have different temperatures. As seen in Figure 103.3 we now obtain a larger radiation energy density and a lower material temperature, in much better agreement with transport theory; indeed for $ct \ge 1$ the material temperature profile is reasonably accurate. The nonequilibrium diffusion results are still in serious error during the earliest instants because (1) the radiation field is assumed to be isotropic and (2) the energy transport is not flux limited. Substantially better results at the earliest times are obtained using flux-limited diffusion (cf. §97).

OTHER SOLUTIONS

Analytical solutions exist for a variety of other nonlinear radiation diffusion problems. For example, Marshak (**M4**) obtained similarity solutions for radiation penetration when T_0 varies as t^α or as exp (αt). Zel'dovich and Raizer give a similarity solution for nonlinear heat conduction (or radiation diffusion) from an instantaneous point source in spherical geometry (**Z3,** 668), and for a slab with a constant *net* (as opposed to *incident*) flux at the boundary (**Z3,** 673). Pomraning (**P3**) solved the non-self-similar problem of the diffusion of radiation from a cavity using a moments method for plane, cylindrical, and spherical geometries. His solution reproduces the exact results in both the early- and late-time limits.

104. Steady Shocks

One of the most interesting phenomena in radiation hydrodynamics is the *radiating shock*, which occurs in a wide variety of astrophysical flows. In this section we consider *steady* radiating shocks, which can arise in steady flows (e.g., stellar winds or accretion flows), and which also can be used as an instantaneous description of propagating shocks when the shock thickness is very small compared to a characteristic length over which the ambient medium changes significantly (e.g., a scale height). As in Chapter 5, we first treat the shock as a discontinuity and derive jump relations that relate the equilibrium states of the upstream and downstream material far from the front; we then consider the structure of the front itself in greater detail.

RANKINE-HUGONIOT RELATIONS FOR A RADIATING SHOCK

Consider a steady, one-dimensional, planar flow of a radiating fluid. Our main goal here is to account for the effects of the radiation energy density, pressure, and energy flux, so we ignore viscosity and thermal conduction and assume the material to be a perfect gas. The flow is then governed by the equation of continuity

$$(\partial\rho/\partial t') + [\partial(\rho v)/\partial x'] = 0, \tag{104.1}$$

momentum [cf. (96.5)]

$$[\partial(\rho v)/\partial t'] + [\partial(\rho v^2 + p + P_0)/\partial x'] = 0, \tag{104.2}$$

and energy [cf. (96.15)]

$$[\partial(\rho e + \tfrac{1}{2}\rho v^2 + E_0)/\partial t'] + \{\partial[\rho v(h + \tfrac{1}{2}v^2) + F_0 + v(E_0 + P_0)]/\partial x'\} = 0, \tag{104.3}$$

where (x', t') denote lab-frame coordinates with respect to which the shock moves with velocity v_S. In (104.2) we have dropped two terms containing F_0, which are $O(\lambda_p v/lc)$ in the diffusion limit, and at most $O(v/c)$ in the streaming limit, compared to $(\partial P_0/\partial x')$, hence are negligible for non-relativistic flow speeds.

It is straightforward to show that (104.1) to (104.3) are invariant under the Galilean transformation

$$x \equiv x' - v_S t', \qquad t' \equiv t, \qquad u \equiv v - v_S \tag{104.4}$$

to the frame in which the shock front is stationary. Thus we obtain shock-frame conservation relations by making the substitutions $(\partial/\partial t') \rightarrow (\partial/\partial t)$, $(\partial/\partial x') \rightarrow (\partial/\partial x)$, and $v \rightarrow u$. Furthermore, in this frame the flow is steady, so we can drop terms in $(\partial/\partial t)$ obtaining

$$d(\rho u)/dx = 0, \tag{104.5}$$

$$d(\rho u^2 + p + P_0)/dx = 0, \tag{104.6}$$

and

$$d[\rho u(h + \tfrac{1}{2}u^2) + F_0 + u(E_0 + P_0)]/dx = 0. \tag{104.7}$$

Integrating (104.5) to (104.7) across the front, and formally taking the limit of vanishing shock thickness as in §56, we obtain the *radiation-modified Rankine–Hugoniot jump conditions*

$$\rho_1 u_1 = \rho_2 u_2 \equiv \dot{m}, \tag{104.8}$$

$$\dot{m}u_1 + p_1 + P_{01} = \dot{m}u_2 + p_2 + P_{02}, \tag{104.9}$$

$$\dot{m}(h_1 + \tfrac{1}{2}u_1^2) + F_{01} + u_1(E_{01} + P_{01}) = \dot{m}(h_2 + \tfrac{1}{2}u_2^2) + F_{02} + u_2(E_{02} + P_{02}), \tag{104.10}$$

results first derived by Marshak (**M4**). As usual the subscripts "1" and "2" denote upstream and downstream conditions, respectively.

Comparing (104.8) to (104.10) with (56.6) to (56.8) we see that the gas pressure and the material enthalpy density are replaced by the total pressure $(p + P_0)$ and total enthalpy density $e + (p + E_0 + P_0)/\rho$. More significant, in contrast to a nonradiating shock [in which the conduction flux, proportional to (dT/dx), necessarily vanishes far from the shock front where there are no gradients] we now may have a nonvanishing radiation flux in both the upstream and downstream flow owing to *nonlocal transport* of radiant energy into optically thin material. Generally speaking, the radiation energy density and pressure are important only at extremely high temperatures (hence for very strong shocks) and/or low densities, whereas radiative energy exchange plays a fundamental role in all radiating shocks.

Note that all radiation quantities in (104.9) and (104.10) are evaluated in the comoving fluid frame on both sides of the jump. By virtue of (91.17) we can rewrite (104.10) in terms of the radiation flux measured in the frame in which the shock is at rest:

$$\dot{m}(h_1 + \tfrac{1}{2}u_1^2) + F_1 = \dot{m}(h_2 + \tfrac{1}{2}u_2^2) + F_2. \tag{104.11}$$

In practice (104.10) is more useful in opaque material where we can use the diffusion approximation, and (104.11) is more useful for transparent material where, if we ignore velocity-dependent terms in the transfer problem, we can write the flux in terms of the Φ operator (cf. §79) operating on the material source function.

It is instructive to analyze the jump conditions (104.8) to (104.10) in two limiting regimes:

(*a*) *Opaque Material* Suppose first that both the upstream and the downstream material is extremely opaque. Then any radiation crossing the front from the hot downstream material into the cooler upstream material will be completely reabsorbed within a thin layer into which it can penetrate by diffusion. To estimate the thickness of this layer we equate the time $t_d = (l/\lambda_p)^2(\lambda_p/c)$ required for the radiation to diffuse a distance l from the front, to the time $t_f = (l/v)$ required for this material to be swept back into the shock front, obtaining

$$l \sim (c/v)\lambda_p. \tag{104.12}$$

The radiation diffusion layer cannot grow much beyond the thickness given by (104.12) because t_d increases quadratically with l, but t_f only linearly.

Outside the diffusion layer conditions again become homogeneous and the radiation flux vanishes. Hence at sufficiently large distances from the shock $F_{01}=F_{02}=0$, and (104.10) simplifies to

$$\dot{m}(h_1+\tfrac{1}{2}u_1^2)+u_1(E_{01}+P_{01})=\dot{m}(h_2+\tfrac{1}{2}u_2^2)+u_2(E_{02}+P_{02}). \quad (104.13)$$

This expression is appropriate for very high-temperature flows (e.g., in a stellar envelope) where the contributions of the radiation pressure and energy density are significant.

Suppose now that the material component of the fluid is a completely ionized plasma, so that $p=\rho kT/\mu_0 m_H \equiv \rho RT$ with μ_0 constant and $\gamma=\frac{5}{3}$, and that the radiation is in equilibrium with the material so that $E_0=3P_0=a_R T^4$. Then (104.8), (104.9), and (104.13) completely determine downstream conditions for given upstream conditions. Defining the compression ratio

$$r \equiv \rho_2/\rho_1 = u_1/u_2 \quad (104.14)$$

we can rewrite (104.9) and (104.13) in nondimensional form:

$$\gamma M_1^2(r-1)/r=(\Pi-1)+\alpha_1[(\Pi/r)^4-1] \quad (104.15)$$

and

$$\tfrac{1}{2}\gamma M_1^2(r^2-1)/r^2=[\gamma/(\gamma-1)][(\Pi/r)-1]+4\alpha_1[(\Pi^4/r^5)-1] \quad (104.16)$$

where

$$\Pi \equiv p_2/p_1, \quad (104.17)$$

$$\alpha_1 \equiv \tfrac{1}{3}a_R T_1^4/p_1, \quad (104.18)$$

and

$$M_1 \equiv u_1/a_1 = u_1(\rho_1/\gamma p_1)^{1/2}. \quad (104.19)$$

Note that for $\alpha_1 \equiv 0$, that is, no radiation pressure, (104.15) and (104.16) reduce to (56.9) and (56.17).

To solve the nonlinear system (104.15) and (104.16) we regard r as the independent variable and α_1 as a given parameter. Eliminating M_1^2 between (104.15) and (104.16) we obtain

$$\alpha_1 r^{-4}(7-r)\Pi^4=(r-r_0)\Pi+\alpha_1(7r-1)+(r_0 r-1) \quad (104.20)$$

where

$$r_0 \equiv (\gamma+1)/(\gamma-1) \quad (104.21)$$

is the maximum compression ratio for a strong shock in a nonradiating gas; for $\gamma=\frac{5}{3}$, $r_0=4$. Equation (104.20) yields $\Pi(r, \alpha_1)$; given Π we find $M_1(r, \alpha_1)$ from either (104.15) or (104.16). These results can then be inverted to find $r(M_1, \alpha_1)$, hence $\Pi(M_1, \alpha_1)$.

In general (104.20) must be solved numerically. However it is obvious that as $r \to 7$, the limiting compression ratio for a gas with $\gamma = \frac{4}{3}$ (e.g., pure radiation), we obtain an infinitely strong shock with $\Pi \propto (7-r)^{-1/3} \to \infty$ and $M_1 \propto \Pi^2 \propto (7-r)^{-2/3} \to \infty$. Because the compression ratio is essentially fixed in very strong shocks, $T_2/T_1 \approx p_2/p_1 = \Pi$, whence we see that in a very strong radiating shock the temperature ratio grows only as $M_1^{1/2}$; in contrast, in a nonradiating shock it rises as M_1^2 [cf. (56.42)].

For a weak shock with $r = 1 + \xi$, where $\xi \ll 1$, one finds from (104.20) that $\not{p} \equiv [(p_2/p_1) - 1]$ is

$$\not{p} = \xi(r_0 + 1 + 32\alpha_1)/(r_0 - 1 + 24\alpha_1) \to \xi(5 + 32\alpha_1)/(3 + 24\alpha_1), \quad (104.22)$$

where the second expression holds for $\gamma = \frac{5}{3}$. Equation (104.22) yields $\not{p} = \Gamma\xi$ [cf. (56.36)] with $\Gamma = \gamma = \frac{5}{3}$ when $\alpha_1 = 0$, and $\Gamma \to \frac{4}{3}$ as $\alpha_1 \to \infty$. Furthermore, one finds from (104.15) and (104.22) that

$$M_1^2 \approx (5 + 40\alpha_1 + 32\alpha_1^2)/(5 + 40\alpha_1). \quad (104.23)$$

This is the expected result because in a weak shock u_1 must equal $[\Gamma_1(1+\alpha_1)p_1/\rho_1]^{1/2}$, the sound speed of the radiating fluid, hence $M_1^2 \equiv u_1^2/a_{\text{gas}}^2 = \Gamma_1(1+\alpha_1)/\gamma$, which, in light of (70.22) for Γ_1, yields (104.23).

An explicit analytical solution can be obtained for the limiting case of a very strong shock propagating into a cold gas (**S2**). We neglect the gas and radiation energy densities and pressures in the upstream material, which is equivalent to dropping unity wherever it appears on the right-hand sides of (104.15) and (104.16). Eliminating M_1^2 between these two equations we obtain

$$\alpha_1 \Pi^3 = r^4(r - r_0)/(7 - r). \quad (104.24)$$

Reverting to dimensional variables we have

$$T_2^3 = (3R\rho_1/a_R)r(r - r_0)/(7 - r) \quad (104.25)$$

which yields $T_2(r)$. Given T_2 we can immediately compute p_2 and P_2. Using (104.24) in the simplified version of (104.15) we find

$$M_1^2 = \Pi r(r - r_0)/\gamma(r - 1)(7 - r). \quad (104.26)$$

Note that (104.24) to (104.26) are valid only for r appreciably larger than r_0.

(b) Optically Thin Upstream Material Suppose now that we have a radiating shock propagating into optically thin upstream material. Focusing mainly on the effects of radiative energy transport across the front, we assume the flow to be cool enough that we can neglect the radiation energy density and pressure. The radiation-modified jump conditions reduce to (56.6), (56.7), and

$$\dot{m}(h_1 + \tfrac{1}{2}u_1^2) + F_{01} = \dot{m}(h_2 + \tfrac{1}{2}u_2^2) + F_{02}. \quad (104.27)$$

We focus on the case of a nonzero net flux across the front, that is,

$|F_{01}-F_{02}|>0$, which will occur if the preshock material is sufficiently transparent that radiation originating in the hot postshock material (which may be so opaque that $F_{02}\approx 0$) can flow freely across the front and escape to infinity upstream (implying a large value for $|F_{01}|$). An example is a strong shock emerging from the photosphere or chromosphere of a star.

By an analysis similar to that leading to (56.20) and (56.21), we find

$$\frac{\rho_2}{\rho_1}=\frac{[(\gamma+1)p_2+(\gamma-1)p_1]u_2+2(\gamma-1)(F_{02}-F_{01})}{[(\gamma+1)p_1+(\gamma-1)p_2]u_2} \tag{104.28}$$

and

$$\frac{T_2}{T_1}=\left(\frac{p_2}{p_1}\right)\frac{[(\gamma+1)p_1+(\gamma-1)p_2]u_1+2(\gamma-1)(F_{01}-F_{02})}{[(\gamma+1)p_2+(\gamma-1)p_1]u_1}. \tag{104.29}$$

For the geometry sketched in Figure 55.2 the shock is moving to the left in the lab frame, hence both u_1 and u_2 are positive in the shock's frame, whereas a net radiation flux into the cooler upstream material implies that $(F_{01}-F_{02})<0$ (i.e., a net flow of energy to the left). Therefore a net flow of radiant energy upstream decreases the temperature jump and increases the density jump. The upstream material is preheated by a *radiation precursor*, and the downstream material is cooled by radiative losses.

In more general terms, the example just discussed illustrates that radiative energy transport across a shock can significantly alter the temperature, density, and velocity profiles in both the upstream and downstream flow over distances determined by the opacity of the material. To analyze these effects in more detail we must now examine the *structure* of radiating shocks.

EQUATIONS OF RADIATING SHOCK STRUCTURE

Adding material viscosity and heat-conduction terms to (104.2) and (104.3), and transforming to the frame of the shock, we obtain general conservation relations which apply throughout the flow. As before, these admit first integrals, that is, (104.8) and

$$\dot{m}u+p+P_0-\mu'(du/dx)=\dot{m}C_1=\text{constant}, \tag{104.30}$$

and

$$\dot{m}(h+\tfrac{1}{2}u^2)+F_0+u(E_0+P_0)-\mu' u(du/dx)-K(dT/dx)=\dot{m}C_2^2=\text{constant}, \tag{104.31}$$

where $\mu'\equiv(\tfrac{4}{3}\mu+\zeta)$ is the effective one-dimensional viscosity.

Equations (104.30) and (104.31) show that in principle the structure of a radiating shock is determined by the combined action of viscosity, thermal conduction, and radiative energy transport. But in practice, photon mean free paths are orders of magnitude larger than particle mean free paths. Hence the viscous-conduction dissipation zone, which is only a few particle mean free paths thick, can be considered to be a mathematical discontinuity across which we allow discrete jumps in temperature, density, pressure, and velocity according to the usual Rankine–Hugoniot relations while the radiation quantities E_0, F_0, and P_0 remain continuous. This discontinuity is embedded in the *radiation exchange zone*, whose thickness is a few to many photon mean free paths, which determines the largest-scale structure of the shock front. We therefore drop the material viscosity and conductivity terms henceforth.

The same remarks apply to the material relaxation zone discussed in §57 as long as the material is assumed to be in LTE, for then the characteristic relaxation length is of the same general size as a particle mean free path, hence is much less than λ_p. As we will see, however, the assumption of LTE is often invalid, particularly in strong shocks where radiation from the hot postshock region is markedly out of equilibrium with the cool preshock material and can therefore drive it out of LTE. Moreover, equilibrium in the downstream material cannot be recovered until the shocked gas has time to recombine and radiate; in fact the size of the material ionization-relaxation zone is of the same order as the size of the radiation-exchange zone. Despite these caveats it is very instructive to analyze shock structure under the assumption of LTE, and we therefore do so in some detail before discussing non-LTE effects, which must be treated numerically.

From a different point of view, we can consider radiation exchange as a mechanism for producing partly or completely dispersed shocks. Unlike the relaxation processes discussed in §57 (but like thermal conduction by electrons), radiation produces not only a tail but also a precursor in the flow. If the shock is not too strong, the tail and precursor can join, giving a completely dispersed continuous solution. With increasing shock strength, a regime is reached in which a continuous solution is not possible; instead, there is a temperature discontinuity at the front, followed by a significant temperature overshoot, as is characteristic of a partly dispersed solution. Beyond a certain critical shock strength, the radiative flux causes this downstream overshoot to collapse to a sharp spike whose thickness is less than one photon mean free path. Finally, for extremely strong shocks (and/or sufficiently hot upstream material) in which the radiation pressure and energy density dominate over the material contributions, continuous solutions are again possible.

APPROXIMATE ANALYSIS OF STRONG SHOCKS WITH NONEQUILIBRIUM RADIATION DIFFUSION

The penetrating phenomenological discussions by Zel'dovich and Raizer (**Z1**), (**Z2**), (**R1**), (**Z3,** Chap. 7) offer considerable insight into the behavior of radiating strong shocks. To make the problem tractable analytically we assume the following. (1) A strong shock propagates into cold material, which implies that we can neglect the upstream pressure and energy density. (2) The material (a perfect gas) remains in LTE with all species of particles at the same kinetic temperature. (3) The radiation field can be treated in the nonequilibrium diffusion approximation. (4) The shock is in an optically thick medium and all radiation emanating from the front is reabsorbed in the upstream material, so that we have a closed thermodynamic system (no open boundaries). (5) It suffices to account for the radiative energy flux, whereas the radiation pressure and energy density can be ignored.

(*a*) *Basic Equations* Under the assumptions stated above the momentum and energy conservation relations (104.30) and (104.31) reduce to

$$\dot{m}u + p = \dot{m}u_1 \tag{104.32}$$

and

$$\dot{m}(h + \tfrac{1}{2}u^2) + F = \tfrac{1}{2}\dot{m}u_1^2. \tag{104.33}$$

The radiation field is governed by

$$(dF/d\tau) = 4\pi B - cE \tag{104.34}$$

and

$$F = -\tfrac{1}{3}c(dE/d\tau) \tag{104.35}$$

where $B = \sigma_R T^4/\pi$, and τ is the optical depth measured in the positive x direction, $d\tau \equiv \kappa\, dx$, with $\tau = 0$ chosen at the shock front. Combining (104.34) and (104.35) we have

$$(d^2F/d\tau^2) = 3F + 16\sigma_R T^3(dT/d\tau). \tag{104.36}$$

We have dropped the subscript "0" on radiation quantities because (104.34) to (104.36) are not correct comoving-frame equations; they are only approximate inasmuch as all velocity-dependent terms (such as the rate of work done by radiation pressure and the advection of radiation energy density) are omitted. These omissions are not serious for our present purposes, and in fact are consistent with the assumption that we can neglect E_0 and P_0 in the fluid conservation relations.

Equations (104.34) to (104.36) are to be solved subject to the boundary conditions $F_1 = E_1 = T_1 = 0$ at $\tau = -\infty$, and $F_2 = 0$, $T = T_2$, $E_2 = 4\pi B/c = a_R T_2^4$ at $\tau = +\infty$, which follow from assumptions (1) and (4) stated above.

In terms of the volume ratio $\eta \equiv \rho_1/\rho$, (104.32) becomes

$$p = \dot{m}u_1(1-\eta). \tag{104.37}$$

Using the perfect gas law for p we have

$$T/T_2 = \eta(1-\eta)/\eta_2(1-\eta_2), \tag{104.38}$$

hence from (104.33) we obtain

$$F = -\dot{m}RT(\eta - \eta_2)/2\eta\eta_2 \tag{104.39a}$$

$$= -\dot{m}RT_2(1-\eta)(\eta-\eta_2)/2\eta_2^2(1-\eta_2). \tag{104.39b}$$

Here $\eta_2 = (\gamma - 1)/(\gamma + 1)$ is the limiting volume ratio for an infinitely strong shock; $\eta_2 = \frac{1}{4}$ for $\gamma = \frac{5}{3}$. $T(\eta)$ and $F(\eta)$ are sketched in Figure 104.2.

(*b*) *Subcritical Shocks* If a shock in cold material is very weak, radiation has negligible influence on the energy balance, and we obtain the usual step discontinuities characteristic of an adiabatic shock in an ideal fluid. With increasing shock strength T_2 rises, and the radiation flux across the front (which can be estimated roughly as the flux $\sigma_R T_2^4$ emitted from an opaque "wall" of postshock material at temperature T_2) increases very rapidly. This radiation is absorbed in the upstream material and heats it to some characteristic temperature T_- immediately in front of the shock; the precursor decays away exponentially upstream as the radiation attenuates in the preshock material. As we will see below, T_- is proportional to the radiation flux incident from the postshock material, and thus rises rapidly as T_2 increases, eventually equaling T_2. Shocks with $T_- < T_2$ are called *subcritical*. Because material entering the shock is preheated, the postshock temperature T_+ overshoots its final equilibrium value T_2. The overshoot decays downstream as the material cools by emitting photons that penetrate across the shock.

In short, radiation acts as a thermodynamic heat-transfer mechanism from hot to cold material in the flow. The resulting shock structure is sketched in Figure 104.1. Preheating produces a small pressure and density rise in the upstream material. Downstream the pressure increases only a small amount from its postshock value; the density shows a larger fractional rise because the downstream material cools while the pressure rises.

From (104.38) one sees that η must always be quite close to unity in the preshock material; indeed, even if T_- is as large as T_2, still $\eta_- = (1 - \eta_2) = 0.75$ for $\gamma = \frac{5}{3}$. Thus in the upstream flow we can set $\eta \approx 1$, and from (104.39) we have

$$F = -\dot{m}RT/(\gamma - 1) = -\dot{m}e, \tag{104.40}$$

which has a straightforward physical interpretation: at any position in front of the shock, the material internal energy flux flowing downstream just equals the radiant flux flowing upstream because all the radiant energy passing that position is absorbed upstream and goes into heating the gas (nominally from zero temperature).

When T_- is appreciably smaller than T_2, we can derive an approximate solution for the structure of the precursor. The thermal energy density $a_R T_-^4$ will be much smaller than the energy density in the radiation field emerging from the shock (which is of order $a_R T_2^4$) hence we can make the simplifying assumptions that we can neglect B in (104.34) and drop the last term from (104.36), obtaining

$$F = F_0 e^{-\sqrt{3}|\tau|} \tag{104.41}$$

and

$$E = -(\sqrt{3}/c)F = -(\sqrt{3}/c)F_0 e^{-\sqrt{3}|\tau|}. \tag{104.42}$$

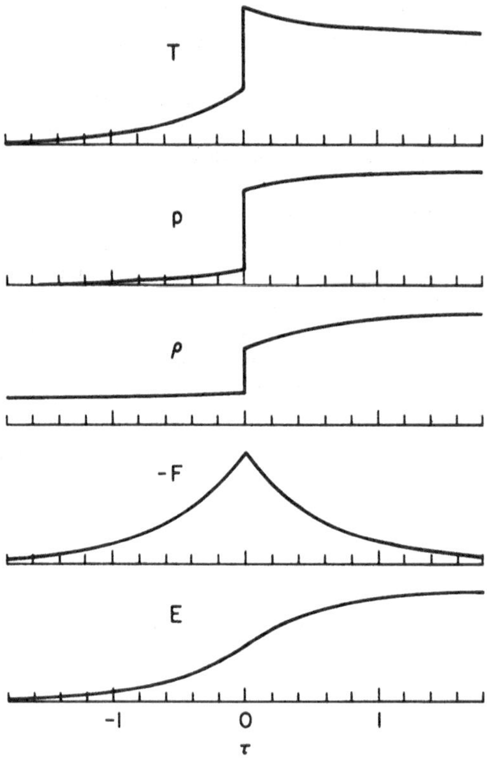

Fig. 104.1 Distribution of temperature, pressure, density, radiation flux, and radiation energy density as a function of optical depth in a subcritical shock.

(Recall that $F<0$ and $\tau<0$ upstream from the shock.) In (104.41) and (104.42) F_0 is the flux at the shock front; we fix its value in terms of T_2 below. Clearly the radiation field in the precursor is severely out of equilibrium.

Using (104.41) in (104.40) we find

$$T = T_- e^{-\sqrt{3}|\tau|}. \tag{104.43}$$

Because the preshock density variation is small, $p \propto T$ hence

$$p = p_- e^{-\sqrt{3}|\tau|}. \tag{104.44}$$

Equations (104.44) and (104.37) imply that

$$\rho - \rho_1 \approx (\rho_- - \rho_1) e^{-\sqrt{3}|\tau|}. \tag{104.45}$$

To develop an approximate solution for the structure of the postshock tail, we first note that both the radiation energy density and the flux must be continuous across the shock front. This result was proved rigorously by Zel'dovich (**Z1**) and is also intuitively obvious from the fact that both E and F are integrals of the source function against well-behaved weight functions. Only if B were *singular* (not merely discontinuous) could we produce discontinuities in E and F, but such singularities are physically inadmissible. Furthermore, (1) if the energy density were discontinuous, (104.35) would imply that the flux is somewhere infinite, whereas by energy conservation it must remain finite. (2) If the flux were discontinuous then (104.34) would imply an infinite net absorption-emission rate at some point in the flow, which is physically nonsensical.

Next we argue that in the postshock material $T \approx T_2$, hence $\pi B \approx \sigma_R T_2^4 = \text{constant}$. From (104.36) we then have

$$F = F_0 e^{-\sqrt{3}\tau}, \tag{104.46}$$

where we invoked continuity of F at $\tau = 0$. To calculate E we eliminate $d\tau$ between (104.34) and (104.35) and use the constancy of B to write

$$F\,dF = \tfrac{1}{3}c^2(E - a_R T_2^4)\,d(E - a_R T_2^4) \tag{104.47}$$

whence we have

$$(c/\sqrt{3})(E - a_R T_2^4) = F = F_0 e^{-\sqrt{3}\tau}. \tag{104.48}$$

Next we note that (104.38) and (104.39) can be combined to yield

$$F = -\dot{m}R(1-\eta)(T - T_2)/2\eta_2(1 - \eta - \eta_2). \tag{104.49}$$

But in the downstream flow $\eta \approx \eta_2$, hence (104.49) reduces to

$$F = -\dot{m}R(T - T_2)/\eta_2(3 - \gamma). \tag{104.50}$$

From (104.48) and (104.50) we have

$$T - T_2 = (T_+ - T_2)e^{-\sqrt{3}\tau}. \tag{104.51}$$

Furthermore, from (104.40) and (104.50) and the continuity of F at $\tau = 0$ we find

$$T_+ - T_2 = [(3-\gamma)/(\gamma+1)]T_-. \tag{104.52}$$

The complete solution implied by (104.41) to (104.52) is sketched in Figure 104.1.

Combining (104.48) with (104.42) at $\tau = 0$ we find

$$E_0 = \tfrac{1}{2}a_R T_2^4 \tag{104.53}$$

and therefore

$$F_0 = -(2/\sqrt{3})\sigma_R T_2^4. \tag{104.54}$$

T_2 is fixed by the upstream flow speed. Thus evaluating (104.33) at $\eta = \eta_2$ (where $F = 0$) we have

$$\gamma R T_2/(\gamma - 1) = \tfrac{1}{2}u_1^2(1 - \eta_2^2) \tag{104.55}$$

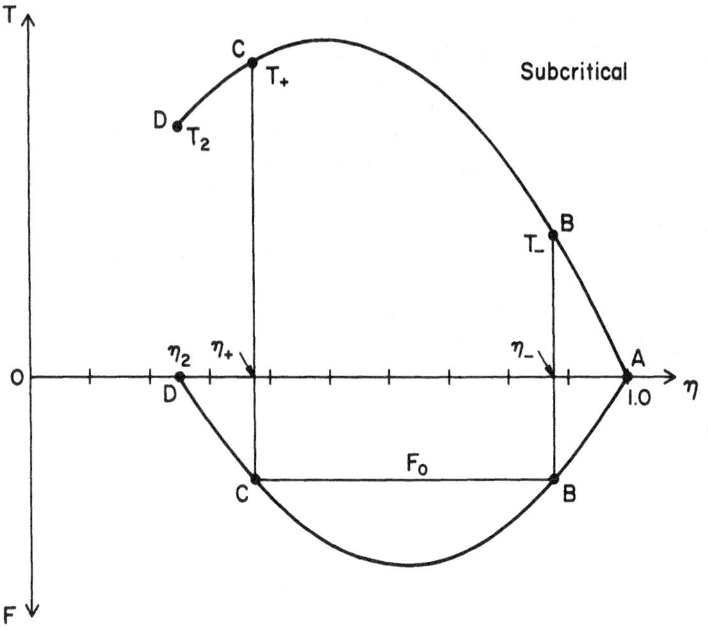

Fig. 104.2 Temperature and radiation flux as a function of volume ratio in a subcritical shock.

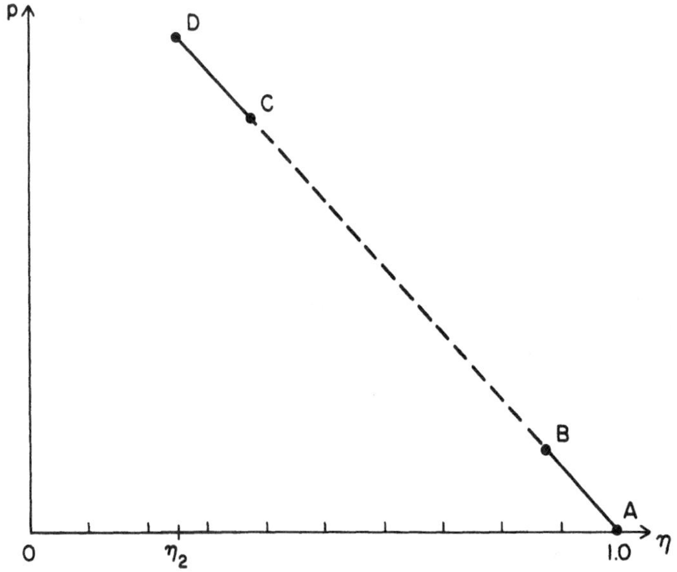

Fig. 104.3 Subcritical shock transition in (p, V) diagram.

or

$$T_2 = 2(\gamma - 1)u_1^2/R(\gamma + 1)^2. \qquad (104.56)$$

It is instructive to interpret the shock transition in terms of $T(\eta)$, $p(\eta)$, and $F(\eta)$ as given by (104.37) to (104.39) and plotted in Figures 104.2 and 104.3. Starting at point A, at $\eta = 1$, the material evolves continuously to point B, at $\eta = \eta_-$, where F rises to F_0, which is fixed by T_2. For F to be continuous the solution must jump discontinuously to point C, at $\eta = \eta_+$, on the left branch of $T(\eta)$ and $F(\eta)$, such that $F(\eta_+) = F(\eta_-)$. As the maximum of $T(\eta)$, at $\eta = \frac{1}{2}$, lies to the left of the minimum of $F(\eta)$, at $\eta = \frac{5}{8}$, T_+ is substantially larger than T_-; thus the temperature at the front is discontinuous. The material then evolves continuously from η_+ to η_2, and $F \to 0$ while $T \to T_2$. As seen in Figure 104.3, p increases monotonically throughout the entire transition, jumping discontinuously between B and C. Evolution of the solution from C to D along the ascending branch of the $T(\eta)$ curve is possible in the present problem (whereas it was forbidden for the thermally conducting shocks studied in §57) because the radiant energy flux is determined from nonequilibrium diffusion theory, hence F is not constrained to be proportional to (dT/dx) as it is for pure thermal conduction (or equilibrium diffusion).

To make the discussion more quantitative, consider a strong shock in a plasma of completely ionized hydrogen. Then (104.56) becomes

$$T_2 = (3m_H/32k)u_1^2, \qquad (104.57)$$

which yields the numerical values listed in Table 104.1. Similarly, using (104.54) in (104.40) we have

$$\dot{m}RT_-/(\gamma - 1) = 2\sigma_R T_2^4/\sqrt{3}, \qquad (104.58)$$

or, for ionized hydrogen,

$$T_- = 4\sigma_R T_2^4/3\sqrt{3}\, kn_{p1}u_1 \qquad (104.59)$$

Table 104.1. Properties of Strong Shocks in Ionized Hydrogen

u_1 (km s^{-1})	T_-	T_+ (K)	T_2
25.0	3,218	8,711	7,102
26.0	4,235	9,799	7,682
27.0	5,515	11,042	8,284
28.0	7,114	12,466	8,909
29.0	9,095	14,105	9,557
29.5	10,251	15,015	9,889

where n_{p1} is the upstream proton density. Taking $n_{p1}=10^{17}\ \mathrm{cm}^{-3}$, a reasonable value for the outer envelope of a star, we obtain the results listed in Table 104.1. Finally we obtain T_+ from T_2 and T_- by using (104.52). The numbers in Table 104.1 show the dramatic increase (as u_1^7) of T_- with increasing flow speed.

(c) *Supercritical Shocks* At some sufficiently large flow speed, u_{crit}, T_- finally equals T_2. We call such shocks *critical shocks* because the structure of the front is quite different in subcritical shocks ($u_1 < u_{\mathrm{crit}}$) and in *supercritical shocks* ($u_1 > u_{\mathrm{crit}}$). We refer to T_- in a critical shock as the *critical temperature.* From Table 104.1 we see that for ionized hydrogen at a density of $10^{17}\ \mathrm{cm}^{-3}$, $u_{\mathrm{crit}} \approx 29.3\ \mathrm{km\ s}^{-1}$ and $T_{\mathrm{crit}} \approx 9750\ \mathrm{K}$.

According to (104.56) and (104.58), T_- can exceed T_2 when u_1 rises above u_{crit}. This conclusion is erroneous for reasons explained below, and it is important to note that the model developed above becomes invalid before T_- actually reaches T_{crit}. In particular, the analysis assumes that the thermal radiation energy density $a_R T^4$ in the precursor is much smaller than the energy density in the radiation penetrating from the shock. But from (104.53) we see that the two will be equal when

$$T_- = (\tfrac{1}{2})^{1/4} T_2 \approx 0.84 T_2 \tag{104.60}$$

at which point the model manifestly fails. Thus we cannot trust the model when T_- is above, say, 70 percent of T_{crit}, and it obviously is not able to predict what happens when T_- equals T_{crit}.

A more thorough analysis (**Z1**), (**R1**) shows that T_- can never exceed T_2. First, one notes that if $T_- > T_2$, the radiation energy density in the precursor would exceed that in the tail, which would imply $(dE/d\tau) < 0$, hence $F > 0$. But we know from (104.39), which follows directly from the basic conservation laws, that $F \leq 0$ everywhere in the flow. Moreover, if T_- were greater than T_2 we would have a closed thermodynamic system in which heat is transferred from low-temperature to high-temperature material, in violation of the second law of thermodynamics. We must therefore conclude that T_- is always $\leq T_2$. A rigorous mathematical analysis of the radiation transport equations leads to the same result as these qualitative physical arguments (**Z1**).

Thus as the mechanical energy of the flow increases above the critical value, the supercritical excess of postshock radiant energy does not force T_- above T_2, but rather drives the radiation precursor more deeply into the upstream flow, producing an extended region with $T \approx T_2$ in front of the shock. The thickness of this zone increases rapidly with increasing T_2.

Viewing the radiant energy transport as a diffusion process we see that in effect a Marshak wave is driven into the preshock material by a radiating "wall" (i.e., the shock) at temperature T_2. In the shock's frame the Marshak wave becomes stationary at the point where material flows into the radiation front at exactly the speed at which the front would otherwise advance into stationary material. Such an accommodation is always possible because radiation from the "wall" initially moves forward at a significant fraction of the speed of light, but asymptotically a Marshak wave moves forward only as $t^{1/2}$, and thus has a velocity decreasing as $t^{-1/2}$. Radiation at the front of the precursor is out of equilibrium as in a subcritical shock, but rapidly comes into equilibrium as the material temperature approaches T_2.

We can develop an approximate model of the structure of the precursor in a supercritical shock by again using semiquantitative arguments. We divide the precursor into a nonequilibrium zone near the radiation front, followed by an equilibrium zone extending back to the shock. We place the boundary separating these zones at an optical depth $|\tau_c|$ in front of the shock where the thermal radiation energy density equals the actual radiation energy density.

In the nonequilibrium zone we recover (104.40) to (104.43) by the same analysis as before, but with $|\tau|$ replaced by $|\tau-\tau_c|$, and with $F_0 = -(c/\sqrt{3})a_R T_c^4$ in (104.41), $E_0 = a_R T_c^4$ in (104.42), and T_- replaced by T_c in (104.43). Here T_c is the temperature at the boundary between the two zones. We fix T_c by using (104.40) at this boundary, which yields

$$T_c^3 = \tfrac{1}{4}\sqrt{3}\, \dot{m}R/(\gamma-1)\sigma_R. \tag{104.61}$$

In the equilibrium zone we take $E = a_R T^4$, hence (104.35) becomes

$$F = -\tfrac{16}{3}\sigma_R T^3(dT/d\tau). \tag{104.62}$$

Combining (104.62) with (104.40) and (104.61) we have

$$T^2\, dT = \tfrac{1}{4}\sqrt{3}\, T_c^3\, d\tau. \tag{104.63}$$

Integrating and demanding continuity at $|\tau| = |\tau_c|$ we obtain

$$\frac{T}{T_c} = \left(\frac{E}{a_R T_c^4}\right)^{1/4} = \frac{\sqrt{3}\,|F|}{4\sigma_R T_c^4} = [1 + (\tfrac{3}{4}\sqrt{3})\,|\tau-\tau_c|]^{1/3}. \tag{104.64}$$

Then using the fact that $T = T_- = T_2$ at $\tau = 0$ we find that τ_c is

$$|\tau_c| = (4/3\sqrt{3})[(T_2/T_c)^3 - 1]. \tag{104.65}$$

From (104.61) and (104.56) one sees that T_2/T_c rises as $u_1^{5/3}$, hence $|\tau_c|$ increases rapidly with increasing shock strength. Because the radiation attenuates exponentially in the nonequilibrium zone, $|\tau_c|$ is essentially the optical thickness of the whole precursor.

Unfortunately, it is not possible to construct a simple, yet realistic, model for the postshock tail in a supercritical shock; nevertheless, a qualitative discussion is worthwhile. From an equilibrium diffusion analysis Prokof'ev (**P5**) argued that radiative exchange would guarantee continuity of the temperature at the shock front and concluded that a supercritical shock comprises a nonlinear radiation-diffusion front within which is imbedded

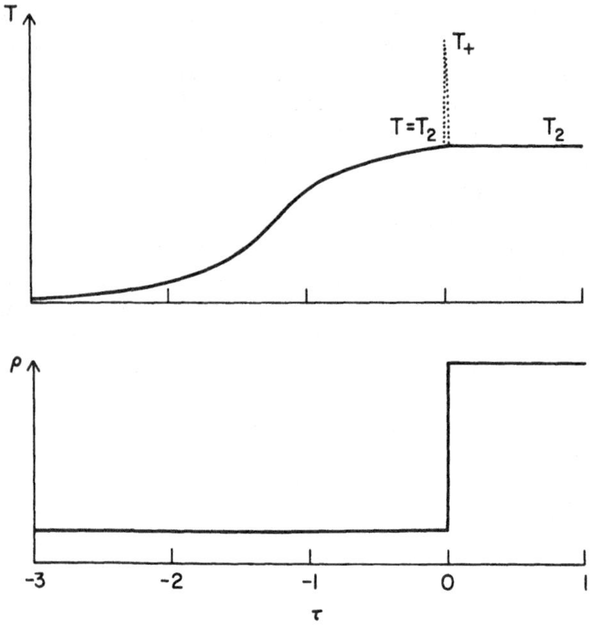

Fig. 104.4 Temperature and density as a function of optical depth in a supercritical shock.

an isothermal shock. At the isothermal shock, the temperature is continuous ($T = T_2$), but other physical variables jump discontinuously, as sketched in Figure 104.4.

While this picture seems reasonable, it poses a puzzle. We have seen that in subcritical shocks there is a postshock region where T rises to $T_+ > T_2$. Moreover, T_+ is a monotone increasing function of T_2. Why should this region suddenly vanish as the shock becomes critical or supercritical? The answer is, it doesn't. Zel'dovich showed (**Z1**) that the temperature distribution in a supercritical shock is *not* continuous. Instead, there is a sharp temperature spike in the downstream flow immediately behind the front (cf. Figure 104.6). As in a subcritical shock, the material in the spike is cooled by radiating into the upstream material. The thickness of the spike turns out to be less than a photon mean free path, and decreases with increasing shock strength; for this reason it is missed by an equilibrium diffusion analysis, which cannot handle properly features on a scale less than λ_p. The spike can be treated correctly only by a detailed transport calculation.

If Prokof'ev's conjecture were true, then in Figure 104.5 the material would evolve continuously from point A to point B, where $T = T_2$, and then jump discontinuously to point D. But, as the curve for $F(\eta)$ shows,

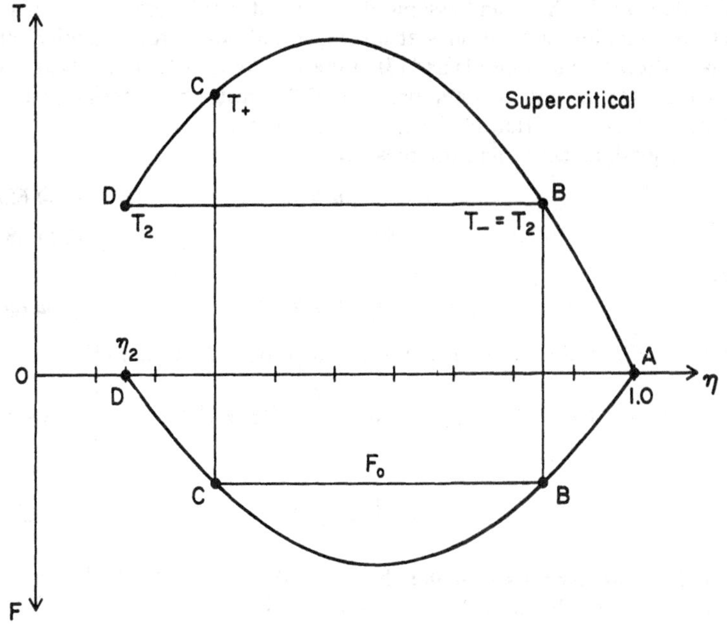

Fig. 104.5 Temperature and radiation flux as a function of volume ratio in a supercritical shock.

the radiation flux would be discontinuous along this path, which is inconsistent with the assumptions underlying the equilibrium diffusion calculation. In reality it is the flux that is continuous (**Z1**). Therefore, in Figure 104.5 the material evolves continuously from A to B, where $F = F_0$, the flux at the front; from (104.64) evaluated at $\tau = 0$ we have

$$F_0 = -(4\sigma_R/\sqrt{3})T_c^3 T_2. \qquad (104.66)$$

The solution then jumps discontinuously, at constant flux, to point C, and finally evolves continuously to point D in a physical distance smaller than a photon mean free path. The temperature jumps discontinuously from T_2 at B to $T_+ > T_2$ at C. We can compute T_+ from (104.38) and (104.39) by demanding that $T_- = T_2$ and $F(\eta_+) = F(\eta_-)$. We find

$$T_+ = (3 - \gamma)T_2. \qquad (104.67)$$

NUMERICAL CALCULATIONS OF RADIATING SHOCK STRUCTURE

The semiquantitative picture of radiating shock structure developed above can be sharpened considerably by recourse to numerical calculation. The classic study was made by Heaslet and Baldwin (**H3**), whose results we summarize here. As before, consider a radiating perfect gas with an imbedded steady shock and assume that the material is sufficiently opaque that the radiation flux vanishes at upstream and downstream infinity. But we now drop assumptions (1) and (3) of the Zel'dovich–Raizer analysis, and thus allow for the enthalpy and pressure of the upstream material, and use a more accurate solution of the transfer equation.

The equations describing the flow are

$$\rho u = \dot{m}, \tag{104.68a}$$

$$\dot{m}u + p = \dot{m}C_1, \tag{104.68b}$$

and

$$\dot{m}(h + \tfrac{1}{2}u^2) + F = \dot{m}C_2^2 \tag{104.68c}$$

where $\dot{m}$, C_1, and C_2 are constants of integration. F is the radiation flux

$$F(\tau) = 2\sigma_R\left[\int_{-\infty}^{\tau} T^4(\tau')E_2\,|\tau'-\tau|\,d\tau' - \int_{\tau}^{\infty} T^4(\tau')E_2\,|\tau'-\tau|\,d\tau'\right] \tag{104.69}$$

where

$$\tau(x) \equiv \int_0^x \kappa(x)\,dx. \tag{104.70}$$

As before, we take $\tau = 0$ at the shock, $\tau < 0$ upstream, and $\tau > 0$ downstream; similarly $F < 0$ in the upstream direction.

Nominally (104.69) is a full transport solution for the lab-frame flux, but it omits velocity-dependent terms (cf. §93); similarly terms in E_0 and P_0 have been ignored in (104.68). All of these terms are small unless the temperature in the flow is very high, and can be neglected in the present context. To simplify the analysis, Heaslet and Baldwin use the exponential approximation (**V6**)

$$E_2(\tau) \approx me^{-n\tau}, \tag{104.71}$$

with $m = \frac{1}{3}n^2$ (which assures recovery of the diffusion limit) and $n = 1.562$.

Introducing a new independent variable $\xi \equiv n\tau$, Heaslet and Baldwin succeeded in reducing (104.68) and (104.69) to a single differential equation:

$$(d^2\theta/d\xi^2) - (\theta - \theta_\infty) = (d\mathcal{F}/d\theta)(d\theta/d\xi), \tag{104.72}$$

where

$$\theta(\xi) \equiv \{[\gamma/(\gamma+1)] - v(\xi)\}^2 \tag{104.73}$$

with $v(\xi) \equiv u(\xi)/C_1$ and

$$C_1 = u[1 + (1/\gamma M^2)] = \text{constant}, \tag{104.74}$$

and where

$$\mathcal{F}(\theta) \equiv \frac{\mathcal{K}}{4}\left[\frac{\gamma}{(\gamma+1)^2} + \left(\frac{\gamma-1}{\gamma+1}\right)\operatorname{sgn}(\xi)\theta^{1/2} - \theta^4\right]. \tag{104.75}$$

The solution depends on the two dimensionless parameters

$$\theta_\infty \equiv \tfrac{1}{4}(v_1 - v_2)^2 \tag{104.76a}$$

and

$$\mathcal{K} \equiv 32m(\gamma-1)\sigma_R C_1^6/(\gamma+1)nR^4\rho_1 u_1 \tag{104.77a}$$

which can be rewritten (**G7**) as

$$\theta_\infty = \tfrac{1}{4}\{[1+(1/\gamma M_1^2)]^{-1} - [1+(1/\gamma M_2^2)]^{-1}\}^2 \tag{104.76b}$$

and

$$\mathcal{K} = \frac{32m\gamma^4}{n(\gamma+1)}\left(M_1 + \frac{1}{\gamma M_1}\right)^6\left(\frac{\sigma_R T_1^4}{\rho_1 u_1 c_p T_1}\right) = \frac{f(M_1)}{\text{Bo}}, \tag{104.77b}$$

where Bo is the Boltzmann number. The parameter θ_∞ is essentially a measure of the shock strength; $\theta_\infty \to 0$ as $M_1 \to 1$, and $\theta_\infty \to (\gamma+1)^{-2}$ as $M_1 \to \infty$ (approaching 0.141 for $\gamma = \frac{5}{3}$ and 0.174 for $\gamma = \frac{7}{5}$). The parameter $\mathcal{K}$, being proportional to the inverse of the Boltzmann number, measures the importance of radiation; $\mathcal{K}$ is large when the upstream gas is hot and/or rarefied. $\mathcal{K}$ and θ_∞ are competing parameters because increasing the shock strength tends to steepen the profile, whereas increasing radiation tends to smear it.

Results from a set of computations for a gas with $\gamma = \frac{7}{5}$ (diatomic molecules) are shown in Figure 104.6. The computations span the range

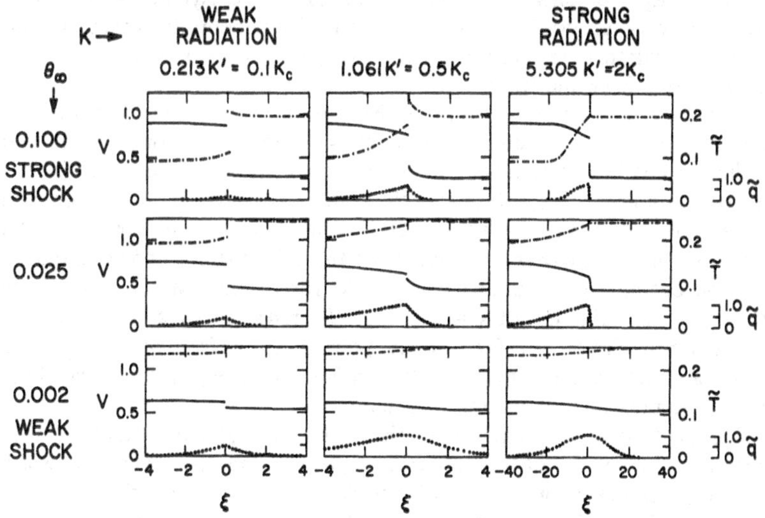

Fig. 104.6 Dimensionless velocity (solid curves), temperature (dash-dot curves), and heat flux (dotted curves) as a function of optical depth in shocks of different strengths and different amounts of radiation. From (**H3**) by permission.

from weak to strong shocks and, at each shock strength, from weak to strong radiation. The plots show the variations of the dimensionless velocity v, temperature

$$\tilde{T} \equiv RT/C_1^2 = v - v^2, \tag{104.78}$$

and radiation flux

$$\tilde{F} \equiv \frac{(\theta_\infty - \theta)}{\theta_\infty} = \frac{4(v_1 - v)(v - v_2)}{(v_1 - v_2)^2} = \frac{-2(\gamma - 1)}{(\gamma + 1)\theta_\infty \dot{m} C_1^2} F. \tag{104.79}$$

The radiation parameter is stated in units of

$$\mathcal{K}' \equiv 2\sqrt{2}\,(\gamma + 1)^7 \theta_\infty^{1/2} / \gamma^3 (\gamma - 1) \tag{104.80}$$

and the "critical" value $\mathcal{K}_c \equiv \frac{3}{2}\sqrt{2}\,\mathcal{K}'$ above which the solution is dispersed by radiation and is therefore continuous unless continuity is precluded by too large a value of θ_∞ (i.e., shock strength).

The bottom row of the figure gives the results for a weak shock ($M_1^2 = 1.20$). For weak radiation v and $\tilde{T}$ are discontinuous, their departures from constancy being antisymmetric about the front, while the radiation flux is symmetric. For moderate and strong radiation the shock is completely dispersed, all variables being continuous; as the radiation parameter $\mathcal{K}$ increases the radiation flux becomes asymmetric, with radiation penetrating deeper into the upstream flow than into the downstream flow.

At intermediate shock strength ($M_1^2 = 2.05$), shown in the middle row, the shock is discontinuous for both weak and moderate radiation, becoming fully dispersed only at the largest value of $\mathcal{K}$. The asymmetry in the flux becomes very pronounced as the radiation parameter increases. In the strong radiation case there is a small temperature peak a short distance downstream from the front.

The top row shows strong shocks ($M_1^2 = 6.4$). In the weak radiation case the results differ but little from a classical inviscid strong shock; the radiation flux is symmetric about the front. The moderate radiation case is a good example of a subcritical shock as described by Zel'dovich and Raizer; note that the present, more accurate, calculation shows that the flux distribution is asymmetric about the front. The strong radiation case is a good example of a supercritical shock; for the largest value of $\mathcal{K}$ the upstream material is hot enough that T_- rises to T_2 and a radiation precursor is driven far (i.e., many photon mean free paths) into the upstream material. The flux distribution is strongly asymmetric, and there is a large postshock temperature spike.

Detailed asymptotic analysis of the equations by Heaslet and Baldwin shows that the width of the precursor is proportional to $\mathcal{K}$ while the width of the postshock temperature spike varies as $\mathcal{K}^{-1}$. As remarked by J. H. Clarke (**G7,** p. 281) the latter result can be understood physically by noting that (104.68c) implies an upper bound to the value of $|F|$, namely,

$$-F = \dot{m}(h - h_2 + \tfrac{1}{2}u^2 - \tfrac{1}{2}u_2^2) < \dot{m}(h + \tfrac{1}{2}u^2) \leq m(h_1 + \tfrac{1}{2}u_1^2) \tag{104.81}$$

which can be attained only in the limiting case that the downstream material is absolutely stationary and cold, all of its internal and kinetic energy being converted into radiation flowing upstream. It then follows from (104.69) that if T in the postshock flow is very large, the only way $|F|$ can remain below its upper bound is for the hot region to be *optically thin*, for then $-F \approx \sigma_R T_+^4 \,\Delta\tau$ where $\Delta\tau \ll 1$ is the optical thickness of the hot region. Thus the postshock temperature overshoot region must collapse to a narrow spike less than one photon mean free path thick. Because T_+ is bounded [cf. (104.67)], the thickness $\Delta\tau$ of the spike does not vanish, but also remains bounded.

Necessary and sufficient conditions for strong shocks to be dispersed by radiative smoothing have been determined by Mitchner and Vinokur (**M12**). They find that in the absence of radiation pressure a sufficient condition for a strong shock in a radiating perfect gas to be discontinuous is that (1) the upstream Mach number exceed the critical value

$$M_1^2 > M_{cr}^2 \equiv (2\gamma - 1)/\gamma(2-\gamma), \qquad (104.82)$$

or that (2) the upstream gas temperature be sufficiently low. For $\gamma = \frac{5}{3}$, $M_{cr}^2 = 4.2$, and for $\gamma = \frac{7}{5}$, $M_{cr}^2 = 2.14$. The intermediate-strength shocks in Figure 104.6 are just below this critical strength, and can be continuous if $\mathcal{K}$ is large, but become discontinuous when $\mathcal{K}$ is small (i.e., the upstream material is cold). Similarly, the strong shocks in that figure are all above the critical strength, hence all are discontinuous.

Mitchner and Vinokur show that the necessary and sufficient conditions for discontinuous shocks can be stated in terms of the upstream Mach number and a dimensionless parameter ψ measuring the influence of radiation. One can show that their $\psi = n\mathcal{K}/4\sqrt{3}\,m$ where $\mathcal{K}$ is defined by (104.77); thus ψ, like $\mathcal{K}$, is proportional to the inverse Boltzmann number. Numerically

$$\psi = 3.39 \times 10^{-12}[(\gamma-1)/(\gamma+1)][1+(\gamma M_1^2)^{-1}]^6(A^3 u_1^5/n_1) \qquad (104.83)$$

where A is the mean molecular weight of the gas in atomic mass units, and n_1 is the upstream particle density.

From numerical integrations Mitchner and Vinokur determine the values $\psi_c(M_1)$ necessary for a continuous shock profile, as shown in Figure 104.7 for various values of γ. For the solution to be continuous ψ must equal or exceed the value implied by the curves. The curve marked "x" indicates the limiting Mach number below which thermal conduction alone can smooth the shock; the curve marked "y" indicates the limiting Mach number given by (104.82), above which all shocks are discontinuous. Necessary *and* sufficient conditions are shown in Figure 104.8; we see that below the critical Mach number M_{cr} continuous solutions are possible if $\psi \geq \psi_c^*$, that is, if the upstream Boltzmann number is sufficiently small. Above that Mach number all solutions are discontinuous.

The effects of viscosity on shock structure in thermally conducting,

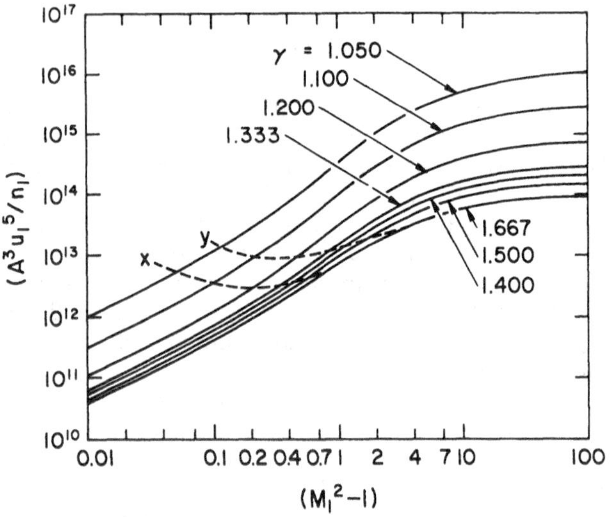

Fig. 104.7 Critical value of radiation parameter necessary for continuous radiating shock transition, as a function of upstream Mach number for various values of γ. From (**M12**), by permission.

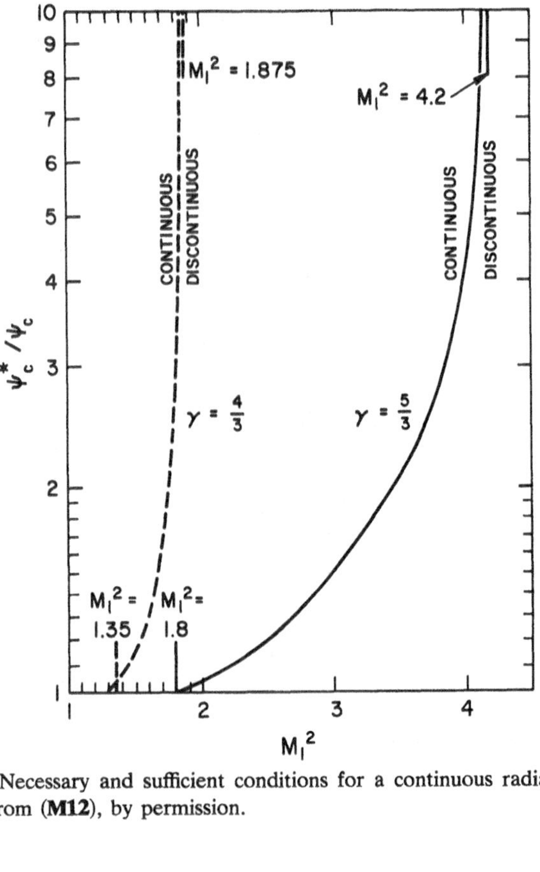

Fig. 104.8 Necessary and sufficient conditions for a continuous radiating shock transition. From (**M12**), by permission.

radiating gases was studied by Traugott. Generally these effects on an isolated shock are small; all the solutions are dispersed by radiation only. The reader can find details in (**T1**).

EFFECTS OF RADIATION PRESSURE AND ENERGY DENSITY ON SHOCK STRUCTURE

The conclusion just stated that shocks above a certain strength are always discontinuous is not strictly correct because we have ignored the pressure and energy density of the radiation. As shown by Belokon (**B3**) and Imshennik (**I1**), (**I2**), when radiation dominates the total fluid pressure and energy density, it is again possible for all variables to remain continuous across the front, even in the absence of viscosity. This remarkable result has direct relevance to astrophysical applications where the upstream material can be both very hot and rarefied, and to very strong shocks that produce enormous postshock temperatures.

To get a physical feeling for the problem, consider an extremely strong shock propagating into cold material. Assume that the radiation and material are in equilibrium so that $E = 3P = a_R T^4$. The shock's structure is described by (104.9) and (104.10), which not only connect initial and final states but also imply that the momentum and energy fluxes are constant throughout the flow. Assume for the moment that $P \gg p$ and $E \gg e$, so that we can neglect p and e. Then (104.9) implies

$$P = \dot{m}u_1(1-\eta) \tag{104.84}$$

and (104.10) implies

$$F = \tfrac{1}{2}\dot{m}u_1^2(1-\eta^2) - 4u_1\eta P = \tfrac{1}{2}\dot{m}u_1^2(1-\eta)(1-7\eta). \tag{104.85}$$

Equation (104.84) shows that P, hence T, is a monotone increasing function of ρ (hence of x), while (104.85) shows that $F \leq 0$ everywhere in the flow, which is consistent with the monotonic increase of $T(x)$.

From our earlier discussion of subcritical shocks we know that we can always obtain a continuous solution if we can assure that $T(\eta)$ is a monotonic decreasing function of η on the range $\eta_2 \leq \eta \leq 1$ instead of passing through a maximum; hence we infer that shocks with $P \gg p$ and $E \gg e$ will be continuous. But (104.84) is an oversimplification; accounting for gas pressure we have

$$\rho RT + \tfrac{1}{3}a_R T^4 = (\rho_1 RT/\eta) + \tfrac{1}{3}a_R T^4 = \dot{m}u_1(1-\eta). \tag{104.86}$$

From (104.86) one sees that $T(\eta = 0) = T(\eta = 1) = 0$, and that $T(\eta)$ actually has a single maximum for some $\eta = \eta_{\max}$, $0 \leq \eta_{\max} \leq 1$. Nevertheless, if the momentum flux on the right-hand side of (104.86) is sufficiently large, we can assure that $P \gg p$ over nearly the whole range $0 < \eta < 1$. In this event we can force $\eta_{\max}$ to be very small because T^4 grows steadily with $(1-\eta)$ until η becomes so small that the term $\rho_1 RT/\eta$ finally becomes competitive with $\tfrac{1}{3}a_R T^4$ and thereafter forces T to decrease. In particular we can force $\eta_{\max}$ to be smaller than the smallest physically realizable value of η (i.e., $\eta_2=\frac{1}{4}$ when $P \ll p$ and $\eta_2=\frac{1}{7}$ when $P \gg p$ in a gas with $\gamma=\frac{5}{3}$). T will then be a monotonic decreasing function of η for $\eta \geq \eta_2$, which is the condition needed to obtain a continuous solution.

To determine the shock strength at which the discontinuity first disappears, we force the point (η_{max}, T_{max}) of (104.86) to coincide exactly with (η_2, T_2) at maximum compression. First rewrite (104.86) as

$$\frac{\rho_1 RT}{\eta}+\tfrac{1}{3}a_R T^4=\left(\frac{\rho_1 RT_2}{\eta_2}+\tfrac{1}{3}a_R T_2^4\right)\left(\frac{1-\eta}{1-\eta_2}\right) \tag{104.87}$$

and adjoin (104.25) rewritten as

$$T_2^3=(3R\rho_1/a_R)(\eta_2-\eta_0)/\eta_2\eta_0(1-7\eta_2) \tag{104.88}$$

where $\eta_0 \equiv (\gamma-1)/(\gamma+1)$. Differentiating (104.87) with respect to η, setting $(dT/d\eta)=0$, and demanding that $(\eta_{max}, T_{max})=(\eta_2, T_2)\equiv(\eta_*, T_*)$ we find

$$T_*^3=(3R\rho_1/a_R)(1-2\eta_*)/\eta_*^2. \tag{104.89}$$

Combining (104.89) with (104.88) we have

$$(14\eta_0-1)\eta_*^2-8\eta_0\eta_*+\eta_0=0, \tag{104.90}$$

which yields the physically relevant root

$$\eta_*=1/[4+(2+\eta_0^{-1})^{1/2}]. \tag{104.91}$$

For $\gamma=\frac{5}{3}$, we find $\eta_*=1/6.45$, close to the limiting value for pure radiation; at this value of η,

$$\alpha_*=(P/p)_*=(a_R/3R\rho_1)\eta_* T_*^3=4.45, \tag{104.92}$$

results obtained by Belokon (**B3**). Shocks sufficiently strong that $\eta_2 \leq \eta_*$ will be continuous.

The analysis above is based on the simplifying assumptions that (1) the gas is a single fluid with all particles in equilibrium at the same temperature, and (2) the upstream material is cold, so that both the radiation and gas pressures are essentially zero. If one relaxes these assumptions one finds (**I1**), (**I2**) that (104.91) yields an upper bound on the compression ratio needed to obtain a continuous solution, and that in fact such solutions can be obtained for a very wide range of upstream flow conditions.

Imshennik considers shocks in a plasma of ions and electrons, with the two species of particles having different temperatures, allowing for radiation pressure and energy density. His results are most conveniently displayed in a plot of the downstream radiation-pressure number $\alpha_2 \equiv (P/p)_2$ as a function of η on a set of Hugoniots corresponding to prechosen values of the upstream radiation-pressure number α_1. In this plot, Figure 104.9, the curve ABC separates continuous and discontinuous solutions. For α_1 less than a certain limiting value $(\alpha_1)_0$, continuous solutions are obtained if the compression ratio r is either less than one critical value or greater than

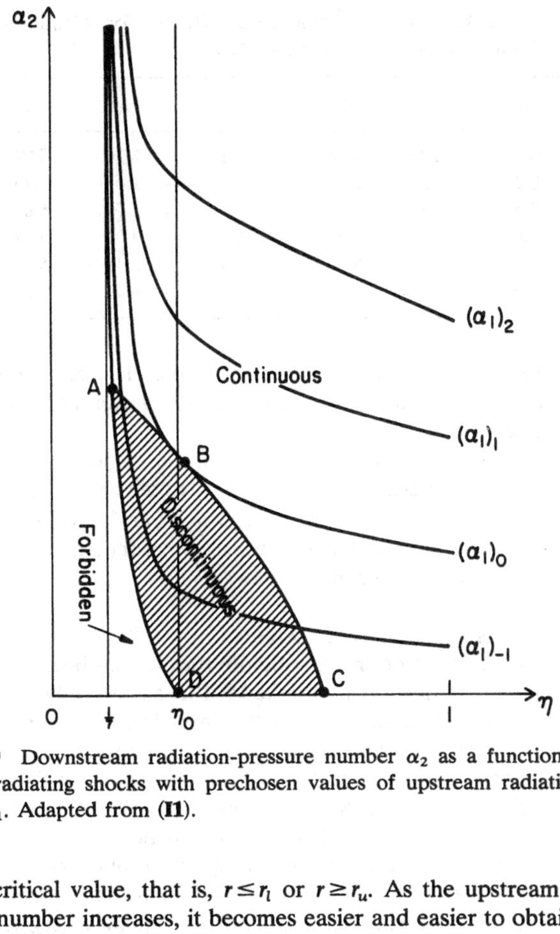

Fig. 104.9 Downstream radiation-pressure number α_2 as a function of volume ratio for radiating shocks with prechosen values of upstream radiation-pressure number α_1. Adapted from **(I1)**.

another critical value, that is, $r \le r_l$ or $r \ge r_u$. As the upstream radiation-pressure number increases, it becomes easier and easier to obtain continuous solutions, that is, r_l increases and r_u decreases. When α_1 just equals the limiting value $(\alpha_1)_0$, $r_l = r_u$, and therefore all shocks with $\alpha_1 \ge (\alpha_1)_0$ are continuous.

Numerical results for r_l and r_u as a function of α_1 in a hydrogen plasma ($Z = 1$) are given in Table 104.2. One sees that $(\alpha_1)_0 = 2.2774$ and $r_l = r_u = 4.33$ (point *B* in Figure 104.9); here $\alpha_2 = 3.8249$. When $\alpha_1 = 0$ we find $r_l = 1.1875$ (point *C* in the figure) corresponding to the critical compression ratio below which shocks are completely dispersed by electron conduction in the absence of radiation (cf. §57), and $r_u = 6.58$ (point *A* in the figure) corresponding to the limiting shock strength in a cold gas beyond which the shock becomes continuous through the action of radiation pressure alone.

Point *A* in the limit as $Z \to \infty$ corresponds to the solution found by Belokon. Imshennik obtains $(Z, r_*, \alpha_*) = (1, 6.58, 6.46)$, $(2, 6.56, 5.79)$,

Table 104.2. Critical Compression Ratios for Continuous Radiating Shocks in a Hydrogen Plasma

α_1	r_l	r_u	α_1	r_l	r_u
0.0	1.19	6.58	1.2	2.26	5.97
0.2	1.31	6.47	1.4	2.48	5.83
0.4	1.48	6.39	1.6	2.72	5.67
0.6	1.66	6.30	1.8	3.00	5.47
0.8	1.85	6.20	2.0	3.44	5.22
1.0	2.05	6.09	2.277...	4.33	4.33

(∞, 6.45, 4.45). As was true for the case of pure electron conduction, the conditions required to guarantee a continuous solution become less stringent with increasing Z (cf. §57).

The curve AD is the Hugoniot for $\alpha_1 = 0$. The region to the left of this curve is "forbidden" because the smallest η to which we can compress a gas with no radiation pressure whatever is $\eta_0 = (\gamma-1)/(\gamma+1)$; if we try to go beyond, we generate a nonzero downstream radiation pressure ($\alpha_2 > 0$) even though $\alpha_1 = 0$ in the upstream flow.

RADIATING SHOCKS IN ISOTHERMAL MATERIAL

Our discussion of radiating shock structure thus far has been predicated on the assumption that the shock is imbedded in an optically thick medium and that all radiation emitted across the front from the hot downstream material is ultimately reabsorbed in the upstream material. The upstream and downstream conditions at large distances from the front are then related by the radiation-modified Rankine–Hugoniot conditions [i.e., (104.8) to (104.10) with $F_{01} = F_{02} = 0$]. But as noted in connection with (104.28) the situation is different if the upstream material is so optically thin that radiation escapes freely to infinity, in which case the temperature jump across the front is smaller, and the density jump is larger, than in an adiabatic shock of the same strength. An extreme case is where the downstream gas radiates away all of the energy of compression across the front and cools back to the original upstream temperature.

In astrophysics the conditions just described actually occur in *gaseous nebulae* where, to a first approximation, the material is essentially in radiative equilibrium in the dilute radiation field of an illuminating star. That is, the temperature of the nebular material is fixed by radiative processes alone, independent of the hydrodynamics, because the radiative heating and cooling rates are orders of magnitude larger than the rate of compressional heating (**O4,** 146), (**S20,** 167). Indeed it is a good approximation to make the idealization that the material remains at *constant temperature* as it passes through the front. More precisely, one is saying that the radiative relaxation zone behind the viscous dissipation zone has a negligible geometric thickness (i.e., is *unresolvable* telescopically) hence the two zones can be lumped together into a composite front that has the net effect of compressing the gas while leaving its temperature unchanged. Such shocks are usually (but loosely) called "isothermal shocks" in astrophysics; we do not use this terminology because it conflicts with that used in Chapter 5 and in the remainder of this chapter.

We can determine the properties of shocks in isothermal material by formally setting $\gamma = 1$ in (56.40) to (56.43), whence we find

$$T_2/T_1 = 1, \tag{104.93}$$

$$\rho_2/\rho_1 = p_2/p_1 = M_1^2, \tag{104.94}$$

and

$$M_2^2 = 1/M_1^2. \tag{104.95}$$

We see that the density jump across a strong shock in isothermal material can be arbitrarily large rather than approaching a finite upper bound as in an adiabatic shock.

Optically thin shocks are continuously damped by radiative energy loss from the thermodynamically open system, hence the steady-flow model provides only an ephemeral snapshot valid over the time required for an element of material to flow through the shock, be compressed and heated (perhaps ionized), and then radiate and cool (perhaps recombine) downstream. The model provides a caricature of the instantaneous behavior of, say, an isolated pulse in the optically thin layers of a stellar atmosphere. Alternatively it could apply to an optically thin shock continuously driven in a laboratory shock tube.

NONEQUILIBRIUM EFFECTS IN OPTICALLY THIN RADIATING SHOCKS

Skalafuris and Whitney have analyzed the nonequilibrium structure of optically thin radiating shocks in hydrogen (**S12**), (**S13**), (**W6**). [A broader discussion of background physics and approximation schemes can be found in (**C18**).] The shocks are assumed to be steady, and to propagate in an infinite homogeneous medium that is optically thick in the Lyman continuum but optically thin in all subordinate continua. Thus all photons emitted by recombination in the subordinate continua escape to infinity without reabsorption, and the gas temperature downstream returns to the original upstream temperature as described above. Their calculation is more realistic than the picture presented above because it allows for (1) different kinetic temperatures for the various particle species, (2) nonequilibrium ionization and recombination, and (3) radiation transport.

The calculations show that the shock can be divided into four zones: (1) a precursor, which is preheated and partially ionized by radiation from behind the front; (2) an external relaxation zone in which translational equilibrium is established for each particle species; (3) an internal relaxation zone in which ionization equilibrates and the particle temperatures

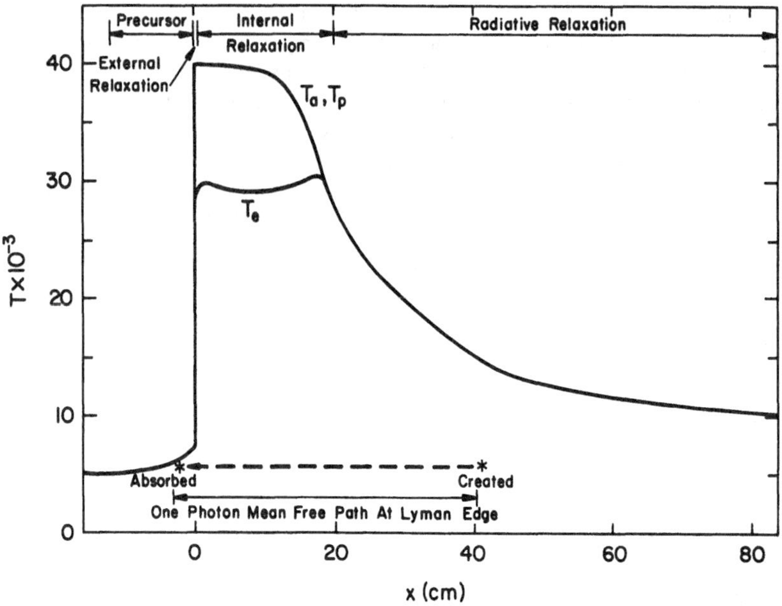

Fig. 104.10 Temperature structure in a nonequilibrium radiating shock. Adapted from (**S12**), (**S13**), and (**G7**).

equalize; and (4) a radiative relaxation zone in which protons and electrons recombine and emit radiation. This structure is sketched in Figure 104.10 for a strong shock propagating at 40 km s^{-1} ($M_1 \approx 5$) into an ambient medium with $T = 5000$ K and $\rho = 2.5\times10^{-9}$ g cm^{-3} ($n_H \approx 1.5\times10^{15}$ cm^{-3}).

The cool material in the precursor absorbs Lyman continuum photons, which slightly heats the gas, and the ionization fraction rises from zero to about 0.01 immediately in front of the shock. The external relaxation zone is of essentially zero thickness both geometrically and optically; as discussed in §57 the atoms and ions are strongly heated by viscous dissipation while the electrons are heated by adiabatic compression.

In the internal relaxation region the proton and atom temperatures are locked together by charge-exchange reactions ($H + H^+ \rightarrow H^+ + H$). The electron gas is cooled by inelastic collisions that ionize atoms (the electrons losing 13.6 eV per ionization from the tail of a Maxwellian distribution with an average energy of only about 2 eV) and are heated by Coulomb collisions with protons. In this region the ionization fraction rises to its equilibrium value (about 0.36) over a distance determined by the collisional ionization rate. The material flows through the entire zone in a time that is short compared to the recombination time, so recombination, leading to photon emission, occurs in a long tail extending far downstream. Indeed, in very strong shocks the postshock plasma is completely ionized and is too hot to recombine efficiently; it must therefore first cool slowly by free–free emission (bremsstrahlung) before recombination can even begin. As predicted by (104.94), the final compression ratio in these shocks is very large.

It is sobering to note that the calculations by Skalafuris and Whitney show that both the internal relaxation and radiative relaxation zones are at most a few meters thick, which is infinitesimal compared to an atmospheric scale height (10^2 to 10^3 km). It is thus virtually impossible to resolve these features in a general time-dependent flow unless adaptive-mesh techniques are used, because otherwise the zoning required would be prohibitively fine.

105. Propagating Shocks

In §104 we concentrated on how radiation affects the *structure* of steady shocks. We now examine how it affects the *propagation* of shocks via radiative energy exchange (particularly radiative losses) and momentum exchange (i.e., the effects of radiation pressure), and by driving the thermodynamic state of the material away from equilibrium.

WEAK SHOCK THEORY

A complete analytical theory of shock propagation can be constructed for weak shocks (cf. §58). In treating radiative effects, attention has been focused exclusively on radiative energy exchange, and both the effects of radiative forces and the dynamics of the radiation field itself have been ignored. The entropy increase across a weak shock front is given by (56.56), which implies that the heat dissipated by the shock is

$$\Delta q = 2\gamma p_0 m^3/3(\gamma+1)^2\rho_0, \tag{105.1}$$

where $m \equiv M_1^2 - 1$. Previously we ignored radiative losses and assumed that Δq went onto raising the temperature of the downstream gas. We now take the opposite extreme view that this energy is all radiated away so that the postshock material ultimately returns to its original temperature.

Different investigators have made differing assumptions about how radiative relaxation behind a weak shock proceeds. For example, in the *Weymann cycle* (**W4**), (**O3**), one assumes that the downstream gas first cools rapidly by radiation at constant density until the specific entropy of the gas returns to its upstream value, and then expands adiabatically back to its original pressure and density. Alternatively, in the *Schatzman cycle* (**S4**) one assumes that the downstream gas first expands adiabatically back to its original upstream pressure (hence to a lower density) and then cools by radiation at constant pressure back to its original density. Both of these cycles are hypothetical and are chosen only because their consequences can be followed analytically; there is no guarantee that either one is an accurate description of reality. For strong shocks, the two cycles convert different amounts of dissipated heat into radiation [see equations (6.185) and (6.194) in (**B6**)]. However for weak shocks they both give equation (105.1) for Δq.

Using (105.1) for the energy radiated per gram, one can proceed as in §58 to deduce equations governing the propagation of pulses or N waves. Accounting for gradients of temperature, hence sound speed, and of the ratio of specific heats γ, one finds (**U1**) that the variation of the Mach number with height for an N wave is given by

$$\frac{dm}{dz} = -\frac{m}{\Lambda_0} + \frac{1}{2H} - \frac{1}{2}\left(\frac{d \ln \gamma}{dz} + \frac{d \ln a}{dz}\right). \tag{105.2}$$

Allowance for refraction of oblique shocks changes the coefficient of $(d \ln a/dz)$ from $-\frac{1}{2}$ to $-\frac{3}{2}$ (**U1**).

Equation (105.2) can be integrated numerically for a given model atmosphere, and radiative losses as a function of height can then be calculated from (105.1). For example, Ulmschneider (**U1**) has made an extensive set of integrations for representative model solar atmospheres. He attempted to determine the properties of shocks that could be responsible for heating the lower solar chromosphere by comparing computed radiative losses with semiempirical estimates obtained from observation. Setting the shock dissipation rate equal to the observed chromospheric radiative flux, one immediately concludes that within the first 1000 km above the photosphere the radiative losses are easily accounted for by weak shocks ($m \lesssim 0.25$). The required mechanical energy flux, of order 2×10^6 ergs cm^{-2} s^{-1}, is only a small fraction of the acoustic flux generated in the subphotospheric convection zone.

The calculations yield an energy dissipation rate that is a sensitive function of the period of the N wave, but that depends only weakly on the input energy flux; hence one can estimate representative wave periods responsible for the heating. Using the best available solar models and estimates of radiative losses one obtains (**U2**) the best fit to the data for periods around 25 to 30 s ($\Lambda \sim 175$ to 200 km). These periods are well below the acoustic-cutoff period in the temperature-minimum region, hence acoustic waves of such periods generated in the convection zone could readily propagate into the chromosphere where they would steepen into sawtooth waves. As remarked in §58, the radiative losses from the waves are severe: over 90 percent of the original wave energy is dissipated at heights below 2000 km.

STRONG LTE SHOCKS

Strong radiating shocks can be studied either with similarity solutions or by direct numerical simulation. We briefly describe here a few representative problems of astrophysical interest.

(*a*) *Similarity Solutions* Consider first the propagation of a strong, *self-similar radiating shock* moving upward in an exponentially stratified atmosphere (**L6**). Both the high temperatures generated in the accelerating shock front and the large photon mean free paths (which increase exponentially with height) at high altitudes imply intense, efficient heat transport by radiation, which will tend to obliterate temperature gradients. To obtain an analytically tractable caricature of these radiative effects, one assumes that the postshock material is *isothermal* (in contrast to the case of adiabatic propagation considered in §60). This extremely rough treatment of radiation manifestly cannot be expected to provide a realistic description of radiative effects on a shock.

As in §60, let Z denote the height of the shock, and let $\zeta \equiv z - Z$ be the distance behind the shock, all in units of scale heights. The speed and position of the shock are again given by (60.15b) and (60.17), where t_∞ is given by (60.18). The numerical value of the similarity exponent α will differ from that found in the adiabatic case; it is found to be a function of the shock strength as measured by the volume ratio η_2 across the shock. Jump conditions at the front may be written

$$\rho_2 = \rho_1/\eta_2, \tag{105.3a}$$

$$v_2 = (1-\eta_2)v_S, \tag{105.3b}$$

and

$$p_2 = (1-\eta_2)\rho_1 v_S^2. \tag{105.3c}$$

These relations suggest using dimensionless variables $\tilde{\rho}$, $\tilde{v}$, and $\tilde{p}$ defined by

$$\rho(\zeta) = [\rho_1(Z)/\eta_2]\tilde{\rho}(\zeta), \tag{105.4a}$$

$$v(\zeta) = [(1-\eta_2)v_{S0}t_\infty/(t_\infty - t)]\tilde{v}(\zeta), \tag{105.4b}$$

and

$$p(\zeta) = [(1-\eta_2)\rho_1(Z)v_{S0}^2 t_\infty^2/(t_\infty - t)^2]\tilde{p}(\zeta). \tag{105.4c}$$

The height variations of these dimensionless variables are found from numerical integration of dimensionless versions of the mass, momentum, and energy conservation equations for isothermal flow, starting from initial conditions $\tilde{\rho}(0) = \tilde{v}(0) = \tilde{p}(0) = 1$. The resulting equations have a singular point; to obtain a unique single-valued solution of the system one must impose a constraint between α and η_2, namely

$$\alpha/(\alpha+1) = [\eta_2(1-\eta_2)]^{1/2}. \tag{105.5}$$

The resulting flow-variable distributions for $\gamma = \frac{5}{3}$ are shown in Figure 105.1.

At high altitudes the preshock density is so low that little radiation is absorbed by the upstream gas, which remains cold. The shock therefore approaches its limiting strength, with $\eta_2 = (\gamma-1)/(\gamma+1)$; hence

$$1/\alpha \to (\gamma+1)[2(\gamma-1)]^{-1/2} - 1, \tag{105.6}$$

from which we find $\alpha = (0.537, 0.885)$ for $\gamma = (\frac{4}{3}, \frac{5}{3})$. The corresponding values for an adiabatic shock are $\alpha = (0.176, 0.204)$. From (60.15b) it then follows that a shock ascends more rapidly in an exponential atmosphere

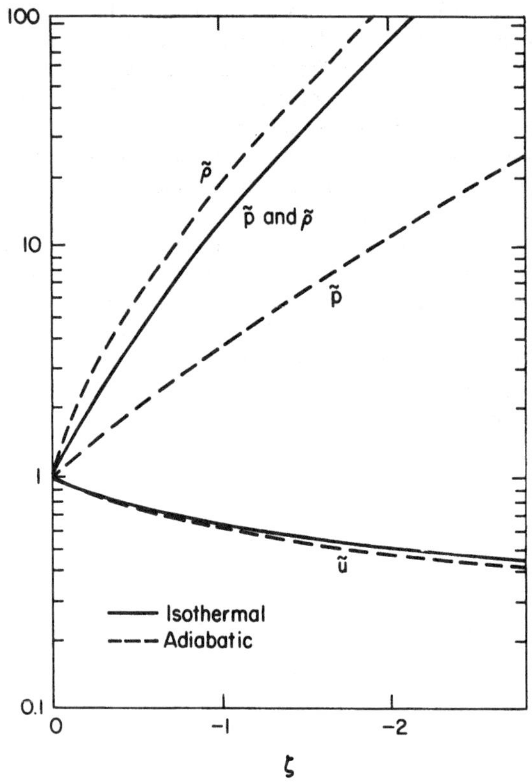

Fig. 105.1 Similarity solutions for radiating shock in an exponential atmosphere. From (**L6**), by permission.

when the downstream gas is isothermal than when it is adiabatic. Higher velocities are reached in the isothermal case because the shock is driven by a larger pressure gradient that develops because the downward flux of radiation from the front raises the temperature, hence pressure, in the dense postshock material, even far from the shock front (**L6**).

Again assuming isothermal downstream material, a similarity solution can also be obtained for power-law density distributions of the form

$$\rho = bx^{\delta}, \tag{105.7}$$

where b and δ are positive constants, and x increases into the medium (**A7**), (**S1**). Equation (105.7) provides a rough caricature of the density distribution in the outer part of a stellar envelope, and thus can be used to study the behavior of a shock emerging from the interior of a star. Taking $t = 0$ to be the instant when the shock arrives at the surface, the shock position is assumed to be given by

$$X = a(-t)^{\alpha}. \tag{105.8}$$

Invoking the jump conditions (105.3), one can obtain a solution in similarity form:

$$x = X\xi(\mu), \tag{105.9a}$$

$$v = (1-\eta_2)\dot{X}\nu(\mu), \tag{105.9b}$$

$$\rho = [\rho_1(X)/\eta_2]g(\mu), \tag{105.9c}$$

and

$$p = (1-\eta_2)\rho_1(X)\dot{X}^2\pi(\mu), \tag{105.9d}$$

where $\mu = 1$ at the position of the shock, and $\nu(1) = g(1) = \pi(1) = 1$. To obtain a physically meaningful solution through the singular point of the differential equations that determine ξ, ν, g, and π, we must constrain the exponent α. An approximate analytical solution (**A7**) gives

$$\alpha = \{1 + \delta[\eta_2/(1-\eta_2)]^{1/2}\}^{-1}. \tag{105.10}$$

The resulting flow-variable distributions for $\gamma = \frac{5}{3}$ are shown in Figure 105.2.

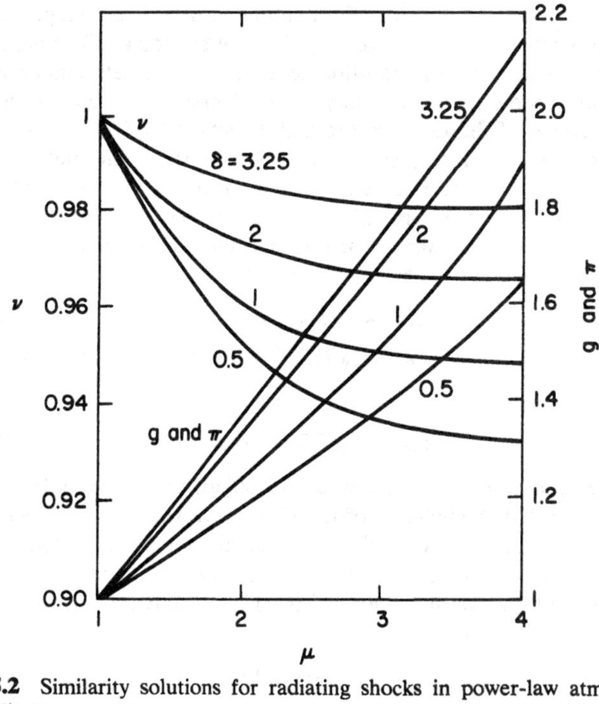

Fig. 105.2 Similarity solutions for radiating shocks in power-law atmospheres. From (**S1**), by permission.

For a stellar envelope in radiative equilibrium $\delta \approx 3.25$ (**C23**). Using this value of δ and taking the limiting value $\eta_2 = (\gamma-1)/(\gamma+1)$ we find $\alpha = 0.348$ for $\gamma = \frac{5}{3}$, which is in good agreement with the more precise value $\alpha = 0.314$ obtained from numerical integration (**S1**). For comparison, the similarity exponent for adiabatic downstream flow (**Z3,** Chap. 12) is $\alpha = 0.590$. From (105.8) one finds that $\dot{X} \propto X^{(1-\alpha)/\alpha}$, hence again the shock velocity is much larger when the downstream material is isothermal than when it is adiabatic.

Similarity solutions for planar radiating shocks driven by a piston are given in (**W1**); these account for radiation diffusion or radiative emission losses in the optically thick and thin limits, respectively. A large number of similarity solutions have been obtained for spherically symmetric radiating blast waves emanating from a point explosion. For example, similarity solutions for very intense explosions in air are given in (**E1**); radiating shocks driven by a piston in planar, cylindrical, and spherical geometry are discussed in (**H4**); a solution for a spherical, radiating, optically thick blast wave in a self-gravitating body such as a star is given in (**O1**); and spherical, radiating shocks in a star with a power-law density distribution are discussed in (**R3**).

While similarity solutions can be used to gain basic insight and to derive scaling rules for the behavior of radiating blast waves, deeper analysis shows that if radiative energy exchange is to be treated at all consistently in either the optically thin or the diffusion limits, then a self-similar solution can be maintained only if the absorption coefficient of the gas varies with a particular power of the temperature and density (**G2**), (**H4**), (**N3**). But the behavior of a material property such as opacity is actually determined by physical laws totally unconnected with the nature of the flow, and usually the requirements for a similarity solution cannot be satisfied. As an example, suppose we assume that the thermal conductivity, Planck mean opacity, and Rosseland mean opacity vary as

$$K = K_0(T/T_0)^{\beta_C}(\rho/\rho_0)^{\delta_C}, \qquad (105.11a)$$

$$\kappa_P = \kappa_{P0}(T/T_0)^{\beta_P}(\rho/\rho_0)^{\delta_P}, \qquad (105.11b)$$

and

$$\kappa_R = \kappa_{R0}(T/T_0)^{\beta_R}(\rho/\rho_0)^{\delta_R}, \qquad (105.11c)$$

where the subscript zero denotes a convenient reference state. Furthermore, assume that the velocity of the blast wave varies as $\xi^{-\lambda/2}$ where ξ is an appropriate similarity variable; for an adiabatic flow $\lambda = 3$ (cf. §60), and for a momentum-conserving shell $\lambda = 6$. Then, in order to obtain a consistent solution one must demand (**G2**) that the temperature exponents in (105.11) be given by

$$\beta_C = \tfrac{1}{2} - (1/\lambda), \qquad (105.12a)$$

$$\beta_P = (1/\lambda) - \tfrac{5}{2}, \qquad (105.12b)$$

and

$$\beta_R = (1/\lambda) + \tfrac{5}{2}. \qquad (105.12c)$$

For an opaque blast wave, the relevant opacity is the Rosseland mean; for a transparent wave it is the Planck mean. Equations (105.12) show that for the values of λ mentioned above β_P must be negative while β_R must be positive. It is extremely unlikely that any real material would have these properties; normally β_P and β_R will have the same sign. Typically β will be positive for a cold gas (e.g., air) that grows more opaque as it becomes excited, dissociates, and ionizes, whereas β will be negative for a hot gas (e.g., in a stellar interior where $\beta_R \approx -3.5$). Thus one can hope to construct a physically consistent solution only for an opaque front in cold material or a transparent front in hot material. An opaque blast in hot material (e.g., a stellar envelope) cannot be treated consistently, for, according to (105.12c), β_R would have the wrong sign.

Thus in many, perhaps most, problems a physically realistic solution will not behave in a self-similar manner. We must therefore turn to numerical modeling; we will discuss a selection of problems of astrophysical interest.

(*b*) *Shock Heating of the Solar Chromosphere* Chromospheric shock heating has been studied extensively by Ulmschneider and his co-workers (**U4**), (**K3**), (**U5**), (**U6**), whose results we summarize briefly here. Numerical modeling allows one to remove the limitations of weak shock theory, and to make a detailed calculation of shock heating in realistic atmospheres. The authors just cited solve the Lagrangean equations of continuity and momentum, and a gas energy equation that allows for radiative absorption and emission terms, for vertically propagating acoustic waves (**U5**). The waves are driven into the atmosphere by a periodic piston at the lower boundary. The radiation field is assumed to be quasi-static and all velocity-dependent terms are neglected (**K3**), hence no distinction is made between lab-frame and comoving-frame radiation quantities, and the dynamical behavior of the radiation field is ignored. Similarly, radiation forces are ignored, which is a good approximation in the solar atmosphere. In solving the transfer equation the Eddington approximation (one angle-point quadrature) is made.

The material is assumed to be in LTE, and to be grey with the opacity taken to be the Rosseland mean. The entropy and sound speed are chosen to be the fundamental thermodynamic variables, and the hydrodynamic equations are solved by the method of characteristics with a shock-finding algorithm. As discussed in §59, this method is a bit complicated to implement well; but it has the great advantage that the viscous dissipation zone (which is only a few particle mean free paths thick, hence always optically thin) is represented by a sharp discontinuity of zero optical thickness. In contrast, the standard Lagrangean pseudoviscosity technique smears the shock over several adjacent zones, which may be optically thick; this method may therefore give a seriously distorted picture of radiative exchange through the shock front. The radiative terms are iterated to consistency with the hydrodynamics at each timestep.

In an initial application of the code, Ulmschneider and Kalkofen (**U4**) studied the propagation and dissipation of short-period waves as perturbations on top of a prescribed (semiempirical) model. This approach does not yield a fully self-consistent final model, but it avoids the necessity of constructing a detailed nongrey, non-LTE, radiative-equilibrium initial model. While the analysis therefore has drawbacks (in particular, it does not handle strong shocks in the fully nonlinear regime quite correctly), it leads to some important conclusions. (1) Heights of shock formation agree with the position of the empirical chromospheric temperature minimum for waves with periods between 25 and 45 s and initial acoustic fluxes between 3×10^7 and 6×10^7 ergs cm^{-2} s^{-1}; both these ranges are in harmony with theoretical predictions of the acoustic spectrum emerging from the convection zone. (2) The mechanical flux in the waves at the height of shock formation agrees well with empirical estimates of chromospheric radiation losses if the waves have periods less than 35 s and initial acoustic fluxes between 2×10^7 and 6×10^7 ergs cm^{-2} s^{-1}. Moreover, such waves explain the observed variation of chromospheric radiation losses with height.

A much more complete ab initio calculation of chromospheric shock heating has been made by Ulmschneider et al. (**U6**). In this analysis the initial model is not fixed as in (**U4**), but is determined from a radiative equilibrium calculation using the hydrodynamic code. The final chromospheric structure is determined from time averages over many wave periods. This procedure has the disadvantage that the initial model represents the solar atmosphere less accurately than the best models that can be produced with a nongrey, multi-angle calculation. However, the disadvantages are outweighed by the fact that we then obtain a self-consistent treatment of the chromosphere's response to shock formation and dissipation. A *differential* comparison of the initial and final models should yield reasonably accurate estimates of the effects of mechanical energy dissipation that are almost independent of the initial model.

The run of physical variables with height is shown in Figure 105.3 for a wave train having a period of 30 s and an initial flux of 5×10^7 ergs cm^{-2} s^{-1}. From Figure 105.3c we see that the average temperature $\bar{T}$ is essentially identical to the radiative equilibrium temperature $T_{\rm RE}$ at heights below about 250 km. Shock dissipation is small at these heights, and in this region of the atmosphere the mean intensity and the temperature merely fluctuate around their radiative equilibrium values. The waves first form shocks at about 590 km, and become fully developed shocks at about 680 km, above which height the atmosphere is strongly heated. At the height of initial shock formation, the waves transport a mechanical flux of 7×10^6 ergs cm^{-2} s^{-1}, having lost much of their original energy by

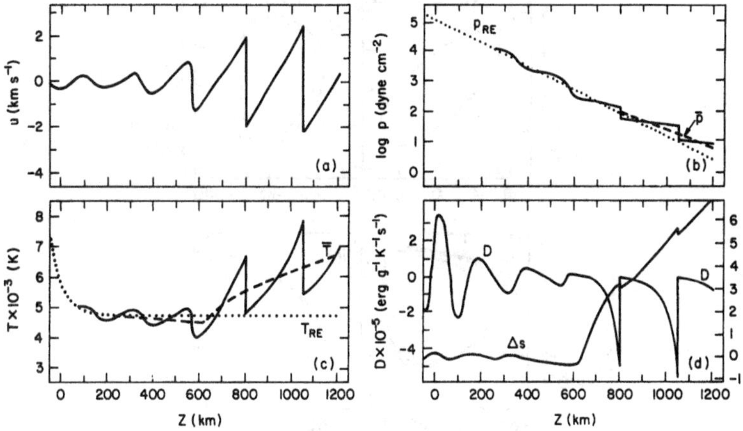

Fig. 105.3 Run of physical variables with height in a model solar atmosphere for a wave train with a period of 30 s. *Dotted curves:* radiative equilibrium temperature and density distributions; *dashed curves:* average temperature and density distributions. From **(U6)**, by permission.

radiative damping in the photosphere. The height of full shock development is lower, for a given period, the greater the initial mechanical flux. Hence the larger the input flux, the lower the point of rapid chromospheric temperature rise, and the greater the gas pressure in the chromosphere.

A surprising result of the calculation is that wave dissipation does not invariably heat the atmosphere, but can actually drive the mean temperature *below* its radiative equilibrium value, producing a temperature minimum of about 4370 K at a height of about 640 km. Radiative gains and losses are crucial in this region of the atmosphere, and because both the opacity and the Planck function are strongly nonlinear and the wave-induced temperature fluctuation is large ($\Delta T \sim 500$ K), it turns out that the net rate at which energy is radiated away at a temperature $T_{\rm RE}+\Delta T$ significantly exceeds the net rate at which it is reabsorbed at $T_{\rm RE}-\Delta T$; therefore $\bar{T}$ must fall below $T_{\rm RE}$. A detailed comparison of the theoretical minimum temperature with empirical models is compromised by the limitations of the calculation (grey material, LTE, monochromatic acoustic waves). Nevertheless, the computations show that the empirical temperature rise in the low chromosphere is easily reproduced for a large range of wave periods and initial mechanical fluxes.

The entropy change Δs of the gas and the net rate of specific entropy generation in the gas by radiation

$$D \equiv (Ds/Dt)_{\rm rad} = 4\pi\kappa(J-B)/\rho T \qquad (105.13)$$

are shown in Figure 105.3d. The thermodynamic behavior of individual

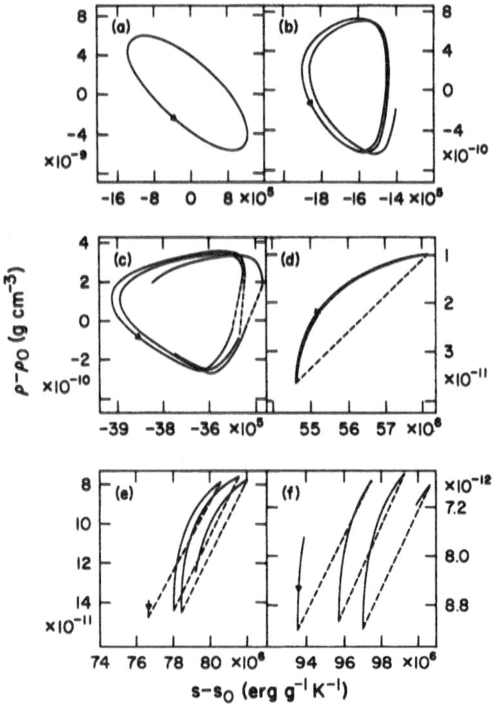

Fig. 105.4 Thermodynamic cycles in (ρ, s) plane. Initial heights: (a) 94 km, (b) 468 km, (c) 624 km, (d) 966 km, (e) 1090 km, (f) 1150 km. From (**U6**), by permission.

fluid elements can be studied in the (ρ, s) and (p, V) planes as shown in Figures 105.4 and 105.5. The fluid elements displayed there are labeled by their initial heights z_0. Both at great depth where the diffusion approximation is valid and $D \propto (d^2B/dt^2)$, and in the optically thin regime where $D \propto -B$, there is a 180° phase shift between D and T, and a 90° shift between T and s. The density depends on s and T in such a way as to have a 135° phase shift relative to s. Thus low in the atmosphere the thermodynamic path in the (ρ, s) plane is an inclined ellipse, as shown in Figure 105.4a. At greater heights (Figure 105.4b) the shape of the cycle is altered by the nonsinusoidal wave form. Near the temperature minimum (Figure 105.4c) weak shocks develop, which produce a very small entropy change for modest density jumps [cf. (56.56)], hence a nearly vertical jump in the (ρ, s) diagram. As the shock strength increases a larger entropy jump is produced; in Figure 105.4d the material approximately follows the Weymann cycle, with an initial drop in entropy at nearly constant density followed by a nearly adiabatic expansion at constant entropy. At yet

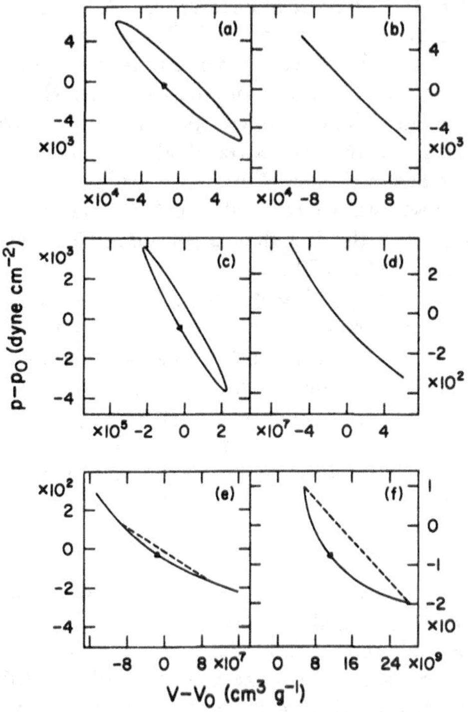

Fig. 105.5 Thermodynamic cycles in (p, V) plane. Initial heights: (a) −44 km, (b) 11 km, (c) 88 km, (d) 550 km, (e) 616 km, (f) 946 km. From (**U6**), by permission.

greater heights (Figures 105.4e and f) a mass element moves continually upward in a mean flow, hence experiences a secular increase in entropy.

From the (p, V) diagrams shown in Figure 105.5 we see that below the photosphere ($\tau > 1$) the cycles perform positive work (Figure 105.5a) indicating amplification of acoustic waves by the κ mechanism (**C23**, §27.6). At the photosphere, $\tau \approx 1$ (Figure 105.5b) the net work is only marginally positive, and for $\tau < 1$ (Figure 105.5c) the net work is negative, indicating radiative damping. At the temperature minimum, where the net radiative exchange is minimal, the cycle lies nearly along an adiabat (Figure 105.5d). At greater heights (Figures 105.5e and f) the material is heated and compressed by shocks and loses energy by radiation, leading to a negative net work, hence wave damping.

As shown in Figure 105.3a, the velocity amplitude of a shock grows only slowly once it is fully developed, despite the exponential density stratification; the growth is hindered by mechanical energy dissipation coupled to radiative energy losses. The Mach number of the shocks shown in the figure is only 1.5 at a height of 1000 km. As the shocks propagate into the outer atmosphere they deposit momentum and loft the material; thus the average pressure gradient outward is shallower than in radiative equilibrium (see Figure 105.3b). Mass elements initially below 600 km are not systematically displaced from their original positions. However, after many cycles mass elements initially above 600 km move to average positions some 100 to 150 km higher. Moreover, fluid elements initially above about 1100 km do not arrive at steady average positions, but rise continually, indicating mass loss into a mean outward flow. The existence of a mean flow is consistent with the fact that large-amplitude waves support a net transport of material (**L2,** 252).

Computations of the type just described have also been made for other stars, and appear to explain some of the observed behavior of stellar chromospheres (**U7**), (**S7**), (**S8**), (**S9**). It should be emphasized, however, that solar observations, which reveal strongly enhanced chromospheric emission in magnetically dominated regions (e.g., *plage* and the *chromospheric network*) clearly indicate that the shock-heating theory discussed above provides only a very incomplete description of real chromospheres, adequate at best to explain only the initial temperature rise in the lower chromosphere.

(*c*) *Supernova Explosions* Type II *supernovae* (**Z4**) are believed to originate when the core of a highly evolved, massive ($\mathcal{M} \gtrsim 5\mathcal{M}_\odot$) star undergoes sudden gravitational collapse and/or a thermonuclear runaway leading to explosive energy release. A total energy of the order of 10^{50} to 10^{51} ergs is released in the star on a time scale short compared to a typical dynamical time scale (e.g., the sound-travel time across the envelope, $\Delta r/a \sim 6$ months), and blows most of the envelope away. A presupernova is expected to have a red supergiant structure comprising (1) a *core* of 1 to $2\mathcal{M}_\odot$ within a radius of 10^7 to 10^8 cm, which collapses to a neutron star or black hole; (2) a *mantle* interior to the helium-burning shell, containing one to several solar masses composed of C, O, Ne, etc. and extending out to 10^{11} cm; and (3) an outer *envelope* composed of He and possibly H (perhaps including an H-burning shell), extending out to 10^{13} to 10^{14} cm. Because supergiants usually show noncatastrophic mass loss in stellar winds, there may also be a circumstellar *shell* containing perhaps $0.01\mathcal{M}_\odot$ and extending out to 10^{15} cm.

As a result of the explosive energy release, and/or because the envelope falls onto the collapsed core and bounces, a very strong shock is driven into the overlying material. The shock is markedly supercritical, hence it drives a radiation front into the upstream gas as discussed in §104. This radiation precursor not only deposits energy and heats the preshock material, but also deposits enough momentum to accelerate the electrons (which then drag along the ions by Coulomb friction) in the upstream gas to high velocities; typically the radiation pressure in the radiative precursor exceeds the gas pressure in the preshock material by one to two orders of magnitude. Thus the velocity jump across the front is sharply reduced and the shock is strongly radiatively mediated. Indeed the shock is fully dispersed by radiation even in the absence of viscosity, and the radiative diffusive effects are so large that a numerical computation can be carried forward without artificial viscosity for several timesteps without becoming unstable. An exhaustive discussion of radiating shock structure in the diffusion regime for conditions appropriate to supernovae has been given by Weaver (**W2**) for a wide range of shock parameters, and accounting for a large variety of physical phenomena that occur at high energy.

The shock strengthens as it moves outward and runs down the density gradient in the stellar envelope; for a typical velocity of a few thousand km s^{-1} it will traverse the envelope in a few hours. An external observer will be unaware that an explosion has occurred until the shock is near optical depth unity at some wavelength; this condition first occurs at high frequencies where the material is more transparent, and the first electromagnetic signal to emerge from the star is a burst of soft X rays lasting about 10^3 s. The luminosity maximum in the visible part of the spectrum does not occur until 4 to 20 days later, depending on whether or not the presupernova is surrounded by a circumstellar shell, when the radiant energy released in the initial event finally escapes by diffusion. As the shock penetrates into optically thin layers the radiation emerging from the front no longer couples efficiently into the upstream material and ceases to accelerate it; the front then steepens into a pure viscous shock that strongly heats the material and drives a hard X-ray burst. Indeed, early calculations (**C19**) assuming compact presupernova envelopes predicted a γ-ray burst and nuclear spallation reactions as the shock unloads in the outer layers. These models did not radiate enough visible luminosity because the material in the blowoff expanded adiabatically and cooled very rapidly; present-day calculations using supergiant presupernovae with extended envelopes easily predict the observed visual luminosities.

As the shocked material expands, it becomes more transparent. The radius of the stellar "photosphere" initially increases as material flows outward, but eventually an external observer will begin to see deeper and deeper into the envelope. When the whole envelope has expanded greatly and becomes transparent the mantle becomes visible; material that has undergone extensive nucleosynthesis is then revealed. Eventually even the mantle becomes transparent as a result of expansion and the compact core remnant (if any) becomes visible.

The radiation hydrodynamics of supernova explosions has been treated by several authors [see e.g., (**C19**), (**L3**), (**C14**), (**F2**)]. One of the most comprehensive discussions is by Falk and Arnett (**F2**), whose results we summarize briefly here. These authors carry out calculations for a variety of initial models, using a one-dimensional spherically symmetric, Lagrangean radiation-hydrodynamics code. The gas is assumed to be a single fluid in LTE. The numerical momentum equation is differenced explicitly as in (98.16), with pressures and radii time centered as in (59.64) and (59.87), and the pseudoviscosity computed as discussed in §59. In optically deep zones the temperature is updated using an implicit difference representation of the equilibrium diffusion equation (97.7). Near the surface ($\tau \lesssim 5$) where radiation transport occurs, the temperature is determined by solving the gas-energy equation (98.20) implicitly with the coupled radiation energy and momentum equations (98.5) and (98.8), which are manipulated into the combined moment equation (98.12) and differenced as in (98.34) to (98.36). Thus both the dynamics of the radiation field itself and its dynamical interaction with matter are taken fully into account. The equilibrium diffusion and radiation transport solutions are joined self-consistently by an iteration procedure. Unfortunately the Eddington approximation is made, which compromises the transport solution in optically thin layers and in the extended shell. The material is assumed to be grey. The overall accuracy of the computation is checked by monitoring the total energy conservation law (98.38).

The results are sensitive to whether or not the presupernova is surrounded by an extended circumstellar shell. Consider first a "compact" model representing a $10\mathcal{M}_\odot$ supergiant with a radius of about 2×10^{14} cm and no circumstellar shell. The light curve is shown in Figure 105.6a; note the very sharp initial pulse at $t=3.4\times10^5$ s, in which the luminosity rises over 8 to 10 orders of magnitude in about a day, followed by a broad plateau. The initial postpeak decline results from adiabatic cooling of the rapidly expanding photospheric layers. The plateau results from a period of diffusive energy release in parallel with expansion, cooling, and recombination of the material. The final turndown occurs when recombination in the envelope is essentially complete and the envelope becomes transparent.

Velocity profiles of the material at various times are shown in Figure 105.6b. The shock is just emerging at $t_6=0.341$, and one sees the "crack of the whip" effect of shock unloading that accelerates the outer layers suddenly to very large velocities. Velocities continue to rise nearly homologously as radiation continues to perform work on the material for the next few days.

Material temperatures are shown in Figure 105.6c. One sees the dramatic heating of the outer layers by shock emergence (shock positions are indicated by tick marks on the curves for $0.094\leq t_6\leq0.341$). After peak luminosity at $t_6=0.341$ the temperature profile is quite flat and the material is continuously cooled by essentially adiabatic expansion. The density structure of the supernova on a mass scale and a radius scale is shown in Figures 105.6d and 105.6e, respectively. The propagation of the shock through the star for $t_6\leq0.341$ shows clearly in Figure 105.6d; the nearly invariant density profile at subsequent times results from the nearly homologous expansion in the explosion. The rapid spatial spreading of the ejected material is shown in Figure 105.6e.

Results for an extended model comprising a $5\mathcal{M}_\odot$ supergiant and an

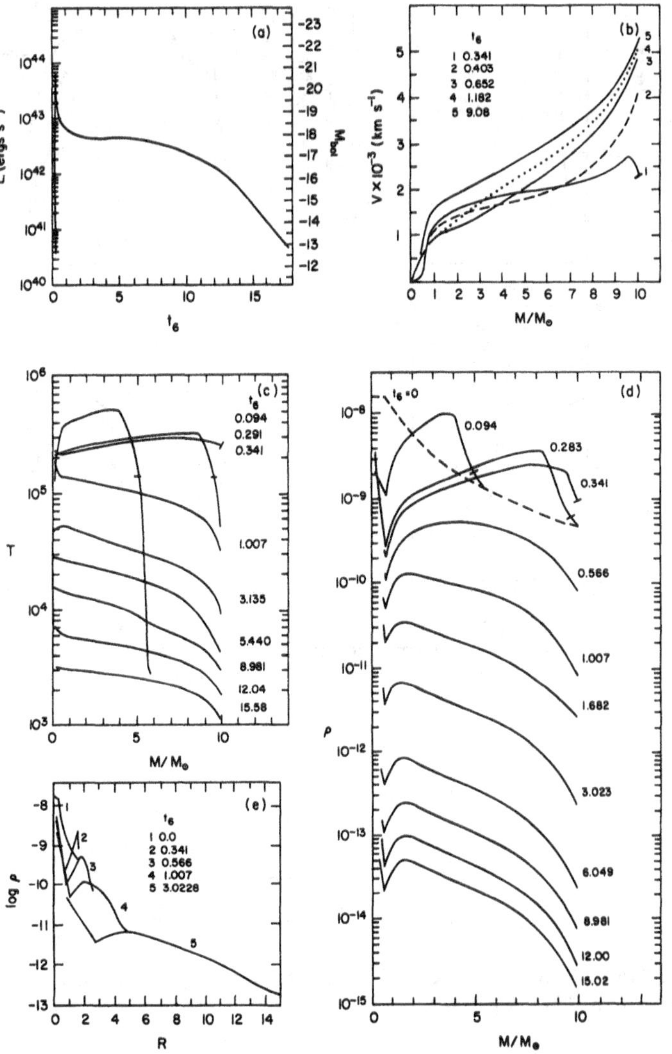

Fig. 105.6 Supernova explosion in a "compact" $10\mathcal{M}_\odot$ supergiant. (a) Light curve. $t_6 = t(s)/10^6$ (b) Velocity as a function of mass. (c) Temperature as a function of mass. (d) Density as a function of mass. (e) Density as a function of radius. From (**F2**), by permission.

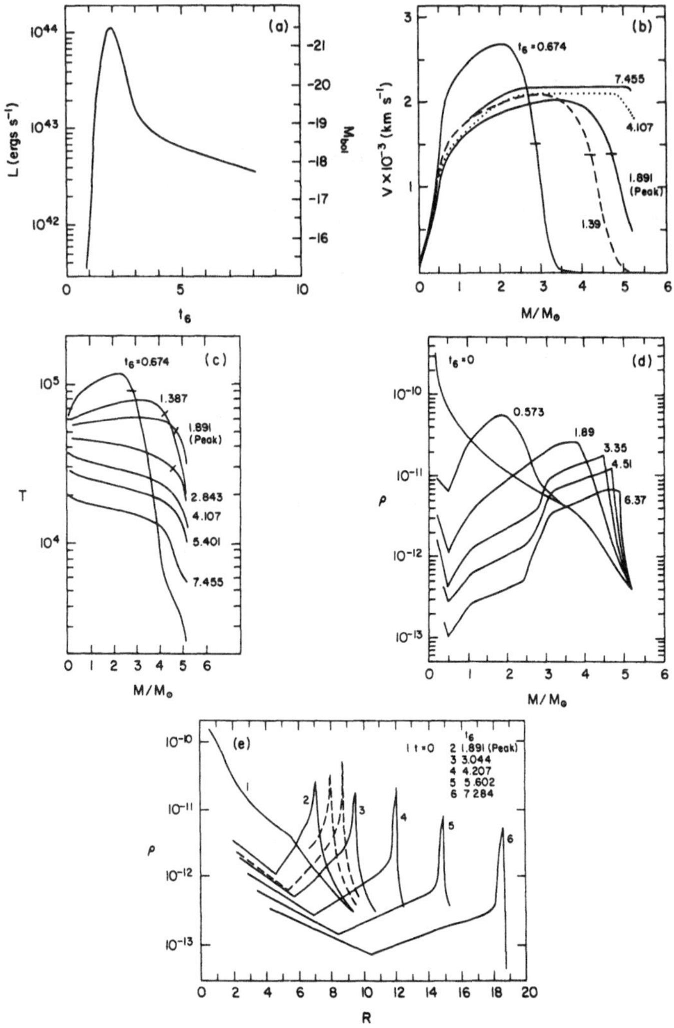

Fig. 105.7 Same as in Fig. 105.6 for extended $5\mathcal{M}_\odot$ model with circumstellar shell. From (**F2**), by permission.

extensive circumstellar shell of radius about 10^{15} cm are shown in Figure 105.7. The light curve, shown in Figure 105.7a, now exhibits a broad peak (width $\approx$20 days), which results when the radiation preceding the shock diffuses through the extended shell. Radiative acceleration and heating of the material is not as efficient as in the compact model, hence the maximum material velocities and peak temperatures are substantially lower. A striking feature of this model appears in the density profiles shown in Figures 105.7d and 105.7e where we see the formation of a dense, thin shell for $M_r \gtrsim 2.7\mathcal{M}_\odot$. This shell (qualitatively similar to the structure in a supernova remnant—see below) develops because photons leak efficiently from the low-density outer layers, and therefore the gas pressure does not rise sufficiently to prevent a density inversion (essentially the gas is thermally unstable). The large density inversion is probably Rayleigh–Taylor unstable, hence it may disintegrate into blobs and filaments, mixing several adjacent mass zones and producing large turbulent motions; the development of violent turbulence would bleed energy from the bulk flow. Growth of Rayleigh–Taylor instabilities in the outer layers of supernova ejecta has been explored by Chevalier and Klein (**C15**) using a two-dimensional hydrodynamics code.

Hard X-ray bursts at the time of shock emergence have been predicted by Falk (**F1**) and by Chevalier and Klein (**C16**), (**K5**) from computations with finely zoned realistic models of the boundary layers. In contrast, Lasher and Chan (**L4**) predict a soft X-ray burst (as do the other authors) but no hard X rays. The essential difference among these calculations is the size of the velocity jump between the emerging shock and the overlying material. In (**F1**), (**C16**), and (**K5**) the jump is large and drives a viscous shock that heats the material to order 10^8 K; in (**L4**) the upstream material is radiatively preaccelerated to almost the shock velocity, hence the velocity jump nearly vanishes and the material is never strongly heated. Falk uses the numerical techniques described above (**F2**), while Chevalier and Klein perform a Lagrangean calculation for a two-fluid (ions and electrons) plasma of ionized hydrogen, treating the radiation dynamics with the nonequilibrium diffusion equation (97.70). A flux limiter (cf. §97) is applied to the flux appearing in the radiative force term in the material momentum equation (96.2), but not to the flux-divergence term in the radiation energy equation (a procedure that is inconsistent). In (**L4**) the radiation is treated by equilibrium diffusion without flux limiting. Chevalier and Klein point out that in optically thin zones the latter approach results in too strong a coupling between the radiation and material, and too large an energy flux, both of which lead to a spuriously large radiative preacceleration of the preshock gas.

Epstein (**E3**) has argued that the equilibrium diffusion results are correct and that Chevalier and Klein's calculation is faulty because of their inconsistent use of flux limiting; he suggests that they obtained about the right energy transport, but seriously underestimated the radiative momentum input to, hence acceleration of, the upstream gas. However Chevalier and Klein report (**C16,** 603) that calculations in which the flux limiter was used consistently in both the momentum and radiation energy equations yielded essentially the same results, with only a minor delay in the time of maximum luminosity (which also increased modestly). The controversy can be settled only by a definitive new calculation using the full radiation energy and momentum equations; but it seems unlikely that equilibrium diffusion with no flux limiting can yield more accurate results than the inherently more complete formulations used in (**F1**), (**F2**), (**K5**), and (**C16**).

(*d*) *Supernova Remnants* The evolution of *supernova remnants*, the interstellar material swept up by the blast wave from a supernova, has been analyzed by several authors using similarity solutions (**P4**), (**C21**), (**C22**), and (**S20,** 200), which do describe many basic features of the flow. Initially the blast wave expands essentially adiabatically, hence its velocity varies as $v_S \propto r^{-3/2}$ [cf. (60.11)]. Eventually the material cools radiatively. A characteristic radiative cooling time is given by the ratio of the material energy density to the rate of radiative energy loss (ignoring reabsorption):

$$t_{rc} \sim \rho c_p T / \sigma_R T^4 \kappa_P, \qquad (105.14)$$

where κ_P is the Planck mean opacity. For $t \gtrsim t_{rc}$ a radiative cooling wave penetrates into the material behind the shock front and the nature of the flow changes markedly. In the model of Poveda and Woltjer (**P4**), it is assumed that the cooling is so efficient that the pressure in the interior of the blast drops essentially to zero; therefore at late times the flow behaves as a momentum-conserving shell that, from simple dimensional arguments, must expand with a front velocity $v_S \sim r^{-3}$. Thus the outer part of the flow collapses into a thin, dense, slowly expanding shell that "snowplows" into the ambient medium, sweeping up a large amount of material.

The treatment of radiative effects in the similarity solutions is highly oversimplified, and a more accurate analysis is needed. Numerical simulations of supernova remnants have been made by Erickson and Olfe (**E4**) and Chevalier (**C13**); we discuss briefly some results from the latter calculation, which is based on more realistic physics. The hydrodynamic code solves the momentum equation including gas pressure and magnetic forces. The gas-energy equation includes approximate optically thin radiative gain and loss terms, but no attempt is made to solve realistic radiation energy and momentum equations. The equation of state allows for ionization, and the time variation of the ionization fraction is calculated with a rate equation that allows for photoionization, collisional ionization, and radiative recombination. Magnetic effects prove to be unimportant except in the dense outer shell, which is mostly supported by magnetic pressure. A rezoning scheme eliminates unnecessary inner zones as the calculation progresses, and the accuracy of the calculation is monitored with a total energy check.

The calculation is started by depositing 3×10^{50} ergs as heat in a small region of a uniform medium having a particle density of 1 cm^{-3}; results are displayed in Figure 105.8. The solution quickly relaxes to an adiabatic Sedov blast wave. As the remnant expands, radiative cooling produces a temperature dip behind the shock front by 4×10^4 years; the postshock pressure likewise drops and a density spike emerges. A dense neutral shell

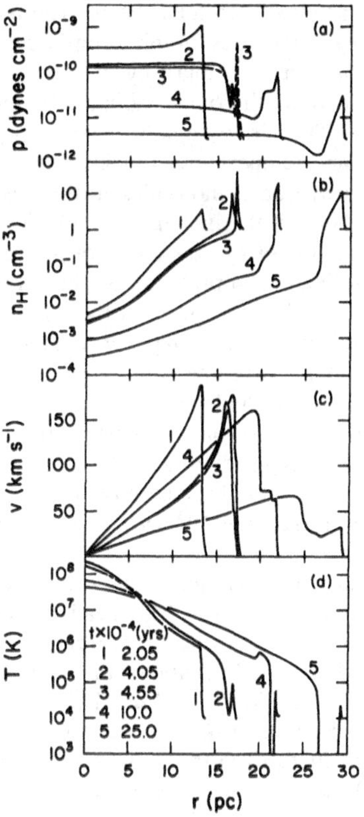

Fig. 105.8 Expansion of a supernova remnant into the interstellar medium. (a) Pressure, (b) hydrogen density, (c) velocity, (d) temperature, all as a function of radius. From (**C13**), by permission.

becomes completely formed by 4.5×10^4 years. Hot, high-pressure inner material accelerates into the pressure low behind the shell, and in fact overtakes the shell and rams into it; shocks then exist on both sides of the shell. The gas is quite cool near the pressure minimum where radiative losses are large, but is reheated to about 10^4 to 10^5 K when it shocks behind the shell. The region where pressure decreases with increasing radius while the density increases is likely to be Rayleigh–Taylor unstable.

The pressure low behind the shell propagates an expansion wave towards the center, which bounces at $t\approx10^5$ years. Similar, but more extreme, dynamical phenomena are shown in (**E4**).

(*e*) *Accretion Flows* Interesting and complex radiation-hydrodynamic phenomena accompany the *accretion flows* that occur in the gravitational collapse of a Jeans-unstable protostellar cloud. The basic scenario is that a hydrogen–helium cloud undergoes an initial rapid *collapse phase* (in about one free-fall time) and forms a quasi-hydrostatic core surrounded by a strong radiating shock where the freely falling envelope slams into the core highly supersonically. The cloud rapidly becomes optically thick, hence radiation is trapped, and temperatures quickly rise in the compressed gas. At about 2000 K the hydrogen molecules in the gas begin to dissociate. This process acts as a sink of thermal energy and triggers a second collapse phase because compression of the gas does not produce a rise in temperature, and a corresponding rise in pressure, because the energy is consumed in further dissociation of the gas. Once all the molecules are destroyed, a second low-mass core forms, and a long *accretion phase* ensues in which all matter in the freely falling envelope is accreted by the core.

The problem is extremely challenging because of the immense range of variation of physical quantities that must be followed and because of the presence of extremely strong radiating shocks. For example, as a result of the collapse the central density of the core rises by 20 orders of magnitude, and pressure jumps of a factor of 10^3 across the accretion shock surrounding the core are typical.

An excellent summary and critique of the literature on the formation of a $1\mathcal{M}_\odot$ star is given in (**W7**). The best available calculation is that by Winkler and Newman (**W8**) who use an accurate equation of state, fairly realistic opacities, and a refined version of the advanced numerical techniques of Tscharnuter and Winkler (**T2**). They solve the equations of radiation hydrodynamics implicitly on an adaptive mesh that automatically resolves all important features in the flow. A tensor artificial viscosity [cf. (59.91) to (59.104)] is used to handle shocks. The adaptive mesh is essential to the success of the calculation because individual fluid elements are first stretched by a factor of about 10^5 and then compressed by a similar factor; neither an Eulerian nor a Lagrangean grid would work well under such circumstances. Furthermore the adaptive mesh allows shock fronts to be resolved in optical depth.

The material is assumed to be a grey gas (single fluid) in LTE. Opacities of the gas and of ice-coated "dust" grains are accounted for. The full Lagrangean radiation energy and momentum equations including *all* velocity-dependent terms (cf. §95) are solved using self-consistent variable Eddington factors determined from a full transport computation. The calculations described in (**T2**), (**W7**), and (**W8**) are thus the most complete and consistent treatment of radiation dynamics available in the literature.

The initial model is a cloud of $1\mathcal{M}_\odot$, contained in a radius of 1.5×10^{17} cm with a constant density of $1.4\times10^{-19}\,\mathrm{g\,cm^{-3}}$, in thermal equilibrium at 10 K with the ambient interstellar medium. A 10 K thermal radiation field is imposed at the outer boundary. The overall temperature and density structure of the protostar is shown in Figure 105.9 at various stages of the main accretion phase. The tenuous envelope, which contains

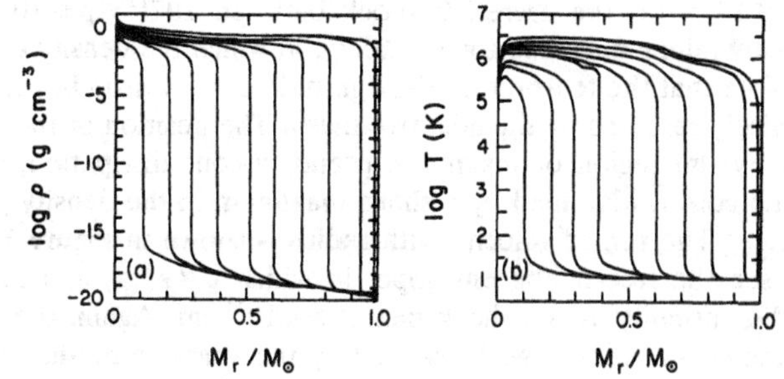

Fig. 105.9 Temperature and density structure in a $1\mathcal{M}$ protostar. Youngest models lie to left, oldest to right. From **(W8)**, by permission.

little mass, but runs over several decades in radius, occupies a narrow region in these plots.

Details of the structure near the end of the main accretion phase are shown in Figure 105.10. In the plot of density, Figure 105.10a, four different structural components can be recognized: (1) a centrally condensed hydrostatic core, (2) an exponential stellar atmosphere between

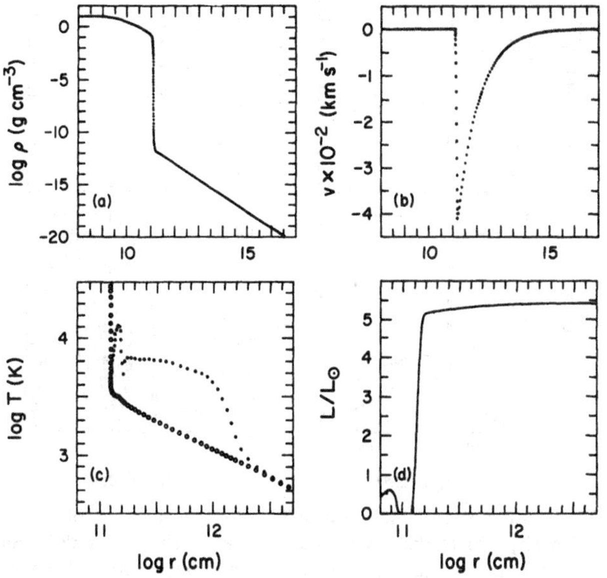

Fig. 105.10 Structure of shock front near end of main accretion phase in a $1\mathcal{M}_\odot$ protostar. (a) Density; (b) velocity; (c) radiation temperature (open circles) and gas temperature (dots); (d) luminosity. From **(W7)** by permission.

$10^{-10} \lesssim \rho \lesssim 10^{-1}$, (3) the accretion shock between $10^{-12} \lesssim \rho \lesssim 10^{-10}$, and (4) the freely falling envelope for $\rho \lesssim 10^{-12}$, in which the density decreases as $r^{-3/2}$. Note that the ten-orders-of-magnitude density drop in the atmosphere is nicely resolved by the adaptive mesh. The position of the shock is identified by the region of maximum pseudoviscous dissipation, and the edge of the core is identified by a sharp maximum in the density gradient $|d \ln \rho / d \ln r|$. The run of velocity with radius is shown in Figure 105.10b. Here we see the free-falling envelope, in which $v \propto r^{-1/2}$, and the sharp velocity discontinuity at the shock near 1.5×10^{11} cm. Again, the velocity jump at the shock front is well resolved by the adaptive mesh.

The radiation temperature, as defined in (97.73), and the material temperature are shown in Figure 105.10c. The gas and radiation are in equilibrium in the core, atmosphere, and outer envelope, but are strongly out of equilibrium in the vicinity of the (smeared-out) accretion shock ($\log r \approx 11.2$) and in the optically thin preshock adiabatic compression zone ($11.3 \lesssim \log r \lesssim 12.3$). As shown in Figure 105.10d, essentially all the kinetic energy of the infalling material is transformed into radiation.

Radiation quantities at various stages in the main accretion phase are shown in Figure 105.11. The run of opacity near the middle of the

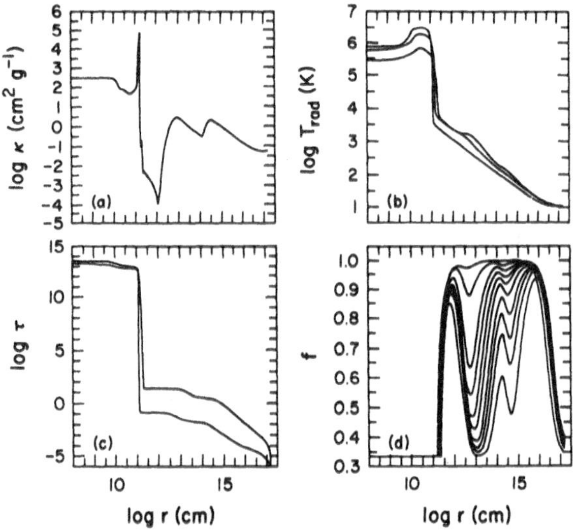

Fig. 105.11 Radiation variables in a $1\mathcal{M}_\odot$ at various stages of main accretion phase. (a) Opacity. (b) Radiation temperature at beginning, middle, and end of accretion phase (bottom to top). (c) Optical depth at beginning and end of accretion phase. Tenuous envelope is optically thick initially and becomes optically thin as it is depleted of material. (d) Eddington factor at various stages of accretion (early at bottom, late at top). From (**W8**), by permission.

accretion phase is shown in Figure 105.11a. Proceeding inward, the opacity first rises to about 1 $cm^2\,g^{-1}$ as the ice mantles on the dust grains absorb, then it drops as these mantles are destroyed at 150 K, rises again as the dust grains themselves absorb, drops again as the dust grains melt at 1600 K, then rises to very large values ($10^5\,cm^2\,g^{-1}$) in the hot shock gas, and finally saturates to a large value in the dense core. Molecular opacities, not accounted for fully in the calculation, could be important in the extensive opacity minimum in the range $11 \lesssim \log r \lesssim 13$. Radiation temperatures at the beginning, middle, and end of the main accretion phase are shown in Figure 105.11b. The run of optical depth near the beginning and end of the accretion phase is shown in Figure 105.11c. One sees a very optically thick core surrounded by an envelope that is initially optically thick but becomes optically thin as essentially all the matter falls into the star.

The run of the variable Eddington factor at various times in the main accretion phase is shown in Figure 105.11d. At the beginning of the accretion phase (lower curve), $f=\frac{1}{3}$ in the core, rises sharply to nearly unity in the optically thin zone at $11 \lesssim \log r \lesssim 13$, drops to $\frac{1}{3}$ again in the opaque zone where the dust grains absorb, rises again at the opacity low near $\log r \approx 14$, drops in the opaque zone where the ice mantles absorb, rises to nearly unity again in the transparent outer envelope, and finally is forced back to $\frac{1}{3}$ again as the stellar radiation field comes into equilibrium with the ambient interstellar field. In contrast, at the end of the accretion phase (upper curve) the Eddington factor rises almost monotonically from $\frac{1}{3}$ in the opaque core to unity in the transparent depleted envelope, and is forced to $\frac{1}{3}$ ultimately only by the imposed radiation at the outer boundary. This complex variation illustrates vividly the essential importance of a *transport* evaluation of the variable Eddington factors, and shows that a non-equilibrium diffusion treatment would be virtually worthless, and that use of an ad hoc flux-limiting procedure would, at best, be of questionable value.

NON-LTE SHOCKS

The solution of the equations of radiation hydrodynamics allowing for departures from LTE for realistic model atoms is quite difficult. One of the most complete efforts of this kind is the work of Klein, Stein, and Kalkofen (**K7**), (**K8**) who consider the propagation of a non-LTE shock driven by a piston that moves with constant velocity into a pure hydrogen atmosphere that is initially in hydrostatic and radiative equilibrium. Here we account for the fact that radiation not only contributes to the energy and momentum balance in the radiating fluid, but also determines the internal excitation and ionization state of the gas.

The computation is performed with a one-dimensional Lagrangean code in planar geometry. The momentum equation is differenced explicitly as in (98.16), (59.64), and (59.87). The energy balance is treated implicitly as in (98.20). In the more refined calculation (**K8**) departures from LTE are allowed in the first two levels ($n = 1$, $n = 2$), while levels $n = 3$ through $n = 10$ are assumed to be in LTE. Radiative bound–bound transitions are ignored, but all other radiative and collisional processes coupling $n = 1$ and 2 to the continuum and to each other are included. Because the rate equations are very stiff on the relevant dynamical time scales, they are differenced fully implicitly as in (85.40), which follows from (85.39) with $\theta = 1$. The radiation field is assumed to be quasi-static, and the time-independent transfer equation (83.58) is solved implicitly at the advanced time level. Thus the dynamical behavior of the radiation field and the effects of radiative forces are ignored. The frequency dependence of the radiation field is, however, treated (with 15 representative frequencies), and variable Eddington factors are obtained from a full transfer solution; hence the effects of nonlocal radiative energy exchange are handled fairly accurately. This work is probably the best non-LTE shock calculation available in the literature.

The initial model atmosphere has an effective temperature $T_{\text{eff}} =$ 11,500 K and a gravity $g = 10^4$, appropriate to a late B-type star. A piston is driven into the bottom of the atmosphere (where $\tau_\nu = 200$ at the most transparent wavelength) with a velocity of 4 km s^{-1} ($= a/6$). The piston generates a shock that heats the gas through which it passes and also drives a radiation diffusion wave into the overlying material. The diffusion wave propagates to the surface in about 35 s, which agrees well with the radiative relaxation time of the atmosphere as computed from (100.17) in the optically thick limit. As shown in Figure 105.12, the radiation diffusion wave produces a large temperature rise in a localized region near $\tau_\lambda = 1$ at the head of the Balmer continuum (i.e., at $\lambda 3648^-$ Å). At the adopted effective temperature, the Balmer continuum acts as a net source of heating to the atmospheric material, which maximizes near the characteristic depth $\tau \approx 1$. As the radiation diffusion wave passes that depth, the mean intensity of the radiation field in the Balmer continuum increases significantly, hence the heating rate of the gas is sharply increased. Furthermore, the ambient conditions are such that a rise in temperature happens to increase the opacity in the Balmer continuum, and therefore the rate at which it absorbs energy, leading to a positive feedback effect reminiscent of thermal instability (cf. §100) and the κ-mechanism in pulsating stars (**C23**, §27.6); see also our earlier discussion of Figure 105.5.

The strong localized radiation heat input near $\tau(\lambda 3648^-) = 1$ raises the temperature, hence pressure, of the gas, and drives a compression wave both upward and downward in the atmosphere (Figure 105.13). The downward propagating wave runs into dense stationary material and is quickly damped. The upward-propagating wave accelerates outward and becomes a weak shock, which passes through the uppermost mass zone of the atmosphere at $t = 550$ s; at that point the Mach number in the shock is 1.7 and the density jump is a factor of 2. Although the secondary shock compresses and heats the gas, it is optically thin in the Balmer and higher

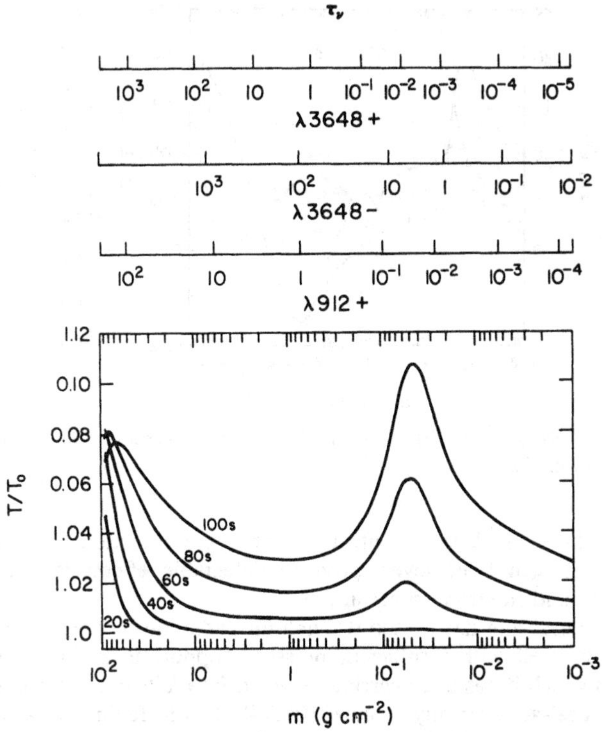

Fig. 105.12 Thermal precursor wave as function of optical depth at various times after piston is set in motion. From (**K8**), by permission.

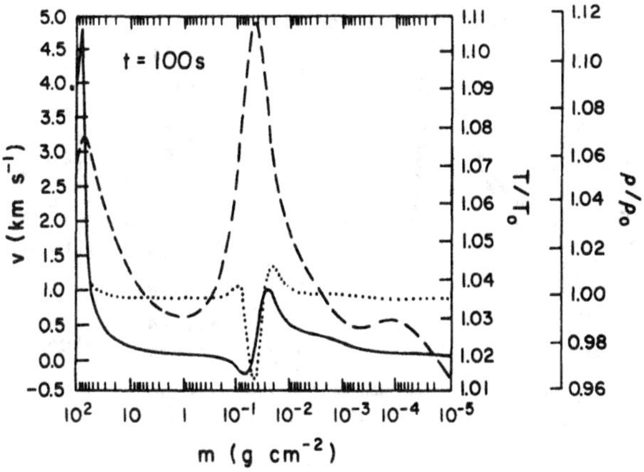

Fig. 105.13 Physical structure of atmosphere at $t = 100$ s. *Solid curve:* velocity; *dashed curve:* temperature; *dotted curve:* density. From (**K8**), by permission.

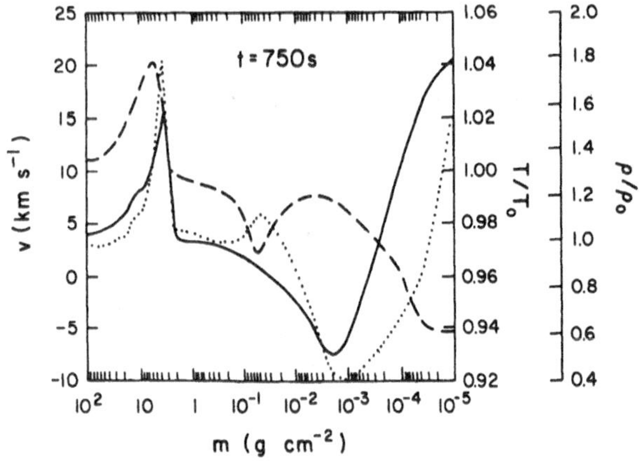

Fig. 105.14 Same as Fig. 105.13 for $t = 750$ s, before accretion shock forms. From (**K8**), by permission.

continua, hence it does not affect the emergent radiation field in visible continua; it would, however, produce detectable effects in the Lyman continuum and in strong spectral lines.

The material that has passed through the radiation-induced compression wave finds itself with insufficient pressure support to maintain it at the height to which it has been carried, hence it free falls back, expanding and cooling quasi-adiabatically (Figure 105.14). The in-falling gas encounters the slowly moving fluid that was not accelerated by the radiation wave and forms an accretion shock at $t = 975$ s (Figure 105.15). This shock produces

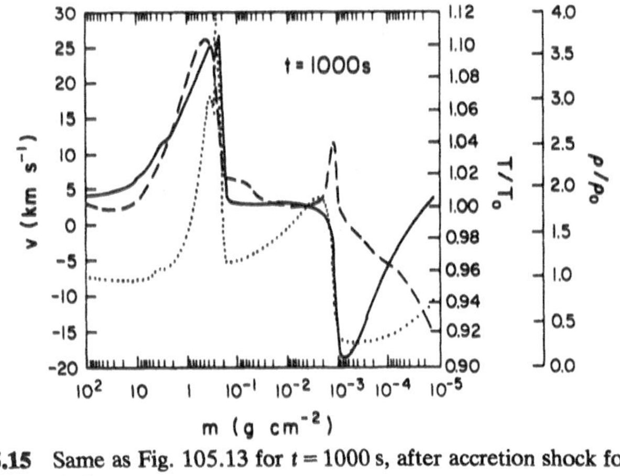

Fig. 105.15 Same as Fig. 105.13 for $t = 1000$ s, after accretion shock forms. From (**K8**), by permission.

a narrow temperature spike that is optically thin, hence incapable of producing an optical signature except in the Lyman continuum or in strong spectral lines.

The primary piston-driven shock finally emerges at $t \approx 1200$ s. Radiation losses have a profound effect on the propagation of this shock. First, the radiating shock's velocity remains nearly constant, in sharp contrast to an adiabatic shock that accelerates rapidly as it rises. Second, the temperature jump in the radiating shock is quite small, $T_{\text{shock}}/T_0 \lesssim 1.2$, in contrast to an adiabatic shock for which $T_{\text{shock}}/T_0 \lesssim 2$ to 10. The compression behind the nearly isothermal radiating shock is about a factor of 2 larger than in the adiabatic shock. When the shock finally reaches small optical depth, it cools rapidly. The emergence of the primary shock is essentially invisible in the emitted radiation field. The reason is that shock heating not only raises the source function S_ν locally, but also raises the opacity of the gas, hence shifts the effective radiating surface near $\tau_\nu \approx 1$ [recall the Eddington–Barbier relation (79.17)] outward into cooler gas. The two effects nearly cancel, and an external observer never actually sees the hot shock front until it is already too optically thin to affect the emergent radiation field significantly; precisely the same phenomenon occurs in fireballs from intense explosions, see (**R2**) and (**Z3,** 598–626).

A related study has been made by Kneer and Nakagawa (**K9**) who compute the time development of a nonequilibrium thermal transient in the solar chromosphere. They formulate the problem in terms of implicit Eulerian difference equations, ignoring all velocity-field effects on the radiation field, which is assumed to be quasi-static. They allow departures from LTE in a two-level hydrogen atom including the $Ly\,\alpha$ transition. They also calculate the response of the emergent $Ly\,\alpha$ radiation field to the thermal pulse.

106. Ionization Fronts

In §§104 and 105 we considered flows in which the radiation is essentially driven by the hydrodynamics, as when radiation is created in the high-temperature downstream gas behind a strong shock. In this and the following section we turn to the opposite case where instead the flow (perhaps including shocks) is driven by radiation. Specifically we examine the physics of *ionization fronts* (or *I-fronts*), which occur when intense radiation from a hot source (e.g., an O-star) eats its way into an ambient cold medium (e.g., the interstellar medium). An I-front is an *interface* only a few photon mean free paths thick, across which the material becomes essentially completely ionized while the temperature and pressure jump nearly discontinuously.

An I-front can produce a wide variety of hydrodynamic phenomena. For example, suppose the material is so rarefied and the incident radiation field is so strong that the photon number density is much larger than the particle number density and recombinations can be ignored. Then the I-front races into the medium at nearly the speed of light, ionizing every atom as it goes. Hydrodynamic motions will develop much more slowly because the speed of sound is much smaller than the speed of light. Thus at the front the upstream and downstream material will coexist at the same density, despite the fact that the temperature, hence pressure, in the downstream material is orders of magnitude larger than in the upstream material, simply because the I-front continually outruns the hydrodynamic motions that would otherwise be driven by the pressure difference. At the other extreme, suppose the medium is very dense. Then the radiation penetrates into the cold material only very slowly by diffusion, essentially in a Marshak wave (cf. §103). The effect of the I-front is to build a radiatively heated pressure reservoir, which drives a shock into the upstream material. Because the radiation front is choked in the dense material and therefore moves slowly, the shock will run ahead of the I-front.

I-FRONT JUMP CONDITIONS

Let us now derive the jump conditions that apply across a steady I-front. We will discuss only the simplest cases, with the goal of providing basic physical orientation, and refer the reader to more comprehensive treatments in the literature for details. Thus, consider an I-front driven by collimated radiation from a steady source incident normally on a planar slab of pure hydrogen. Assume that the material is completely neutral upstream and completely ionized downstream. Furthermore, for simplicity, assume that no recombinations occur in the ionized material so that radiation from the source always arrives unattenuated at the current position of the I-front. We can then have a steady flow, and it is convenient to transform to the frame moving with the front, using the geometric and sign conventions indicated in Figure 106.1. As usual, subscripts "1" and "2" refer to upstream and downstream quantities, respectively.

The density of the material is

$$\rho = (n_{\mathrm{H}} + n_p) m_{\mathrm{H}}, \tag{106.1}$$

the gas pressure is

$$p = (n_{\mathrm{H}} + n_p + n_e) kT, \tag{106.2}$$

and the internal energy of the material is

$$\rho e = \tfrac{3}{2} kT (n_{\mathrm{H}} + n_p + n_e) + \varepsilon_{\mathrm{H}} n_p. \tag{106.3}$$

Here n_{H} is the density of neutral hydrogen atoms, n_p is the proton density, n_e is the electron density, and ε_{H} is the ionization potential of hydrogen. From charge conservation we have $n_e \equiv n_p$.

The radiation field has a specific intensity $I(\mu, \nu) = I_\nu\, \delta(\mu + 1)$, that is, it is nonzero only for $\mu = -1$. The number flux of ionizing photons (photons

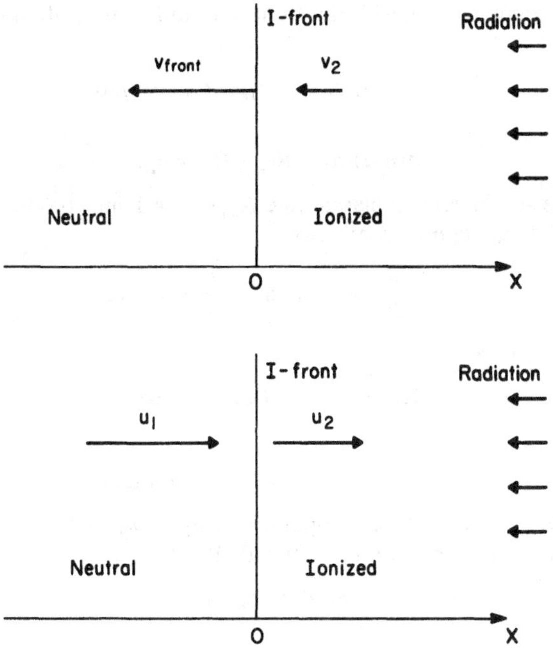

Fig. 106.1 Fluid velocities near ionization front, as measured in lab frame (top) and I-front's frame (bottom).

$\text{cm}^{-2}\,\text{s}^{-1}$) is

$$\phi = \int_{\nu_H}^{\infty} (I_\nu/h\nu)\, d\nu \tag{106.4}$$

where ν_H is the threshold frequency for hydrogen ionization. Because we neglect recombinations the transfer equation simplifies to

$$(dI_\nu/dx) = n_H \alpha_\nu I_\nu \tag{106.5}$$

(recall $\mu = -1$), which implies that

$$(d\phi/dx) = n_H \bar{\alpha} \phi, \tag{106.6}$$

where we have defined a mean cross section

$$\bar{\alpha} \equiv \left[\int_{\nu_H}^{\infty} (\alpha_\nu I_\nu/h\nu)\, d\nu\right] \Big/ \phi. \tag{106.7}$$

Similarly, we define a mean photon energy

$$\bar{\varepsilon} = h\bar{\nu} \equiv \left(\int_{\nu_H}^{\infty} \alpha_\nu I_\nu \, d\nu\right) \Big/ \bar{\alpha}\phi. \tag{106.8}$$

In the frame moving with the front, the statistical equilibrium equations become

$$d(n_{\rm H}u)/dx = -R_{1\kappa} + R_{\kappa 1} = -n_{\rm H}\bar{\alpha}\phi \tag{106.9}$$

and

$$d(n_{\rm p}u)/dx = R_{1\kappa} - R_{\kappa 1} = n_{\rm H}\bar{\alpha}\phi, \tag{106.10}$$

where we set the recombination rate $R_{\kappa 1} \equiv 0$. Adding (106.9) and (106.10) we get the equation of continuity

$$\frac{d}{dx}[(n_{\rm H} + n_{\rm p})u] = \frac{d}{dx}(Nu) = 0, \tag{106.11}$$

whence we have

$$Nu = (n_{\rm H}u)_1 = (n_{\rm p}u)_2 = \text{constant} \tag{106.12a}$$

or

$$\rho_1 u_1 = \rho_2 u_2 = \text{constant}, \tag{106.12b}$$

where we have used the assumption that $n_{\rm H,2} = n_{\rm p,1} \equiv 0$.

Combining (106.6) with (106.9) and (106.10) we find

$$(n_{\rm H}u)_1 = (n_{\rm p}u)_2 = \phi_2 \tag{106.13a}$$

or

$$\rho_1 u_1 = \rho_2 u_2 = m_{\rm H}\phi_2, \tag{106.13b}$$

which make the physically obvious statement that each ionizing photon incident at the front converts one hydrogen atom to a proton-electron pair. We have set $\phi_1 \equiv 0$ because all photons are absorbed in the front.

The fluid momentum equation in the frame of the I-front yields

$$\rho_1 u_1^2 + p_1 = \rho_2 u_2^2 + p_2 \tag{106.14}$$

where $p_1 = k(n_{\rm H}T)_1$ and $p_2 = k[(n_p + n_e)T]_2 = 2k(n_p T)_2$. The material energy equation in the frame of the front is

$$\frac{d}{dx}\{u[\rho(e + \tfrac{1}{2}u^2) + p]\} = n_{\rm H}\bar{\alpha}\bar{\varepsilon}\phi. \tag{106.15}$$

Integrating across the discontinuity with the aid of (106.6) we find

$$\tfrac{5}{2}(p_2/\rho_2) + \tfrac{1}{2}u_2^2 = \tfrac{5}{2}(p_1/\rho_1) + \tfrac{1}{2}u_1^2 + (\bar{\varepsilon} - \varepsilon_{\rm H})/m_{\rm H}. \tag{106.16}$$

Equations (106.12), (106.14), and (106.16) uniquely determine the downstream conditions in the flow for given upstream conditions and a specified photon flux (**K1**). However in astrophysical applications it is usually the case that radiative relaxation times are orders of magnitude smaller than typical flow times through the I-front. Hence, as was true for shocks (cf. §104), we can often make the simplifying assumption that both the upstream and downstream material remains isothermal at temperatures appropriate to the radiative heating and cooling mechanisms occurring in the neutral and ionized gases, respectively. We can thus replace (106.16) by the conditions

$$p_1/\rho_1 = kT_1/\mu_1 m_H = a_1^2 \tag{106.17a}$$

and

$$p_2/\rho_2 = kT_2/\mu_2 m_H = a_2^2, \tag{106.17b}$$

where a denotes the isothermal sound speed. In the interstellar medium, typical temperatures are $T_1 \approx 10^2$ K in the neutral gas, where $\mu_1 = 1$, and $T_2 \approx 10^4$ K in the ionized gas, where $\mu_2 = \frac{1}{2}$ (**S20**), (**S21**), (**S22**). Therefore typical sound speeds are $a_1 \approx 0.9 \text{ km s}^{-1}$ and $a_2 \approx 13 \text{ km s}^{-1}$.

TYPES OF I-FRONTS

Solutions for the jump conditions written above are described in a fundamental paper by Kahn (**K1**), who developed a comprehensive classification scheme for I-fronts, discussed their basic physical properties, and delineated the conditions under which they can occur in nature. An exhaustive treatment of these questions was later given by Axford (**A8**), who also analyzed the structure of I-fronts in great detail.

Combining (106.13), (106.14), and (106.17) we find

$$\rho_2/\rho_1 = \{a_1^2 + u_1^2 \pm [(a_1^2 + u_1^2)^2 - 4a_2^2 u_1^2]^{1/2}\}/2a_2^2. \tag{106.18}$$

The restriction that ρ_2/ρ_1 be real implies that u_1 must satisfy the inequalities

$$u_1 \geq u_R \equiv a_2 + (a_2^2 - a_1^2)^{1/2} \approx 2a_2, \tag{106.19}$$

or

$$u_1 \leq u_D \equiv a_2 - (a_2^2 - a_1^2)^{1/2} \approx a_1^2/2a_2 \ll a_1, \tag{106.20}$$

where the approximations apply when $a_2 \gg a_1$.

When u_1 exceeds the critical velocity u_R we have an *R-type* ionization front; "R" stands for "rarefied" because such fronts occur when $\rho_1 < \rho_R \equiv m_H \phi/u_R$. For fixed ϕ, $u_1 \to \infty$ (more precisely $u_1 \to c$) as $\rho_1 \to 0$. Fronts for which $u_1 = u_R$ and $\rho_1 = \rho_R$ are called *R-critical.* Similarly, when $u_1 < u_D$ and $\rho_1 > \rho_D \equiv m_H \phi/u_D$ we have a *D-type* ionization front; "D" stands for "dense". Fronts for which $u_1 = u_D$ and $\rho_1 = \rho_D$ are called *D-critical.* Fronts for which $u_D < u_1 < u_R$ and $\rho_R < \rho_1 < \rho_D$ are called *M-type.*

In an R-front $u_1 \geq u_R > a_2 > a_1$. Hence R-fronts always advance supersonically into the neutral gas, and thus cannot be preceded by a hydrodynamic disturbance of the upstream material. D-fronts always advance subsonically into the neutral gas, and thus can be preceded by hydrodynamic disturbances (e.g., a shock or a rarefication). A steady I-front cannot advance into the neutral gas when conditions ahead of it are M-type.

R-FRONTS

We obtain simple expressions for the density jump and downstream velocity in R-fronts in the limiting case that $u_1 \gg u_R$. Expanding (106.18) and choosing the negative root we find

$$\rho_2/\rho_1 \approx 1+(a_2/u_1)^2 \approx 1. \tag{106.21}$$

Such fronts are called *weak R-fronts* because the material is only slightly compressed as it passes through the front. Choosing the positive root we find

$$\rho_2/\rho_1 \approx (u_1/a_2)^2[1-(a_2/u_1)^2] \gg 1. \tag{106.22}$$

Such fronts are called *strong R-fronts* because the material is greatly compressed.

The downstream velocity in a weak R-front is

$$u_2 \approx u_1[1-(a_2/u_1)^2] \gg a_2, \tag{106.23}$$

and in a strong R-front

$$u_2 \approx a_2^2/u_1 \ll a_2. \tag{106.24}$$

Thus a weak R-front moves supersonically with respect to both the neutral and the ionized gas. In the lab frame the neutral material is at rest, $v_1 = u_1 + v_f = 0$, hence the front moves to the left (cf. Figure 106.1) with a speed $v_f = -u_1$, and the ionized gas moves subsonically to the left with a speed

$$v_2 = u_2 + v_f = u_2 - u_1 \approx -a_2^2/u_1. \tag{106.25}$$

In contrast, a strong R-front moves supersonically with respect to the neutral gas but only subsonically with respect to the ionized gas. Therefore in the lab frame the ionized gas moves to the left supersonically, almost with the speed of the I-front:

$$v_2 \approx -u_1[1-(a_2/u_1)^2]. \tag{106.26}$$

For an R-critical front one finds

$$u_2 = a_2 \tag{106.27}$$

and

$$\rho_2/\rho_1 \approx 2-\tfrac{1}{2}(a_1/a_2)^2 \approx 2. \tag{106.28}$$

Thus an R-critical front moves exactly sonically with respect to the ionized gas, and produces a moderate density jump across the front. In the lab frame the ionized gas moves to the left nearly sonically.

D-FRONTS

In contrast to an R-front, in which ρ_2 always exceeds ρ_1, the gas passing through a D-front undergoes *expansion*. We can obtain simple expressions for ρ_2/ρ_1 and u_2 by assuming that $u_1 \ll u_D \ll a_1$. Then by choosing the positive root in (106.18) we find that in a *weak D-front*

$$\rho_2/\rho_1 = (a_1^2/2a_2^2)(1+\delta) \approx (u_D/a_2)(1+\delta) \ll 1, \qquad (106.29)$$

where

$$\delta \equiv [1-(2u_1a_2/a_1^2)^2]^{1/2} \qquad (106.30)$$

increases from zero to one as u_1 decreases from u_D to zero. In a *strong D-front* we find

$$\rho_2/\rho_1 \approx (a_1^2/4a_2^2)(u_1/u_D)^2 \approx \tfrac{1}{2}(u_1/u_D)(u_D/a_2) \ll 1, \quad (106.31)$$

which is smaller than the density ratio in a weak D-front by an additional factor of (u_1/u_D).

The downstream velocities in weak and strong D-fronts are, respectively,

$$u_2/a_2 \approx (u_1/u_D)/(1+\delta) \ll 1 \qquad (106.32)$$

and

$$u_2/a_2 \approx 2(u_D/u_1) \gg 1. \qquad (106.33)$$

Thus weak D-fronts move subsonically with respect to both the neutral and ionized gas, whereas strong D-fronts move subsonically into the neutral gas but supersonically with respect to the ionized gas. For a D-critical front one finds

$$u_2 = a_2 \qquad (106.34)$$

and

$$\rho_2/\rho_1 = a_1^2/2a_2^2 = (u_D/a_2) \ll 1. \qquad (106.35)$$

Thus a D-critical front moves exactly sonically with respect to the ionized gas.

In the lab frame the ionized gas behind a D-front advancing into neutral material at rest always moves to the right, subsonically for weak fronts, nearly sonically for critical fronts, and supersonically for strong fronts.

A more detailed and complete discussion of the properties of steady I-fronts can be found in (**A8**).

RELATION TO COMBUSTION WAVES

Ionization fronts resemble *combustion waves* (**L2,** Chap. 14), (**C20,** Chap. 3, Sec. E) in many respects. In both cases the "chemical composition" of the gas changes across a sharp interface as a result of energy input into the gas: from exothermic chemical reactions (which *burn* the gas) in the case of combustion waves, and from an external radiation source in the case of I-fronts (which "dissociate" atoms into ions and electrons). In general terms D-fronts resemble *deflagrations* (or *flame fronts*) and R-fronts resemble *detonations.*

A significant difference between the two theories is that, according to the *Chapman–Jouguet hypothesis,* only weak deflagrations and strong detonations are possible. In the former case the front propagates subsonically with respect to both the unburnt (upstream) and burnt (downstream) gas; in the latter it is driven by a shock that propagates supersonically into the unburnt gas, but subsonically with respect to the combination products. Strong deflagrations and weak detonations are forbidden.

In contrast, for I-fronts, weak R-fronts (corresponding to weak detonations) are not only possible, but, as we will see below, play a central role in the dynamics of gaseous nebulae and ablation fronts. Similarly, whereas the D-fronts in gaseous nebulae are usually D-critical or weak-D, transient strong D-fronts can arise.

Indeed it is strong R-fronts (analogous to strong detonations) that are not expected to occur in nature because some additional (i.e., nonradiative) mechanism would be required to maintain the large velocity of the compressed, ionized gas behind the front. Moreover, sonic disturbances in the hot gas behind the front can catch up with the front and can continually weaken it. An essential reason for this difference is that the exothermic chemical reaction that powers a detonation wave is actually *driven* by the wave itself; that is, the high temperatures in the shock cause the upstream gas to ignite spontaneously as it passes through the front, while the energy thus released propels the shock forward. Thus the strong detonation (supersonic upstream, subsonic downstream) is the only natural solution. However an I-front will propagate naturally at the speed of light (a signal speed that is independent of the hydrodynamic state of the material) until the density, hence absorption coefficient, of the upstream material becomes large enough to slow the radiation front to a diffusion wave. Thus the weak R-front is a natural solution that reflects the properties of the externally imposed energy source.

A much more penetrating analysis of the differences between combustion waves and I-fronts was carried out by Axford (**A8**), who demonstrates that the Chapman–Jouguet hypothesis is invalid for I-fronts.

ABLATION FRONTS

When intense radiation from a hot source (e.g., an O-star) penetrates into optically thick cold material (e.g., an interstellar cloud) bounded by vacuum, an *ablation front* is driven into the medium, and hot ionized material expands rapidly away from the boundary surface in a *blowoff*. To gain insight we first assume planar geometry, and, following Kahn (**K1**), we consider what happens at the vacuum-cloud interface as the intensity of the incident radiation is progressively increased from a very low to a very high level.

When the incident photon flux is very small, conditions in the neutral gas are of extreme D-type, hence the radiation produces a very weak D-front which propagates only very slowly into the neutral material. The ionized gas expands gently into the vacuum, essentially as a mild *evaporation*. This loss of material induces a weak rarefaction (or *expansion wave*) to propagate at the speed of sound into the cold medium ahead of the I-front,

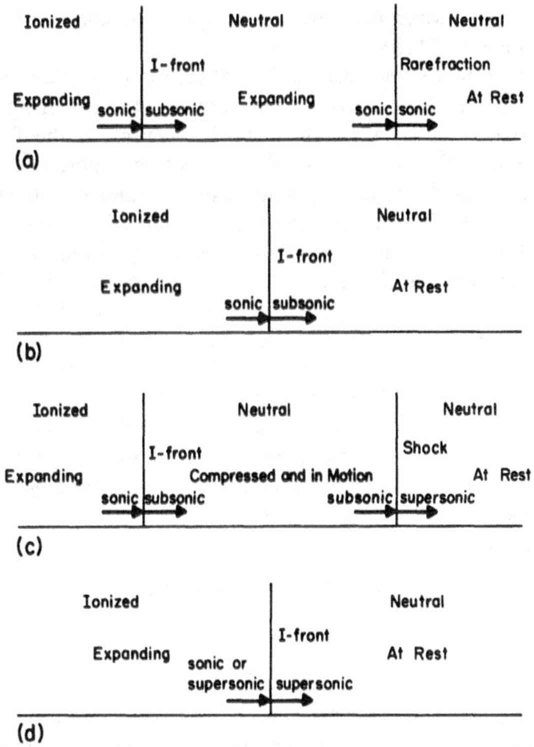

Fig. 106.2 State of flow in an I-front for (a) a very weak D-front; (b) a D-critical front; (c) a D-critical front preceded by a shock; and (d) an R-critical front.

which penetrates only subsonically. The resulting flow pattern is sketched in Figure 106.2a.

As the photon flux is increased, the rate of ablation increases, and the pressure in the ionized gas in the blowoff rises. Eventually the back pressure becomes large enough to prevent any expansion of the neutral gas, at which point conditions in the flow are D-critical, as sketched in Figure 106.2b.

A further increase in the photon flux would raise the I-front velocity above the D-critical limit u_D, hence the conditions in the neutral gas would become M-type, and the front could not propagate directly into the quiet material. But at the same time the pressure in the blowoff rises above the value necessary to just stop the backward expansion of the neutral gas, and thus drives a shock that moves supersonically into the quiet gas ahead of the I-front. This shock compresses the material passing through it enough that the postshock density rises to the value ρ_D required to permit continued D-critical propagation of the I-front at the specified value of ϕ.

The resulting flow, comprising an I-front and an *ablation-driven shock*, is sketched in Figure 106.2c.

As the photon flux is increased still further, the ionization rate becomes larger and larger, and the I-front travels through the gas at a speed closer and closer to that of its antecedent shock. Eventually the flux becomes large enough that the I-front and shock have the same speed and thus merge into a single front propagating into the quiet gas. Conditions are then R-critical, and the flow pattern is as sketched in Figure 106.2d.

If the photon flux is made even greater, the I-front moves into the cold material so fast that the shock can no longer keep up with it. The I-front then propagates directly into the neutral gas as a weak R-front.

Newton's third law of motion implies that the rapid loss of high-velocity material in the blowoff from an ablation front will accelerate the neutral medium in the opposite direction. Oort and Spitzer (**O2**) have suggested that this *rocket effect* can accelerate neutral interstellar clouds near O- and B-stars to very large velocities. Thus if radiation on a cloud of mass $\mathcal{M}$ forces it to lose mass at a rate $d\mathcal{M}/dt$ in material expanding into vacuum with a velocity V at the ablation front, then the velocity v of the cloud can be determined from

$$\mathcal{M}(dv/dt) = -V(d\mathcal{M}/dt), \tag{106.36}$$

which yields the standard rocket equation

$$v = V \ln (\mathcal{M}_0/\mathcal{M}). \tag{106.37}$$

Here $\mathcal{M}_0$ is the initial mass of the cloud, which is assumed to be initially at rest. The mass-loss rate is related to the photon number flux by

$$(d\mathcal{M}/dt) = -\pi R^2 \phi m_{\rm H}, \tag{106.38}$$

where R characterizes the projected cross section of the cloud to the stellar radiation. Here we have assumed that each photon ionizes an atom, and that the ionized material all leaves in the blowoff.

To trace the time history of a cloud in detail, one must make a variety of additional assumptions. But with reasonable, if simplified, models one can show that there is a critical initial mass $\mathcal{M}_{\rm crit}$ at which clouds engulfed in the region of ionized gas surrounding an O-star just evaporate as the cloud remnant reaches the edge of that region (**O2**), (**S20**), (**S21**). Clouds with $\mathcal{M}_0 < \mathcal{M}_{\rm crit}$ evaporate completely; clouds with $\mathcal{M}_0 > \mathcal{M}_{\rm crit}$ can survive and escape from the ionized region, sometimes with large velocities.

The efficiency of the rocket effect is somewhat reduced by recombination of ions and electrons in the blowoff. Neutral atoms formed by recombination attenuate the radiation from the external source before it reaches the ablation front; they thus produce an *insulating layer* that decreases the rate of energy deposition into the front. This decrease can be quite significant in planar geometry (**K1**), but is less serious for a uniformly irradiated spherical medium because the geometrical divergence of a radial blowoff leads to a rapid reduction of the density, hence recombination rate, in the expanding material. In this case we can get rather efficient energy deposition into a spherical ablation front that drives a *converging spherical shock* into the cold medium. The converging shock can collapse the core of the original sphere to very high densities. In fact it is just such *radiation-driven implosions* that are used in laser-fusion experiments to compress pellets containing appropriate isotopes of hydrogen to the high densities and temperatures needed to ignite thermonuclear reactions (**M2**).

Similar effects occur in interstellar cloud complexes near or around young clusters containing O- and B-stars. In particular the *bright rims* sometimes observed to surround dark clouds near very hot stars may be insulating layers formed by blowoff from the clouds. Furthermore, radiation-driven implosion of interstellar clouds may provide an effective mechanism of star formation (**E2**), (**S3**).

Thus suppose a single *seed* O-star "lights" deep inside a massive (10^5 to $10^6 \mathcal{M}_\odot$) interstellar molecular cloud complex that has a very inhomogeneous structure consisting of dense condensations surrounded by a more rarefied medium. Radiation from the seed star will preferentially burn through the less-dense interstices in the cloud complex, and can produce ablation fronts around several nearby condensations in the original cloud. Each of these fronts may implode a condensation to the point where it becomes *Jeans unstable*, collapses gravitationally, and forms a new star. Radiation from these new stars may then implode still more stars, and one can imagine the possibility of a multiplicative runaway leading to a violent burst of star formation in the cloud.

There is, of course, a competition between the loss of material into the blowoff and the effects of the converging shock. If the implosion proceeds too slowly, an initial condensation will evaporate before it can collapse gravitationally. Thus the radiation-driven implosion mechanisms may produce mainly high-mass stars, though recent work (**K6**) suggests that irradiation of a condensation by *multiple* driving stars may produce low-mass stars as well.

The multiplicative star-formation mechanism described above implies a rapid building of radiation inside the cloud complex. Ultraviolet stellar radiation will continually ionize the less-dense regions between condensations in the complex. In due course, each ionization is followed by a recombination, which results in the emission of one or more photons in subordinate continua and in spectral lines (including Lyman α). This recombination radiation scatters around within the transparent ionized interstices between condensations and steadily accumulates, as in a reservoir, because new ionizing photons are continually emitted by stars. The level of this reservoir, representing the time-integrated luminous output from all stars embedded in the cloud complex, can become quite high. Eventually the ionizing radiation burns through at some position on the outer boundary surface of the cloud complex. It is interesting to speculate whether one might observe an intense, nonequilibrium burst of radiation at this instant, as the radiation reservoir stored in the cloud pours out. At a somewhat later time one might also expect to observe an energetic hydrodynamic flow through the site of the radiative burn-out.

RADIATION-DRIVEN EXPLOSIONS

A radiation-driven explosion is produced when a large amount of radiant energy is released nearly instantaneously from a point source in a cold gas. The radiation both ionizes and strongly heats the gas and can thus drive violent hydrodynamic phenomena. Good examples are *H II regions*, which are regions of ionized hydrogen in the interstellar medium surrounding O- and B-stars, and *fireballs* produced by extremely strong explosions in the Earth's atmosphere.

The dynamics of fireballs is significantly influenced by gravity (stratification of the ambient atmosphere) and by reflected shocks (in explosions near the ground). As fireballs are discussed extensively in (**Z3,** Chap. 9) and the references cited therein, they will not be considered further here. Rather, we discuss qualitatively the dynamical behavior of an H II region as it expands into the surrounding H I region (i.e., the neutral interstellar medium) until it comes into equilibrium with its surroundings and forms a static *Strömgren sphere* (**S26**) around the exciting star. We assume that the H I region is initially homogeneous, and neglect gravitational forces.

Numerous studies have been made of the dynamics of H II regions. Simple analytical considerations are summarized in (**S20**), (**S21**). A similarity solution was constructed by Goldsworthy (**G6**), but unfortunately it is valid only for a particular initial density distribution ($\rho \propto r^{-3/2}$ in spherical geometry and $\rho \propto r^{-1}$ in cylindrical geometry). Moreover, for a steady photon flux the solution requires that the gas temperature must vanish at the origin in spherical geometry, which is unphysical, hence one is forced to cylindrical geometry, which is unrealistic. Thus these solutions have only limited value. Vandervoort (**V1**), (**V2**), (**V3**) discussed the early phases of evolution of H II regions using the method of characteristics. The effects of various physical processes on the structure of H II regions in the steady-flow approximation were analyzed by Hjellming (**H5**).

The most realistic models have been constructed by Mathews (**M6**) and Lasker (**L5**) using numerical methods. We will discuss these shortly, but first, following Mathews and O'Dell (**M7**), it is instructive to study the evolution of an H II region by an analysis of the behavior of the gas in the (p, V) diagram. As we saw in §56, conservation of mass and momentum across a front imply that

$$p_2 - p_1 = -\dot{m}^2(V_2 - V_1), \tag{106.39}$$

where $\dot{m}$ is the mass flux and $V \equiv 1/\rho$. This result, which applies for both shocks and I-fronts, shows that the initial and final states of the gas must be connected by a straight line of negative slope in the (p, V) diagram. For the

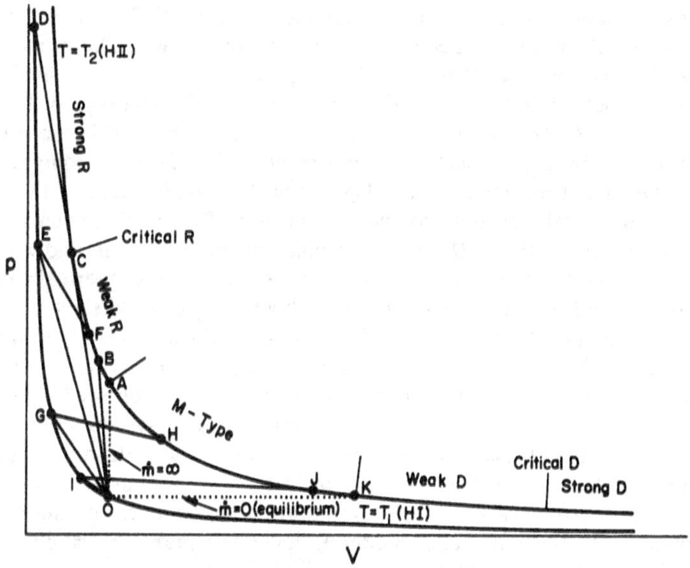

Fig. 106.3 Evolution of shocks and ionization fronts in an H II region, shown in (p, V) diagram. Adapted with permission, from (**M7**) in *Annual Review of Astronomy and Astrophysics,* Vol. **7,** © 1966 by Annual Reviews, Inc.

problem under consideration, the initial and final states must lie on the isotherms $T \equiv T_1$ and $T \equiv T_2$ which, as sketched in Figure 106.3, are hyperbolae in the (p, V) plane. Shocks are represented by straight lines joining two points on the same isotherm; I-fronts by straight lines with negative slope joining points on the two isotherms.

If the initial state of the gas is represented by point O at (p_1, V_1) on the H I isotherm, various types of I-fronts *with respect to this point* are shown on the H II isotherm. Thus the separation between M-type conditions and R-fronts occurs at infinite photon flux (which implies $\dot{m} = \infty$), and between M-type conditions and D-fronts at zero photon flux (which implies that $\dot{m} = 0$). A transition via an I-front from point O to any point in the M-type region is clearly impossible because the slope of the line joining the initial and final states would be positive. R-critical and D-critical fronts occur when the line from O is just tangent to the H II isotherm, for this is the condition that the front propagate sonically at $T = T_2$.

Suppose now that an O-star begins to radiate essentially instantaneously in a large, low-density H I cloud having $(p, V) = (p_1, V_1)$. Then initially $\dot{m}$ is very large and a weak R-front moves rapidly [$u_1 \sim 800 \text{ km s}^{-1}$ (**M6**)] outward, producing a transition such as OB in Figure 106.3. As the front moves outward, ϕ, hence $\dot{m}$, decreases owing to spatial dilution and to absorptions resulting from recombinations in the H II region. Eventually the front becomes R-critical, producing the transition *OC* in Figure 106.3, and moves at the sound speed relative to the ionized gas. The pressure in the H II region exceeds that in the H I region by two orders of magnitude, hence it must expand, and will drive a shock into the H I region. In fact, the transition *OC* can also be viewed as a strong shock (*OD*) in the neutral gas followed by a D-critical ionization front (*DC*) *relative to point D*.

As the I-front continues outward ϕ and $\dot{m}$ decrease further and the front moves subsonically relative to the ionized gas. The shock driven by the excess pressure of the H II region can thus outrun the I-front, and we now have transitions of the type *OEF* in which a shock in the isothermal neutral gas is followed by a weak D-front. The shock progressively slows and the I-front progressively weakens, passing from transitions like *OGH* to transitions like *OIJ*, in which the shock is very weak (*OI* is nearly tangent to the isotherm, hence the Mach number is near unity) and is followed by a very weak D-front ($\dot{m}$ is nearly zero). Ultimately, the system approaches equilibrium in the transition *OK*, with $\dot{m}=0$, and a static Strömgren sphere is formed. Note, however, that all early type stars embedded in H II regions have strong winds (cf. §107) that are a major source of energy and momentum to the interstellar medium, hence the purely radiation-driven flow discussed above ceases to provide a realistic description at late times when the dynamical effects of the stellar wind dominate.

If the sequence just described is reversed, we recover the scenario discussed earlier for ablation fronts. An important difference is that in Kahn's analysis (**K1**) of an I-front backed by vacuum, rarefaction waves running ahead of a D-front tend to maintain it in a D-critical condition until ϕ becomes large enough to force the front to be R-critical. Much of the early analytical work on the propagation of I-fronts in H II regions was based on the simplifying assumption that the I-front following remained exactly D-critical. But numerical calculations show (**L5**) that in reality this approximation is not at all appropriate for H II regions (because the large pressure in the ionized gas drives a strong shock, which must be followed by a weak D-front) until the H II region expands nearly to its equilibrium position and $\phi \rightarrow 0$.

The numerical models (**M6**), (**L5**) of H II regions employ the one-dimensional Lagrangean hydrodynamics schemes discussed in §59. One uses the equation of continuity to relate radii to a Lagrangean mass or space variable as in (59.84); Euler's equation of motion with a pseudo-viscous pressure and zero gravity [cf. (59.82)]; and an energy equation of the general form

$$(De/Dt)+p[D(1/\rho)/Dt]=\mathcal{G}-\mathcal{L}, \tag{106.40}$$

where $\mathcal{G}$ and $\mathcal{L}$ are radiative gains and losses per unit mass. The ionization state of the material is determined by the rate equation

$$(Dx/Dt)=(1-x)\bar{\alpha}\phi-(\rho/m_{\rm H})x^2\beta(T), \tag{106.41}$$

where x is the ionization fraction and $\beta(T)$ is a recombination coefficient, while the photon flux follows from the (quasi-static) transfer equation

$$\frac{1}{r^2}\frac{\partial}{\partial r}(r^2\phi)=\frac{-\rho}{m_{\rm H}}(1-x)\bar{\alpha}\phi. \qquad (106.42)$$

These equations are discretized and solved using basically the techniques described in §59. The I-front, however, requires special treatment. Mathews (**M6**) used a special integration scheme suggested by Henyey to handle the numerical stiffness of (106.41) as the ionization fraction approaches equilibrium. In addition he used extremely fine zones in the I-front along with a rezoning scheme that added new zones in the upstream gas as it entered the I-front and discarded unnecessary zones in the downstream flow. In contrast, Lasker (**L5**) used an algorithm that smears the discontinuous I-front over a few zones, a method analogous to using pseudoviscosity to smear shocks. In present-day computations it would be preferable to handle the I-front with an adaptive-mesh technique.

Lasker's calculations follow the evolution of an H II region well beyond the R-critical stage studied by Vandervoort and by Mathews. The exciting

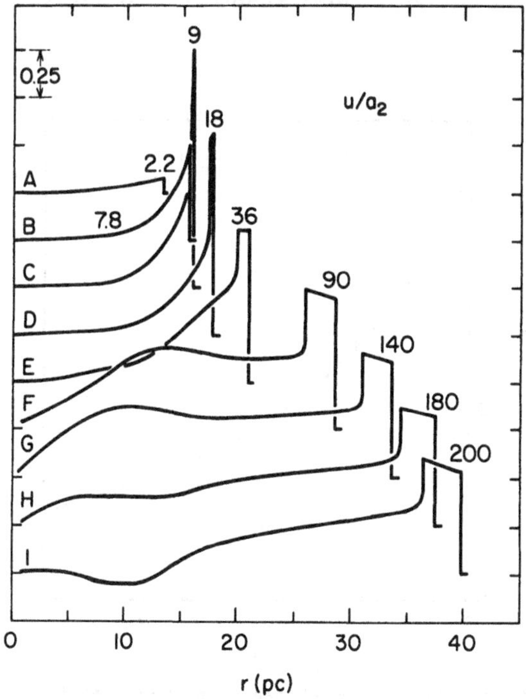

Fig. 106.4 Time evolution of velocity structure of an H II region. From (**L5**), by permission.

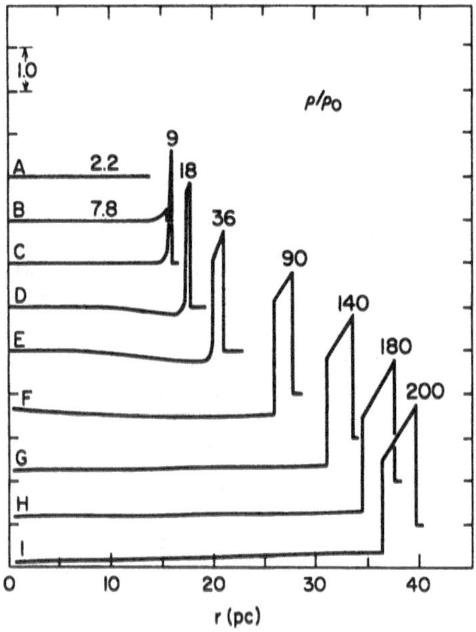

Fig. 106.5 Time evolution of density structure of an H II region. From (**L5**), by permission.

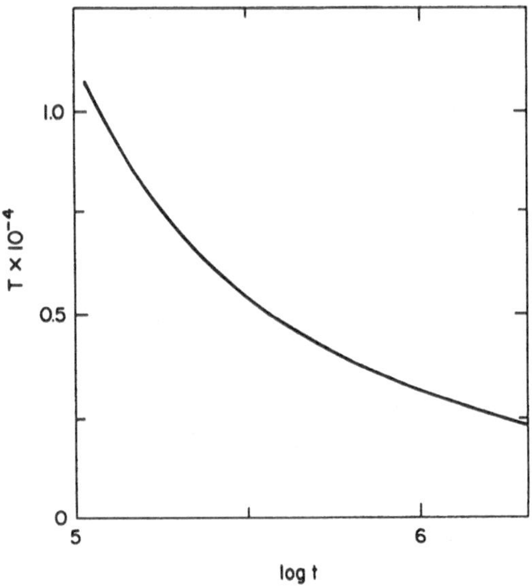

Fig. 106.6 Temperature immediately behind shock in an H II region. From (**L5**), by permission.

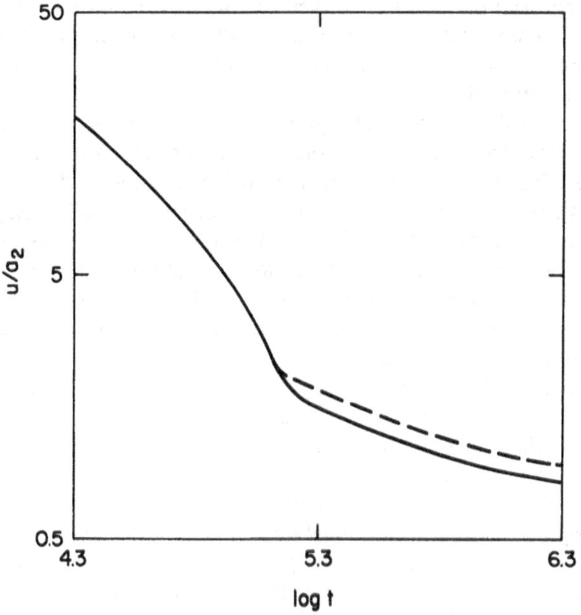

Fig. 106.7 Velocity of shock (upper curve) and I-front (lower curve) in an H II region. From (**L5**), by permission.

star is assumed to start radiating instantaneously in an infinite homogeneous cloud with an initial temperature $T_0 = 100$ K. The results shown in Figures 106.4 to 106.7 apply to a model with an initial number density $N_0 = 6.4\ \text{cm}^{-3}$. Models A to C cover the initial stages of R-front propagation; in model C the shock has just formed and is slightly separated from the I-front. The shock compresses the neutral material as it passes over it, and in model D a distinct shell of compressed neutral gas is evident. This shell, which is driven outward by excess pressure in the H II region, becomes thicker and thicker in subsequent models as the shock progressively moves away from the I-front. The strength of the shock decreases in time both because of geometrical divergence and because the pressure and density drop in the H II region as it expands. The temperature immediately behind the shock and the velocities of the I-front and shock as functions of time are shown in Figures 106.6 and 106.7.

107. Radiation-Driven Winds

The main effect of radiation in the radiation-driven flows we have considered so far is to provide an energy input to the gas, which heats it, hence raises its pressure, and thus drives a flow, perhaps explosively. We now consider an example of a flow driven by radiative *momentum* input to the gas, that is, a flow that results from the work done by the *radiation force* on the material, even in the absence of net energy exchange between the gas and the radiation field.

In recent years a variety of observations from spacecraft and ground-based observatories have shown that hot, luminous, early type stars have massive stellar winds. Analyses of line profiles and infrared emission (**A4**), (**A5**), (**G1**), (**L1**) imply mass-loss rates $\dot{\mathcal{M}}$ of order 10^{-6} to $10^{-5}\mathcal{M}_\odot$/year for O-stars, and perhaps up to $10^{-4}\mathcal{M}_\odot$/year for Wolf–Rayet stars; recall from §61 that the mass-loss rate in the solar wind is only $10^{-14}\mathcal{M}_\odot$/year. The observations indicate transonic winds, with flow velocities that rise from near zero in the stellar photosphere to highly supersonic values within one stellar radius from the surface. For O-stars the observed terminal velocities v_∞ are typically about three times the escape velocity v_{esc}, which is about 1000 to 1500 km s^{-1} for a main-sequence O-star and 600 to 900 km s^{-1} for an O-supergiant (**A1**), (**G1**). The sound velocity in the atmospheres of these stars is about 25 km s^{-1}.

Lucy and Solomon (**L10**) recognized that these flows cannot be explained by the thermal wind model described in §61 because if the specific enthalpy at the critical point is to provide the observed terminal kinetic energy flux, then the critical-point temperature would have to be $T_c \sim 3\times10^7$ K for $v_\infty \approx 3000$ km s^{-1}. This high value is excluded because lines from ions that would be destroyed by collisional ionization at temperatures greater than about 3×10^5 K are present throughout most of the flow. (However, soft X rays are also observed from O-stars, hence there must be at least some material at coronal temperatures embedded in the flow; nevertheless the bulk of the flow is too cool to be a thermal wind.)

The most natural way to explain the flow is that it is driven by momentum input to the gas from the intense radiation fields of these extremely luminous stars, in particular by the radiation force exerted on strong spectrum lines. We emphasize that, as in §61, our goal here is to elucidate some of the underlying physics of radiation-driven winds, not to develop realistic models of the winds of particular stars.

THE EDDINGTON-LIMIT LUMINOSITY

Radiative momentum input to the gas results when photons are absorbed from the anisotropic (indeed almost purely radially streaming at large distances from the star) stellar radiation field, and then scattered isotropically. The absorbed photons deposit all their outward-directed momentum into the material, but because the scattering process is isotropic, the reemitted photons produce no net change in the momentum of the material, which therefore experiences a net gain of outward momentum. The ions scattering the radiation are thus accelerated radially and they drag along the rest of the plasma through momentum exchange in Coulomb collisions.

Therefore the outward acceleration of the gas is

$$g_R = \int_0^\infty \chi_\nu F_\nu \, d\nu/\rho c, \tag{107.1}$$

where χ_ν is the total extinction from all sources (continua, electron scattering, and lines) and F_ν is the radiation flux. Note that a photon is not destroyed when it is scattered, but is merely redshifted by at most $\Delta\nu = \nu_0 v_\infty/c$; because $\Delta\nu/\nu_0 \ll 1$, each photon can in principle be scattered many times before it is extinguished.

The outward *radiative acceleration* g_R is to be compared with the inward acceleration of gravity, $g = G\mathcal{M}/r^2$; if g is everywhere greater than g_R then the atmosphere remains in hydrostatic equilibrium and does not expand. For convenience define the *force ratio*

$$\Gamma \equiv g_R/g. \tag{107.2}$$

In O-stars the continuous opacity is dominated by electron scattering in those spectral regions where most of the flux emerges. We therefore obtain a reasonable lower bound for Γ if we assume that the opacity is pure Thomson scattering, namely

$$\Gamma_e = s_e L/4\pi c G\mathcal{M}, \tag{107.3}$$

where $s_e \equiv n_e\sigma_e/\rho$ is the electron scattering coefficient per gram.

Consider now a spherically symmetric steady flow from a star. Parameterizing the radiation force as in (107.2), we write the momentum equation (96.2) as

$$\rho v(dv/dr) = -(dp/dr) - G\mathcal{M}(1-\Gamma)\rho/r. \tag{107.4}$$

The pressure can be expressed as $p = a^2\rho$, where a is the isothermal sound speed, assumed to be, in general, a function of r. [For brevity we drop the subscript "T" used in (51.26); no confusion should result because we will not be referring to the adiabatic sound speed in this section.] From the equation of state and the continuity equation we find

$$\rho^{-1}(dp/dr) = (da^2/dr) - (2a^2/r) - (a^2/v)(dv/dr), \tag{107.5}$$

whence we can rewrite (107.4) as

$$[1-(a^2/v^2)]v(dv/dr) = (2a^2/r) - (da^2/dr) - G\mathcal{M}(1-\Gamma)/r^2. \tag{107.6}$$

We now ask under what conditions one can have a continuous transonic flow under the combined action of gravity and radiation (**M3**). For simplicity, assume the envelope is isothermal and drop (da^2/dr). It is then evident that to obtain a smooth transition from subsonic flow at small r to supersonic flow at large r, the right-hand side of (107.6) must (1) vanish at the sonic radius r_s where $v(r_s) \equiv v_s = a$; (2) be negative for $r < r_s$; and (3) be positive for $r > r_s$. The condition for $r < r_s$ can be met only if $\Gamma < 1$ in that region; that is, in the subsonic flow region the radiation force must be less than that of gravity if a steady flow is to accelerate outward. In contrast, in the supersonic flow region ($r > r_s$) Γ may become arbitrarily large; indeed the larger it is, the greater is the momentum input to the gas, and the larger (dv/dr), hence v_∞, will be.

If Γ is greater than unity everywhere in a stellar envelope, steady transonic flow is impossible; one must have either an initially subsonic flow that decelerates outward, an initially supersonic flow that accelerates outward, or (most likely) a time-dependent flow. As Eddington pointed out, if Γ_e (which always underestimates the radiation force because $\chi_\nu \geq n_e\sigma_e$) rises to unity at some point in the envelope, one can expect $\Gamma \geq 1$ throughout the remainder of the stellar interior because both the radiation flux and the force of gravity scale as r^{-2}. In this event the material is unbound gravitationally, so the star becomes unstable, and can freely expand homologously on a short time scale. The critical luminosity

$$L_E \equiv 4\pi c G\mathcal{M}/s_e \tag{107.7}$$

is called the *Eddington-limit luminosity*. Objects of radius R having $L \gg L_E$ can be expected to be blown apart by radiation pressure on a dynamical time scale of order

$$t_L \sim (4\pi c R^3/s_e L)^{1/2}. \tag{107.8}$$

Numerically (107.3) gives $\Gamma_e \approx 2.5\times 10^{-5}(L/L_\odot)(\mathcal{M}_\odot/\mathcal{M})$; for an O-star $L \approx 10^6 L_\odot$ and $\mathcal{M} \approx 60\mathcal{M}_\odot$, hence $\Gamma_e \approx 0.4$. Thus the radiation force from continuum opacity alone does not exceed gravity, which implies (1) that normal O-stars are stable against radiative disruption, and (2) that the continuum radiation force cannot drive a transonic wind by itself. We must therefore look to spectral lines to provide the required force.

THE RADIATION FORCE ON SPECTRAL LINES

In order to focus on the momentum (as opposed to energy) transfer from radiation to the material, we assume pure conservative scattering lines. At great optical depth where the diffusion approximation is valid, $F_\nu \propto \chi_\nu^{-1}$, hence in this regime the product $\chi_\nu F_\nu$ in (107.1) is independent of the value of χ_ν, and lines are no more effective than the continuum in delivering momentum to the gas. Therefore at depth Γ remains essentially equal to Γ_e. However in optically thin material the situation is quite different. Near the surface of a star F_ν can rise far above its diffusion-limit value because intense radiation emerges from the material below, and none is incident from above.

To estimate the maximum force that can result from a single line, assume that some optically thin material is irradiated from below with unattenuated continuum radiation, that is, $F_\nu = F_c = \pi B_\nu(T_{\text{eff}})$. Then an upper limit to the acceleration of the gas produced by a single line of an atom of chemical species k, in excitation state i of ionization state j, is

$$g_R^0 = (\pi^2 e^2/mc^2) f B_\nu(T_{\text{eff}})(n_{ijk}/N_{jk})(N_{jk}/N_k)(\alpha_k/X m_H), \tag{107.9}$$

where n_{ijk} is the population of the particular level, N_{jk} is the total number of ions in all excitation states of ionization stage j, N_k is the total number density in all ionization stages of species k, α_k is the abundance of species k relative to hydrogen, and X is the mass fraction of the stellar material that is hydrogen. For example, Lucy and Solomon (**L10**) considered the C IV resonance line at $\lambda 1548$ Å; adopting an oscillator strength $f=0.2$, $T_{\text{eff}}=25{,}000$ K (to maximize B_ν), $\alpha_C=3\times10^{-4}$, $X=1$, and $(n_{ijC}/N_{jC})=1$ they found

$$\log (g_R^0)_{\lambda 1548} = 5.47 + \log (N_{jC}/N_C) \tag{107.10}$$

For an O-supergiant $\log g \approx 3$; hence in the outer layers of such a star the upper limit for the radiation force from even this one line exceeds the force of gravity by a factor of 300!

The estimate just derived is (purposely) a gross upper limit because ions in the underlying stellar photosphere produce a dark absorption line in which $F_\nu \ll F_c$. To account for this effect, Lucy and Solomon solved the line transfer equation in detail and found that above a certain level in the atmosphere the radiation force given by (107.1) for the C IV line above still exceeded gravity. Similar results are also obtained from model atmosphere calculations for early type stars, where the radiation force from a realistic line spectrum is often found to exceed gravity at the surface of the model. Thus for O-stars the radiation force obtained when the atmosphere is assumed to be static is incompatible with that assumption; hence hydrostatic equilibrium in the outermost layers is not possible, and an outflow of material must inevitably occur.

To understand how the flow develops, consider the following scenario. Once the gas in the uppermost layer begins to move outward, its spectrum lines will be Doppler shifted away from their rest wavelengths and will therefore begin to intercept the intense photospheric flux in the adjacent continuum, which enhances the momentum input to the material, hence increases its outward acceleration. The underlying layers must expand to fill the rarefaction left by the outward motion of the upper layers. Furthermore, the absorption lines in these lower layers begin to desaturate because the lines in overlying layers have been Doppler shifted, hence the underlying layers also begin to experience a radiative force that exceeds gravity, and behave, in turn, in the manner just described. Clearly a flow can be initiated by this mechanism; we now must inquire whether (1) the rate of mass loss so produced is significant, and (2) the variation of the radiation force with depth will be consistent with the requirements for transonic flow.

In connection with the latter point, one should note that the radiation force on the continuum plus lines has precisely the right behavior to produce a transonic wind. That is, Γ is less than unity in the diffusion regime inside the star, approaches unity in the atmosphere as some lines begin to desaturate and the gas begins to flow, and reaches very large values in the supersonic flow region where the lines are sufficiently displaced from their rest frequencies to absorb continuum radiation from the underlying photosphere. Moreover, we will shortly see that to a good approximation the radiation force on lines varies as a power of the velocity gradient in the flow; this dependence allows the force and the flow it drives to accommodate to one another, so that a steady transonic flow can be attained.

The first attempt to obtain quantitative results was made by Lucy and Solomon, who evaluated (107.1) numerically for scattering lines formed in an expanding envelope above a hydrostatic photosphere. In solving the transfer equation, re-emissions can be ignored because they contribute nothing to the net force exerted by radiation on the material. Therefore, the incident photospheric intensity is simply attenuated exponentially as it scatters in a line, hence $I_\nu(\tau_\nu) = I_\nu(0)\exp(-\tau_\nu/\mu)$, where τ_ν is the optical depth, at lab-frame frequency ν, from the base of the envelope to the test point, allowing for Doppler shifting of the line profile along the path. Then

$$g_{R,l} = (2\pi/c\rho)\sum_l \int_0^1 d\mu \int_0^\infty d\nu \chi_l(\nu) I_\nu(0)\mu e^{-\tau_\nu/\mu} \tag{107.11}$$

where the sum extends over all lines considered.

Lucy and Solomon coupled (107.11) to the equations of steady flow to construct radiatively driven wind models for O- and B-stars. They assumed planar geometry (adequate for the flow inside the sonic point), isothermal material, a simple nebular photoionization-recombination ionization equilibrium, and that the radiation force results from absorption in the resonance lines of a few abundant ions. The solution was obtained by an iteration procedure that yields the mass flux as an eigenvalue. A large number of models were constructed for a wide range of stellar parameters. These models successfully produced transonic flows having reasonable terminal velocities, $v_\infty \sim 3000$ km s^{-1}, but the computed mass-loss rates were only $10^{-8}\mathcal{M}_\odot$/year or less, which is two orders of magnitude smaller than the observed values.

The source of this discrepancy was identified by Castor, Abbott, and Klein (**C11**) who pointed out that *hundreds* of lines in the spectrum make important contributions to the total radiation force, so that Lucy and Solomon's estimate, based on only a few lines, is roughly a hundredfold too small (**C12**). As it would be hopeless to calculate the aggregate radiation force from hundreds of lines by a direct numerical solution of the transfer equation, recourse must be had to an approximate analytical method. The essential point is to account for saturation in the lines, so that the correct transition is made between the optically thick and thin limits. This problem was solved in detail by Castor (**C9**); here we make only a simple heuristic argument to recover the main result.

We assume that the incident photospheric radiation field on the lines is essentially radial, and approximate the momentum absorbed, per unit mass, by a line of opacity χ_l and width $\Delta\nu_D$ from the unattenuated continuum flux F_c as $g_{R,l}(0) = \chi_l \, \Delta\nu_D F_c/c\rho$. We ignore re-emissions, as before; then the incident flux is attenuated as $e^{-\tau_l}$ where, as in (107.11), τ_l is the line optical depth in a layer, allowing for Doppler shifts. The average rate of momentum input to a layer of optical depth τ_l is then

$$\langle g_{R,l} \rangle = g_{R,l}(0)\tau_l^{-1} \int_0^{\tau_l} e^{-\tau'} \, d\tau' \tag{107.12}$$

or

$$\langle g_{R,l} \rangle = (\chi_l \, \Delta\nu_D F_c/c\rho)(1 - e^{-\tau_l})/\tau_l. \tag{107.13}$$

In their work Castor, Abbott, and Klein approximate $\tau_l^{-1}(1-e^{-\tau_l})$ by $\min(1, \tau_l^{-1})$. In the optically thin limit, (107.13) reduces to

$$\langle g_{R,l} \rangle_{\text{thin}} = \chi_l \, \Delta\nu_D F_c/c\rho, \tag{107.14}$$

so that, as in (107.9), the force on a line is proportional to its opacity, hence strong lines are more important than weak lines. In the optically thick limit (107.13) reduces to

$$\langle g_{R,l} \rangle_{\text{thick}} = \chi_l \, \Delta\nu_D F_c/c\rho\tau_l, \tag{107.15}$$

which shows that the force on a line is independent of its strength (because τ_l scales as χ_l), hence all lines are of equal importance, as expected in the diffusion limit.

We must now specify the effective optical thickness of the envelope. For a static medium

$$\tau_l = \int_R^\infty \chi_l \, dr, \tag{107.16}$$

hence τ_l is determined by the strength of the line and the amount of material in the line-forming layers.

For an expanding medium the situation is quite different. Here photons emitted at one position are always redshifted when they arrive at some other position in the flow by an amount proportional to the average velocity gradient times the distance between the two positions. Therefore photons emitted at line center at some point can interact with the material only within a localized *resonance region;* beyond this region they fall too far in the wing of the line profile of the material at the remote position to be absorbed effectively, hence they escape without further interactions. Line transport in such a flow regime is described by *Sobolev theory* (**S14**), (**S15**), (**C8**). We cannot discuss this theory in detail here, but from mere dimensional considerations one can see that for an idealized square-topped line profile of width $\Delta\nu = \nu_0 v_{\text{th}}/c$ (where v_{th} is the thermal speed of the absorbing atoms), the characteristic distance within which radiative interactions can occur must be of order $l \sim v_{\text{th}}/|\nabla v|$. Hence for radially streaming radiation in an expanding medium we can take

$$\tau_1 \equiv \chi_1 v_{\text{th}}/(dv/dr). \tag{107.17}$$

The important difference between (107.16) and (107.17) is that a large velocity gradient serves to reduce τ_l from its static value, hence to desaturate the line, and thus to increase the radiation force on that line, perhaps by orders of magnitude.

In estimating the total radiation force from an ensemble of lines, we will use (107.17) throughout the entire wind, even though it becomes invalid in the nearly hydrostatic photosphere (because the line radiation force is unimportant there anyway). It is convenient to use a depth variable that is independent of line strength, so we define $\beta_l \equiv n_e\sigma_e/\chi_l$ and introduce an equivalent electron optical depth scale

$$t \equiv \beta_l \tau_l = n_e \sigma_e v_{\text{th}}/(dv/dr). \tag{107.18}$$

The total radiation force is obtained by summing (107.13) over all lines, which gives

$$g_{R,l} = (s_e F/c)M(t) = (s_e L/4\pi c r^2)M(t) \tag{107.19}$$

where

$$M(t) \equiv \frac{1}{F}\sum_l \frac{F_c(\nu_l)\,\Delta\nu_{D,l}(1-e^{-t/\beta_l})}{t} \approx \frac{1}{F}\sum_l F_c(\nu_l)\,\Delta\nu_{D,l}\min\left(\frac{1}{\beta_l},\frac{1}{t}\right) \tag{107.20}$$

is the *line force multiplier*. The calculation of the radiation force is thus reduced to the evaluation of $M(t)$ which, for a specified temperature and density and a given set of lines, is a function of only the one parameter t. The local excitation and ionization equilibrium enters through the parameter β_l.

It is important to note that (107.20) has only limited accuracy because two important approximations have been made in deriving it. (1) We assumed radially streaming radiation, and ignored the angular integration over the finite solid angle subtended by the stellar photosphere. The effect of this omission is to overestimate the radiation force close to the photosphere. (2) We assumed that each photon is scattered only once in one resonance region, and ignored the possibility of multiple scatterings in several (perhaps overlapping) lines. The effect of this omission is to underestimate the total amount of momentum that a photon can deposit in the gas in the high flow-velocity region. We return to these issues later.

LINE-DRIVEN WINDS

A comprehensive and internally consistent analytical theory for line-driven winds was first developed by Castor, Abbott, and Klein (**C11**), which, for brevity, we call the *CAK theory*. They assumed that the flow is steady and spherically symmetric, that the gas is a single fluid, and that conduction and viscosity can be neglected; these assumptions are justified in detail in (**C12**). The flow is calculated for a given temperature distribution $T(r)$, which ultimately is determined in an iteration procedure by imposing radiative equilibrium. The latter assumption is reasonable because the thermal relaxation time of the gas is much shorter than a characteristic flow time, but may lead to an unrealistic temperature distribution (e.g., if the flow is unstable and disintegrates into shocks) and to an unrealistic predicted spectrum. But we emphasize that the temperature distribution can have essentially no influence on the gross dynamics of the wind unless temperatures rise to order 10^7 K, and/or there are extreme temperature gradients. Gas pressure, hence temperature, is important only in the subsonic flow regime, which is also the part of the flow where radiative equilibrium is most likely to be a good approximation; it is inconsequential in the supersonic flow regime.

Castor, Abbott, and Klein evaluated the line force multiplier $M(\ell)$ for the spectrum of the representative ion C^{++}, and assuming that those results were typical they scaled them to account for the total abundance of C, N, and O. The occupation numbers were computed from LTE. Their results are well fitted by the formula

$$M(\ell) = k\ell^{-\alpha} \tag{107.21}$$

with $k \approx \frac{1}{30}$ and $\alpha = 0.7$. An exhaustive analysis (**A3**) based on a complete line list for all relevant ions of the elements H to Zn yields a more accurate expression valid for $10^4 \text{ K} \le T_{\text{eff}} \le 5 \times 10^4$ K, namely,

$$M(\ell) = 0.28(N_{11})^{0.09}\ell^{-0.56}. \tag{107.22}$$

Here $N_{11} \equiv (n_e/W) \times 10^{-11}$ and W is the *dilution factor* of the radiation field [the fraction of 4π steradians subtended by photospheric radiation, cf. (**M10,** 120)]. Substituting (107.21) and (107.18) into (107.19) we have

$$g_{R,l} = \left(\frac{s_e L k}{4\pi c r^2}\right)\left(\frac{1}{n_e \sigma_e v_{\text{th}}}\frac{dv}{dr}\right)^\alpha = \frac{C}{r^2}\left(r^2 v \frac{dv}{dr}\right)^\alpha. \tag{107.23}$$

The second equality follows from the equation of continuity, and the constant is

$$C = (s_e L k/4\pi c)(4\pi/s_e v_{\text{th}}\dot{\mathcal{M}})^\alpha. \tag{107.24}$$

Using (107.23) for the line radiation force we can rewrite the equation of motion (107.6) as

$$\left(1 - \frac{a^2}{v^2}\right)v\frac{dv}{dr} = \frac{2a^2}{r} - \frac{da^2}{dr} - \frac{G\mathcal{M}(1-\Gamma_e)}{r^2} + \frac{C}{r^2}\left(r^2 v \frac{dv}{dr}\right)^\alpha. \tag{107.25}$$

Unlike (61.13) for thermal winds, (107.25) is *nonlinear* in (dv/dr); as a result it has quite different mathematical properties. In particular, notice that the sonic point $(v = a)$ is not the critical point of (107.25) because when the left-hand side vanishes, the right-hand side can be made to vanish as well with a suitable choice of (dv/dr), which need not (1) vanish, or (2) become infinite, or (3) be discontinuous. This difference from thermal wind theory results from our use of a force law that has an *explicit* dependence on (dv/dr). Had we used some generic $g_{R,l}$ (perhaps obtained from a numerical line transfer computation) which depends explicitly on r but only *implicitly* on (dv/dr), we would again conclude that the sonic point r_s is the critical point. The solution would then proceed as in thermal wind theory, but at the cost, as we shortly see, of losing important physical insight (and possibly of poor numerical convergence as well).

Equation (107.25) is equivalent to

$$F(u, w, w') \equiv [1 - \tfrac{1}{2}(a^2/w)]w' - h(u) - C(w')^\alpha = 0, \qquad (107.26)$$

where $w \equiv \frac{1}{2}v^2$, $u \equiv -1/r$, $w' \equiv (dw/du)$, and

$$h(u) \equiv -G\mathcal{M}(1-\Gamma_e) - 2(a^2/u) - (da^2/du). \qquad (107.27)$$

The differential equation (107.26) has a *singular point* at which solutions terminate, have cusps, or show other discontinuities; it is defined by the condition

$$\partial F(u, w, w')/\partial w' = 1 - \tfrac{1}{2}(a^2/w) - \alpha C(w')^{\alpha-1} = 0. \qquad (107.28)$$

One may eliminate w' between (107.26) and (107.28), and for a given value of C thus determine the *locus of singular points* $w(u, C)$. To guarantee that the solution passes smoothly through the singular point we demand that w' be continuous there; this requirement can be met only if the solution is *tangent* to the singular locus at its point of contact, which is guaranteed by imposing the *regularity condition*

$$(dF/du)_c = [(\partial F/\partial u) + w'(\partial F/\partial w)]_c = 0. \qquad (107.29)$$

Equations (107.26), (107.28), and (107.29) uniquely determine the *critical point* u_c (or r_c) for a given C, or, conversely, C for a specified r_c.

A detailed analysis (**C11**), (**A2**) of the behavior of (107.26) and (107.27) shows that the (u, w) plane is divided into five regions, in each of which there are zero (regions IV and V), one (regions I and III), or two (region II) mathematically valid solutions. An example is shown in Figure 107.1 for a case with $v_{esc} = 4.9a$ and $\Gamma_{R,l} = g_{R,l}/g = 0.76(w')^{1/2}$; the photospheric radius is denoted as R. We see that there is a unique transonic solution in which the subcritical and supercritical branches join smoothly at tangency with the locus of singular points.

Figure 107.1 shows that a line-driven wind is already supersonic at the critical point. It thus appears that the critical point is located beyond the position in the flow where information can still be propagated upstream, and it is not obvious how conditions at r_c are able to determine conditions in the entire flow. We must therefore examine the physical significance of the critical point carefully. Abbott (**A2**) developed a physical interpretation of the critical point by examining the behavior of small-amplitude disturbances of the flow in its vicinity. Thus consider a time-dependent planar flow with velocity $v(z')$ and radiation force $f_l(z', v, dv/dz')$ directed along the z' axis. For simplicity ignore stratification effects. In the neighborhood

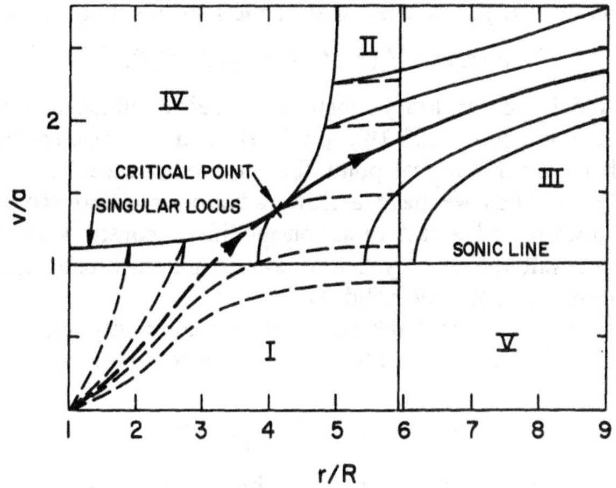

Fig. 107.1 Topology of radiatively driven wind solutions, showing singular locus, critical point, and regions with zero, one, or two mathematically valid solutions. From (**A2**), by permission.

of some point z'_0 make a Galilean transformation to a frame moving with a uniform velocity $v_0 \equiv v(z'_0)$, so that $z = z' - v_0 t$. Write the perturbed velocity as $v = v_0 + v_1$. Then the linearized continuity equation is

$$(\partial \rho_1/\partial t) + \rho_0(\partial v_1/\partial z) = 0, \qquad (107.30)$$

and the vertical component of the momentum equation is

$$\rho_0(\partial v_1/\partial t) = -(\partial p_1/\partial z) + \rho_0 f'_l(\partial v_1/\partial z), \qquad (107.31)$$

where we assumed $f_l = f_l(\ell) = f_l[\rho^{-1}(dv/dz)]$, and f'_l denotes the derivative of f_l with respect to (dv_1/dz). We restrict attention to vertically propagating disturbances; Abbott analyzes obliquely propagating disturbances as well.

Using the isothermal equation of state to eliminate the pressure perturbation $p_1 = a^2\rho_1$, we can combine (107.30) and (107.31) into

$$(\partial^2 v_1/\partial t^2) - a^2(\partial^2 v_1/\partial z^2) - f'_l(\partial^2 v_1/\partial t\, \partial z) = 0. \qquad (107.32)$$

Equation (107.32) is a wave equation that can be recast as

$$(\partial^2 v_1/\partial p\, \partial q) = 0 \qquad (107.33)$$

by transforming from (z, t) to new coordinates $p \equiv z + C_- t$ and $q \equiv z - C_+ t$, where

$$C_- = \tfrac{1}{2}f'_l + [(\tfrac{1}{2}f'_l)^2 + a^2]^{1/2} \qquad (107.34a)$$

and

$$C_+ = -\tfrac{1}{2}f'_l + [(\tfrac{1}{2}f'_l)^2 + a^2]^{1/2}. \qquad (107.34b)$$

The solution of (107.33) is composed of the two traveling waves

$$v_1(z, t) = V_1(z + C_-t) + V_2(z - C_+t), \qquad (107.35)$$

where V_1 and V_2 are arbitrary functions of their arguments. In the CAK model it follows from (107.18), (107.21), and the planar version of (107.28) that near the critical point $f'_l/a \sim v/a \gg 1$, hence $C_- \approx f'_l \gg a$, and $C_+ \approx a(a/f'_l) \ll a$. Thus we have a slow radiation-modified acoustic wave traveling outward and a fast radiation-modified acoustic wave traveling inward. This result applies only to long wavelength disturbances, for which the CAK force law could be valid (**O6**).

Abbott showed (**A1**) that the full nonlinear continuity and momentum equations for a line-driven wind have characteristics in the (r, t) plane given by

$$(dr/dt) = (v - \tfrac{1}{2}f'_l) \pm [(\tfrac{1}{2}f'_l)^2 + a^2]^{1/2}. \qquad (107.36)$$

These characteristics define the speed at which a disturbance will propagate in the flow. Transforming (107.35) back to the rest frame we have

$$v_1(z', t) = V_1[z' - (v_0 - C_-)t] + V_2[z' - (v_0 + C_+)t], \qquad (107.37)$$

which represents disturbances propagating outward with speeds $v_0 + C_+$ and $v_0 - C_-$; from (107.34) we see that these speeds are in exact agreement with (107.36).

Now at the critical point, the CAK singularity condition (107.28) implies $[1 - (a^2/v_c^2)]v_c = f'_l$. When this relation is substituted into (107.36) we find that the velocity of the radiation-modified acoustic waves as seen by an observer in the rest frame is

$$v_\pm = \tfrac{1}{2}(v_c \pm v_c)[1 + (a^2/v_c^2)], \qquad (107.38)$$

or

$$v_+ = v_c[1 + (a^2/v_c^2)] \qquad (107.39a)$$

and

$$v_- = 0. \qquad (107.39b)$$

Thus *for the adopted force law* the flow speed at the critical point of a line-driven wind just equals the inward propagation speed of small disturbances, hence beyond this point information can no longer propagate upstream in the flow. Therefore, as is true for thermal winds, the critical point is, in fact, the point farthest downstream that can still communicate with all other points on a streamline. The main difference between the two theories is the characteristic signal speed: in the absence of radiation the signal and sound speeds are the same, hence the sonic and critical points coincide, whereas in a radiating fluid they differ, hence the sonic and critical points are distinct. These conclusions depend sensitively, however, on the force law adopted, and may not be valid in general (**O6**).

MOMENTUM TRANSPORT IN THE WIND

A line-driven wind deposits momentum (originally photon momentum) in the interstellar medium at a rate $\dot{\mathcal{M}}v_\infty$. If we assume that every photon emitted by the star scatters exactly once in the wind, then an upper bound on the mass-loss rate is

$$\dot{\mathcal{M}} \le L/v_\infty c = 7\times 10^{-12}(L/L_\odot)(3000/v_\infty) \tag{107.40}$$

where $\mathcal{M}$ is measured in $\mathcal{M}_\odot$/year and v_∞ in km s^{-1}. For a typical O-star $L\approx 10^6 L_\odot$ and $v_\infty \approx 3000$ km s^{-1}, hence $\dot{\mathcal{M}}\approx 7\times 10^{-6}\mathcal{M}_\odot$/year, which is, in fact, a typical observed value. The parameter

$$\varepsilon \equiv c v_\infty \dot{\mathcal{M}}/L \tag{107.41}$$

provides a measure of the efficiency with which matter is radiatively ejected in a wind; for single scattering of all photons ε cannot exceed unity.

A more complete picture of the momentum distribution in a wind emerges from integrating the momentum equation (107.4) over all mass in the envelope (**A2**). For a general force law $f_l(r, v, dv/dr)$ we obtain

$$\int_0^{v_\infty} 4\pi r^2 \rho v\, dv + \int_R^\infty 4\pi r^2 \rho\left[\frac{G\mathcal{M}(1-\Gamma_e)}{r^2} + \frac{1}{\rho}\frac{dp}{dr}\right] dr = \int_R^\infty 4\pi r^2 \rho f_l\, dr. \tag{107.42}$$

The first integral in (107.42) is simply $\dot{\mathcal{M}}v_\infty$. To evaluate the second integral we argue that inside the sonic radius the gas is very nearly in hydrostatic equilibrium, in which case the integrand vanishes, whereas outside the sonic radius the gas pressure gradient is negligible compared to gravity because the line force dominates. Using (107.3) we can then approximate the second integral as

$$[L(1-\Gamma_e)/c\Gamma_e]\int_{r_s}^\infty n_e\sigma_e\, dr = L(1-\Gamma_e)\tau_e/c\Gamma_e, \tag{107.43}$$

where τ_e is the electron-scattering optical depth exterior to the sonic point. Finally, using (107.18) and (107.19), we can write the third integral in (107.42) as $\beta L/c$ where

$$\beta \equiv v_{\rm th}^{-1}\int_0^{v_\infty} M(t)t\, dv \tag{107.44}$$

is essentially the line optical depth of the envelope, and equals the equivalent number of strong lines a photon encounters as it traverses the wind. For a single-scattering model, $\beta \le 1$.

Thus momentum conservation in the wind implies that

$$\dot{\mathcal{M}}v_\infty + [\tau_e(1-\Gamma_e)/\Gamma_e](L/c) = \beta L/c, \tag{107.45}$$

which shows that the momentum transferred from photons to the gas goes partly into the momentum lost in the wind and partly into supporting the extended envelope against gravity. One sees that the parameter ε defined in (107.41) underestimates the total photon momentum consumed in driving a wind of a given $\dot{\mathcal{M}}v_\infty$ because it omits the momentum transfer rate required to support the envelope.

RESULTS FROM CAK THEORY

For the CAK radiation-force law, explicit analytical expressions can be obtained for the mass-loss rate, the velocity law, and the critical radius (**C11**). Assuming that $v_{esc} \gg a$ and taking the radius at which the velocity vanishes to be approximately the sonic radius r_s (which in turn is nearly the same as the photospheric radius R) one finds

$$\dot{\mathcal{M}} = \left(\frac{4\pi G\mathcal{M}}{s_e v_{th}}\right)\alpha\left(\frac{1-\alpha}{1-\Gamma_e}\right)^{(1-\alpha)/\alpha}(k\Gamma_e)^{1/\alpha}, \tag{107.46}$$

$$v^2 = \frac{2G\mathcal{M}(1-\Gamma_e)\alpha}{(1-\alpha)}\left(\frac{1}{r_s}-\frac{1}{r}\right), \tag{107.47}$$

and

$$r_c/r_s = 1+\{-\tfrac{1}{2}n+[\tfrac{1}{4}n^2+4-2n(n+1)]^{1/2}\}^{-1}. \tag{107.48}$$

Equation (107.48) is based on the assumption that $a^2 \propto T \propto r^{-n}$; likely values for n lie between 0 (isothermal) and $\frac{1}{2}$ (radiative equilibrium), hence $1.5 \lesssim (r_c/r_s) \lesssim 1.74$. From (107.47) we have

$$v_\infty/v_{esc} = [\alpha/(1-\alpha)]^{1/2}; \tag{107.49}$$

thus for $0.5 \lesssim \alpha \lesssim 0.7$, CAK theory predicts $1 \lesssim v_\infty/v_{esc} \lesssim 1.5$.

For the CAK model one can also evaluate τ_e in (107.45) analytically (**A1**), obtaining

$$\tau_e = [(1-\alpha)\Gamma_e/\alpha(1-\Gamma_e)](\dot{\mathcal{M}}v_\infty c/L). \tag{107.50}$$

Hence a momentum transfer rate $(1-\alpha)\dot{\mathcal{M}}v_\infty/\alpha$ is needed just to support the envelope in the CAK model. Combining (107.50) and (107.45) we find

$$(\dot{\mathcal{M}}v_\infty)_{CAK} = \alpha\beta(L/c), \tag{107.51}$$

or $\varepsilon_{CAK} = \alpha\beta$; inasmuch as $0.5 \lesssim \alpha \lesssim 0.7$ and $\beta \lesssim 1$ for single scattering, ε_{CAK} can never exceed unity, and is more likely of order 0.5.

To construct a complete stellar wind model with CAK theory one chooses L, $\mathcal{M}$, R, and an assumed temperature distribution $T(r)$; for a given choice of k and α in (107.21) the mass-loss rate is determined almost entirely by $\mathcal{M}$ and L via Γ_e. One next makes an initial guess for r_c from (107.48) with $r_s = R$; equation (107.25) is then integrated numerically, and the run of optical depth with radius is computed. The value of r_c is adjusted until an optical depth of about $\frac{2}{3}$ is reached at the photospheric radius R. Having constructed a dynamical model, one may use the resulting density structure in a spherical model atmosphere code and calculate the temperature structure by enforcing radiative equilibrium. This new temperature distribution can then be used to reconstruct the dynamical model, and the procedure iterated. Because the dynamics is insensitive to the temperature structure, the iteration process converges rapidly.

Castor, Abbott, and Klein published a solution for parameters appropriate to an O5 star: $\mathcal{M} = 60\mathcal{M}_\odot$, $L = 9.7 \times 10^5 L_\odot$, $R = 9.6 \times 10^{11}$ cm $= 13.8 R_\odot$, $T_{\text{eff}} = 49{,}300$ K, $\log g = 3.94$, and $\Gamma_e = 0.4$. The resulting mass-loss rate is $\dot{\mathcal{M}} = 6.6 \times 10^{-6} \mathcal{M}_\odot$/year, a reasonable value for a star like ζ Puppis. The terminal velocity is $v_\infty = 1515\ \text{km s}^{-1}$, so $\dot{\mathcal{M}} \approx \frac{1}{2} L/v_\infty c$, which shows that about one half of the momentum originally carried by radiation is transferred to the flow. Stellar evolution theory gives main-sequence lifetimes of about 3×10^6 years at this mass, which implies a total mass-loss in the wind of about one-third the original mass of the star. Thus the stellar winds from O-stars may have very significant effects on their evolution.

Some results for this model are shown in Figure 107.2. The letters *P*, *S*, and *C* designate the photosphere, sonic point, and critical point respectively. The velocity variation (Figure 107.2a) is quite abrupt, with highly

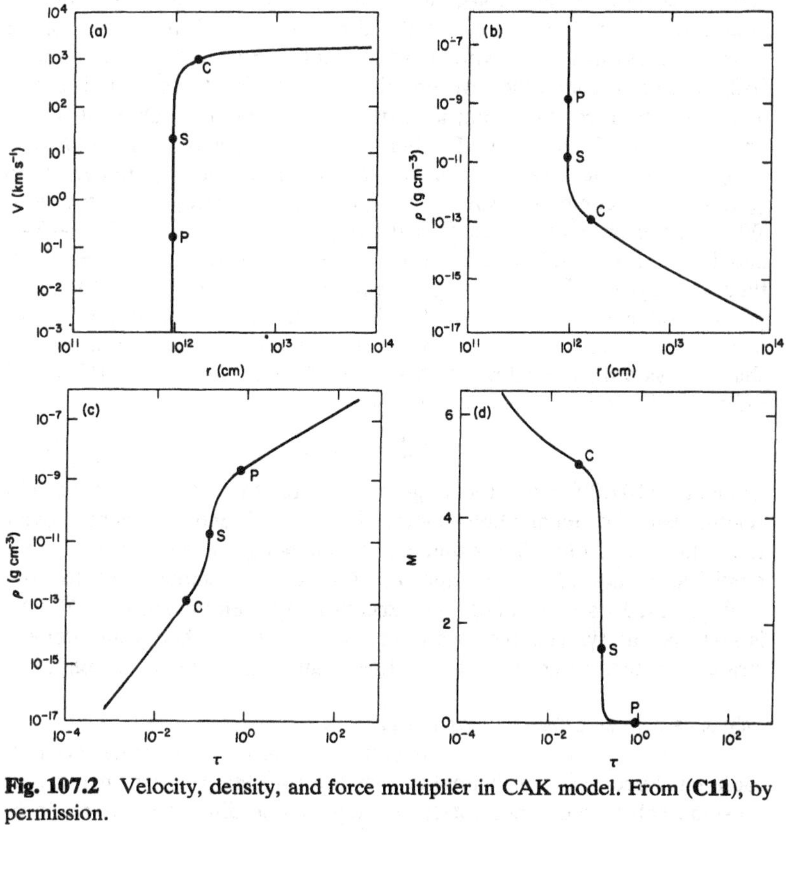

Fig. 107.2 Velocity, density, and force multiplier in CAK model. From (**C11**), by permission.

supersonic flow being achieved within a fraction of a stellar radius above the photosphere. The density distribution (Figure 107.2b) has a decided "core-halo" nature: inside the sonic point the density gradient is nearly hydrostatic, while outside the critical point the velocity is essentially constant at v_∞, hence $\rho \propto r^{-2}$. As seen in Figure 107.2c the halo is transparent in the continuum, and radiates mainly in strong spectral lines. The run of the radiation-force multiplier is shown in Figure 107.2d. In the outer envelope $M \approx 5$, which implies that the radiation force on the lines is about twice the force of gravity (recall that $\Gamma_e = 0.4$); adding the radiation force on the electrons and subtracting the force of gravity we find that the gas outside the critical point experiences a net outward acceleration of about 1.5 times gravity.

COMPARISON WITH OBSERVATIONS

On the whole, CAK gives a coherent and satisfying account of the basic dynamics of line-driven winds. Nevertheless it also shows significant discrepancies with observations, an analysis of which leads to a deeper understanding of the physics of the flow. (1) A critical comparison of observed mass-loss rates with those computed from CAK theory shows (**A3**) that for a comprehensive line list the radiation force is sufficient to drive the observed mass flux; if anything the computed values of $\dot{\mathcal{M}}$ are about a factor of 2 too large. Furthermore, the predicted scaling of $\dot{\mathcal{M}}$ with L agrees with observation over about four orders of magnitude. (2) In contrast, the computed values of $v_\infty/v_{\rm esc}$ are systematically too low. Whereas the observations show that $v_\infty/v_{\rm esc}$ is about 1 to 1.5 for early-A and late-B stars, and rises to about 3.0 for early-B and O-stars, CAK theory always predicts $1 \lesssim v_\infty/v_{\rm esc} \lesssim 1.5$ [cf. (107.49)]. Thus (107.20) fails to provide sufficient radiative acceleration in the high-velocity part of the flow. (3) the CAK velocity distribution (107.47) likewise rises much too sharply inside the critical point. A variety of observations (**B2**), (**C7**), (**L1**) indicate a "softer" velocity law, rising like

$$v = v_\infty[1-(R/r)]. \tag{107.52}$$

Evidently (107.20) gives too large a radiation force in the low-velocity regime near the stellar photosphere. (4) The CAK model cannot provide the total momentum flux observed in the winds of some stars. Using empirical values of v_∞, τ_e, and Γ_e, Abbott (**A2**) shows that for two well-observed stars, β in (107.45) exceeds unity even though ε in (107.41) is less than unity. This result is in conflict with CAK theory and demonstrates the need to account for multiple scattering of the stellar photons.

TRANSFER AND MULTIPLE-SCATTERING EFFECTS

To improve upon the CAK models one must use a more accurate radiation-force law, which implies that the transfer problem in the lines must be solved more accurately. A step in this direction was made by

Weber (**W3**) who calculated self-consistent, line-driven, steady-flow models by solving the comoving-frame line transfer equations numerically [cf. (**M10,** Chap. 14), (**M11**)] for a prechosen distribution of line strengths. This approach is expected to yield better results because: (1) it removes the Sobolev approximation inherent in (107.17) and (107.20) (which surely breaks down near the photosphere because the velocity gradient becomes small and continuum sources and sinks become increasingly important), and (2) it accounts accurately for the angular distribution of the radiation field instead of assuming radial streaming (again, an effect that is important near the stellar surface).

Weber's results are encouraging. First, the velocity rise is softer, mimicking (107.52) fairly closely near the surface of the star, and shifting towards a relation like (107.47) at large distances. Second, the terminal velocities are larger, by about a factor of 4, than the CAK results for the same line strengths. Weber finds that these improvements result mainly from accounting for the radiation field's angular distribution, in particular for the finite solid angle subtended by the stellar photosphere. This result is in harmony with the analysis by Castor (**C10**) who showed that neglect of the angular distribution causes CAK theory to overestimate the line force by about a factor of 2 near the stellar photosphere. In fact, the radiation force calculated from the CAK formula (with an optimized choice of k and α) using the wind structure obtained from the transfer solution agrees closely with the force given by the transfer calculation. Therefore the velocity distribution in the flow is quite sensitive to even small departures from the CAK force law; Abbott shows (**A2**) that this sensitivity to small changes is a peculiarity of the CAK model and is not a general property of line-driven winds.

In Weber's approach each line is modeled in detail. It is hopeless to use such a method to obtain the force law for a realistic line spectrum having hundreds to perhaps thousands of important lines, each of which may have a distinctive response to variations of temperature and density. It is therefore necessary to develop a simpler theory that still accounts for the important physics. The problem has been addressed by Castor and Friend (**C10**), (**F3**) who calculate the line radiation force allowing for multiple scattering in an ensemble of lines described by a statistical model for the distribution of lines over frequency and line strength, and also accounting for the angular distribution of the radiation field. They perform a consistent solution of the dynamical equations and random-line transfer equations to evaluate a correction factor to the force computed for radially streaming radiation.

As predicted by Castor (**C10**) the resulting force is substantially smaller near the star and much larger at large distances. Most of the discrepancies between theory and observation are removed by Friend and Castor's work (FC). For example, for a model that closely resembles the one published by CAK, $v_\infty(\text{FC}) = 3900 \text{ km s}^{-1}$ instead of $v_\infty(\text{CAK}) = 1515 \text{ km s}^{-1}$, while near the star the velocity behaves like (107.52) instead of (107.47), as desired. The mass-loss rate is nearly unchanged: $\dot{\mathcal{M}}(\mathrm{FC}) = 8.6 \times 10^{-6} \mathcal{M}_\odot$ /year, $\dot{\mathcal{M}}(\mathrm{CAK}) = 6.6 \times 10^{-6} \mathcal{M}_\odot$ /year. The velocity at the critical point drops from $v_c(\mathrm{CAK}) = 950 \text{ km s}^{-1}$ to $v_c(\mathrm{FC}) = 275 \text{ km s}^{-1}$, and the critical point moves inward from $r_c(\mathrm{CAK}) = 1.5R$ to $r_c(\mathrm{FC}) = 1.06R$. The radiation force at r_c in the FC model is only two thirds as large as the force in the CAK model, but the total momentum flux in the wind is much larger, $\varepsilon(\mathrm{FC}) = 1.71$ compared to $\varepsilon(\mathrm{CAK}) = 0.51$. Likewise, the effective number of scatterings is $= 1.93$; both of these results vividly illustrate the effect of multiple scattering.

Recent models of line-driven winds include the effects of rotation and magnetic fields; see (**C10**), (**F4**), (**N2**).

ALTERNATIVE WIND THEORIES

The cool (i.e., radiative equilibrium), line-driven wind model appears to provide a good basic picture of the dynamics of the flow, but yields little, if any, information about the temperature structure and excitation-ionization equilibrium of the material. The latter are quantities of considerable interest because observations show spectrum lines from "anomalously" high ions such as N V and O VI, as well as soft X rays, all of which indicate gas temperatures far in excess of T_{eff} of the star.

A variety of models have been suggested to explain these observations including (1) the *modified cool wind model* in which the gas temperature is about 6×10^4 K and the wind is optically thick in the He II resonance continuum; (2) the *warm wind model* in which the gas temperature is of order 2×10^5 K; and (3) the *hybrid corona plus cool wind model* in which a thin ($\sim 0.1R$) hot corona with $T \sim 5 \times 10^6$ K is surrounded by a cool ($T \approx 0.8 T_{\mathrm{eff}}$) envelope. All of these models require a source of nonradiative energy input, such as heating by shocks that grow from instabilities; their relative merits are discussed in (**C4**), (**C5**), (**C6**) and the references cited therein.

All of the models mentioned so far have difficulty in explaining the soft X-ray data. Models developed by Lucy and White (**L9**), (**L11**) to explain the X-ray data invoke the growth of instabilities into the nonlinear regime. In the more recent version of the theory it is argued that small flow perturbations are radiatively amplified into shocks, which survive until "shadowing" by following shocks deprives them of the radiation force that drives them, thus allowing them to dissipate and decay.

Observations show variations in the spectra produced by winds on time-scales from hours to years. They strongly suggest that the winds may in fact be unstable. The stability of line-driven winds has been examined theoretically by several authors (**A2**), (**C2**), (**K2**), (**M1**), (**M5**), (**N1**), (**O5**). The results obtained depend sensitively on the radiation force law adopted.

For example, in an optically thin disturbance, a velocity-induced Doppler shift from the rest position of a saturated line produces a net radiation force $\delta g_{R,l} = Aw_1$, where A is positive. This force is like that for a damped harmonic oscillator, but with a negative "damping coefficient", hence one expects the perturbation to be unstable; this can indeed be the case. Nelson and Hearn (**N1**) and Martens (**M5**) showed that under certain conditions an initial disturbance varying as $e^{i\omega t}$ is *absolutely unstable* (i.e., ω is complex with a negative real part), and grows exponentially. Under similar assumptions MacGregor et al. (**M1**) showed that a driven disturbance (real ω, complex k) is subject to a *drift instability*, and grows in amplitude as it propagates outward in the wind. These results can be understood intuitively by noting that in this case w_1 and $\delta g_{R,l}$ are in phase, hence the work done by the radiation force, which is proportional to $\langle w_1 \, \delta g_{R,l} \rangle$ is necessarily positive (**O5**). In contrast, Abbott (**A2**) considered optically thick disturbances and assumed that the line radiation force depends on the velocity gradient, not the velocity perturbation. As discussed earlier, he found stable radiation-modified acoustic waves. His results can also be understood intuitively by noting that in this case w_1 and $\delta g_{R,l}$ are 90° out of phase, hence $\langle w_1 \, \delta g_{R,l} \rangle \equiv 0$, so that the radiation force does no net work on the perturbation (**O5**). However Abbott's result applies only to long-wavelength disturbances, assuming the validity of the Sobolev force law, and may not be achieved in real stellar winds. A more complete theory that recovers these two limiting cases and works at intermediate optical thicknesses as well has been constructed by Owocki and Rybicki (**O6**). They conclude that the wind must inevitably be unstable to short-wavelength disturbances.

References

(A1) Abbott, D. C. (1978) *Astrophys. J.*, **225,** 893.

(A2) Abbott, D. C. (1980) *Astrophys. J.*, **242,** 1183.

(A3) Abbott, D. C. (1982) *Astrophys. J.*, **259,** 282.

(A4) Abbott, D. C., Bieging, J. H., and Churchwell, E. (1981) *Astrophys. J.*, **250,** 645.

(A5) Abbott, D. C., Bieging, J. H., Churchwell, E., and Cassinelli, J. P. (1980) *Astrophys. J.*, **238,** 196.

(A6) Anderson, D. (1973) *Astron. Astrophys.*, **29,** 23.

(A7) Ashraf, S. and Ahmad, Z. (1974) *Indian J. Pure Appl. Math.*, **6,** 1090.

(A8) Axford, W. I. (1961) *Phil. Trans. Roy. Soc.* (*London*), **A253,** 301.

(B1) Barfield, W. D., von Holdt, R., and Zachariasen, F. (1954) *Los Alamos Scientific Laboratory Report No. LA-1709.* Los Alamos: University of California.

(B2) Barlow, M. J. and Cohen, M. (1977) *Astrophys. J.*, **213,** 737.

(B3) Belokon, V. A. (1959) *Soviet Phys. J.E.T.P.*, **9,** 235.

(B4) Berthomieu, G., Provost, J., and Rocca, A. (1976) *Astron. Astrophys.*, **47,** 413.

(B5) Bond, J. W., Watson, K. M., and Welch, J. A. (1965) *Atomic Theory of Gas Dynamics.* Reading: Addison-Wesley.

(B6) Bray, R. J. and Loughhead, R. E. (1974) *The Solar Chromosphere.* London: Chapman and Hall.
(C1) Campbell, P. M. and Nelson, R. G. (1964) *Lawrence Radiation Laboratory Report No. UCRL-7838.* Livermore: University of California.
(C2) Carlberg, R. G. (1980) *Astrophys. J.*, **241,** 1131.
(C3) Carslaw, H. S. and Jaeger, J. C. (1959) *Conduction of Heat in Solids.* (2nd ed.) Oxford: Oxford University Press.
(C4) Cassinelli, J. P. (1979) *Ann. Rev. Astron. Astrophys.*, **17,** 275.
(C5) Cassinelli, J. P. (1979) in *Mass Loss and Evolution of O-Type Stars,* ed. P. S. Conti and C. W. H. de Loore, p. 201. Dordrecht: Reidel.
(C6) Cassinelli, J. P., Castor, J. I., and Lamers, H. J. G. L. M. (1976) *Pub. Astron. Soc. Pacific,* **90,** 496.
(C7) Cassinelli, J. P., Olson, G. L., and Stalio, R. (1978) *Astrophys. J.*, **220,** 573.
(C8) Castor, J. I. (1970) *Mon. Not. Roy. Astron. Soc.*, **149,** 111.
(C9) Castor, J. I. (1974) *Mon. Not. Roy. Astron. Soc.*, **169,** 279.
(C10) Castor, J. I. (1979) in *Mass Loss and Evolution of O-Type Stars,* ed P. S. Conti and C. W. H. de Loore, p. 175. Dordrecht: Reidel.
(C11) Castor, J. I., Abbott, D. C., and Klein, R. I. (1975) *Astrophys. J.*, **195,** 157.
(C12) Castor, J. I., Abbott, D. C., and Klein, R. I. (1976) in *Physique des Mouvements dans les Atmospheres Stellaires,* ed. R. Cayrel and M. Steinberg, p. 363. Paris: Centre National de la Recherche Scientifique.
(C13) Chevalier, R. A. (1974) *Astrophys. J.*, **188,** 501.
(C14) Chevalier, R. A. (1976) *Astrophys. J.*, **207,** 872.
(C15) Chevalier, R. A. and Klein, R. I. (1978) *Astrophys. J.*, **219,** 994.
(C16) Chevalier, R. A. and Klein, R. I. (1979) *Astrophys. J.*, **234,** 597.
(C17) Clark, P. A. and Clark, A. (1973) *Solar Phys.*, **30,** 319.
(C18) Clarke, J. H. and Ferrari, C. (1965) *Phys. Fluids,* **8,** 2121.
(C19) Colgate, S. A. and White, R. H. (1966) *Astrophys. J.*, **143,** 626.
(C20) Courant, R. and Friedrichs, K. O. (1976) *Supersonic Flow and Shock Waves.* New York: Springer.
(C21) Cox, D. P. (1972) *Astrophys. J.*, **178,** 159.
(C22) Cox, D. P. (1972) *Astrophys. J.*, **178,** 169.
(C23) Cox, J. P. and Giuli, R. T. (1968) *Principles of Stellar Structure.* New York: Gordon and Breach.
(C24) Cram, L. E. (1977) *Astron. Astrophys.*, **59,** 151.
(D1) Delache, P. and Froeschle, C. (1972) *Astron. Astrophys.*, **16,** 348.
(E1) Elliot, L. A. (1960) *Proc. Roy. Soc. (London),* **A258,** 287.
(E2) Elmegreen, B. G. and Lada, C. J. (1977) *Astrophys. J.*, **214,** 725.
(E3) Epstein, R. I. (1981) *Astrophys. J. Letters,* **244,** L89.
(E4) Erickson, G. G. and Olfe, D. B. (1973) *Phys. Fluids,* **16,** 2121.
(F1) Falk, S. W. (1978) *Astrophys. J. Letters,* **226,** L113.
(F2) Falk, S. W. and Arnett, W. D. (1977) *Astrophys. J. Suppl.*, **33,** 515.
(F3) Friend, D. B. and Castor, J. I. (1983) *Astrophys. J.*, **272,** 259.
(F4) Friend, D. B. and MacGregor, K. B. (1984) *Astrophys. J.*, in press.
(F5) Froeschle, C. (1973) *Astron. Astrophys.*, **26,** 229.
(F6) Froeschle, C. (1977) *Astron. Astrophys.*, **55,** 45.
(G1) Garmany, C. D., Olson, G. L., Conti, P. S., and van Steenberg, M. E. (1981) *Astrophys. J.*, **250,** 660.
(G2) Ghoniem, A. F., Kamel, M. M., Berger, S. A., and Oppenheim, A. K. (1982) *J. Fluid Mech.*, **117,** 473.

(G3) Gingerich, O., Noyes, R. W., and Kalkofen, W. (1971) *Solar Phys.*, **18,** 347.
(G4) Giovanelli, R. G. (1978) *Solar Phys.*, **59,** 293.
(G5) Giovanelli, R. G. (1979) *Solar Phys.*, **62,** 253.
(G6) Goldsworthy, F. A. (1961) *Phil. Trans. Roy. Soc.* (*London*), **A253,** 277.
(G7) Goulard, R. (1966) in *Aerodynamic Phenomena in Stellar Atmospheres*, ed. R. N. Thomas, p. 247. London: Academic Press.
(H1) Hearn, A. G. (1972) *Astron. Astrophys.*, **19,** 417.
(H2) Hearn, A. G. (1973) *Astron. Astrophys.*, **23,** 97.
(H3) Heaslet, M. A. and Baldwin, B. S. (1963) *Phys. Fluids*, **6,** 781.
(H4) Helliwell, J. B. (1969) *J. Fluid Mech.*, **37,** 497.
(H5) Hjellming, R. M. (1966) *Astrophys. J.*, **143,** 420.
(I1) Imshennik, V. S. (1962) *Soviet Physics J.E.T.P.*, **15,** 167.
(I2) Imshennik, V. S. (1975) *Sov. J. Plasma Phys.*, **1,** 108.
(K1) Kahn, F. D. (1954) *Bull. Astron. Inst. Netherlands*, **12,** 187.
(K2) Kahn, F. D. (1981) *Mon. Not. Roy. Astron. Soc.*, **196,** 641.
(K3) Kalkofen, W. and Ulmschneider, P. (1977) *Astron. Astrophys.*, **57,** 193.
(K4) Kass, W. and O'Keeffe, M. (1966) *J. Appl. Phys.*, **37,** 2377.
(K5) Klein, R. I. and Chevalier, R. A. (1978) *Astrophys. J.*, **223,** L109.
(K6) Klein, R. I., Sandford, M. T., and Whitaker, R. W. (1983) *Astrophys. J. Letters*, **271,** L69.
(K7) Klein, R. I., Stein, R. F., and Kalkofen, W. (1976) *Astrophys. J.*, **205,** 499.
(K8) Klein, R. I., Stein, R. F., and Kalkofen, W. (1978) *Astrophys. J.*, **220,** 1024.
(K9) Kneer, F. and Nakagawa, Y. (1976) *Astron. Astrophys.*, **46,** 65.
(L1) Lamers, H. J. G. L. M. and Morton, D. C. (1976) *Astrophys. J. Suppl.*, **32,** 715.
(L2) Landau, L. D. and Lifschitz, E. M. (1959) *Fluid Mechanics.* Reading: Addison-Wesley.
(L3) Lasher, G. (1975) *Astrophys. J.*, **201,** 194.
(L4) Lasher, G. J. and Chan, K. L. (1979) *Astrophys. J.*, **230,** 742.
(L5) Lasker, B. M. (1966) *Astrophys. J.*, **143,** 700.
(L6) Laumbach, D. D. and Probstein, R. F. (1970) *Phys. Fluids*, **13,** 1178.
(L7) Le Guet, F. (1972) *Astron. Astrophys.*, **16,** 356.
(L8) Levy, M. (1974) *Astron. Astrophys.*, **31,** 451.
(L9) Lucy, L. B. (1982) *Astrophys. J.*, **255,** 286.
(L10) Lucy, L. B. and Solomon, P. M. (1970) *Astrophys. J.*, **159,** 879.
(L11) Lucy, L. B. and White, R. L. (1980) *Astrophys. J.*, **241,** 300.
(M1) MacGregor, K. B., Hartmann, L., and Raymond, J. C. (1979) *Astrophys. J.*, **231,** 514.
(M2) Manheimer, W. M., Colombant, D. G., and Gardner, J. H. (1982) *Phys. Fluids*, **25,** 1644.
(M3) Marlborough, J. M. and Roy, J. R. (1970) *Astrophys. J.*, **160,** 221.
(M4) Marshak, R. E. (1958) *Phys. Fluids*, **1,** 24.
(M5) Martens, P. C. H. (1979) *Astron. and Astrophys.*, **75,** L7.
(M6) Mathews, W. G. (1965) *Astrophys. J.*, **142,** 1120.
(M7) Mathews, W. G. and O'Dell, C. R. (1969) *Ann. Rev. Astron. Astrophys.*, **7,** 67.
(M8) Mihalas, B. R. W. (1979) Ph.D. Thesis, University of Colorado.
(M9) Mihalas, B. W. and Toomre, J. (1982) *Astrophys. J.*, **263,** 386.

(M10) Mihalas, D. (1978) *Stellar Atmospheres.* (2nd ed.) San Francisco: Freeman.

(M11) Mihalas, D., Kunasz, P. B., and Hummer, D. G. (1975) *Astrophys. J.*, **202,** 465.

(M12) Mitchner, M. and Vinokur, M. (1963) *Phys. Fluids*, **6,** 1682.

(M13) Morse, P. M. and Feshbach, H. (1953) *Methods of Theoretical Physics.* New York: McGraw-Hill.

(N1) Nelson, G. D. and Hearn, A. G. (1978) *Astron. and Astrophys.*, **65,** 223.

(N2) Nerney, S. (1980) *Astrophys. J.*, **242,** 723.

(N3) NiCastro, J. R. A. J. (1970) *Phys. Fluids*, **13,** 2000.

(O1) Ojha, S. N. (1972) *Acta Phys. Acad. Sci. Hungar.*, **31,** 375.

(O2) Oort, J. H. and Spitzer, L. (1955) *Astrophys. J.*, **121,** 6.

(O3) Osterbrock, D. E. (1961) *Astrophys. J.*, **134,** 347.

(O4) Osterbrock, D. E. (1974) *Astrophysics of Gaseous Nebulae.* San Francisco: Freeman.

(O5) Owocki, S. P. and Rybicki, G. B. (1983) Paper presented at the 161st Meeting of the American Astronomical Society. Cambridge, Massachusetts.

(O6) Owocki, S. P. and Rybicki, G. B. (1984) *Astrophys. J.*, in press.

(P1) Pai, S.-I. (1966) *Radiation Gas Dynamics.* New York: Springer.

(P2) Petschek, A. G., Williamson, R. E., and Wooten, J. K. (1960) *Los Alamos Scientific Laboratory Report No. LAMS-2421.* Los Alamos: University of California.

(P3) Pomraning, G. C. (1967) *J. Appl. Phys.*, **38,** 3845.

(P4) Poveda, A. and Woltjer, L. (1968) *Astron. J.*, **73,** 65.

(P5) Prokof'ev, V. A. (1952) *Uch. Zap. Mos. Gos. Univ. Mekh.*, **172,** 79.

(R1) Raizer, Yu. P. (1957) *Soviet Phys. J.E.T.P.*, **5,** 1242.

(R2) Raizer, Yu. P. (1958) *Soviet Phys. J.E.T.P.*, **6,** 77.

(R3) Ray, G. D. and Bhowmick, J. B. (1975) *Indian J. Pure Appl. Math.*, **7,** 96.

(S1) Sachdev, P. L. and Ashraf, S. (1971) *Phys. Fluids*, **14,** 2107.

(S2) Sachs, R. G. (1946) *Phys. Rev.* **69,** 514.

(S3) Sandford, M., Whitaker, R., and Klein, R. I. (1982) *Astrophys. J.*, **260,** 183.

(S4) Schatzman, E. (1949) *Ann. d'Astrophys.*, **12,** 203.

(S5) Schmieder, B. (1977) *Solar Phys.*, **54,** 269.

(S6) Schmieder, B. (1978) *Solar Phys.*, **57,** 245.

(S7) Schmitz, F. and Ulmschneider, P. (1980) *Astron. Astrophys.*, **84,** 93.

(S8) Schmitz, F. and Ulmschneider, P. (1980) *Astron. Astrophys.*, **84,** 191.

(S9) Schmitz, F. and Ulmschneider, P. (1981) *Astron. Astrophys.*, **93,** 178.

(S10) Schwarzschild, M. (1958) *Structure and Evolution of the Stars.* Princeton: Princeton University Press.

(S11) Shafranov, V. D. (1957) *Soviet Phys. J.E.T.P.*, **5,** 1183.

(S12) Skalafuris, A. J. (1965) *Astrophys. J.*, **142,** 351.

(S13) Skalafuris, A. J. (1969) *Astrophys. Space Sci.*, **2,** 258.

(S14) Sobolev, V. V. (1958) in *Theoretical Astrophysics*, ed. V. A. Ambartsumian, Chap. 29. London: Pergamon.

(S15) Sobolev, V. V. (1960) *Moving Envelopes of Stars.* Cambridge: Harvard University Press.

(S16) Souffrin, P. (1966) *Ann. d'Astrophys.*, **29,** 55.

(S17) Souffrin, P. (1972) *Astron. Astrophys.*, **17,** 458.

(S18) Spiegel, E. A. (1957) *Astrophys. J.*, **126,** 202.

(S19) Spiegel, E. A. (1976) in *Physique des Mouvements dans les Atmospheres*

Stellaires, ed. R. Cayrel and M. Steinberg, p. 19. Paris: Centre National de la Recherche Scientifique.
(S20) Spitzer, L. (1968) *Diffuse Matter in Space.* New York: Interscience.
(S21) Spitzer, L. (1978) *Physical Processes in the Interstellar Medium.* New York: Wiley.
(S22) Spitzer, L. and Savedoff, M. P. (1950) *Astrophys. J.*, **111,** 593.
(S23) Stein, R. F. and Spiegel, E. A. (1967) *J. Acous. Soc. America*, **42,** 866.
(S24) Stix, M. (1970) *Astron. Astrophys.*, **4,** 189.
(S25) Stokes, G. G. (1851) *Phil. Mag.*, **1,** 305.
(S26) Strömgren, B. (1939) *Astrophys. J.*, **89,** 526.
(T1) Traugott, S. C. (1965) *Phys. Fluids*, **8,** 834.
(T2) Tscharnuter, W. M. and Winkler, K.-H. (1979) *Comp. Phys. Comm.*, **18,** 171.
(U1) Ulmschneider, P. (1970) *Solar Phys.*, **12,** 403.
(U2) Ulmschneider, P. (1971) *Astron. Astrophys.*, **12,** 297.
(U3) Ulmschneider, P. (1971) *Astron. Astrophys.*, **14,** 275.
(U4) Ulmschneider, P. and Kalkofen, W. (1977) *Astron. Astrophys.*, **57,** 199.
(U5) Ulmschneider, P., Kalkofen, W., Nowak, T., and Bohn, H. U. (1977) *Astron. Astrophys.*, **54,** 61.
(U6) Ulmschneider, P., Schmitz, F., Kalkofen, W., and Bohn, H. U. (1978) *Astron. Astrophys.*, **70,** 487.
(U7) Ulmschneider, P., Schmitz, F., Renzini, A., Cacciari, C., Kalkofen, W., and Kurucz, R. (1977) *Astron. Astrophys.*, **70,** 487.
(U8) Ulrich, R. K. (1970) *Astrophys. J.*, **162,** 993.
(U9) Unno, W. and Spiegel, E. A. (1966) *Pub. Astron. Soc. Japan*, **18,** 85.
(V1) Vandervoort, P. O. (1963) *Astrophys. J.*, **137,** 381.
(V2) Vandervoort, P. O. (1963) *Astrophys. J.*, **138,** 426.
(V3) Vandervoort, P. O. (1964) *Astrophys. J.*, **139,** 889.
(V4) Vernazza, J. E., Avrett, E. H., and Loeser, R. (1973) *Astrophys. J.*, **184,** 605.
(V5) Vernazza, J. E., Avrett, E. H., and Loeser, R. (1976) *Astrophys. J. Suppl.*, **30,** 1.
(V6) Vincenti, W. G. and Baldwin, B. S. (1962) *J. Fluid Mech.*, **12,** 449.
(V7) Vincenti, W. G. and Kruger, C. H. (1965) *Introduction to Physical Gas Dynamics.* New York: Wiley.
(W1) Wang, K. C. (1964) *J. Fluid Mech.*, **20,** 447.
(W2) Weaver, T. A. (1976) *Astrophys. J. Suppl.*, **32,** 233.
(W3) Weber, S. V. (1981) *Astrophys. J.*, **243,** 954.
(W4) Weymann, R. (1960) *Astrophys. J.*, **132,** 452.
(W5) Whitney, C. A. (1963) *Astrophys. J.*, **138,** 537.
(W6) Whitney, C. A. and Skalafuris, A. J. (1963) *Astrophys. J.*, **138,** 200.
(W7) Winkler, K.-H. A. and Newman, M. J. (1980) *Astrophys. J.*, **236,** 201.
(W8) Winkler, K.-H. A. and Newman, M. J. (1980) *Astrophys. J.*, **238,** 311.
(Z1) Zel'dovich, Ya. B. (1957) *Soviet Phys. J.E.T.P.*, **5,** 919.
(Z2) Zel'dovich, Ya. B. and Raizer, Yu. P. (1957) *Usp. Fiz. Nauk*, **63,** 613.
(Z3) Zel'dovich, Ya. B. and Raizer, Yu. P. (1966) *Physics of Shock Waves and High-Temperature Hydrodynamic Phenomena.* New York: Academic.
(Z4) Zwicky, F. (1965) in *Stellar Structure*, ed. L. H. Aller and D. B. McLaughlin, Chap. 7. Chicago: University of Chicago.

APPENDIX

Elements of Tensor Calculus

The equations of radiation hydrodynamics are most naturally expressed in terms of vectors and tensors. We summarize here the concepts used elsewhere in this book. While reasonably complete derivations are given, no attempt at mathematical rigor is made; the reader should consult the references listed at the end of §A3 for further details.

A1 Notation

The three types of geometrical objects with which we will deal are *scalars*, *vectors*, and *tensors* (of the second rank). Scalars will be written as italic or Greek symbols, usually without an affix. Suffixes may be used in some instances to denote a quantity evaluated at a particular position or time, or in a particular reference frame. Vectors and tensors will be distinguished by the use of a special type font or by indices that denote components. Vectors will be written in boldface type (e.g., $\mathbf{v}$); tensors will be written in Gothic type (e.g., $\mathfrak{R}$). Individual components of vectors and tensors will be denoted by italic or Greek symbols with one more suffixes (e.g., v^i, $R^{\alpha\beta}$). In the text, Roman indices range from 1 to 3, and denote components in a three-dimensional Euclidian space, while Greek indices range from 0 to 3, and denote components in the four-dimensional spacetime of special relativity, 0 indicating time. To avoid confusion with powers of scalars, specific components of vectors and tensors with definite numerical (or symbolic) values assigned to their indices may be written e.g., $V^{(k)}$ or $R^{(\gamma)(\delta)}$. Finally, *matrices*, two-dimensional rectangular arrays such as appear in a transformation of coordinates (e.g., rotation or Lorentz transformation) will also be written in boldface type. These may be of arbitrarily large dimensionality, depending on the use to which they are put. The distinction we make between a matrix and a tensor (which sometimes is represented by a matrix of its components) is that the latter is a *physical* or *geometrical* entity whose components transform, under a change of coordinate systems, according to particular transformation laws, while the former is merely an array of numbers defined in such a way as to systematize algebraic manipulations involving systems of equations or coordinate transformations.

As we will see below, in curvilinear coordinates vectors and tensors can be described by *abstract components* of two different kinds, called *contravariant* and *covariant*, which have different transformation properties under a change of coordinates. Contravariant components will be denoted with superscripts (e.g., v^i, $T^{\alpha\beta}$) and covariant components with subscripts (e.g., V_α, R_{ij}). In general these abstract components differ from the *physical components*, which give the values of the components in physical units along the directions of the coordinate curves. Physical components will be labeled with subscripts that indicate the relevant coordinate [e.g., v_r, v_θ, v_ϕ for the spherical polar coordinates (r, θ, ϕ)].

In Cartesian coordinates, all three kinds of components (contravariant, covariant, and physical) are identical, and usually no distinction is made among them by changes in the positions of component labels. We will often depart from this practice, however, and write even Cartesian tensors with subscripts and superscripts when it serves our purposes to do so (in particular we always write the coordinates themselves as contravariant quantities x^i). An advantage is gained by this device because one can then see by inspection the invariance and transformation properties of an equation under a change of coordinates. As we will see, the power of tensor notation is that it allows us to write equations in a *covariant form*, which means that the equation has the *same form in all coordinate systems.* This formalism is thus responsive to the demands of relativity, which insists that equations expressing genuine physical laws must remain valid in all coordinate systems.

The Einstein summation convention, by which repeated indices imply sums over the appropriate range, will be used throughout. For example

$$a^ib_i \equiv a^1b_1 + a^2b_2 + a^3b_3, \tag{A1.1}$$

and similarly for Greek indices. Summed indices are *dummy* and may be replaced by any other symbol without changing the meaning of the expression (e.g., $a_ib^i \equiv a_kb^k$, etc.). In cases where repeated indices appear but summation is not implied we will write the indices in parentheses [e.g., $g_{(i)(i)}$ denotes that particular tensor component].

Ordinary *partial derivatives* $\partial/\partial x^i$ will often be abbreviated to the notation $_{,i}$ thus: $(\partial v^j/\partial x^i) = v^j_{,i}$. When convenient to do so we will sometimes abbreviate $\partial/\partial t$ to $_{,t}$. *Covariant derivatives* (cf. §A3.10) will be written $_{;\alpha}$ thus: $T^{\alpha\beta}_{;\beta}$.

A2 Cartesian Tensors

Let us now consider vectors and tensors in a three-dimensional space with orthogonal Cartesian coordinates. It is straightforward to generalize most of the results obtained to n dimensions, but we will not pursue this matter.

A2.1. Vectors and Their Algebra

Choose an origin O and three mutually perpendicular coordinate axes with a right-handed orientation. A vector $\mathbf{a}$ is a directed line segment drawn

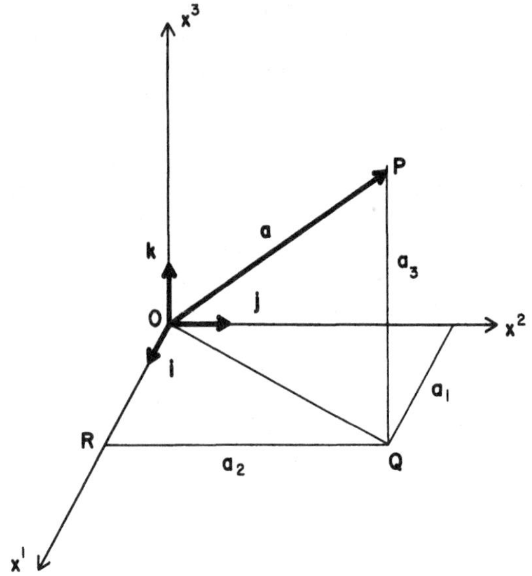

Fig. A1 Right-handed coordinate system.

from O to some point P whose coordinates are (a_1, a_2, a_3). This number triple completely specifies the vector by giving the components along each coordinate axis, that is, the length of the projection of the vector onto that axis (see Figure A1).

By applying the Pythagorean theorem to triangles OPQ and OQR in Figure A1 we see that the *length* (or *magnitude*) of $\mathbf{a}$ is

$$a = |\mathbf{a}| = (a_i a_i)^{1/2}. \tag{A2.1}$$

Unit vectors are vectors of unit length. In particular we may choose *basis vectors*

$$\mathbf{e}_{(1)} = \mathbf{i} = (1, 0, 0); \qquad \mathbf{e}_{(2)} = \mathbf{j} = (0, 1, 0); \qquad \text{and} \qquad \mathbf{e}_{(3)} = \mathbf{k} = (0, 0, 1). \tag{A2.2}$$

Then

$$\mathbf{a} = a_1\mathbf{i} + a_2\mathbf{j} + a_3\mathbf{k}. \tag{A2.3}$$

If we multiply a vector $\mathbf{A}$ by a scalar α we obtain a new vector $\mathbf{B} = \alpha\mathbf{A}$ with components $B_i = \alpha A_i$. $\mathbf{B}$ lies along $\mathbf{A}$, has magnitude $|\mathbf{B}| = \alpha\,|\mathbf{A}|$, and points in the same direction (or opposite to) as $\mathbf{A}$ according to whether α is greater than (or less than) zero. Vectors may be added and subtracted; thus $\mathbf{C} = \mathbf{A} \pm \mathbf{B}$ has components $C_i = A_i \pm B_i$. Furthermore, $\mathbf{A} + \mathbf{B} = \mathbf{B} + \mathbf{A}$; $(\mathbf{A} + \mathbf{B}) + \mathbf{C} = \mathbf{A} + (\mathbf{B} + \mathbf{C})$; $\mathbf{A} - (-\mathbf{B}) = \mathbf{A} + \mathbf{B}$; and $\alpha(\mathbf{A} + \mathbf{B}) = \alpha\mathbf{A} + \alpha\mathbf{B}$.

If some vector quantity, say velocity $\mathbf{v}$, can be assigned a definite value at each point $\mathbf{x}=(x^{(1)}, x^{(2)}, x^{(3)})$ within some region during a definite time interval, then $\mathbf{v}$ is a *vector field.* Similarly we may have *scalar fields* [e.g., pressure $p(x^{(1)}, x^{(2)}, x^{(3)}, t)$] and *tensor fields* (e.g., the radiation stress-energy tensor R).

A2.2. Scalar Product

Consider two vectors $\mathbf{a}$ and $\mathbf{b}$ and their difference $\mathbf{c}=\mathbf{a}-\mathbf{b}$, as shown in Figure A2. Then from the familar law of cosines we know that $c^2=a^2+b^2-2ab\cos\theta$, hence

$$2ab\cos\theta = a_i a_i + b_i b_i - (a_i - b_i)(a_i - b_i) = 2a_i b_i. \tag{A2.4}$$

The quantity

$$\mathbf{a}\cdot\mathbf{b} \equiv a_i b_i \tag{A2.5}$$

is called the *scalar* (or *inner*, or *dot*) *product* of $\mathbf{a}$ and $\mathbf{b}$. If $\mathbf{m}$ and $\mathbf{n}$ are unit vectors along $\mathbf{a}$ and $\mathbf{b}$, we see from (A2.4) that

$$\cos\theta = \mathbf{m}\cdot\mathbf{n} = a_i b_i/(ab). \tag{A2.6}$$

This is a convenient way to determine the angle between any two vectors. Notice that when two vectors $\mathbf{a}$ and $\mathbf{b}$ are *orthogonal*, $\theta=\pi/2$, hence $\mathbf{a}\cdot\mathbf{b}=0$; in particular $\mathbf{i}\cdot\mathbf{j}=\mathbf{i}\cdot\mathbf{k}=\mathbf{j}\cdot\mathbf{k}=0$ as would be expected from (A2.2). If $\mathbf{b}\equiv\mathbf{a}$ then $\theta=0$ and (A2.4) yields (A2.1) for the length of a vector. In general the scalar product gives the length of one vector times the projection of the length of another vector onto the first.

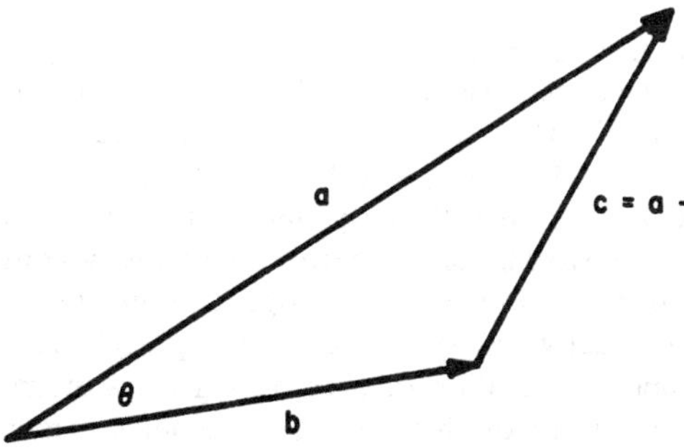

Fig. A2 Vector subtraction.

A2.3. Orthogonal Transformations

Let us now inquire how vectors are affected by changes in the coordinate system. Having fixed the origin O, the only significant change we can make is to perform a rigid rotation of the three axes around O. (We could also reflect, that is, reverse the direction of, axes, but we will not consider that case in this book.) We then obtain new basis vectors $\bar{\mathbf{e}}_j$. Let l_{ij} be the cosine of the angle between $\mathbf{e}_i$ and $\bar{\mathbf{e}}_j$. Then $\bar{\mathbf{e}}_j$ can be resolved along the old basis set and expressed as

$$\bar{\mathbf{e}}_i = l_{ji}\mathbf{e}_j. \tag{A2.7}$$

Similarly, we can resolve $\mathbf{e}_i$ along the new basis set to find

$$\mathbf{e}_i = l_{ij}\bar{\mathbf{e}}_j. \tag{A2.8}$$

Now choose some vector $\mathbf{a}$; we can express $\mathbf{a}$ in terms of its components in either system. Noting that we get the same vector in either case, we see that

$$\mathbf{a} = a_i\mathbf{e}_i \equiv \bar{\mathbf{a}} = \bar{a}_j\bar{\mathbf{e}}_j = \bar{a}_j l_{ij}\mathbf{e}_i, \tag{A2.9}$$

hence

$$a_i = l_{ij}\bar{a}_j. \tag{A2.10}$$

By reversing the argument we find

$$\bar{a}_i = l_{ji}a_j. \tag{A2.11}$$

The matrix $\mathbf{L}$ whose element in the ith row and jth column is l_{ij} is the *transformation matrix* from basis set $\bar{\mathbf{e}}_i$ to set $\mathbf{e}_i$. From (A2.7) and (A2.8) or (A2.10) and (A2.11) we see that the *inverse transformation* $\mathbf{L}^{-1}$ has a matrix $\mathbf{L}^t$ that is the *transpose* of $\mathbf{L}$. This implies that $\mathbf{L}$ must be an *orthogonal* matrix. It is easy to prove that this is so. Let the *Kronecker δ symbol* be defined such that $\delta_{ij} = 0$ if $i \neq j$, and $\delta_{(i)(i)} = 1$. Then $\mathbf{e}_i \cdot \mathbf{e}_j = \delta_{ij} = \bar{\mathbf{e}}_i \cdot \bar{\mathbf{e}}_j$. Hence

$$\delta_{ij} = \mathbf{e}_i \cdot \mathbf{e}_j = (l_{ik}\bar{\mathbf{e}}_k)\cdot(l_{jm}\bar{\mathbf{e}}_m) = l_{ik}l_{jm}(\bar{\mathbf{e}}_k \cdot \bar{\mathbf{e}}_m) = l_{ik}l_{jm}\,\delta_{km} = l_{ik}l_{jk}. \tag{A2.12}$$

Thus the row vectors of $\mathbf{L}$ are *orthonormal* (i.e., of unit length and mutually orthogonal). Starting from $\bar{\mathbf{e}}_i \cdot \bar{\mathbf{e}}_j$ one can show that the column vectors of $\mathbf{L}$ are also orthonormal. Therefore $\mathbf{L}$ is in fact orthogonal, and $(\mathbf{L}\mathbf{L}^t) = (\mathbf{L})_{ik}(\mathbf{L}^t)_{kj} = l_{ik}l_{jk} = \delta_{ij} = (\mathbf{I})_{ij}$ where $\mathbf{I}$ is the *identity matrix*.

Although here we started with the geometrical notion of a vector and then deduced its transformation properties, a perfectly consistent set of results is obtained if one proceeds in the opposite direction and *defines* a vector $\mathbf{a}$ to be an object whose components (a_1, a_2, a_3) become $(\bar{a}_1, \bar{a}_2, \bar{a}_3)$, where $\bar{a}_i = l_{ji}a_j$, under a rotation of axes having a transformation matrix $(\mathbf{L})_{ij} = l_{ij}$. The naturalness of this approach becomes evident when one considers general vectors and tensors in curvilinear coordinates.

A2.4. Transformation Properties and Algebra of Tensors

We define a Cartesian tensor of *rank n* to be a geometrical object with n indices, which transforms according to the rule

$$\bar{A}_{ab\ldots n} = l_{pa}l_{qb}\ldots l_{tn}A_{pq\ldots t}. \tag{A2.13}$$

Vectors as defined above are obviously tensors of rank one. Scalars do not change their value under coordinate transformation ($\bar{\alpha} = \alpha$) and thus can be considered to be tensors of rank zero. Aside from scalars and vectors, the tensors we shall most frequently encounter in this book are of rank two (e.g., A_{ij} or $T_{\alpha\beta}$, having 9 or 16 components in three- and four-dimensional spaces, respectively). For example, the Kronecker δ symbol δ_{ij} is a tensor of the second rank, which has the property of being invariant under coordinate transformation:

$$\bar{\delta}_{ij} = l_{ki} l_{mj}\, \delta_{km} = l_{ki} l_{kj} = \delta_{ij}. \tag{A2.14}$$

For this reason it is sometimes called the *isotropic tensor.*

Tensors obey simple algebraic rules. Thus if $\mathsf{B} = \alpha \mathsf{A}$ then $B_{ij} = \alpha A_{ij}$. Tensors of identical rank may be added and subtracted; thus if $\mathsf{C} = \mathsf{A} \pm \mathsf{B}$, $C_{ij} = A_{ij} \pm B_{ij}$. Similarly, $\mathsf{A} + \mathsf{B} = \mathsf{B} + \mathsf{A}$; $\mathsf{A} + (\mathsf{B} + \mathsf{C}) = (\mathsf{A} + \mathsf{B}) + \mathsf{C}$; and $\alpha(\mathsf{A} + \mathsf{B}) = \alpha \mathsf{A} + \alpha \mathsf{B}$. Tensors may also be multiplied. Thus if $A_{ab\ldots m}$ and $B_{pq\ldots n}$ are tensors of rank m and n, respectively, then the set of products $A_{ab\ldots m} B_{pq\ldots n}$ are the components of a tensor of rank $m+n$. In particular if we form the *outer* (or *tensor*) *product* of two vectors a_i and b_j we obtain a second-rank tensor $T_{ij} = a_i b_j$.

A2.5. Symmetry

A tensor is *symmetric* with respect to two indices, say i and j, if interchange of the indices does not change the value of the tensor component (e.g., if $A_{ab\ldots i\ldots j\ldots n} = A_{ab\ldots j\ldots i\ldots n}$). A tensor is *antisymmetric* (or *skew symmetric*) with respect to two indices if their interchange produces a component of the same magnitude but opposite sign.

Any tensor T_{ij} of rank two can be uniquely decomposed into a symmetric part S_{ij} and an antisymmetric part A_{ij}. Thus defining

$$S_{ij} \equiv \tfrac{1}{2}(T_{ij} + T_{ji}) \tag{A2.15}$$

and

$$A_{ij} \equiv \tfrac{1}{2}(T_{ij} - T_{ji}) \tag{A2.16}$$

we have

$$T_{ij} = S_{ij} + A_{ij}. \tag{A2.17}$$

In three dimensions a second-rank symmetric tensor has only six distinct components and a second-rank antisymmetric tensor has only three distinct nonzero components.

A2.6 Contraction

Given a tensor of order n, we may form a new tensor of order $n-2$ by *contraction* in which we set two indices to the same value and sum over their range. The rank of a tensor is thus equal to the number of *free* indices. For example, if we contract the second-rank tensor $T_{ij} = a_i b_j$ we get the scalar (tensor of rank zero) $T_{ii} = a_i b_i$, the usual inner product of **a** and **b**. We can verify directly that the inner product is in fact a scalar invariant under coordinate transformation:

$$\bar{a}_i \bar{b}_i = l_{ji} a_j l_{ki} b_k = (l_{ji} l_{ki}) a_j b_k = \delta_{jk} a_j b_k = a_j b_j. \tag{A2.18}$$

By contracting a tensor A_{ij} of rank two we can form a unique scalar A_{ii}, called the *trace*, the sum of the diagonal elements. From an argument similar to (A2.18) we can show that A_{ii} is invariant, a fact we exploit in our discussion of fluid kinematics in §21. Notice that for the Kronecker δ tensor, $\delta_{ii} = n$ where n is the dimensionality of the space.

Contraction of the fourth-order tensor formed by the multiplication of two second-order tensors, e.g., $A_{ijkl} = B_{ij} C_{kl}$, yields four distinct "inner products," each of which is a second-order tensor, namely $B_{ij} C_{il}$, $B_{ij} C_{kj}$, $B_{ij} C_{ki}$, and $B_{ij} C_{jl}$. Thus while $\mathbf{a} \cdot \mathbf{b}$ has a unique meaning for vectors, a similar notation for tensors is ambiguous, and we will avoid it, preferring instead to use component notation, which is explicit.

A2.7. The Permutation Symbol

In a space of three dimensions we define the *permutation symbol* e_{ijk} such that

$$e_{ijk} = \begin{cases} +1 \text{ if } ijk \text{ is an even permutation of } 123, \\ -1 \text{ if } ijk \text{ is an odd permutation of } 123, \\ 0 \text{ if any two indices are the same.} \end{cases} \tag{A2.19}$$

The generalization to n dimensions is obvious. This symbol proves to be extraordinarily useful in a variety of contexts. A result of particular importance is the statement

$$e_{ijk} e_{ilm} \equiv \delta_{jl} \delta_{km} - \delta_{jm} \delta_{kl}, \tag{A2.20}$$

which follows immediately from a direct enumeration of cases. From (A2.20) we easily have $e_{ijk} e_{ijl} = 2\delta_{kl}$. We show below that e_{ijk} is a tensor of rank three whose value is invariant under coordinate rotation.

A2.8. Determinants

The *determinant* of the $(n \times n)$ matrix **a** with components a_{ij} is defined to be the sum of the $n!$ distinct products composed of one element from each row (in order) and column, each given a positive or negative sign according to whether an even or odd number of permutations is required to restore the column indices to ascending numerical order. The same definition with the words "row" and "column" interchanged also holds. Thus

$$|\mathbf{a}| = |a_{ij}| = \begin{vmatrix} a_{11} a_{12} \dots a_{1n} \\ a_{21} a_{22} \dots a_{2n} \\ \vdots \quad \vdots \qquad \vdots \\ a_{n1} a_{n2} \dots a_{nn} \end{vmatrix} \equiv a \tag{A2.21}$$

can be written compactly as

$$a = e_{j_1 j_2 \ldots j_n} a_{1j_1} a_{2j_2} \ldots a_{nj_n} = e_{i_1} e_{i_2} \ldots e_{i_n} a_{i_1 1} a_{i_2 2} \ldots a_{i_n n}. \qquad \text{(A2.22)}$$

For simplicity in constructing proofs, let us temporarily set $n = 3$, so that

$$a = e_{ijk} a_{1i} a_{2j} a_{3k} = e_{ijk} a_{i1} a_{j2} a_{k3}. \qquad \text{(A2.23)}$$

From (A2.23) one sees immediately that the determinant of the transpose of a matrix equals the determinant of the matrix itself. Let us now show that the sum $e_{ijk} a_{pi} a_{qj} a_{rk}$ is skew symmetric under interchange of two rows, say p and q:

$$e_{ijk} a_{pi} a_{qj} a_{rk} = e_{jik} a_{pj} a_{qi} a_{rk} = e_{jik} a_{qi} a_{pj} a_{rk} = -e_{ijk} a_{qi} a_{pj} a_{rk}. \qquad \text{(A2.24)}$$

A similar result is obtained for interchanges of the other indices. It follows that

$$e_{ijk} a_{ip} a_{jq} a_{kr} = e_{pqr} a. \qquad \text{(A2.25)}$$

By a similar analysis we find

$$e_{ijk} a_{pi} a_{qj} a_{rk} = e_{pqr} a. \qquad \text{(A2.26)}$$

Equations (A2.25) and (A2.26) show that if any two rows or columns are identical (or even scalar multiples of one another) the determinant is zero.

Determinants can also be expanded in *cofactors*. For example, expanding along the first row we have

$$a = |a_{ij}| = a_{1j_1} e_{j_1 j_2 \ldots j_n} a_{2j_2} \ldots a_{nj_n} = a_{(1)k} A^k_{(1)}, \qquad \text{(A2.27)}$$

where the cofactor

$$A^k_{(1)} \equiv e_{kj_2 \ldots j_n} a_{2j_2} \ldots a_{nj_n}. \qquad \text{(A2.28)}$$

From (A2.28) we immediately see that $a_{(i)k} A^k_{(j)} = \delta_{ij} a$.

Consider now the determinant of the matrix $\mathbf{c}$, which is the product of two matrices $\mathbf{a}$ and $\mathbf{b}$, so that $c_{ij} = a_{ik} b_{kj}$. Then

$$\begin{aligned} c = |c_{ij}| &= e_{pq\ldots t} c_{p1} c_{q2} \ldots c_{tn} = e_{pq\ldots t} (a_{pi} b_{i1})(a_{qj} b_{j2}) \ldots (a_{tl} b_{ln}) \\ &= (e_{pq\ldots t} a_{pi} a_{qj} \ldots a_{tl}) b_{i1} b_{j2} \ldots b_{ln} = |a_{ij}|\, e_{ij\ldots l} b_{i1} b_{j2} \ldots b_{ln} \qquad \text{(A2.29)} \\ &= |a_{ij}|\, |b_{ij}|, \end{aligned}$$

where we have used (A2.25). We thus recover the familiar rule that the determinant of the product of two matrices equals the product of the determinants of those matrices.

Applying this result to the transformation matrix $\mathbf{L}$ introduced in §A2.3 we find

$$L^2 = |\mathbf{L}^t|\,|\mathbf{L}| = |\mathbf{L}^t \mathbf{L}| = |\mathbf{L}^{-1} \mathbf{L}| = |\mathbf{I}| = 1, \qquad \text{(A2.30)}$$

so that $L = \pm 1$. The case $L = +1$ applies to *rotations* of coordinates, and $L = -1$ applies if an odd number of the coordinate axes undergoes *reflection*, thereby changing the system from right handed to left handed; we consider only rotations. We can now see that the permutation symbol is invariant under rotations:

$$\bar{e}_{ijk} = l_{pi}l_{qj}l_{rk}e_{pqr} = e_{ijk}\,|l_{ij}| = Le_{ijk} = e_{ijk}. \tag{A2.31}$$

A2.9. Cross Products; Triple Products

We define the *cross* (or *vector*) *product* of $\mathbf{a}$ and $\mathbf{b}$ to be the vector $\mathbf{c}=\mathbf{a}\times\mathbf{b}$, whose components are

$$c_i = e_{ijk}a_jb_k. \tag{A2.32}$$

Note then that $\mathbf{b}\times\mathbf{a}=-(\mathbf{a}\times\mathbf{b})$ and that $\mathbf{a}\times\mathbf{a}=0$ for any $\mathbf{a}$. Using (A2.2) in (A2.32) one finds $\mathbf{i}\times\mathbf{j}=\mathbf{k}$; $\mathbf{j}\times\mathbf{k}=\mathbf{i}$; $\mathbf{k}\times\mathbf{i}=\mathbf{j}$; $\mathbf{i}\times\mathbf{i}=\mathbf{j}\times\mathbf{j}=\mathbf{k}\times\mathbf{k}=0$. From (A2.32) we see that $\mathbf{c}$ can be written symbolically as the determinant

$$\mathbf{c}=\begin{vmatrix} \mathbf{i} & \mathbf{j} & \mathbf{k} \\ a_1 & a_2 & a_3 \\ b_1 & b_2 & b_3 \end{vmatrix}. \tag{A2.33}$$

Consider now the geometrical interpretation of $\mathbf{a}\times\mathbf{b}$. Without changing $\mathbf{a}$, $\mathbf{b}$, or $\mathbf{c}$ we can rotate the coordinate axes so that $\mathbf{i}'$ lies along $\mathbf{a}$, and $\mathbf{i}'$ and $\mathbf{j}'$ lie in the plane defined by $\mathbf{a}$ and $\mathbf{b}$. We can then write $\mathbf{a}=a\mathbf{i}'$ and $\mathbf{b}=b\cos\theta\mathbf{i}'+b\sin\theta\mathbf{j}'$ where θ is the angle between $\mathbf{a}$ and $\mathbf{b}$. Therefore

$$\mathbf{c}=a\mathbf{i}'\times b(\cos\theta\mathbf{i}'+\sin\theta\mathbf{j}')=ab\sin\theta\mathbf{k}'. \tag{A2.34}$$

Thus $\mathbf{c}$ is a vector perpendicular to the plane of $\mathbf{a}$ and $\mathbf{b}$, whose magnitude is $ab\sin\theta$; this is the area of the parallelogram generated by $\mathbf{a}$ and $\mathbf{b}$ (i.e., $\mathbf{a}$ and $\mathbf{b}$ along two of its sides).

We can define two kinds of triple products of vectors. The *scalar triple product* is

$$\mathbf{a}\cdot(\mathbf{b}\times\mathbf{c})=e_{ijk}a_ib_jc_k. \tag{A2.35}$$

This product has a simple geometrical interpretation: it is the length of $\mathbf{a}$ projected onto a vector perpendicular to the plane of $\mathbf{b}$ and $\mathbf{c}$, times the area of the parallelogram generated by $\mathbf{b}$ and $\mathbf{c}$, and hence is the volume of the parallelepiped whose sides are $\mathbf{a}$, $\mathbf{b}$, and $\mathbf{c}$. Notice that (A2.35) is unaltered by *cyclic permutation* (i.e., $i\to j\to k\to i$), hence

$$\mathbf{a}\cdot(\mathbf{b}\times\mathbf{c})=\mathbf{b}\cdot(\mathbf{c}\times\mathbf{a})=\mathbf{c}\cdot(\mathbf{a}\times\mathbf{b}), \tag{A2.36}$$

which is also self-evident from the geometrical meaning of the scalar triple product.

The *vector triple product* is $\mathbf{d}=\mathbf{a}\times(\mathbf{b}\times\mathbf{c})$. Because $(\mathbf{b}\times\mathbf{c})$ is perpendicular to $\mathbf{b}$ and $\mathbf{c}$, while $\mathbf{d}$ is perpendicular to $(\mathbf{b}\times\mathbf{c})$, it follows that $\mathbf{d}$ lies in the plane of $\mathbf{b}$ and $\mathbf{c}$. We see this explicitly by using (A2.20) to show that

$$d_i=e_{ijk}a_j(e_{klm}b_lc_m)=(\delta_{il}\,\delta_{jm}-\delta_{im}\,\delta_{jl})a_jb_lc_m=b_i(a_jc_j)-c_i(a_jb_j), \tag{A2.37}$$

and hence

$$\mathbf{a}\times(\mathbf{b}\times\mathbf{c})=(\mathbf{a}\cdot\mathbf{c})\mathbf{b}-(\mathbf{a}\cdot\mathbf{b})\mathbf{c}. \qquad \text{(A2.38)}$$

By similar use of (A2.20) it is easy to prove the useful relations

$$(\mathbf{a}\times\mathbf{b})\cdot(\mathbf{c}\times\mathbf{d})=(\mathbf{a}\cdot\mathbf{c})(\mathbf{b}\cdot\mathbf{d})-(\mathbf{a}\cdot\mathbf{d})(\mathbf{b}\cdot\mathbf{c}) \qquad \text{(A2.39)}$$

and

$$\begin{aligned}(\mathbf{a}\times\mathbf{b})\times(\mathbf{c}\times\mathbf{d})&=\mathbf{c}\cdot(\mathbf{d}\times\mathbf{a})\mathbf{b}-\mathbf{c}\cdot(\mathbf{d}\times\mathbf{b})\mathbf{a}\\&=\mathbf{a}\cdot(\mathbf{b}\times\mathbf{d})\mathbf{c}-\mathbf{a}\cdot(\mathbf{b}\times\mathbf{c})\mathbf{d}.\end{aligned} \qquad \text{(A2.40)}$$

A2.10. Gradient, Divergence, Laplacian, and Curl

Thus far we have dealt with the algebra of individual vectors. We now turn to the calculus of (continuous and differentiable) vector fields. First, notice that (A2.10) and (A2.11) can be applied to the position vector of a point, from which it follows that

$$\frac{\partial x^i}{\partial \bar{x}^j}=l_{ij} \qquad \text{(A2.41a)}$$

and

$$\frac{\partial \bar{x}^i}{\partial x^j}=l_{ji}. \qquad \text{(A2.41b)}$$

Starting with the scalar field $f=f(x, y, z)$, form the vector

$$\nabla f=\frac{\partial f}{\partial x}\mathbf{i}+\frac{\partial f}{\partial y}\mathbf{j}+\frac{\partial f}{\partial z}\mathbf{k}, \qquad \text{(A2.42)}$$

which is called the *gradient* of f. We can verify that ∇f is, in fact, a vector in the sense of §A2.3 by noting that in a new coordinate system the component $(\nabla f)_i=(\partial f/\partial x^i)=f_{,i}$ becomes

$$\overline{(\nabla f)}_i=(\partial f/\partial \bar{x}^i)=(\partial f/\partial x^j)(\partial x^j/\partial \bar{x}^i)=l_{ji}(\partial f/\partial x_j)=l_{ji}(\nabla f)_j \qquad \text{(A2.43)}$$

which is consistent with (A2.10). Now choose a *level surface* on which $f(x, y, z)\equiv$ a constant; then for any $d\mathbf{r}=(dx, dy, dz)$ lying in this surface

$$df=\frac{\partial f}{\partial x}dx+\frac{\partial f}{\partial y}dy+\frac{\partial f}{\partial z}dz=(\nabla f)\cdot d\mathbf{r}\equiv 0. \qquad \text{(A2.44)}$$

Thus geometrically ∇f is a vector field perpendicular to level surfaces of f, and $(\nabla f)\cdot d\mathbf{r}$, for arbitrary $d\mathbf{r}$, measures the change in the value of f along the increment $d\mathbf{r}$.

In Cartesian coordinates, n successive differentiations of a tensor of rank n yield a new tensor of rank $m+n$. For example, consider $A_{ij,pq}$, which we see is in fact a tensor of rank four because

$$\bar{A}_{ab,cd} = \frac{\partial^2 \bar{A}_{ab}}{\partial \bar{x}^c \, \partial \bar{x}^d} = \frac{\partial^2}{\partial \bar{x}^c \, \partial \bar{x}^d}(l_{ia} l_{jb} A_{ij}) = \frac{\partial x^k}{\partial \bar{x}^c} \frac{\partial}{\partial x^k}\left[\frac{\partial x^l}{\partial \bar{x}^d} \frac{\partial}{\partial x^l}(l_{ia} l_{jb} A_{ij})\right]$$
$$= l_{ia} l_{jb} l_{kc} l_{ld} A_{ij,kl} \tag{A2.45}$$

Here we have used (A2.41a) and the constancy of the l_{ij}'s under a given rotation of coordinates.

We may regard

$$\boldsymbol{\nabla} = \frac{\partial}{\partial x}\mathbf{i} + \frac{\partial}{\partial y}\mathbf{j} + \frac{\partial}{\partial z}\mathbf{k}, \tag{A2.46}$$

called *del*, as a symbolic vector. The dot product of del with a vector field $\mathbf{a}$ yields the *divergence* of $\mathbf{a}$:

$$\boldsymbol{\nabla} \cdot \mathbf{a} = (\partial a_1/\partial x) + (\partial a_2/\partial y) + (\partial a_3/\partial z) = a_{i,i} \tag{A2.47}$$

The divergence of a vector is obviously a scalar, a fact also indicated by the notation $a_{i,i}$ which shows that it is the contraction of the second-order tensor $a_{i,j}$. By direct calculation it is easy to see that

$$\boldsymbol{\nabla} \cdot (\alpha \mathbf{a}) = (\alpha a_i)_{,i} = \alpha_{,i} a_i + \alpha a_{i,i} = \mathbf{a} \cdot (\boldsymbol{\nabla}\alpha) + \alpha \boldsymbol{\nabla} \cdot \mathbf{a}. \tag{A2.48}$$

If we calculate the divergence of the gradient of a scalar field f we obtain the *Laplacian* of f:

$$\nabla^2 f = \boldsymbol{\nabla} \cdot (\boldsymbol{\nabla} f) = (f_{,i})_{,i} = f_{,ii} = (\partial^2 f/\partial x^2) + (\partial^2 f/\partial y^2) + (\partial^2 f/\partial z^2) \tag{A2.49}$$

where the summation convention holds. The Laplacian of a vector $\mathbf{a}$ is a new vector $\mathbf{b} = \nabla^2 \mathbf{a}$ whose components are $b_i = a_{i,jj}$ (sum on j).

The symbolic cross product of $\boldsymbol{\nabla}$ with a vector field $\mathbf{a}$ yields a new vector field called the *curl* of $\mathbf{a}$. It has components

$$b_i = (\boldsymbol{\nabla} \times \mathbf{a})_i = e_{ijk}(\partial/\partial x^j) a_k = e_{ijk} a_{k,j}. \tag{A2.50}$$

Thus $b_1 = (a_{3,2} - a_{2,3})$, etc. The curl is sometimes written as the symbolic determinant

$$\nabla \times \mathbf{a} = \begin{vmatrix} \mathbf{i} & \mathbf{j} & \mathbf{k} \\ \partial/\partial x & \partial/\partial y & \partial/\partial z \\ a_1 & a_2 & a_3 \end{vmatrix}, \tag{A2.51}$$

but in practice (A2.50) is more useful in establishing vector identities. For example, to calculate the curl of the curl of a vector we write

$$[\boldsymbol{\nabla} \times (\boldsymbol{\nabla} \times \mathbf{a})]_i = e_{ijk}(e_{klm} a_{m,l})_{,j} = e_{kij} e_{klm} a_{m,jl} = (\delta_{il}\,\delta_{jm} - \delta_{im}\,\delta_{jl}) a_{m,jl}$$
$$= a_{j,ji} - a_{i,jj} = [\boldsymbol{\nabla}(\boldsymbol{\nabla} \cdot \mathbf{a})]_i - (\nabla^2 \mathbf{a})_i, \tag{A2.52}$$

hence

$$\boldsymbol{\nabla} \times (\boldsymbol{\nabla} \times \mathbf{a}) = \boldsymbol{\nabla}(\boldsymbol{\nabla} \cdot \mathbf{a}) - \nabla^2 \mathbf{a}. \tag{A2.53}$$

By similar reasoning it is easy to prove the useful relations

$$\nabla \cdot (\nabla \times \mathbf{a}) = 0, \tag{A2.54}$$

$$\nabla \cdot (\mathbf{a} \times \mathbf{b}) = \mathbf{b} \cdot (\nabla \times \mathbf{a}) - \mathbf{a} \cdot (\nabla \times \mathbf{b}), \tag{A2.55}$$

$$\nabla \times (\nabla f) = 0, \tag{A2.56}$$

$$\nabla \times (\alpha \mathbf{a}) = (\nabla \alpha) \times \mathbf{a} + \alpha (\nabla \times \mathbf{a}), \tag{A2.57}$$

$$\nabla \times (\mathbf{a} \times \mathbf{b}) = (\nabla \cdot \mathbf{b})\mathbf{a} - (\nabla \cdot \mathbf{a})\mathbf{b} + (\mathbf{b} \cdot \nabla)\mathbf{a} - (\mathbf{a} \cdot \nabla)\mathbf{b}, \tag{A2.58}$$

and

$$\nabla(\mathbf{a} \cdot \mathbf{b}) = (\mathbf{a} \cdot \nabla)\mathbf{b} + (\mathbf{b} \cdot \nabla)\mathbf{a} + \mathbf{a} \times (\nabla \times \mathbf{b}) + \mathbf{b} \times (\nabla \times \mathbf{a}). \tag{A2.59}$$

A2.11. Duals

Consider an antisymmetric second-rank tensor Ω_{ij} in three-space. With any such tensor we may associate a vector by the definition

$$\omega_i = \tfrac{1}{2} e_{ijk} \Omega_{jk}. \tag{A2.60}$$

The vector ω_i is called the *dual* of Ω_{jk} because of the reciprocal relation that

$$\Omega_{ij} = e_{ijk} \omega_k, \tag{A2.61}$$

which can easily be verified by substitution from (A2.60) and use of (A2.20). Thus

$$\mathbf{\Omega} = \begin{pmatrix} 0 & \omega_3 & -\omega_2 \\ -\omega_3 & 0 & \omega_1 \\ \omega_2 & -\omega_1 & 0 \end{pmatrix}. \tag{A2.62}$$

The dual concept can be generalized to spaces of higher dimension and tensors of higher rank, see (**S2,** 134–135) and (**S2,** 245–247).

A result of considerable importance is that

$$\Omega_{jk} a_j = e_{ijk} \omega_i a_j = (\boldsymbol{\omega} \times \mathbf{a})_k, \tag{A2.63}$$

which shows that this particular sum of a vector against an antisymmetric tensor is identical to the cross product of the vector with the vector dual of the tensor. We exploit this result in our discussion of fluid kinematics in §21.

Vectors of the type described above are called *axial vectors* (or *pseudovectors* because they are "really" tensors of rank two). Important examples of axial vectors are the cross product, $\mathbf{c} = \mathbf{a} \times \mathbf{b}$, for which the associated tensor has components $C_{ij} = a_i b_j - a_j b_i$, and the curl, $\mathbf{b} = \nabla \times \mathbf{a}$, for which the associated tensor has components $B_{ij} = a_{j,i} - a_{i,j}$. An interesting distinction between axial vectors and vectors of the type defined in §A2.1, called *polar vectors,* is that under reversal of the directions of the coordinate axes, the components of polar vectors change sign whereas the components of axial vectors are unaltered. This statement is obviously true for the two examples given above. As discussed in §21, the angular velocity $\boldsymbol{\omega}$ of a rigid body or an infinitesimal element of fluid may be considered to be an axial vector.

A2.12. The Divergence Theorem

The *divergence theorem* (also known as *Gauss's theorem* or *Green's theorem*) is one of the most useful tools of tensor calculus, and is employed frequently in almost all branches of theoretical physics. Let V be a simple convex volume with surface S. Let $\mathbf{n}$ be the outward-pointing normal at any point on S. Then at each position on S we can write an *oriented surface element* as $d\mathbf{S} = \mathbf{n}\, dS$. It is easy to see that the projection of $d\mathbf{S}$ onto a plane perpendicular to any particular direction $\mathbf{l}$ is $\mathbf{l} \cdot d\mathbf{S} = \mathbf{l} \cdot \mathbf{n}\, dS$.

The divergence theorem states that for any differentiable function f,

$$\int_V f_{,i}\, dV = \int_S f n_i\, dS. \tag{A2.64}$$

To prove this theorem, choose $i = 3$ and partition S into upper and lower surfaces S^+ and S^- with respect to the $(x^{(1)}, x^{(2)})$ plane (see Figure A3). Consider an elementary vertical rectangular tube within V, having a volume δV and a projection Σ on the $(x^{(1)}, x^{(2)})$ plane. Let S^+ be given by $x^{(3)} = g^+(x^{(1)}, x^{(2)})$ and S^- by $x^{(3)} = g^-(x^{(1)}, x^{(2)})$. Then carrying out the integration over δV we have

$$\int_{\delta V} f_{,3}\, dx^{(1)}\, dx^{(2)}\, dx^{(3)} = \int_\Sigma \{f[x^{(1)}, x^{(2)}, g^+(x^{(1)}, x^{(2)})] - f[x^{(1)}, x^{(2)}, g^-(x^{(1)}, x^{(2)})]\}\, dx^{(1)}\, dx^{(2)}. \tag{A2.65}$$

But from the definition of the oriented surface element we see that on S^+

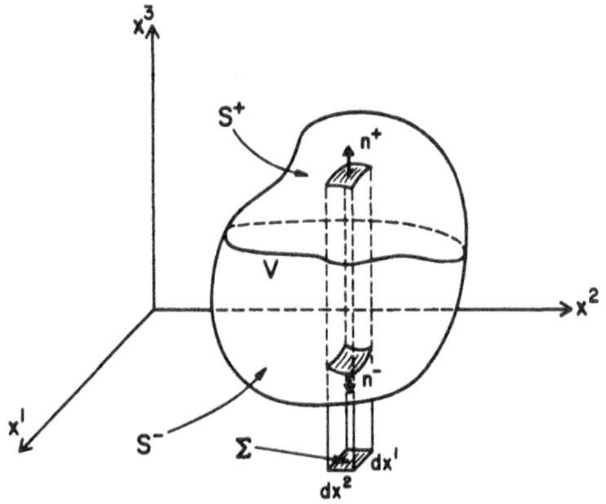

Fig. A3 Geometry of surface and volume integrals.

we have $dx^{(1)}\,dx^{(2)} = n_3^+\,dS^+$, and on S^-, $dx^{(1)}\,dx^{(2)} = -n_3^-\,dS^-$. Hence

$$\int_{\delta V} f_{,3}\,dV = \int_{\delta S^+} f[x^{(1)}, x^{(2)}, g^+(x^{(1)}, x^{(2)})]n_3^+\,dS^+$$
$$+\int_{\delta S^-} f[x^{(1)}, x^{(2)}, g^-(x^{(1)}, x^{(2)})]n_3^-\,dS^- \qquad \text{(A2.66)}$$
$$= \int_{\delta S^+} f^+ n_3^+\,dS^+ + \int_{\delta S^-} f^- n_3^-\,dS^-,$$

where now f^+ and f^- denote the value of f on S^+ and S^-, respectively. Finally, by summing over all elementary tubes, and recognizing that S is the union of S^+ and S^-, we recover (A2.64) for $i = 3$; the choice of i is arbitrary, hence the theorem holds for all i.

Perhaps the most familiar form of the divergence theorem is that for a vector, say $\mathbf{F}$. Writing $f = F^i$ in (A2.64) and summing we have

$$\int_V F^i_{,i}\,dV = \int_S F^i n_i\,dS, \qquad \text{(A2.67)}$$

or

$$\int_V \nabla \cdot \mathbf{F}\,dV = \int_S \mathbf{F} \cdot \mathbf{n}\,dS. \qquad \text{(A2.68)}$$

As another example, if we choose $f = e_{ijk}a_j$, then $f_{,i} = e_{kij}a_{j,i} = (\nabla \times \mathbf{a})_k$, while $fn_i = e_{ijk}n_i a_j = (\mathbf{n} \times \mathbf{a})_k$, so that

$$\int_V (\nabla \times \mathbf{a})\,dV = \int_S (\mathbf{n} \times \mathbf{a})\,dS. \qquad \text{(A2.69)}$$

It is very important to note that (A2.64) is quite general, and holds for *any* differentiable f, whether scalar, vector, or tensor (the latter usually being considered one component at a time, or as the contraction of a set of components against the derivative). In fact the theorem is actually a result from analysis, and has no roots in vector or tensor analysis per se.

A2.13. Stokes's Theorem

If S is a caplike surface bounded by a closed curve C, *Stokes's theorem* states that for a differentiable vector field $\mathbf{a}$,

$$\int_S (\nabla \times \mathbf{a}) \cdot \mathbf{n}\,dS = \oint_C \mathbf{a} \cdot \mathbf{t}\,ds, \qquad \text{(A2.70)}$$

where $\mathbf{t}$ is the unit tangent to C. To prove (A2.70), cover S with a rectilinear coordinate mesh (u, v) so that S is a collection of points $\mathbf{r}(u, v)$. Consider the integrals in (A2.70) for an element $\alpha\beta\gamma\delta$ bounded by the curve Γ, where $\alpha=(u, v)$, $\beta=(u+du, v)$, $\gamma=(u+du, v+dv)$, $\delta=(u, v+dv)$. Then to first order

$$\mathbf{a}(u+du, v)=\mathbf{a}(u, v)+\left[\left(\frac{\partial \mathbf{r}}{\partial u}\, du\right)\cdot \boldsymbol{\nabla}\right]\mathbf{a}, \tag{A2.71}$$

and similarly for $\mathbf{a}(u, v+dv)$. Then, calculating the line integral to first order we have

$$\oint_\Gamma \mathbf{a}\cdot\mathbf{t}\, ds=\left\{\left[\left(\frac{\partial \mathbf{r}}{\partial u}\cdot\boldsymbol{\nabla}\right)\mathbf{a}\right]\cdot\frac{\partial \mathbf{r}}{\partial v}-\left[\left(\frac{\partial \mathbf{r}}{\partial v}\cdot\boldsymbol{\nabla}\right)\mathbf{a}\right]\cdot\frac{\partial \mathbf{r}}{\partial u}\right\} du\, dv. \tag{A2.72}$$

But by using component notation and (A2.20), (A2.32), and (A2.50) we see that

$$\begin{aligned}\frac{\partial r_j}{\partial u} a_{k,j}\frac{\partial r_k}{\partial v}-\frac{\partial r_j}{\partial v} a_{k,j}\frac{\partial r_k}{\partial u}&=(\delta_{jl}\,\delta_{km}-\delta_{jm}\,\delta_{kl})a_{k,j}\frac{\partial r_l}{\partial u}\frac{\partial r_m}{\partial v}\\&=e_{ijk}a_{k,j}e_{ilm}\frac{\partial r_l}{\partial u}\frac{\partial r_m}{\partial v}\\&=(\boldsymbol{\nabla}\times\mathbf{a})\cdot\left(\frac{\partial \mathbf{r}}{\partial u}\times\frac{\partial \mathbf{r}}{\partial v}\right).\end{aligned} \tag{A2.73}$$

From the geometrical meaning of the cross product we know that $[(\partial\mathbf{r}/\partial u)\times(\partial\mathbf{r}/\partial v)]\, du\, dv$ is just the oriented area $\delta\mathbf{S}$ of $\alpha\beta\gamma\delta$. We therefore find that (A2.70) holds for the element $\alpha\beta\gamma\delta$. Now sum over all elements. The line integrals on the interior mesh lines cancel in pairs, leaving only the line integral around the bounding curve C; the surface integrals sum to the integral over the whole surface. Thus (A2.70) is valid as stated.

As was true for the divergence theorem, Stokes's theorem is quite general, and can be written

$$\int_S e_{ijk}a_{k,j}n_i\, dS=\oint_C a_i t_i\, ds \tag{A2.74}$$

where a_k may be the components of any differentiable tensor (e.g., T_{klm} with lm fixed).

A3 General Tensors

We now consider general tensors in curvilinear coordinates. We will not usually specify the dimensionality of the space, and most of the results are valid in n dimensions. In this section, unless specified otherwise, both roman and Greek indices are assumed to run from 1 to n.

A3.1. Transformation Properties

In order to deal with vectors and tensors in curvilinear coordinates, we must now consider transformations of a quite general, but not arbitrary, kind. We restrict attention to what we shall call *admissible transformations*, which have the following properties.

1. They are real, single-valued transformations of the form

$$\bar{x}^i = f^i(x^{(1)}, \ldots, x^{(n)}), \qquad (i = 1, \ldots, n). \tag{A3.1}$$

2. They are *reversible* so that

$$x^i = g^i(\bar{x}^{(1)}, \ldots, \bar{x}^{(n)}), \qquad (i = 1, \ldots, n). \tag{A3.2}$$

3. The g^i are single valued so that the direct and inverse transformations are one to one.

To guarantee these properties it is sufficient to demand that the f^i and g^i be continuous and have continuous derivatives, and that the Jacobian determinant

$$J \equiv \left| \frac{\partial x^i}{\partial \bar{x}^\gamma} \right| \tag{A3.3}$$

be nonzero everywhere in the domain of the transformation.

Under general transformations, vectors and tensors are represented by two different kinds of abstract components, called *contravariant* and *covariant*, each of which will in general differ from the *physical* components of the tensor. We emphasize that all three sets of components represent the *same physical quantity*, and all are related by definite rules (cf §§A3.5 and A3.7); the three representations can be used interchangeably as convenient.

A *contravariant tensor* of rank m has components $A^{ab\ldots m}$ that transform according to the rule

$$\bar{A}^{\alpha\beta\ldots\mu} = \frac{\partial \bar{x}^\alpha}{\partial x^a} \frac{\partial \bar{x}^\beta}{\partial x^b} \cdots \frac{\partial \bar{x}^\mu}{\partial x^m} A^{ab\ldots m}. \tag{A3.4}$$

In particular a contravariant vector transforms as

$$\bar{A}^\alpha = (\partial \bar{x}^\alpha / \partial x^a) A^a. \tag{A3.5}$$

The archetype for a contravariant vector is the set of coordinate differentials dx^i for which we obviously have $d\bar{x}^\alpha = (\partial \bar{x}^\alpha / \partial x^a)\, dx^a$.

A *covariant tensor* of rank m has components $A_{ab\ldots m}$ that transform according to the rule

$$A_{\alpha\beta\ldots\mu} = \frac{\partial x^a}{\partial \bar{x}^\alpha} \frac{\partial x^b}{\partial \bar{x}^\beta} \cdots \frac{\partial x^m}{\partial \bar{x}^\mu} A_{ab\ldots m}. \tag{A3.6}$$

In particular a covariant vector transforms as

$$\bar{A}_\alpha = (\partial x^a / \partial \bar{x}^\alpha) A_a. \tag{A3.7}$$

The archetype for a covariant vector is the gradient $\phi_{,a}$ for which we obviously have $\bar{\phi}_{,\alpha} = (\partial x^a / \partial \bar{x}^\alpha) \phi_{,a}$.

A *mixed tensor* of contravariant rank m and covariant rank n has components $A^{ab\ldots m}_{kl\ldots n}$ that transform according to the rule

$$\bar{A}^{\alpha\beta\ldots\mu}_{\kappa\lambda\ldots\nu} = \frac{\partial\bar{x}^\alpha}{\partial x^a}\frac{\partial\bar{x}^\beta}{\partial x^b}\cdots\frac{\partial\bar{x}^\mu}{\partial x^m}\frac{\partial x^k}{\partial\bar{x}^\kappa}\frac{\partial x^l}{\partial\bar{x}^\lambda}\cdots\frac{\partial x^n}{\partial\bar{x}^\nu}A^{ab\ldots m}_{kl\ldots n}. \tag{A3.8}$$

In general we suppose that tensors of the kinds defined above can exist throughout a finite region of space, and thereby constitute a tensor field.

A3.2. Tensor Algebra

General tensors obey simple rules of algebra. We may multiply a tensor whose components are $A^{ab\ldots c}_{ij\ldots k}$. by a scalar α to obtain a tensor whose components are $\alpha A^{ab\ldots c}_{ij\ldots k}$. We can add and subtract tensors of identical contravariant and covariant ranks (which are shown by the number of *free*, that is, unsummed, indices of the appropriate kinds); these operations are associative. For a tensor $A^{ab\ldots c}_{ij\ldots k}$ of contravariant rank c and covariant rank k, and a tensor $B^{lm\ldots n}_{pq\ldots r}$ of contravariant rank n and covariant rank r, the outer product

$$C^{ab\ldots clm\ldots n}_{ij\ldots kpq\ldots r} = A^{ab\ldots c}_{ij\ldots k}B^{lm\ldots n}_{pq\ldots r} \tag{A3.9}$$

is a tensor of contravariant rank $(c+n)$ and covariant rank $(k+r)$, as can be verified immediately by application of (A3.8). The outer product is distributive.

From a tensor $A^{ab\ldots k}_{mn\ldots t}$ of contravariant rank k and covariant rank t one may construct a new tensor of contravariant rank $(k-1)$ and covariant rank $(t-1)$ by the operation of contraction, in which one covariant and one contravariant index are set to the same value and summed. For example, contract A^{ab}_{lm} to form A^{ab}_{lb}. Then

$$\bar{A}^{\alpha\beta}_{\lambda\mu} = \frac{\partial\bar{x}^\alpha}{\partial x^a}\frac{\partial\bar{x}^\beta}{\partial x^b}\frac{\partial x^l}{\partial\bar{x}^\lambda}\frac{\partial x^m}{\partial\bar{x}^\mu}A^{ab}_{lm}. \tag{A3.10}$$

implies that

$$\bar{A}^{\alpha\beta}_{\lambda\beta} = \frac{\partial\bar{x}^\alpha}{\partial x^a}\frac{\partial x^l}{\partial\bar{x}^\lambda}\left(\frac{\partial\bar{x}^\beta}{\partial x^b}\frac{\partial x^m}{\partial\bar{x}^\beta}\right)A^{ab}_{lm}. \tag{A3.11}$$

But

$$\frac{\partial\bar{x}^\beta}{\partial x^b}\frac{\partial x^m}{\partial\bar{x}^\beta} \equiv \frac{\partial x^m}{\partial x^b} = \delta^m_b \tag{A3.12}$$

where δ^m_b is the mixed tensor that represents the Kronecker δ symbol. Note in passing that δ^i_j is the isotropic tensor whose value is the same in all coordinate systems (which is trivial to prove). Using (A3.12) in (A3.11) we have

$$\bar{A}^{\alpha\beta}_{\lambda\beta} = \frac{\partial\bar{x}^\alpha}{\partial x^a}\frac{\partial x^l}{\partial\bar{x}^\lambda}A^{ab}_{lb}, \tag{A3.13}$$

which is the correct transformation for tensor whose contravariant and covariant ranks are unity (and hence reduced by one from those of the original tensor, as claimed above).

The *null tensor* is that tensor whose components are all zero in some coordinate system. It follows from (A3.4), (A3.6), and (A3.8) that its components must remain zero in all other admissible coordinate systems. This result is of great importance in physics, for it implies that if we can express a physical law as a tensor equation in some frame, say $A^{ab\ldots k}_{lm\ldots p} = B^{ab\ldots k}_{lm\ldots p}$, then this equation remains true in *all* coordinate systems (hence the physical law is covariant) because $(A^{ab\ldots k}_{lm\ldots p} - B^{ab\ldots k}_{lm\ldots p}) \equiv 0$ in the first frame, and hence in every frame.

If the interchange of two contravariant (or covariant) indices of a tensor does not alter the value of any of its components, the tensor is symmetric with respect to those indices. A tensor is antisymmetric with respect to a pair of indices if their interchange changes the sign but not the magnitude of the tensor's components. The symmetry properties of pure contravariant or pure covariant tensors are intrinsic (i.e., they remain the same in all coordinate frames). Symmetry (or antisymmetry) is *not* intrinsic to mixed tensors, however, because the relationship $A^i_j = A^j_i$, for example, in one coordinate system will not in general carry over to another. These statements may be proved directly by application of (A3.8).

A3.3. Relative Tensors

The tensors described above are *absolute* tensors. *Relative* tensors transform according to a more general law: a relative tensor of contravariant rank m, covariant rank n, and *weight* W transforms as

$$\bar{A}^{\alpha\beta\ldots\mu}_{\kappa\lambda\ldots\nu} = J^W \frac{\partial \bar{x}^\alpha}{\partial x^a} \cdots \frac{\partial \bar{x}^\mu}{\partial x^m} \frac{\partial x^k}{\partial \bar{x}^\kappa} \cdots \frac{\partial x^n}{\partial \bar{x}^\nu} A^{ab\ldots m}_{kl\ldots n}, \tag{A3.14}$$

where J is the Jacobian of the transformation, $J = |\partial x^i/\partial \bar{x}^\gamma|$. Absolute tensors are obviously relative tensors of weight zero; similarly a relative scalar of weight zero is an absolute scalar. Scalars and tensors of weight one are often given the special names scalar and tensor *density* for reasons indicated in §A3.4.

A3.4. The Line Element and the Metric Tensor

In a Euclidian three-space E_3 the length ds of the line element corresponding to an infinitesimal displacement vector dy^k in orthogonal Cartesian coordinates is given by

$$ds^2 = dy^k\, dy^k \tag{A3.15}$$

where k is summed. Generalizing, we adopt (A3.15) as the definition of ds^2 in E_n, where k now runs from 1 to n. Suppose now we transform to a curvilinear coordinate system x^i, and that we can express $y^k = y^k(x^{(1)}, x^{(2)}, \ldots, x^{(n)})$ and hence $dy^k = (\partial y^k/\partial x^i)\,dx^i$. Then in terms of the new coordinates the line element is

$$ds^2 = \left(\frac{\partial y^k}{\partial x^i}\right)\left(\frac{\partial y^k}{\partial x^j}\right) dx^i\,dx^j \equiv g_{ij}\,dx^i\,dx^j, \tag{A3.16}$$

where k is summed from 1 to n.

The tensor g_{ij} is the *metric tensor*; any space characterized by a metric tensor is called a *Riemannian space.* The metric tensor is obviously symmetric in its indices, and is an absolute covariant tensor of rank two, which implies that the line element is a scalar. We verify these statements directly by transforming to a new coordinate system $\bar{x}^i$; then

$$\bar{g}_{ab} \equiv \frac{\partial y^k}{\partial \bar{x}^a}\frac{\partial y^k}{\partial \bar{x}^b} = \frac{\partial x^i}{\partial \bar{x}^a}\frac{\partial y^k}{\partial x^i}\frac{\partial x^j}{\partial \bar{x}^b}\frac{\partial y^k}{\partial x^j} = \frac{\partial x^i}{\partial \bar{x}^a}\frac{\partial x^j}{\partial \bar{x}^b}\,g_{ij}, \tag{A3.17}$$

which is the correct transformation law for a second-rank covariant tensor. The fact that ds is a scalar is then obvious. In Cartesian coordinates the elements of the metric tensor are $g_{ij} = \delta_{ij}$ (the Kronecker δ), and hence are everywhere constant. Any coordinate system in which the elements of g_{ij} are constant, but not necessarily δ_{ij}, may also be considered to be Cartesian because in this case one can reduce g_{ij} to δ_{ij} by a suitable linear transformation.

As an example of a metric in curvilinear coordinates, consider spherical polar coordinates $(x^{(1)}, x^{(2)}, x^{(3)}) = (r, \theta, \phi)$, for which $y^{(1)} = r \sin\theta \cos\phi$, $y^{(2)} = r \sin\theta \sin\phi$, and $y^{(3)} = r\cos\theta$. Then from (A3.16) we find

$$ds^2 = dr^2 + r^2\,d\theta^2 + r^2 \sin^2\theta\,d\phi^2, \tag{A3.18}$$

so that $g_{11} = 1$, $g_{22} = r^2$, $g_{33} = r^2\sin^2\theta$, and $g_{ij} = 0$ for $i \neq j$.

Given the covariant tensor g_{ij}, we can construct a second-order contravariant tensor g^{ij}, which is called the *reciprocal* (or *conjugate*) *tensor,* defined such that

$$g_{ij}g^{jk} \equiv \delta_i^k. \tag{A3.19}$$

Equation (A3.19) states that the components of g^{ij} are the elements of the inverse of the matrix whose components are g_{ij}. As long as $\mathbf{g}$ is nonsingular (i.e., $g = |g_{ij}| \neq 0$), its inverse is unique; therefore (A3.19) uniquely determines g^{ij}, and specifically implies that $g^{ij} = G^{ij}/g$ where G^{ij} is the cofactor of element g_{ij} in $|g_{ij}|$.

The determinant g appears in many tensor formulae. We can show that g is a relative scalar by taking determinants in (A3.17) and using (A2.29) to find

$$\bar{g} = |\bar{g}_{ij}| = \left|\frac{\partial x^a}{\partial \bar{x}^i}\frac{\partial x^b}{\partial \bar{x}^j}\,g_{ab}\right| = \left|\frac{\partial x^a}{\partial \bar{x}^i}\right|\left|\frac{\partial x^b}{\partial \bar{x}^j}\right||g_{ab}| = J^2 g \tag{A3.20}$$

where J is the Jacobian of the transformation; this is the transformation law for a relative scalar of weight two. It follows that $g^{1/2}$ is a relative scalar of weight one.

The factor g also appears in definitions of volume elements via the formula

$$dV = dy^{(1)}\,dy^{(2)}\dots dy^{(n)} = \begin{vmatrix} \dfrac{\partial y^{(1)}}{\partial x^{(1)}} & \dfrac{\partial y^{(1)}}{\partial x^{(2)}} \cdots & \dfrac{\partial y^{(1)}}{\partial x^{(n)}} \\ \vdots & & \vdots \\ \dfrac{\partial y^{(n)}}{\partial x^{(1)}} & \dfrac{\partial y^{(n)}}{\partial x^{(2)}} \cdots & \dfrac{\partial y^{(n)}}{\partial x^{(n)}} \end{vmatrix} dx^{(1)}\,dx^{(2)}\dots dx^{(n)}$$

$$= g^{1/2}\,dx^{(1)}\,dx^{(2)}\dots dx^{(n)} \qquad \text{(A3.21)}$$

where the y^i again denote orthogonal Cartesian coordinates. We can justify (A3.21) in two different ways. First, we can regard it as a result from analysis obtained by a direct evaluation of an n-dimensional iterated integral (**J1**, 183 *et seq.*). By direct transformation $y^i \to x^i$ one finds

$$I = \int dy^{(n)} \int dy^{(n-1)} \dots \int dy^{(1)} f[y^{(1)}, y^{(2)}, \dots, y^{(n)}]$$

$$\equiv \int dx^{(n)} \int dx^{(n-1)} \dots \int dx^{(1)} f[y^{(1)}, y^{(2)}, \dots, y^{(n)}] J(y^{(1)}, \dots, y^{(n)}/x^{(1)}, \dots, x^{(n)})$$

$$\text{(A3.22)}$$

where in the second integral y^i is regarded as $y^i(x^{(1)}, \dots, x^{(n)})$, and J denotes the Jacobian of the transformation. Alternatively we can recognize that (A3.21) is the natural generalization of (A2.35) to an n-dimensional space, and view it as giving the volume of an n-dimensional parallelepiped whose sides are spanned by the elementary vectors $(dx^{(1)}, 0, \dots, 0)$, $\dots, (0, 0, \dots, dx^{(n)})$. Under the transformation $x^i \to y^i$ these vectors become $\left(\dfrac{\partial y^{(1)}}{\partial x^{(1)}}, \dfrac{\partial y^{(2)}}{\partial x^{(1)}}, \dots, \dfrac{\partial y^{(n)}}{\partial x^{(1)}}\right) dx^1, \dots, \left(\dfrac{\partial y^{(1)}}{\partial x^{(n)}}, \dfrac{\partial y^{(2)}}{\partial x^{(n)}}, \dots, \dfrac{\partial y^{(n)}}{\partial x^{(n)}}\right) dx^{(n)}$, which again leads to (A3.21). We thus identify the rightmost member of (A3.21) as the *invariant volume element*; the *volume* of any finite region is then

$$V = \iint \dots \int g^{1/2}\,dx^{(1)}\,dx^{(2)}\dots dx^{(n)}. \qquad \text{(A3.23)}$$

Finally, suppose we calculate the mass within some volume containing fluid of density ρ; then

$$M = \iint \dots \int \rho\,dy^{(1)}\dots dy^{(n)} = \iint \dots \int \rho g^{1/2}\,dx^{(1)}\dots dx^{(n)}$$

$$\equiv \iint \dots \int \bar{\rho}\,dx^{(1)}\dots dx^{(n)}. \qquad \text{(A3.24)}$$

Thus in terms of $\bar{\rho} \equiv g^{1/2}\rho$, a relative scalar of weight one, the integral giving the mass assumes an invariant form. It is this result that motivates the name scalar "density" for relative scalars of weight one.

A3.5. Associated Tensors

Having at our disposal the metric tensor and its reciprocal we can carry out the operation of *raising* and *lowering indices* to construct new tensors *associated* with any given tensor. To lower a contravariant index (say j) we multiply the tensor by g_{ij} and sum against j; for example

$$g_{ij}T^{kj}_{\;\;..ab} = T^{k}_{\;.iab}. \tag{A3.25a}$$

Notice that it may be necessary to use a notation that shows explicitly which index is affected, as was done here by filling vacant positions with dots, because in general the tensors $g_{ij}T^{jk} = T_i^{\;k}$ and $g_{ij}T^{kj} = T^k_{\;.i}$ will be different. The operation of raising covariant indices proceeds similarly; for example

$$g^{ij}T^{k}_{\;.iab} = T^{kj}_{\;\;..ab}, \tag{A3.25b}$$

which shows explicitly that the operation of raising is the direct inverse of lowering. These operations can also be performed on relative tensors.

In the case of vectors the notation is unambiguous, and we can write

$$A_i = g_{ij}A^j \tag{A3.26a}$$

and

$$A^i = g^{ij}A_j. \tag{A3.26b}$$

Moreover we have

$$A_i = g_{ij}A^j = g_{ij}g^{jk}A_k = \delta_i^k A_k \equiv A_i \tag{A3.27}$$

which shows the complete reciprocity of contravariant and covariant components.

The fact that we can raise and lower indices at will shows convincingly that, as mentioned before, the contravariant, covariant, and physical components of a tensor are all merely different representations of the same physical entity. A direct geometrical interpretation of this relationship can be most easily provided for vectors. Choose a set of basis vectors along coordinate curves:

$$\mathbf{a}_i \equiv \mathbf{r}_{,i}. \tag{A3.28}$$

Then

$$ds^2 = d\mathbf{r} \cdot d\mathbf{r} = (\mathbf{r}_{,i}\,dx^i) \cdot (\mathbf{r}_{,j}\,dx^j) = (\mathbf{a}_i \cdot \mathbf{a}_j)\,dx^i\,dx^j = g_{ij}\,dx^i\,dx^j \tag{A3.29}$$

shows that $\mathbf{a}_i \cdot \mathbf{a}_j = g_{ij}$. Furthermore we see that these vectors are not in general unit vectors because $\mathbf{a}_{(i)} \cdot \mathbf{a}_{(i)} = g_{(i)(i)}$ is not necessarily unity. Now resolve $\mathbf{A}$ along this basis: $\mathbf{A} = A^i\mathbf{a}_i$. Then we can see that the geometrical interpretation of the contravariant components of $\mathbf{A}$ is that $(g_{(i)(i)})^{1/2}A^i$ is the length, along the unit vector $\mathbf{e}_i \equiv \mathbf{a}_i/(g_{(i)(i)})^{1/2}$, of the ith edge of the parallelepiped whose diagonal is $\mathbf{A}$.

Alternatively, define the reciprocal basis set

$$\mathbf{a}^i \equiv (\mathbf{a}_j \times \mathbf{a}_k)/g^{1/2} \tag{A3.30}$$

where (ijk) is a cyclic permutation of (123). Then clearly $\mathbf{a}_i \cdot \mathbf{a}^j = \delta_i^j$, and by using (A3.30) in (A2.40) it is easy to show that

$$\mathbf{a}_i = g^{1/2}(\mathbf{a}^j \times \mathbf{a}^k), \tag{A3.31}$$

where again (ijk) is a cyclic permutation of (123). Moreover, because

$$ds^2 = d\mathbf{r} \cdot d\mathbf{r} = (\mathbf{a}^i\, dx_i) \cdot (\mathbf{a}^j\, dx_j) = (\mathbf{a}^i \cdot \mathbf{a}^j) g_{il} g_{jm}\, dx^l\, dx^m \equiv g_{lm}\, dx^l\, dx^m, \tag{A3.32}$$

we see that $g_{lm} = g_{il}g_{jm}(\mathbf{a}^i \cdot \mathbf{a}^j)$. Contracting both sides of this equality against $g^{l\beta}g^{m\alpha}$ we find

$$g^{l\beta}g^{m\alpha}g_{lm} = g^{l\beta}\,\delta_l^\alpha = g^{\alpha\beta} = g^{l\beta}g^{m\alpha}g_{il}g_{jm}(\mathbf{a}^i \cdot \mathbf{a}^j) = \delta_i^\beta\,\delta_j^\alpha(\mathbf{a}^i \cdot \mathbf{a}^j) = \mathbf{a}^\alpha \cdot \mathbf{a}^\beta, \tag{A3.33}$$

so that we must have $\mathbf{a}^i \cdot \mathbf{a}^j = g^{ij}$.

Now resolve $\mathbf{A}$ along the $\mathbf{a}^i$ as $\mathbf{A} = A_i\mathbf{a}^i$. Then

$$\mathbf{A} \cdot \mathbf{a}_i = (A_k\mathbf{a}^k) \cdot \mathbf{a}_i = A_k\,\delta_i^k = A_i \equiv (A^k\mathbf{a}_k) \cdot \mathbf{a}_i = (\mathbf{a}_i \cdot \mathbf{a}_k)A^k = g_{ik}A^k \tag{A3.34}$$

which shows that the A_i are in fact the covariant components associated with the contravariant components A^i. In addition we can now see that $A_i/(g_{(i)(i)})^{1/2} = \mathbf{A} \cdot \mathbf{e}_i$ so that the geometrical interpretation of the covariant components A_i is that $A_i/(g_{(i)(i)})^{1/2}$ is the length of the orthogonal projection of $\mathbf{A}$ onto the unit vector that is tangent to the x^i coordinate curve. [See (**A2,** §§7.22 and 7.35) and (**S1,** §45) for further details.]

The geometrical interpretations given above show that in orthogonal Cartesian coordinates, for which the $g_{(i)(i)} = 1$, the contravariant and covariant coordinates of a vector are identical. But in curvilinear coordinates one sees [e.g., from (A3.18) and (A3.26)] that the two types of abstract components can be quite different, and moreover, from the discussion above, that individual components of a given type do not necessarily even have the same physical units (cf. §A3.7).

A3.6. Scalar Product

The natural covariant generalization of (A2.5) is

$$\mathbf{a} \cdot \mathbf{b} = a_jb^j = g_{ij}a^ib^j = g^{ij}a_ib_j. \tag{A3.35}$$

This expression is manifestly invariant under coordinate transformation:

$$\bar{a}_i\bar{b}^i = (\partial x^p/\partial \bar{x}^i)a_p(\partial \bar{x}^i/\partial x^q)b^q = \delta_q^p a_p b^q = a_q b^q. \tag{A3.36}$$

The natural covariant generalization of (A2.1) for the magnitude of a vector is

$$|\mathbf{a}| = (a_i a^i)^{1/2} = (g_{ij} a^i a^j)^{1/2} = (g^{ij} a_i a_j)^{1/2}. \qquad \text{(A3.37)}$$

Furthermore, this suggests that we take as the covariant generalization of (A2.6) the expression

$$\cos\theta = (a_i b^i)/[(a_i a^i)^{1/2}(b_i b^i)^{1/2}] = (g_{ij} a^i b^j)/[(g_{ij} a^i a^j)^{1/2}(g_{ij} b^i b^j)^{1/2}]. \qquad \text{(A3.38)}$$

As before, two vectors are considered to be orthogonal if $\cos\theta = 0$.

Choosing displacement vectors $(dx^{(1)}, 0, 0)$, $(0, dx^{(2)}, 0)$, and $(0, 0, dx^{(3)})$ along the coordinate curves of a three space we find from (A3.38) that the angles θ_{12}, θ_{13}, and θ_{23} between these curves are

$$\cos\theta_{12} = g_{12}/(g_{11}g_{22})^{1/2} \qquad \text{(A3.39a)}$$

$$\cos\theta_{13} = g_{13}/(g_{11}g_{33})^{1/2} \qquad \text{(A3.39b)}$$

and

$$\cos\theta_{23} = g_{23}/(g_{22}g_{33})^{2}. \qquad \text{(A3.39c)}$$

From (A3.39) it immediately follows that the necessary and sufficient condition for a curvilinear coordinate system to be orthogonal is that $g_{ij} \equiv 0$ for $i \neq j$. In this important case, which is the only one we consider in our work, the metric tensor is diagonal and its reciprocal is simply $g^{(i)(i)} = 1/g_{(i)(i)}$. For example, in spherical polar coordinates $g^{11} = 1$, $g^{22} = 1/r^2$, $g^{33} = 1/r^2 \sin^2\theta$.

A3.7. *Physical Components*

Consider an orthogonal coordinate system in three-space. The metric is diagonal and the line element can be written

$$ds^2 = (h_1\, dx^{(1)})^2 + (h_2\, dx^{(2)})^2 + (h_3\, dx^{(3)})^2, \qquad \text{(A3.40)}$$

where $h_i \equiv (g_{(i)(i)})^{1/2}$. It is clear that the increment of path length associated with a coordinate increment dx^i is not dx^i itself, but $ds^{(i)} = h_{(i)}\, dx^{(i)}$. More generally, using (A3.37) to calculate the length of a vector we have $a^2 = (h_1 a^{(1)})^2 + (h_2 a^{(2)})^2 + (h_3 a^{(3)})^2$. To obtain consistency with the Pythagorean theorem, we find that with the abstract contravariant component $a^{(i)}$ we must associate the *physical component*

$$a(i) = h_{(i)} a^{(i)}. \qquad \text{(A3.41)}$$

Using (A3.37) again, now for covariant components, and noting that $g^{(i)(i)} = (1/h_i)^2$, we have $a^2 = (a_{(1)}/h_1)^2 + (a_{(2)}/h_2)^2 + (a_{(3)}/h_3)^2$ which shows that the physical components are related to covariant components by the expression

$$a(i) = a_{(i)}/h_{(i)}. \qquad \text{(A3.42)}$$

To compute the physical components of a tensor T we notice that in Cartesian coordinates if λ^i and μ^i are unit vectors along some coordinate axes, then the expression

$$c = T_{ij}\lambda^i\mu^j \tag{A3.43}$$

gives the physical component of the tensor along those axes. But this expression is an invariant, and can be applied in curvilinear coordinates as well. If $\boldsymbol{\lambda}$ is to be a unit vector, we must have $h_i^2(\lambda^i)^2 = 1$. Thus if we choose three unit vectors along the $x^{(1)}$, $x^{(2)}$, and $x^{(3)}$ coordinate curves, we must have $\boldsymbol{\lambda}_{(1)} = (1/h_1, 0, 0)$, $\boldsymbol{\lambda}_{(2)} = (0, 1/h_2, 0)$, and $\boldsymbol{\lambda}_{(3)} = (0, 0, 1/h_3)$, respectively. Using these vectors in (A3.43) we find that the physical components of T in terms of its covariant components are

$$T(i, j) = T_{(i)(j)}/(h_{(i)}h_{(j)}). \tag{A3.44}$$

Carrying out the same analysis for contravariant components we have

$$T(i, j) = h_{(i)}h_{(j)}T^{(i)(j)}. \tag{A3.45}$$

Generalization of these expressions to nonothogonal systems is discussed in (**A2,** §§7.42 and 7.43).

As a specific example, the relations between the abstract and physical components of a vector in spherical coordinates are:

$$v^{(1)} = v_r,\ v^{(2)} = v_\theta/r,\ v^{(3)} = v_\phi/(r\sin\theta); \tag{A3.46a}$$

and

$$v_1 = v_r, \qquad v_2 = rv_\theta, \qquad v_3 = (r\sin\theta)v_\phi; \tag{A3.46b}$$

For a symmetric tensor in spherical coordinates we have

$$\begin{aligned} &T^{11} = T_{rr},\ T^{12} = T_{r\theta}/r,\ T^{13} = T_{r\phi}/(r\sin\theta),\ T^{22} = T_{\theta\theta}/r^2,\\ &T^{23} = T_{\theta\phi}/(r^2\sin\theta), \qquad \text{and} \qquad T^{33} = T_{\phi\phi}/(r\sin\theta)^2, \end{aligned} \tag{A3.47}$$

with analogous formulae for covariant components.

A3.8. The Levi-Civita Tensor

In curvilinear coordinates where index position is significant, the appropriate generalizations of (A2.25) and (A2.26) are

$$e_{pq\ldots t}\,|a^b_c| = e_{ij\ldots n}a^i_p a^j_q \ldots a^n_t, \tag{A3.48}$$

and

$$e^{pq\ldots t}\,|a^b_c| = e^{ij\ldots n}a^p_i a^q_j \ldots a^t_n, \tag{A3.49}$$

where a^b_c is the element in the bth row and cth column. In particular, if we set $a^b_c = (\partial x^b/\partial \bar{x}^c)$, (A3.48) becomes

$$\bar{e}_{pq\ldots t}J = \left(\frac{\partial x^i}{\partial \bar{x}^p}\right)\left(\frac{\partial x^j}{\partial \bar{x}^q}\right)\cdots\left(\frac{\partial x^n}{\partial \bar{x}^t}\right)e_{ij\ldots n}, \tag{A3.50}$$

which shows that the covariant permutation symbol is a relative tensor of weight −1. By a similar analysis one finds that $e^{pq\ldots t}$ is a relative tensor of weight +1. Recalling from (A3.20) that $J=(\bar{g}/g)^{1/2}$ and that g is a relative scalar of weight two, we then see that

$$\varepsilon_{ij\ldots k} \equiv g^{1/2} e_{ij\ldots k} \tag{A3.51}$$

and

$$\varepsilon^{ij\ldots k} \equiv g^{-1/2} e^{ij\ldots k} \tag{A3.52}$$

are of weight zero, and hence are absolute tensors. These are the covariant and contravariant components of the *Levi-Civita tensor*, which is skew symmetric in all indices.

Using the Levi-Civita tensor we can write a covariant generalization of the cross product (A2.32) as

$$c_i = \varepsilon_{ijk} a^j b^k \tag{A3.53a}$$

or

$$c^i = \varepsilon^{ijk} a_j b_k. \tag{A3.53b}$$

Similarly the covariant generalization of (A2.60) for the vector dual associated with an antisymmetric tensor in three-space is

$$\omega_i = \tfrac{1}{2}\varepsilon_{ijk}\Omega^{jk}, \tag{A3.54}$$

and

$$\Omega^{jk} = \varepsilon^{ijk}\omega_i. \tag{A3.55}$$

A3.9. Christoffel Symbols

As will be seen in §A3.10, certain combinations of partial derivatives of the metric tensor appear when we attempt to construct a covariant generalization of the operation of differentiation. Thus we define the *Christoffel symbol of the first kind* to be

$$[ij, k] \equiv \tfrac{1}{2}(g_{ik,j} + g_{jk,i} - g_{ij,k}), \tag{A3.56}$$

and the *Christoffel symbol of the second kind* as

$$\left\{ {i \atop j\ k} \right\} \equiv g^{il}[jk, l]. \tag{A3.57}$$

The rather cumbersome (and customary) notation employed here emphasizes that the *Christoffel symbols are not tensors* (see below). By inspection of (A3.56) it is obvious that $[ij, k] \equiv [ji, k]$, and hence that $\{ {i \atop j\ k} \} \equiv \{ {i \atop k\ j} \}$. Notice that *in Cartesian coordinates all Christoffel symbols (of both kinds) are identically zero.*

From (A3.56) we easily find the useful result

$$g_{ij,k} = [ik, j] + [jk, i] \tag{A3.58}$$

and hence

$$g_{ij,k} = g_{jl}\begin{Bmatrix} l \\ i\ k \end{Bmatrix} + g_{il}\begin{Bmatrix} l \\ j\ k \end{Bmatrix}. \tag{A3.59}$$

We can also write an extremely important formula for the derivative of the determinant g in terms of Christoffel symbols of the second kind. From the fact that $g = g_{(i)j}G^{(i)j}$, where G^{ij} is the cofactor of g_{ij}, and recalling that $G^{ij} = gg^{ij}$, we see from (A3.59) that

$$g_{,k} = (\partial g/\partial g_{ij})g_{ij,k} = G^{ij}g_{ij,k} = gg^{ij}\left(g_{jl}\begin{Bmatrix} l \\ i\ k \end{Bmatrix} + g_{il}\begin{Bmatrix} l \\ j\ k \end{Bmatrix}\right)$$

$$= g\left(\begin{Bmatrix} i \\ i\ k \end{Bmatrix} + \begin{Bmatrix} j \\ j\ k \end{Bmatrix}\right) = 2g\begin{Bmatrix} i \\ i\ k \end{Bmatrix}. \tag{A3.60}$$

Therefore

$$\begin{Bmatrix} i \\ i\ k \end{Bmatrix} = (\ln g^{1/2})_{,k}. \tag{A3.61}$$

For orthogonal coordinate systems, the Christoffel symbols can be written in a very compact form that is useful for computation. If $g_{ij} = 0$ when $j \neq i$ one easily finds from (A3.56) and (A3.57) that

$$\begin{Bmatrix} i \\ i\ i \end{Bmatrix} = \tfrac{1}{2}(\ln g_{ii})_{,i}, \tag{A3.62a}$$

$$\begin{Bmatrix} i \\ i\ j \end{Bmatrix} = \tfrac{1}{2}(\ln g_{ii})_{,j}, \tag{A3.62b}$$

$$\begin{Bmatrix} i \\ j\ j \end{Bmatrix} = -\tfrac{1}{2}(g_{jj})_{,i}/g_{ii}, \tag{A3.62c}$$

and

$$\begin{Bmatrix} i \\ j\ k \end{Bmatrix} = 0. \tag{A3.62d}$$

In (A3.62), i, j, and k are distinct, and there is no sum on repeated indices. In particular, for spherical coordinates we find, using (A3.18) and (A3.62) that the nonzero Christoffel symbols are

$$\begin{Bmatrix} 1 \\ 2\ 2 \end{Bmatrix} = -r, \qquad \begin{Bmatrix} 1 \\ 3\ 3 \end{Bmatrix} = -r\sin^2\theta,$$

$$\begin{Bmatrix} 2 \\ 3\ 3 \end{Bmatrix} = -\sin\theta\cos\theta, \qquad \text{and} \qquad \begin{Bmatrix} 2 \\ 1\ 2 \end{Bmatrix} = 1/r, \tag{A3.63}$$

$$\begin{Bmatrix} 3 \\ 2\ 3 \end{Bmatrix} = \cot\theta, \qquad \begin{Bmatrix} 3 \\ 1\ 3 \end{Bmatrix} = 1/r.$$

Last, we must develop the transformation law for Christoffel symbols. Consider two curvilinear coordinate systems y^i and x^α with metric tensors

h_{ij} and $g_{\alpha\beta}$, respectively. Then

$$h_{ij} = (\partial x^\alpha/\partial y^i)(\partial x^\beta/\partial y^j)g_{\alpha\beta} \equiv x^\alpha_{,i}x^\beta_{,j}g_{\alpha\beta} \tag{A3.64}$$

and

$$h^{ij} = (\partial y^i/\partial x^\alpha)(\partial y^j/\partial x^\beta)g^{\alpha\beta} \equiv y^i_{,\alpha}y^j_{,\beta}g^{\alpha\beta}. \tag{A3.65}$$

Differentiating h_{ij} we have

$$\begin{aligned} h_{ij,k} &= g_{\alpha\beta}(x^\alpha_{,ik}x^\beta_{,j} + x^\alpha_{,i}x^\beta_{,kj}) + x^\alpha_{,i}x^\beta_{,j}x^\gamma_{,k}g_{\alpha\beta,\gamma} \\ &= g_{\alpha\beta}(x^\alpha_{,ik}x^\beta_{,j} + x^\alpha_{,kj}x^\beta_{,i}) + x^\alpha_{,i}x^\beta_{,j}x^\gamma_{,k}g_{\alpha\beta,\gamma}. \end{aligned} \tag{A3.66}$$

The second step follows because α and β are dummy and $g_{\alpha\beta}$ is symmetric. Now permuting $i \to j \to k \to i$ in (A3.66) and adding, we find that

$$[ij, k] = \tfrac{1}{2}(h_{jk,i} + h_{ki,j} - h_{ij,k}) = x^\alpha_{,i}x^\beta_{,j}x^\gamma_{,k}[\alpha\beta, \gamma] + x^\alpha_{,ij}x^\beta_{,k}g_{\alpha\beta}, \tag{A3.67}$$

which shows that $[\alpha\beta, \gamma]$ is, in general, *not* a tensor. It would be a tensor only if the second term on the right-hand side were to vanish identically, which happens to be true if the coordinate transformation is linear (i.e., $x^\alpha = c^\alpha_i y^i$ where the c's are constants), but not in general. Using (A3.57) and (A3.65) in (A3.67) we find

$$\begin{aligned} \begin{Bmatrix} l \\ i\ j \end{Bmatrix} &= y^l_{,\varepsilon}y^k_{,\zeta}g^{\varepsilon\zeta}(x^\alpha_{,i}x^\beta_{,j}x^\gamma_{,k}[\alpha\beta, \gamma] + x^\alpha_{,ij}x^\beta_{,k}g_{\alpha\beta}) \\ &= y^l_{,\varepsilon}x^\alpha_{,i}x^\beta_{,j}g^{\varepsilon\gamma}[\alpha\beta, \gamma] + y^l_{,\varepsilon}x^\alpha_{,ij}g^{\varepsilon\beta}g_{\alpha\beta}, \end{aligned} \tag{A3.68}$$

which simplifies to

$$\begin{Bmatrix} k \\ i\ j \end{Bmatrix} = y^k_{,\gamma}x^\alpha_{,i}x^\beta_{,j}\begin{Bmatrix} \gamma \\ \alpha\ \beta \end{Bmatrix} + y^k_{,\alpha}x^\alpha_{,ij}. \tag{A3.69}$$

Equation (A3.69) shows that $\{^{\ k}_{i\ j}\}$ is also not in general a tensor, although it would be if the coordinate transformation were linear. Finally, by contracting (A3.69) against $x^\lambda_{,k}$ we obtain the useful result

$$x^\lambda_{,ij} = x^\lambda_{,k}\begin{Bmatrix} k \\ i\ j \end{Bmatrix} - x^\alpha_{,i}x^\beta_{,j}\begin{Bmatrix} \lambda \\ \alpha\ \beta \end{Bmatrix}. \tag{A3.70}$$

A3.10. Covariant Differentiation

We are now in a position to generalize the notion of differentiation into a covariant form. Suppose we differentiate the covariant vector

$$B_i \equiv (\partial x^\alpha/\partial y^i)A_\alpha \equiv x^\alpha_{,i}A_\alpha. \tag{A3.71}$$

We obtain

$$B_{i,j} = x^\alpha_{,i}x^\beta_{,j}A_{\alpha,\beta} + x^\alpha_{,ij}A_\alpha. \tag{A3.72}$$

It is obvious from (A3.72) that $B_{i,j}$ is not a tensor. But if we use (A3.70) in

(A3.72) we can rewrite the equation as

$$B_{i,j} = x^{\alpha}_{,i}x^{\beta}_{,j}\left(A_{\alpha,\beta} - \begin{Bmatrix} \lambda \\ \alpha\ \beta \end{Bmatrix} A_{\lambda}\right) + \begin{Bmatrix} k \\ i\ j \end{Bmatrix} x^{\lambda}_{,k}A_{\lambda}, \tag{A3.73}$$

or

$$B_{i,j} - \begin{Bmatrix} k \\ i\ j \end{Bmatrix} B_k = x^{\alpha}_{,i}x^{\beta}_{,j}\left(A_{\alpha,\beta} - \begin{Bmatrix} \lambda \\ \alpha\ \beta \end{Bmatrix} A_{\lambda}\right). \tag{A3.74}$$

Thus the combination

$$B_{i;j} \equiv B_{i,j} - \begin{Bmatrix} k \\ i\ j \end{Bmatrix} B_k \tag{A3.75}$$

is a covariant second-rank tensor, and reduces to the ordinary partial derivative of B_i in Cartesian coordinates. We therefore take (A3.75) as the definition of the *covariant derivative* of the vector B_i.

By a similar analysis, one can show that the covariant derivative of a contravariant vector is a mixed tensor of the second rank:

$$B^i_{;j} \equiv B^i_{,j} + \begin{Bmatrix} i \\ j\ k \end{Bmatrix} B^k. \tag{A3.76}$$

The extra terms containing Christoffel symbols that appear in equations (A3.75) and (A3.76) account for the effects of curvature of the coordinate system [see (**S1**, §46) for a detailed discussion].

These formulae are easily extended to mixed tensors of arbitrary rank. We find that the covariant derivative of the mixed tensor $A^{ab\ldots c}_{ij\ldots k}$ is

$$\begin{aligned} A^{ab\ldots c}_{ij\ldots k;q} = A^{ab\ldots c}_{ij\ldots k,q} &+ \begin{Bmatrix} a \\ \alpha\ q \end{Bmatrix} A^{\alpha b\ldots c}_{ij\ldots k} + \begin{Bmatrix} b \\ \alpha\ q \end{Bmatrix} A^{a\alpha\ldots c}_{ij\ldots k} + \ldots + \begin{Bmatrix} c \\ \alpha\ q \end{Bmatrix} A^{ab\ldots\alpha}_{ij\ldots k} \\ &- \begin{Bmatrix} \alpha \\ i\ q \end{Bmatrix} A^{ab\ldots c}_{\alpha j\ldots k} - \begin{Bmatrix} \alpha \\ j\ q \end{Bmatrix} A^{ab\ldots c}_{i\alpha\ldots k} - \ldots - \begin{Bmatrix} \alpha \\ k\ q \end{Bmatrix} A^{ab\ldots c}_{ij\ldots\alpha}. \end{aligned} \tag{A3.77}$$

In particular, for a contravariant tensor of the second rank,

$$T^{ij}_{;k} = T^{ij}_{,k} + \begin{Bmatrix} i \\ l\ k \end{Bmatrix} T^{lj} + \begin{Bmatrix} j \\ l\ k \end{Bmatrix} T^{il}. \tag{A3.78}$$

Note in passing that the covariant derivative of a scalar is identical to its ordinary partial derivative, and that the operation of covariant differentiation increases the rank of the resulting tensor by one relative to the original tensor. Furthermore, it is straightforward to generalize (A3.77) to relative tensors [see, e.g., (**L1,** §36) or (**S2,** §7.2)], but we will not require this result in our work.

The derivation of formulae for covariant differentiation given above proceeds by direct analysis. While this approach has the merit of brevity, it fails to communicate the deeper geometrical significance of the process, which raises questions concerning parallel transport and transplantation of vectors in a curved space. Discussions of these important and interesting matters, and generalizations of covariant differentiation to nonmetrical spaces, can be found in (**A1,** §§2.1, 2.2, and 3.1), (**L2,** §§33–36 and 39–41), and (**S2,** Chap. 8).

Let us now calculate the covariant derivative of the metric tensor; from (A3.77) we have

$$g_{ij;k} = g_{ij,k} - \begin{Bmatrix} l \\ i\ k \end{Bmatrix} g_{lj} - \begin{Bmatrix} l \\ j\ k \end{Bmatrix} g_{il}. \tag{A3.79}$$

But from (A3.59) the right-hand side is identically zero. Thus we have *Ricci's theorem*: the covariant derivative of the metric tensor (or its reciprocal) is identically zero in any coordinate system. This implies that in tensor equations we may freely interchange the operations of raising and lowering indices and of covariant differentiation. By a similar calculation we find that the Kronecker delta also behaves like a constant under covariant differentiation:

$$\delta^i_{j;k} = \delta^i_{j,k} + \begin{Bmatrix} i \\ k\ l \end{Bmatrix} \delta^l_j - \begin{Bmatrix} l \\ j\ k \end{Bmatrix} \delta^i_l = 0 + \begin{Bmatrix} i \\ k\ j \end{Bmatrix} - \begin{Bmatrix} i \\ j\ k \end{Bmatrix} \equiv 0. \tag{A3.80}$$

Suppose now that through some region in which a tensor field $A^{ab\ldots c}_{ij\ldots k}$ is defined, we choose a specific path, parameterized in terms of a path-length variable s as $x^i(s)$. Then we define the *intrinsic* (or *absolute*) *derivative* of the tensor field along this path to be

$$\begin{aligned} \frac{\delta A^{ab\ldots c}_{ij\ldots k}}{\delta s} &\equiv A^{ab\ldots c}_{ij\ldots k;q}\frac{dx^q}{ds} \\ &= \frac{dA^{ab\ldots c}_{ij\ldots k}}{ds} + \left(\begin{Bmatrix} a \\ m\ q \end{Bmatrix} A^{mb\ldots c}_{ij\ldots k} + \ldots - \begin{Bmatrix} m \\ i\ q \end{Bmatrix} A^{ab\ldots c}_{mj\ldots k} - \ldots \right) \frac{dx^q}{ds}. \end{aligned} \tag{A3.81}$$

Here we have written $(dA^{ab\ldots c}_{ij\ldots k}/ds) \equiv (\partial A^{ab\ldots c}_{ij\ldots k}/\partial x^q)(dx^q/ds)$. In particular, for a contravariant vector A^i,

$$\frac{\delta A^i}{\delta s} = \frac{dA^i}{ds} + \begin{Bmatrix} i \\ j\ k \end{Bmatrix} A^j \frac{dx^k}{ds}. \tag{A3.82}$$

In (A3.81) and (A3.82), s is an arbitrary path-length parameter. But in problems of fluid flow, it is natural to describe the path followed by a fluid element in terms of the time t, so that $x^i = x^i(t)$, and $(dx^i/dt) \equiv x^i_{,t} = v^i$, the velocity of the fluid element. In addition we must then allow for the possibility that any vector or tensor field may be an explicit function of time as well as of position, say $A^i = A^i(\mathbf{r}, t)$. Suppose now we choose time as the independent variable; then the intrinsic derivative with respect to time is the derivative with respect to time along the path followed by the fluid, that is, as measured in a frame moving with the fluid. In physical terms it is therefore identical to the *Lagrangean derivative* employed in descriptions of fluid kinematics and dynamics (cf. §15). For example

$$(\delta A^i/\delta t) \equiv (DA^i/Dt) = A^i_{,t} + A^i_{,j}v^j + \left\{ {i \atop j\ k} \right\} A^j v^k. \tag{A3.83}$$

Equation (A3.83) is the covariant generalization of the customary Lagrangean derivative for the vector field A^i; similar formulae can be written for tensors.

A3.11. Gradient, Divergence, Laplacian, and Curl

Covariant generalizations of the various operations with the symbolic operator ∇ discussed in §A2.10 can in most instances be obtained simply by replacing the partial derivatives with covariant derivatives. In the case of the gradient of a scalar field, the two derivatives are identical $f_{,i} \equiv f_{;i}$, and we obtain a covariant vector, say $\mathbf{F}$. For instance, in spherical coordinates the covariant components of $\mathbf{F}$ are $(\partial f/\partial r,\ \partial f/\partial \theta,\ \partial f/\partial \phi)$. Then from (A3.42) we find physical components

$$\nabla f = \left(\frac{\partial f}{\partial r}, \frac{1}{r}\frac{\partial f}{\partial \theta}, \frac{1}{r \sin\theta}\frac{\partial f}{\partial \phi} \right). \tag{A3.84}$$

One can of course also form the gradient of vector and tensor fields, for example, $(\nabla v^i)_j = v^i_{;j}$, etc.

The covariant generalization of the divergence of a vector is

$$\nabla \cdot \mathbf{v} = v^i_{;i} = v^i_{,i} + \left\{ {i \atop i\ j} \right\} v^j. \tag{A3.85}$$

In view of (A3.61) we can rewrite (A3.85) as

$$v^i_{;i} = g^{-1/2}(g^{1/2} v^i)_{,i} \tag{A3.86}$$

which is a convenient form for calculating $\nabla \cdot \mathbf{v}$ in curvilinear coordinates. For example, in spherical coordinates $g^{1/2} = r^2 \sin\theta$, and we find

$$v^i_{;i} = \frac{\partial v^{(1)}}{\partial r} + \frac{\partial v^{(2)}}{\partial \theta} + \frac{\partial v^{(3)}}{\partial \phi} + \frac{2v^{(1)}}{r} + \cot\theta\, v^{(2)}. \tag{A3.87}$$

Converting to physical components via (A3.46a) we recover the familiar result

$$v^i_{;i} = \frac{1}{r^2}\frac{\partial (r^2 v_r)}{\partial r} + \frac{1}{r\sin\theta}\frac{\partial(\sin\theta\, v_\theta)}{\partial \theta} + \frac{1}{r\sin\theta}\frac{\partial v_\phi}{\partial \phi}. \tag{A3.88}$$

For an arbitrary tensor we can form a divergence by contracting the covariant derivative index against any contravariant index, for example, $A^{i_1\cdots i_k\cdots i_n}_{j_1\cdots j_m;i_k}$. In our work we have occasion to deal only with second-rank tensors, which, in view of (A2.17), we can assume to have a definite symmetry. Applying (A3.78) to a symmetric tensor we find, using (A3.61),

$$S^{ij}_{;j} = S^{ij}_{,j} + \left\{ {i \atop k\ j} \right\} S^{kj} + \left\{ {j \atop k\ j} \right\} S^{ik} = g^{-1/2}(g^{1/2}S^{ij})_{,j} + \left\{ {i \atop j\ k} \right\} S^{jk}. \tag{A3.89}$$

For an antisymmetric tensor, individual terms in the last sum in (A3.89) cancel in pairs because the Christoffel symbols are symmetric in j and k, and we obtain the simpler result

$$A^{ij}_{;j} = g^{-1/2}(g^{1/2}A^{ij})_{,j}. \tag{A3.90}$$

Using (A3.89) and (A3.63) for a symmetric tensor T^{ij} in spherical coordinates and converting to physical components via (A3.47) we obtain the useful results:

$$T^{1j}_{;j} = \frac{1}{r^2}\frac{\partial(r^2 T_{rr})}{\partial r} + \frac{1}{r\sin\theta}\frac{\partial(\sin\theta T_{r\theta})}{\partial\theta} + \frac{1}{r\sin\theta}\frac{\partial T_{r\phi}}{\partial\phi} - \frac{1}{r}(T_{\theta\theta} + T_{\phi\phi}), \tag{A3.91a}$$

$$T^{2j}_{;j} = \frac{1}{r}\left[\frac{1}{r^2}\frac{\partial(r^2 T_{r\theta})}{\partial r} + \frac{1}{r\sin\theta}\frac{\partial(\sin\theta T_{\theta\theta})}{\partial\theta} + \frac{1}{r\sin\theta}\frac{\partial T_{\theta\phi}}{\partial\phi} + \frac{1}{r}(T_{r\theta} - \cot\theta T_{\phi\phi})\right] \tag{A3.91b}$$

and

$$T^{3j}_{;j} = \frac{1}{r\sin\theta}\left[\frac{1}{r^2}\frac{\partial(r^2 T_{r\phi})}{\partial r} + \frac{1}{r\sin\theta}\frac{\partial(\sin\theta T_{\theta\phi})}{\partial\theta} + \frac{1}{r\sin\theta}\frac{\partial T_{\phi\phi}}{\partial\phi} + \frac{1}{r}(T_{r\phi} + \cot\theta T_{\theta\phi})\right]. \tag{A3.91c}$$

The additional factors outside the square brackets account for the fact that the quantities on the left-hand side of (A3.91) are contravariant components, not physical components.

The easiest way to find a covariant expression for the Laplacian of a scalar is to follow (A2.49) and take the divergence of the vector obtained by forming the gradient of f. Thus $g^{ij}f_{,i}$ is a contravariant representation of ∇f, hence

$$\nabla^2 f = (g^{ij}f_{,i})_{;j} = g^{-1/2}(g^{1/2}g^{ij}f_{,i})_{,j} \tag{A3.92}$$

where we have used (A3.86). Similarly, for a vector we could write $\nabla^2 a^k = (g^{ij}a^k_{;i})_{;j}$.

As an example, the Laplacian of a scalar in spherical coordinates is, from

(A3.92),

$$\nabla^2 f = \frac{1}{r^2 \sin\theta}\left[\frac{\partial}{\partial r}\left(r^2 \sin\theta \frac{\partial f}{\partial r}\right) + \frac{\partial}{\partial \theta}\left(\sin\theta \frac{\partial f}{\partial \theta}\right) + \frac{\partial}{\partial \phi}\left(\frac{1}{\sin\theta}\frac{\partial f}{\partial \phi}\right)\right]$$

$$= \frac{1}{r^2}\frac{\partial}{\partial r}\left(r^2 \frac{\partial f}{\partial r}\right) + \frac{1}{r^2 \sin\theta}\frac{\partial}{\partial \theta}\left(\sin\theta \frac{\partial f}{\partial \theta}\right) + \frac{1}{r^2 \sin^2\theta}\frac{\partial^2 f}{\partial \phi^2}. \tag{A3.93}$$

For the curl of a vector we can obtain a covariant generalization of (A2.50) by replacing the permutation symbol with the Levi-Civita tensor and partial derivatives with covariant derivatives. Then

$$(\nabla \times \mathbf{a})^i = \varepsilon^{ijk} a_{k;j} = g^{-1/2} e^{ijk} a_{k;j} = g^{-1/2}(a_{k;j} - a_{j;k})$$

$$= g^{-1/2}\left(a_{k,j} - \left\{ {l \atop k\ j} \right\} a_l - a_{j,k} + \left\{ {l \atop j\ k} \right\} a_l\right) \tag{A3.94}$$

where (i, j, k) are distinct and are a cyclic permutation of $(1, 2, 3)$. But the Christoffel symbols are symmetric in their lower indices, so that (A3.94) reduces to

$$(\nabla \times \mathbf{a})^i = g^{-1/2}(a_{k,j} - a_{j,k}), \tag{A3.95}$$

which is easy to evaluate. For example, in spherical coordinates $g^{1/2} = r^2 \sin\theta$, and converting to physical components via (A3.46) one easily finds

$$(\nabla \times \mathbf{a})_r = \frac{1}{r \sin\theta}\left[\frac{\partial(\sin\theta\, a_\phi)}{\partial \theta} - \frac{\partial a_\theta}{\partial \phi}\right], \tag{A3.96a}$$

$$(\nabla \times \mathbf{a})_\theta = \frac{1}{r}\left[\frac{1}{\sin\theta}\frac{\partial a_r}{\partial \phi} - \frac{\partial (r a_\phi)}{\partial r}\right], \tag{A3.96b}$$

and

$$(\nabla \times \mathbf{a})_\phi = \frac{1}{r}\left[\frac{\partial (r a_\theta)}{\partial r} - \frac{\partial a_r}{\partial \theta}\right]. \tag{A3.96c}$$

Finally, note in passing that (A3.94) shows that $(\nabla \times \mathbf{a})$ is the dual of the antisymmetric tensor $A_{ij} = a_{j,i} - a_{i,j}$.

A3.12. Geodesics

Suppose that a constant vector $\mathbf{A}$ in Cartesian coordinates is moved parallel to itself through a displacement dy^i; we know that all components A^i remain unchanged, so that $dA^i \equiv 0$. Now consider the same operation transformed to curvilinear coordinates in which $B^\alpha = (\partial x^\alpha / \partial y^i) A^i = x^\alpha_{,i} A^i$. Then

$$dB^\alpha = x^\alpha_{,ij} A^i\, dy^j + x^\alpha_{,i}\, dA^i = x^\alpha_{,ij} A^i y^j_{,\beta}\, dx^\beta \tag{A3.97}$$

because $dA^i = 0$. Thus for *parallel displacement*,

$$dB^\alpha = x^\alpha_{,ij} y^j_{,\beta} y^i_{,\gamma} B^\gamma \, dx^\beta. \tag{A3.98}$$

Now using (A3.70) with $\{^{\,k}_{i\ j}\} = 0$ for Cartesian coordinates, (A3.98) becomes

$$dB^\alpha = -B^\gamma x^\rho_{,i} x^\sigma_{,j} \begin{Bmatrix} \alpha \\ \rho \ \sigma \end{Bmatrix} y^i_{,\gamma} y^j_{,\beta} \, dx^\beta = -\delta^\rho_\gamma \delta^\sigma_\beta \begin{Bmatrix} \alpha \\ \rho \ \sigma \end{Bmatrix} B^\gamma \, dx^\beta = -\begin{Bmatrix} \alpha \\ \beta \ \gamma \end{Bmatrix} B^\beta \, dx^\gamma. \tag{A3.99}$$

If we parameterize a path as $x^i(s)$, then (A3.99) shows that for parallel displacement of **B** along that path the intrinsic derivative will be identically zero:

$$\frac{\delta B^\alpha}{\delta s} = \frac{dB^\alpha}{ds} + \begin{Bmatrix} \alpha \\ \beta \ \gamma \end{Bmatrix} B^\beta \frac{dx^\gamma}{ds} = B^\alpha_{;\beta} \frac{dx^\beta}{ds} \equiv 0. \tag{A3.100}$$

In a Euclidian space we can construct a *straight line* by choosing the curve that has the property that an arbitrary vector displaced along it always remains parallel to itself. We generalize the notion of a straight line in a Riemannian space to that of a *geodesic*, which is the curve generated by parallel displacement of its unit tangent vector; that is, along a geodesic the tangents at all points are parallel, so that the curve's direction remains "constant" in the curved space. The equations describing a geodesic follow immediately by substituting $B^i = \lambda^i = (dx^i/ds)$ into (A3.100), which yields

$$\frac{d^2x^i}{ds^2} + \begin{Bmatrix} i \\ j \ k \end{Bmatrix} \frac{dx^j}{ds} \frac{dx^k}{ds} = 0. \tag{A3.101}$$

Other forms are discussed in (**S2,** §2.4).

Another property of straight lines in Euclidian space is that they are the shortest distance between two points. If one requires that a geodesic have the property of minimizing the path length between two points and therefore that

$$\delta \int_{P_1}^{P_2} ds = \delta \int_{P_1}^{P_2} \left(g_{ij} \frac{dx^i}{ds} \frac{dx^j}{ds} \right)^{1/2} ds = 0, \tag{A3.102}$$

then a variational analysis leads again to (A3.101) [see e.g. (**A1,** 55–57), (**L1,** §128), or (**S1,** §58)].

A3.13. Integral Theorems

It is possible to generalize the divergence theorem and Stokes's theorem to curvilinear coordinates in n dimensions, and also to nonmetrical spaces. We will not develop these generalizations here because we do not require them in our work; the reader may pursue these matters in (**S2,** Chap. 7).

References

(A1) Adler, R., Bazin, M., and Schiffer, M. (1975) *Introduction to General Relativity*. New York: McGraw-Hill.

(A2) Aris, R. (1962) *Vectors, Tensors, and the Basic Equations of Fluid Mechanics.* Englewood Cliffs: Prentice-Hall.

(J1) Jeffreys, H. and Jeffreys, B. D. (1950) *Methods of Mathematical Physics.* Cambridge: Cambridge University Press.

(L1) Lass, H. (1950) *Vector and Tensor Analysis.* New York: McGraw-Hill.

(L2) Lawden, D. F. (1975) *An Introduction to Tensor Calculus and Relativity.* London: Chapman and Hall.

(S1) Sokolnikoff, I. S. (1964) *Tensor Analysis.* New York: Wiley.

(S2) Synge, J. L. and Schild, A. (1949) *Tensor Calculus.* Toronto: University of Toronto Press.

Glossary of Physical Symbols

a	Sound speed
a_n	Frequency quadrature weight
a_R	Radiation density constant
a_T	Isothermal sound speed
a_x, a_y, a_z	Components of acceleration in (x, y, z) directions
a_0	Bohr radius
$\mathbf{a}$	Acceleration
$\mathscr{a}$	Invariant opacity
A	Atomic weight
A	Surface area
A_i	Vector part of Chapman–Enskog solution of Boltzmann equation
A_{ji}	Einstein spontaneous emission probability
A_k	Amplitude of kth Fourier component
A^α	Four acceleration
$\mathscr{A}_0$	Avogadro's number
b	Impact parameter in collision
b_i	Non-LTE departure coefficient, n_i/n_i^*
b_m	Angle quadrature weight
$\mathscr{b}$	Dimensionless impact parameter
B_{ij}	Tensor part of Chapman–Enskog solution of Boltzmann equation
B_{ij}	Einstein absorption probability
B_{ji}	Einstein stimulated emission probability
$B(T), B$	Integrated Planck function
$B_\nu(T), B(\nu, T), B_\nu$	Planck function
Bo	Boltzmann number
c	Speed of light
c_p	Specific heat at constant pressure
$\tilde{c}_p$	Specific heat at constant pressure per heavy particle
c_v	Specific heat at constant volume
$\tilde{c}_v$	Specific heat at constant volume per heavy particle
C	Heat capacity
C_{AB}	Bimolecular collision frequency
C_I	Numerical constant in Saha ionization formula
C_{ij}	Collision rate from level i to level j
$C_{i\kappa}$	Collisional ionization rate from level i

C_v	Heat capacity at constant volume
C_α	Coefficient in power-law potential
d	Diameter of rigid elastic sphere
d_0	Average distance between particles
D	Debye length
D_{ij}	Traceless rate of strain tensor
$D_{\alpha\beta}$	Covariant traceless shear tensor
(D/Dt)	Lagrangean time derivative (fluid element fixed)
$(Df/Dt)_{\text{coll}}$	Collisional source term in Boltzmann equation
$\mathscr{D}f$	Differential operator in Boltzmann equation
D	Traceless shear tensor
e	Electron charge
e	Specific internal energy (per gram)
e_{dissoc}	Internal energy of molecular dissociation
e_{exc}	Internal energy of atomic excitation
e_{ion}	Internal energy of atomic ionization
e_{rot}	Internal energy of molecular rotation
e_{trans}	Translational internal energy
e_{vib}	Internal energy of molecular vibration
e_{ijk}, e^{ijk}	Permutation symbol
$\hat{e}$	Internal energy per unit volume
$\tilde{e}$	Internal energy per particle
$\bar{e}$	Total specific internal energy of radiating fluid, $e_{\text{gas}}+(E/\rho)$
$\mathscr{e}$	Invariant emissivity
$\hat{\mathscr{e}}$	Total internal energy per unit volume, including rest energy, $\rho_0(c^2+e)$
$\tilde{\mathscr{e}}$	Total energy of particle including rest energy
$\mathscr{e}(\nu)$	Radiation energy spectral profile
$\mathscr{e}(\infty)$	Energy flux per particle at infinity in stellar wind
$\mathscr{e}_c(\infty)$	Heat-conduction flux per particle at infinity in stellar wind
E_{ij}	Rate of strain tensor
E_w	Energy transported in wave per period
$E_{\alpha\beta}$	Covariant shear tensor
$E^*(T)$, E^*	Radiation energy density in thermal equilibrium at temperature T, $4\pi B/c$
$E^*(\nu, T)$, E_ν^*	Monochromatic radiation energy density in thermal equilibrium at temperature T, $4\pi B_\nu/c$
$E(\mathbf{x}, t)$, E	Radiation energy density
$E(\mathbf{x}, t; \nu)$, E_ν	Monochromatic radiation energy density
$E(z)$	Wave-amplitude scale factor
$\mathscr{E}$	Energy flux in spherical flow
$\mathscr{E}$	Explosion energy
$\mathscr{E}$	Internal energy in a volume V
$\mathscr{E}$	Scalar part of Eckart decomposition of radiation stress-energy tensor
$\mathscr{E}^{n+1}$	Total energy in flow at time t^{n+1}
$\mathscr{E}_c(\infty)$	Heat-conduction flux at infinity in stellar wind
E	Rate of strain tensor
f	Oscillator strength of spectral line

f_l	Line radiation force
$f_\nu, f(\mathbf{x}, t; \nu)$	Monochromatic variable Eddington factor
f_0	Maxwellian velocity distribution
f_1	First-order term in Chapman–Enskog solution of Boltzmann equation
$\hat{f}$	Dimensionless distribution function
$f(\mathbf{x}, \mathbf{u}, t)$	Particle distribution function
$f_R(\mathbf{x}, t; \mathbf{n}, p), f_R$	Photon distribution function
$\mathbf{f}$	Newtonian force density
$\mathbf{f}_R$	Radiation force
$\not{f}_\nu$	Radiation flux spectral profile
F_{BB}	Radiation flux from black body
F^α	Four-force density
$F(v)$	Spontaneous recombination probability for electrons of speed v
$\mathbf{F}$	Force
$\mathbf{F}(\mathbf{x}, t), \mathbf{F}$	Radiation flux
$\mathbf{F}(\mathbf{x}, t; \nu), \mathbf{F}_\nu$	Monochromatic radiation flux
$\mathscr{F}$	Particle flux in spherical flow
$\mathscr{F}^\alpha$	Vector part of Eckart decomposition of radiation stress-energy tensor
g	Acceleration of gravity (planar geometry)
g	Determinant of metric tensor
g	Relative speed of collision partners
$g_{electron}$	Statistical weight of free electron
g_i	Statistical weight of state i
$g_{ij}, g_{\alpha\beta}$	Metric tensor
g_R	Radiative acceleration
$g_{R,l}$	Radiative acceleration from spectral lines
$\overset{0}{g}_R$	Invariant photon distribution function for blackbody radiation at rest relative to observer
$\mathbf{g}$	Gravitational acceleration
$\mathbf{g}$	Relative velocity of collision partners
G	Newtonian gravitation constant
G^α	Radiation four-force density
$G(v)$	Induced recombination probability for electrons of speed v
$\mathbf{G}$	Center of mass velocity
$\mathbf{G}$	Space components of radiation four-force density
$\mathbf{G}(\Delta t, k)$	Amplification matrix of system of difference equations
$\mathscr{G}$	Radiative gain rate per unit mass
$\boldsymbol{\mathscr{G}}$	Total radiation momentum density, $\mathbf{F}/c^2$
$\boldsymbol{\mathscr{G}}_\nu$	Monochromatic radiation momentum density, $\mathbf{F}_\nu/c^2$
h	Planck constant
$\hbar$	$h/2\pi$
h	Specific enthalpy
h_1	Normalized flux perturbation, H_1/B_0
$\tilde{h}$	Enthalpy per particle
$h(\mu, \nu), h_{\mu\nu}, h$	Antisymmetric part of specific intensity, $\frac{1}{2}[I(+\mu, \nu) - I(-\mu, \nu)]$
$\tilde{\hbar}$	Total enthalpy per particle, including rest energy, $m_0(c^2 + e) + (p/N)$

H	Scale height, density scale height
H_p	Pressure scale height
$\mathbf{H}(\mathbf{x}, t)$, $\mathbf{H}$	Integrated Eddington flux, $\mathbf{F}/4\pi$
$\mathbf{H}(\mathbf{x}, t; \nu)$, $\mathbf{H}_\nu$	Monochromatic Eddington flux, $F_\nu/4\pi$
$\mathbf{i}$	Unit vector along x axis
I	Boltzmann collision integral
$I(\mathbf{x}, t; \mathbf{n}, \nu)$, $I(\mu, \nu)$, I_ν, I	Specific intensity
$I(\mathbf{x}, t; +\mathbf{n}, \nu)$, $I(+\mu, \nu)$, I^+	Outward directed intensity
$I(\mathbf{x}, t; -\mathbf{n}, \nu)$, $I(-\mu, \nu)$, I^-	Inward directed intensity
$\mathscr{I}$	Boltzmann collision integral
$\mathscr{I}(\mu, \nu)$, $\mathscr{I}$	Invariant intensity
I	Unit tensor
j_1	Normalized mean intensity perturbation, J_1/B_0
$j(\mu, \nu)$, $j_{\mu\nu}$, j	Symmetric part of specific intensity, $\frac{1}{2}[I(+\mu, \nu)+I(-\mu, \nu)]$
$\mathbf{j}$	Unit vector along y axis
J	Jacobian determinant of transformation
$\bar{J}$	Mean intensity averaged over line profile, $\int \phi_\nu J_\nu \, d\nu$
$J(f_i, f_j)$	Boltzmann collision integral for functions f_i and f_j
$J(\mathbf{x}, t; \nu)$, J_ν	Monochromatic mean intensity
k	Boltzmann constant
k	Wavenumber
k_H	χ_H/χ_R
k_J	κ_J/κ_P
k_Q	Dimensionless coefficient in pseudoviscosity
k_x, k_y, k_z	Components of wave vector in (x, y, z) directions
$\mathbf{k}$	Unit vector along z axis
$\mathbf{k}$	Wave vector
K	Coefficient of thermal conductivity
K_e	Electron thermal conduction coefficient
K_L	Conduction coefficient of a Lorentz gas
K_p	Proton thermal conduction coefficient
K_R	Radiative conductivity
K^α	Photon propagation four-vector
K_ν	$(c/4\pi)P_\nu$
$\bar{K}$	Total conductivity of radiating fluid
Kn	Knudsen number
l	Characteristic length
ℓ	Affine path-length variable
L	Wave damping length
L_E	Eddington luminosity
$L(r, t)$, $L(r)$, L_r, L	Luminosity passing through sphere of radius r
$\mathbf{L}$	Linear difference operator
$\mathbf{L}$, $L_\beta^{\alpha'}$	Lorentz transformation matrix
$\mathscr{L}$	Radiative loss rate per unit mass
$\mathscr{L}$	Thermalization length
m	Mass
m	Column mass

m	Electron mass
m	Relativistic mass
m_e	Electron mass
m_{H}	Mass of hydrogen atom
m_p	Proton mass
m_0	Rest mass
$\dot{m}$	Mass flux (planar geometry)
$\tilde{m}$	Reduced mass of collision partners
$\mathfrak{m}$	M_1^2-1, where M_1 is upstream Mach number in shock
M	Total mass of collision partners
M_{cr}	Critical Mach number
M_r	Mass contained within sphere of radius r
M^{α}	Photon four-momentum
$M^{\alpha\beta}$	Material stress-energy tensor
$M(\ell)$	Radiation force multiplier
M	Mach number
$\mathcal{M}$	Mass of star
$\dot{\mathcal{M}}$	Radial mass flux (spherical geometry)
M	Material stress-energy tensor
n	Number of moles of gas
n_e	Number density of free electrons
n_H	Number density of neutral hydrogen atoms
n_i^*	Number density in level i in thermal equilibrium
n_p	Number density of protons
n_s	Number density of particles of species s
n_x, n_y, n_z	Components of photon propagation vector in (x, y, z) directions
n_κ	Number density of ions
$n(k)$	Inverse radiative relaxation time for disturbance of wavenumber k
$n_{\mathrm{E}}(k)$	Inverse radiative relaxation time in Eddington approximation for disturbance of wavenumber k
$\mathbf{n}$	Direction of photon propagation
$\mathbf{n}$	Unit normal to surface
N	Total number density of particles
N_{H}	Number density of all (neutral plus ionized) hydrogen atoms
N_j	Number density of ion state j in all excitation states
Nu	Nusselt number
$\mathcal{N}$	Total number of particles in volume V
p	Impact parameter
p	Pressure
p_g, p_{gas}	Gas pressure
p_ν	Photoionization probability
p_1	Pressure perturbation in wave
$\bar{p}$	Total pressure in radiating fluid, $p_{\mathrm{gas}}+p_{\mathrm{rad}}$
$\mathbf{p}$	Momentum
$\mathfrak{p}$	Fractional pressure jump across shock front, $(p_2-p_1)/p_1$
P	Complex amplitude of pressure perturbation in a wave
P	zz or rr component of radiation pressure tensor

P^*	Thermal radiation pressure
P_c	Average collision probability per unit length
P_d	Photon destruction probability
P_{ij}	Probability of transition from level i to level j
$P_{rr}, P_{\theta\theta}, P_{\phi\phi}$	Diagonal components of radiation pressure tensor in spherical geometry
P_{xx}, P_{yy}, P_{zz}	Diagonal components of radiation pressure tensor in planar geometry
P^{α}	Four-momentum
$P^{\alpha\beta}$	Projection tensor
$P^{\alpha\beta}, P^{ij}$	Radiation pressure tensor
P_{ν}	zz or rr components of monochromatic radiation pressure tensor
P_{ν}^*	Monochromatic thermal radiation pressure
$P_e(\tau)$	Photon escape probability
$\bar{P}(\mathbf{x}, t; \nu), \bar{P}_{\nu}$	Mean monochromatic radiation pressure
Pe	Peclet number
Pr	Prandtl number
$\mathcal{P}^{\alpha\beta}$	Tensor part of Eckart decomposition of radiation stress-energy tensor
P	Projection tensor
P	Radiation pressure tensor
$\mathsf{P}(\mathbf{x}, t; \nu), \mathsf{P}_{\nu}$	Monochromatic radiation pressure tensor
q	Heat transferred to unit mass of gas
$\dot{q}$	Rate of heat input to gas
q_{ν}	Sphericality factor
$q(\tau)$	Hopf function
$\mathbf{q}$	Thermal conduction flux
$\mathcal{q}^{n+1}$	Integrated heat input to flow at time t^{n+1}
Q	Heat gained or lost by gas
Q	Viscous pressure, pseudoviscous pressure
Q	$-(\partial \ln \rho/\partial \ln T)_p$
Q^{α}	Heat-flux four-vector
Q	Pseudoviscosity tensor
r	Radial coordinate in spherical coordinate system
r	Ratio of continuum to line opacity, κ_c/χ_l
r	a/cBo
r_c	Core radius
r_c	Critical radius in stellar wind
r_s	Radial position of shock
r_s	Sonic radius in stellar wind
r_0	Maximum compression ratio in shock $(\gamma+1)/(\gamma-1)$
$\hat{\mathbf{r}}$	Radial unit vector in spherical coordinate system
$\mathcal{r}$	Generalized Lagrangean radial coordinate
R	Complex amplitude of density perturbation in a wave
R	Gas constant for particular gas
R	Spectral radius of matrix
R	Stellar radius
R_{ij}	Radiative rate from level i to level j

$R_{i\kappa}$	Photoionization rate from level i
$R^{\alpha\beta}$	Radiation stress-energy tensor
$R(\mathbf{n}', \nu'; \mathbf{n}, \nu)$	Redistribution function
Re	Reynolds number
Re_R	Radiative Reynolds number
$\mathcal{R}$	Universal gas constant
R	Radiation stress-energy tensor
s	Path length
s	Spacetime interval
s	Specific entropy
s_e	Electron scattering coefficients per gram, $n_e\sigma_e/\rho$
s_{elec}	Entropy of electronic excitation
s_{trans}	Translational entropy
S	Entropy in volume V
S	Surface, surface area
S_l	Line source function
S_{rad}	Entropy of thermal radiation
S^{α}	Entropy-flux density four-vector
S^{α}_{β}	Projection tensor
$S^{\alpha\beta}$	Total radiating-fluid stress-energy tensor, $M^{\alpha\beta}+R^{\alpha\beta}$
S_ν	Source function
t	Time
t_c	Self-collision time
t_d	Radiation diffusion time, $l^2/c\lambda_p$
t_D	Deflection time
t_E	Energy-exchange time
t_{ee}	Electron self-collision time
t_{ep}	Electron-proton energy-exchange time
t_f	Fluid flow time, l/v
t_{ic}	Inelastic collision time
t_L	Radiation pressure disruption time
t_N	Nuclear time scale
t_{pp}	Proton self-collision time
t_R	Radiation flow time, l/c
t_{rc}	Radiative cooling time
t_{relax}	Relaxation time
t_{rr}	Radiative recombination time
$t_{RR}(k)$, t_{RR}	Radiative relaxation time
t_s	Strong-collision time
t_λ	Photon flight time, λ_p/c
$\mathbf{t}$	Surface force on fluid element
$\mathbf{t}_{\mathrm{visc}}$	Viscous drag force
$\not{t}$	Equivalent electron optical depth
T	Temperature
T_e	Electron temperature
T_{eff}	Effective temperature
T_{ij}	Stress tensor
T_{ion}	Ion temperature
T_k	Kinetic temperature

T_p	Proton temperature
T_R	Radiation temperature
T_1	Temperature perturbation in wave
T_-	Temperature immediately in front of shock
T_+	Temperature immediately behind shock
T	Stress tensor
u	Fluid speed relative to shock front
u	Speed
u	Velocity component along x axis
u_D	Critical speed of D-type ionization front
u_g	Horizontal component of group velocity
u_R	Critical speed of R-type ionization front
$\mathbf{u}$	Particle velocity
U	Complex amplitude of horizontal component of wave velocity
U	Random particle speed
U_x, U_y, U_z	Components of particle random velocity in (x, y, z) directions
$\mathbf{U}$	Random velocity of particle
$\mathcal{U}$	Dimensionless random particle speed
$\mathcal{U}_i$	ith component of dimensionless random particle velocity
$\mathcal{U}_l^\circ \mathcal{U}_m$	Traceless outer product of dimensionless random velocity components
$\boldsymbol{\mathcal{U}}$	Dimensionless random particle velocity
v	Specific volume $(1/\rho)$
v	Speed
v	Velocity component along y axis
v_c	Critical velocity in stellar wind
v_{esc}	Escape velocity
v_g	Group speed
$v_{i,j}$	Velocity gradient tensor
v_p	Phase speed
v_r	Radial velocity in spherical coordinate system
v_s	Shock speed
v_t	Phase trace speed
v_x, v_y, v_z	Components of velocity in (x, y, z) directions
v_θ	Tangential velocity in spherical coordinate system
v_ϕ	Azimuthal velocity in spherical coordinate system
v_∞	Terminal flow speed in stellar wind
$\mathbf{v}$	Fluid velocity
$\mathbf{v}_g$	Group velocity
$\mathbf{v}_1$	Velocity perturbation in wave
V	Specific volume in shock theory $(1/\rho)$
V	Volume
V^α	Four-velocity
$\mathcal{V}$	Material volume (fixed in fluid)
w	Velocity component along z axis
w_g	Vertical component of group velocity
w_k	Quadrature weight
w_1	Vertical component of wave velocity
W	Complex amplitude of vertical component of wave velocity

W	Rate of energy input from nonmechanical sources
W	Thermodynamic probability
W	Work done by gas
$\mathcal{W}^{n+1}$	Boundary work term in flow at time t^{n+1}
x	Cartesian coordinate
x	Degree of ionization [e.g., $n_p/(n_H+n_p)$]
x	Dimensionless frequency displacement from line center, $(\nu-\nu_0)/\Delta\nu_D$
x_f	Position of a front
$\mathbf{x}$	Position vector
$\mathbf{x}_1$	Fluid displacement in wave
X	Complex amplitude of horizontal component of wave displacement
$X_r[f(x)]$	X operator
y	Cartesian coordinate
z	Cartesian coordinate
z	Dimensionless radiative relaxation rate, nt_λ
z	Dimensionless wavenumber, ak/ω
Z	Charge number
Z	Complex amplitude of vertical component of wave displacement
Z	Partition function
Z_{elec}	Partition function for electronic excitation
Z_{rot}	Partition function for molecular rotation
Z_{trans}	Partition function for translational motions
Z_{vib}	Partition function for molecular vibration
α	Angle of wave phase propagation relative to horizontal plane
α	Exponent in power-law potential
α	Exponent in radiation force law
α	Ratio of radiation pressure to gas pressure in thermal equilibrium, P^*/p_g
$\bar{\alpha}$	Mean photoionization cross section
$\alpha_{ij}(\nu)$	Bound–bound absorption cross section
$\alpha_{i\kappa}(\nu)$	Bound–free absorption cross section
$\alpha_{\kappa\kappa}(\nu, T)$	Free–free absorption cross section
α_ν	Photoionization cross section
β	Coefficient of thermal expansion
β	Effective line optical depth in stellar wind
β	$1/kT$
β	v/c
β_l	$(n_e\sigma_e/\chi_l)=\ell/\tau_l$
γ	Ratio of specific heats
γ	$(1-v^2/c^2)^{-1/2}$
$\boldsymbol{\gamma}$	Dimensionless relative velocity of collision partners
Γ	Circulation
Γ	Force ratio g_R/g
$\Gamma, \Gamma_1, \Gamma_2, \Gamma_3$	Generalized adiabatic exponents
Γ^b_{ac}	Ricci rotation coefficient
$\boldsymbol{\Gamma}$	Dimensionless center of mass velocity

δ	Continuum destruction probability
δ	Width of shock front
δ_{AB}	Phase shift between quantities A and B in a wave
δ_{ij}, δ^{ij}	Kronecker delta symbol
$\delta(x)$	Dirac delta function
$(\delta/\delta t)$	Intrinsic derivative with respect to time
Δ	Thickness of temperature relaxation layer
$\Delta\nu_D$	Line Doppler width
ε	Momentum flux in stellar wind normalized to momentum in radiation field
ε	Rate of thermonuclear energy release (per gram)
ε	Thermalization parameter
ε^{α}_{a}, ε^{a}_{α}	Transformation between coordinate and tetrad frames
ε_H	Ionization potential (from ground state) of hydrogen
ε_i	Energy above ground of state i
ε_{iH}	Excitation energy of state i of hydrogen
ε_{Ii}	Ionization potential above state i
ε_{ion}	Ionization energy
ε_w	Wave energy density
ε_∞	Residual energy at infinity per particle in stellar wind
$\bar{\varepsilon}$	Mean energy per ionizing photon
$\boldsymbol{\varepsilon}_a$	Basis vectors in orthonormal tetrad frame
$\boldsymbol{\varepsilon}_\alpha$	Basis vectors of coordinate system
ζ	Coefficient of bulk viscosity
ζ_R	Bulk viscosity coefficient for radiation
ζ_1	Vertical displacement of fluid element in a wave
η	Photoionization sink term in line source function
η	Volume ratio in shock, $V/V_1=\rho_1/\rho$
η_l	Line emission coefficient
$\eta_{\alpha\beta}$	Lorentz metric
η_*	Maximum compression ratio in radiation-dominated shock
$\eta(\mathbf{x}, t; \mathbf{n}, \nu)$, η_ν	Emission coefficient
$\eta^s(\mathbf{x}, t; \mathbf{n}, \nu)$, η^s_ν	Scattering emission coefficient
$\boldsymbol{\eta}$	Lorentz metric tensor
θ	Fluid expansion
θ	Polar angle in spherical coordinate system
θ	Recombination source term in line source function
θ	Time-centering coefficient in implicit difference equation, $0\leq\theta\leq 1$
θ_1	Normalized temperature perturbation, T_1/T_0
$\hat{\boldsymbol{\theta}}$	Unit vector in direction of increasing polar angle in spherical coordinate system
Θ	Complex amplitude of temperature perturbation in a wave
Θ	Polar angle of radiation propagation vector relative to local outward normal in planar or spherical geometry
κ_c	Continuum opacity
κ_E	Absorption mean opacity
κ_J	Absorption mean opacity
κ_P	Planck mean opacity

$\kappa_{P,g}$	Group Planck mean
κ_s	Coefficient of adiabatic compressibility
κ_T	Coefficient of isothermal compressibility
$\kappa(\mathbf{x}, t; \mathbf{n}, \nu)$, κ_ν, κ	True absorption coefficient
λ	Coefficient of dilatational viscosity (second coefficient of viscosity)
λ	Dimensionless potential energy in stellar wind
λ	Eigenvalue
λ	Particle mean free path
λ_p	Photon mean free path
λ_t	Free-flight distance of photon in time Δt, $c\,\Delta t$
λ_R	Rosseland mean free path, χ_R^{-1}
λ_ν	Photon mean free path at frequency ν, χ_ν^{-1}
Λ	Ratio of maximum to minimum impact parameter
Λ	Wavelength
$\Lambda_\tau[f(x)]$, Λ_τ	Lambda operator
$\mathbf{\Lambda}$	Lorentz transformation in Minkowski metric
μ	Angle cosine of photon propagation vector relative to outward normal, $\mu \equiv \mathbf{n} \cdot \mathbf{k}$ or $\mathbf{n} \cdot \hat{\mathbf{r}}$
μ	Coefficient of dynamical viscosity
μ	Mean molecular weight
μ_e	Electron viscosity coefficient
μ_m	Angle-quadrature point
μ_p	Proton viscosity coefficient
μ_Q	Artificial viscosity coefficient
μ_R	Coefficient of radiative viscosity
$\mu_{\alpha\beta}$	Minkowski metric
μ'	Effective viscosity coefficient for one-dimensional flows, $\mu' = \mu + \frac{3}{4}\zeta$
$\boldsymbol{\mu}$	Momentum density
$\boldsymbol{\mu}_w$	Wave momentum density
ν	Frequency
ν	Kinematic viscosity coefficient (μ/ρ)
ν	Inverse radiative relaxation time for optically thin disturbance
ν_i	Occupation number of state i (number of particles in state i in volume V)
ν_0	Line-center frequency
ξ	Similarity variable
$\bar{\xi}$	Average thermal coupling parameter
ξ_k	Amplification factor of kth Fourier component
ξ_ν	Monochromatic thermalization parameter, $(r + \varepsilon\phi_\nu)/(r + \phi_\nu)$
ξ_1	Horizontal displacement of fluid element in a wave
Π	Gas pressure ratio in radiating shock
Π_{ij}	Momentum flux-density tensor
$\mathbf{\Pi}$	Momentum flux-density tensor
ρ	Newtonian density
ρ	Lab density of proper mass, $\gamma\rho_0$
ρ'	Lab density of relative mass, $\gamma^2\rho_0$
ρ_0	Proper density of proper mass

ρ_{00}	Mass density of fluid including internal energy, $\rho_0(1+e/c^2)$
ρ_{000}	Mass density of fluid including enthalpy, $\rho_0(1+e/c^2)+p/c^2$
ρ_1	$\gamma^2\rho_{000}$
ρ_1	Density perturbation in wave
ρ_*	$\gamma\rho_{000}$
σ_e	Thomson electron scattering cross section
σ_{ij}	Viscous stress tensor
σ_l	Line scattering coefficient
σ_R	Stefan–Boltzmann constant
σ_T, $\sigma_{(0)}$	Total collision cross section
$\sigma_{(2)}$	Collision cross section in transport coefficient
$\sigma(g, \chi)$; $\sigma(\mathbf{u}_1, \mathbf{u}_2; \mathbf{u}_1', \mathbf{u}_2')$	Collision cross section
$\sigma(\mathbf{x}, t; \mathbf{n}, \nu)$, σ_ν	Scattering coefficient
$\boldsymbol{\sigma}$	Viscous stress tensor
Σ	Strength of vortex tube
τ	Average collision time
τ	Dimensionless temperature in stellar wind
τ	Optical depth
τ	Proper time
τ	Scaled temperature in thermal front
τ	Wave period
τ_a	Optical thickness of disturbance of frequency ω traveling with speed of sound, $a\kappa/\omega$
τ_{ac}	Acoustic-cutoff period
τ_c	Optical thickness of disturbance of frequency ω traveling with speed of light, $c\kappa/\omega$
τ_e	Electron optical depth of stellar wind
τ_l	Effective line optical depth
τ_R	Rosseland optical depth
τ_Λ	Optical thickness of disturbance of wavelength Λ
$\tau_\nu(\mathbf{x}, \mathbf{x}')$, τ_ν	Monochromatic optical depth
ϕ	Azimuthal angle in spherical coordinate system
ϕ	Photon number flux
ϕ	Velocity potential
ϕ	Wave phase
ϕ_ν, $\phi(\nu)$	Line profile function
$\phi(r)$, ϕ	Potential
$\boldsymbol{\phi}$	Newtonian three-force
$\boldsymbol{\phi}_w$	Wave energy flux
$\hat{\boldsymbol{\phi}}$	Unit vector in direction of increasing azimuthal angle in spherical coordinate system
Φ	Azimuthal angle of radiation propagation vector around local outward normal in planar or spherical geometry
Φ	Potential
Φ	Viscous dissipation function
Φ_1	Scaled first-order term in Chapman–Enskog solution of Boltzmann equation, f_1/f_0
Φ^α	Four-force

$\Phi_{ij}(T)$	Saha–Boltzmann factor of bound state i of ion state j relative to ground state of ion $j+1$. $[n_{ij}^* = n_e n_{0,j+1} \Phi_{ij}(T)]$
$\Phi_\tau[f(x)]$	Phi operator
χ	Angle of deflection in collision
χ	Thermal diffusivity, $K/\rho c_p$
χ_H	Flux mean opacity
$\chi_l, \chi_l(\nu)$	Line absorption coefficient
χ_R	Rosseland mean opacity
$\chi_{R,g}$	Group Rosseland mean
$\chi(\mathbf{x}, t; \mathbf{n}, \nu), \chi_\nu$	Opacity coefficient (per unit volume)
ψ	Dimensionless kinetic energy in stellar wind
ψ	Wave phase
$\psi(\mathbf{x}, t; \mathbf{n}, \nu), \psi_\nu, \psi$	Photon number density
$\boldsymbol{\psi}$	Solution vector in complete linearization method
ω	Angular frequency
ω	Solid angle
ω_a	Acoustic cutoff frequency
ω_{BV}	Brunt–Väisälä frequency
ω_g	Brunt–Väisälä frequency in isothermal medium
$\omega_r, \omega_\theta, \omega_\phi$	Radial, tangential, and azimuthal components of vorticity
$\omega_x, \omega_y, \omega_z$	Components of vorticity in (x, y, z) directions
$\omega(\nu), \omega_\nu$	Opacity per gram, χ_ν/ρ
$\omega_{aN}(t_{RR})$	Effective acoustic-cutoff frequency in Newtonian cooling approximation
$\omega_{gN}(t_{RR})$	Effective maximum gravity-wave frequency in Newtonian cooling approximation
$\boldsymbol{\omega}$	Vorticity
Ω	Solid angle
Ω_{ij}	Vorticity tensor
$\Omega_{\alpha\beta}$	Covariant rotation tensor
$,i$	Partial derivative with respect to coordinate x^i
$,t$	Partial derivative with respect to time
$;\alpha$	Covariant derivative with respect to coordinate x^α
$(\partial/\partial t)$	Eulerian time derivative (space coordinates fixed)
∂_a	Pfaffian derivative
$\boldsymbol{\nabla}_k$	Gradient with respect to wave-vector components
$\oplus$	Earth symbol
$\odot$	Sun symbol

Index

Astronomy

BURNHAM'S CELESTIAL HANDBOOK, Robert Burnham, Jr. Thorough guide to the stars beyond our solar system. Exhaustive treatment. Alphabetical by constellation: Andromeda to Cetus in Vol. 1; Chamaeleon to Orion in Vol. 2; and Pavo to Vulpecula in Vol. 3. Hundreds of illustrations. Index in Vol. 3. 2,000pp. 6$^1/_8$ x 9$^1/_4$.
Vol. I: 0-486-23567-X
Vol. II: 0-486-23568-8
Vol. III: 0-486-23673-0

EXPLORING THE MOON THROUGH BINOCULARS AND SMALL TELESCOPES, Ernest H. Cherrington, Jr. Informative, profusely illustrated guide to locating and identifying craters, rills, seas, mountains, other lunar features. Newly revised and updated with special section of new photos. Over 100 photos and diagrams. 240pp. 8$^1/_4$ x 11. 0-486-24491-1

THE EXTRATERRESTRIAL LIFE DEBATE, 1750–1900, Michael J. Crowe. First detailed, scholarly study in English of the many ideas that developed from 1750 to 1900 regarding the existence of intelligent extraterrestrial life. Examines ideas of Kant, Herschel, Voltaire, Percival Lowell, many other scientists and thinkers. 16 illustrations. 704pp. 5$^3/_8$ x 8$^1/_2$. 0-486-40675-X

THEORIES OF THE WORLD FROM ANTIQUITY TO THE COPERNICAN REVOLUTION, Michael J. Crowe. Newly revised edition of an accessible, enlightening book re-creates the change from an earth-centered to a sun-centered conception of the solar system. 242pp. 5$^3/_8$ x 8$^1/_2$. 0-486-41444-2

ARISTARCHUS OF SAMOS: The Ancient Copernicus, Sir Thomas Heath. Heath's history of astronomy ranges from Homer and Hesiod to Aristarchus and includes quotes from numerous thinkers, compilers, and scholasticists from Thales and Anaximander through Pythagoras, Plato, Aristotle, and Heraclides. 34 figures. 448pp. 5$^3/_8$ x 8$^1/_2$.
0-486-43886-4

A COMPLETE MANUAL OF AMATEUR ASTRONOMY: TOOLS AND TECHNIQUES FOR ASTRONOMICAL OBSERVATIONS, P. Clay Sherrod with Thomas L. Koed. Concise, highly readable book discusses: selecting, setting up and maintaining a telescope; amateur studies of the sun; lunar topography and occultations; observations of Mars, Jupiter, Saturn, the minor planets and the stars; an introduction to photoelectric photometry; more. 1981 ed. 124 figures. 25 halftones. 37 tables. 335pp. 6$^1/_2$ x 9$^1/_4$. 0-486-42820-8

AMATEUR ASTRONOMER'S HANDBOOK, J. B. Sidgwick. Timeless, comprehensive coverage of telescopes, mirrors, lenses, mountings, telescope drives, micrometers, spectroscopes, more. 189 illustrations. 576pp. 5$^5/_8$ x 8$^1/_4$. (Available in U.S. only.)
0-486-24034-7

STAR LORE: Myths, Legends, and Facts, William Tyler Olcott. Captivating retellings of the origins and histories of ancient star groups include Pegasus, Ursa Major, Pleiades, signs of the zodiac, and other constellations. "Classic."—Sky & Telescope. 58 illustrations. 544pp. 5$^3/_8$ x 8$^1/_2$. 0-486-43581-4

Chemistry

THE SCEPTICAL CHYMIST: THE CLASSIC 1661 TEXT, Robert Boyle. Boyle defines the term "element," asserting that all natural phenomena can be explained by the motion and organization of primary particles. 1911 ed. viii+232pp. 5³/₈ x 8¹/₂. 0-486-42825-7

RADIOACTIVE SUBSTANCES, Marie Curie. Here is the celebrated scientist's doctoral thesis, the prelude to her receipt of the 1903 Nobel Prize. Curie discusses establishing atomic character of radioactivity found in compounds of uranium and thorium; extraction from pitchblende of polonium and radium; isolation of pure radium chloride; determination of atomic weight of radium; plus electric, photographic, luminous, heat, color effects of radioactivity. ii+94pp. 5³/₈ x 8¹/₂. 0-486-42550-9

CHEMICAL MAGIC, Leonard A. Ford. Second Edition, Revised by E. Winston Grundmeier. Over 100 unusual stunts demonstrating cold fire, dust explosions, much more. Text explains scientific principles and stresses safety precautions. 128pp. 5³/₈ x 8¹/₂. 0-486-67628-5

MOLECULAR THEORY OF CAPILLARITY, J. S. Rowlinson and B. Widom. History of surface phenomena offers critical and detailed examination and assessment of modern theories, focusing on statistical mechanics and application of results in mean-field approximation to model systems. 1989 edition. 352pp. 5³/₈ x 8¹/₂. 0-486-42544-4

CHEMICAL AND CATALYTIC REACTION ENGINEERING, James J. Carberry. Designed to offer background for managing chemical reactions, this text examines behavior of chemical reactions and reactors; fluid-fluid and fluid-solid reaction systems; heterogeneous catalysis and catalytic kinetics; more. 1976 edition. 672pp. 6¹/₈ x 9¹/₄. 0-486-41736-0 $31.95

ELEMENTS OF CHEMISTRY, Antoine Lavoisier. Monumental classic by founder of modern chemistry in remarkable reprint of rare 1790 Kerr translation. A must for every student of chemistry or the history of science. 539pp. 5³/₈ x 8¹/₂. 0-486-64624-6

MOLECULES AND RADIATION: An Introduction to Modern Molecular Spectroscopy. Second Edition, Jeffrey I. Steinfeld. This unified treatment introduces upper-level undergraduates and graduate students to the concepts and the methods of molecular spectroscopy and applications to quantum electronics, lasers, and related optical phenomena. 1985 edition. 512pp. 5³/₈ x 8¹/₂. 0-486-44152-0

A SHORT HISTORY OF CHEMISTRY, J. R. Partington. Classic exposition explores origins of chemistry, alchemy, early medical chemistry, nature of atmosphere, theory of valency, laws and structure of atomic theory, much more. 428pp. 5³/₈ x 8¹/₂. (Available in U.S. only.) 0-486-65977-1

GENERAL CHEMISTRY, Linus Pauling. Revised 3rd edition of classic first-year text by Nobel laureate. Atomic and molecular structure, quantum mechanics, statistical mechanics, thermodynamics correlated with descriptive chemistry. Problems. 992pp. 5³/₈ x 8¹/₂. 0-486-65622-5

ELECTRON CORRELATION IN MOLECULES, S. Wilson. This text addresses one of theoretical chemistry's central problems. Topics include molecular electronic structure, independent electron models, electron correlation, the linked diagram theorem, and related topics. 1984 edition. 304pp. 5³/₈ x 8¹/₂. 0-486-45879-2

Engineering

DE RE METALLICA, Georgius Agricola. The famous Hoover translation of greatest treatise on technological chemistry, engineering, geology, mining of early modern times (1556). All 289 original woodcuts. 638pp. $6^3/_4$ x 11. 0-486-60006-8

FUNDAMENTALS OF ASTRODYNAMICS, Roger Bate et al. Modern approach developed by U.S. Air Force Academy. Designed as a first course. Problems, exercises. Numerous illustrations. 455pp. $5^3/_8$ x $8^1/_2$. 0-486-60061-0

DYNAMICS OF FLUIDS IN POROUS MEDIA, Jacob Bear. For advanced students of ground water hydrology, soil mechanics and physics, drainage and irrigation engineering and more. 335 illustrations. Exercises, with answers. 784pp. $6^1/_8$ x $9^1/_4$. 0-486-65675-6

THEORY OF VISCOELASTICITY (SECOND EDITION), Richard M. Christensen. Complete consistent description of the linear theory of the viscoelastic behavior of materials. Problem-solving techniques discussed. 1982 edition. 29 figures. xiv+364pp. $6^1/_8$ x $9^1/_4$. 0-486-42880-X

MECHANICS, J. P. Den Hartog. A classic introductory text or refresher. Hundreds of applications and design problems illuminate fundamentals of trusses, loaded beams and cables, etc. 334 answered problems. 462pp. $5^3/_8$ x $8^1/_2$. 0-486-60754-2

MECHANICAL VIBRATIONS, J. P. Den Hartog. Classic textbook offers lucid explanations and illustrative models, applying theories of vibrations to a variety of practical industrial engineering problems. Numerous figures. 233 problems, solutions. Appendix. Index. Preface. 436pp. $5^3/_8$ x $8^1/_2$. 0-486-64785-4

STRENGTH OF MATERIALS, J. P. Den Hartog. Full, clear treatment of basic material (tension, torsion, bending, etc.) plus advanced material on engineering methods, applications. 350 answered problems. 323pp. $5^3/_8$ x $8^1/_2$. 0-486-60755-0

A HISTORY OF MECHANICS, René Dugas. Monumental study of mechanical principles from antiquity to quantum mechanics. Contributions of ancient Greeks, Galileo, Leonardo, Kepler, Lagrange, many others. 671pp. $5^3/_8$ x $8^1/_2$. 0-486-65632-2

STABILITY THEORY AND ITS APPLICATIONS TO STRUCTURAL MECHANICS, Clive L. Dym. Self-contained text focuses on Koiter postbuckling analyses, with mathematical notions of stability of motion. Basing minimum energy principles for static stability upon dynamic concepts of stability of motion, it develops asymptotic buckling and postbuckling analyses from potential energy considerations, with applications to columns, plates, and arches. 1974 ed. 208pp. $5^3/_8$ x $8^1/_2$. 0-486-42541-X

BASIC ELECTRICITY, U.S. Bureau of Naval Personnel. Originally a training course; best nontechnical coverage. Topics include batteries, circuits, conductors, AC and DC, inductance and capacitance, generators, motors, transformers, amplifiers, etc. Many questions with answers. 349 illustrations. 1969 edition. 448pp. $6^1/_2$ x $9^1/_4$. 0-486-20973-3

Mathematics

FUNCTIONAL ANALYSIS (Second Corrected Edition), George Bachman and Lawrence Narici. Excellent treatment of subject geared toward students with background in linear algebra, advanced calculus, physics and engineering. Text covers introduction to inner-product spaces, normed, metric spaces, and topological spaces; complete orthonormal sets, the Hahn-Banach Theorem and its consequences, and many other related subjects. 1966 ed. 544pp. $6^1/_8$ x $9^1/_4$. 0-486-40251-7

DIFFERENTIAL MANIFOLDS, Antoni A. Kosinski. Introductory text for advanced undergraduates and graduate students presents systematic study of the topological structure of smooth manifolds, starting with elements of theory and concluding with method of surgery. 1993 edition. 288pp. $5^3/_8$ x $8^1/_2$. 0-486-46244-7

VECTOR AND TENSOR ANALYSIS WITH APPLICATIONS, A. I. Borisenko and I. E. Tarapov. Concise introduction. Worked-out problems, solutions, exercises. 257pp. $5^5/_8$ x $8^1/_4$. 0-486-63833-2

AN INTRODUCTION TO ORDINARY DIFFERENTIAL EQUATIONS, Earl A. Coddington. A thorough and systematic first course in elementary differential equations for undergraduates in mathematics and science, with many exercises and problems (with answers). Index. 304pp. $5^3/_8$ x $8^1/_2$. 0-486-65942-9

FOURIER SERIES AND ORTHOGONAL FUNCTIONS, Harry F. Davis. An incisive text combining theory and practical example to introduce Fourier series, orthogonal functions and applications of the Fourier method to boundary-value problems. 570 exercises. Answers and notes. 416pp. $5^3/_8$ x $8^1/_2$. 0-486-65973-9

COMPUTABILITY AND UNSOLVABILITY, Martin Davis. Classic graduate-level introduction to theory of computability, usually referred to as theory of recurrent functions. New preface and appendix. 288pp. $5^3/_8$ x $8^1/_2$. 0-486-61471-9

AN INTRODUCTION TO MATHEMATICAL ANALYSIS, Robert A. Rankin. Dealing chiefly with functions of a single real variable, this text by a distinguished educator introduces limits, continuity, differentiability, integration, convergence of infinite series, double series, and infinite products. 1963 edition. 624pp. $5^3/_8$ x $8^1/_2$. 0-486-46251-X

METHODS OF NUMERICAL INTEGRATION (SECOND EDITION), Philip J. Davis and Philip Rabinowitz. Requiring only a background in calculus, this text covers approximate integration over finite and infinite intervals, error analysis, approximate integration in two or more dimensions, and automatic integration. 1984 edition. 624pp. $5^3/_8$ x $8^1/_2$. 0-486-45339-1

INTRODUCTION TO LINEAR ALGEBRA AND DIFFERENTIAL EQUATIONS, John W. Dettman. Excellent text covers complex numbers, determinants, orthonormal bases, Laplace transforms, much more. Exercises with solutions. Undergraduate level. 416pp. $5^3/_8$ x $8^1/_2$. 0-486-65191-6

RIEMANN'S ZETA FUNCTION, H. M. Edwards. Superb, high-level study of landmark 1859 publication entitled "On the Number of Primes Less Than a Given Magnitude" traces developments in mathematical theory that it inspired. xiv+315pp. $5^3/_8$ x $8^1/_2$. 0-486-41740-9

Math—Decision Theory, Statistics, Probability

INTRODUCTION TO PROBABILITY, John E. Freund. Featured topics include permutations and factorials, probabilities and odds, frequency interpretation, mathematical expectation, decision-making, postulates of probability, rule of elimination, much more. Exercises with some solutions. Summary. 1973 edition. 247pp. 5³/₈ x 8¹/₂.
0-486-67549-1

STATISTICAL AND INDUCTIVE PROBABILITIES, Hugues Leblanc. This treatment addresses a decades-old dispute among probability theorists, asserting that both statistical and inductive probabilities may be treated as sentence-theoretic measurements, and that the latter qualify as estimates of the former. 1962 edition. 160pp. 5³/₈ x 8¹/₂.
0-486-44980-7

APPLIED MULTIVARIATE ANALYSIS: Using Bayesian and Frequentist Methods of Inference, Second Edition, S. James Press. This two-part treatment deals with foundations as well as models and applications. Topics include continuous multivariate distributions; regression and analysis of variance; factor analysis and latent structure analysis; and structuring multivariate populations. 1982 edition. 692pp. 5³/₈ x 8¹/₂. 0-486-44236-5

LINEAR PROGRAMMING AND ECONOMIC ANALYSIS, Robert Dorfman, Paul A. Samuelson and Robert M. Solow. First comprehensive treatment of linear programming in standard economic analysis. Game theory, modern welfare economics, Leontief input-output, more. 525pp. 5³/₈ x 8¹/₂. 0-486-65491-5

PROBABILITY: AN INTRODUCTION, Samuel Goldberg. Excellent basic text covers set theory, probability theory for finite sample spaces, binomial theorem, much more. 360 problems. Bibliographies. 322pp. 5³/₈ x 8¹/₂. 0-486-65252-1

GAMES AND DECISIONS: INTRODUCTION AND CRITICAL SURVEY, R. Duncan Luce and Howard Raiffa. Superb nontechnical introduction to game theory, primarily applied to social sciences. Utility theory, zero-sum games, n-person games, decision-making, much more. Bibliography. 509pp. 5³/₈ x 8¹/₂. 0-486-65943-7

INTRODUCTION TO THE THEORY OF GAMES, J. C. C. McKinsey. This comprehensive overview of the mathematical theory of games illustrates applications to situations involving conflicts of interest, including economic, social, political, and military contexts. Appropriate for advanced undergraduate and graduate courses; advanced calculus a prerequisite. 1952 ed. x+372pp. 5³/₈ x 8¹/₂. 0-486-42811-7

FIFTY CHALLENGING PROBLEMS IN PROBABILITY WITH SOLUTIONS, Frederick Mosteller. Remarkable puzzlers, graded in difficulty, illustrate elementary and advanced aspects of probability. Detailed solutions. 88pp. 5³/₈ x 8¹/₂. 0-486-65355-2

PROBABILITY THEORY: A CONCISE COURSE, Y. A. Rozanov. Highly readable, self-contained introduction covers combination of events, dependent events, Bernoulli trials, etc. 148pp. 5³/₈ x 8¹/₄. 0-486-63544-9

THE STATISTICAL ANALYSIS OF EXPERIMENTAL DATA, John Mandel. First half of book presents fundamental mathematical definitions, concepts and facts while remaining half deals with statistics primarily as an interpretive tool. Well-written text, numerous worked examples with step-by-step presentation. Includes 116 tables. 448pp. 5³/₈ x 8¹/₂. 0-486-64666-1

History of Math

THE WORKS OF ARCHIMEDES, Archimedes (T. L. Heath, ed.). Topics include the famous problems of the ratio of the areas of a cylinder and an inscribed sphere; the measurement of a circle; the properties of conoids, spheroids, and spirals; and the quadrature of the parabola. Informative introduction. clxxxvi+326pp. $5^3/_8 \times 8^1/_2$. 0-486-42084-1

A SHORT ACCOUNT OF THE HISTORY OF MATHEMATICS, W. W. Rouse Ball. One of clearest, most authoritative surveys from the Egyptians and Phoenicians through 19th-century figures such as Grassman, Galois, Riemann. Fourth edition. 522pp. $5^3/_8 \times 8^1/_2$. 0-486-20630-0

THE HISTORY OF THE CALCULUS AND ITS CONCEPTUAL DEVELOPMENT, Carl B. Boyer. Origins in antiquity, medieval contributions, work of Newton, Leibniz, rigorous formulation. Treatment is verbal. 346pp. $5^3/_8 \times 8^1/_2$. 0-486-60509-4

THE HISTORICAL ROOTS OF ELEMENTARY MATHEMATICS, Lucas N. H. Bunt, Phillip S. Jones, and Jack D. Bedient. Fundamental underpinnings of modern arithmetic, algebra, geometry and number systems derived from ancient civilizations. 320pp. $5^3/_8 \times 8^1/_2$. 0-486-25563-8

THE HISTORY OF THE CALCULUS AND ITS CONCEPTUAL DEVELOPMENT, Carl B. Boyer. Fluent description of the development of both the integral and differential calculus—its early beginnings in antiquity, medieval contributions, and a consideration of Newton and Leibniz. 368pp. $5^3/_8 \times 8^1/_2$. 0-486-60509-4

GAMES, GODS & GAMBLING: A HISTORY OF PROBABILITY AND STATISTICAL IDEAS, F. N. David. Episodes from the lives of Galileo, Fermat, Pascal, and others illustrate this fascinating account of the roots of mathematics. Features thought-provoking references to classics, archaeology, biography, poetry. 1962 edition. 304pp. $5^3/_8 \times 8^1/_2$. (Available in U.S. only.) 0-486-40023-9

OF MEN AND NUMBERS: THE STORY OF THE GREAT MATHEMATICIANS, Jane Muir. Fascinating accounts of the lives and accomplishments of history's greatest mathematical minds—Pythagoras, Descartes, Euler, Pascal, Cantor, many more. Anecdotal, illuminating. 30 diagrams. Bibliography. 256pp. $5^3/_8 \times 8^1/_2$. 0-486-28973-7

HISTORY OF MATHEMATICS, David E. Smith. Nontechnical survey from ancient Greece and Orient to late 19th century; evolution of arithmetic, geometry, trigonometry, calculating devices, algebra, the calculus. 362 illustrations. 1,355pp. $5^3/_8 \times 8^1/_2$. Two-vol. set. Vol. I: 0-486-20429-4 Vol. II: 0-486-20430-8

A CONCISE HISTORY OF MATHEMATICS, Dirk J. Struik. The best brief history of mathematics. Stresses origins and covers every major figure from ancient Near East to 19th century. 41 illustrations. 195pp. $5^3/_8 \times 8^1/_2$. 0-486-60255-9

Physics

OPTICAL RESONANCE AND TWO-LEVEL ATOMS, L. Allen and J. H. Eberly. Clear, comprehensive introduction to basic principles behind all quantum optical resonance phenomena. 53 illustrations. Preface. Index. 256pp. 5³/₈ x 8¹/₂. 0-486-65533-4

QUANTUM THEORY, David Bohm. This advanced undergraduate-level text presents the quantum theory in terms of qualitative and imaginative concepts, followed by specific applications worked out in mathematical detail. Preface. Index. 655pp. 5³/₈ x 8¹/₂. 0-486-65969-0

ATOMIC PHYSICS (8th EDITION), Max Born. Nobel laureate's lucid treatment of kinetic theory of gases, elementary particles, nuclear atom, wave-corpuscles, atomic structure and spectral lines, much more. Over 40 appendices, bibliography. 495pp. 5³/₈ x 8¹/₂. 0-486-65984-4

A SOPHISTICATE'S PRIMER OF RELATIVITY, P. W. Bridgman. Geared toward readers already acquainted with special relativity, this book transcends the view of theory as a working tool to answer natural questions: What is a frame of reference? What is a "law of nature"? What is the role of the "observer"? Extensive treatment, written in terms accessible to those without a scientific background. 1983 ed. xlviii+172pp. 5³/₈ x 8¹/₂. 0-486-42549-5

AN INTRODUCTION TO HAMILTONIAN OPTICS, H. A. Buchdahl. Detailed account of the Hamiltonian treatment of aberration theory in geometrical optics. Many classes of optical systems defined in terms of the symmetries they possess. Problems with detailed solutions. 1970 edition. xv + 360pp. 5³/₈ x 8¹/₂. 0-486-67597-1

PRIMER OF QUANTUM MECHANICS, Marvin Chester. Introductory text examines the classical quantum bead on a track: its state and representations; operator eigenvalues; harmonic oscillator and bound bead in a symmetric force field; and bead in a spherical shell. Other topics include spin, matrices, and the structure of quantum mechanics; the simplest atom; indistinguishable particles; and stationary-state perturbation theory. 1992 ed. xiv+314pp. 6¹/₈ x 9¹/₄. 0-486-42878-8

LECTURES ON QUANTUM MECHANICS, Paul A. M. Dirac. Four concise, brilliant lectures on mathematical methods in quantum mechanics from Nobel Prize-winning quantum pioneer build on idea of visualizing quantum theory through the use of classical mechanics. 96pp. 5³/₈ x 8¹/₂. 0-486-41713-1

THIRTY YEARS THAT SHOOK PHYSICS: THE STORY OF QUANTUM THEORY, George Gamow. Lucid, accessible introduction to influential theory of energy and matter. Careful explanations of Dirac's anti-particles, Bohr's model of the atom, much more. 12 plates. Numerous drawings. 240pp. 5³/₈ x 8¹/₂. 0-486-24895-X

ELECTRONIC STRUCTURE AND THE PROPERTIES OF SOLIDS: THE PHYSICS OF THE CHEMICAL BOND, Walter A. Harrison. Innovative text offers basic understanding of the electronic structure of covalent and ionic solids, simple metals, transition metals and their compounds. Problems. 1980 edition. 582pp. 6¹/₈ x 9¹/₄. 0-486-66021-4

HYDRODYNAMIC AND HYDROMAGNETIC STABILITY, S. Chandrasekhar. Lucid examination of the Rayleigh-Benard problem; clear coverage of the theory of instabilities causing convection. 704pp. $5^3/_8 \times 8^1/_4$. 0-486-64071-X

INVESTIGATIONS ON THE THEORY OF THE BROWNIAN MOVEMENT, Albert Einstein. Five papers (1905–8) investigating dynamics of Brownian motion and evolving elementary theory. Notes by R. Fürth. 122pp. $5^3/_8 \times 8^1/_2$. 0-486-60304-0

THE PHYSICS OF WAVES, William C. Elmore and Mark A. Heald. Unique overview of classical wave theory. Acoustics, optics, electromagnetic radiation, more. Ideal as classroom text or for self-study. Problems. 477pp. $5^3/_8 \times 8^1/_2$. 0-486-64926-1

GRAVITY, George Gamow. Distinguished physicist and teacher takes reader-friendly look at three scientists whose work unlocked many of the mysteries behind the laws of physics: Galileo, Newton, and Einstein. Most of the book focuses on Newton's ideas, with a concluding chapter on post-Einsteinian speculations concerning the relationship between gravity and other physical phenomena. 160pp. $5^3/_8 \times 8^1/_2$. 0-486-42563-0

PHYSICAL PRINCIPLES OF THE QUANTUM THEORY, Werner Heisenberg. Nobel Laureate discusses quantum theory, uncertainty, wave mechanics, work of Dirac, Schroedinger, Compton, Wilson, Einstein, etc. 184pp. $5^3/_8 \times 8^1/_2$. 0-486-60113-7

ATOMIC SPECTRA AND ATOMIC STRUCTURE, Gerhard Herzberg. One of best introductions; especially for specialist in other fields. Treatment is physical rather than mathematical. 80 illustrations. 257pp. $5^3/_8 \times 8^1/_2$. 0-486-60115-3

AN INTRODUCTION TO STATISTICAL THERMODYNAMICS, Terrell L. Hill. Excellent basic text offers wide-ranging coverage of quantum statistical mechanics, systems of interacting molecules, quantum statistics, more. 523pp. $5^3/_8 \times 8^1/_2$. 0-486-65242-4

THEORETICAL PHYSICS, Georg Joos, with Ira M. Freeman. Classic overview covers essential math, mechanics, electromagnetic theory, thermodynamics, quantum mechanics, nuclear physics, other topics. First paperback edition. xxiii + 885pp. $5^3/_8 \times 8^1/_2$. 0-486-65227-0

PROBLEMS AND SOLUTIONS IN QUANTUM CHEMISTRY AND PHYSICS, Charles S. Johnson, Jr. and Lee G. Pedersen. Unusually varied problems, detailed solutions in coverage of quantum mechanics, wave mechanics, angular momentum, molecular spectroscopy, more. 280 problems plus 139 supplementary exercises. 430pp. $6^1/_2 \times 9^1/_4$. 0-486-65236-X

THEORETICAL SOLID STATE PHYSICS, Vol. 1: Perfect Lattices in Equilibrium; Vol. II: Non-Equilibrium and Disorder, William Jones and Norman H. March. Monumental reference work covers fundamental theory of equilibrium properties of perfect crystalline solids, non-equilibrium properties, defects and disordered systems. Appendices. Problems. Preface. Diagrams. Index. Bibliography. Total of 1,301pp. $5^3/_8 \times 8^1/_2$. Two volumes. Vol. I: 0-486-65015-4 Vol. II: 0-486-65016-2

WHAT IS RELATIVITY? L. D. Landau and G. B. Rumer. Written by a Nobel Prize physicist and his distinguished colleague, this compelling book explains the special theory of relativity to readers with no scientific background, using such familiar objects as trains, rulers, and clocks. 1960 ed. vi+72pp. $5^3/_8 \times 8^1/_2$. 0-486-42806-0

CATALOG OF DOVER BOOKS